GOVERNORS STATE UNIVERSITY LIBRARY
3 1611 00233 1814

Undergraduate Texts in Mathematics

Undergraduate Texts in Mathematics

Abbott: Understanding Analysis.
Anglin: Mathematics: A Concise History and Philosophy.
Readings in Mathematics.
Anglin/Lambek: The Heritage of Thales.
Readings in Mathematics.
Apostol: Introduction to Analytic Number Theory. Second edition.
Armstrong: Basic Topology.
Armstrong: Groups and Symmetry.
Axler: Linear Algebra Done Right. Second edition.
Beardon: Limits: A New Approach to Real Analysis.
Bak/Newman: Complex Analysis. Second edition.
Banchoff/Wermer: Linear Algebra Through Geometry. Second edition.
Berberian: A First Course in Real Analysis.
Bix: Conics and Cubics: A Concrete Introduction to Algebraic Curves.
Brémaud: An Introduction to Probabilistic Modeling.
Bressoud: Factorization and Primality Testing.
Bressoud: Second Year Calculus.
Readings in Mathematics.
Brickman: Mathematical Introduction to Linear Programming and Game Theory.
Browder: Mathematical Analysis: An Introduction.
Buchmann: Introduction to Cryptography.
Buskes/van Rooij: Topological Spaces: From Distance to Neighborhood.
Callahan: The Geometry of Spacetime: An Introduction to Special and General Relavitity.
Carter/van Brunt: The Lebesgue–Stieltjes Integral: A Practical Introduction.
Cederberg: A Course in Modern Geometries. Second edition.
Chambert-Loir: A Field Guide to Algebra
Childs: A Concrete Introduction to Higher Algebra. Second edition.
Chung/AitSahlia: Elementary Probability Theory: With Stochastic Processes and an Introduction to Mathematical Finance. Fourth edition.
Cox/Little/O'Shea: Ideals, Varieties, and Algorithms. Second edition.
Croom: Basic Concepts of Algebraic Topology.
Curtis: Linear Algebra: An Introductory Approach. Fourth edition.
Daepp/Gorkin: Reading, Writing, and Proving: A Closer Look at Mathematics.
Devlin: The Joy of Sets: Fundamentals of Contemporary Set Theory. Second edition.
Dixmier: General Topology.
Driver: Why Math?
Ebbinghaus/Flum/Thomas: Mathematical Logic. Second edition.
Edgar: Measure, Topology, and Fractal Geometry.
Elaydi: An Introduction to Difference Equations. Third edition.
Erdős/Surányi: Topics in the Theory of Numbers.
Estep: Practical Analysis in One Variable.
Exner: An Accompaniment to Higher Mathematics.
Exner: Inside Calculus.
Fine/Rosenberger: The Fundamental Theory of Algebra.
Fischer: Intermediate Real Analysis.
Flanigan/Kazdan: Calculus Two: Linear and Nonlinear Functions. Second edition.
Fleming: Functions of Several Variables. Second edition.
Foulds: Combinatorial Optimization for Undergraduates.
Foulds: Optimization Techniques: An Introduction.
Franklin: Methods of Mathematical Economics.

(continued after index)

J. David Logan

A First Course in Differential Equations

With 55 Figures

J. David Logan
Willa Cather Professor of Mathematics
Department of Mathematics
University of Nebraska at Lincoln
Lincoln, NE 68588-0130
USA
dlogan@math.unl.edu

Mathematics Subject Classification (2000): 34-xx, 15-xx

Library of Congress Control Number: 2005926697 (hardcover);
Library of Congress Control Number: 2005926698 (softcover)

ISBN-10: 0-387-25963-5 (hardcover)
ISBN-13: 978-0387-25963-5

ISBN-10: 0-387-25964-3 (softcover)
ISBN-13: 978-0387-25964-2

Printed in the United States of America. (SBA)

9 8 7 6 5 4 3 2 1

springeronline.com

Dedicated to—

Reece Charles Logan,
Jaren Logan Golightly

Contents

Preface

There are many excellent texts on elementary differential equations designed for the standard sophomore course. However, in spite of the fact that most courses are one semester in length, the texts have evolved into calculus-like presentations that include a large collection of methods and applications, packaged with student manuals, and Web-based notes, projects, and supplements. All of this comes in several hundred pages of text with busy formats. Most students do not have the time or desire to read voluminous texts and explore internet supplements. The format of this differential equations book is different; it is a one-semester, brief treatment of the basic ideas, models, and solution methods. Its limited coverage places it somewhere between an outline and a detailed textbook. I have tried to write concisely, to the point, and in plain language. Many worked examples and exercises are included. A student who works through this primer will have the tools to go to the next level in applying differential equations to problems in engineering, science, and applied mathematics. It can give some instructors, who want more concise coverage, an alternative to existing texts.

The numerical solution of differential equations is a central activity in science and engineering, and it is absolutely necessary to teach students some aspects of scientific computation as early as possible. I tried to build in flexibility regarding a computer environment. The text allows students to use a calculator or a computer algebra system to solve some problems numerically and symbolically, and templates of MATLAB and Maple programs and commands are given in an appendix. The instructor can include as much of this, or as little of this, as he or she desires.

For many years I have taught this material to students who have had a standard three-semester calculus sequence. It was well received by those who appreciated having a small, definitive parcel of material to learn. Moreover, this text gives students the opportunity to start reading mathematics at a slightly higher level than experienced in pre-calculus and calculus. Therefore the book can be a bridge in their progress to study more advanced material at the junior–senior level, where books leave a lot to the reader and are not packaged in elementary formats.

Chapters 1, 2, 3, 5, and 6 should be covered in order. They provide a route to geometric understanding, the phase plane, and the qualitative ideas that are important in differential equations. Included are the usual treatments of separable and linear first-order equations, along with second-order linear homogeneous and nonhomogeneous equations. There are many applications to ecology, physics, engineering, and other areas. These topics will give students key skills in the subject. Chapter 4, on Laplace transforms, can be covered at any time after Chapter 3, or even omitted. Always an issue in teaching differential equations is how much linear algebra to cover. In two extended sections in Chapter 5 we introduce a moderate amount of matrix theory, including solving linear systems, determinants, and the eigenvalue problem. In spite of the book's brevity, it still contains slightly more material than can be comfortably covered in a single three-hour semester course. Generally, I assign most of the exercises; hints and solutions for selected problems are given in Appendix D.

I welcome suggestions, comments, and corrections. Contact information is on my Web site: `http://www.math.unl.edu/~dlogan`, where additional items may be found.

I would like to thank John Polking at Rice University for permitting me to use his MATLAB program pplane7 to draw some of the phase plane diagrams and Mark Spencer at Springer for his enthusiastic support of this project. Finally, I would like to thank Tess for her continual encouragement and support for my work.

David Logan
Lincoln, Nebraska

To the Student

What is a course in differential equations about? Here are some informal, preparatory remarks to give you some sense of the subject before we take it up seriously.

You are familiar with algebra problems and solving algebraic equations. For example, the solutions to the quadratic equation

$$x^2 - x = 0$$

are easily found to be $x = 0$ and $x = 1$, which are numbers. A differential equation (sometimes abbreviated DE) is another type of equation where the unknown is not a number, but a function. We will call it $u(t)$ and think of it as a function of time. A DE also contains derivatives of the unknown function, which are also not known. So a DE is an equation that relates an unknown function to some of its derivatives. A simple example of a DE is

$$u'(t) = u(t),$$

where $u'(t)$ denotes the derivative of $u(t)$. We ask what function $u(t)$ solves this equation. That is, what function $u(t)$ has a derivative that is equal to itself? From calculus you know that one such function is $u(t) = e^t$, the exponential function. We say this function is a solution of the DE, or it solves the DE. Is it the only one? If you try $u(t) = Ce^t$, where C is any constant whatsoever, you will also find it is a solution. So differential equations have lots of solutions (fortunately we will see they are quite similar, and the fact that there are many allows some flexibility in imposing other desired conditions).

This DE was very simple and we could guess the answer from our calculus knowledge. But, unfortunately (or, fortunately!), differential equations are usually more complicated. Consider, for example, the DE

$$u''(t) + 2u'(t) + 2u(t) = 0.$$

This equation involves the unknown function and both its first and second derivatives. We seek a function for which its second derivative, plus twice its first derivative, plus twice the function itself, is zero. Now can you quickly guess a function $u(t)$ that solves this equation? It is not likely. An answer is

$$u(t) = e^{-t} \cos t.$$

And,

$$u(t) = e^{-t} \sin t$$

works as well. Let's check this last one by using the product rule and calculating its derivatives:

$$\begin{aligned} u(t) &= e^{-t} \sin t, \\ u'(t) &= e^{-t} \cos t - e^{-t} \sin t, \\ u''(t) &= -e^{-t} \sin t - 2e^{-t} \cos t + e^{-t} \sin t. \end{aligned}$$

Then

$$\begin{aligned} & u''(t) + 2u'(t) + 2u(t) \\ = \; & -e^{-t} \sin t - 2e^{-t} \cos t + e^{-t} \sin t + 2(e^{-t} \cos t - e^{-t} \sin t) + 2e^{-t} \sin t \\ = \; & 0. \end{aligned}$$

So it works! The function $u(t) = e^{-t} \sin t$ solves the equation $u''(t) + 2u'(t) + 2u(t) = 0$. In fact,

$$u(t) = Ae^{-t} \sin t + Be^{-t} \cos t$$

is a solution regardless of the values of the constants A and B. So, again, differential equations have lots of solutions.

Partly, the subject of differential equations is about developing methods for finding solutions.

Why differential equations? Why are they so important to deserve a course of study? Well, differential equations arise naturally as *models* in areas of science, engineering, economics, and lots of other subjects. Physical systems, biological systems, economic systems—all these are marked by change. Differential equations model real-world systems by describing how they change. The unknown function $u(t)$ could be the current in an electrical circuit, the concentration of a chemical undergoing reaction, the population of an animal species in an ecosystem, or the demand for a commodity in a micro-economy. Differential equations are laws that dictate change, and the unknown $u(t)$, for which we solve, describes exactly how the changes occur. In fact, much of the reason that the calculus was developed by Isaac Newton was to describe motion and to solve differential equations.

For example, suppose a particle of mass m moves along a line with constant velocity V_0. Suddenly, say at time $t = 0$, there is imposed an external resistive force F on the particle that is proportional to its velocity $v = v(t)$ for times $t > 0$. Notice that the particle will slow down and its velocity will change. From this information can we predict the velocity $v(t)$ of the particle at any time $t > 0$? We learned in calculus that Newton's second law of motion states that the mass of the particle times its acceleration equals the force, or $ma = F$. We also learned that the derivative of velocity is acceleration, so $a = v'(t)$. Therefore, if we write the force as $F = -kv(t)$, where k is the proportionality constant and the minus sign indicates the force opposes the motion, then

$$mv'(t) = -kv(t).$$

This is a differential equation for the unknown velocity $v(t)$. If we can find a function $v(t)$ that "works" in the equation, and also satisfies $v(0) = V_0$, then we will have determined the velocity of the particle. Can you guess a solution? After a little practice in Chapter 1 we will be able to solve the equation and find that the velocity decays exponentially; it is given by

$$v(t) = V_0 e^{-kt/m}, \quad t \geq 0.$$

Let's check that it works:

$$mv'(t) = mV_0 \left(-\frac{k}{m}\right) e^{-kt/m} = -kV_0 e^{-kt/m} = -kv(t).$$

Moreover, $v(0) = V_0$. So it does check. The differential equation itself is a *model* that governs the dynamics of the particle. We set it up using Newton's second law, and it contains the unknown function $v(t)$, along with its derivative $v'(t)$. The solution $v(t)$ dictates how the system evolves.

In this text we study differential equations and their applications. We address two principal questions. (1) How do we find an appropriate DE to model a physical problem? (2) How do we understand or solve the DE after we obtain it? We learn modeling by examining models that others have studied (such as Newton's second law), and we try to create some of our own through exercises. We gain understanding and learn solution techniques by practice.

Now we are ready. Read the text carefully with pencil and paper in hand, and work through all the examples. Make a commitment to solve most of the exercises. You will be rewarded with a knowledge of one of the monuments of mathematics and science.

1 Differential Equations and Models

In science, engineering, economics, and in most areas where there is a quantitative component, we are greatly interested in describing how systems evolve in time, that is, in describing a system's **dynamics**. In the simplest one-dimensional case the state of a system at any time t is denoted by a function, which we generically write as $u = u(t)$. We think of the dependent variable u as the state variable of a system that is varying with time t, which is the independent variable. Thus, knowing u is tantamount to knowing what state the system is in at time t. For example, $u(t)$ could be the population of an animal species in an ecosystem, the concentration of a chemical substance in the blood, the number of infected individuals in a flu epidemic, the current in an electrical circuit, the speed of a spacecraft, the mass of a decaying isotope, or the monthly sales of an advertised item. Knowledge of $u(t)$ for a given system tells us exactly how the state of the system is changing in time. Figure 1.1 shows a **time series plot** of a generic state function. We always use the variable u for a generic state; but if the state is "population", then we may use p or N; if the state is voltage, we may use V. For mechanical systems we often use $x = x(t)$ for the position.

One way to obtain the state $u(t)$ for a given system is to take measurements at different times and fit the data to obtain a nice formula for $u(t)$. Or we might read $u(t)$ off an oscilloscope or some other gauge or monitor. Such curves or formulas may tell us *how* a system behaves in time, but they do not give us insight into *why* a system behaves in the way we observe. Therefore we try to formulate explanatory models that underpin the understanding we seek. Often these models are dynamic equations that relate the state $u(t)$ to its rates of

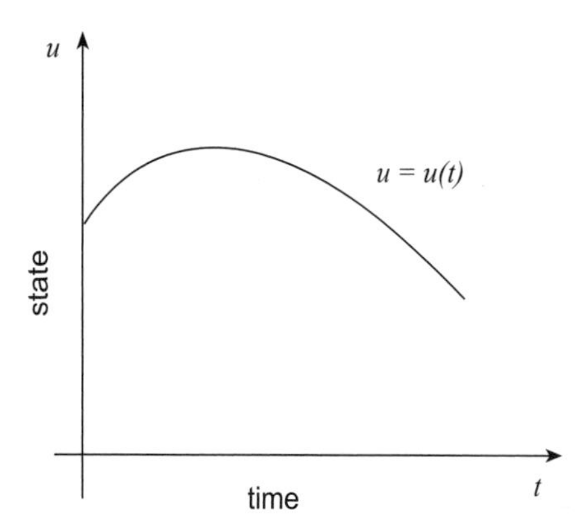

Figure 1.1 Time series plot of a generic state function $u = u(t)$ for a system.

change, as expressed by its derivatives $u'(t), u''(t), ...$, and so on. Such equations are called **differential equations** and many laws of nature take the form of such equations. For example, Newton's second law for the motion for a mass acted upon by external forces can be expressed as a differential equation for the unknown position $x = x(t)$ of the mass.

In summary, a differential equation is an equation that describes how a state $u(t)$ changes. A common strategy in science, engineering, economics, etc., is to formulate a basic principle in terms of a differential equation for an unknown state that characterizes a system and then solve the equation to determine the state, thereby determining how the system evolves in time.

1.1 Differential Equations

1.1.1 Equations and Solutions

A **differential equation** (abbreviated **DE**) is simply an equation for an unknown state function $u = u(t)$ that connects the state function and some of its derivatives. Several notations are used for the derivative, including

$$u', \frac{du}{dt}, \dot{u}, ...$$

The *overdot* notation is common in physics and engineering; mostly we use the simple *prime* notation. The reader should be familiar with the definition of the derivative:

$$u'(t) = \lim_{h \to 0} \frac{u(t+h) - u(t)}{h}.$$

For small h, the difference quotient on the right side is often taken as an approximation for the derivative. Similarly, the second derivative is denoted by

$$u'', \frac{d^2u}{dt^2}, \ddot{u}, \ldots$$

and so forth; the nth derivative is denoted by $u^{(n)}$. The first derivative of a quantity is the "rate of change of the quantity" measuring how fast the quantity is changing, and the second derivative measures how fast the rate is changing. For example, if the state of a mechanical system is position, then its first derivative is velocity and its second derivative is acceleration, or the rate of change of velocity. Differential equations are equations that relate states to their rates of change, and many natural laws are expressed in this manner. The order of the highest derivative that occurs in the DE is called the **order** of the equation.

Example 1.1

Three examples of differential equations are

$$\begin{aligned}
\theta'' + \sqrt{\frac{g}{l}} \sin\theta &= 0, \\
Lq'' + Rq' + \frac{1}{C}q &= \sin\omega t, \\
p' &= rp(1 - \frac{p}{K}).
\end{aligned}$$

The first equation models the angular deflections $\theta = \theta(t)$ of a pendulum of length l; the second models the charge $q = q(t)$ on a capacitor in an electrical circuit containing an inductor, resistor, and a capacitor, where the current is driven by a sinusoidal electromotive force operating at frequency ω; in the last equation, called the logistics equation, the state function $p = p(t)$ represents the population of an animal species in a closed ecosystem; r is the population growth rate and K represents the capacity of the ecosystem to support the population. The unspecified constants in the various equations, l, L, R, C, ω, r, and K are called **parameters**, and they can take any value we choose. Most differential equations that model physical processes contain such parameters. The constant g in the pendulum equation is a **fixed parameter** representing the acceleration of gravity on earth. In mks units, $g = 9.8$ meters per second-squared. The unknown in each equation, $\theta(t)$, $q(t)$, and $p(t)$, is the state function. The first two equations are *second-order* and the third equation is *first-order*. Note that all the state variables in all these equations depend on time t. Because time dependence is understood we often save space and drop that dependence when writing differential equations. So, for example, in the first equation θ means $\theta(t)$ and θ'' means $\theta''(t)$.

In this chapter we focus on first-order differential equations and their origins. We write a generic first-order equation for an unknown state $u = u(t)$ in the form

$$u' = f(t, u). \tag{1.1}$$

When we have solved for the derivative, we say the equation is in **normal** form. There are several words we use to classify DEs, and the reader should learn them. If f does not depend *explicitly* on t (i.e., the DE has the form $u' = f(u)$), then we call the DE **autonomous**. Otherwise it is **nonautonomous**. For example, the equation $u' = -3u^2 + 2$ is autonomous, but $u' = -3u^2 + \cos t$ is nonautonomous. If f is a linear function in the variable u, then we say (1.1) is **linear**; else it is **nonlinear**. For example, the equation $u' = -3u^2 + 2$ is nonlinear because $f(t, u) = -3u^2 + 2$ is a quadratic function of u, not a linear one. The general form of a **first-order linear equation** is

$$u' = p(t)u + q(t),$$

where p and q are known functions. Note that in a linear equation both u and u' occur alone and to the first power, but the time variable t can occur in any manner. Linear equations occur often in theory and applications, and their study forms a significant part of the subject of differential equations.

A function $u = u(t)$ is a **solution**[1] of the DE (1.1) on an interval $I : a < t < b$ if it is differentiable on I and, when substituted into the equation, it satisfies the equation identically for all $t \in I$; that is,

$$u'(t) = f(t, u(t)), \quad t \in I.$$

Therefore, a function is a solution if, when substituted into the equation, every term cancels out. In a differential equation the solution is an unknown state function to be found. For example, in $u' = -u + e^{-t}$, the unknown is a function $u = u(t)$; we ask what function $u(t)$ has the property that its derivative is the same as the negative of the function, plus e^{-t}.

Example 1.2

This example illustrates what we might expect from a first-order linear DE. Consider the DE

$$u' = -u + e^{-t}.$$

[1] We are overburdening the notation by using the same symbol u to denote both a variable and a function. It would be more precise to write "$u = \varphi(t)$ is a solution," but we choose to stick to the common use, and abuse, of a single letter.

The state function $u(t) = te^{-t}$ is a solution to the DE on the interval $I: -\infty < t < \infty$. (Later, we learn how to find this solution). In fact, for any constant C the function $u(t) = (t + C)e^{-t}$ is a solution. We can verify this by direct substitution of u and u' into the DE; using the product rule for differentiation,

$$u' = (t + C)(-e^{-t}) + e^{-t} = -u + e^{-t}.$$

Therefore $u(t)$ satisfies the DE regardless of the value of C. We say that this expression $u(t) = (t + C)e^{-t}$ represents a **one-parameter family** of solutions (one solution for each value of C). This example illustrates the usual state of affairs for any first-order linear DE—there is a one-parameter family of solutions depending upon an arbitrary constant C. This family of solutions is called a **general solution**. The fact that there are many solutions to first-order differential equations turns out to be fortunate because we can adjust the constant C to obtain a specific solution that satisfies other conditions that might apply in a physical problem (e.g., a requirement that the system be in some known state at time $t = 0$). For example, if we require $u(0) = 1$, then $C = 1$ and we obtain a **particular solution** $u(t) = (t+1)e^{-t}$. Figure 1.2 shows a plot of the one-parameter family of solutions for several values of C. Here, we are using the word parameter in a different way from that in Example 1.1; there, the word parameter refers to a physical number in the equation itself that is fixed, yet arbitrary (like resistance in a circuit).

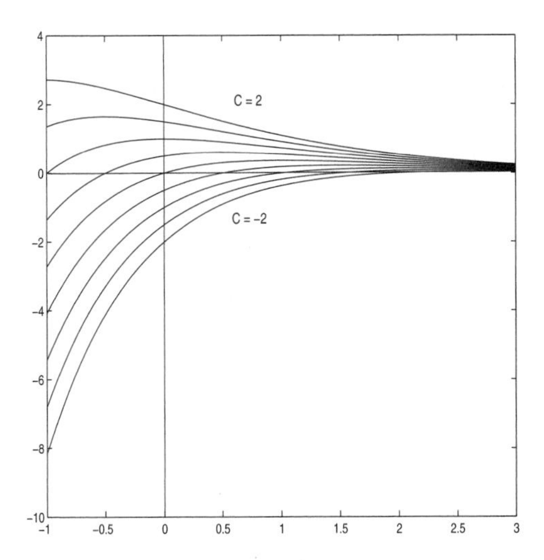

Figure 1.2 Time series plots of several solutions to $u' = e^{-t} - u$ on the interval $-1 \le t \le 3$. The solution curves, or the one-parameter family of solutions, are $u(t) = (t + C)e^{-t}$, where C is an arbitrary constant, here taking several values between -2 and 2.

An **initial value problem** (abbreviated **IVP**) for a first-order DE is the problem of finding a solution $u = u(t)$ to (1.1) that satisfies an **initial condition** $u(t_0) = u_0$, where t_0 is some fixed value of time and u_0 is a fixed state. We write the IVP concisely as

$$\text{(IVP)} \qquad \begin{cases} u' = f(t,u), \\ u(t_0) = u_0. \end{cases} \tag{1.2}$$

The initial condition usually picks out a specific value of the arbitrary constant C that appears in the general solution of the equation. So, it selects one of the many possible states that satisfy the differential equation. The accompanying graph (figure 1.3) depicts a solution to an IVP.

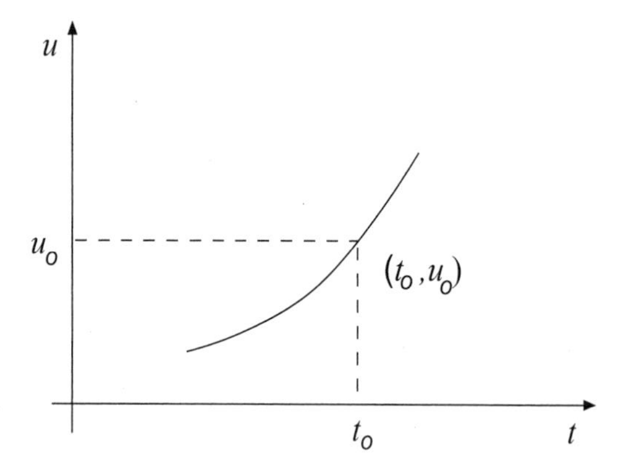

Figure 1.3 Solution to an initial value problem. The fundamental questions are: (a) is there a solution curve passing through the given point, (b) is the curve the only one, and (c) what is the interval (α, β) on which the solution exists.

Geometrically, solving an initial value problem means to find a solution to the DE that passes through a specified point (t_0, u_0) in the plane. Referring to Example 1.2, the IVP

$$u' = -u + e^{-t}, \quad u(0) = 1$$

has solution $u(t) = (t+1)e^{-t}$, which is valid for all times t. The solution curve passes through the point $(0, 1)$, corresponding to the initial condition $u(0) = 1$. Again, the initial condition selects one of the many solutions of the DE; it fixes the value of the arbitrary constant C.

There are many interesting mathematical questions about initial value problems:

1. **(Existence)** Given an initial value problem, is there a solution? This is the question of existence. Note that there may be a solution even if we cannot find a formula for it.
2. **(Uniqueness)** If there is a solution, is the solution unique? That is, is it the only solution? This is the question of uniqueness.
3. **(Interval of existence)** For which times t does the solution to the initial value problem exist?

Obtaining resolution of these theoretical issues is an interesting and worthwhile endeavor, and it is the subject of advanced courses and books in differential equations. In this text we only briefly discuss these matters. The next three examples illustrate why these are reasonable questions.

Example 1.3

Consider the initial value problem

$$u' = u\sqrt{t-3}, \quad u(1) = 2.$$

This problem has no solution because the derivative of u is not defined in an interval containing the initial time $t = 1$. There cannot be a solution curve passing through the point $(1, 2)$.

Example 1.4

Consider the initial value problem

$$u' = 2u^{1/2}, \quad u(0) = 0.$$

The reader should verify that both $u(t) = 0$ and $u(t) = t^2$ are solutions to this initial value problem on $t > 0$. Thus, it does not have a unique solution. More than one state evolves from the initial state.

Example 1.5

Consider the two similar initial value problems

$$\begin{aligned} u' &= 1 - u^2, \quad u(0) = 0, \\ u' &= 1 + u^2, \quad u(0) = 0. \end{aligned}$$

The first has solution

$$u(t) = \frac{e^{2t} - 1}{e^{2t} + 1},$$

which exists for every value of t. Yet the second has solution

$$u(t) = \tan t,$$

which exists only on the interval $\frac{-\pi}{2} < t < \frac{\pi}{2}$. So the solution to the first initial value problem is defined for all times, but the solution to the second "blows up" in finite time. These two problems are quite similar, yet the times for which their solutions exist are quite different.

The following theorem, which is proved in advanced books, provides partial answers to the questions raised above. The theorem basically states that if the right side $f(t, u)$ of the differential equation is nice enough, then there is a unique solution in a neighborhood the initial value.

Theorem 1.6

Let the function f be continuous on the open rectangle $R : a < t < b, c < u < d$ in the tu-plane and consider the initial value problem

$$\begin{cases} u' = f(t, u), \\ u(t_0) = u_0, \end{cases} \tag{1.3}$$

where (t_0, u_0) lies in the rectangle R. Then the IVP (1.3) has a solution $u = u(t)$ on some interval (α, β) containing t_0, where $(\alpha, \beta) \subset (a, b)$. If, in addition, the partial derivative[2] $f_u(t, u)$ is continuous on R, then (1.3) has a unique solution.

The **interval of existence** is the set of time values for which the solution to the initial value problem exists. Theorem 1.6 is called a *local* existence theorem because it guarantees a solution only in a small neighborhood of the initial time t_0; the theorem does not state how large the interval of existence is. Observe that the rectangle R mentioned in the theorem is open, and hence the initial point cannot lie on its boundary. In Example 1.5 both right sides of the equations, $f(t, u) = 1 - u^2$ and $f(t, u) = 1 + u^2$, are continuous in the plane, and their partial derivatives, $f_u = -2u$ and $f_u = 2u$, are continuous in in the plane. So the initial value problem for each would have a unique solution regardless of the initial condition.

In addition to theoretical questions, there are central issues from the viewpoint of *modeling* and *applications*; these are the questions we mentioned in the "To the Student" section.

1. How do we determine a differential equation that models, or governs, a given physical observation or phenomenon?

[2] We use subscripts to denote partial derivatives, and so $f_u = \frac{\partial f}{\partial u}$.

2. How do we find a solution (either analytically, approximately, graphically, or numerically) $u = u(t)$ of a differential equation?

The first question is addressed throughout this book by formulating model equations for systems in particles dynamics, circuit theory, biology, and in other areas. We learn some basic principles that sharpen our ability to invent explanatory models given by differential equations. The second question is one of developing methods, and our approach is to illustrate some standard analytic techniques that have become part of the subject. By an **analytic method** we mean manipulations that lead to a formula for the solution; such formulas are called **analytic solutions** or **closed-form** solutions. For most real-world problems it is difficult or impossible to obtain an analytic solution. By a **numerical solution** we mean an approximate solution that is obtained by some computer algorithm; a numerical solution can be represented by a data set (table of numbers) or by a graph. In real physical problems, numerical methods are the ones most often used. **Approximate solutions** can be formulas that approximate the actual solution (e.g., a polynomial formula) or they can be numerical solutions. Almost always we are interested in obtaining a graphical representation of the solution. Often we apply **qualitative methods**. These are methods designed to obtain important information from the DE without actually solving it either numerically or analytically. For a simple example, consider the DE $u' = u^2 + t^2$. Because $u' > 0$ we know that all solution curves are increasing. Or, for the DE $u' = u^2 - t^2$, we know solution curves have a horizontal tangent as they cross the straight lines $u = \pm t$. Quantitative methods emphasize understanding the underlying model, recognizing properties of the DE, interpreting the various terms, and using graphical properties to our benefit in interpreting the equation and plotting the solutions; often these aspects are more important than actually learning specialized methods for obtaining a solution formula.

Many of the methods, both analytic and numerical, can be performed easily on computer algebra systems such as Maple, Mathematica, or MATLAB, and some can be performed on advanced calculators that have a built-in computer algebra system. Although we often use a computer algebra system to our advantage, especially to perform tedious calculations, our goal is to understand concepts and develop technique. Appendix B contains information on using MATLAB and Maple.

1.1.2 Geometrical Interpretation

What does a differential equation $u' = f(t, u)$ tell us geometrically? At each point (t, u) of the tu-plane, the value of $f(t, u)$ is the slope u' of the solution curve $u = u(t)$ that goes through that point. This is because

$$u'(t) = f(t, u(t)).$$

This fact suggests a simple graphical method for constructing approximate solution curves for a differential equation. Through each point of a selected set of points (t, u) in some rectangular region (or window) of the tu-plane we draw a short line segment with slope $f(t, u)$. The collection of all these line segments, or mini-tangents, form the **direction field**, or **slope field**, for the equation. We may then sketch solution curves that fit this direction field; the curves must have the property that at each point the tangent line has the same slope as the slope of the direction field. For example, the slope field for the differential equation $u' = -u + 2t$ is defined by the right side of the differential equation, $f(t, u) = -u + 2t$. The slope field at the point $(2, 4)$ is $f(2, 3) = -3 + 2 \cdot 4 = 5$. This means the solution curve that passes through the point $(2, 4)$ has slope 5. Because it is tedious to calculate several mini-tangents, simple programs have been developed for calculators and computer algebra systems that perform this task automatically for us. Figure 1.4 shows a slope field and several solution curves that have been fit into the field.

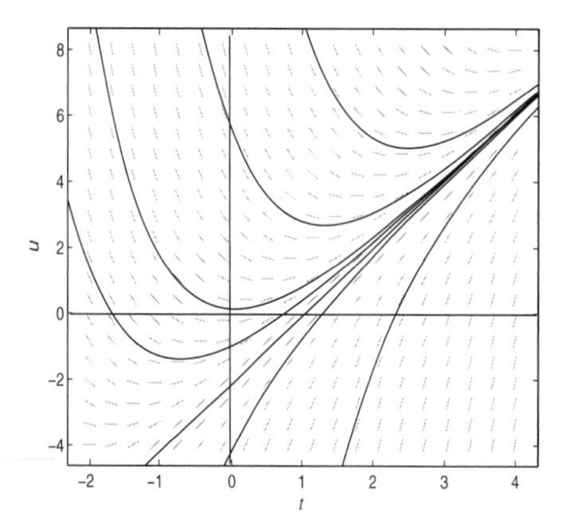

Figure 1.4 The slope field in the window $-2 \leq t \leq 4$, $-4 \leq u \leq 8$, with several approximate solution curves for the DE $u' = -u + 2t$.

Notice that a problem in differential equations is just opposite of that in differential calculus. In calculus we know the function (curve) and are asked to find the derivative (slope); in differential equations we know the slopes and try to find the state function that fits them.

Also observe that the simplicity of autonomous equations (no time t dependence on the right side)

$$u' = f(u)$$

shows itself in the slope field. In this case the slope field is independent of time, so on each horizontal line in the tu plane, where u has the same value, the slope field is the same. For example, the DE $u' = 3u(5-u)$ is autonomous, and along the horizontal line $u = 2$ the slope field has value 18. This means solution curves cross the line $u = 2$ with a relatively steep slope $u' = 18$.

EXERCISES

1. Verify that the two differential equations in Example 1.5 have solutions as stated.

2. From the set of commonly encountered functions, guess a nonzero solution $u = u(t)$ to the DE $u' = u^2$.

3. Show that $u(t) = \ln(t + C)$ is a one-parameter family of solutions of the DE $u' = e^{-u}$, where C is an arbitrary constant. Plot several members of this family. Find and plot a particular solution that satisfies the initial condition $u(0) = 0$.

4. Find a solution $u = u(t)$ of $u' + 2u = t^2 + 4t + 7$ in the form of a quadratic function of t.

5. Find value(s) of m such that $u = t^m$ is a solution to $2tu' = u$.

6. Plot the one-parameter family of curves $u(t) = (t^2 - C)e^{3t}$, and find a differential equation whose solution is this family.

7. Show that the one-parameter family of straight lines $u = Ct + f(C)$ is a solution to the differential equation $tu' - u + f(u') = 0$ for any value of the constant C.

8. Classify the first-order equations as linear or nonlinear, autonomous or nonautonomous.

   a) $u' = 2t^3u - 6$.

   b) $(\cos t)u' - 2u \sin u = 0$.

   c) $u' = \sqrt{1 - u^2}$.

   d) $7u' - 3u = 0$.

9. Explain Example 1.4 in the context of Theorem 1.6. In particular, explain why the theorem does not apply to this initial value problem. Which hypothesis fails?

10. Verify that the initial value problem $u' = \sqrt{u}$, $u(0) = 0$, has infinitely many solutions of the form

$$u(t) = \begin{cases} 0, & t \leq a \\ \frac{1}{4}(t-a)^2, & t > a, \end{cases}$$

where $a > 0$. Sketch these solutions for different values of a. What hypothesis fails in Theorem 1.6?

11. Consider the linear differential equation $u' = p(t)u + q(t)$. Is it true that the sum of two solutions is again a solution? Is a constant times a solution again a solution? Answer these same questions if $q(t) = 0$. Show that if u_1 is a solution to $u' = p(t)u$ and u_2 is a solution to $u' = p(t)u + q(t)$, then $u_1 + u_2$ is a solution to $u' = p(t)u + q(t)$.

12. By hand, sketch the slope field for the DE $u' = u(1 - \frac{u}{4})$ in the window $0 \leq t \leq 8$, $0 \leq u \leq 8$ at integer points. What is the value of the slope field along the lines $u = 0$ and $u = 4$? Show that $u(t) = 0$ and $u(t) = 4$ are constant solutions to the DE. On your slope field plot, draw in several solution curves.

13. Using a software package, sketch the slope field in the window $-4 \leq t \leq 4$, $-2 \leq u \leq 2$ for the equation $u' = 1 - u^2$ and draw several approximate solution curves. Lines and curves in the tu plane where the slope field is zero are called **nullclines**. For the given DE, find the nullclines. Graph the locus of points where the slope field is equal to -3.

14. Repeat Exercise 13 for the equation $u' = t - u^2$.

15. In the tu plane, plot the nullclines of the differential equation $u' = 2u^2(u - 4\sqrt{t})$.

16. Using concavity, show that the second-order DE $u'' - u = 0$ cannot have a solution (other than the $u = 0$ solution) that takes the value zero more than once. (Hint: construct a contradiction argument—if it takes the value zero twice, it must have a negative minimum or positive maximum.)

17. For any solution $u = u(t)$ of the DE $u'' - u = 0$, show that $(u')^2 - u^2 = C$, where C is a constant. Plot this one parameter-family of curves on a uu' set of axes.

18. Show that if u_1 and u_2 are both solutions to the DE $u' + p(t)u = 0$, then u_1/u_2 is constant.

19. Show that the linear initial value problem

$$u' = \frac{2(u-1)}{t}, \quad u(0) = 1,$$

has a continuously differentiable solution (i.e., a solution whose first derivative is continuous) given by

$$u(t) = \begin{cases} at^2 + 1, & t < 0, \\ bt^2 + 1, & t > 0, \end{cases}$$

for any constants a and b. Yet, there is no solution if $u(0) \neq 1$. Do these facts contradict Theorem 1.6?

1.2 Pure Time Equations

In this section we solve the simplest type of differential equation. First we need to recall the fundamental theorem of calculus, which is basic and used regularly in differential equations. For reference, we state the two standard forms of the theorem. They show that differentiation and integration are inverse processes.

Fundamental Theorem of Calculus I. If u is a differentiable function, the integral of its derivative is

$$\int_a^b \frac{d}{dt} u(t)dt = u(b) - u(a).$$

Fundamental Theorem of Calculus II. If g is a continuous function, the derivative of an integral with variable upper limit is

$$\frac{d}{dt} \int_a^t g(s)ds = g(t),$$

where the lower limit a is any number.

This last expression states that the function $\int_a^t g(s)ds$ is an antiderivative of g (i.e., a function whose derivative is g). Notice that $\int_a^t g(s)ds + C$ is also an antiderivative for any value of C.

The simplest differential equation is one of the form

$$u' = g(t), \tag{1.4}$$

where the right side of the differential equation is a given, known function $g(t)$. This equation is called a **pure time equation**. Thus, we seek a state function u whose derivative is $g(t)$. The fundamental theorem of calculus II, u must be an **antiderivative** of g. We can write this fact as $u(t) = \int_a^t g(s)ds + C$, or using the indefinite integral notation, as

$$u(t) = \int g(t)dt + C, \tag{1.5}$$

where C is an arbitrary constant, called the **constant of integration**. Recall that antiderivatives of a function differ by an additive constant. Thus, all solutions of (1.4) are given by (1.5), and (1.5) is called the general solution. The fact that (1.5) solves (1.4) follows from the fundamental theorem of calculus II.

Example 1.7

Find the general solution to the differential equation

$$u' = t^2 - 1.$$

Because the right side depends only on t, the solution u is an antiderivative of the right side, or

$$u(t) = \frac{1}{3}t^3 - t + C,$$

where C is an arbitrary constant. This is the general solution and it graphs as a family of cubic curves in the tu plane, one curve for each value of C. A particular antiderivative, or solution, can be determined by imposing an initial condition that picks out a specific value of the constant C, and hence a specific curve. For example, if $u(1) = 2$, then $\frac{1}{3}(1)^3 - 1 + C = 2$, giving $C = \frac{8}{3}$. The solution to the initial value problem is then $u(t) = \frac{1}{3}t^3 - t + \frac{8}{3}$.

Example 1.8

For equations of the form $u'' = g(t)$ we can take two successive antiderivatives to find the general solution. The following sequence of calculations shows how. Consider the DE

$$u'' = t + 2.$$

Then

$$\begin{aligned} u' &= \frac{1}{2}t^2 + 2t + C_1; \\ u &= \frac{1}{6}t^3 + t^2 + C_1 t + C_2. \end{aligned}$$

Here C_1 and C_2 are two arbitrary constants. For second-order equations we always expect two arbitrary constants, or a two-parameter family of solutions. It takes two auxiliary conditions to determine the arbitrary constants. In this example, if $u(0) = 1$ and if $u'(0) = 0$, then $c_1 = 1$ and $c_2 = 1$, and we obtain the particular solution $u = \frac{1}{6}t^3 + t^2 + 1$.

Example 1.9

The autonomous equation

$$u' = f(u)$$

cannot be solved by direct integration because the right side is not a known function of t; it depends on u, which is the unknown in the problem. Equations with the unknown u on the right side are not pure time equations.

Often it is not possible to find a simple expression for the antiderivative, or indefinite integral. For example, the functions $\frac{\sin t}{t}$ and e^{-t^2} have no simple analytic expressions for their antiderivatives. In these cases we must represent the antiderivative of g as

$$u(t) = \int_a^t g(s)ds + C$$

with a variable upper limit. Here, a is any fixed value of time and C is an arbitrary constant. We have used the dummy variable of integration s to avoid confusion with the upper limit of integration, the independent time variable t. It is really not advisable to write $u(t) = \int_a^t g(t)dt$.

Example 1.10

Solve the initial value problem

$$\begin{aligned} u' &= e^{-t^2}, \quad t > 0 \\ u(0) &= 2. \end{aligned}$$

The right side of the differential equation has no simple expression for its antiderivative. Therefore we write the antiderivative in the form

$$u(t) = \int_0^t e^{-s^2} ds + C.$$

The common strategy is to take the lower limit of integration to be the initial value of t, here zero. Then $u(0) = 2$ implies $C = 2$ and we obtain the solution to the initial value problem in the form of an integral,

$$u(t) = \int_0^t e^{-s^2} ds + 2. \tag{1.6}$$

If we had written the solution of the differential equation as

$$u(t) = \int e^{-t^2} dt + C,$$

in terms of an indefinite integral, then there would be no way to use the initial condition to evaluate the constant of integration, or evaluate the solution at a particular value of t.

We emphasize that integrals with a variable upper limit of integration define a function. Referring to Example 1.10, we can define the special function "erf" (called the **error function**) by

$$\operatorname{erf}(t) = \frac{2}{\sqrt{\pi}} \int_0^t e^{-s^2}\, ds.$$

The factor $\frac{2}{\sqrt{\pi}}$ in front of the integral normalizes the function to force $\operatorname{erf}(+\infty) = 1$. Up to this constant multiple, the erf function gives the area under a bell-shaped curve $\exp(-s^2)$ from 0 to t. In terms of this special function, the solution (1.6) can be written

$$u(t) = 2 + \frac{\sqrt{\pi}}{2} \operatorname{erf}(t).$$

The erf function, which is plotted in figure 1.5, is an important function in probability and statistics, and in diffusion processes. Its values are tabulated in computer algebra systems and mathematical handbooks.

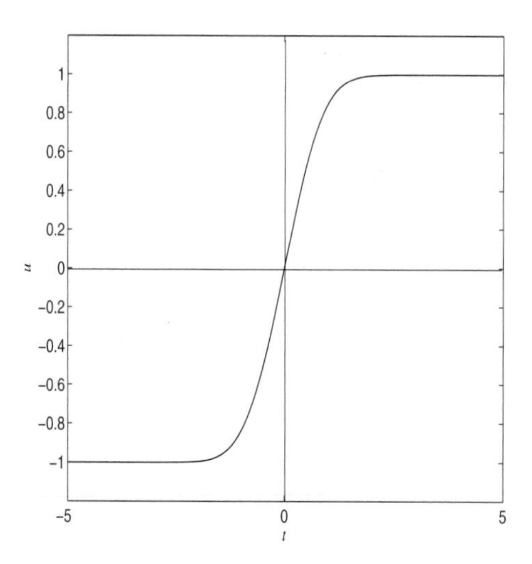

Figure 1.5 Graph of the erf function.

Functions defined by integrals are common in the applied sciences and are equally important as functions defined by simple algebraic formulas. To the point, the reader should recall that the natural logarithm can be defined by the integral

$$\ln t = \int_1^t \frac{1}{s} ds, \quad t > 0.$$

One important viewpoint is that differential equations often define special functions. For example, the initial value problem

$$u' = \frac{1}{t}, \quad u(1) = 0,$$

can be used to define the natural logarithm function $\ln t$. Other special functions of mathematical physics and engineering, for example, Bessel functions, Legendre polynomials, and so on, are usually defined as solutions to differential equations. By solving the differential equation numerically we can obtain values of the special functions more efficiently than looking those values up in tabulated form.

We end this section with the observation that one can find solution formulas using computer algebra systems like Maple, MATLAB, Mathematica, etc., and calculators equipped with computer algebra systems. Computer algebra systems do symbolic computation. Below we show the basic syntax in Maple, Mathematica, and on a TI-89 that returns the general solution to a differential equation or the solution to an initial value problem. MATLAB has a special add-on symbolic package that has similar commands. Our interest in this text is to use MATLAB for scientific computation, rather than symbolic calculation. Additional information on computing environments is in Appendix B.

The general solution of the first-order differential equation $u' = f(t, u)$ can be obtained as follows:

```
deSolve(u'=f(t,u),t,u)                        (TI-89)
dsolve(diff(u(t),t)=f(t,u(t)),u(t));          (Maple)
DSolve[u'[t]==f[t,u[t]], u[t], t]             (Mathematica)
```

To solve the initial value problem $u' = f(t, u)$, $u(a) = b$, the syntax is.

```
deSolve(u'= f(t,u) and u(a)=b, t, u)                    (TI-89)
dsolve(diff(u(t),t) = f(t,u(t)), u(a)=b, u(t));         (Maple)
DSolve[u'[t]==f[t,u[t]], u[a]==b, u[t], t]              (Mathematica)
```

EXERCISES

1. Using antiderivatives, find the general solution to the pure time equation $u' = t\cos(t^2)$, and then find the particular solution satisfying the initial condition $u(0) = 1$. Graph the particular solution on the interval $[-5, 5]$.

2. Solve the initial value problem $u' = \frac{t+1}{\sqrt{t}}, \quad u(1) = 4$.

3. Find a function $u(t)$ that satisfies the initial value problem $u'' = -3\sqrt{t}$, $u(1) = 1$, $u'(1) = 2$.

4. Find all state functions that solve the differential equation $u' = te^{-2t}$.

5. Find the solution to the initial value problem $u' = \frac{e^{-t}}{\sqrt{t}}$, $u(1) = 0$, in terms of an integral. Graph the solution on the interval $[1, 4]$ by using numerical integration to calculate values of the integral.

6. The differential equation $u' = 3u + e^{-t}$ can be converted into a pure time equation for a new dependent variable y using the transformation $u = ye^{3t}$. Find the pure time equation for y, solve it, and then determine the general solution u of the original equation.

7. Generalize the method of Exercise 6 by devising an algorithm to solve $u' = au + q(t)$, where a is any constant and q is a given function. In fact, show that
$$u(t) = Ce^{at} + e^{at} \int_0^t e^{-as} q(s) ds.$$
Using the fundamental theorem of calculus, verify that this function does solve $u' = au + q(t)$.

8. Use the chain rule and the fundamental theorem of calculus to compute the derivative of $\text{erf}(\sin t)$.

9. **Exact equations**. Consider a differential equation written in the (non-normal) form $f(t, u) + g(t, u)u' = 0$. If there is a function $h = h(t, u)$ for which $h_t = f$ and $h_u = g$, then the differential equation becomes $h_t + h_u u' = 0$, or, by the chain rule, just $\frac{d}{dt} h(t, u) = 0$. Such equations are called **exact** equations because the left side is (exactly) a total derivative of the function $h = h(t, u)$. The general solution to the equation is therefore given implicitly by $h(t, u) = C$, where C is an arbitrary constant.

   a) Show that $f(t, u) + g(t, u)u' = 0$ is exact if, and only if, $f_u = g_t$.

   b) Use part (a) to check if the following equations are exact. If the equation is exact, find the general solution by solving $h_t = f$ and $h_u = g$ for h (you may want to review the method of finding potential functions associated with a conservative force field from your multivariable calculus course).

      i. $u^3 + 3tu^2 u' = 0$.

      ii. $t^3 + \frac{u}{t} + (u^2 + \ln t)u' = 0$.

      iii. $u' = -\frac{\sin u - u \sin t}{t \cos u + \cos t}$.

10. An **integral equation** is an equation where the unknown $u(t)$ appears under an integral sign. Use the fundamental theorem of calculus to show that the integral equation

$$u(t) + \int_0^t e^{-p(t-s)} u(s)ds = A; \qquad p, A \quad \text{constants},$$

can be transformed into an initial value problem for $u(t)$.

11. Show that the integral equation

$$u(t) = e^{-2t} + \int_0^t su(s)ds$$

can be transformed into an initial value problem for $u(t)$.

12. Show, by integration, that the initial value problem (1.3) can be transformed into the integral equation

$$u(t) = u_0 + \int_0^t f(s, u(s))ds.$$

13. From the definition of the derivative, a difference quotient approximation to the first derivative is $u'(t) \cong \frac{u(t+h)-u(t)}{h}$. Use Taylor's theorem to show that an approximation for the second derivative is

$$u''(t) \cong \frac{u(t+h) - 2u(t) + u(t-h)}{h^2}.$$

(Recall that Taylor's expansion for a function u about the point t with increment h is

$$u(t+h) = u(t) + u'(t)h + \frac{1}{2}u''(t)h^2 + \cdots.$$

Use this and and a similar formula for $u(t-h)$.)

1.3 Mathematical Models

By a **mathematical model** we mean an equation, or set of equations, that describes some physical problem or phenomenon that has its origin in science, engineering, or some other area. Here we are interested in differential equation models. By **mathematical modeling** we mean the process by which we obtain and analyze the model. This process includes introducing the important and relevant quantities or variables involved in the model, making model-specific assumptions about those quantities, solving the model equations by some method, and then comparing the solutions to real data and interpreting the results. Often the solution method involves computer simulation. This comparison may lead to revision and refinement until we are satisfied that the model accurately describes the physical situation and is predictive of other similar observations. Therefore the subject of mathematical modeling involves physical intuition, formulation of equations, solution methods, and analysis. Overall, in mathematical modeling the overarching objective is to make sense of the world as we observe it, often by inventing caricatures of reality. Scientific exactness is sometimes sacrificed for mathematical tractability. Model predictions depend strongly on the assumptions, and changing the assumptions changes the model. If some assumptions are less critical than others, we say the model is robust to those assumptions.

The best strategy to learn modeling is to begin with simple examples and then graduate to more difficult ones. The reader is already familiar with some models. In an elementary science or calculus course we learn that Newton's second law, force equals mass times acceleration, governs mechanical systems such as falling bodies; Newton's inverse-square law of gravitation describes the motion of the planets; Ohm's law in circuit theory dictates the voltage drop across a resistor in terms of the current; or the law of mass action in chemistry describes how fast chemical reactions occur. In this course we learn new models based on differential equations. The importance of differential equations, as a subject matter, lies in the fact that differential equations describe many physical phenomena and laws in many areas of application. In this section we introduce some simple problems and develop differential equations that model the physical processes involved.

The first step in modeling is to select the relevant variables (independent and dependent) and parameters that describe the problem. Physical quantities have **dimensions** such as time, distance, degrees, and so on, or corresponding **units** such as seconds, meters, and degrees Celsius. The equations we write down as models must be dimensionally correct. Apples cannot equal oranges. Verifying that each term in our model has the same dimensions is the first task in obtaining a correct equation. Also, checking dimensions can often give us insight into what a term in the model might be. We always should be aware of the dimensions of the quantities, both variables and parameters, in a model, and we should always try to identify the physical meaning of the terms in the equations we obtain.

All of these comments about modeling are perhaps best summarized in a quote attributed to the famous psychologist, Carl Jung: "Science is the art of creating suitable illusions which the fool believes or argues against, but the wise man enjoys their beauty and ingenuity without being blind to the fact they are human veils and curtains concealing the abysmal darkness of the unknowable."

When one begins to feel too confident in the correctness of the model, he or she should recall this quote.

1.3.1 Particle Dynamics

In the late 16th and early 17th centuries scientists were beginning to quantitatively understand the basic laws of motion. Galileo, for example, rolled balls down inclined planes and dropped them from different heights in an effort to understand dynamical laws. But it was Isaac Newton in the mid-1600s (who developed calculus and the theory of gravitation) who finally wrote down a basic law of motion, known now as **Newton's second law**, that is in reality a differential equation for the state of the dynamical system. For a particle of mass m moving along a straight line under the influence of a specified external force F, the law dictates that "mass times acceleration equals the force on the particle," or

$$mx'' = F(t, x, x') \quad \text{(Newton's second law)}.$$

This is a second-order differential equation for the unknown location or position $x = x(t)$ of the particle. The force F may depend on time t, position $x = x(t)$, or velocity $x' = x'(t)$. This DE is called the **equation of motion** or the **dynamical equation** for the system. For second-order differential equations we impose *two* initial conditions, $x(0) = x_0$ and $x'(0) = v_0$, which fix the initial position and initial velocity of the particle, respectively. We expect that if the initial position and velocity are known, then the equation of motion should determine the state for all times $t > 0$.

Example 1.11

Suppose a particle of mass m is falling downward through a viscous fluid and the fluid exerts a resistive force on the particle proportional to the square of its velocity. We measure positive distance downward from the top of the fluid surface. There are two forces on the particle, gravity and fluid resistance. The gravitational force is mg and is positive because it tends to move the mass in a positive downward direction; the resistive force is $-ax'^2$, and it is negative because it opposes positive downward motion. The net force is then $F = mg - ax'^2$, and the equation of motion is $mx'' = mg - a(x'^2)^2$. This second-order equation can immediately be reformulated as a first-order differential equation for the velocity $v = x'$. Clearly

$$v' = g - \frac{a}{m}v^2.$$

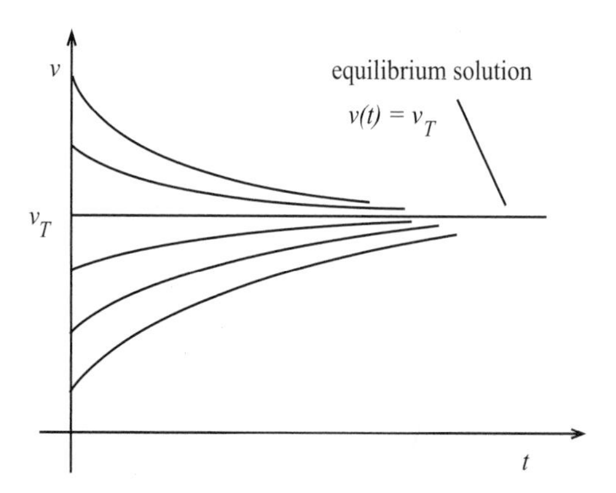

Figure 1.6 Generic solution curves, or time series plots, for the model $v' = g - (a/m)v^2$. For $v < v_T$ the solution curves are increasing because $v' > 0$; for $v > v_T$ the solution curves are decreasing because $v' < 0$. All the solution curves approach the constant terminal velocity solution $v(t) = v_T$.

If we impose an initial velocity, $v(0) = v_0$, then this equation and the initial condition gives an initial value problem for $v = v(t)$. Without solving the DE we can obtain important qualitative information from the DE itself. Over a long time, if the fluid were deep, we would observe that the falling mass would approach a constant, terminal velocity v_T. Physically, the terminal velocity occurs when the two forces, the gravitational force and resistive force, balance. Thus $0 = g - (av_T^2/m)$, or

$$v_T = \sqrt{\frac{mg}{a}}.$$

By direct substitution, we note that $v(t) = v_T$ is a constant solution of the differential equation with initial condition $v(0) = v_T$. We call such a constant solution an **equilibrium**, or **steady-state**, solution. It is clear that, regardless of the initial velocity, the system approaches this equilibrium state. This supposition is supported by the observation that $v' > 0$ when $v < v_T$ and $v' < 0$ when $v > v_T$. Figure 1.6 shows what we expect, illustrating several generic solution curves (time series plots) for different initial velocities. To find the position $x(t)$ of the object we would integrate the velocity $v(t)$, once it is determined; that is, $x(t) = \int_0^t v(s)ds$.

Example 1.12

A ball of mass m is tossed upward from a building of height h with initial velocity v_0. If we ignore air resistance, then the only force is that due to gravity, having magnitude mg, directed downward. Taking the positive direction upward with $x = 0$ at the ground, the model that governs the motion (i.e., the height $x = x(t)$ of the ball), is the initial value problem

$$mx'' = -mg, \quad x(0) = h, \quad x'(0) = v_0.$$

Note that the force is negative because the positive direction is upward. Because the right side is a known function (a constant in this case), the differential equation is a pure time equation and can be solved directly by integration (antiderivatives). If $x''(t) = -g$ (i.e., the second derivative is the constant $-g$), then the first derivative must be $x'(t) = -gt + c_1$, where c_1 is some constant (the constant of integration). We can evaluate c_1 using the initial condition $x'(0) = v_0$. We have $x'(0) = -g \times 0 + c_1 = v_0$, giving $c_1 = v_0$. Therefore, at any time the velocity is given by

$$x'(t) = -gt + v_0.$$

Repeating, we take another antiderivative. Then

$$x(t) = -\frac{1}{2}gt^2 + v_0 t + c_2,$$

where c_2 is some constant. Using $x(0) = h$ we find that $c_2 = h$. Therefore the height of the ball at any time t is given by the familiar physics formula

$$x(t) = -\frac{1}{2}gt^2 + v_0 t + h.$$

Example 1.13

Imagine a mass m lying on a table and connected to a spring, which is in turn attached to a rigid wall (figure 1.7). At time $t = 0$ we displace the mass a positive distance x_0 to the right of equilibrium and then release it. If we ignore friction on the table then the mass executes simple harmonic motion; that is, it oscillates back and forth at a fixed frequency. To set up a model for the motion we follow the doctrine of mechanics and write down Newton's second law of motion, $mx'' = F$, where the state function $x = x(t)$ is the position of the mass at time t (we take $x = 0$ to be the equilibrium position and $x > 0$ to the right), and F is the external force. All that is required is to impose the form of the force. Experiments confirm that if the displacement is not too large (which we assume), then the force exerted by the spring is proportional to its displacement from equilibrium. That is,

$$F = -kx. \tag{1.7}$$

The minus sign appears because the force opposes positive motion. The proportionality constant k (having dimensions of force per unit distance) is called the **spring constant**, or **stiffness** of the spring, and equation (1.7) is called **Hooke's law**. Not every spring behaves in this manner, but Hooke's law is used as a model for some springs; it is an example of what in engineering is called a **constitutive relation**. It is an empirical result rather than a law of nature. To give a little more justification for Hooke's law, suppose the force F depends on the displacement x through $F = F(x)$, with $F(0) = 0$. Then by Taylor's theorem,

$$\begin{aligned} F(x) &= F(0) + F'(0)x + \frac{1}{2}F''(0)x^2 + \cdots \\ &= -kx + \frac{1}{2}F''(0)x^2 + \cdots, \end{aligned}$$

where we have defined $F'(0) = -k$. So Hooke's law has a general validity if the displacement is small, allowing the higher-order terms in the series to be neglected. We can measure the stiffness k of a spring by letting it hang from a ceiling without the mass attached; then attach the mass m and measure the elongation L after it comes to rest. The force of gravity mg must balance the restoring force kx of the spring, so $k = mg/L$. Therefore, assuming a Hookean spring, we have the equation of motion

$$mx'' = -kx \tag{1.8}$$

which is the **spring-mass equation**. The initial conditions (released at time zero at position x_0) are

$$x(0) = x_0, \quad x'(0) = 0.$$

We expect oscillatory motion. If we attempt a solution of (1.8) of the form $x(t) = A\cos\omega t$ for some frequency ω and amplitude A, we find upon substitution that $\omega = \sqrt{k/m}$ and $A = x_0$. Therefore the displacement of the mass is given by

$$x(t) = x_0 \cos\sqrt{k/m}t.$$

This solution represents an oscillation of amplitude x_0, frequency $\sqrt{k/m}$, and period $2\pi/\sqrt{k/m}$.

Example 1.14

Continuing with Example 1.13, if there is damping (caused, for example, by friction or submerging the system in a liquid), then the spring-mass equation

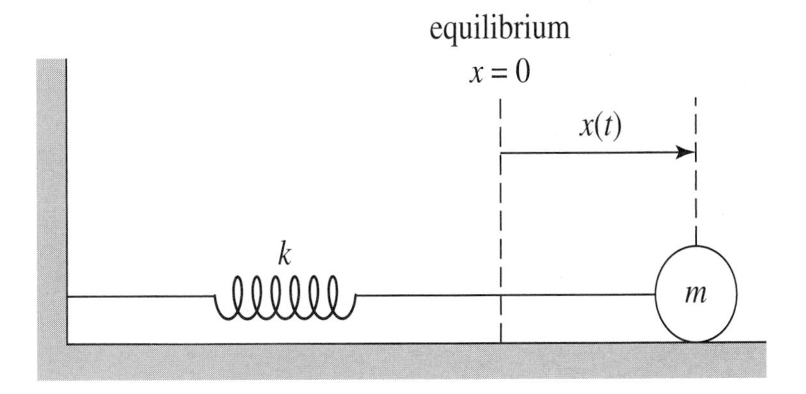

Figure 1.7 Spring-mass oscillator.

must be modified to account for the damping force. The simplest assumption, again a constitutive relation, is to take the resistive force F_{r} to be proportional to the velocity of the mass. Thus, also assuming Hooke's law for the spring force F_{s}, we have the **damped spring-mass equation**

$$mx'' = F_{\mathrm{r}} + F_{\mathrm{s}} = -cx' - kx.$$

The positive constant c is the damping constant. Both forces have negative signs because both oppose positive (to the right) motion. For this case we expect some sort of oscillatory behavior with the amplitude decreasing during each oscillation. In Exercise 1 you will show that solutions representing decaying oscillations do, in fact, occur.

Example 1.15

For conservative mechanical systems, another technique for obtaining the equation of motion is to apply the conservation of energy law: the kinetic energy plus the potential energy remain constant. We illustrate this method by finding the equation governing a frictionless pendulum of length l whose bob has mass m. See figure 1.8. As a state variable we choose the angle θ that it makes with the vertical. As time passes, the bob traces out an arc on a circle of radius l; we let s denote the arclength measured from rest ($\theta = 0$) along the arc. By geometry, $s = l\theta$. As the bob moves, its kinetic energy is one-half its mass times the velocity-squared; its potential energy is mgh, where h is the height above zero-potential energy level, taken where the pendulum is at rest. Therefore $\frac{1}{2}m(s')^2 + mgl(1 - \cos\theta) = E$, where E is the constant energy. In terms of the angle θ,

$$\frac{1}{2}l(\theta')^2 + g(1 - \cos\theta) = C, \tag{1.9}$$

where $C = E/ml$. The initial conditions are $\theta(0) = \theta_0$ and $\theta'(0) = \omega_0$, where θ_0 and ω_0 are the initial angular displacement and angular velocity, respectively. As it stands, the differential equation (1.9) is first-order; the constant C can be determined by evaluating the differential equation at $t = 0$. We get $C = \frac{1}{2}l\omega_0^2 + g(1 - \cos\theta_0)$. By differentiation with respect to t, we can write (1.9) as

$$\theta'' + \frac{g}{l}\sin\theta = 0. \tag{1.10}$$

This is a second-order nonlinear DE in $\theta(t)$ called the **pendulum equation**. It can also be derived directly from Newton's second law by determining the forces, which we leave as an exercise (Exercise 6). We summarize by stating that for a conservative mechanical system the equation of motion can be found either by determining the energies and applying the conservation of energy law, or by finding the forces and using Newton's second law of motion.

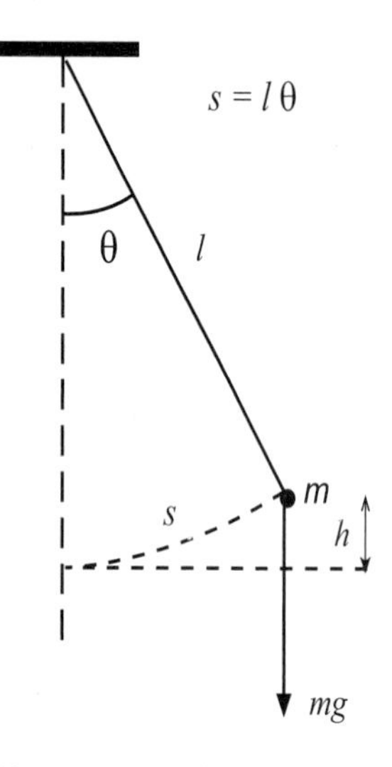

Figure 1.8 A pendulum consisting of a mass m attached to a rigid, weightless, rod of length l. The force of gravity is mg, directed downward. The potential energy is mgh where h is the height of the mass above the equilibrium position.

EXERCISES

1. When a mass of 0.3 kg is placed on a spring hanging from the ceiling, it elongates the spring 15 cm. What is the stiffness k of the spring?

2. Consider a damped spring-mass system whose position $x(t)$ is governed by the equation $mx'' = -cx' - kx$. Show that this equation can have a "decaying-oscillation" solution of the form $x(t) = e^{-\lambda t}\cos \omega t$. (Hint: By substituting into the differential equations, show that the decay constant λ and frequency ω can be determined in terms of the given parameters m, c, and k.)

3. A car of mass m is moving at speed V when it has to brake. The brakes apply a constant force F until the car comes to rest. How long does it take the car to stop? How far does the car go before stopping? Now, with specific data, compare the time and distance it takes to stop if you are going 30 mph vs. 35 mph. Take $m = 1000$ kg and $F = 6500$ N. Write a short paragraph on recommended speed limits in a residential areas.

4. Derive the pendulum equation (1.10) from the conservation of energy law (1.9) by differentiation.

5. A pendulum of length 0.5 meters has a bob of mass 0.1 kg. If the pendulum is released from rest at an angle of 15 degrees, find the total energy in the system.

6. Derive the pendulum equation (1.10) by resolving the gravitational force on the bob in the tangential and normal directions along the arc of motion and then applying Newton's second law. Note that only the tangential component affects the motion.

7. If the amplitude of the oscillations of a pendulum is small, then $\sin\theta$ is nearly equal to θ (why?) and the nonlinear equation (1.10) is approximated by the linear equation $\theta'' + (g/l)\theta = 0$.

    a) Show that the approximate linear equation has a solution of the form $\theta(t) = A\cos\omega t$ for some value of ω, which also satisfies the initial conditions $\theta(0) = A$, $\theta'(0) = 0$. What is the period of the oscillation?

    b) A 650 lb wrecking ball is suspended on a 20 m cord from the top of a crane. The ball, hanging vertically at rest against the building, is pulled back a small distance and then released. How soon does it strike the building?

8. An enemy cannon at distance L from a fort can fire a cannon ball from the top of a hill at height H above the ground level with a muzzle velocity v. How high should the wall of the fort be to guarantee that a cannon ball

will not go over the wall? Observe that the enemy can adjust the angle of its shot. (Hint: Ignoring air resistance, the governing equations follow from resolving Newton's second law for the horizontal and vertical components of the force: $mx'' = 0$ and $my'' = -mg$.)

1.3.2 Autonomous Differential Equations

In this section we introduce some simple qualitative methods to understand the dynamics of an autonomous differential equation

$$u' = f(u).$$

We introduce the methods in the context of population ecology, as well as in some other areas in the life sciences.

Models in biology often have a different character from fundamental laws in the physical sciences, such as Newton's second law of motion in mechanics or Maxwell's equations in electrodynamics. Ecological systems are highly complex and it is often impossible to include every possible factor in a model; the chore of modeling often comes in knowing what effects are important, and what effects are minor. Many models in ecology are often not based on physical law, but rather on observation, experiment, and reasoning.

Ecology is the study of how organisms interact with their environment. A fundamental problem in population ecology is to determine what mechanisms operate to regulate animal populations. Let $p = p(t)$ denote the population of an animal species at time t. For the human population, T. Malthus (in the late 1700s) proposed the model

$$\frac{p'}{p} = r,$$

which states that the "*per capita* growth rate is constant," where the constant $r > 0$ is the **growth rate** given in dimensions of time^{-1}. We can regard r as the birth rate minus the death rate, or $r = b - d$. This *per capita* law is same as

$$p' = rp,$$

which says that the growth rate is proportional to the population. It is easily verified (check this!) that a one-parameter family of solutions is given by

$$p(t) = Ce^{rt},$$

where C is any constant. If there is an initial condition imposed, that is, $p(0) = p_0$, then $C = p_0$ and we have picked out a particular solution $p(t) = p_0e^{rt}$ of the DE, that is, the one that satisfies the initial condition. Therefore, the Malthus model predicts exponential population growth (figure 1.9).

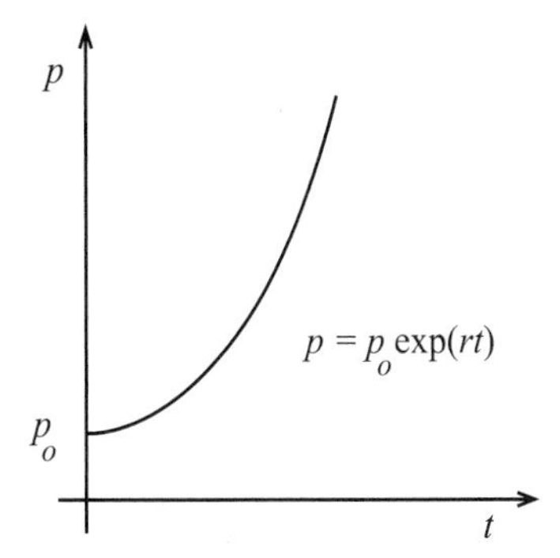

Figure 1.9 The Malthus model for population growth: $p(t) = p_0 e^{rt}$.

The reader should note a difference between the phrases "*per capita* growth rate" and "growth rate." To say that the *per capita* growth rate is 2% (per time) is to say that $p'/p = 0.02$, which gives exponential growth; to say that the growth rate is 2% (animals per time) is to say $p' = 0.02$, which forces $p(t)$ to be of the form $p(t) = 0.02t + K$, (K constant), which is linear growth.

In animal populations, for fairly obvious reasons, we do not expect exponential growth over long times. Environmental factors and competition for resources limit the population when it gets large. Therefore we might expect the *per capita* growth rate r (which is constant in the Malthus model) to decrease as the population increases. The simplest assumption is a linearly decreasing *per capita* growth rate where the rate becomes zero at some maximum carrying capacity K. See figure 1.10. This gives the **logistics model** of population growth (developed by P. Verhulst in the 1800s) by

$$\frac{p'}{p} = r(1 - \frac{p}{K}) \quad \text{or} \quad p' = rp(1 - \frac{p}{K}). \tag{1.11}$$

Clearly we may write this autonomous equation in the form

$$p' = rp - \frac{r}{K}p^2.$$

The first term is a positive **growth term**, which is just the Malthus term. The second term, which is quadratic in p, decreases the population growth rate and is the **competition term**. Note that if there were p animals, then there would be about p^2 encounters among them. So the competition term is proportional to the number of possible encounters, which is a reasonable model. Exercise 11 presents an alternate derivation of the logistics model based on food supply.

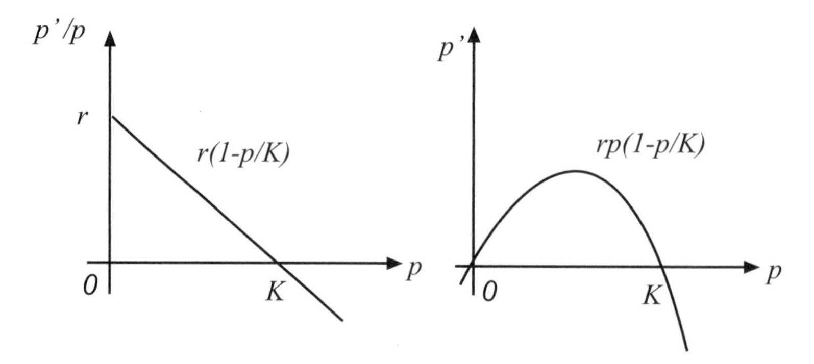

Figure 1.10 Plots of the logistics model of population growth. The left plot shows the *per capita* growth rate vs. population, and the right plot shows the growth rate vs. population. Both plots give important interpretations of the model.

For any initial condition $p(0) = p_0$ we can find the formula for the solution to the logistics equation (1.11). (You will solve the logistics equation in Exercise 8.) But, there are qualitative properties of solutions that can be exposed without actually finding the solution. Often, all we may want are qualitative features of a model. First, we note that there are two constant solutions to (1.11), $p(t) = 0$ and $p(t) = K$, corresponding to no animals (extinction) and to the number of animals represented by the carrying capacity, respectively. These constant solutions are found by setting the right side of the equation equal to zero (because that forces $p' = 0$, or $p =$ constant). The constant solutions are called steady-state, or **equilibrium**, solutions. If the population is between $p = 0$ and $p = K$ the right side of (1.11) is positive, giving $p' > 0$; for these population numbers the population is increasing. If the population is larger than the carrying capacity K, then the right side of (1.11) is negative and the population is decreasing. These facts can also be observed from the growth rate plot in figure 1.10. These observations can be represented conveniently on a **phase line** plot as shown in figure 1.11. We first plot the growth rate p' vs. p, which in this case is a parabola opening downward. The points of intersection on the p axis are the equilibrium solutions 0 and K. We then indicate by a directional arrow on the p axis those values of p where the solution $p(t)$ is increasing (where $p' > 0$) or decreasing ($p' < 0$). Thus the arrow points to the right when the graph of the growth rate is above the axis, and it points to the left when the graph is below the axis. In this context we call the p axis a phase line. We can regard the phase line as a one-dimensional, parametric solution space with the population $p = p(t)$ tracing out points on that line as t increases. In the range $0 < p < K$ the arrow points right because $p' > 0$. So

$p(t)$ increases in this range. For $p > K$ the arrow points left because $p' < 0$. The population $p(t)$ decreases in this range. These qualitative features can be easily transferred to time series plots (figure 1.12) showing $p(t)$ vs. t for different initial conditions.

Both the phase line and the time series plots imply that, regardless of the initial population (if nonzero), the population approaches the carrying capacity K. This equilibrium population $p = K$ is called an **attractor**. The zero population is also an equilibrium population. But, near zero we have $p' > 0$, and so the population diverges away from zero. We say the equilibrium population $p = 0$ is a **repeller**. (We are considering only positive populations, so we ignore the fact that $p = 0$ could be approached on the left side). In summary, our analysis has determined the complete qualitative behavior of the logistics population model.

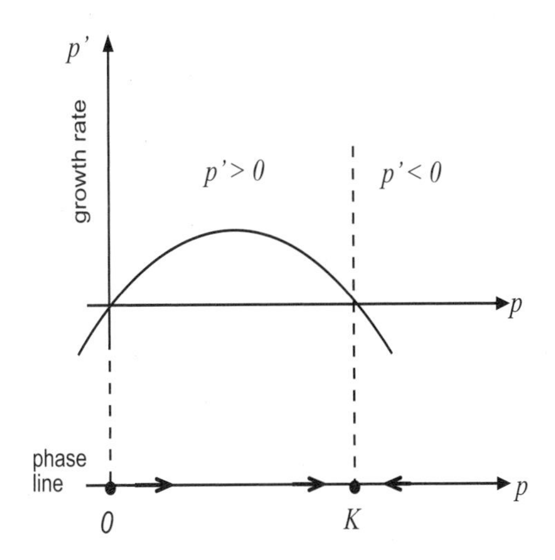

Figure 1.11 The p axis is the phase line, on which arrows indicate an increasing or decreasing population for certain ranges of p.

This qualitative method used to analyze the logistics model is applicable to any autonomous equation

$$u' = f(u). \tag{1.12}$$

The **equilibrium solutions** are the constant solutions, which are roots of the algebraic equation $f(u) = 0$. Thus, if u^* is an equilibrium, then $f(u^*) = 0$.

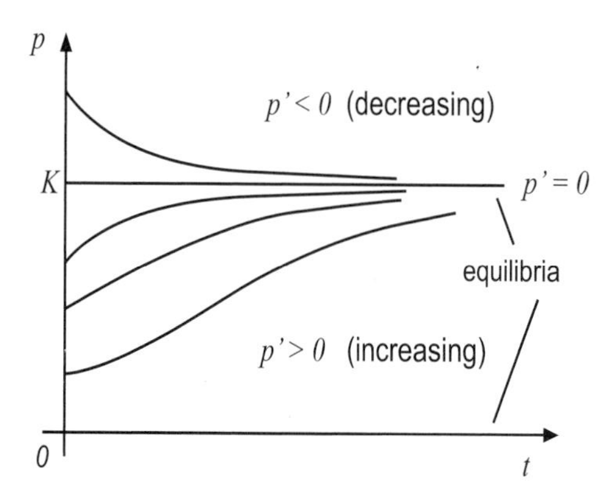

Figure 1.12 Time series plots of solutions to the logistics equation for various initial conditions. For $0 < p < K$ the population increases and approaches K, whereas for $p > K$ the population decreases to K. If $p(0) = K$, then $p(t) = K$ for all times $t > 0$; this is the equilibrium solution.

These are the values where the graph of $f(u)$ vs. u intersects the u-axis. We always assume the equilibria are **isolated**; that is, if u^* is an equilibrium, then there is an open interval containing u^* that contains no other equilibria. Figure 1.13 shows a generic plot where the equilibria are $u^* = a,\ b,\ c$. In between the equilibria we can observe the values of u for which the population is increasing ($f(u) > 0$) or decreasing ($f(u) < 0$). We can then place arrows on the phase line, or the u-axis, in between the equilibria showing direction of the movement (increasing or decreasing) as time increases. If desired, the information from the phase line can be translated into time series plots of $u(t)$ vs. t (figure 1.14). In between the constant, equilibrium solutions, the other solution curves increase or decrease; oscillations are not possible. Moreover, assuming f is a well-behaved function ($f'(u)$ is continuous), solution curves actually approach the equilibria, getting closer and closer as time increases. By uniqueness, the curves never intersect the constant equilibrium solutions.

On the phase line, if arrows on both sides of an equilibrium point toward that equilibrium point, then we say the equilibrium point is an **attractor**. If both of the arrows point away, the equilibrium is called a **repeller**. Attractors are called **asymptotically stable** because if the system is in that constant equilibrium state and then it is given a small **perturbation** (i.e., a change or "bump") to a nearby state, then it just returns to that state as $t \to +\infty$. It is clear that real systems will seek out the stable states. Repellers are **unstable** because a small perturbation can cause the system to go to a different equilib-

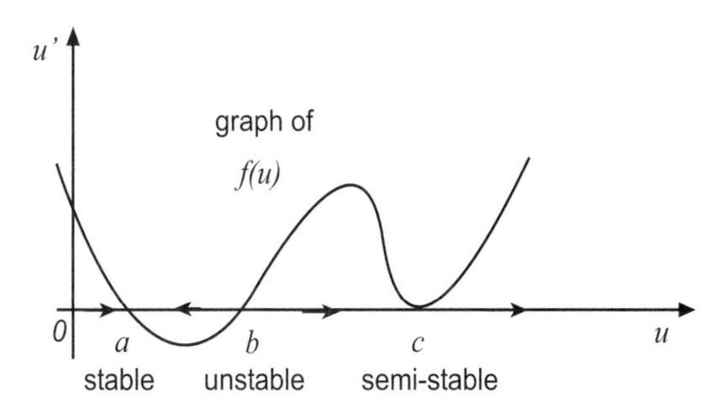

Figure 1.13 A generic plot showing $f(u)$, which is u' vs. u. The points of intersection, a, b, c, on the u-axis are the equilibria. The arrows on the u-axis, or phase line, show how the state u changes with time between the equilibria. The direction of the arrows is read from the plot of $f(u)$. They are to the right when $f(u) > 0$ and to the left when $f(u) < 0$. The phase line can either be drawn as a separate line with arrows, as in figure 1.11, or the arrows can be drawn directly on the u-axis of the plot, as is done here.

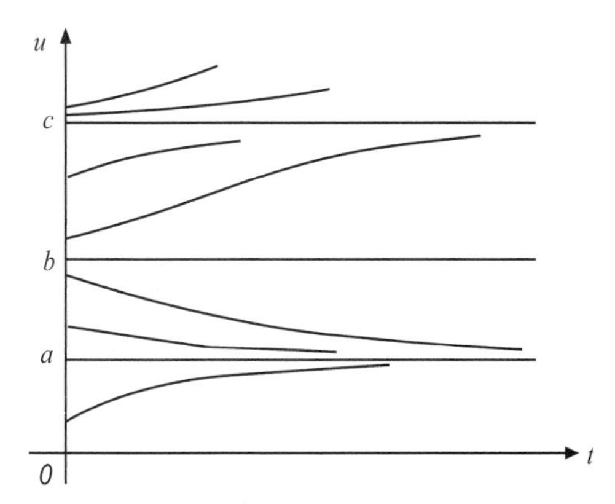

Figure 1.14 Time series plots of (1.12) for different initial conditions. The constant solutions are the equilibria.

rium or even go off to infinity. In the logistics model for population growth we observe (figure 1.11) that the equilibrium $u = K$ is an asymptotically stable attractor, and the zero population $u = 0$ is unstable; all solutions approach the carrying capacity $u = K$ at $t \to +\infty$. Finally, if one of the arrows points toward the equilibrium and one points away, we say the equilibrium is **semi-stable**. Semi-stable equilibria are not stable.

We emphasize that when we say an equilibrium u^* is asymptotically stable, our understanding is that this is with respect to *small* perturbations. To fix the idea, consider a population of fish in a lake that is in an asymptotically stable state u^*. A small death event, say caused by some toxic chemical that is dumped into the lake, will cause the population to drop. Asymptotic stability means that the system will return the original state u^* over time. We call this **local asymptotic stability**. If many fish are killed by the pollution, then the perturbation is not small and there is no guarantee that the fish population will return to the original state u^*. For example, a catastrophe or bonanza could cause the population to jump beyond some other equilibrium. If the population returns to the state u^* for all perturbations, no matter how large, then the state u^* is called **globally asymptotically stable**. A more precise definition of local asymptotic stability can be given as follows. An isolated equilibrium state u^* of (1.12) is locally asymptotically stable if there is an open interval I containing u^* with $\lim_{t\to+\infty} u(t) = u^*$ for any solution $u = u(t)$ of (1.12) with $u(0)$ in I. That is, each solution starting in I converges to u^*.

Note that a semi-stable point is not asymptotically stable; such points are, in fact, not stable.

Example 1.16

(Dimensionless Models) When we formulate a mathematical model we sometimes trade in the dimensioned quantities in our equations for dimensionless ones. In doing so we obtain a *dimensionless* model, often containing fewer parameters than the original model. The idea is simple. If, for example, time t is the independent variable in a model of population growth and a constant r, with dimension time^{-1}, representing the *per capita* growth rate appears in the model, then the variable $\tau = t/r^{-1} = rt$ has no dimensions, that is, it is dimensionless (time divided by time). It can serve as a new independent variable in the model representing "dimensionless time", or time measured relative to the inverse growth rate. We say r^{-1} is a **time scale** in the problem. Every *variable* in a model has a natural **scale** with which we can measure its relative value; these scales are found from the parameters in the problem. The population p of an animal species in a geographical region can be scaled by the carrying capacity K of the region, which is the number of animals the region can support. Then the variable $P = p/K$ is dimensionless (animals divided by animals) and represents the fraction of the region's capacity that is filled. If the carrying capacity is large, the actual population p could be large, requiring us to work with and plot big numbers. However, the dimensionless population P is represented by smaller numbers which are easier to deal with and plot. For some models selecting dimensionless dependent and independent variables can pay off in great benefits—it can help us understand the magnitude of various terms in the equations, and it can reduce the number of parameters in a problem, thus giving simplification. We illustrate this procedure for the initial value problem for the logistics model,

$$p' = rp(1 - \frac{p}{K}), \quad p(0) = p_0. \tag{1.13}$$

There are two variables in the problem, the independent variable t, measured in *time*, and the dependent variable p, measured in *animals*. There are three parameters in the problem: the carrying capacity K and initial population p_0, both measured in animals, and the growth rate r measured in 1/time. Let us define new dimensionless variables $\tau = rt = t/r^{-1}$ and $P = p/K$. These represent a "dimensionless time" and a "dimensionless population"; P is measured relative to the carrying capacity and t is measured relative to the growth rate; the values K and r^{-1} are called scales. Now we transform the DE into the new dimensionless variables. First, we transform the derivative:

$$\frac{dp}{dt} = \frac{d(KP)}{d(\tau/r)} = rK\frac{dP}{d\tau}.$$

Then the logistics DE in (1.13) becomes

$$rK\frac{dP}{d\tau} = r(KP)(1 - \frac{KP}{K}),$$

or

$$\frac{dP}{d\tau} = P(1 - P).$$

In dimensionless variables τ and P, the parameters in the DE disappeared! Next, the initial condition becomes $KP(0) = p_0$, or

$$P(0) = \alpha,$$

where $\alpha = p_0/K$ is a dimensionless parameter (animals divided by animals). In summary, the dimensioned model (1.13), with three parameters, can be replaced by the dimensionless model with only a single dimensionless parameter α:

$$\frac{dP}{d\tau} = P(1 - P), \quad P(0) = \alpha. \tag{1.14}$$

What this tells us is that although three parameters appear in the original problem, only a single combination of those parameters is relevant. We may as well work with the simpler, equivalent, dimensionless model (1.14) where populations are measured relative to the carrying capacity and time is measured relative to how fast the population is growing. For example, if the carrying capacity is $K = 300,000$, and the dimensioned p varies between $0 < p < 300,000$, it is much simpler to have dimensionless populations P with $0 < P < 1$. Furthermore, in the simplified form (1.14) it is easy to see that the equilibria are $P = 0$ and $P = 1$, the latter corresponding to the carrying capacity $p = K$.

We have pointed out that an autonomous model can be easily analyzed qualitatively without ever finding the solution. In this paragraph we introduce a simple method for solving a general autonomous equation

$$u' = f(u). \tag{1.15}$$

The method is called **separation of variables**. If we divide both sides of the equation by $f(u)$, we get

$$\frac{1}{f(u)}u' = 1.$$

Now, remembering that u is a function of t, we integrate both sides with respect to t to obtain

$$\int \frac{1}{f(u)}u'dt = \int 1dt + C = t + C,$$

where C is an arbitrary constant. A substitution $u = u(t)$, $du = u'(t)dt$ reduces the integral on the left and we obtain

$$\int \frac{1}{f(u)}du = t + C. \tag{1.16}$$

This equation, once the integral is calculated, defines the general solution $u = u(t)$ of (1.15) implicitly. We may or may not be able to actually calculate the integral and solve for u in terms of t to determine an explicit solution $u = u(t)$. This method of separating the variables (putting all the terms with u on the left side) is a basic technique in differential equations; it is adapted to more general equations in Chapter 2.

Example 1.17

Consider the **growth–decay** model

$$u' = ru, \tag{1.17}$$

where r is a given constant. If $r < 0$ then the equation models **exponential decay**; if $r > 0$ then the equation models **exponential growth** (e.g., population growth, as in the Malthus model). We apply the separation of variables method. Dividing by u (we could divide by ru, but we choose to leave the constant the right side) and taking antiderivatives gives

$$\int \frac{1}{u} u' dt = \int r dt + C.$$

Because $u'dt = du$, we can write

$$\int \frac{1}{u} du = rt + C.$$

Integrating gives

$$\ln |u| = rt + C \quad \text{or} \quad |u| = e^{rt+C} = e^C e^{rt}.$$

This means either $u = e^C e^{rt}$ or $u = -e^C e^{rt}$. Therefore the general solution of the growth–decay equation can be written compactly as

$$u(t) = C_1 e^{rt},$$

where C_1 has been written for $\pm e^C$, and is an arbitrary constant. If an initial condition

$$u(0) = u_0 \tag{1.18}$$

is prescribed on (1.17), it is straightforward to show that $C_1 = u_0$ and the solution to the initial value problem (1.17)–(1.18) is

$$u(t) = u_0 e^{rt}.$$

The growth–decay equation and its solution given in Example 1.16 occur often enough in applications that they are worthy of memorization. The equation models processes like growth of a population, mortality (death), growth of principal in a money account where the interest is compounded continuously at rate r, and radioactive decay, like the decay of Carbon-14 used in carbon dating.

EXERCISES

1. (The Allee effect) At low population densities it may be difficult for an animal to reproduce because of a limited number of suitable mates. A population model that predicts this behavior is the Allee model (W. C. Allee, 1885–1955)

$$p' = rp\left(\frac{p}{a} - 1\right)\left(1 - \frac{p}{K}\right), \quad 0 < a < K.$$

Find the *per capita* growth rate and plot the *per capita* rate vs. p. Graph p' vs. p, determine the equilibrium populations, and draw the phase line. Which equilibria are attractors and which are repellers? Which are asymptotically stable? From the phase line plot, describe the long time behavior of the system for different initial populations, and sketch generic time series plots for different initial conditions.

2. Modify the logistics model to include harvesting. That is, assume that the animal population grows logistically while, at the same time, animals are being removed (by hunting, fishing, or whatever) at a constant rate of h animals per unit time. What is the governing DE? Determine the equilibria. Which are asymptotically stable? Explain how the system will behave for different initial conditions. Does the population ever become extinct?

3. The **Ricker population law** is

$$p' = rpe^{-ap},$$

where r and a are constants. Determine the dimensions of r and a. At what population is the growth rate maximum? Make a generic sketch of the *per capita* growth rate and write a brief explanation of how a population behaves under this law. Is it possible to use the separation of variables method to find a simple formula for $p(t)$?

4. In this exercise we introduce a simple model of growth of an individual organism over time. For simplicity, we assume it is shaped like a cube having sides equal to $L = L(t)$. Organisms grow because they assimilate nutrients and then use those nutrients in their energy budget for maintenance and to build structure. It is conjectured that the organism's growth rate in volume equals the assimilation rate minus the rate food is used. Food is assimilated at a rate proportional to its surface area because food must ultimately pass across the cell walls; food is used at a rate proportional to its volume because ultimately cells are three-dimensional. Show that the differential equation governing its size $L(t)$ can be written

$$L'(t) = a - bL,$$

where a and b are positive parameters. What is the maximum length the organism can reach? Using separation of variables, show that if the length of the organism at time $t = 0$ is $L(0) = 0$ (it is very small), then the length is given by $L(t) = \frac{a}{b}(1 - e^{-bt})$. Does this function seem like a reasonable model for growth?

5. In a classical ecological study of budworm outbreaks in Canadian fir forests, researchers proposed that the budworm population N was governed by the

law

$$N' = rN\left(1 - \frac{N}{K}\right) - P(N),$$

where the first term on the right represents logistics growth, and where $P(N)$ is a *bird-predation* rate given by

$$P(N) = \frac{aN^2}{N^2 + b^2}.$$

Sketch a graph of the bird-predation rate vs. N and discuss its meaning. What are the dimensions of all the constants and variables in the model? Select new dimensionless independent and dependent variables by

$$\tau = \frac{t}{b/a}, \quad n = \frac{N}{b}$$

and reformulate the model in dimensionless variables and dimensionless constants. Working with the dimensionless model, show that there is at least one and at most three positive equilibrium populations. What can be said about their stability?

6. Use the method of separation of variables to find the general solution to the following autonomous differential equations.

   a) $u' = \sqrt{u}$.

   b) $u' = e^{-2u}$.

   c) $u' = 1 + u^2$.

   d) $u' = 3u - a$, where a is a constant.

   e) $u' = \frac{u}{4+u^2}$.

   f) $u' = e^{u^2}$.

7. In Exercises 6 (a)–(f) find the solution to the resulting IVP when $u(0) = 1$.

8. Find the general solution to the logistics equation $u' = ru(1 - u/K)$ using the separation of variables method. Hint: use the partial fractions decomposition

$$\frac{1}{u(K-u)} = \frac{1/K}{u} + \frac{1/K}{K-u}.$$

Show that a solution curve that crosses the line $u = K/2$ has an inflection point at that position.

9. (Carbon dating) The half-life of Carbon-14 is 5730 years. That is, it takes this many years for half of a sample of Carbon-14 to decay. If the decay of Carbon-14 is modeled by the DE $u' = -ku$, where u is the amount of Carbon-14, find the decay constant k. (Answer: 0.000121 yr^{-1}). In an artifact the percentage of the original Carbon-14 remaining at the present day was measured to be 20 percent. How old is the artifact?

10. In 1950, charcoal from the Lascaux Cave in France gave an average count of 0.09 disintegrations of C^{14} (per minute per gram). Living wood gives 6.68 disintegrations. Estimate the date that individuals lived in the cave.

11. In the usual Malthus growth law $N' = rN$ for a population of size N, assume the growth rate is a linear function of food availability F; that is, $r = bF$, where b is the conversion factor of food into newborns. Assume that F_T is the total, constant food in the system with $F_T = F + cN$, where cN is amount of food already consumed. Write down a differential equation for the population N. What is the carrying capacity? What is the population as t gets large?

12. One model of tumor growth is the Gompertz equation

$$R' = -aR\ln\left(\frac{R}{k}\right),$$

where $R = R(t)$ is the tumor radius, and a and k are positive constants. Find the equilibria and analyze their stability. Can you solve this differential equation for $R(t)$?

13. A population model is given by $p' = rP(P-m)$, where r and m are positive constants. State reasons for calling this the **explosion–extinction** model.

14. In a fixed population of N individuals let I be the number of individuals infected by a certain disease and let S be the number susceptible to the disease with $I+S = N$. Assume that the rate that individuals are becoming infected is proportional to the number of infectives times the number of susceptibles, or $I' = aSI$, where the positive constant a is the transmission coefficient. Assume no individual gets over the disease once it is contracted. If $I(0) = I_0$ is a small number of individuals infected at $t = 0$, find an initial value problem for the number infected at time t. Explain how the disease evolves. Over a long time, how many contract the disease?

15. In Example 1.11 we modeled the velocity of an object falling in a fluid by the equation $mv' = mg - av^2$. If $v(0) = 0$, find an analytic formula for $v(t)$.

1.3.3 Stability and Bifurcation

Differential equations coming from modeling physical phenomena almost always contain one or more parameters. It is of great interest to determine how equilibrium solutions depend upon those parameters. For example, the logistics growth equation

$$p' = rp(1 - \frac{p}{K})$$

has two parameters: the growth rate r and the carrying capacity K. Let us add **harvesting**; that is, we remove animals at a constant rate $H > 0$. We can think of a fish population where fish are caught at a given rate H. Then we have the model

$$p' = rp(1 - \frac{p}{K}) - H. \tag{1.19}$$

We now ask how possible equilibrium solutions and their stability depend upon the rate of harvesting H. Because there are three parameters in the problem, we can nondimensionalize to simplify it. We introduce new dimensionless variables by

$$u = \frac{p}{K}, \quad \tau = rt.$$

That is, we measure populations relative to the carrying capacity and time relative to the inverse growth rate. In terms of these dimensionless variables, (1.19) simplifies to (check this!)

$$u' = u(1 - u) - h,$$

where $h = H/rK$ is a single dimensionless parameter representing the ratio of the harvesting rate to the product of the growth rate and carrying capacity. We can now study the effects of changing h to see how harvesting influences the steady-state fish populations in the model. In dimensionless form, we think of h as the harvesting parameter; information about changing h will give us information about changing H.

The equilibrium solutions of the dimensionless model are roots of the quadratic equation

$$f(u) = u(1 - u) - h = 0,$$

which are

$$u^* = \frac{1}{2} \pm \frac{1}{2}\sqrt{1 - 4h}.$$

The growth rate $f(u)$ is plotted in figure 1.15 for different values of h. For $h < 1/4$ there are two positive equilibrium populations. The graph of $f(u)$ in this case is concave down and the phase line shows that the smaller one is unstable, and the larger one is asymptotically stable. As h increases these populations begin to come together, and at $h = 1/4$ there is only a single unstable equilibrium. For $h > 1/4$ the equilibrium populations cease to exist. So, when harvesting is small, there are two equilibria, one being stable; as harvesting increases the equilibrium disappears. We say that a **bifurcation** (bifurcation means "dividing") occurs at the value $h = 1/4$. This is the value where there is a significant change in the character of the equilibria. For $h \geq 1/4$ the population will become extinct, regardless of the initial condition (because $f(u) < 0$ for all u). All these facts can be conveniently represented on a **bifurcation diagram**. See figure 1.16. In a bifurcation diagram we plot the equilibrium solutions u^* vs. the parameter h. In this context, h is called the **bifurcation parameter**. The plot is a parabola opening to the left. We observe that the upper branch of the parabola corresponds to the larger equilibrium, and all solutions represented by that branch are asymptotically stable; the lower branch, corresponding to the smaller solution, is unstable.

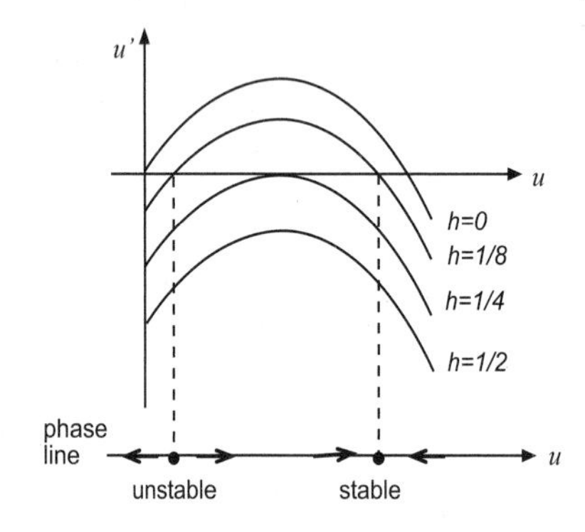

Figure 1.15 Plots of $f(u) = u(1-u) - h$ for different values of h. The phase line is plotted in the case $h = 1/8$.

Finally, we give an analytic criterion that allows us to determine stability of an equilibrium solution by simple calculus. Let

$$u' = f(u) \tag{1.20}$$

be a given autonomous systems and u^* an isolated equilibrium solution, so that $f(u^*) = 0$. We observe from figure 1.13 that when the slope of the graph of $f(u)$ at the equilibrium point is negative, the graph falls from left to right and both arrows on the phase line point toward the equilibrium point. Therefore, a condition that guarantees the equilibrium point u^* is asymptotically stable is $f'(u^*) < 0$. Similarly, if the graph of $f(u)$ strictly increases as it passes through the equilibrium, then $f'(u^*) > 0$ and the equilibrium is unstable. If the slope of $f(u)$ is zero at the equilibrium, then any pattern of arrows is possible and there is no information about stability. If $f'(u^*) = 0$, then u^* is a critical point of

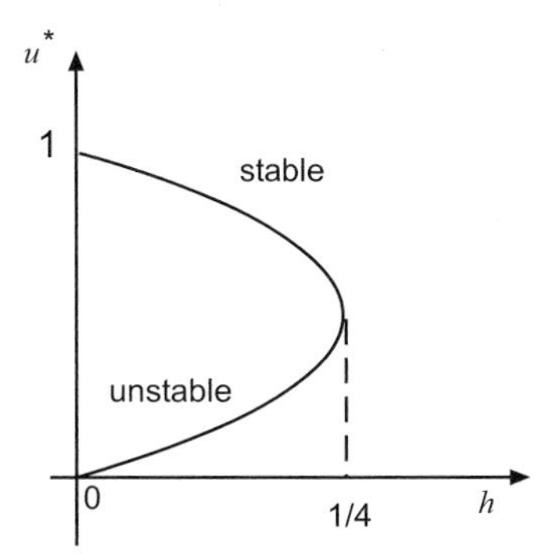

Figure 1.16 Bifurcation diagram: plot of the equilibrium solution as a function of the bifurcation parameter h, $u^* = \frac{1}{2} \pm \frac{1}{2}\sqrt{1-4h}$. For $h > \frac{1}{4}$ there are no equilibria and for $h < \frac{1}{4}$ there are two, with the larger one being stable. A bifurcation occurs at $h = \frac{1}{4}$.

f and could be a local maximum, local minimum, or have an inflection point. If there is a local maximum or local minimum, then u^* is semi-stable (which is not stable). If there is an inflection point, then f changes sign at u^* and we obtain either a repeller or an attractor, depending on how the concavity changes, negative to positive, or positive to negative.

Theorem 1.18

Let u^* be an isolated equilibrium for the autonomous system (1.20). If $f'(u^*) < 0$, then u^* is asymptotically stable; if $f'(u^*) > 0$, then u^* is unstable. If $f'(u^*) = 0$, then there is no information about stability.

Example 1.19

Consider the logistics equation $u' = f(u) = ru(1 - u/K)$. The equilibria are $u^* = 0$ and $u^* = K$. The derivative of $f(u)$ is $f'(u) = r - 2ru/K$. Evaluating the derivative at the equilibria gives

$$f'(0) = r > 0, \quad f'(K) = -r < 0.$$

Therefore $u^* = 0$ is unstable and $u^* = K$ is asymptotically stable.

EXERCISES

1. A fish population in a lake is harvested at a constant rate, and it grows logistically. The growth rate is 0.2 per month, the carrying capacity is 40

(thousand), and the harvesting rate is 1.5 (thousand per month). Write down the model equation, find the equilibria, and classify them as stable or unstable. Will the fish population ever become extinct? What is the most likely long-term fish population?

2. For the following equations, find the equilibria and sketch the phase line. Determine the type of stability of all the equilibria. Use Theorem 1.18 to confirm stability or instability.

   a) $u' = u^2(3 - u)$.

   b) $u' = 2u(1 - u) - \frac{1}{2}u$.

   c) $u' = (4 - u)(2 - u)^3$.

3. For the following models, which contain a parameter h, find the equilibria in terms of h and determine their stability. Construct a bifurcation diagram showing how the equilibria depend upon h, and label the branches of the curves in the diagram as unstable or stable.

   a) $u' = hu - u^2$.

   b) $u' = (1 - u)(u^2 - h)$.

4. Consider the model $u' = (\lambda - b)u - au^3$, where a and b are fixed positive constants and λ is a parameter that may vary.

   a) If $\lambda < b$ show that there is a single equilibrium and that it is asymptotically stable.

   b) If $\lambda > b$ find all the equilibria and determine their stability.

   c) Sketch the bifurcation diagram showing how the equilibria vary with λ. Label each branch of the curves shown in the bifurcation diagram as stable or unstable.

5. The biomass P of a plant grows logistically with intrinsic growth rate r and carrying capacity K. At the same time it is consumed by herbivores at a rate

$$\frac{aP}{b + P},$$

per herbivore, where a and b are positive constants. The model is

$$P' = rP(1 - \frac{P}{K}) - \frac{aPH}{b + P},$$

where H is the density of herbivores. Assume $aH > br$, and assume r, K, a, and b are fixed. Plot, as a function of P, the growth rate and the consumption rate for several values of H on the same set of axes, and identify the values of P that give equilibria. What happens to the equilibria as the herbivory H is steadily increased from a small value to a large value? Draw a bifurcation diagram showing this effect. That is, plot equilibrium solutions vs. the parameter H. If herbivory is slowly increased so that the plants become extinct, and then it is decreased slowly back to a low level, do the plants return?

6. A deer population grows logistically and is harvested at a rate proportional to its population size. The dynamics of population growth is modeled by

$$P' = rP(1 - \frac{P}{K}) - \lambda P,$$

where λ is the *per capita* harvesting rate. Use a bifurcation diagram to explain the effects on the equilibrium deer population when λ is slowly increased from a small value to a large value.

7. Draw a bifurcation diagram for the model $u' = u^3 - u + h$, where h is the bifurcation parameter. Label branches of the curves as stable or unstable.

8. Consider the model $u' = u(u - e^{\lambda u})$, where λ is a parameter. Draw the bifurcation diagram, plotting the equilibrium solution(s) u^* vs. λ. Label each curve on the diagram as stable or unstable.

1.3.4 Heat Transfer

An object of uniform temperature T_0 (e.g., a potato) is placed in an oven of temperature T_e. It is observed that over time the potato heats up and eventually its temperature becomes that of the oven environment, T_e. We want a model that governs the temperature $T(t)$ of the potato at any time t. **Newton's law of cooling** (heating), a constitutive model inferred from experiment, dictates that the rate of change of the temperature of the object is proportional to the difference between the temperature of the object and the environmental temperature. That is,

$$T' = -h(T - T_e). \tag{1.21}$$

The positive proportionality constant h is the **heat loss coefficient**. There is a fundamental assumption here that the heat is instantly and uniformly distributed throughout the body and there are no temperature gradients, or spatial variations, in the body itself. From the DE we observe that $T = T_e$ is an equilibrium solution. If $T > T_e$ then $T' < 0$, and the temperature decreases; if $T < T_e$ then $T' > 0$, and the temperature increases. Plotting the phase line easily shows that this equilibrium is stable (Exercise!).

We can find a formula for the temperature $T(t)$ satisfying (1.21) using the separation of variables method introduced in the last section. Here, for variety, we illustrate another simple method that uses a **change of variables.** Let $u = T - T_e$. Then $u' = T'$ and (1.21) may be written $u' = -hu$. This is the decay equation and we have memorized its general solution $u = Ce^{-ht}$. Therefore $T - T_e = Ce^{-ht}$, or

$$T(t) = T_e + Ce^{-ht}.$$

This is the general solution of (1.21). If we impose an initial condition $T(0) = T_0$, then one finds $C = T_0 - T_e$, giving

$$T(t) = T_e + (T_0 - T_e)e^{-ht}.$$

We can now see clearly that $T(t) \to T_e$ as $t \to \infty$. A plot of the solution showing how an object heats up is given in figure 1.17.

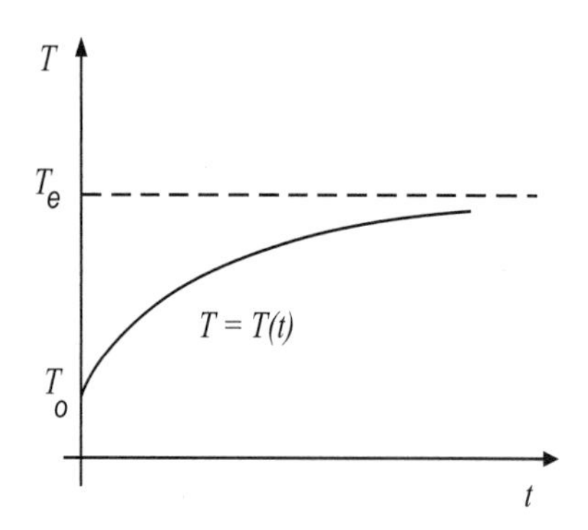

Figure 1.17 Temperature history in Newton's law of cooling.

If the environmental, or ambient, temperature fluctuates, then T_e is not constant but rather a function of time $T_e(t)$. The governing equation becomes

$$T' = -h(T - T_e(t)).$$

In this case there are no constant, or equilibrium, solutions. Writing this model in a different way,

$$T' = -hT + hT_e(t).$$

The first term on the right is internal to the system (the body being heated) and, considered alone with zero ambient temperature, leads to an exponentially decaying temperature (recall that $T' = -hT$ has solution $T = Ce^{-ht}$). Therefore, there is a transient governed by the natural system that decays away. The external, environmental temperature $T_e(t)$ gives rise to time-dependent dynamics and eventually takes over to drive the system; we say the system is "driven", or forced, by the environmental temperature. In Chapter 2 we develop methods to solve this equation with time dependence in the environmental temperature function.

EXERCISES

1. A small solid initially of temperature 22°C is placed in an ice bath of 0°C. It is found experimentally, by measuring temperatures at different times, that the natural logarithm of the temperature $T(t)$ of the solid plots as a linear function of time t; that is,

$$\ln T = -at + b.$$

Show that this equation is consistent with Newton's law of cooling. If the temperature of the object is 8°C degrees after two hours, what is the heat loss coefficient? When will the solid be 2°C?

2. A small turkey at room temperature 70°F is places into an oven at 350°F. If $h = 0.42$ per hour is the heat loss coefficient for turkey meat, how long should you cook the turkey so that it is uniformly 200°F? Comment on the validity of the assumptions being made in this model?

3. The body of a murder victim was discovered at 11:00 A.M. The medical examiner arrived at 11:30 A.M. and found the temperature of the body was 94.6°F. The temperature of the room was 70°F. One hour later, in the same room, he took the body temperature again and found that it was 93.4°F. Estimate the time of death.

4. Suppose the temperature inside your winter home is 68°F at 1:00 P.M. and your furnace then fails. If the outside temperature is 10°F and you notice that by 10:00 P.M. the inside temperature is 57°F, what will be the temperature in your home the next morning at 6:00 A.M.?

5. The temperature $T(t)$ of an exothermic, chemically reacting sample placed in a furnace is governed by the initial value problem

$$T' = -k(T - T_e) + qe^{-\theta/T}, \quad T(0) = T_0,$$

where the term $qe^{-\theta/T}$ is the rate heat is generated by the reaction. What are the dimensions of all the constants (k, T_e, q, T_0, and θ) in the problem? Scale time by k^{-1} and temperature by T_e to obtain the dimensionless model

$$\frac{d\psi}{d\tau} = -(\psi - 1) + ae^{-b/\psi}, \quad \psi(0) = \gamma,$$

for appropriately chosen dimensionless parameters a, b, and c. Fix $a = 1$. How many positive equilibria are possible, depending upon the value of b? (Hint: Graph the heat loss term and the heat generation term vs. ψ on the same set of axes for different values of b).

1.3.5 Chemical Reactors

A **continuously stirred tank reactor** (also called a chemostat, or compartment) is a basic unit of many physical, chemical, and biological processes. A continuously stirred tank reactor is a well-defined geometric volume or entity where substances enter, react, and are then discharged. A chemostat could be an organ in our body, a polluted lake, an industrial chemical reactor, or even an ecosystem. See figure 1.18.

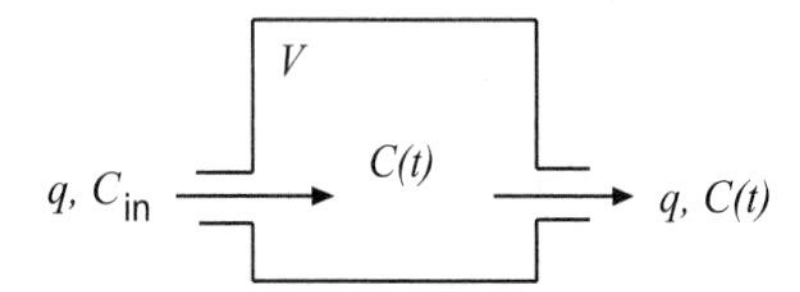

Figure 1.18 A chemostat, or continuously stirred tank reactor.

We illustrate a reactor model with a specific example. Consider an industrial pond with constant volume V cubic meters. Suppose that polluted water containing a toxic chemical of concentration C_{in} grams per cubic meter is dumped into the pond at a constant volumetric flow rate of q cubic meters per day. At the same time the continuously mixed solution in the pond is drained off at the same flow rate q. If the pond is initially at concentration C_0, what is the concentration $C(t)$ of the chemical in the pond at any time t?

The key idea in all chemical mixture problems is to obtain a model by conserving mass: the rate of change of mass in the pond must equal the rate mass flows in minus the rate mass flows out. The total mass in the pond at any time is VC, and the mass flow rate is the volumetric flow rate times the mass concentration; thus mass balance dictates

$$(VC)' = qC_{\text{in}} - qC.$$

Hence, the initial value problem for the chemical concentration is

$$VC' = qC_{\text{in}} - qC, \quad C(0) = C_0, \tag{1.22}$$

where C_0 is the initial concentration in the tank. This initial value problem can be solved by the separation of variables method or the change of variables method (Section 1.3.4). See Exercise 1.

Now suppose we add degradation of the chemical while it is in the pond, assuming that it degrades to inert products at a rate proportional to the amount present. We represent this decay rate as rC gm per cubic meter per day, where r is constant. Then the model equation becomes

$$VC' = qC_{\text{in}} - qC - rVC.$$

Notice that we include a factor V in the last term to make the model dimensionally correct. A similar model holds when the volumetric flow rates in and out are different, which gives a changing volume $V(t)$. Letting q_{in} and q_{out} denote those flow rates, respectively, we have

$$(V(t)C)' = q_{\text{in}}C_{\text{in}} - q_{\text{out}}C - rV(t)C,$$

where $V(t) = V_0 + (q_{\text{in}} - q_{\text{out}})t$, and where V_0 is the initial volume. Methods developed in Chapter 2 show how this equation is solved.

EXERCISES

1. Solve the initial value problem (1.22) and obtain a formula for the concentration in the reactor at time t.

2. An industrial pond having volume 100 m^3 is full of pure water. Contaminated water containing a toxic chemical of concentration 0.0002 kg per m^3 is then is pumped into the pond with a volumetric flow rate of 0.5 m^3 per minute. The contents are well-mixed and pumped out at the same flow rate. Write down an initial value problem for the contaminant concentration $C(t)$ in the pond at any time t. Determine the equilibrium concentration and its stability. Find a formula for the concentration $C(t)$.

3. In the preceding problem, change the flow rate out of the pond to 0.6 m^3 per minute. How long will it take the pond to empty? Write down a revised initial value problem.

4. A vat of volume 1000 gallons initially contains 5 lbs of salt. For $t > 0$ a salt brine of concentration 0.1 lbs per gallon is pumped into the tank at the rate of 2 gallons per minute; the perfectly stirred mixture is pumped out at the same flow rate. Derive a formula for the concentration of salt in the tank at any time t. Check your answer on a computer algebra system, and sketch a graph of the concentration vs. time.

5. Consider a chemostat of constant volume where a chemical **C** is pumped into the reactor at constant concentration and constant flow rate. While in the reactor it reacts according to $\mathbf{C} + \mathbf{C} \to$ products. From the law of mass action the rate of the reaction is $r = kC^2$, where k is the rate constant. If the concentration of **C** in the reactor is given by $C(t)$, then

mass balance leads the governing equation $(VC)' = qC_{\text{in}} - qC - kVC^2$. Find the equilibrium state(s) and analyze their stability. Redo this problem after nondimensionalizing the equation (pick time scale V/q and concentration scale C_{in}).

6. Work Exercise 5 if the rate of reaction is given by **Michaelis–Menten kinetics**
$$r = \frac{aC}{b+C},$$
where a and b are positive constants.

7. A **batch reactor** is a reactor of volume V where there are no in and out flow rates. Reactants are loaded instantaneously and then allowed to react over a time T, called the residence time. Then the contents are expelled instantaneously. Fermentation reactors and even sacular stomachs of some animals can be modeled as batch reactors. If a chemical is loaded in a batch reactor and it degrades with rate $r(C) = kC$, given in mass per unit time, per unit volume, what is the residence time required for 90 percent of the chemical to degrade?

8. The **Monod equation** for conversion of a chemical substrate of concentration C into its products is
$$\frac{dC}{dt} = -\frac{aC}{b+C},$$
where a and b are positive constants. This equation, with Michaelis–Menten kinetics, describes how the substrate is being used up through chemical reaction. If, in addition to reaction, the substrate is added to the solution at a constant rate R, write down a differential equation for C. Find the equilibrium solution and explain how the substrate concentration evolves for various initial conditions.

9. Consider the chemical reaction $\mathrm{A} + \mathrm{B} \xrightarrow{k} \mathrm{C}$, where one molecule of A reacts with one molecule of B to produce one molecule of C, and the rate of the reaction is k, the rate constant. By the law of mass action in chemistry, the reaction rate is $r = kab$, where a and b represent the time-dependent concentrations of the reactants A and B. Thus, the rates of change of the reactants and product are governed by the three equations
$$a' = -kab, \quad b' = -kab, \quad c' = kab.$$
If, initially, $a(0) = a_0$, $b(0) = b_0$, and $c(0) = 0$, with $a_0 > b_0$, find a single, first-order differential equation that involves only the concentration $a = a(t)$. What is the limiting concentration $\lim_{t\to\infty} a(t)$? What are the other two limiting concentrations?

10. Digestion in the stomach (gut) in some organisms can be modeled as a chemical reactor of volume V, where food enters and is broken down into nutrient products, which are then absorbed across the gut lining; the food–product mixture in the stomach is perfectly stirred and exits at the same rate as it entered. Let S_0 be the concentration of a substrate (food) consumed at rate q (volume per time). In the gut the rate of substrate breakdown into the nutrient product, $S \to P$, is given by kVS, where k is the rate constant and $S = S(t)$ is the substrate concentration. The nutrient product, of concentration $P = P(t)$, is then absorbed across the gut boundary at a rate aVP, where a is the absorption constant. At all times the contents are thoroughly stirred and leave the gut at the flow rate q.

    a) Show that the model equations are

    $$\begin{aligned} VS' &= qS_0 - qS - kVS, \\ VP' &= kVS - aVP - qP. \end{aligned}$$

    b) Suppose the organism eats continuously, in a steady-state mode, where the concentrations become constant. Find the steady-steady, or equilibrium, concentrations S_e and P_e.

    c) Some ecologists believe that animals regulate their consumption rate in order to maximize the absorption rate of nutrients. Show that the maximum nutrient concentration P_e occurs when the consumption rate is $q = \sqrt{ak}V$.

    d) Show that the maximum absorption rate is therefore $\frac{akS_0V}{(\sqrt{a}+\sqrt{k})^2}$.

1.3.6 Electric Circuits

Our modern, technological society is filled with electronic devices of all types. At the basis of these are electrical circuits. The simplest circuit unit is the loop in figure 1.19 that contains an electromotive force (emf) $E(t)$ (a battery or generator that supplies energy), a resistor, an inductor, and a capacitor, all connected in series. A capacitor stores electrical energy on its two plates, a resistor dissipates energy, usually in the form of heat, and an inductor acts as a "choke" that resists changes in current. A basic law in electricity, **Kirchhoff's law,** tells us that the sum of the voltage drops across the circuit elements (as measured, e.g., by a voltmeter) in a loop must equal the applied emf. In symbols,

$$V_L + V_R + V_C = E(t).$$

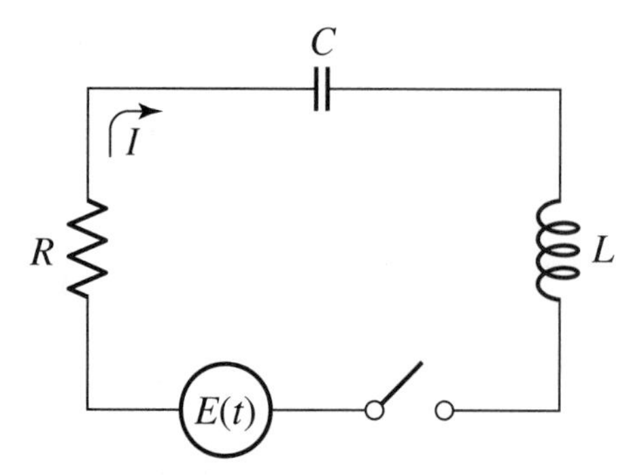

Figure 1.19 An RCL circuit with an electromotive force $E(t)$ supplying the electrical energy.

This law comes from conservation of energy in a current loop, and it is derived in elementary physics texts. A voltage drop across an element is an energy potential that equals the amount of work required to move a charge across that element.

Let $I = I(t)$ denote the current (in amperes, or charge per second) in the circuit, and let $q = q(t)$ denote the charge (in coulombs) on the capacitor. These quantities are related by

$$q' = I.$$

There are several choices of state variables to describe the response of the circuit: charge on the capacitor q, current I, or voltage V_C across the capacitor. Let us write Kirchhoff's law in terms of charge. By Ohm's law the voltage drop across the resistor is proportional to the current, or

$$V_R = RI,$$

where the proportionality constant R is called the resistance (measured in ohms). The voltage drop across a capacitor is proportional to the charge on the capacitor, or

$$V_C = \frac{1}{C}q,$$

where C is the capacitance (measured in farads). Finally, the voltage drop across an inductor is proportional to how fast the current is changing, or

$$V_L = LI',$$

where L is the inductance (measured in henrys). Substituting these voltage drops into Kirchhoff's law gives

$$LI' + RI + \frac{1}{C}q = E(t),$$

or, using $q' = I$,

$$Lq'' + Rq' + \frac{1}{C}q = E(t).$$

This is the **RCL circuit equation**, which is a second-order DE for the charge q. The initial conditions are

$$q(0) = q_0, \quad q'(0) = I(0) = I_0.$$

These express the initial charge on the capacitor and the initial current in the circuit. Here, $E(t)$ may be a given constant (e.g., $E(t) = 12$ for a 12-volt battery) or may be a oscillating function of time t (e.g., $E(t) = A\cos\omega t$ for an alternating voltage potential of amplitude A and frequency ω).

If there is no inductor, then the resulting RC circuit is modeled by the first-order equation

$$Rq' + \frac{1}{C}q = E(t).$$

If $E(t)$ is constant, this equation can be solved using separation of variables or the change of variables method (Exercise 2). We show how to solve second-order differential equations in Chapter 3.

EXERCISES

1. Write down the equation that governs an RC circuit with a 12-volt battery, taking $R = 1$ and $C = \frac{1}{2}$. Determine the equilibrium solution and its stability. If $q(0) = 5$, find a formula for $q(t)$. Find the current $I(t)$. Plot the charge and the current on the same set of axes.

2. In an arbitrary RC circuit with constant emf E, use the method of separation of variables to derive the formula

$$q(t) = Ke^{-t/RC} + EC$$

for the charge on the capacitor, where K is an arbitrary constant. If $q(0) = q_0$, what is K?

3. An RCL circuit with an applied emf given by $E(t)$ has initial charge $q(0) = q_0$ and initial current $I(0) = I_0$. What is $I'(0)$? Write down the circuit equation and the initial conditions in terms of current $I(t)$.

4. Formulate the governing equation of an RCL circuit in terms of the current $I(t)$ when the circuit has an emf given by $E(t) = A\cos\omega t$. What are the appropriate initial conditions?

5. Find the DE model for the charge in an LC circuit with no emf. Show that the response of the circuit may have the form $q(t) = A\cos\omega t$ for some amplitude A and frequency ω.

6. Consider a standard RCL circuit with no emf, but with a voltage drop across the resistor given by a nonlinear function of current,

$$V_R = \frac{1}{2}\left(\frac{1}{3}I^3 - I\right)$$

(This replaces Ohm's law.) If $C = L = 1$, find a differential equation for the current $I(t)$ in the circuit.

7. Write the RCL circuit equation with the voltage $V_c(t)$ as the unknown state function.

2
Analytic Solutions and Approximations

In the last chapter we studied several first-order DE models and a few elementary techniques to help understand the qualitative behavior of the models. In this chapter we introduce analytic solution techniques for first-order equations and some general methods of approximation, including numerical methods.

2.1 Separation of Variables

In Section 1.3.2 we presented a simple algorithm to obtain an analytic solution to an autonomous equation $u' = f(u)$ called **separation of variables**. Now we show that this method is applicable to a more general class of equations. A **separable equation** is a first-order differential where the right side can be factored into a product of a function of t and a function of u. That is, a separable equation has the form

$$u' = g(t)h(u). \tag{2.1}$$

To solve separable equations we take the expression involving u to the left side and then integrate with respect to t, remembering that $u = u(t)$. Therefore, dividing by $h(u)$ and taking the antiderivatives of both sides with respect to t gives

$$\int \frac{1}{h(u)} u' dt = \int g(t) dt + C,$$

where C is an arbitrary constant of integration. (Both antiderivatives generate an arbitrary constant, but we have combined them into a single constant C). Next we change variables in the integral on the left by letting $u = u(t)$, so that $du = u'(t)dt$. Hence,

$$\int \frac{1}{h(u)} du = \int g(t)dt + C.$$

This equation, once the integrations are performed, yields an equation of the form

$$H(u) = G(t) + C, \tag{2.2}$$

which defines the general solution u implicitly as a function of t. We call (2.2) the **implicit solution**. To obtain an **explicit solution** $u = u(t)$ we must solve (2.2) for u in terms of t; this may or may not be possible. As an aside, we recall that if the antiderivatives have no simple expressions, then we write the antiderivatives with limits on the integrals.

Example 2.1

Solve the initial value problem

$$u' = \frac{t+1}{2u}, \quad u(0) = 1.$$

We recognize the differential equation as separable because the right side is $\frac{1}{2u}(t+1)$. Bringing the $2u$ term to the left side and integrating gives

$$\int 2uu'dt = \int (t+1)dt + C,$$

or

$$\int 2udu = \frac{1}{2}t^2 + t + C.$$

Therefore

$$u^2 = \frac{1}{2}t^2 + t + C.$$

This equation is the general **implicit solution**. We can solve for u to obtain two forms for **explicit solutions**,

$$u = \pm\sqrt{\frac{1}{2}t^2 + t + C}.$$

Which sign do we take? The initial condition requires that u be positive. Thus, we take the plus sign and apply $u(0) = 1$ to get $C = 1$. The solution to the initial value problem is therefore

$$u = \sqrt{\frac{1}{2}t^2 + t + 1}.$$

This solution is valid as long as the expression under the radical is not negative. In the present case the solution is defined for all times $t \in \mathbf{R}$ and so the interval of existence is the entire real line.

Example 2.2

Solve the initial value problem

$$u' = \frac{2\sqrt{u}e^{-t}}{t}, \quad u(1) = 4.$$

Note that we might expect trouble at $t = 0$ because the derivative is undefined there. The equation is separable so we separate variables and integrate with respect to t:

$$\frac{1}{2}\int \frac{u'}{\sqrt{u}}dt = \int \frac{e^{-t}}{t}dt + C.$$

We can integrate the left side exactly, but the integral on the right cannot be resolved in closed form. Hence we write it with variable limits and we have

$$\sqrt{u} = \int_1^t \frac{e^{-t}}{t}dt + C.$$

Judiciously we chose the lower limit as $t = 1$ so that the initial condition would be easy to apply. Clearly we get $C = 2$. Therefore

$$\sqrt{u} = \int_1^t \frac{e^{-t}}{t}dt + 2,$$

or

$$u(t) = \left(\int_1^t \frac{e^{-t}}{t}dt + 2\right)^2.$$

This solution is valid on $0 < t < \infty$. In spite of the apparent complicated form of the solution, which contains an integral, it is not difficult to plot using a computer algebra system. The plot is shown in figure 2.1.

The method of separation of variables is a key technique in differential equations. Many important models turn out to be separable, not the least of which is the autonomous equation.

EXERCISES

1. Find the general solution in explicit form of the following equations.

   a) $u' = \frac{2u}{t+1}$.

   b) $u' = \frac{t\sqrt{t^2+1}}{\cos u}$.

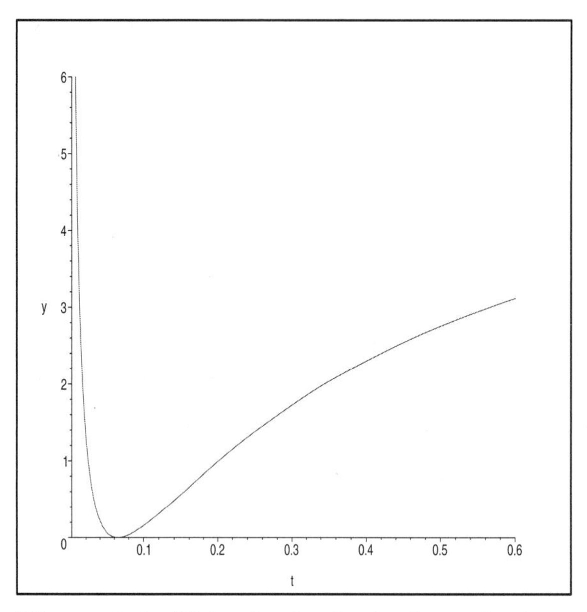

Figure 2.1 This plot is obtained on the interval (0, 0.6] using the Maple command: plot((evalf(2+int(exp(-s)/s, s=1..t)))∧2, t=0..0.6, y=0..6);.

c) $u' = (t+1)(u^2+1)$.

d) $u' + u + \frac{1}{u} = 0$.

2. Find the solution to the initial value problem

$$u' = t^2 e^{-u}, \quad u(0) = \ln 2,$$

and determine the interval of existence.

3. Draw the phase line associated with the DE $u' = u(4-u^2)$ and then solve the DE subject to the initial condition $u(0) = 1$. (Hint: for the integration you will need a partial fractions expansion

$$\frac{1}{u(4-u^2)} = \frac{a}{u} + \frac{b}{2+u} + \frac{c}{2-u},$$

where a, b, and c are to be determined.)

4. Find the general solution in implicit form to the equation

$$u' = \frac{4-2t}{3u^2-5}.$$

Find the solution when $u(1) = 3$ and plot the solution. What is its interval of existence?

5. Solve the initial value problem $u' = \frac{2tu^2}{1+t^2}$, $u(t_0) = u_0$, and find the interval of existence when $u_0 < 0$, when $u_0 > 0$, and when $u_0 = 0$.

6. Find the general solution of the DE

$$u' = 6t(u-1)^{2/3}.$$

Show that there is no value of the arbitrary constant that gives the solution $u = 1$. (A solution to a DE that cannot be obtained from the general solution by fixing a value of the arbitrary constant is called a **singular solution**).

7. Find the general solution of the DE

$$(T^2 - t^2)u' + tu = 0,$$

where T is a fixed, positive parameter. Find the solution to the initial value problem when $u(T/2) = 1$. What is the interval of existence?

8. Allometric growth describes temporal relationships between sizes of different parts of organisms as they grow (e.g., the leaf area and the stem diameter of a plant). We say two sizes u_1 and u_2 are *allometrically* related if their relative growth rates are proportional, or

$$\frac{u_1'}{u_1} = a\frac{u_2'}{u_2}.$$

Show that if u_1 and u_2 are allometrically related, then $u_1 = Cu_2^a$, for some constant C.

9. A differential equation of the form

$$u' = F\left(\frac{u}{t}\right),$$

where the right depends only on the ratio of u and t, is called a **homogeneous**. Show that the substitution $u = ty$ transforms a homogeneous equation into a first-order separable equation for $y = y(t)$. Use this method to solve the equation

$$u' = \frac{4t^2 + 3u^2}{2tu}.$$

10. Solve the initial value problem

$$\frac{d}{dt}\left(u(t)e^{2t}\right) = e^{-t}, \quad u(0) = 3.$$

11. Find the general solution $u = u(r)$ of the DE

$$\frac{1}{r}\frac{d}{dr}(ru'(r)) = -p,$$

where p is a positive constant.

12. A population of u_0 individuals all has HIV, but none has the symptoms of AIDS. Let $u(t)$ denote the number that does not have AIDS at time $t > 0$. If $r(t)$ is the *per capita* rate of individuals showing AIDS symptoms (the conversion rate from HIV to AIDS), then $u'/u = -r(t)$. In the simplest case we can take r to be a linear function of time, or $r(t) = at$. Find $u(t)$ and sketch the solution when $a = 0.2$ and $u_0 = 100$. At what time is the rate of conversion maximum?

13. An arrow of mass m is shot vertically upward with initial velocity 160 ft/sec. It experiences both the deceleration of gravity and a deceleration of magnitude $mv^2/800$ due to air resistance. How high does the arrow go?

14. In very cold weather the thickness of ice on a pond increases at a rate inversely proportional to its thickness. If the ice initially is 0.05 inches thick and 4 hours later it is 0.075 inches thick, how thick will it be in 10 hours?

15. Write the solution to the initial value problem

$$u' = -u^2e^{-t^2}, \quad u(0) = \frac{1}{2}$$

in terms of the erf function, $\text{erf}(t) = \frac{2}{\sqrt{\pi}}\int_0^t e^{-s^2}ds$.

16. Use separation of variables to solve the following problems. Write the solution explicitly when possible.

    a) $u' = p(t)u$, where $p(t)$ is a given continuous function.

    b) $u' = -2tu, \quad u(1) = 2$. Plot the solution on $0 \le t \le 2$.

    c) $u' = \begin{cases} -2u, & 0 < t < 1 \\ -u^2, & 1 \le t \le 2 \end{cases}, \quad u(0) = 5.$

    Find a continuous solution on the interval $0 \le t \le 3$ and plot the solution.

17. A certain patch is populated with a cohort of newly hatched grasshoppers numbering h_0. As time proceeds they die of natural causes at *per capita* rate m, and they are eaten by spiders at the rate $aH/(1+bH)$ per spider, where

H is the population of grasshoppers, and a and b are positive constants. Thus, the dynamics is given by

$$H' = -mH - \frac{aH}{1+bH}S,$$

where S is the spider population, and time is given in days.

a) Determine the units on the constants m, a, and b.

b) Choose new dimensionless variables $\tau = mt$ and $h = bH$, and reformulate the differential equation and initial condition in a dimensionless problem for $h = h(\tau)$. In your differential equation you should have a single dimensionless constant given by $\lambda = aS/m$.

c) Solve the dimensionless initial value problem to obtain a formula for $h(\tau)$. What is $\lim_{\tau\to\infty} h(\tau)$?

18. Let N_0 be the number of individuals in a cohort at time $t = 0$ and $N = N(t)$ be the number of those individuals alive at time t. If m is the constant *per capita* mortality rate, then $N'/N = -m$, which gives $N(t) = N_0 e^{-mt}$. The **survivorship function** is defined by $S(t) = N(t)/N_0$, and $S(T)$ therefore gives the probability of an individual living to age T. In the case of a constant *per capita* mortality the survivorship curve is a decaying exponential.

a) What fraction die before age T? Calculate the fraction of individuals that die between age a and age b.

b) If the *per capita* death rate depends on time, or $m = m(t)$, find a formula for the survivorship function (your answer will contain an integral).

c) What do you think the human survivorship curve looks like?

2.2 First-Order Linear Equations

A differential equation of the form

$$u' = p(t)u + q(t). \tag{2.3}$$

is called a **first-order linear equation**. The given functions p and q are assumed to be continuous. If $q(t) = 0$, then the equation is called **homogeneous**;

otherwise it is called **nonhomogeneous**. Linear equations have a nice structure to their solution set, and we are able to derive the general solution. The homogeneous equation

$$u' = p(t)u, \tag{2.4}$$

without the nonhomogeneous term $q(t)$, can readily be solved by separation of variables to obtain

$$u_h(t) = Ce^{P(t)}, \quad \text{where} \quad P(t) = \int p(t)dt, \tag{2.5}$$

where C is an arbitrary constant. We have placed a subscript h on this solution to distinguish it from the solution of the nonhomogeneous equation (2.3). (The solution $u_h(t)$ to the homogeneous equation is sometimes called the complementary solution; we just refer to it as the homogeneous solution.) Also, note that we have used the indefinite integral notation for the function $P(t)$; in some cases we have to represent $P(t)$ in the form

$$P(t) = \int_a^t p(s)ds,$$

with limits of integration. We always choose the lower limit a to be the value of time where the initial condition is prescribed.

We now describe a standard technique to solve the nonhomogeneous equation (2.3). The idea is to try a solution of the form (2.5) where we let the constant C in the homogeneous solution vary as a function of t; we then substitute this form into (2.3) to determine the $C = C(t)$. The method is, for obvious reasons, called **variation of parameters**.[1] Thus, assume a solution to (2.3) of the form

$$u(t) = C(t)e^{P(t)}.$$

Then, plugging in,

$$C'(t)e^{P(t)} + C(t)e^{P(t)}P'(t) = p(t)C(t)e^{P(t)} + q(t).$$

But $P' = p$ and therefore two of the terms cancel, giving

$$C'(t)e^{P(t)} = q(t),$$

or

$$C'(t) = e^{-P(t)}q(t).$$

Integration yields

$$C(t) = \int e^{-P(t)}q(t)dt + K,$$

[1] Another method using *integrating factors* is presented in the Exercises.

where K is a constant of integration. So we have

$$\begin{aligned} u(t) &= \left(\int e^{-P(t)}q(t)dt + K\right)e^{P(t)} \\ &= Ke^{P(t)} + e^{P(t)}\int e^{-P(t)}q(t)dt, \end{aligned} \tag{2.6}$$

which is the general solution to the general linear, nonhomogeneous equation (2.3). If the antiderivative in the last equation cannot be calculated explicitly, then we write the solution as

$$u(t) = Ke^{P(t)} + e^{P(t)}\int_a^t e^{-P(\tau)}q(\tau)d\tau.$$

We urge the reader not to memorize these formulas; rather, remember the *method* and apply it to each problem as you solve it.

Example 2.3

Find the general solution to

$$u' = \frac{1}{t}u + t^3.$$

The homogeneous equation is $u' = \frac{1}{t}u$ and has solution

$$u_h(t) = Ce^{\int(1/t)dt} = Ce^{\ln t} = Ct.$$

Therefore we vary the parameter C and assume a solution of the original non-homogeneous equation of the form

$$u(t) = C(t)t.$$

Substituting into the equation, we get

$$u' = C(t) + C'(t)t = \frac{1}{t}C(t)t + t^3,$$

or

$$C'(t) = t^2.$$

Therefore $C(t) = \int t^2 dt = \frac{1}{3}t^3 + K$ and the general solution to the original equation is

$$u(t) = \left(\frac{1}{3}t^3 + K\right)t = \frac{1}{3}t^4 + Kt.$$

The arbitrary constant K can be determined by an initial condition.

Example 2.4

Consider the DE

$$u' = 2u + t. \tag{2.7}$$

The associated homogeneous equation is

$$u' = 2u,$$

which has solution $u_h = Ce^{2t}$. Therefore we assume the solution of (2.7) is of the form

$$u(t) = C(t)e^{2t}.$$

Substituting into the original equation gives

$$C(t)2e^{2t} + C'(t)e^{2t} = 2C(t)e^{2t} + t,$$

or

$$C'(t) = te^{-2t}.$$

Integrating,

$$C(t) = \int te^{-2t}dt + K = -\frac{1}{4}e^{-2t}(2t+1) + K.$$

The integral was calculated analytically using integration by parts. Therefore the general solution of (2.7) is

$$\begin{aligned} u(t) &= \left(-\frac{1}{4}e^{-2t}(2t+1) + K\right)e^{2t} \\ &= Ke^{2t} - \frac{1}{4}(2t+1). \end{aligned}$$

Notice that the general solution is composed of two terms, $u_h(t)$ and $u_p(t)$, defined by

$$u_h = Ke^{2t}, \quad u_p = -\frac{1}{4}(2t+1).$$

We know u_h is the general solution to the homogeneous equation, and it is easy to show that u_p is a particular solution to the nonhomogeneous equation (2.7). So, the general solution to the nonhomogeneous equation (2.7) is the sum of the general solution to the associated homogeneous equation and any particular solution to the nonhomogeneous equation (2.7).

Example 2.4 illustrates a general principle that reveals the structure of the solution to a first-order linear DE. The general solution can be written as the sum of the solution to the homogeneous equation and any particular solution of the nonhomogeneous equation. Precisely, the basic structure theorem for first-order linear equations states:

Theorem 2.5

(Structure Theorem) The general solution of the nonhomogeneous equation

$$u' = p(t)u + q(t)$$

is the sum of the general solution u_h of the homogeneous equation $u' = p(t)u$ and a particular solution u_p to the nonhomogeneous equation. In symbols,

$$u(t) = u_h(t) + u_p(t),$$

where $u_h(t) = Ke^{P(t)}$ and $u_p = e^{P(t)} \int e^{-P(t)}q(t)dt$, and where $P(t) = \int p(t)dt$.

Example 2.6

Consider an RC electrical circuit where the resistance is $R = 1$ and the capacitance is $C = 0.5$. Initially the charge on the capacitor is $q(0) = 5$. The current is driven by an emf that generates a variable voltage of $\sin t$. How does the circuit respond? The governing DE for the charge $q(t)$ on the capacitor is

$$Rq' + \frac{1}{C}q = \sin t,$$

or, substituting the given parameters,

$$q' = -2q + \sin t. \qquad (2.8)$$

The homogeneous equation $q' = -2q$ has solution $q_h = Ce^{-2t}$. We assume the solution to the nonhomogeneous equation has the form $q = C(t)e^{-2t}$. Substituting into (2.8) gives

$$C(t)(-2q'e^{-2t}) + C'(t)qe^{-2t} = -2C(t)e^{-2t} + \sin t,$$

or

$$C'(t) = e^{2t} \sin t.$$

Integrating,

$$C(t) = \int e^{2t} \sin t dt + K = e^{2t}(\frac{2}{5} \sin t - \frac{1}{5} \cos t) + K,$$

where K is a constant of integration. The integral was calculated using software (or, one can use integration by parts). Therefore the general solution of (2.8) is

$$q(t) = C(t)e^{-2t} = \frac{2}{5} \sin t - \frac{1}{5} \cos t + Ke^{-2t}.$$

Next we apply the initial condition $q(0) = 5$ to obtain $K = 26/5$. Therefore the solution to the initial value problem is

$$q(t) = \frac{2}{5}\sin t - \frac{1}{5}\cos t + \frac{26}{5}e^{-2t}.$$

The solution is consistent with Theorem 2.5. Also, there is an important physical interpretation of the solution. The homogeneous solution is the **transient response** $q_h(t) = \frac{26}{5}e^{-2t}$ that depends upon the initial charge and decays over a time; what remains over a long time is the particular solution, which is regarded as the **steady-state response** $q_p(t) = \frac{26}{5}\sin t - \frac{1}{5}\cos t$. The homogeneous solution ignores the forcing term (the emf), whereas the particular solution arises from the forcing term. After a long time the applied emf drives the response of the system. This behavior is characteristic of forced linear equations coming from circuit theory and mechanics. The solution is a sum of two terms, a contribution due to the internal system and initial data (the decaying transient), and a contribution due to the external forcing term (the steady response). Figure 2.2 shows a plot of the solution.

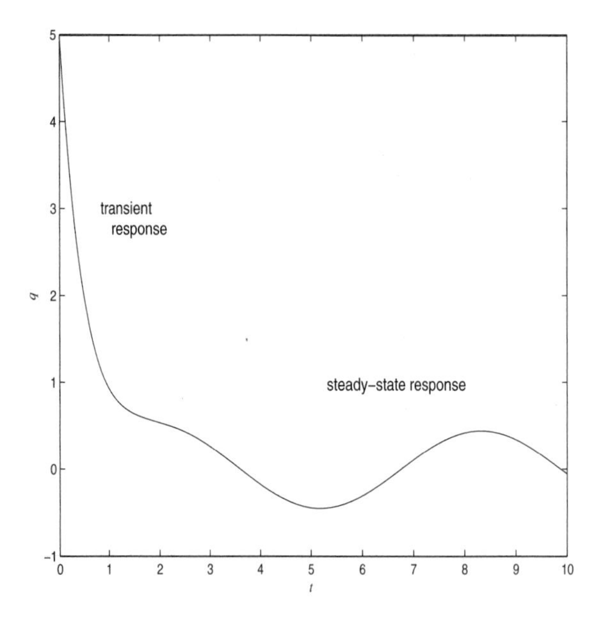

Figure 2.2 Response of the circuit in Example 2.6 showing the initial transient and the long-time steady-state.

Example 2.7

(Sales Response to Advertising) The field of economics has always been a source of interesting phenomena modeled by differential equations. In this example we set up a simple model that allows management to assess the effectiveness of an advertising campaign. Let $S = S(t)$ be the monthly sales of an item. In the absence of advertising it is observed from sales history data that the logarithm of the monthly sales decreases linearly in time, or $\ln S = -at + b$. Thus $S' = -aS$, and sales are modeled by exponential decay. To keep sales up, advertising is required. If there is a lot of advertising, then sales tend to saturate at some maximum value $S = M$; this is because there are only finitely many consumers. The rate of increase in sales due to advertising is jointly proportional to the advertising rate $A(t)$ and to the degree the market is not saturated; that is,

$$rA(t)\left(\frac{M-S}{M}\right).$$

The constant r measures the effectiveness of the advertising campaign. The term $\frac{M-S}{M}$ is a measure of the market share that has still not purchased the product. Then, combining both natural sales decay and advertising, we obtain the model

$$S' = -aS + rA(t)\left(\frac{M-S}{M}\right).$$

The first term on the right is the natural decay rate, and the second term is the rate of sales increase due to advertising, which drives the sales. As it stands, because the advertising rate A is not constant, there are no equilibria (constant solutions). We can rearrange the terms and write the equation in the form

$$S' = -\left(a + \frac{rA(t)}{M}\right)S + rA(t). \tag{2.9}$$

Now we recognize that the sales are governed by a first-order linear DE. The Exercises request some solutions for different advertising strategies.

EXERCISES

1. Find the general solution of $u' = -\frac{1}{t}u + t$.
2. Find the general solution of $u' = -u + e^t$.
3. Show that the general solution to the DE $u' + au = \sqrt{1+t}$ is given by

$$u(t) = Ce^{-at} + \int_0^t e^{-a(t-s)}\sqrt{1+s}ds.$$

4. A decaying battery generating $200e^{-5t}$ volts is connected in series with a 20 ohm resistor, and a 0.01 farad capacitor. Assuming $q = 0$ at $t = 0$, find the charge and current for all $t > 0$. Show that the charge reaches a maximum and find the time it is reached.

5. Solve $u'' + u' = 3t$ by introducing $y = u'$.

6. Solve $u' = (t + u)^2$ by letting $y = t + u$.

7. Express the general solution of the equation $u' = 2tu + 1$ in terms of the erf function.

8. Find the solution to the initial value problem $u' = pu + q, \quad u(0) = u_0$, where p and q are constants.

9. Find a formula for the general solution to the DE $u' = pu + q(t)$, where p is constant. Find the solution satisfying $u(t_0) = u_0$.

10. A differential equation of the form

$$u' = a(t)u + g(t)u^n$$

is called a **Bernoulli equation**, and it arises in many applications. Show that the Bernoulli equation can be reduced to the linear equation

$$y' = (1 - n)a(t)y + (1 - n)g(t)$$

by changing the dependent variable from u to y via $y = u^{1-n}$.

11. Solve the Bernoulli equations (see Exercise 10).

    a) $u' = \frac{2}{3t}u + \frac{2t}{u}$.

    b) $u' = u(1 + ue^t)$.

    c) $u' = -\frac{1}{t}u + \frac{1}{tu^2}$.

12. Initially, a tank contains 60 gal of pure water. Then brine containing 1 lb of salt per gallon enters the tank at 2 gal/min. The perfectly mixed solution is drained off at 3 gal/min. Determine the amount (in lbs) of salt in the tank up until the time it empties.

13. A large industrial retention pond of volume V, initially free of pollutants, was subject to the inflow of a contaminant produced in the factory's processing plant. Over a period of b days the EPA found that the inflow concentration of the contaminant decreased linearly (in time) to zero from its initial value of a (grams per volume), its flow rate q (volume per day) being constant. During the b days the spillage to the local stream was also

q. What is the concentration in the pond after b days? Do a numerical experiment using a computer algebra system where $V = 6000$ cubic meters, $b = 20$ days, $a = 0.03$ grams per cubic meter, and $q = 50$ cubic meters per day. With this data, how long would it take for the concentration in the pond to get below the required EPA level of 0.00001 grams per cubic meter if fresh water is pumped into the pond at the same flow rate, with the same spillover?

14. Determine the dimensions of the various quantities in the sales–advertising model (2.9). If A is constant, what is the equilibrium?

15. (Technology Transfer) Suppose a new innovation is introduced at time $t = 0$ in a community of N possible users (e.g., a new pesticide introduced to a community of farmers). Let $x(t)$ be the number of users who have adopted the innovation at time t. If the rate of adoption of the innovation is jointly proportional to the number of adoptions and the number of those who have not adopted, write down a DE model for $x(t)$. Describe, qualitatively, how $x(t)$ changes in time. Find a formula for $x(t)$.

16. A house is initially at 12 degrees Celsius when its heating–cooling system fails. The outside temperature varies according to $T_e = 9 + 10\cos 2\pi t$, where time is given in days. The heat loss coefficient is $h = 3$ degrees per day. Find a formula for the temperature variation in the house and plot it along with T_e on the same set of axes. What is the time lag between the maximum inside and outside temperature?

17. In the sales response to advertising model (2.9), assume $S(0) = S_0$ and that advertising is constant over a fixed time period T, and is then removed. That is,

$$A(t) = \begin{cases} a, & 0 \leq t \leq T \\ 0, & t > T \end{cases}$$

Find a formula for the sales $S(t)$. (Hint: solve the problem on two intervals and piece together the solutions in a continuous way).

18. In a community having a fixed population N, the rate that people hear a rumor is proportional to the number of people who have not yet heard the rumor. Write down a DE for the number of people P who have heard the rumor. Over a long time, how many will hear the rumor? Is this a believable model?

19. An object of mass $m = 1$ is dropped from rest at a large height, and as it falls it experiences the force of gravity mg and a time-dependent resistive force of magnitude $F_r = \frac{2}{t+1}v$, where v is its velocity. Write down an initial value problem that governs its velocity and find a formula for the solution.

20. The **MacArthur–Wilson model** of the dynamics of species (e.g., bird species) that inhabit an island located near a mainland was developed in the 1960s. Let P be the number of species in the source pool on the mainland, and let $S = S(t)$ be the number of species on the island. Assume that the rate of change of the number of species is

$$S' = \chi - \mu,$$

where χ is the colonization rate and μ is the extinction rate. In the MacArthur–Wilson model,

$$\chi = I(1 - \frac{S}{P}) \quad \text{and} \quad \mu = \frac{E}{P}S,$$

where I and E are the maximum colonization and extinction rates, respectively.

 a) Over a long time, what is the expected equilibrium for the number of species inhabiting the island? Is this equilibrium stable?

 b) Given $S(0) = S_0$, find an analytic formula for $S(t)$.

 c) Suppose there are two islands, one large and one small, with the larger island having the smaller maximum extinction rate. Both have the same colonization rate. Show that the smaller island will eventually have fewer species.

21. (**Integrating Factor Method**) There is another popular method, called the integrating factor method, for solving first-order linear equations written in the form

$$u' - p(t)u = q(t).$$

If this equation is multiplied by $e^{-\int p(t)dt}$, called an **integrating factor**, show that

$$\left(ue^{-\int p(t)dt}\right)' = q(t)e^{-\int p(t)dt}.$$

(Note that the left side is a total derivative). Next, integrate both sides and show that you obtain (2.6). Use this method to solve Exercises 1 and 2 above.

2.3 Approximation

The fact is that most differential equations cannot be solved with simple analytic formulas. Therefore we are interested in developing methods to approximate solutions to differential equations. Approximations can come in the form of a formula or a data set obtained by computer methods. The latter forms the basis of modern scientific computation.

2.3.1 Picard Iteration

We first introduce an iterative procedure, called Picard iteration (E. Picard, 1856-1941), that is adapted from the classical fixed point method to approximate solutions of nonlinear algebraic equations. In Picard iteration we begin with an assumed first approximation of the solution to an initial value problem and then calculate successively better approximations by an iterative, or recursive, procedure. The result is a set of recursive analytic formulas that approximate the solution. We first review the standard fixed point method for algebraic equations.

Example 2.8

Consider the problem of solving the nonlinear algebraic equation

$$x = \cos x.$$

Graphically, it is clear that there is a unique solution because the curves $y = x$ and $y = \cos x$ cross at a single point. Analytically we can approximate the root by making an initial guess x_0 and then successively calculate better approximations via

$$x_{k+1} = \cos x_k \quad \text{for} \quad k = 0, 1, 2, ...$$

For example, if we choose $x_0 = 0.9$, then $x_1 = \cos x_0 = \cos(0.9) = 0.622$, $x_2 = \cos x_1 = \cos(0.622) = 0.813$, $x_3 = \cos x_2 = \cos(0.813) = 0.687$, $x_4 = \cos x_3 = \cos(0.687) = 0.773$, $x_5 = \cos x_4 = \cos(0.773) = 0.716, ...$ Thus we have generated a sequence of approximations 0.9, 0.622, 0.813, 0.687, 0.773, 0.716, ... If we continue the process, the sequence converges to $x^* = 0.739$, which is the solution to $x = \cos x$ (to three decimal places). This method, called **fixed point iteration**, can be applied to general algebraic equations of the form

$$x = g(x).$$

The iterative procedure

$$x_{k+1} = g(x_k), \quad k = 0, 1, 2, ...$$

will converge to a root x^* provided $|g'(x^*)| < 1$ and the initial guess x_0 is sufficiently close to x^*. The conditions stipulate that the graph of g is not too steep (its absolute slope at the root must be bounded by one), and the initial guess is close to the root.

We pick up on this iteration idea for algebraic equations to obtain an approximation method for solving the initial value problem

$$\text{(IVP)} \qquad \begin{cases} u' = f(t,u), \\ u(t_0) = u_0. \end{cases} \tag{2.10}$$

First, we turn this initial value problem into an equivalent integral equation by integrating the DE from t_0 to t:

$$u(t) = u_0 + \int_{t_0}^{t} f(s, u(s))ds.$$

Now we define a type of fixed point iteration, called **Picard iteration**, that is based on this integral equation formulation. We define the iteration scheme

$$u_{k+1}(t) = u_0 + \int_{t_0}^{t} f(s, u_k(s))ds, \quad k = 0, 1, 2, ..., \tag{2.11}$$

where $u_0(t)$ is an initial approximation (we often take the initial approximation to be the constant function $u_0(t) = u_0$). Proceeding in this manner, we generate a sequence $u_1(t)$, $u_2(t)$, $u_3(t)$,... of iterates, called **Picard iterates**, that under certain conditions converge to the solution of the original initial value problem (2.10).

Example 2.9

Consider the linear initial value problem

$$u' = 2t(1+u), \quad u(0) = 0.$$

Then the iteration scheme is

$$u_{k+1}(t) = \int_0^t 2s(1 + u_k(s))ds, \quad k = 0, 1, 2, ...,$$

Take $u_0 = 0$, then

$$u_1(t) = \int_0^t 2s(1+0)ds = t^2.$$

Then

$$u_2(t) = \int_0^t 2s(1 + u_1(s))ds = \int_0^t 2s(1+s^2)ds = t^2 + \frac{1}{2}t^4.$$

Next,

$$u_3(t) = \int_0^t 2s(1+u_2(s))ds = u_{k+1}(t) = \int_0^t 2s(1+s^2+\frac{1}{2}s^4)ds = t^2+\frac{1}{2}t^4+\frac{1}{6}t^6.$$

In this manner we generate a sequence of approximations to the solution to the IVP. In the present case, one can verify that the analytic solution to the IVP is

$$u(t) = e^{t^2} - 1.$$

The Taylor series expansion of this function is

$$u(t) = e^{t^2} - 1 = t^2 + \frac{1}{2}t^4 + \frac{1}{6}t^6 + \cdots + \frac{1}{n!}t^{2n} + \cdots,$$

and it converges for all t. Therefore the successive approximations generated by Picard iteration are the partial sums of this series, and they converge to the exact solution.

The Picard procedure (2.11) is especially important from a theoretical viewpoint. The method forms the basis of an existence proof for the solution to a general *nonlinear* initial value problem; the idea is to show that there is a limit to the sequence of approximations, and that limit is the solution to the initial value problem. This topic is discussed in advanced texts on differential equations. Practically, however, Picard iteration is not especially useful for problems in science and engineering. There are other methods, based upon numerical algorithms, that give highly accurate approximations. We discuss these methods in the next section.

Finally, we point out that Picard iteration is guaranteed to converge if the right side of the equation $f(t, u)$ is regular enough; specifically, the first partial derivatives of f must be continuous in an open rectangle of the tu plane containing the initial point. However, convergence is only guaranteed locally, in a small interval about t_0.

EXERCISES

1. Consider the initial value problem

$$u' = 1 + u^2, \quad u(0) = 0.$$

Apply Picard iteration with $u_0 = 0$ and compute four terms. If the process continues, to what function will the resulting series converge?

2. Apply Picard iteration to the initial value problem

$$u' = t - u, \quad u(0) = 1,$$

to obtain three Picard iterates, taking $u_0 = 1$. Plot each iterate and the exact solution on the same set of axes.

2.3.2 Numerical Methods

As already emphasized, most differential equations cannot be solved analytically by a simple formula. In this section we develop a class of methods that solve an initial value problem numerically, using a computer algorithm. In industry and science, differential equations are almost always solved numerically because most real-world problems are too complicated to solve analytically. And, even if the problem can be solved analytically, often the solution is in the form of a complicated integral that has to be resolved by a computer calculation anyway. So why not just begin with a computational approach in the first place?

We study numerical approximations by a method belonging to a class called **finite difference methods**. Here is the basic idea. Suppose we want to solve the following initial value problem on the interval $0 \leq t \leq T$:

$$u' = f(t, u), \quad u(0) = u_0. \tag{2.12}$$

Rather than seek a continuous solution defined at each time t, we develop a strategy of discretizing the problem to determine an approximation at discrete times in the interval of interest. Therefore, the plan is to replace the continuous time model (2.12) with an approximate discrete time model that is amenable to computer solution.

To this end, we divide the interval $0 \leq t \leq T$ into N segments of constant length h, called the **stepsize**. Thus the stepsize is $h = T/N$. This defines a set of equally spaced discrete times $0 = t_0, t_1, t_2, ..., t_N = T$, where $t_n = nh$, $n = 0, 1, 2, ..., N$. Now, suppose we know the solution $u(t_n)$ of the initial value problem at time t_n. How could we estimate the solution at time t_{n+1}? Let us integrate the DE (2.12) from t_n to t_{n+1} and use the fundamental theorem of calculus. We get the equation

$$u(t_{n+1}) - u(t_n) = \int_{t_n}^{t_{n+1}} f(t, u)dt. \tag{2.13}$$

The integral can be approximated using the left-hand rule, giving

$$u(t_{n+1}) - u(t_n) \approx hf(t_n, u(t_n)).$$

If we denote by u_n the approximation of the solution $u(t_n)$ at $t = t_n$, then this last formula suggests the recursion formula

$$u_{n+1} = u_n + hf(t_n, u_n). \tag{2.14}$$

If $u(0) = u_0$, then (2.14) provides an algorithm for calculating approximations u_1, u_2, u_3, etc., recursively, at times t_1, t_2, t_3,... This method is called the **Euler method**, named after the Swiss mathematician L. Euler (1707–1783). The discrete approximation consisting of the values u_0, u_1, u_2, u_3, etc. is called a **numerical solution** to the initial value problem. The discrete values approximate the graph of the exact solution, and often they are connected with line segments to obtain a continuous curve. It seems evident that the smaller the stepsize h, the better the approximation. One can show that the cumulative error over an interval $0 \leq t \leq T$ is bounded by the stepsize h; thus, the Euler method is said to be of **order** h.

Example 2.10

Consider the initial value problem

$$u' = 1 + tu, \quad u(0) = 0.25.$$

Here $f(t, u) = 1 + tu$ and the Euler difference equation (2.14) with stepsize h is

$$\begin{aligned} u_{n+1} &= u_n + h(1 + t_n u_n) \\ &= u_n + h(1 + nhu_n), \quad n = 0, 1, 2, 3, ... \end{aligned}$$

We take $h = 0.1$. Beginning with $u_0 = 0.25$ we have

$$u_1 = u_0 + (0.1)(1 + (0)(0.1)u_0) = 0.25 + (0.1)(1) = 0.350.$$

Then

$$u_2 = u_1 + (0.1)(1 + (1)(0.1)u_1) = 0.35 + (0.1)(1 + (1)(0.1)(0.35)) = 0.454.$$

Next

$$u_3 = u_2 + (0.1)(1 + (2)(0.1)u_2) = 0.454 + (0.1)(1 + (2)(0.1)(0.454)) = 0.563.$$

Continuing in this manner we generate a sequence of numbers at all the discrete time points. We often connect the approximations by straight line segments to generate a continuous curve. In figure 2.3 we compare the discrete solution to the exact solution $u(t) = e^{t^2/2}(0.25 + \int_0^t e^{-s^2/2}ds)$. Because it is tedious to do numerical calculations by hand, one can program a calculator or write a simple set of instructions for a computer algebra system to do the work for us. Most calculators and computer algebra systems have built-in programs that implement the Euler algorithm automatically. Below is a MATLAB m-file to perform the calculations in Example 2.10 and plot the approximate solution on the interval [0,1]. We take 10 steps, so the stepsize is $h = 1/10 = 0.1$.

```
function euler1D
Tmax=1; N=10; h=Tmax/N;
u=0.25; uhistory=0.25;
for n=1:N;
u=u+h*(1+n*h*u);
uhistory=[uhistory, u];
end
T=0:h:Tmax;
plot(T,uhistory)
xlabel('time t'), ylabel('u')
```

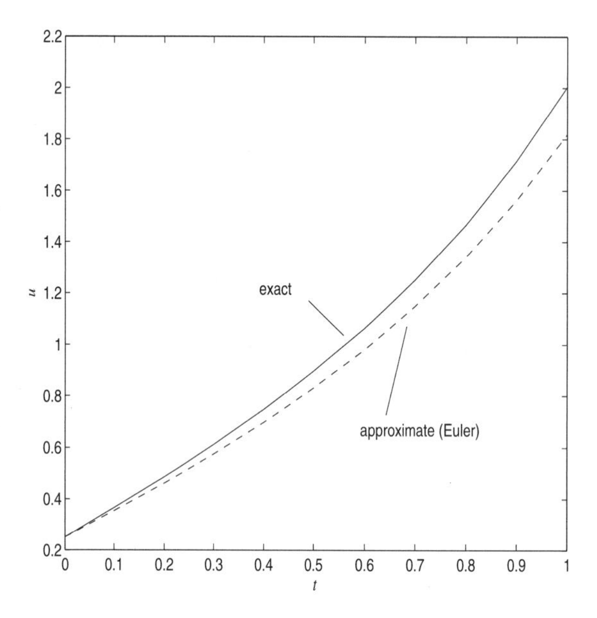

Figure 2.3 The numerical solution (fit with a continuous curve) and exact solution in Example 2.10. Here, $h = 0.1$. A closer approximation can be obtained with a smaller stepsize.

In science and engineering we often write simple programs that implement recursive algorithms; that way we know the skeleton of our calculations, which is often preferred to plugging into an unknown black box containing a canned program.

There is another insightful way to understand the Euler algorithm using the direction field. Beginning at the initial value, we take $u_0 = u(0)$. To find u_1, the approximation at t_1, we march from (t_0, u_0) along the direction field segment with slope $f(t_0, u_0)$ until we reach the point (t_1, u_1) on the vertical line $t = t_1$.

Then, from (t_1, u_1) we march along the direction field segment with slope $f(t_1, u_1)$ until we reach (t_2, u_2). From (t_2, u_2) we march along the direction field segment with slope $f(t_2, u_2)$ until we reach (t_3, u_3). We continue in this manner until we reach $t_N = T$. So how do we calculate the u_n? Inductively, let us assume we are at (t_n, u_n) and want to calculate u_{n+1}. We march along the straight line segment with slope $f(t_n, u_n)$ to (t_{n+1}, u_{n+1}). Thus, writing the slope of this segment in two different ways,

$$\frac{u_{n+1} - u_n}{t_{n+1} - t_n} = f(t_n, u_n).$$

But $t_{n+1} - t_n = h$, and therefore we obtain

$$u_{n+1} = u_n + hf(t_n, u_n),$$

which is again the Euler formula. In summary, the Euler method computes approximate values by moving in the direction of the slope field at each point. This explains why the the numerical solution in Example 2.10 (figure 2.3) lags behind the increasing exact solution.

The Euler algorithm is the simplest method for numerically approximating the solution to a differential equation. To obtain a more accurate method, we can approximate the integral on the right side of (2.13) by the trapezoidal rule, giving

$$u_{n+1} - u_n = \frac{h}{2}[f(t_n, u_n) + f(t_{n+1}, u_{n+1})]. \tag{2.15}$$

This difference equation is not as simple as it may first appear. It does not give the u_{n+1} explicitly in terms of the u_n because the u_{n+1} is tied up in a possibly nonlinear term on the right side. Such a difference equation is called an **implicit equation**. At each step we would have to solve a nonlinear algebraic equation for the u_{n+1}; we can do this numerically, which would be time consuming. Does it pay off in more accuracy? The answer is yes. The Euler algorithm makes a cumulative error over an interval proportional to the stepsize h, whereas the implicit method makes an error of order h^2. Observe that $h^2 < h$ when h is small.

A better approach, which avoids having to solve a nonlinear algebraic equation at each step, is to replace the u_{n+1} on the right side of (2.15) by the u_{n+1} calculated by the simple Euler method in (2.14). That is, we compute a "predictor"

$$\widetilde{u}_{n+1} = u_n + hf(t_n, u_n), \tag{2.16}$$

and then use that to calculate a "corrector"

$$u_{n+1} = u_n + \frac{1}{2}h[f(t_n, u_n) + f(t_{n+1}, \widetilde{u}_{n+1})]. \tag{2.17}$$

This algorithm is an example of a **predictor–corrector method**, and again the cumulative error is proportional to h^2, an improvement to the Euler method. This method is called the **modified Euler method** (also, Heun's method and the second-order Runge–Kutta method).

The Euler and modified Euler methods are two of many numerical constructs to solve differential equations. Because solving differential equations is so important in science and engineering, and because real-world models are usually quite complicated, great efforts have gone into developing accurate, efficient methods. The most popular algorithm is the highly accurate fourth-order **Runge–Kutta method**, where the cumulative error over a bounded interval is proportional to h^4. The Runge–Kutta update formula is

$$u_{n+1} = u_n + \frac{h}{6}(k_1 + k_2 + k_3 + k_4),$$

where

$$\begin{aligned} k_1 &= f(t_n, u_n), \\ k_2 &= f(t_n + \frac{h}{2}, u_n + \frac{h}{2}k_1), \\ k_3 &= f(t_n + \frac{h}{2}, u_n + \frac{h}{2}k_2), \\ k_4 &= f(t_n + h, u_n + hk_3). \end{aligned}$$

We do not derive the formulas here, but they follow by approximating the integral in (2.13) by Simpson's rule, and then averaging. The Runge–Kutta method is built in on computer algebra systems and on scientific calculators.

Note that the order of the error makes a big difference in the accuracy. If $h = 0.1$ then the cumulative errors over an interval for the Euler, modified Euler, and Runge–Kutta methods are proportional to 0.1, 0.01, and 0.0001, respectively.

2.3.3 Error Analysis

Readers who want a detailed account of the errors involved in numerical algorithms should consult a text on numerical analysis or on numerical solution of differential equations. In this section we give only a brief elaboration of the comments made in the last section on the order of the error involved in Euler's method.

Consider again the initial value problem

$$u' = f(t, u), \quad u(0) = u_0 \tag{2.18}$$

on the interval $0 \le t \le T$, with solution $u(t)$. For our argument we assume u has a continuous second derivative on the interval (which implies that the second derivative is bounded). The Euler method, which gives approximations u_n at the discrete points $t_n = nh$, $n = 1, 2, ..., N$, is the recursive algorithm

$$u_{n+1} = u_n + hf(t_n, u_n). \tag{2.19}$$

We want to calculate the error made in performing one step of the Euler algorithm. Suppose at the point t_n the approximation u_n is exact; that is, $u_n = u(t_n)$. Then we calculate the error at the next step. Let $E_{n+1} = u(t_{n+1}) - u_{n+1}$ denote the error at the $(n+1)$st step. Evaluating the DE at $t = t_n$, we get

$$u'(t_n) = f(t_n, u(t_n)).$$

Recall from calculus (Taylor's theorem with remainder) that if u has two continuous derivatives then

$$\begin{aligned} u(t_{n+1}) &= u(t_n + h) = u(t_n) + u'(t_n)h + \frac{1}{2}u''(\tau_n)h^2 \\ &= u(t_n) + hf(t_n, u(t_n)) + \frac{1}{2}u''(\tau_n)h^2 \\ &= u_n + hf(t_n, u_n) + \frac{1}{2}u''(\tau_n)h^2, \end{aligned} \tag{2.20}$$

where the second derivative is evaluated at some point τ_n in the interval (t_n, t_{n+1}). Subtracting (2.19) from (2.20) gives

$$E_{n+1} = \frac{1}{2}u''(\tau_n)h^2.$$

So, if u_n is exact, the Euler algorithm makes an error proportional to h^2 in computing u_{n+1}. So, at each step the Euler algorithm gives an error of order h^2. This is called the **local error**. Notice that the absolute error is $|E_{n+1}| = \frac{1}{2}|u''(\tau_n)|h^2 \le \frac{1}{2}Ch^2$, where C is an absolute bound for the second derivative of u on the entire interval; that is, $|u''(t)| \le C$ for $0 \le t \le T$.

If we apply the Euler method over an entire interval of length T, where $T = Nh$ and N the number of steps, then we expect to make a cumulative error of N times the local error, or an error bounded by a constant times h. This is why we say the cumulative error in Euler's method is order h.

An example will confirm this calculation. Consider the initial value problem for the growth–decay equation:

$$u' = ku, \quad u(0) = u_0,$$

with exact solution $u(t) = u_0 e^{kt}$. The Euler method is

$$u_{n+1} = u_n + hku_n = (1 + hk)u_n.$$

We can iterate to find the exact formula for the sequence of approximations:

$$\begin{aligned} u_1 &= (1+hk)u_0, \\ u_2 &= (1+hk)u_1 = (1+hk)^2 u_0, \\ u_3 &= (1+hk)u_2 = (1+hk)^3 u_0, \\ &\cdots \\ u_n &= (1+hk)^n u_0. \end{aligned}$$

One can calculate the cumulative error in applying the method over an interval $0 \le t \le T$ with $T = Nh$, where N is the total number of steps. We have

$$E_N = u(T) - u_N = u_0[e^{kT} - (1+hk)^N].$$

The exponential term in the parentheses can be expressed in its Taylor series, $e^{kT} = 1 + kT + \frac{1}{2}(kT)^2 + \cdots$, and the second term can be expanded using the binomial theorem, $(1+hk)^N = 1 + Nhk + \frac{N(N+1)}{2}(hk)^2 + \cdots + (hk)^N$. Using $T = Nh$,

$$\begin{aligned} E_N &= u_0[1 + kT + \frac{1}{2}(kT)^2 + \cdots - 1 - Nhk - \frac{N(N+1)}{2}(hk)^2 - \cdots - (hk)^N] \\ &= -\frac{u_0 T k^2}{2} h + \text{ terms containing at least } h^2. \end{aligned}$$

So the cumulative error is the order of the stepsize h.

EXERCISES

1. Use the Euler method and the modified Euler method to numerically solve the initial value problem $u' = 0.25u - t^2$, $u(0) = 2$, on the interval $0 \le t \le 2$ using a stepsize $h = 0.25$. Find the exact solution and compare it, both graphically and in tabular form, to the numerical solutions. Perform the same calculation with $h = 0.1$, $h = 0.01$, and $h = 0.001$. Confirm that the cumulative error at $t = 2$ is roughly order h for the Euler method and order h^2 for the modified Euler method.

2. Use the Euler method to solve the initial value problem $u' = u \cos t$, $u(0) = 1$ on the interval $0 \le t \le 20$ with 50, 100, 200, and 400 steps. Compare to the exact solution and comment on the accuracy of the numerical algorithm.

3. A population of bacteria, given in millions of organisms, is governed by the law
$$u' = 0.6u\left(1 - \frac{u}{K(t)}\right), \quad u(0) = 0.2,$$
where in a periodically varying environment the carrying capacity is $K(t) = 10 + 0.9 \sin t$, and time is given in days. Plot the bacteria population for 40 days.

4. Consider the initial value problem for the decay equation,

$$u' = -ru, \quad u(0) = u_0.$$

Here, r is a given positive decay constant. Find the exact solution to the initial value problem and the exact solution to the sequence of difference approximations $u_{n+1} = u_n - hru_n$ defined by the Euler method. Does the discrete solution give a good approximation to the exact solution for all stepsizes h? What are the constraints on h?

5. Suppose the temperature inside your winter home is 68 degrees at 1:00 P.M. and your furnace then fails. If the outside temperature has an hourly variation over each day given by $15 + 10\cos\frac{\pi t}{12}$ degrees (where $t = 0$ represents 2:00 P.M.), and you notice that by 10:00 P.M. the inside temperature is 57 degrees, what will be the temperature in your home the next morning at 6:00 A.M.? Sketch a plot showing the temperature inside your home and the outside air temperature.

6. Write a program in your computer algebra system that uses the Runge-Kutta method for solving the initial value problem (2.12), and use the program to numerically solve the problem

$$u' = -u^2 + 2t, \quad u(0) = 1.$$

7. Consider the initial value problem $u' = 5u - 6e^{-t}$, $u(0) = 1$. Find the exact solution and plot it on the interval $0 \le t \le 3$. Next use the Euler method with $h = 0.1$ to obtain a numerical solution. Explain the results of this numerical experiment.

8. Consider the initial value problem

$$u' = -u + (15 - u)e^{-a/(u+1)}, \quad u(0) = 1,$$

where a is a parameter. This model arises in the study of a chemically reacting fluid passing through a continuously stirred tank reactor, where the reaction gives off heat. The variable u is related to the temperature in the reactor (Logan 1997, pp. 430–434). Plot the solution for $a = 5.2$ and for $a = 5.3$ to show that the model is sensitive to small changes in the parameter a (this sensitivity is called **structural instability**). Can you explain why this occurs? (Plot the bifurcation diagram with bifurcation parameter a.)

3
Second-Order Differential Equations

Second-order differential equations are one of the most widely studied classes of differential equations in mathematics, physical science, and engineering. One sure reason is that Newton's second law of motion is expressed as a law that involves acceleration of a particle, which is the second derivative of position. Thus, one-dimensional mechanical systems are governed naturally by a second-order equations.

There are really two strategies in dealing with a second-order differential equation. We can always turn a single, second-order differential equation into a system of two simultaneous first-order equations and study the system. Or, we can deal with the equation itself, as it stands. For example, consider the damped spring-mass equation

$$mx'' = -kx - cx'.$$

From Section 1.3 we recall that this equation models the decaying oscillations of a mass m under the action of two forces, a restoring force $-kx$ caused by the spring, and a frictional force $-cx'$ caused by the damping mechanism. This equation is nothing more than a statement of Newton's second law of motion. We can easily transform this equation into a system of two first-order equations with two unknowns by selecting a second unknown state function $y = y(t)$ defined by $y(t) = x'(t)$; thus y is the velocity. Then $my' = -kx - cy$. So the second-order equation is equivalent to

$$\begin{aligned} x' &= y, \\ y' &= -\frac{k}{m}x - \frac{c}{m}y. \end{aligned}$$

This is a simultaneous system of two equations in two unknowns, the position $x(t)$ and the velocity $y(t)$; both equations are first-order. Why do this? Have we gained advantage? Is the system easier to solve than the single equation? The answers to these questions emerge as we study both types of equations in the sequel. Here we just make some general remarks that you may find cryptic. It is probably easier to find the solution formula to the second-order equation directly. But the first-order system embodies a geometrical structure that reveals the underlying dynamics in a far superior way. And, first-order systems arise just as naturally as second-order equations in many other areas of application. Ultimately, it comes down to one's perspective and what information one wants to get from the physical system. Both viewpoints are important.

In this chapter we develop some methods for understanding and solving a single second-order equation. In Chapters 5 and 6 we examine systems of first-order equations.

3.1 Particle Mechanics

Some second-order differential equations can be reduced essentially to a single first-order equation that can be handled by methods from Chapter 2. We place the discussion in the context of particle mechanics to illustrate some of the standard techniques. The general form of Newton's law is

$$mx'' = F(t, x, x'), \tag{3.1}$$

where $x = x(t)$ is the displacement from equilibrium.

(a) If the force does not depend on the position x, then (3.1) is

$$mx'' = F(t, x').$$

We can make the velocity substitution $y = x'$ to obtain

$$my' = F(t, y),$$

which is a first-order differential equation that can be solved with the methods of the preceding chapter. Once the velocity $y = y(t)$ is found, then the position $x(t)$ can be recovered by anti-differentiation, or $x(t) = \int y(t)dt + C$.

(b) If the force does not depend explicitly on time t, then (3.1) becomes

$$mx'' = F(x, x').$$

Again we introduce $y = x'$. Using the chain rule to compute the second derivative (acceleration),

$$x'' = \frac{dy}{dt} = \frac{dy}{dx}\frac{dx}{dt} = y\frac{dy}{dx}.$$

Then

$$my\frac{dy}{dx} = F(x, y),$$

which is a first-order differential equation for the velocity y in terms of the position x. If we solve this equation to obtain $y = y(x)$, then we can recover $x(t)$ by solving the equation $x' = y(x)$ by separation of variables.

(c) In the important, special case where the force F depends only on the position x we say F is a **conservative force**. See also Example 1.5. Then, using the same calculation as in (b), Newton's law becomes

$$my\frac{dy}{dx} = F(x),$$

which is a separable equation. We may integrate both sides with respect to x to get

$$m\int y\frac{dy}{dx}dx = \int F(x)dx + E,$$

or

$$\frac{1}{2}my^2 = \int F(x)dx + E.$$

Note that the left side is the kinetic energy, one-half the mass times the velocity-squared. We use the symbol E for the constant of integration because it must have dimensions of energy. We recall from calculus that the **potential energy** function $V(x)$ is defined by $-dV/dx = F(x)$, or the "force is the negative gradient of the potential."Then $\int F(x)dx = -V(x)$ and we have

$$\frac{1}{2}my^2 + V(x) = E, \tag{3.2}$$

which is the **energy conservation theorem**: the kinetic plus potential energy for a conservative system is constant. The constant E, which represents the total energy in the system, can be computed from knowledge of the initial position $x(0) = x_0$ and initial velocity $y(0) = y_0$, or $E = \frac{1}{2}y_0^2 + V(x_0)$. We regard the conservation of energy law as a reduction of Newton's second law; the latter is a second-order equation, whereas (3.2) is a first-order equation if we replace the position y by dx/dt. It may be recast into

$$\frac{dx}{dt} = \pm\sqrt{\frac{2}{m}}\sqrt{E - V(x)}. \tag{3.3}$$

This equation is separable, and its solution would give $x = x(t)$. The appropriate sign is taken depending upon whether the velocity is positive or negative during a certain phase of the motion.

Usually we analyze conservative systems in phase space (xy-space, or the **phase plane**) by plotting y vs. x from equation (3.2) for different values of the parameter E. The result is a one-parameter family of curves, or **orbits**, in the xy plane along which the motion occurs. The set of these curves forms the **phase diagram** for the system. On these orbits we do not know how x and y depend upon time t unless we solve (3.3). But we do know how velocity relates to position.

Example 3.1

Consider a spring-mass system without damping. The governing equation is

$$mx'' = -kx,$$

where k is the spring constant. The force is $-kx$ and the potential energy $V(x)$ is given by

$$V(x) = -\int -kx\,dx = \frac{k}{2}x^2.$$

We have picked the constant of integration to be zero, which automatically sets the zero level of potential energy at $x = 0$ (i.e., $V(0) = 0$). Conservation of energy is expressed by (3.2), or

$$\frac{1}{2}my^2 + \frac{k}{2}x^2 = E,$$

which plots as a family of concentric ellipses in the xy phase plane, one ellipse for each value of E. These curves represent oscillations, and the mass tracks on one of these orbits in the phase plane, continually cycling as time passes; the position and velocity cycle back and forth. At this point we could attempt to solve (3.3) to determine how x varies in time, but in the next section we find an easier method to solve second-order linear equations for $x(t)$ directly.

EXERCISES

1. Consider a dynamical system governed by the equation $x'' = -x + x^3$. Hence, $m = 1$. Find the potential energy $V(x)$ with $V(0) = 0$. How much total energy E is in the system if $x(0) = 2$ and $x'(0) = 1$? Plot the orbit in the xy phase plane of a particle having this amount of total energy.

2. In a conservative system show that the conservation of energy law (3.2) can be obtained by multiplying the governing equation $mx'' = F(x)$ by x' and noting that $\frac{d}{dt}(x'^2) = 2x'x''$.

3. In a conservative system derive the relation

$$t = \pm \int \frac{dx}{\sqrt{2(E - V(x))}} + C,$$

which gives time as an antiderivative of an expression that is a function of position.

4. A bullet is shot from a gun with muzzle velocity 700 meters per second horizontally at a point target 100 meters away. Neglecting air resistance, by how much does the bullet miss its target?

5. Solve the following differential equations by reducing them to first-order equations.

   a) $x'' = -\frac{2}{t}x'$.

   b) $x'' = xx'$.

   c) $x'' = -4x$.

   d) $x'' = (x')^2$.

   e) $tx'' + x' = 4t$.

6. In a nonlinear spring-mass system the equation governing displacement is $x'' = -2x^3$. Show that conservation of energy for the system can be expressed as $y^2 = C - x^4$, where C is a constant. Plot this set of orbits in the phase plane for different values of C. If $x(0) = x_0 > 0$ and $x'(0) = 0$, show that the period of oscillations is

$$T = \frac{4}{x_0} \int_0^1 \frac{dr}{\sqrt{1 - r^4}}.$$

   Sketch a graph of the period T vs. x_0. (Hint: in (3.3) separate variables and integrate over one-fourth of a period.)

3.2 Linear Equations with Constant Coefficients

We recall two models from Chapter 1. For a spring-mass system with damping the displacement $x(t)$ satisfies

$$mx'' + cx' + kx = 0.$$

The current $I(t)$ in an RCL circuit with no emf satisfies

$$LI'' + RI' + \frac{1}{C}I = 0.$$

The similarity between these two models is called the **mechanical-electrical analogy**. The spring constant k is analogous to the inverse capacitance $1/C$; both a spring and a capacitor store energy. The damping constant c is analogous to the resistance R; both friction in a mechanical system and a resistor in an electrical system dissipate energy. The mass m is analogous to the inductance

L; both represent "inertia" in the system. All of the equations we examine in the next few sections can be regarded as either circuit equations or mechanical problems.

After dividing by the leading coefficient, both equations above have the form

$$u'' + pu' + qu = 0, \tag{3.4}$$

where p and q are constants. An equation of the form (3.4) is called a **second-order linear equation with constant coefficients**. Because zero is on the right side (physically, there is no external force or emf), the equation is **homogeneous**. Often the equation is accompanied by initial data of the form

$$u(0) = A, \quad u'(0) = B. \tag{3.5}$$

The problem of solving (3.4) subject to (3.5) is called the **initial value problem** (IVP). Here the initial conditions are given at $t = 0$, but they could be given at any time $t = t_0$. Fundamental to our discussion is the following existence-uniqueness theorem, which we assume to be true. It is proved in advanced texts.

Theorem 3.2

The initial value problem (3.4)–(3.5) has a unique solution that exists on $-\infty < t < \infty$.

The plan is this. We first note that the DE (3.4) always has two independent solutions $u_1(t)$ and $u_2(t)$ (by **independent** we mean one is not a multiple of the other). We prove this fact by actually exhibiting the solutions explicitly. If we multiply each by an arbitrary constant and form the combination

$$u(t) = c_1 u_1(t) + c_2 u_1(t),$$

where c_1 and c_2 are the arbitrary constants, then we can easily check that $u(t)$ is also a solution to (3.4). This combination is called the **general solution** to (3.4). We prove at the end of this section that all solutions to (3.4) are contained in this combination. Finally, to solve the initial value problem we use the initial conditions (3.5) to uniquely determine the constants c_1 and c_2.

We try a solution to (3.4) of the form $u = e^{\lambda t}$, where λ is to be determined. We suspect something like this might work because every term in (3.4) has to be the same type of function in order for cancellation to occur; thus u, u', and u'' must be the same form, which suggests an exponential for u. Substitution of $u = e^{\lambda t}$ into (3.4) instantly leads to

$$\lambda^2 + p\lambda + q = 0, \tag{3.6}$$

which is a quadratic equation for the unknown λ. Equation (3.6) is called the **characteristic equation**. Solving, we obtain roots

$$\lambda = \frac{1}{2}(-p \pm \sqrt{p^2 - 4q}).$$

These roots of the characteristic equation are called the **characteristic values** (or **roots**) corresponding to the differential equation (3.4). There are three cases, depending upon whether the discriminant $p^2 - 4q$ is positive, zero, or negative. The reader should memorize these three cases and the forms of the solution.

Case 1. If $p^2 - 4q > 0$ then there are two real unequal characteristic values λ_1 and λ_2. Hence, there are two independent, exponential-type solutions

$$u_1(t) = e^{\lambda_1 t}, \quad u_2(t) = e^{\lambda_2 t},$$

and the general solution to (3.4) is

$$u(t) = c_1 e^{\lambda_1 t} + c_2 e^{\lambda_2 t}. \tag{3.7}$$

Case 2. If $p^2 - 4q = 0$ then there is a double root $\lambda = -p/2$. Then one solution is $u_1 = e^{\lambda t}$. A second independent solution in this case is $u_2 = te^{\lambda t}$. Therefore the general solution to (3.4) in this case is

$$u(t) = c_1 e^{\lambda t} + c_2 t e^{\lambda t}. \tag{3.8}$$

Case 3. If $p^2 - 4q < 0$ then the roots of the characteristic equation are complex conjugates having the form

$$\lambda = \alpha \pm i\beta.$$

Therefore two *complex* solutions of (3.4) are

$$e^{(\alpha + i\beta)t}, \quad e^{(\alpha - i\beta)t}.$$

To manufacture *real* solutions we use a fundamental result that holds for all linear, homogeneous equations.

Theorem 3.3

If $u = g(t) + ih(t)$ is a complex solution to the differential equation (3.4), then its real and imaginary parts, $g(t)$ and $h(t)$, are real solutions.

The simple proof is requested in the Exercises.

Let us take the first of the complex solutions given above and expand it into its real and imaginary parts using **Euler's formula:** $e^{i\beta t} = \cos\beta t + i\sin\beta t$. We have

$$e^{(\alpha+i\beta)t} = e^{\alpha t}e^{i\beta t} = e^{\alpha t}(\cos\beta t + i\sin\beta t) = e^{\alpha t}\cos\beta t + ie^{\alpha t}\sin\beta t.$$

Therefore, by Theorem 3.3, $u_1 = e^{\alpha t}\cos\beta t$ and $u_2 = e^{\alpha t}\sin\beta t$ are two real, independent solutions to equation (3.4). If we take the second of the complex solutions, $e^{(\alpha-i\beta)t}$ instead of $e^{(\alpha+i\beta)t}$, then we get the same two real solutions. Consequently, in the case that the characteristic values are complex $\lambda = \alpha \pm i\beta$, the general solution to DE (3.4) is

$$u(t) = c_1e^{\alpha t}\cos\beta t + c_2e^{\alpha t}\sin\beta t. \tag{3.9}$$

In the case of complex eigenvalues, we recall from trigonometry that (3.9) can be written differently as

$$u(t) = e^{\alpha t}(c_1\cos\beta t + c_2\sin\beta t) = e^{\alpha t}A\cos(\beta t - \varphi),$$

where A is called the **amplitude** and φ is the **phase**. This latter form is called the **phase–amplitude form** of the general solution. Written in this form, A and φ play the role of the two arbitrary constants, instead of c_1 and c_2. One can show that that all these constants are related by

$$A = \sqrt{c_1^2 + c_2^2}, \quad \varphi = \arctan\frac{c_2}{c_1}.$$

This is because the cosine of difference expands to

$$A\cos(\beta t - \varphi) = A\cos(\beta t)\cos\varphi + A\sin(\beta t)\sin\varphi.$$

Comparing this expression to $c_1\cos\beta t + c_2\sin\beta t$, gives

$$A\cos\varphi = c_1, \quad A\sin\varphi = c_2.$$

Squaring and adding this last set of equations determines A, and dividing the set of equations determines φ.

Observe that the solution in the complex case is oscillatory in nature with $e^{\alpha t}$ multiplying the amplitude A. If $\alpha < 0$ then the solution will be a decaying oscillation and if $\alpha > 0$ the solution will be a growing oscillation. If $\alpha = 0$ then the solution is

$$u(t) = c_1\cos\beta t + c_2\sin\beta t = A\cos(\beta t - \varphi),$$

and it oscillates with constant amplitude A and period $2\pi/\beta$. The frequency β is called the **natural frequency** of the system.

There is some useful terminology used in engineering to describe the motion of a spring-mass system with damping, governed by the equation

$$mx'' + cx' + kx = 0.$$

The characteristic equation is

$$m\lambda^2 + c\lambda + k = 0,$$

with roots

$$\lambda = \frac{-c \pm \sqrt{c^2 - 4mk}}{2m}.$$

If the roots are complex ($c^2 < 4mk$) then the system is **under-damped** (representing a decaying oscillation); if the roots are real and equal ($c^2 = 4mk$) then the system is **critically damped** (decay, no oscillations, and at most one pass through equilibrium $x = 0$); if the roots are real and distinct ($c^2 > 4mk$) then the system is **over-damped** (a strong decay toward $x = 0$). The same terminology can be applied to an RCL circuit.

Example 3.4

The differential equation $u'' - u' - 12u = 0$ has characteristic equation $\lambda^2 - \lambda - 12 = 0$ with roots $\lambda = -3, 4$. These are real and distinct and so the general solution to the DE is $u = c_1 e^{-3t} + c_2 e^{4t}$. Over a long time the contribution e^{-3t} decays and the solution is dominated by the e^{4t} term. Thus, eventually the solution grows exponentially.

Example 3.5

The differential equation $u'' + 4u' + 4u = 0$ has characteristic equation $\lambda^2 + 4\lambda + 4 = 0$, with roots $\lambda = -2, -2$. Thus the eigenvalues are real and equal, and the general solution is $u = c_1 e^{-2t} + c_2 te^{-2t}$. This solution decays as time gets large (recall that a decaying exponential dominates the linear growth term t so that te^{-2t} goes to zero).

Example 3.6

The differential equation $u''+2u'+2u = 0$ models a damped spring-mass system with $m = 1$, $c = 2$, and $k = 2$. It has characteristic equation $\lambda^2 + 2\lambda + 1 = 0$. The quadratic formula gives complex roots $\lambda = -1 \pm 2i$. Therefore the general solution is

$$u = c_1 e^{-t} \cos 2t + c_2 e^{-t} \sin 2t,$$

representing a decaying oscillation. Here, the natural frequency of the undamped oscillation is 2. In phase–amplitude form we can write

$$u = Ae^{-t}\cos(2t - \varphi).$$

Let us assume that the mass is given an initial velocity of 3 from an initial position of 1. Then the initial conditions are $u(0) = 1$, $u'(0) = 3$. We can use these conditions directly to determine either c_1 and c_2 in the first form of the solution, or A and φ in the phase-amplitude form. Going the latter route, we apply the first condition to get

$$u(0) = Ae^{-0}\cos(2(0) - \varphi) = A\cos\varphi = 1.$$

To apply the other initial condition we need the derivative. We get

$$u' = -2Ae^{-t}\sin(2t - \varphi).$$

Then

$$u'(0) = -2Ae^{-0}\sin(2(0) - \varphi) = 2A\sin\varphi = 3.$$

Therefore we have

$$A\cos\varphi = 1, \quad A\sin\varphi = \frac{3}{2}.$$

Squaring both equations and summing gives $A^2 = 13/4$, so the amplitude is $A = \sqrt{13/4}$. Note that the cosine is positive and the sine is positive, so the phase angle lies in the first quadrant. The phase is

$$\varphi = \arctan(\frac{3}{2}) \doteq 0.983 \text{ radians}.$$

Therefore the solution to the initial value problem is

$$u = \sqrt{\frac{13}{4}}e^{-t}\cos(2t - 0.983).$$

This solution represents a decaying oscillation. The oscillatory part has natural frequency 2 and the period is π. See figure 3.1. The phase has the effect of translating the $\cos 2t$ term by $0.983/2$, which is called the phase shift.

To summarize, we have observed that the differential equation (3.4) always has two independent solutions $u_1(t)$ and $u_2(t)$, and that the combination

$$u(t) = c_1u_1(t) + c_2u_1(t)$$

is also a solution, called the general solution. Now, as promised, we show that the general solution contains all possible solutions to (3.4). To see this let $u_1(t)$ and $u_2(t)$ be special solutions that satisfy the initial conditions

$$u_1(0) = 1, \quad u_1'(0) = 0,$$

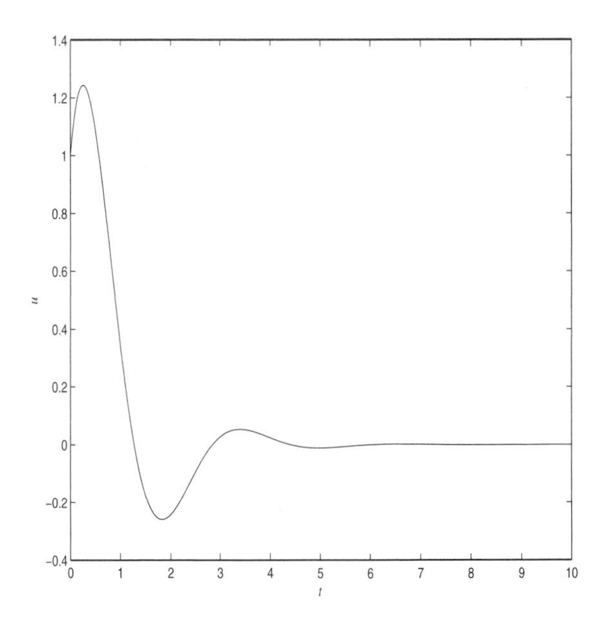

Figure 3.1 Plot of the solution.

and

$$u_2(0) = 0, \quad u_2'(0) = 1,$$

respectively. Theorem 3.2 implies these two solutions exist. Now let $v(t)$ be any solution of (3.4). It will satisfy some conditions at $t = 0$, say, $v(0) = a$ and $v'(0) = b$. But the function

$$u(t) = au_1(t) + bu_1(t)$$

satisfies those same initial conditions, $u(0) = a$ and $u'(0) = b$. Must $u(t)$ therefore equal $v(t)$? Yes, by the uniqueness theorem, Theorem 3.2. Therefore $v(t) = au_1(t)+bu_1(t)$, and the solution $v(t)$ is contained in the general solution.

Two equations occur so frequently that it is worthwhile to memorize them along with their solutions. The pure oscillatory equation

$$u'' + k^2u = 0$$

has characteristic roots $\lambda = \pm ki$, and the general solution is

$$u = c_1 \cos kt + c_2 \sin kt.$$

On the other hand, the equation

$$u'' - k^2u = 0$$

has characteristic roots $\lambda = \pm k$, and thus the general solution is

$$u = c_1 e^{kt} + c_2 e^{-kt}.$$

This latter equation can also be written in terms of the the hyperbolic functions cosh and sinh as

$$u = C_1 \cosh kt + C_2 \sinh kt,$$

where

$$\cosh kt = \frac{e^{kt} + e^{-kt}}{2}, \quad \sinh kt = \frac{e^{kt} - e^{-kt}}{2}.$$

Sometimes the hyperbolic form of the general solution is easier to work with.

EXERCISES

1. Find the general solution of the following equations:

   a) $u'' - 4u' + 4u = 0$.

   b) $u'' + u' + 4u = 0$.

   c) $u'' - 5u' + 6u = 0$.

   d) $u'' + 9u = 0$.

   e) $u'' - 2u' = 0$.

   f) $u'' - 12u = 0$.

2. Find the solution to the initial value problem $u'' + u' + u = 0$, $u(0) = u'(0) = 1$, and write it in phase-amplitude form.

3. A damped spring-mass system is modeled by the initial value problem

   $$u'' + 0.125u' + u = 0, \quad u(0) = 2, \quad u'(0) = 0.$$

   Find the solution and sketch its graph over the time interval $0 \le t \le 50$. If the solution is written in the form $u(t) = Ae^{-t/16}\cos(\omega t - \varphi)$, find A, ω, and φ.

4. For which values of the parameters a and b (if any) will the solutions to $u'' - 2au' + bu = 0$ oscillate with no decay (i.e., be periodic)? Oscillate with decay? Decay without oscillations?

5. An RCL circuit has equation $LI'' + I' + I = 0$. Characterize the types of current responses that are possible, depending upon the value of the inductance L.

6. An oscillator with damping is governed by the equation $x'' + 3ax' + bx = 0$, where a and b are positive parameters. Plot the set of points in the ab plane (i.e., ab parameter space) where the system is critically damped.

7. Find a DE that has general solution $u(t) = c_1e^{4t} + c_2e^{-6t}$.

8. Find a DE that has solution $u(t) = e^{-3t} + 2te^{-3t}$. What are the initial conditions?

9. Find a DE that has solution $u(t) = \sin 4t + 3\cos 4t$.

10. Find a DE that has general solution $u(t) = A\cosh 5t + B\sinh 5t$, where A and B are arbitrary constants. Find the arbitrary constants when $u(0) = 2$ and $u'(0) = 0$.

11. Find a DE that has solution $u(t) = e^{-2t}(\sin 4t + 3\cos 4t)$. What are the initial conditions?

12. Describe the current response $I(t)$ of a LC circuit with $L = 5$ henrys, $C = 2$ farads, with $I(0) = 0$, $I'(0) = 1$.

13. Prove Theorem 3.3 by substituting u into the equation and separating real and imaginary parts, using linearity. Then use the fact that a complex quantity is zero if, and only if, its real and imaginary parts are zero.

3.3 The Nonhomogeneous Equation

In the last section we solved the **homogeneous equation**

$$u'' + pu' + qu = 0. \tag{3.10}$$

Now we consider the **nonhomogeneous equation**

$$u'' + pu' + qu = g(t), \tag{3.11}$$

where a known term $g(t)$, called a **source term** or **forcing term**, is included on the right side. In mechanics it represents an applied, time-dependent force; in a circuit it represents an applied voltage (an emf, such as a battery or generator). There is a general structure theorem, analogous to Theorem 1.2 for first-order linear equations, that dictates the form of the solution to the nonhomogeneous equation.

Theorem 3.7

All solutions of the nonhomogeneous equation (3.11) are given by the sum of the general solution to the homogeneous equation (3.10) and any particular solution to the nonhomogeneous equation. That is, the general solution to (3.11) is

$$u(t) = c_1u_1(t) + c_2u_1(t) + u_p(t),$$

where u_1 and u_2 are independent solutions to (3.10) and u_p is any solution to (3.11).

This result is very easy to show. If $u(t)$ is any solution whatsoever of (3.11), and $u_p(t)$ is a particular solution, then $u(t)-u_p(t)$ must satisfy the homogeneous equation (3.10). Therefore, by the results in the last section we must have $u(t) - u_p(t) = c_1u_1(t) + c_2u_1(t)$.

3.3.1 Undetermined Coefficients

We already know how to find the solution of the homogeneous equation, so we need techniques to find a particular solution u_p to (3.11). One method that works for many equations is to simply make a judicious guess depending upon the form of the source term $g(t)$. Officially, this method is called the **method of undetermined coefficients** because we eventually have to find numerical coefficients in our guess. This works because all the terms on the left side of (3.11) must add up to give $g(t)$. So the particular solution cannot be too wild if $g(t)$ is not too wild; in fact, it nearly must have the same form as $g(t)$. The method is successful for forcing terms that are exponential functions, sines and cosines, polynomials, and sums and products of these common functions. Here are some basic rules without some caveats, which come later. The capital letters in the list below denote known constants in the source term $g(t)$, and the lowercase letters denote coefficients to be determined in the trial form of the particular solution when it is substituted into the differential equation.

1. If $g(t) = Ae^{\gamma t}$ is an exponential, then we try an exponential $u_p = ae^{\gamma t}$.
2. If $g(t) = A\sin\omega t$ or $g(t) = A\cos\omega t$, then we try a combination $u_p = a\sin\omega t$ $+b\cos\omega t$.
3. If $g(t) = A_nt^n + A_{n-1}t^{n-1} + \cdots + A_0$ is a polynomial of degree n, then we try $u_p = a_nt^n + a_{n-1}t^{n-1} + \cdots + a_0$, a polynomial of degree n.
4. If $g(t) = (A_nt^n + A_{n-1}t^{n-1} + \cdots + A_1t + A_0)e^{\gamma t}$, then we try $u_p = (a_nt^n + a_{n-1}t^{n-1} + \cdots a_1t + a_0)e^{\gamma t}$.
5. If $g(t) = Ae^{\gamma t}\sin\omega t$ or $g(t) = Ae^{\gamma t}\cos\omega t$, then we try $u_p = ae^{\gamma t}\sin\omega t + be^{\gamma t}\cos\omega t$.

If the source term $g(t)$ is a sum of two different types, we take the net guess to be a sum of the two individual guesses. For example, if $g(t) = 3t-1+7e^{-2t}$, a polynomial plus an exponential, then a good guess would be $u_p = at+b+ce^{-2t}$. The following examples show how the method works.

Example 3.8

Find a particular solution to the differential equation

$$u'' - u' + 7u = 5t - 3.$$

The right side, $g(t) = 5t - 3$, is a polynomial of degree 1 so we try $u_p = at + b$. Substituting, $-a + 7(at + b) = 5t - 3$. Equating like terms (constant term and terms involving t) gives $-a + 7b = -3$ and $7a = 5$. Therefore $a = 5/7$ and $b = -16/49$. A particular solution to the equation is therefore

$$u_p(t) = \frac{5}{7} - \frac{16}{49}t.$$

Example 3.9

Consider the equation

$$u'' + 3u' + 3u = 6e^{-2t}.$$

The homogeneous equation has characteristic polynomial $\lambda^2+3\lambda+3 = 0$, which has roots $\lambda = -\frac{3}{2} \pm \frac{\sqrt{3}}{2}i$. Thus the solution to the homogeneous equation is

$$u_h(t) = c_1 e^{-3t/2} \cos \frac{\sqrt{3}}{2}t + c_2 e^{-3t/2} \sin \frac{\sqrt{3}}{2}t.$$

To find a particular solution to the nonhomogeneous equation note that $g(t) = 6e^{-2t}$. Therefore we guess $u_p = ae^{-2t}$. Substituting this trial function into the nonhomogeneous equation gives, after canceling e^{-2t}, the equation $4a - 6a + 3a = 6$. Thus $a = 1$ and a particular solution to the nonhomogeneous equation is $u_p = e^{-2t}$. The general solution to the original nonhomogeneous equation is

$$u(t) = c_1 e^{-3t/2} \cos \frac{\sqrt{3}}{2}t + c_2 e^{-3t/2} \sin \frac{\sqrt{3}}{2}t + e^{-2t}.$$

Example 3.10

Find a particular solution to the DE

$$u'' + 2u = \sin 3t.$$

Our basic rule above dictates we try a solution of the form $u_p = a \sin 3t + b \cos 3t$. Then, upon substituting,

$$-9a \sin 3t - 9b \cos 3t + 2a \sin 3t + 2b \cos 3t = \sin 3t.$$

Equating like terms gives $-9a + 2a = 1$ and $b = 0$ (there are no cosine terms on the right side). Hence $a = -1/7$ and a particular solution is $u_p = -\frac{1}{7} \sin 3t$.

For this equation, because there is no first derivative, we did not need a cosine term in the guess. If there were a first derivative, a cosine would have been required.

Example 3.11

Next we modify Example 3.10 and consider

$$u'' + 9u = \sin 3t.$$

The rules dictate the trial function $u_p = a \sin 3t + b \cos 3t$. Substituting into (3.15) yields

$$-9a \sin 3t - 9b \cos 3t + 9a \sin 3t + 9b \cos 3t = \sin 3t.$$

But the terms on the left cancel completely and we get $0 = \sin 3t$, an absurdity. The method failed! This is because the homogeneous equation $u'' + 9u = 0$ has eigenvalues $\lambda = \pm 3i$, which lead to independent solutions $u_1 = \sin 3t$ and $u_2 = \cos 3t$. The forcing term $g(t) = \sin 3t$ is not independent from those two basic solutions; it duplicates one of them, and in this case the method as presented above fails. The fact that we get 0 when we substitute our trial function into the equation is no surprise—it is a solution to the homogeneous equation. To remedy this problem, we can modify our original guess by multiplying it by t. That is, we attempt a particular solution of the form

$$u_p = t(a \sin 3t + b \cos 3t).$$

Calculating the second derivative u_p'' and substituting, along with u_p, into the original equation leads to (show this!)

$$6a \cos 3t - 6b \sin 3t = \sin 3t.$$

Hence $a = 0$ and $b = -1/6$. We have found a particular solution

$$u_p = -\frac{1}{6} t \cos 3t.$$

Therefore the general solution of the original nonhomogeneous equation is the homogeneous solution plus the particular solution,

$$u(t) = c_1 \cos 3t + c_2 \sin 3t - \frac{1}{6} t \cos 3t.$$

Notice that the solution to the homogeneous equation is oscillatory and remains bounded; the particular solution oscillates without bound because of the increasing time factor t multiplying that term.

The technique for finding the form of the particular solution that we used in the preceding example works in general; this is the main caveat in the set of rules listed above.

Caveat. *If a term in the initial trial guess for a particular solution* u_p *duplicates one of the basic solutions for the homogeneous equation, then modify the guess by multiplying by the smallest power of* t *that eliminates the duplication.*

Example 3.12

Consider the DE

$$u'' - 4u' + u = 5te^{2t}.$$

The initial guess for a particular solution is $u_p = (at + b)e^{2t}$. But, as you can check, e^{2t} and te^{2t} are basic solutions to the homogeneous equation $u'' - 4u' + u = 0$. Multiplying the first guess by t gives $u_p = (at^2 + bt)e^{2t}$, which still does not eliminate the duplication because of the te^{2t} term. So, multiply by another t to get $u_p = (at^3 + bt^2)e^{2t}$. Now no term in the guess duplicates one of the basic homogeneous solutions and so this is the correct form of the particular solution. If desired, we can substitute this form into the differential equation to determine the exact values of the coefficients a and b. But, without actually finding the coefficients, the form of the general solution is

$$u(t) = c_1e^{2t} + c_2te^{2t} + (at^3 + bt^2)e^{2t}.$$

The constants c_1 and c_2 could be determined at this point by initial conditions, if given. Sometimes knowing the form of the solution is enough.

Example 3.13

Consider an RCL circuit where $R = 2$, $L = C = 1$, and the current is driven by an electromotive force of $2\sin 3t$. The circuit equation for the voltage $V(t)$ across the capacitor is

$$V'' + 2V' + V = 2\sin 3t.$$

For initial data we take

$$V(0) = 4, \quad V'(0) = 0.$$

We recognize this as a nonhomogeneous linear equation with constant coefficients. So the general solution will be the sum of the general solution to the homogeneous equation

$$V'' + 2V' + V = 0$$

plus any particular solution to the nonhomogeneous equation. The homogeneous equation has characteristic equation $\lambda^2 + 2\lambda + 1 = 0$ with a double root $\lambda = -1$. Thus the homogeneous solution is

$$V_h = e^{-t}(c_1 + c_2 t).$$

Notice that this solution, regardless of the values of the constants, will decay away in time; this part of the solution is called the **transient response** of the circuit. To find a particular solution we use undetermined coefficients and assume it has the form

$$V_p = a \sin 3t + b \cos 3t.$$

Substituting this into the nonhomogeneous equation gives a pair of linear equations for a and b,

$$-4a - 3b = 1, \quad 7a - 9b = 0.$$

We find $a = -0.158$ and $b = -0.123$. Therefore the general solution is

$$V(t) = e^{-t}(c_1 + c_2 t) - 0.158 \sin 3t - 0.123 \cos 3t.$$

Now we apply the initial conditions. Easily $V(0) = 4$ implies $c_1 = 4.123$. Next we find $V'(t)$ so that we can apply the condition $V'(0) = 0$. Leaving this as an exercise, we find $c_2 = 4.597$. Therefore the voltage on the capacitor is

$$V(t) = e^{-t}(4.123 + 4.597t) - 0.158 \sin 3t - 0.123 \cos 3t.$$

As we observed, the first term always decays as time increases. Therefore we are left with only the particular solution $-0.158 \sin 3t - 0.123 \cos 3t$, which takes over in time. It is called the **steady-state response** of the circuit (figure 3.2).

The method of undetermined coefficients works for nonhomogeneous *first-order* linear equations as well, provided the equation has constant coefficients.

Example 3.14

Consider the equation

$$u' + qu = g(t).$$

The homogeneous solution is $u_h(t) = Ce^{-qt}$. Provided $g(t)$ has the right form, a particular solution $u_p(t)$ can be found by the method of undetermined coefficients exactly as for second-order equations: make a trial guess and substitute into the equation to determine the coefficients in the guess. The general solution to the nonhomogeneous equation is then $u(t) = u_h(t) + u_p(t)$. For example, consider the equation

$$u' - 3u = t - 2.$$

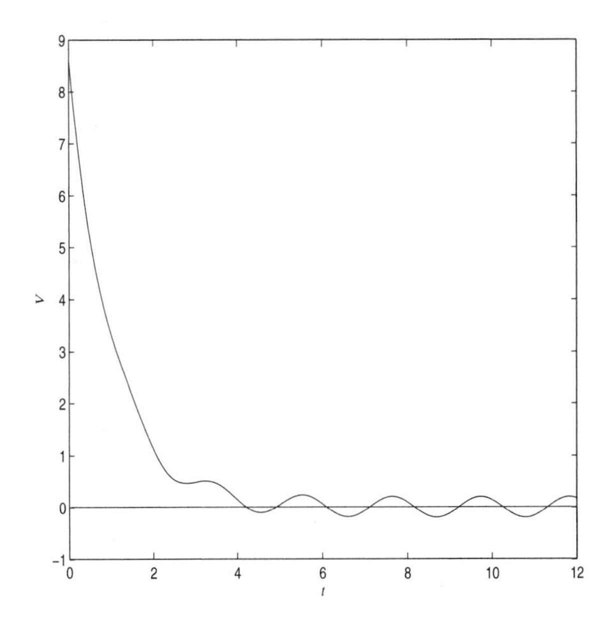

Figure 3.2 A plot of the voltage $V(t)$ in Example 3.13. Initially there is a transient caused by the initial conditions. It decays away and is replaced by a steady-state response, an oscillation, that is caused by the forcing term.

The homogeneous solution is $u_h = Ce^{3t}$. To find a particular solution make the trial guess

$$u_p = at + b.$$

Substituting this into the equation gives $a = -\frac{1}{3}$ and $b = \frac{5}{3}$. Consequently, the general solution is

$$u(t) = Ce^{3t} - \frac{1}{3}t + \frac{5}{3}.$$

EXERCISES

1. Each of the following functions represents $g(t)$, the right side of a nonhomogeneous equation. State the form of an initial trial guess for a particular solution $u_p(t)$.

   a) $3t^3 - 1$.

   b) 12.

   c) t^2e^{3t}.

   d) $5\sin 7t$.

   e) $e^{2t}\cos t + t^2$.

f) $te^{-t} \sin \pi t$.

2. Find the general solution of the following nonhomogeneous equations:

   a) $u'' + 7u = te^{3t}$.

   b) $u'' - u' = 6 + e^{2t}$.

   c) $u' + u = t^2$.

   d) $u'' - 3u' - 4u = 2t^2$.

   e) $u'' + u = 9e^{-t}$.

   f) $u' + u = 4e^{-t}$.

   g) $u'' - 4u = \cos 2t$.

   h) $u'' + u' + 2u = t \sin 2t$

3. Solve the initial value problem $u'' - 3u' - 40u = 2e^{-t}$, $u(0) = 0$, $u'(0) = 1$.

4. Find the solution of $u'' - 2u' = 4$, $u(0) = 1$, $u'(0) = 0$.

5. Find the particular solution to the equation $u'' + u' + 2u = \sin^2 t$? (Hint: use a double angle formula to rewrite the right side.)

6. An RL circuit contains a 2 ohm resistor and a 5 henrys inductor connected in series with a 10 volt battery. If the open circuit is suddenly closed at time zero, find the current for all times $t > 0$. Plot the current vs. time and identify the steady-state response.

7. A circuit contains a 10^{-3} farad capacitor in series with a 20 volt battery and an inductor of 0.4 henrys. At $t = 0$ both $q = 0$ and $I = 0$. Find the charge $q(t)$ on the capacitor and describe the response of the circuit in terms of transients and steady-states.

8. An RCL circuit contains a battery generating 110 volts. The resistance is 16 ohms, the inductance is 2 henrys, and the capacitance is 0.02 farads. If $q(0) = 5$ and $I(0) = 0$, find the charge $q(t)$ current response of the circuit. Identify the transient solution and the steady-state response.

3.3.2 Resonance

The phenomenon of resonance is a key characteristic of oscillating systems. Resonance occurs when the frequency of a forcing term has the same frequency as the natural oscillations in the system; resonance gives rise to large amplitude oscillations. To give an example, consider a pendulum that is oscillating at its natural frequency. What happens when we deliberately force the pendulum (say, by giving it a tap with our finger in the positive angular direction) at a frequency near this natural frequency? So, every time the bob passes through $\theta = 0$ with a positive direction, we give it a positive tap. We will clearly increase its amplitude. This is the phenomenon of resonance. It can occur in circuits where we force (by a generator) the system at its natural frequency, and it can occur in mechanical systems and structures where an external periodic force is applied at the same frequency as the system would naturally oscillate. The results could be disastrous, such as a blown circuit or a fallen building; a building or bridge could have a natural frequency of oscillation, and the wind could provide the forcing function. Another imagined example is a company of soldiers marching in cadence across a suspension bridge at the same frequency as the natural frequency of the structure.

We consider a model problem illustrating this phenomenon, an LC circuit that is forced with a sinusoidal voltage source of frequency β. If $L = 1$ the governing equation for the charge on the capacitor will have the form

$$u'' + \omega^2 u = \sin \beta t, \tag{3.12}$$

where $\omega^2 = 1/C$. Assume first that $\beta \neq \omega$ and take initial conditions

$$u(0) = 0, \quad u'(0) = 1.$$

The homogeneous equation has general solution

$$u_h = c_1 \cos \omega t + c_2 \sin \omega t,$$

which gives natural oscillations of frequency ω. A particular solution has the form $u_p = a \sin \beta t$. Substituting into the DE gives $a = 1/(\omega^2 - \beta^2)$. So the general solution of (3.12) is

$$u = c_1 \cos \omega t + c_2 \sin \omega t + \frac{1}{\omega^2 - \beta^2} \sin \beta t. \tag{3.13}$$

At $t = 0$ we have $u = 0$ and so $c_1 = 0$. Also $u'(0) = 1$ gives $c_2 = -\frac{\beta + \omega(\omega^2 - \beta^2)}{\omega^2 - \beta^2}$. Therefore the solution to the initial value problem is

$$u = -\frac{\beta + \omega(\omega^2 - \beta^2)}{\omega^2 - \beta^2} \sin \omega t + \frac{1}{\omega^2 - \beta^2} \sin \beta t. \tag{3.14}$$

This solution shows that the charge response is a sum of two oscillations of different frequencies. If the forcing frequency β is close to the natural frequency ω, then the amplitude is bounded, but it is obviously large because of the factor $\omega^2 - \beta^2$ in the denominator. Thus the system has large oscillations when β is close to ω.

What happens if $\beta = \omega$? Then the general solution in (3.13) is not valid because there is division by zero, and we have to re-solve the problem. The circuit equation is

$$u'' + \omega^2 u = \sin \omega t, \tag{3.15}$$

where the circuit is forced at the same frequency as its natural frequency. The homogeneous solution is the same as before, but the particular solution will now have the form

$$u_p = t(a \sin \omega t + b \cos \omega t),$$

with a factor of t multiplying the terms. Therefore the general solution of (3.15) has the form

$$u(t) = c_1 \cos \omega t + c_2 \sin \omega t + t(a \sin \omega t + b \cos \omega t).$$

Without actually determining the constants, we can see the nature of the response. Because of the t factor in the particular solution, the amplitude of the oscillatory response $u(t)$ will grow in time. This is the phenomenon of **pure resonance**. It occurs when the frequency of the external force is the same as the natural frequency of the system.

What happens if we include damping in the circuit (i.e., a resistor) and still force it at the natural frequency? Consider

$$u'' + 2\sigma u' + 2u = \sin \sqrt{2}t,$$

where 2σ is a small $(0 < \sigma)$ damping coefficient, for example, resistance. The homogeneous equation $u'' + 2\sigma u' + 2u = 0$ has solution $u = e^{-\sigma t}(c_1 \cos \sqrt{2 - \sigma^2}t + c_2 \sin \sqrt{2 - \sigma^2}t)$. Now the particular solution has the form $u_p = a \cos \sqrt{2}t + b \sin \sqrt{2}t$, where a and b are constants (found by substituting into the DE). So, the response of the circuit is

$$u = e^{-\sigma t}(c_1 \cos \sqrt{2 - \sigma^2}t + c_2 \sin \sqrt{2 - \sigma^2}t) + a \cos \sqrt{2}t + b \sin \sqrt{2}t.$$

The transient is a decaying oscillation of frequency $\sqrt{2 - \sigma^2}$, and the steady-state response is periodic of frequency $\sqrt{2}$. The solution will remain bounded, but its amplitude will be large if σ is very small.

EXERCISES

1. Graph the solution (3.14) for several different values of β and ω. Include values where these two frequencies are close.

2. Find the general solution of the equation $u'' + 16u = \cos 4t$.

3. Consider a general LC circuit with input voltage $V_0 \sin \beta t$. If β and the capacitance C are known, what value of the inductance L would cause resonance?

4. Consider the equation
$$u'' + \omega^2 u = \cos \beta t.$$

   a) Find the solution when the initial conditions are $u(0) = u'(0) = 0$ when $\omega \neq \beta$.

   b) Use the trigonometric identity $2 \sin A \sin B = \cos(A - B) - \cos(A + B)$ to write the solution as a product of sines.

   c) Take $\omega = 55$ and $\beta = 45$ and plot the solution in part (b).

   d) Show that the solution in (c) can be interpreted as a high-frequency response contained in a low-frequency amplitude envelope. (We say the high frequency is *modulated* by the low frequency.) This is the phenomenon of **beats**.

3.4 Variable Coefficients

Next we consider second-order, linear equations with given variable coefficients $p(t)$ and $q(t)$:
$$u'' + p(t)u' + q(t)u = g(t). \tag{3.16}$$
Except for a few cases, these equations cannot be solved in analytic form using familiar functions. Even the simplest equation of this form,
$$u'' - tu = 0$$
(where $p(t) = g(t) = 0$ and $q(t) = -t$), which is called **Airy's equation**, requires the definition of a new class of functions (Airy functions) to characterize the solutions. Nevertheless, there is a well-developed theory for these equations, and we list some of the main results. We require that the coefficients $p(t)$ and $q(t)$, as well as the forcing term $g(t)$, be continuous functions on the interval I of interest. We list some basic properties of these equations; the reader will observe that these are the same properties shared by second-order, constant coefficient equations studied in Section 3.2.

1. (**Existence-Uniqueness**) If I is an open interval and t_0 belongs to I, then the initial value problem
$$u'' + p(t)u' + q(t)u = g(t), \tag{3.17}$$
$$u(t_0) = a, \quad u'(t_0) = b, \tag{3.18}$$
has a unique solution on I.

2. (**Superposition of Solutions**) If u_1 and u_2 are independent solutions of the associated homogeneous equation

$$u'' + p(t)u' + q(t)u = 0 \tag{3.19}$$

on an interval I, then $u(t) = c_1u_1 + c_2u_2$ is a solution on the interval I for any constants c_1 and c_2. Moreover, all solutions of the homogeneous equation are contained in the general solution.

3. (**Nonhomogeneous Equation**) All solutions to the nonhomogeneous equation (3.17) can be represented as the sum of the general solution to the homogeneous equation (3.19) and any particular solution to the nonhomogeneous equation (3.17). In symbols,

$$u(t) = c_1u_1(t) + c_2u_2(t) + u_p(t),$$

which is called the general solution to (3.17)

The difficulty, of course, is to find two independent solutions u_1 and u_2 to the homogeneous equation, and to find a particular solution. As we remarked, this task is difficult for equations with variable coefficients. The method of writing down the characteristic polynomial, as we did for constant coefficient equations, *does not work.*

3.4.1 Cauchy–Euler Equation

One equation that can be solved analytically is an equation of the form

$$u'' + \frac{b}{t}u' + \frac{c}{t^2}u = 0,$$

or

$$t^2u'' + btu' + cu = 0,$$

which is called a **Cauchy–Euler equation**. In each term the exponent on t coincides with the order of the derivative. Observe that we must avoid $t = 0$ in our interval of solution, because $p(t) = b/t$ and $q(t) = c/t^2$ are not continuous at $t = 0$. We try to find a solution of the form of a power function $u = t^m$. (Think about why this might work). Substituting gives the **characteristic equation**

$$m(m-1) + bm + c = 0,$$

which is a quadratic equation for m. There are three cases. If there are two distinct real roots m_1 and m_2, then we obtain two independent solutions t^{m_1} and t^{m_1}. Therefore the general solution is

$$u = c_1t^{m_1} + c_2t^{m_2}.$$

If the characteristic equation has two equal roots $m_1 = m_2 = m$, then t^m and $t^m \ln t$ are two independent solutions; in this case the general solution is

$$u = c_1 t^m + c_2 t^m \ln t.$$

When the characteristic equation has complex conjugate roots $m = \alpha \pm i\beta$, we note, using the properties of logarithms, exponentials, and Euler's formula, that a complex solution is

$$t^m = t^{\alpha+i\beta} = t^\alpha t^{i\beta} = t^\alpha e^{\ln t^{i\beta}} = t^\alpha e^{i\beta \ln t} = t^\alpha[\cos(\beta \ln t) + i \sin(\beta \ln t)].$$

The real and imaginary parts of this complex function are therefore real solutions (Theorem 3.3). So the general solution in the complex case is

$$u = c_1 t^\alpha \cos(\beta \ln t) + c_2 t^\alpha \sin(\beta \ln t).$$

Figure 3.3 shows a graph of the function $\sin(5 \ln t)$, which is a function of the type that appears in this solution. Note that this function oscillates less and less as t gets large because $\ln t$ grows very slowly. As t nears zero it oscillates infinitely many times. Because of the scale, these oscillations are not apparent on the plot.

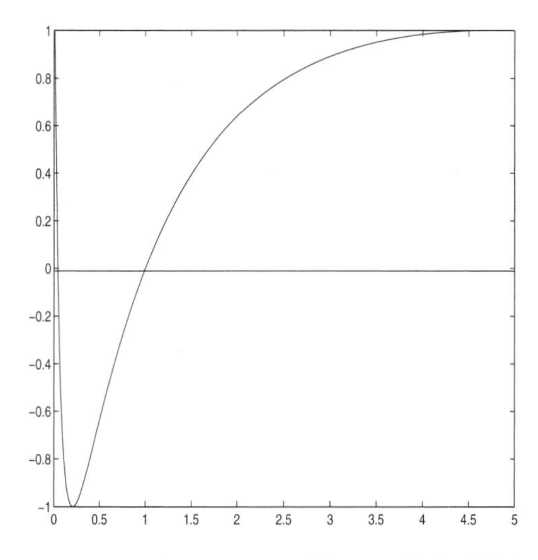

Figure 3.3 Plot of $\sin(5 \ln t)$.

Example 3.15

Consider the equation

$$t^2 u'' + t u' + 9u = 0.$$

The characteristic equation is $m(m-1)+m+9=0$, which has roots $m=\pm 3i$. The general solution is therefore

$$u = c_1 \cos(3\ln t) + c_2 \sin(3\ln t).$$

Example 3.16

Consider the equation

$$u'' = \frac{2}{t}u'.$$

We can write this in Cauchy–Euler form as

$$t^2u'' - 2tu' = 0,$$

which has characteristic equation $m(m-1)-2m=0$. The roots are $m=0$ and $m=3$. Therefore the general solution is

$$u(t) = c_1 + c_2t^3.$$

Example 3.17

Solve the initial value problem

$$t^2u'' + 3tu' + u = 0, \quad u(1) = 0, \quad u'(1) = 2.$$

The DE is Cauchy–Euler type with characteristic equation $m(m-1)+3m+1=0$. This has a double root $m=-1$, and so the general solution is

$$u(t) = \frac{c_1}{t} + \frac{c_2}{t}\ln t.$$

Now, $u(1) = c_1 = 0$ and so $u(t) = \frac{c_2}{t}\ln t$. Taking the derivative, $u'(t) = \frac{c_2}{t^2}(1-\ln t)$. Then $u'(1) = c_2 = 2$. Hence, the solution to the initial value problem is

$$u(t) = \frac{2}{t}\ln t.$$

A. Cauchy (1789–1857) and L. Euler (1707–1783) were great mathematicians who left an indelible mark on the history of mathematics and science. Their names are encountered often in advanced course in mathematics and engineering.

3.4.2 Power Series Solutions

In general, how are we to solve variable coefficient equations? Some equations can be transformed into the Cauchy–Euler equation, but that is only a small class. If we enter the equation in a computer algebra system such as Maple or Mathematica, the system will often return a general solution that is expressed in terms of so-called special functions (such as Bessel functions, Airy functions, and so on). We could define these special functions by the differential equations that we cannot solve. This is much like defining the natural logarithm function $\ln t$ as the solution to the initial value problem $u' = \frac{1}{t}$, $u(1) = 0$, as in Chapter 1. For example, we could define functions $Ai(t)$ and $Bi(t)$, the Airy functions, as two independent solutions of the DE $u'' - tu = 0$. Many of the properties of these special functions could be derived directly from the differential equation itself. But how could we get a "formula" for those functions? One way to get a representation of solutions to equations with variable coefficients is to use power series.

Let p and q be continuous on an open interval I containing t_0 and also have continuous derivatives of all orders on I. Solutions to the second-order equation with variable coefficients,

$$u'' + p(t)u' + q(t)u = 0, \tag{3.20}$$

can be approximated near $t = t_0$ by assuming a power series solution of the form

$$u(t) = \sum_{n=0}^{\infty} a_n (t - t_0)^n = a_0 + a_1(t - t_0) + a_2(t - t_0)^2 + a_3(t - t_0)^3 + \cdots.$$

The idea is to simply substitute the series and its derivatives into the differential equation and collect like terms, thereby determining the coefficients a_n. We recall from calculus that a power series converges only at $t = t_0$, for all t, or in an interval $(t_0 - R, t_0 + R)$, where R is the radius of convergence. Within its radius of convergence the power series represents a function, and the power series may be differentiated term by term to obtain derivatives of the function.

Example 3.18

Consider the DE

$$u'' - (1 + t)u = 0$$

on an interval containing $t_0 = 0$. We have

$$\begin{aligned} u(t) &= a_0 + a_1 t + a_2 t^2 + a_3 t^3 + a_4 t^4 + \cdots, \\ u'(t) &= a_1 + 2a_2 t + 3a_3 t^2 + 4a_4 t^3 + \cdots, \\ u''(t) &= 2a_2 + 6a_3 t + 12a_4 t^2 + \cdots. \end{aligned}$$

Substituting into the differential equation gives

$$2a_2 + 6a_3t + 12a_4t^2 + \cdots - (1+t)(a_0 + a_1t + a_2t^2 + a_3t^3 + \cdots) = 0.$$

Collecting like terms,

$$(-a_0 + 2a_2) + (-a_0 - a_1 + 6a_3)t + (-a_2 - a_1 + 12a_4)t^2 + \cdots = 0.$$

Therefore

$$\begin{aligned} -a_0 + 2a_2 &= 0, \\ -a_0 - a_1 + 6a_3 &= 0, \\ -a_2 - a_1 + 12a_4 &= 0, \ldots. \end{aligned}$$

Notice that all the coefficients can be determined in terms of a_0 and a_1. We have

$$a_2 = \frac{1}{2}a_0, \quad a_3 = \frac{1}{6}(a_0 + a_1), \quad a_4 = \frac{1}{12}(a_1 + a_2) = \frac{1}{12}(a_1 + \frac{1}{2}a_0), \ldots.$$

Therefore the power series for the solution $u(t)$ can be written

$$\begin{aligned} u(t) &= a_0 + a_1t + \frac{1}{2}a_0t^2 + \frac{1}{6}(a_0 + a_1)t^3 + \frac{1}{12}(a_1 + \frac{1}{2}a_0)t^4 + \cdots \\ &= a_0(1 + \frac{1}{2}t^2 + \frac{1}{6}t^3 + \frac{1}{24}t^4 + \cdots) + a_1(t + \frac{1}{6}t^3 + \frac{1}{12}t^4 + \cdots), \end{aligned}$$

which gives the general solution as a linear combination of two independent power series solutions

$$\begin{aligned} u_1(t) &= 1 + \frac{1}{2}t^2 + \frac{1}{6}t^3 + \frac{1}{24}t^4 + \cdots, \\ u_2(t) &= t + \frac{1}{6}t^3 + \frac{1}{12}t^4 + \cdots. \end{aligned}$$

The two coefficients a_0 and a_1 can be determined from initial conditions. For example, if

$$u(0) = 1, \quad u'(0) = 3,$$

then $a_0 = 1$ and $a_1 = 3$, which gives the power series solution

$$\begin{aligned} u(t) &= (1 + \frac{1}{2}t^2 + \frac{1}{6}t^3 + \frac{1}{24}t^4 + \cdots) + 3(t + \frac{1}{6}t^3 + \frac{1}{12}t^4 + \cdots) \\ &= 1 + 3t + \frac{1}{2}t^2 + \frac{2}{3}t^3 + \frac{7}{24}t^4 + \cdots. \end{aligned}$$

In this example, the power series converges for all t. We have only calculated five terms, and our truncated power series is an approximation to the actual solution to the initial value problem in a neighborhood of $t = 0$. Figure 3.4 shows the polynomial approximations by taking the first term, the first two terms, the first three, and so on.

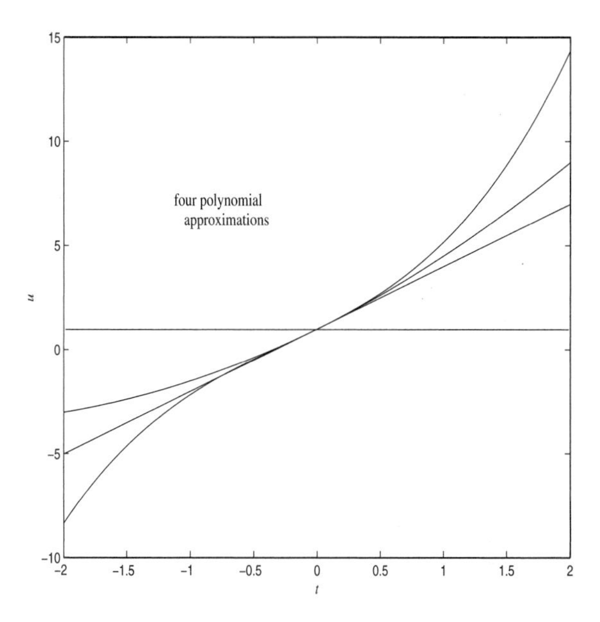

Figure 3.4 Successive polynomial approximations.

There are many important equations of the form (3.20) where the coefficients p and q do not satisfy the regularity properties (having continuous derivatives of all order) mentioned at the beginning of this subsection. However, if p and q are not too ill-behaved at t_0, we can still seek a series solution. In particular, if $(t-t_0)p(t)$ and $(t-t_0)^2q(t)$ have convergent power series expansions in an interval about t_0, then we say t_0 is a **regular singular point** for (3.20), and we attempt a series solution of the form

$$u(t) = t^r \sum_{n=0}^{\infty} a_n (t-t_0)^n,$$

where r is some number. Substitution of this form into (3.20) leads to equations for both r and the coefficients a_n. This technique, which is called the **Frobenius method**, is explored in the Exercises.

3.4.3 Reduction of Order

If one solution $u_1(t)$ of the DE

$$u'' + p(t)u' + q(t)u = 0$$

happens to be known, then a second, linearly independent solution $u_2(t)$ can be found of the form $u_2(t) = v(t)u_1(t)$, for some $v(t)$ to be determined. To find

$v(t)$ we substitute this form for $u_2(t)$ into the differential equation to obtain a first-order equation for $v(t)$. This method is called **reduction of order**, and we illustrate it with an example.

Example 3.19

Consider the DE

$$u'' - \frac{1}{t}u' + \frac{1}{t^2}u = 0.$$

An obvious solution is $u_1(t) = t$. So let $u_2 = v(t)t$. Substituting, we get

$$(2v' + tv'') - \frac{1}{t}(v + tv') + \frac{1}{t^2}vt = 0,$$

which simplifies to

$$tv'' + v' = 0.$$

Letting $w = v'$, we get the first-order equation

$$tw' + w = 0.$$

By separating variables and integrating we get $w = 1/t$. Hence $v = \int \frac{1}{t}dt = \ln t$, and the second independent solution is $u_2(t) = t \ln t$. Consequently, the general solution of the equation is

$$u(t) = c_1 t + c_2 t \ln t.$$

Note that this example is a Cauchy–Euler equation; but the method works on general linear second-order equations.

3.4.4 Variation of Parameters

There is a general formula for the particular solution to the nonhomogeneous equation

$$u'' + p(t)u' + q(t)u = g(t), \tag{3.21}$$

called the variation of parameters formula.

Recall how we attacked the first-order linear equation

$$u' + p(t)u = g(t)$$

in Chapter 1. We first found the solution of the associated homogeneous equation $u' + p(t)u = 0$, as $Ce^{-P(t)}$, where $P(t)$ is an antiderivative of $p(t)$. Then we found the solution to the nonhomogeneous equation by varying the constant

C (i.e., by assuming $u(t) = C(t)e^{-P(t)}$). Substitution of this into the equation yielded $C(t)$ and therefore the solution.

The same method works for second-order equations, but the calculations are more involved. Let u_1 and u_2 be independent solutions to the homogeneous equation

$$u'' + p(t)u' + q(t)u = 0.$$

Then

$$u_h(t) = c_1u_1(t) + c_2u_2(t)$$

is the general solution of the homogeneous equation. To find a particular solution we vary both parameters c_1 and c_2 and take

$$u_p(t) = c_1(t)u_1(t) + c_2(t)u_2(t). \tag{3.22}$$

Now we substitute this expression into the nonhomogeneous equation to get expressions for $c_1(t)$ and $c_2(t)$. This is a tedious task in calculus and algebra, and we leave most of the details to the interested reader. But here is how the argument goes. We calculate u_p' and u_p'' so that we can substitute into the equation. For notational simplicity, we drop the t variable in all of the functions. We have

$$u_p' = c_1u_1' + c_2u_2' + c_1'u_1 + c_2'u_2.$$

There is flexibility in our answer so let us set

$$c_1'u_1 + c_2'u_2 = 0. \tag{3.23}$$

Then

$$\begin{aligned} u_p' &= c_1u_1' + c_2u_2', \\ u_p'' &= c_1u_1'' + c_2u_2'' + c_1'u_1' + c_2'u_2'. \end{aligned}$$

Substituting these into the nonhomogeneous DE gives

$$c_1u_1'' + c_2u_2'' + c_1'u_1' + c_2'u_2' + p(t)[c_1u_1' + c_2u_2'] + q(t)[c_1u_1 + c_2u_2] = g(t).$$

Now we observe that u_1 and u_2 satisfy the homogeneous equation, and this simplifies the last equation to

$$c_1'u_1' + c_2'u_2' = g(t). \tag{3.24}$$

Equations (3.23) and (3.24) form a system of two linear algebraic equations in the two unknowns c_1' and c_2'. If we solve these equations and integrate we finally obtain (readers should fill in the details)

$$c_1(t) = -\int \frac{u_2(t)g(t)}{W(t)}dt, \quad c_2(t) = \int \frac{u_1(t)g(t)}{W(t)}dt, \tag{3.25}$$

where

$$W(t) = u_1(t)u_2'(t) - u_1'(t)u_2(t). \tag{3.26}$$

This expression $W(t)$ is called the **Wronskian**. Combining the previous expressions gives the **variation of parameters formula** for the particular solution of (3.21):

$$u_p(t) = -u_1(t)\int \frac{u_2(t)g(t)}{W(t)}dt + u_2(t)\int \frac{u_1(t)g(t)}{W(t)}dt.$$

The general solution of (3.21) is the homogeneous solution $u_h(t)$ plus this particular solution. If the antiderivatives in (3.25) cannot be computed explicitly, then the integrals should be written with a variable limit of integration.

Example 3.20

Find a particular solution to the DE

$$u'' + 9u = 3\sec 3t.$$

Here the homogeneous equation $u'' + 9u = 0$ has two independent solutions $u_1 = \cos 3t$ and $u_2 = \sin 3t$. The Wronskian is

$$W(t) = 3\cos^2 t + 3\sin^2 3t = 3.$$

Therefore

$$c_1(t) = -\int \frac{\sin 3t \cdot 3\sec 3t}{3}dt, \quad c_2(t) = \int \frac{\cos 3t \cdot 3\sec 3t}{3}dt.$$

Simplifying,

$$c_1(t) = -\int \tan 3t dt = \frac{1}{3}\ln(\cos 3t), \quad c_2(t) = \int 1 dt = t.$$

We do not need constants of integration because we seek only the particular solution. Therefore the particular solution is

$$u_p(t) = \frac{1}{3}\ln(\cos 3t) + t\sin 3t.$$

The general solution is

$$u(t) = c_1\cos 3t + c_2\sin 3t + \frac{1}{3}\ln(\cos 3t) + t\sin 3t.$$

The constants may be determined by initial data, if given.

When the second-order equation has constant coefficients and the forcing term is a polynomial, exponential, sine, or cosine, then the method of undetermined coefficients works more easily than the variation of parameters formula. For other cases we use the formula or Laplace transform methods, which are the subject of Chapter 4. Of course, the easiest method of all is to use a computer algebra system. When you have paid your dues by using analytic methods on several problems, then you have your license and you may use a computer algebra system. The variation of parameters formula is important because it is often used in the theoretical analysis of problems in advanced differential equations.

EXERCISES

1. Solve the following initial value problems:

   a) $t^2u'' + 3tu' - 8u = 0$, $u(1) = 0$, $u'(1) = 2$.

   b) $t^2u'' + tu' = 0$, $u(1) = 0$, $u'(1) = 2$.

   c) $t^2u'' - tu' + 2u = 0$, $u(1) = 0$, $u'(1) = 1$.

2. For what value(s) of β is $u = t^\beta$ a solution to the equation $(1 - t^2)u'' - 2tu' + 2u = 0$?

3. This exercise presents a transformation method for solving a Cauchy–Euler equation. Show that the transformation $x = \ln t$ to a new independent variable x transforms the Cauchy–Euler equation $at^2u'' + btu' + cu = 0$ into an linear equation with constant coefficients. Use this method to solve Exercise 1a.

4. Use the power series method to obtain two independent, power series solutions to $u'' + u = 0$ about $t_0 = 0$ and verify that the series are the expansions of $\cos t$ and $\sin t$ about $t = 0$.

5. Use the power series method to find the first three terms of two independent power series solutions to Airy's equation $u'' - tu = 0$, centered at $t_0 = 0$.

6. Find the first three terms of two independent power series solutions to the equation $(1 + t^2)u'' + u = 0$ near $t_0 = 0$.

7. Solve the first-order nonlinear initial value problem $u' = 1 + u^2$, $u(0) = 1$, using a power series method. Compare the accuracy of the partial sums to the exact solution. (Hint: you will have to square out a power series.)

8. Consider the equation $u'' - 2tu' + 2nu = 0$, which is Hermite's differential equation, an important equation in quantum theory. Show that if n is a nonnegative integer, then there is a polynomial solution $H_n(t)$ of degree n,

which is called a Hermite polynomial of degree n. Find $H_0(t), ..., H_5(t)$ up to a constant multiple.

9. Consider the equation $u'' - 2au' + a^2u = 0$, which has solution $u = e^{at}$. Use reduction of order to find a second independent solution.

10. One solution of
$$u'' - \frac{t+2}{t}u' + \frac{t+2}{t^2}u = 0$$
is $u_1(t) = t$. Find a second independent solution.

11. One solution of
$$t^2u'' + tu' + (t^2 - \frac{1}{4})u = 0$$
is $u_1(t) = \frac{1}{\sqrt{t}} \cos t$. Find a second independent solution.

12. Let $y(t)$ be one solution of the equation $u'' + p(t)u' + q(t)u = 0$. Show that the reduction of order method with $u(t) = v(t)y(t)$ leads to the first-order linear equation
$$yz' + (2y' + py)z = 0, \quad z = v'.$$
Show that
$$z(t) = \frac{Ce^{-\int p(t)dt}}{y(t)^2},$$
and then find a second linear independent solution of the equation in the form of an integral.

13. Use ideas from the last exercise to find a second-order linear equation that has independent solutions e^t and $\cos t$.

14. Let u_1 and u_2 be independent solutions of the linear equation $u'' + p(t)u' + q(t)u = 0$ on an interval I and let $W(t)$ be the Wronskian of u_1 and u_2. Show that
$$W'(t) = -p(t)W(t),$$
and then prove that $W(t) = 0$ for all $t \in I$, or $W(t)$ is never zero on I.

15. Find the general solution of $u'' + tu' + u = 0$ given that $u = e^{-t^2/2}$ is one solution.

16. Use the transformation $u = \exp\left(\int y(t)dt\right)$ to convert the second-order equation $u'' + p(t)u' + q(t)u = 0$ to a **Riccati equation** $y' + y^2 + p(t)y + q(t) = 0$. Conversely, show that the Riccati equation can be reduced to the second-order equation in u using the transformation $y = u'/u$. Solve the first-order nonautonomous equation
$$y' = -y^2 + \frac{3}{t}y.$$

17. Use the variation of parameters formula to find a particular solution to the following equations.

    a) $u'' + \frac{1}{t}u = a$, where a is a constant. Note that 1 and $\ln t$ are two independent solutions to the homogeneous equation.

    b) $u'' + u = \tan t$.

    c) $u'' - u = te^t$.

    d) $u'' - u = \frac{1}{t}$.

    e) $t^2u'' - 2u = t^3$.

18. (Frobenius method) Consider the differential equation (**Bessel's equation** of order k)

$$u'' + \frac{1}{t}u' + \left(1 - \frac{k^2}{t^2}\right)u = 0,$$

where k is a real number.

    a) Show that $t_0 = 0$ is a regular singular point for the equation.

    b) Assuming a solution of the form $u(t) = t^r \sum_{n=0}^{\infty} a_n t^n$, show that $r = \pm k$.

    c) In the case that $k = \frac{1}{3}$, find the first three terms of two independent series solutions to the DE.

    d) Show that if $k = 0$ then the Frobenius method leads to only one series solution, and find the first three terms. (The entire series, which converges for all t, is denoted by $J_0(t)$ and is called a **Bessel function** of the first kind of order zero. Finding a second independent solution is beyond the scope of our treatment.)

3.5 Boundary Value Problems and Heat Flow

Let us consider the following problem in steady-state heat conduction. A cylindrical, uniform, metallic bar of length L and cross-sectional area A is insulated on its lateral side. We assume the left face at $x = 0$ is maintained at T_0 degrees and that the right face at $x = L$ is held at T_L degrees. What is the temperature distribution $u = u(x)$ in the bar after it comes to equilibrium? Here $u(x)$ represents the temperature of the entire cross section of the bar at position x, where $0 < x < L$. We are assuming that heat flows only in the axial direction along the bar, and we are assuming that any transients caused by initial temperatures in the bar have decayed away. In other words, we have waited long enough for the temperature to reach a steady state. One might conjecture that the temperature distribution is a linear function of x along the bar; that is, $u(x) = T_0 + \frac{T_L - T_0}{L}x$. This is indeed the case, which we show below. But also we want to consider a more complicated problems where the bar has both a variable conductivity and an internal heat source along its length. An internal heat source, for example, could be resistive heating produced by a current running through the medium.

The physical law that provides the basic model is conservation of energy. If $[x, x + dx]$ is any small section of the bar, then the rate that heat flows in at x, minus the rate that heat flows out at $x + dx$, plus the rate that heat is generated by sources, must equal zero, because the system is in a steady state. See figure 3.5.

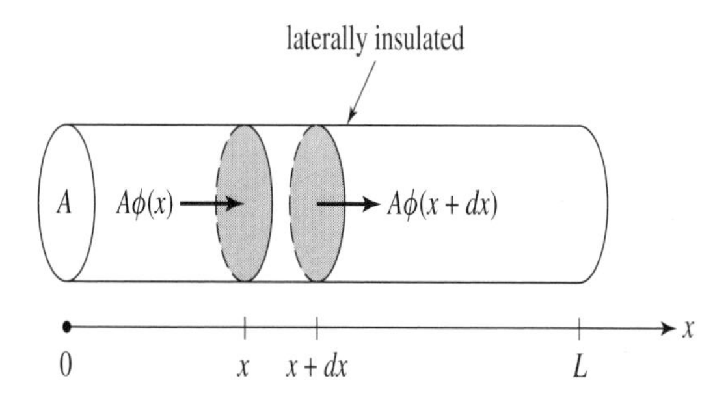

Figure 3.5 Cylindrical bar, laterally insulated, through which heat is flowing in the x-direction. The temperature is uniform in a fixed cross-section.

If we denote by $\phi(x)$ the rate that heat flows to the right at any section x (measured in calories/(area · time), and we let $f(x)$ denote the rate that heat is internally produced at x, measured in calories/(volume · time), then

$$A\phi(x) - A\phi(x + dx) + f(x)A\,dx = 0.$$

Canceling A, dividing by dx, and rearranging gives

$$\frac{\phi(x + dx) - \phi(x)}{dx} = f(x).$$

Taking the limit as $dx \to 0$ yields

$$\phi'(x) = f(x). \tag{3.27}$$

This is an expression of energy conservation in terms of flux. But what about temperature? Empirically, the flux $\phi(x)$ at a section x is found to be proportional to the negative temperature gradient $-u'(x)$ (which measures the steepness of the temperature distribution, or profile, at that point), or

$$\phi(x) = -K(x)u'(x). \tag{3.28}$$

This is **Fourier's heat conduction law**. The given proportionality factor $K(x)$ is called the **thermal conductivity**, in units of energy/(length · degrees · time), which is a measure of how well the bar conducts heat at location x. For a uniform bar K is constant. The minus sign in (3.28) means that heat flows from higher temperatures to lower temperatures. Fourier's law seems intuitively correct and it conforms with the second law of thermodynamics; the larger the temperature gradient, the faster heat flows from high to low temperatures. Combining (3.27) and (3.28) leads to the equation

$$-(K(x)u'(x))' = f(x), \quad 0 < x < L, \tag{3.29}$$

which is the **steady-state heat conduction equation**. When the **boundary conditions**

$$u(0) = T_0, \quad u(L) = T_1, \tag{3.30}$$

are appended to (3.29), we obtain a **boundary value problem** for the temperature $u(x)$. Boundary conditions are conditions imposed on the unknown state u given at different values of the independent variable x, unlike initial conditions that are imposed at a single value. For boundary value problems we usually use x as the independent variable because boundary conditions usually refer to the boundary of a spatial domain.

Note that we could expand the heat conduction equation to

$$-K(x)u''(x) - K'(x)u'(x) = f(x), \tag{3.31}$$

but there is little advantage in doing so.

Example 3.21

If there are no sources ($f(x) = 0$) and if the thermal conductivity $K(x) = K$ is constant, then the boundary value problem reduces to

$$\begin{aligned} u'' &= 0, \quad 0 < x < L, \\ u(0) &= T_0, \quad u(L) = T_1. \end{aligned}$$

Thus the bar is homogeneous and can be characterized by a constant conductivity. The general solution of $u'' = 0$ is $u(x) = c_1 x + c_2$; applying the boundary conditions determines the constants c_1 and c_2 and gives the linear temperature distribution $u(x) = T_0 + \frac{T_L - T_0}{L}x$, as we previously conjectured.

In nonuniform systems the thermal conductivity K depends upon location x in the system. And, K may depend upon the temperature u as well. Moreover, the heat source term f could depend on location and temperature. In these cases the steady-state heat conduction equation (3.29) takes the more general form

$$-(K(x,u)u')' = f(x,u),$$

which is a nonlinear second-order equation for the steady temperature distribution $u = u(x)$.

Boundary conditions at the ends of the bar may also specify the flux rather than the temperature. For example, in a homogeneous system, if heat is injected at $x = 0$ at a rate of N calories per area per time, then the left boundary condition takes the form $\phi(0) = N$, or

$$-Ku'(0) = N.$$

Thus, a flux condition at an endpoint imposes a condition on the derivative of the temperature at that endpoint. In the case that the end at $x = L$, say, is insulated, so that no heat passes through that end, then the boundary condition is

$$u'(L) = 0,$$

which is called an **insulated boundary condition**. As the reader can see, there are a myriad of interesting boundary value problems associated with heat flow. Similar equations arise in diffusion processes in biology and chemistry, for example, in the diffusion of toxic substances where the unknown is the chemical concentration.

Boundary value problems are much different from initial value problems in that they may have no solution, or they may have infinitely many solutions. Consider the following.

Example 3.22

When $K = 1$ and the heat source term is $f(u) = 9u$ and both ends of a bar of length $L = 2$ are held at $u = 0$ degrees, the boundary value problem becomes

$$\begin{aligned} -u'' &= 9u, \quad 0 < x < 2. \\ u(0) &= 0, \quad u(2) = 0. \end{aligned}$$

The general solution to the DE is $u(x) = c_1 \sin 3x + c_2 \cos 3x$, where c_1 and c_2 are arbitrary constants. Applying the boundary condition at $x = 0$ gives $u(0) = c_1 \sin(3 \cdot 0) + c_2 \cos(3 \cdot 0) = c_2 = 0$. So the solution must have the form $u(x) = c_1 \sin 3x$. Next apply the boundary condition at $x = 2$. Then $u(2) = c_1 \sin(6) = 0$, to obtain $c_1 = 0$. We have shown that the only solution is $u(x) = 0$. There is no nontrivial steady state. But if we make the bar length π, then we obtain the boundary value problem

$$\begin{aligned} -u'' &= 9u, \quad 0 < x < \pi. \\ u(0) &= 0, \quad u(\pi) = 0. \end{aligned}$$

The reader should check that this boundary value problem has infinitely many solutions $u(x) = c_1 \sin 3x$, where c_1 is any number. If we change the right boundary condition, one can check that the boundary value problem

$$\begin{aligned} -u'' &= 9u, \quad 0 < x < \pi. \\ u(0) &= 0, \quad u(\pi) = 1, \end{aligned}$$

has no solution at all.

Example 3.23

Find all real values of λ for which the boundary value problem

$$-u'' = \lambda u, \quad 0 < x < \pi. \tag{3.32}$$

$$u(0) = 0, \quad u'(\pi) = 0, \tag{3.33}$$

has a nontrivial solution. These values are called the **eigenvalues**, and the corresponding nontrivial solutions are called the **eigenfunctions**. Interpreted in the heat flow context, the left boundary is held at zero degrees and the right end is insulated. The heat source is $f(u) = \lambda u$. We are trying to find which linear heat sources lead to nontrivial steady states. To solve this problem we consider different cases because the form of the solution will be different for $\lambda = 0$, $\lambda < 0$, $\lambda > 0$. If $\lambda = 0$ then the general solution of $u'' = 0$ is $u(x) = ax+b$. Then $u'(x) = a$. The boundary condition $u(0) = 0$ implies $b = 0$ and the boundary condition $u'(\pi) = 0$ implies $a = 0$. Therefore, when $\lambda = 0$, we get only a trivial solution. Next consider the case $\lambda < 0$ so that the general solution will have the form

$$u(t) = a \sinh \sqrt{-\lambda}x + b \cosh \sqrt{-\lambda}x.$$

The condition $u(0) = 0$ forces $b = 0$. Then $u'(t) = a\sqrt{-\lambda} \cosh \sqrt{-\lambda}x$. The right boundary condition becomes $u'(\pi) = a\sqrt{-\lambda}\cosh(\sqrt{-\lambda} \cdot 0) = 0$, giving $a = 0$. Recall that $\cosh 0 = 1$. Again there is only the trivial solution. Finally assume $\lambda > 0$. Then the general solution takes the form

$$u(t) = a \sin \sqrt{\lambda}x + b \cos \sqrt{\lambda}x.$$

The boundary condition $u(0) = 0$ forces $b = 0$. Then $u(t) = a \sin \sqrt{\lambda} x$ and $u'(x) = a\sqrt{\lambda} \cos \sqrt{\lambda} x$. Applying the right boundary condition gives

$$u'(\pi) = a\sqrt{\lambda} \cos \sqrt{\lambda}\pi = 0.$$

Now we do not have to choose $a = 0$ (which would again give the trivial solution) because we can satisfy this last condition with

$$\cos \sqrt{\lambda}\pi = 0.$$

The cosine function is zero at the values $\pi/2 \pm n\pi$, $n = 0, 1, 2, 3, ...$Therefore

$$\sqrt{\lambda}\pi = \pi/2 + n\pi, \qquad n = 0, 1, 2, 3, ...$$

Solving for λ yields

$$\lambda = \left(\frac{2n+1}{2}\right)^2, \qquad n = 0, 1, 2, 3, ...$$

Consequently, the values of λ for which the original boundary value problem has a nontrivial solution are $\frac{1}{4}, \frac{9}{4}, \frac{25}{4}, ...$ These are the eigenvalues. The corresponding solutions are

$$u(x) = a \sin\left(\frac{2n+1}{2}\right)x, \qquad n = 0, 1, 2, 3, ...$$

These are the eigenfunctions. Notice that the eigenfunctions are unique only up to a constant multiple. In terms of heat flow, the eigenfunctions represent possible steady-state temperature profiles in the bar. The eigenvalues are those values λ for which the boundary value problem will have steady-state profiles.

Boundary value problems are of great interest in applied mathematics, science, and engineering. They arise in many contexts other than heat flow, including wave motion, quantum mechanics, and the solution of partial differential equations.

EXERCISES

1. A homogeneous bar of length 40 cm has its left and right ends held at 30°C and 10°C, respectively. If the temperature in the bar is in steady-state, what is the temperature in the cross section 12 cm from the left end? If the thermal conductivity is K, what is the rate that heat is leaving the bar at its right face?

2. The thermal conductivity of a bar of length $L = 20$ and cross-sectional area $A = 2$ is $K(x) = 1$, and an internal heat source is given by $f(x) = 0.5x(L - x)$. If both ends of the bar are maintained at zero degrees, what is the steady state temperature distribution in the bar? Sketch a graph of $u(x)$. What is the rate that heat is leaving the bar at $x = 20$?

3. For a metal bar of length L with no heat source and thermal conductivity $K(x)$, show that the steady temperature in the bar has the form

$$u(x) = c_1 \int_0^x \frac{dy}{K(y)} + c_2,$$

where c_1 and c_2 are constants. What is the temperature distribution if both ends of the bar are held at zero degrees? Find an analytic formula and plot the temperature distribution in the case that $K(x) = 1 + x$. If the left end is held at zero degrees and the right end is insulated, find the temperature distribution and plot it.

4. Determine the values of λ for which the boundary value problem

$$\begin{aligned} -u'' &= \lambda u, \quad 0 < x < 1, \\ u(0) &= 0, \quad u(1) = 0, \end{aligned}$$

has a nontrivial solution.

5. Consider the nonlinear heat flow problem

$$\begin{aligned} (uu')' &= 0, \quad 0 < x < \pi, \\ u(0) &= 0, \quad u'(\pi) = 1, \end{aligned}$$

where the thermal conductivity depends on temperature and is given by $K(u) = u$. Find the steady-state temperature distribution.

6. Show that if there is a solution $u = u(x)$ to the boundary value problem (3.29)–(3.30), then the following condition must hold:

$$-K(L)u'(L) + K(0)u'(0) = \int_0^L f(x)dx.$$

Interpret this condition physically.

7. Consider the boundary value problem

$$u'' + \omega^2 u = 0, \quad u(0) = a, \quad u(L) = b.$$

When does a unique solution exist?

8. Find all values of λ for which the boundary value problem

$$\begin{aligned} -u'' - 2u' &= \lambda u, \quad 0 < x < 1, \\ u(0) &= 0, \quad u(1) = 0, \end{aligned}$$

has a nontrivial solution.

9. Show that the eigenvalues of the boundary value problem

$$\begin{aligned} -u'' &= \lambda u, \quad 0 < x < 1, \\ u'(0) &= 0, \quad u(1) + u'(1) = 0, \end{aligned}$$

are given by the numbers $\lambda_n = p_n^2$, $n = 1, 2, 3, \ldots$, where the p_n are roots of the equation $\tan p = 1/p$. Plot graphs of $\tan p$ and $1/p$ and indicate graphically the locations of the values p_n. Numerically calculate the first four eigenvalues.

10. Find the values of λ (eigenvalues) for which the boundary value problem

$$\begin{aligned} -x^2u'' - xu' &= \lambda u, \quad 1 < x < e^{\pi}, \\ u(1) &= 0, \quad u(e^{\pi}) = 0, \end{aligned}$$

has a nontrivial solution.

3.6 Higher-Order Equations

So far we have dealt with first- and second-order equations. Higher-order equations occur in some applications. For example, in solid mechanics the vertical deflection $y = y(x)$ of a beam from its equilibrium satisfies a fourth-order equation. However, the applications of higher-order equations are not as extensive as those for their first- and second-order counterparts.

Here, we outline the basic results for a homogeneous, nth-order linear DE with constant coefficients:

$$u^{(n)} + p_{n-1}u^{(n-1)} + \cdots + p_1 u' + p_0 u = 0. \tag{3.34}$$

The p_i, $i = 0, 1, \ldots, n-1$, are specified constants. The **general solution** of (3.34) has the form

$$u(t) = c_1 u_1(t) + c_2 u_2(t) + \cdots + c_n u_n(t),$$

where $u_1(t), u_2(t), \ldots, u_n(t)$ are independent solutions, and where $c_1, c_2, \ldots, c_n$ are arbitrary constants. In different words, the general solution is a linear combination of n different basic solutions. To find these basic solutions we try the same strategy that worked for a second-order equation, namely assume a solution of the form of an exponential function

$$u(t) = e^{\lambda t},$$

where λ is to be determined. Substituting into the equation gives

$$\lambda^n + p_{n-1}\lambda^{n-1} + \cdots + p_1\lambda + p_0 = 0, \tag{3.35}$$

which is an nth degree polynomial equation for λ. Equation (3.35) is the **characteristic equation**. From algebra we know that there are n roots $\lambda_1, \lambda_2, ..., \lambda_n$. Here we are counting multiple roots and complex roots (the latter will always occur in complex conjugate pairs $a \pm bi$). A root $\lambda = a$ has **multiplicity** K if $(\lambda - a)^K$ appears in the factorization of the characteristic polynomial.

If the characteristic roots are all *real and distinct*, we will obtain n different basic solutions $u_1(t) = e^{\lambda_1 t}$, $u_2(t) = e^{\lambda_2 t}, ..., u_n(t) = e^{\lambda_n t}$. In this case the general solution of (3.34) will be a linear combination of these,

$$u(t) = c_1e^{\lambda_1 t} + c_2e^{\lambda_2 t} + \cdots + c_ne^{\lambda_n t}. \tag{3.36}$$

If the roots of (3.35) are not real and distinct then we proceed as might be expected from our study of second-order equations. A complex conjugate pair, $\lambda = a \pm ib$ gives rise to two real solutions $e^{at}\cos bt$ and $e^{at}\sin bt$. A double root λ (multiplicity 2) leads to two solutions $e^{\lambda t}$ and $te^{\lambda t}$. A triple root λ (multiplicity 3) leads to three independent solutions $e^{\lambda t}$, $te^{\lambda t}$, $t^2e^{\lambda t}$, and so on. In this way we can build up from the factorization of the characteristic polynomial a set of n independent, basic solutions of (3.34). The hardest part of the problem is to find the characteristic roots; computer algebra systems are often useful for this task.

As may be expected from our study of second-order equations, an nth-order nonhomogeneous equation of the form

$$u^{(n)} + p_{n-1}u^{(n-1)} + \cdots + p_1u' + p_0u = g(t), \tag{3.37}$$

has a general solution that is the sum of the general solution (3.36) of the homogeneous equation and a particular solution to the equation (3.37). This result is true even if the coefficients p_i are functions of t.

Example 3.24

If the characteristic equation for a 6th-order equation has roots $\lambda = -2 \pm 3i, 4, 4, 4, -1$, the general solution will be

$$u(t) = c_1e^{-2t}\cos 3t + c_2e^{-2t}\sin 3t + c_3e^{4t} + c_4te^{4t} + c_5t^2e^{4t} + c_6e^{-t}.$$

Initial conditions for an nth order equation (3.34) at $t = 0$ take the form

$$u(0) = \alpha_1, \quad u'(0) = \alpha_2, ..., u^{(n-1)}(0) = \alpha_{n-1},$$

where the α_i are given constants. Thus, for an nth-order initial value problem we specify the value of the function and all of its derivatives up to the $(n-1)$st-order, at the initial time. These initial conditions determine the n arbitrary constants in the general solution and select out a unique solution to the initial value problem.

Example 3.25

Consider

$$u''' - 2u'' - 3u' = 5e^{4t}.$$

The characteristic equation for the homogeneous equation is

$$\lambda^3 - 2\lambda^2 - 3\lambda = 0,$$

or

$$\lambda(\lambda - 3)(\lambda + 1) = 0.$$

The characteristic roots are $\lambda = 0, -1, 3$, and therefore the homogeneous equation has solution

$$u_h(t) = c_1 + c_2 e^{-t} + c_3 e^{3t}.$$

The particular solution will have the form $u_p(t) = ae^{4t}$. Substituting into the original nonhomogeneous equation gives $a = 1/4$. Therefore the general solution to the equation is

$$u(t) = c_1 + c_2 e^{-t} + c_3 e^{3t} + \frac{1}{4}e^{4t}.$$

The three constants can be determined from initial conditions. For example, for a third-order equation the initial conditions at time $t = 0$ have the form

$$u(0) = \alpha, \quad u'(0) = \beta, \quad u''(0) = \gamma,$$

for some given constants α, β, γ. Of course, initial conditions can be prescribed at any other time t_0.

EXERCISES

1. Find the general solution of the following differential equations:

   a) $u''' + u' = 0$.

   b) $u'''' + u' = 1$.

c) $u'''' + u'' = 0$.

d) $u''' - u' - 8u = 0$.

e) $u''' + u'' = 2e^t + 3t^2$.

2. Solve the initial value problem $u''' - u'' - 4u' - 4u = 0$, $u(0) = 2$, $u'(0) = -1$, $u''(0) = 5$.

3. Write down a linear, fifth-order differential equation whose general solution is

$$u = c_1 + c_2 t + c_3 e^{-4t} + e^{5t}(c_4 \cos 2t + c_5 \sin 5t).$$

4. Show that the third-order equation $u''' + 2u'' - 5u' - u = 0$ can be written as an equivalent system of three first-order equations in the variables u, v, and w, where $v = u'$ and $w = u''$.

5. What is the general solution of a fourth-order differential equation if the four characteristic roots are $\lambda = 3 \pm i,\ 3 \pm i$? What is the differential equation?

3.7 Summary and Review

One way to think about learning and solving differential equations is in terms of pattern recognition. Although this is a very "compartmentalized" way of thinking, it does help our learning process. When faced with a differential equation, what do we do? The first step is to recognize what type it is. It is like a pianist recognizing a certain set of notes in a complicated musical piece and then playing those notes easily because of long hours of practice. In differential equations we must practice to recognize an equation and learn the solution technique that works for that equation. At this point in your study, what kinds of equations should you surely be able to recognize and solve?

The simplest is the **pure time equation**

$$u' = g(t).$$

Here u is the antiderivative of $g(t)$, and we sometimes have to write the solution as an integral when we cannot find a simple form for the antiderivative. The next simplest equation is the **separable equation**

$$u' = g(t)f(u),$$

where the right side is a product of functions of the dependent and independent variables. These are easy: just separate variables and integrate. **Autonomous equations** have the form

$$u' = f(u),$$

where the right side depends only on u. These equations are separable, should we want to attempt a solution. But often, for autonomous equations, we apply qualitative methods to understand the behavior of solutions. This includes graphing $f(u)$ vs. u, finding the equilibrium solutions, and then drawing arrows on the phase line to determine stability of the equilibrium solutions and whether u is increasing or decreasing. Nearly always these qualitative methods are superior to having an actual solution formula. First-order autonomous equations cannot have oscillatory solutions. Finally, the first-order **linear equation** is

$$u' = p(t)u + g(t).$$

Here we use variation of parameters or integrating factors. Sometimes an equation can be solved by multiple methods; for example, $u' = 2u - 7$ is separable, linear, and autonomous.

There are other first-order nonlinear equations that can be solved, and some of these were introduced in the Exercises. The **Bernoulli equation**

$$u' = p(t)u + g(t)u^n$$

can be transformed into a linear equation for the variable $y = u^{1-n}$, and the **homogeneous equation**

$$u' = f(\frac{u}{t})$$

can be transformed into a separable equation for the variable $y = u/t$. Solutions to special and unusual equations can sometimes be found in mathematical handbooks or in computer algebra systems.

There are really only two second-order linear equations that can be solved simply. These are the **equation with constant coefficients**

$$au'' + bu' + cu = 0,$$

where we have solutions of the form $u = e^{\lambda t}$, with λ satisfying the characteristic equation $a\lambda^2 + b\lambda + c = 0$, and the **Cauchy–Euler equation**

$$at^2u'' + btu' + cu = 0,$$

where we have solutions of the form $u = t^m$, where m satisfies the characteristic equation $am(m-1) + bm + c = 0$. For these two problems we must distinguish when the roots of the characteristic equation are real and unequal, real and equal, or complex. When the right side of either of these equations is nonzero, then the equation is nonhomogeneous. Then we can find particular solutions using the **variation of parameters** method, which works for all linear equations, or use **undetermined coefficients**, which works only for constant coefficient equations with special right sides. Nonhomogeneous linear equations with constant coefficients can also be handled by Laplace transforms, which are discussed in the next chapter. All these methods extend to higher-order equations.

Generally, we cannot easily solve homogeneous second-order linear equations with variable coefficients, or equations having the form

$$u'' + p(t)u' + q(t)u = 0.$$

Many of these equations have solutions that can be written as power series. These power series solutions define **special functions** in mathematics, such as Bessel functions, Hermite polynomials, and so forth. In any case, you can *not* solve these variable coefficient equations using the characteristic polynomial, and nonhomogeneous equations are not amenable to the methods of undetermined coefficients. If you are fortunate enough to find one solution, you can determine a second by reduction of order. If you are lucky enough to find two independent solutions to the homogeneous equation, the method of variation of parameters gives a particular solution.

The basic structure theorem holds for all linear nonhomogeneous equations: the general solution is the sum of the general solution to the homogeneous equation and a particular solution. This result is fundamental.

Second-order equations coming from Newton's second law have the form $x'' = F(t, x, x')$. These can be reduced to first-order equations when t or x is missing from the force F, or when $F = F(x)$, which is the conservative case.

The Exercises give you review and practice in identifying and solving differential equations.

EXERCISES

1. Identify each of the differential equations and find the general solution. Some of the solutions may contain an integral.

   a) $2u'' + 5u' - 3u = 0$.

   b) $u' - Ru = 0$, where R is a parameter.

   c) $u' = \cos t - u \cos t$.

   d) $u' - 6u = e^t$.

   e) $u'' = -\frac{2}{t^2}u$.

   f) $u'' + 6u' + 9u = 5 \sin t$.

g) $u' = -8t + 6$.

h) $u'' + u = t^2 - 2t + 2$

i) $u' + u - tu^3 = 0$.

j) $2u'' + u' + 3u = 0$.

k) $x'' = (x')^3$.

l) $tu' + u = t^2u^2$.

m) $u'' = -3u^2$.

n) $tu' = u - \frac{t}{2}\cos^2\left(\frac{2u}{t}\right)$.

o) $u''' + 5u'' - 6u' = 9e^{3t}$.

p) $(6tu - u^3) + (4u + 3t^2 - 3tu^2)u' = 0$.

2. Solve the initial value problem $u' = u^2 \cos t$, $u(0) = 2$, and find the interval of existence.

3. Solve the initial value problem $u' = \frac{2}{t}u + t$, $u(1) = 2$, and find the interval of existence.

4. Use the power series method to find the first three terms of two independent solutions to $u'' + tu' + tu = 0$ valid near $t = 0$.

5. For all cases, find the equilibrium solutions for $u' = (u - a)(u^2 - a)$, where a is a real parameter, and determine their stability. Summarize the information on a bifurcation diagram.

6. A spherical water droplet loses volume by evaporation at a rate proportional to its surface area. Find its radius $r = r(t)$ in terms of the proportionality constant and its initial radius r_0.

7. A population is governed by the law $p' = rp\left(\frac{K-p}{K+ap}\right)$, where r, K, and a are positive constants. Find the equilibria and their stability. Describe, in words, the dynamics of the population.

8. Use the variation of parameters method to find a particular solution to $u'' - u' - 2u = \cosh t$.

9. If e^{-t^2} is one solution to the differential equation $u''+4tu'+2(2t^2+1)u = 0$, find the solution satisfying the conditions $u(0) = 3$, $u'(0) = 1$.

10. Sketch the slope field for the differential equation $u' = -t^2 + \sin(u)$ in the window $-3 \leq t \leq 3$, $-3 \leq t \leq 3$, and then superimpose on the field the two solution curves that satisfy $u(-2) = 1$ and $u(-1) = 1$, respectively.

11. Solve $u' = 4tu - \frac{2u}{t}\ln u$ by making the substitution $y = \ln u$.

12. Adapt your knowledge about solution methods for Cauchy–Euler equations to solve the third-order initial value problem:

$$t^3u''' - t^2u'' + 2tu' - 2u = 0$$

with $u(1) = 3$, $u'(1) = 2$, $u''(1) = 1$.

4 Laplace Transforms

The Laplace method for solving linear differential equations with constant coefficients is based upon transforming the differential equation into an algebraic equation. It is especially applicable to models containing a nonhomogeneous forcing term (such as the electrical generator in a circuit) that is either discontinuous or is applied only at a single instant of time (an impulse).

This method can be regarded as another tool, in addition to variation of parameters and undetermined coefficients, for solving nonhomogeneous equations. It is often a key topic in engineering where the stability properties of linear systems are addressed.

The material in this chapter is not needed for the remaining chapters, so it may be read at any time.

4.1 Definition and Basic Properties

A successful strategy for many problems is to transform them into simpler ones that can be solved more easily. For example, some problems in rectangular coordinates are better understood and handled in polar coordinates, so we make the usual coordinate transformation $x = r\cos\theta$ and $y = r\sin\theta$. After solving the problem in polar coordinates, we can return to rectangular coordinates by the inverse transformation $r = \sqrt{x^2 + y^2}$, $\theta = \arctan\frac{y}{x}$. A similar technique holds true for many differential equations using **integral transform methods**. In this chapter we introduce the Laplace transformation which has the effect of turning a differential equation with state function $u(t)$ into an algebra problem for an associated transformed function $U(s)$; we can easily solve the algebra problem for $U(s)$ and then return to $u(t)$ via an inverse transformation. The technique is applicable to both homogeneous and nonhomogeneous linear differential equations with constant coefficients, and it is a standard method for engineers and applied mathematicians. It is particularly useful for differential equations that contain piecewise continuous forcing functions or functions that act as an impulse. The transform goes back to the late 1700s and is named for the great French mathematician and scientist Pierre de Laplace, although the basic integral goes back earlier to L. Euler. The English engineer O. Heaviside developed much of the operational calculus for transform methods in the early 1900s.

Let $u = u(t)$ be a given function defined on $0 \leq t < \infty$. The **Laplace transform** of $u(t)$ is the function $U(s)$ defined by

$$U(s) = \int_0^\infty u(t)e^{-st}dt, \tag{4.1}$$

provided the improper integral exists. The integrand is a function of t and s, and we integrate on t, leaving a function of s. Often we represent the Laplace transform in function notation,

$$\mathcal{L}[u(t)](s) = U(s) \quad \text{or just} \quad \mathcal{L}[u] = U(s).$$

$\mathcal{L}$ represents a function-like operation, called an operator or transform, whose domain and range are sets of functions; $\mathcal{L}$ takes a function $u(t)$ and transforms it into a new function $U(s)$ (see figure 4.1). In the context of Laplace transformations, t and u are called the **time domain** variables, and s and U are called the **transform domain** variables. In summary, the Laplace transform maps functions $u(t)$ to functions $U(s)$ and is somewhat like mappings we consider in calculus, such as $y = f(x) = x^2$, which maps numbers x to numbers y.

We can compute the Laplace transform of many common functions directly from the definition (4.1).

Example 4.1

Let $u(t) = e^{at}$. Then

$$U(s) = \int_0^\infty e^{at}e^{-st}dt = \int_0^\infty e^{(a-s)t}dt = \frac{1}{a-s}e^{(a-s)t}\Big|_{t=0}^{t=\infty} = \frac{1}{s-a}, \qquad s > a.$$

In different notation, $\mathcal{L}[e^{at}] = \frac{1}{s-a}$. Observe that this transform exists only for $s > a$ (otherwise the integral does not exist). Sometimes we indicate the values of s for which the transformed function $U(s)$ is defined.

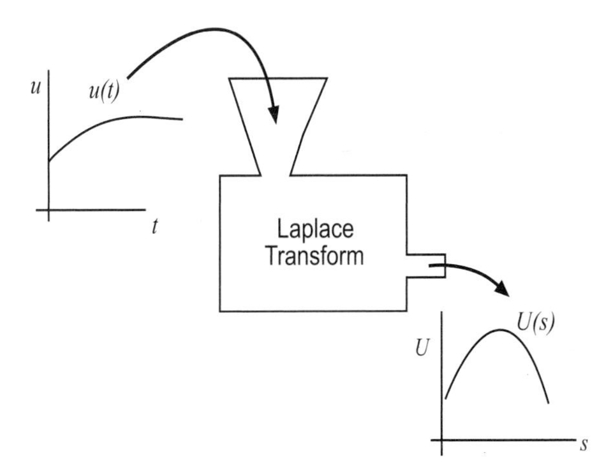

Figure 4.1 The Laplace transform as a machine that transforms functions $u(t)$ to functions $U(s)$.

Example 4.2

Let $u(t) = t$. Then, using integration by parts,

$$U(s) = \int_0^\infty te^{-st}dt = \left[t\frac{e^{-st}}{-s}\right]_{t=0}^{t=\infty} - \frac{1}{s}\int_0^\infty 1 \cdot e^{-st}dt = \frac{1}{s^2}, \quad s > 0.$$

Example 4.3

The unit switching function $h_a(t)$ is defined by $h_a(t) = 0$ if $t < a$ and $h_a(t) = 1$ if $t \geq a$. The switch is off if $t < a$, and it is on when $t \geq a$. Therefore the function $h_a(t)$ is a step function where the step from 0 to 1 occurs at $t = a$. The switching function is also called the **Heaviside function**. The Laplace transform of $h_a(t)$ is

$$\begin{aligned}
\mathcal{L}[h_a(t)] &= \int_0^\infty h_a(t)e^{-st}dt \\
&= \int_0^a h_a(t)e^{-st}dt + \int_a^\infty h_a(t)e^{-st}dt \\
&= \int_0^a 0 \cdot e^{-st}dt + \int_a^\infty 1 \cdot e^{-st}dt \\
&= -\frac{1}{s}e^{-st}\Big|_{t=a}^{t=\infty} = \frac{1}{s}e^{-as}, \quad s > 0.
\end{aligned}$$

Example 4.4

The Heaviside function is useful for expressing multi-lined functions in a single formula. For example, let

$$f(t) = \begin{cases} \frac{1}{2}, & 0 \leq t < 2 \\ t-1, & 2 \leq t \leq 3 \\ 5-t^2, & 3 < t \leq 6 \\ 0, & t > 6 \end{cases}$$

(The reader should plot this function). This can be written in one line as

$$f(t) = \frac{1}{2}h_0(t) + (t - 1 - \frac{1}{2})h_2(t) + (5 - t^2 - (t-1))h_3(t) - (5-t^2)h_6(t).$$

The first term switches on the function $1/2$ at $t = 0$; the second term switches off $1/2$ and switches on $t-1$ at time $t = 2$; the third term switches off $t-1$ and switches on $5 - t^2$ at $t = 3$; finally, the last term switches off $5 - t^2$ at $t = 6$. Later we show how to find Laplace transforms of such functions.

As you may have already concluded, calculating Laplace transforms may be tedious business. Fortunately, generations of mathematicians, scientists, and engineers have computed the Laplace transforms of many, many functions, and the results have been catalogued in tables and in software systems. Some of the tables are extensive, but here we require only a short table, which is given at the end of the chapter. The table lists a function $u(t)$ in the first column, and its transform $U(s)$, or $\mathcal{L}u$, in the second. The various functions in the first column are discussed in the sequel. Computer algebra systems also have commands that calculate the Laplace transform (see Appendix B).

Therefore, given $u(t)$, the Laplace transform $U(s)$ can be computed by the definition, given in formula (4.1). We can also think of the opposite problem: given $U(s)$, find a function $u(t)$ whose Laplace transform is $U(s)$. This is the inverse problem. Unfortunately, there is no elementary formula that we can write down that computes $u(t)$ in terms of $U(s)$ (there is a formula, but it involves a contour integration in the complex plane). In elementary treatments we are satisfied with using tables. For example, if $U(s) = \frac{1}{s-2}$, then the table gives $u(t) = e^{2t}$ as the function that has $U(s)$ as its transform. When we think of it this way, we say $u(t) = e^{2t}$ is the "inverse transform" of $U(s) = \frac{1}{s-2}$, and we write

$$e^{2t} = \mathcal{L}^{-1}\left[\frac{1}{s-2}\right].$$

In general we use the notation

$$U = \mathcal{L}(u), \quad u = \mathcal{L}^{-1}[U].$$

We think of $\mathcal{L}$ as an operator (**transform**) and $\mathcal{L}^{-1}$ as the inverse operation (**inverse transform**). The functions $u(t)$ and $U(s)$ form a transform pair, and they are listed together in two columns of a table. Computer algebra systems also supply inverse transforms.

One question that should be addressed concerns the *existence* of the transform. That is, which functions have Laplace transforms? Clearly if a function grows too quickly as t gets large, then the improper integral will not exist and there will be no transform. There are two conditions that guarantee existence, and these are reasonable conditions for most problems in science and engineering. First, we require that $u(t)$ not grow too fast; a way of stating this mathematically is to require that there exist constants $M > 0$ and α for which

$$|u(t)| \leq Me^{\alpha t}$$

is valid for all $t > t_0$, where t_0 is some value of time. That is, beyond the value t_0 the function is bounded above and below by an exponential function. Such functions are said to be of **exponential order**. Second, we require that $u(t)$ be **piecewise continuous** on $0 \leq t < \infty$. In other words, the interval $0 \leq t < \infty$ can be divided into intervals on which u is continuous, and at any point of discontinuity u has finite left and right limits, except possibly at $t = +\infty$. One can prove that if u is piecewise continuous on $0 \leq t < \infty$ and of exponential order, then the Laplace transform $U(s)$ exists for all $s > \alpha$.

What makes the Laplace transform so useful for differential equations is that it turns derivative operations in the time domain into multiplication operations in the transform domain. The following theorem gives the crucial operational formulas stating how the derivatives transform.

Theorem 4.5

Let $u(t)$ be a function and $U(s)$ its transform. Then

$$\mathcal{L}[u'] = sU(s) - u(0), \tag{4.2}$$

$$\mathcal{L}[u''] = s^2U(s) - su(0) - u'(0). \tag{4.3}$$

Proof. These facts are easily proved using integration by parts. We have

$$\begin{aligned} \mathcal{L}[u'] &= \int_0^\infty u'(t)e^{-st}dt = \left[u(t)e^{-st}\right]_{t=0}^{t=\infty} - \int_0^\infty -su(t)e^{-st}dt \\ &= -u(0) + sU(s), \quad s > 0. \end{aligned}$$

The second operational formula (4.3) is derived using two successive integration by parts, and we leave the calculation to the reader.

These formulas allow us to transform a differential equation with unknown $u(t)$ into an algebraic problem with unknown $U(s)$. We solve for $U(s)$ and then find $u(t)$ using the inverse transform $u = \mathcal{L}^{-1}[U]$. We elaborate on this method in the next section.

Before tackling the solution of differential equations, we present additional important and useful properties.

(a) (**Linearity**) The Laplace transform is a linear operation; that is, the Laplace transform of a sum of two functions is the sum of the Laplace transforms of each, and the Laplace transform of a constant times a function is the constant times the transform of the function. We can express these rules in symbols by a single formula:

$$\mathcal{L}[c_1 u + c_2 v] = c_1\mathcal{L}[u] + c_2\mathcal{L}[v]. \tag{4.4}$$

Here, u and v are functions and c_1 and c_2 are any constants. Similarly, the inverse Laplace transform is a linear operation:

$$\mathcal{L}^{-1}[c_1 u + c_2 v] = c_1\mathcal{L}^{-1}[u] + c_2\mathcal{L}^{-1}[v]. \tag{4.5}$$

(b) (**Shift Property**) The Laplace transform of a function times an exponential, $u(t)e^{at}$, shifts the transform of U; that is,

$$\mathcal{L}[u(t)e^{at}] = U(s-a). \tag{4.6}$$

(c) (**Switching Property**) The Laplace transform of a function that switches on at $t = a$ is given by

$$\mathcal{L}[h_a(t)u(t-a)] = U(s)e^{-as}. \tag{4.7}$$

Proofs of some of these relations follow directly from the definition of the Laplace transform, and they are requested in the Exercises.

EXERCISES

1. Use the definition of the Laplace transform to compute the transform of the square pulse function $u(t) = 1$, $1 \le t \le 2$; $u(t) = 0$, otherwise.

2. Derive the operational formula (4.3).

3. Sketch the graphs of $\sin t$, $\sin(t - \pi/2)$, and $h_{\pi/2}(t)\sin(t-\pi/2)$. Find the Laplace transform of each.

4. Find the Laplace transform of $t^2 e^{-3t}$.

5. Find $\mathcal{L}[\sinh kt]$ and $\mathcal{L}[\cosh kt]$ using the fact that $\mathcal{L}\left[e^{kt}\right] = \frac{1}{s-k}$.

6. Find $\mathcal{L}\left[e^{-3t} + 4\sin kt\right]$ using the table. Find $\mathcal{L}\left[e^{-3t}\sin 2t\right]$ using the shift property (4.6).

7. Using the switching property (4.7), find the Laplace transform of the function

$$u(t) = \begin{cases} 0 & t < 2 \\ e^{-t}, & t > 2. \end{cases}$$

8. From the definition (4.1), find $\mathcal{L}\left[1/\sqrt{t}\right]$ using the integral substitution $st = r^2$ and then noting $\int_0^\infty \exp(-r^2)dr = \sqrt{\pi}/2$.

9. Does the function $u(t) = e^{t^2}$ have a Laplace transform? What about $u(t) = 1/t$? Comment.

10. Derive the operational formulas (4.6) and (4.7).

11. Plot the *square-wave* function

$$f(t) = \sum_{n=0}^{\infty}(-1)^n h_n(t)$$

on the interval $t > 0$ and find its transform $F(s)$. (Hint: use the geometric series $1 + x + x^2 + \cdots = \frac{1}{1-x}$.)

12. Show that

$$\mathcal{L}\left[\int_0^t u(r)dr\right] = \frac{U(s)}{s}.$$

13. Derive the formulas

$$\mathcal{L}\left[tu(t)\right] = -U'(s), \quad \mathcal{L}^{-1}[U'(s)] = -tu(t).$$

Use these formulas to find the inverse transform of $\arctan \frac{a}{s}$.

14. Show that

$$\mathcal{L}\left[\frac{u(t)}{t}\right] = \int_s^\infty U(r)dr,$$

and use the result to find

$$\mathcal{L}\left[\frac{\sinh t}{t}\right].$$

15. Show that

$$\mathcal{L}\left[f(t)h_a(t)\right] = e^{-as}\mathcal{L}[f(t+a)],$$

and use this formula to compute $\mathcal{L}[t^2 h_1(t)]$.

16. Find the Laplace transform of the function in Example 4.4.

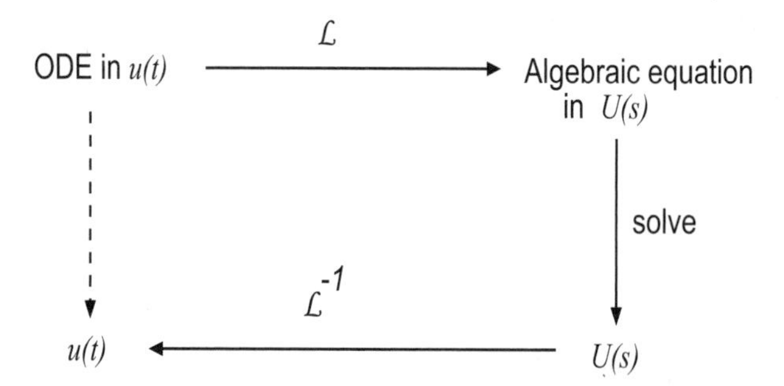

Figure 4.2 A DE for an unknown function $u(t)$ is transformed to an algebraic equation for its transform $U(s)$. The algebraic problem is solved for $U(s)$ in the transform domain, and the solution is returned to the original time domain via the inverse transform.

17. The **gamma function** is defined by

$$\Gamma(x) = \int_0^\infty e^{-t} t^{x-1} dt, \quad x > -1.$$

a) Show that $\Gamma(n+1) = n\Gamma(n)$ and $\Gamma(n+1) = n!$ for nonnegative integers n. Show that $\Gamma(\frac{1}{2}) = \sqrt{\pi}$.

b) Show that $\mathcal{L}\left[t^a\right] = \frac{\Gamma(a+1)}{s^{a+1}}, \quad s > 0$.

4.2 Initial Value Problems

The following examples illustrate how Laplace transforms are used to solve initial value problems for linear differential equations with constant coefficients. The method works on equations of all orders and on systems of several equations in several unknowns. We assume $u(t)$ is the unknown state function. The idea is to take the transform of each term in the equation, using the linearity property. Then, using Theorem 4.5, reduce all of the derivative terms to algebraic expressions and solve for the transformed state function $U(s)$. Finally, invert $U(s)$ to recover the solution $u(t)$. Figure 4.2 illustrates this three-step method.

Example 4.6

Consider the second-order initial value problem

$$u'' + \omega^2 u = 0, \quad u(0) = 0, \quad u'(0) = 1.$$

Taking transforms of both sides and using the linearity property gives

$$\mathcal{L}[u''] + \omega^2 \mathcal{L}[u] = \mathcal{L}[0].$$

Then Theorem 4.5 gives

$$s^2 U(s) - su(0) - u'(0) + \omega^2 U(s) = 0,$$

which is an algebraic equation for the transformed state $U(s)$. Using the initial conditions, we get

$$s^2 U(s) - 1 + \omega^2 U(s) = 0.$$

Solving for the transform function $U(s)$ gives

$$U(s) = \frac{1}{\omega^2 + s^2} = \frac{1}{\omega} \frac{\omega}{\omega^2 + s^2},$$

which is the solution in the transform domain. Therefore, from the table, the inverse transform is

$$u(t) = \frac{1}{\omega} \sin \omega t,$$

which is the solution to the original initial value problem.

Example 4.7

Solve the first-order nonhomogeneous equation

$$u' + 2u = e^{-t}, \quad u(0) = 0.$$

Taking Laplace transforms of each term

$$\mathcal{L}[u'] + \mathcal{L}[2u] = \mathcal{L}[e^{-t}],$$

or

$$sU(s) - u(0) + 2U(s) = \frac{1}{s+1}.$$

Solving for the transformed function $U(s)$ gives

$$U(s) = \frac{1}{(s+1)(s+2)}.$$

Now we can look up the inverse transform in the table. We find

$$u(t) = \mathcal{L}^{-1}\left[\frac{1}{(s+1)(s+2)}\right] = e^{-t} - e^{-2t}.$$

Example 4.8

(Partial Fractions, I) Sometimes the table may not include an entry for the inverse transform that we seek, and so we may have to algebraically manipulate or simplify our expression so that it can be reduced to table entries. A standard technique is to expand complex fractions into their "partial fraction" expansion. In the last example we had

$$U(s) = \frac{1}{(s+1)(s+2)}.$$

We can expand $U(s)$ as

$$\frac{1}{(s+1)(s+2)} = \frac{a}{(s+1)} + \frac{b}{(s+2)},$$

for some constants a and b to be determined. Combining terms on the right side gives

$$\begin{aligned} \frac{1}{(s+1)(s+2)} &= \frac{a(s+2)+b(s+1)}{(s+1)(s+2)} \\ &= \frac{(a+b)s+2a+b}{(s+1)(s+2)}. \end{aligned}$$

Comparing numerators on the left and right force $a+b=0$ and $2a+b=1$. Hence $a=-b=1$ and we have

$$U(s) = \frac{1}{(s+1)(s+2)} = \frac{1}{(s+1)} + \frac{-1}{(s+2)}.$$

We have reduced the complex fraction to the sum of two simple, easily identifiable, fractions that are easily found in the table. Using the linearity property of the inverse transform,

$$\begin{aligned} \mathcal{L}^{-1}[U(s)] &= \mathcal{L}^{-1}\left[\frac{1}{(s+1)}\right] - \mathcal{L}^{-1}\left[\frac{1}{(s+2)}\right] \\ &= e^{-t} - e^{-2t}. \end{aligned}$$

Example 4.9

(Partial Fractions, II) A common expression in the transform domain that requires inversion is a fraction of the form

$$U(s) = \frac{1}{s^2+bs+c}.$$

If the denominator has two distinct real roots, then it factors and we can proceed as in the previous example. If the denominator has complex roots then the following "complete the square" technique may be used. For example, consider

$$U(s) = \frac{1}{s^2 + 3s + 6}.$$

Then, completing the square in the denominator,

$$\begin{aligned} U(s) &= \frac{1}{s^2 + 3s + \left(\frac{3}{2}\right)^2 - \left(\frac{3}{2}\right)^2 + 6} \\ &= \frac{1}{\left(s + \frac{3}{2}\right)^2 + \left(\frac{\sqrt{15}}{2}\right)^2}. \end{aligned}$$

This entry is in the table, up to a factor of $\frac{\sqrt{15}}{2}$. Therefore we multiply and divide by this factor and locate the inverse transform in the table as

$$u(t) = \frac{2}{\sqrt{15}} e^{-3t/2} \sin \frac{\sqrt{15}}{2} t.$$

Example 4.10

In this example we calculate the response of an RC circuit when the emf is a discontinuous function. These types of problems occur frequently in engineering, especially electrical engineering, where discontinuous inputs to circuits are commonplace. Therefore, consider an RC circuit containing a 1 volt battery, and with zero initial charge on the capacitor. Take $R = 1$ and $C = 1/3$. Assume the switch is turned on from $1 \le t \le 2$, and is otherwise switched off, giving a square pulse. The governing equation for the charge on the capacitor is

$$q' + 3q = h_1(t) - h_2(t), \quad q(0) = 0.$$

We apply the basic technique. Taking the Laplace transform gives

$$sQ(s) - q(0) + 3Q(s) = \frac{1}{s}(e^{-s} - e^{-2s}).$$

Solving for $Q(s)$ yields

$$\begin{aligned} Q(s) &= \frac{1}{s(s+3)}(e^{-s} - e^{-2s}) \\ &= \frac{1}{s(s+3)} e^{-s} - \frac{1}{s(s+3)} e^{-2s}. \end{aligned}$$

Now we have to invert, which is usually the hardest part. Each term on the right has the form $U(s)e^{-as}$, and therefore we can apply the switching property (4.7). From the table we have

$$\mathcal{L}^{-1}\left[\frac{1}{s(s+3)}\right] = \frac{1}{3}(1 - e^{-3t}).$$

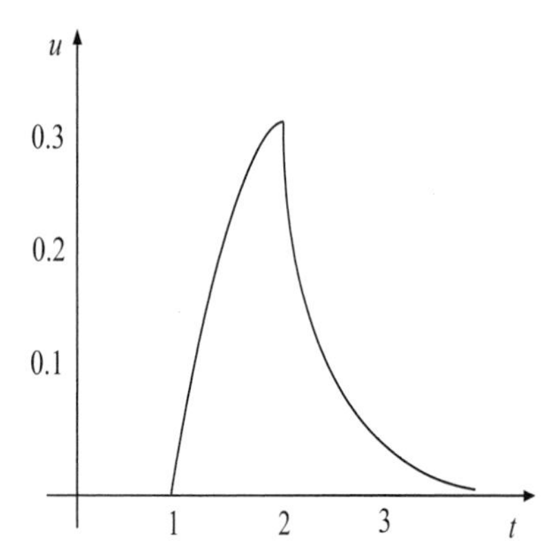

Figure 4.3 The charge response is zero up to time $t = 1$, when the switch is closed. The charge increases until $t = 2$, when the switch is again opened. The charge then decays to zero.

Therefore, by the shift property,

$$\mathcal{L}^{-1}\left[\frac{1}{s(s+3)}e^{-s}\right] = \frac{1}{3}(1 - e^{-3(t-1)})h_1(t).$$

Similarly,

$$\mathcal{L}^{-1}\left[\frac{1}{s(s+3)}e^{-2s}\right] = \frac{1}{3}(1 - e^{-3(t-2)})h_2(t).$$

Putting these two results together gives

$$q(t) = \frac{1}{3}(1 - e^{-3(t-1)})h_1(t) - \frac{1}{3}(1 - e^{-3(t-2)})h_2(t).$$

We can use software to plot the charge response. See figure 4.3.

Because there are extensive tables and computer algebra systems containing large numbers of inverse transforms, the partial fractions technique for inversion is not used as often as in the past.

EXERCISES

1. Find A, B, and C for which

$$\frac{1}{s^2(s-1)} = \frac{As+B}{s^2} + \frac{C}{s-1}.$$

   Then find the inverse Laplace transform of $\frac{1}{s^2(s-1)}$.

2. Find the inverse transform of the following functions.

a) $U(s) = \frac{s}{s^2+7s-8}$.

b) $U(s) = \frac{3-2s}{s^2+2s+10}$.

c) $\frac{2}{(s-5)^4}$.

d) $\frac{7}{s}e^{-4s}$.

3. Solve the following initial value problems using Laplace transforms.

a) $u' + 5u = h_2(t), \quad u(0) = 1.$

b) $u' + u = \sin 2t, \quad u(0) = 0.$

c) $u'' - u' - 6u = 0, \quad u(0) = 2,\ u'(0) = -1$

d) $u'' - 2u' + 2u = 0, \quad u(0) = 0,\ u'(0) = 1.$

e) $u'' - 2u' + 2u = e^{-t}, \quad u(0) = 0,\ u'(0) = 1.$

f) $u'' - u' = 0, \quad u(0) = 1,\ u'(0) = 0.$

g) $u'' + 0.4u' + 2u = 1 - h_5(t), \quad u(0) = 0,\ u'(0) = 0.$

h) $u'' + 9u = \sin 3t, \quad u(0) = 0,\ u'(0) = 0.$

i) $u'' - 2u = 1, \quad u(0) = 1,\ u'(0) = 0.$

4. Use Laplace transforms to solve the two simultaneous differential equations

$$\begin{aligned} x' &= x - 2y - t \\ y' &= 3x + y, \end{aligned}$$

with $x(0) = y(0) = 0$. (Hint: use what you know about solving single equations, letting $\mathcal{L}[x] = X(s)$ and $\mathcal{L}[y] = Y(s)$.)

5. Show that

$$L[t^n u(t)] = (-1)^n U^{(n)}(s)$$

for $n = 1, 2, 3, ...$

4.3 The Convolution Property

The additivity property of Laplace transforms is stated in (4.4): the Laplace transform of a sum is the sum of the transforms. But what can we say about the Laplace transform of a product of two functions? It is *not* the product of the two Laplace transforms. That is, if $u = u(t)$ and $v = v(t)$ with $\mathcal{L}[u] = U(s)$ and $\mathcal{L}[v] = V(s)$, then $\mathcal{L}[uv] \neq U(s)V(s)$. If this is not true, then what is true? We ask it this way. What function has transform $U(s)V(s)$? Or, differently, what is the inverse transform of $U(s)V(s)$. The answer may surprise you because it is nothing one would easily guess. The function whose transform is $U(s)V(s)$ is the convolution of the two functions $u(t)$ and $v(t)$. It is defined as follows. If u and v are two functions defined on $[0, \infty)$, the **convolution** of u and v, denoted by $u * v$, is the function defined by

$$(u * v)(t) = \int_0^t u(\tau)v(t-\tau)d\tau.$$

Sometimes it is convenient to write the convolution as $u(t) * v(t)$. The **convolution property** of Laplace transforms states that

$$\mathcal{L}[u * v] = U(s)V(s).$$

It can be stated in terms of the inverse transform as well:

$$\mathcal{L}^{-1}[U(s)V(s)] = u * v.$$

This property is useful because when solving a DE we often end up with a product of transforms; we may use this last expression to invert the product.

The convolution property is straightforward to verify using a multi-variable calculus technique, interchanging the order of integration. The reader should check the following steps.

$$\begin{aligned}
\mathcal{L}\left(\int_0^t u(\tau)v(t-\tau)d\tau\right) &= \int_0^\infty \left(\int_0^t u(\tau)v(t-\tau)d\tau\right) e^{-st}dt \\
&= \int_0^\infty \left(\int_0^t u(\tau)v(t-\tau)e^{-st}d\tau\right) dt \\
&= \int_0^\infty \left(\int_\tau^\infty u(\tau)v(t-\tau)e^{-st}dt\right) d\tau \\
&= \int_0^\infty \left(\int_\tau^\infty v(t-\tau)e^{-st}dt\right) u(\tau)d\tau \\
&= \int_0^\infty \left(\int_0^\infty v(r)e^{-s(r+\tau)}dr\right) u(\tau)d\tau \\
&= \int_0^\infty \left(\int_0^\infty v(r)e^{-sr}dr\right) e^{-s\tau}u(\tau)d\tau \\
&= \left(\int_0^\infty e^{-s\tau}u(\tau)d\tau\right)\left(\int_0^\infty v(r)e^{-sr}dr\right).
\end{aligned}$$

This last expression is $U(s)V(s)$.

Example 4.11

Find the convolution of 1 and t^2. We have

$$\begin{aligned} 1 * t^2 &= \int_0^t 1 \cdot (t-\tau)^2 d\tau = \int_0^t (t^2 - 2t\tau + \tau^2) d\tau \\ &= t^2 \cdot t - 2t(\frac{t^2}{2}) + \frac{t^3}{3} = \frac{t^3}{3}. \end{aligned}$$

Notice also that the convolution of t^2 and 1 is

$$t^2 * 1 = \int_0^t \tau^2 \cdot 1 d\tau = \frac{t^3}{3}.$$

In the exercises you are asked to show that $u * v = v * u$, so the order of the two functions under convolution does not matter.

Example 4.12

Find the inverse of $U(s) = \frac{3}{s(s^2+9)}$. We can do this by partial fractions, but here we use convolution. We have

$$\begin{aligned} \mathcal{L}^{-1}\left[\frac{3}{s(s^2+9)}\right] &= \mathcal{L}^{-1}\left[\frac{1}{s}\frac{3}{(s^2+9)}\right] \\ &= 1 * \sin 3t = \int_0^t \sin 3\tau d\tau \\ &= \frac{1}{3}(1 - \cos 3t). \end{aligned}$$

Example 4.13

Solve the nonhomogeneous DE

$$u'' + k^2 u = f(t),$$

where f is any given input function, and where $u(0)$ and $u'(0)$ are specified initial conditions. Taking the Laplace transform,

$$s^2 U(s) - su(0) - u'(0) + k^2 U(s) = F(s).$$

Then

$$U(s) = u(0)\frac{s}{s^2+k^2} + u'(0)\frac{1}{s^2+k^2} + \frac{F(s)}{s^2+k^2}.$$

Now we can invert each term, using the convolution property on the last term, to get the solution formula

$$u(s) = u(0)\cos kt + \frac{u'(0)}{k}\sin kt + \frac{1}{k}\int_0^t f(\tau)\sin k(t-\tau) dr.$$

EXERCISES

1. Compute the convolution of $\sin t$ and $\cos t$.

2. Compute the convolution of t and t^2.

3. Use the convolution property to find the general solution of the differential equation $u' = au + q(t)$ using Laplace transforms

4. Use a change of variables to show that the order of the functions used in the definition of the convolution does not matter. That is,
$$(u * v)(t) = (v * u)(t).$$

5. Solve the initial value problem
$$u'' - \omega^2 u = f(t), \quad u(0) = u'(0) = 0.$$

6. Use Exercise 5 to find the solution to
$$u'' - 4u = 1 - h_1(t), \quad u(0) = u'(0) = 0.$$

7. Write an integral expression for the inverse transform of $U(s) = \frac{1}{s}e^{-3s}F(s)$, where $\mathcal{L}[f] = F$.

8. Find a formula for the solution to the initial value problem
$$u'' - u' = f(t), \quad u(0) = u'(0) = 0.$$

9. An integral equation is an equation where the unknown function $u(t)$ appears under an integral sign (see also the exercises in Section 1.2). Consider the integral equation
$$u(t) = f(t) + \int_0^t k(t-\tau)u(\tau)d\tau,$$
where f and k are given functions. Using convolution, find a formula for $U(s)$ in terms of the transforms of F and K of f and k, respectively.

10. Using the idea in the preceding exercise, solve the following integral equations.

    a) $u(t) = t - \int_0^t (t-\tau)u(\tau)d\tau$.

    b) $u(t) = \int_0^t e^{t-\tau}u(\tau)d\tau$.

11. Solve the integral equation for $u(t)$:
$$f(t) = \frac{1}{\sqrt{\pi}} \int_0^t \frac{u(\tau)}{\sqrt{t-\tau}} d\tau.$$
(Hint: use the gamma function from the Exercise 17 in Section 4.1.)

4.4 Discontinuous Sources

The problems we are solving have the general form

$$\begin{aligned} u'' + bu' + cu &= f(t), \quad t > 0 \\ u(0) &= u_1, \; u'(0) = u_2. \end{aligned}$$

If f is a continuous function, then we can use variation of parameters to find the particular solution; if f has the special form of a polynomial, exponential, sine, or cosine, or sums and products of these forms, we can use the method of undetermined coefficients (judicious guessing) to find the particular solution. If, however, f is a piecewise continuous source with different forms on different intervals, then we would have to find the general solution on each interval and determine the arbitrary constants to match up the solutions at the endpoints of the intervals. This is an algebraically difficult task. However, using Laplace transforms, the task is not so tedious. In this section we present additional examples on how to deal with discontinuous forcing functions.

Example 4.14

As we noted earlier, the Heaviside function is very useful for writing piecewise, or multi-lined, functions in a single line. For example,

$$\begin{aligned} f(t) &= \begin{cases} t, & 0 < t < 1 \\ 2, & 1 \le t \le 3 \\ 0, & t > 3 \end{cases} \\ &= t + (2-t)h_1(t) - 2h_3(t). \end{aligned}$$

The first term switches on the function t at $t = 0$; the second term switches on the function 2 and switches off the function t at $t = 1$; and the last term switches off the function 2 at $t = 3$. By linearity, the Laplace transform of $f(t)$ is given by

$$F(s) = \mathcal{L}[t] + 2\mathcal{L}[h_1(t)] - \mathcal{L}[th_1(t)] - 2\mathcal{L}[h_3(t)].$$

The second and fourth terms are straightforward from Example 4.3, and $\mathcal{L}[t] = 1/s^2$. The third term can be calculated using $\mathcal{L}\left[f(t)h_a(t)\right] = e^{-as}\mathcal{L}[f(t+a)]$. With $f(t) = t$ we have

$$\mathcal{L}\left[th_1(t)\right] = e^{-s}\mathcal{L}[t+1] = \frac{1}{s^2}e^{-s} + \frac{1}{s}e^{-s}.$$

Putting all these results together gives

$$F(s) = \frac{1}{s^2} + \frac{2}{s}e^{-s} - \left(\frac{1}{s^2}e^{-s} + \frac{1}{s}e^{-s}\right) - \frac{2}{s}e^{-3s}.$$

Example 4.15

Solve the initial value problem

$$u'' + 9u = e^{-0.5t} h_4(t), \quad u(0) = u'(0) = 0,$$

where the forcing term is an exponential decaying term that switches on at time $t = 4$. The Laplace transform of the forcing term is

$$\mathcal{L}[e^{-0.5t} h_4(t)] = e^{-4s} \mathcal{L}[e^{-0.5(t+4)}] = e^{-2} \frac{1}{s+0.5} e^{-4s}.$$

Then, taking the transform of the the equation,

$$s^2 U(s) + 9U(s) = e^{-2} \frac{1}{s+0.5} e^{-4s}.$$

Whence

$$U(s) = e^{-2} \frac{1}{(s+0.5)(s^2+9)} e^{-4s}.$$

Now we need the shift theorem. But first we find the inverse transform of $\frac{1}{(s+0.5)(s^2+9)}$. Here we leave it as an exercise (partial fractions) to show

$$\mathcal{L}^{-1}\left[\frac{1}{(s+0.5)(s^2+9)}\right] = \frac{3e^{-0.5t} - 3\cos 3t + 0.5 \sin 3t}{27.75}.$$

Therefore, by the shift property,

$$\begin{aligned} u(t) &= e^{-2} \mathcal{L}^{-1}\left[\frac{e^{-4s}}{(s+0.5)(s^2+9)}\right] \\ &= h_4(t) \frac{3e^{-0.5(t-4)} - 3\cos 3(t-4) + 0.5 \sin 3(t-4)}{27.75 e^2}, \end{aligned}$$

which is the solution. Notice that the solution does not switch on until $t = 4$. At that time the forcing term turns on, producing a transient; eventually its effects decay away and an oscillating steady-state takes over.

EXERCISES

1. Sketch the function $f(t) = 2h_3(t) - 2h_4(t)$ and find its Laplace transform.
2. Find the Laplace transform of $f(t) = t^2 h_3(t)$.
3. Invert $F(s) = (s-2)^{-4}$.
4. Sketch the following function, write it as a single expression, and then find its transform:

$$f(t) = \begin{cases} 3, & 0 \le t < 2 \\ 2, & 2 \le t < \pi \\ 6, & \pi \le t \le 7 \\ 0, & t > 7. \end{cases}$$

5. Find the inverse transform of
$$U(s) = \frac{1-e^{-4s}}{s^2}.$$

6. Solve the initial value problem
$$u'' + 4u = \begin{cases} \cos 2t, & 0 \le t \le 2\pi, \\ 0, & t > 2\pi, \end{cases}$$
where $u(0) = u'(0) = 0$. Sketch the solution.

7. Consider the initial value problem $u' = u + f(t)$, $u(0) = 1$, where $f(t)$ is given by
$$f(t) = \begin{cases} 0, & 0 < t \le 1 \\ -2, & t > 1. \end{cases}$$
Solve this problem in two ways: (a) by solving the problem on two intervals and pasting together the solutions in a continuous way, and (b) by Laplace transforms.

8. An LC circuit with $L = C = 1$ is "ramped-up" with an applied voltage
$$e(t) = \begin{cases} t, & 0 \le t \le 9 \\ 9, & t > 9. \end{cases}$$
Initially there is no charge on the capacitor and no current. Find and sketch a graph of the voltage response on the capacitor.

9. Solve $u' = -u + h_1(t) - h_2(t)$, $u(0) = 1$.

10. Solve the initial value problem
$$u'' + \pi^2 u = \begin{cases} \pi^2, & 0 < t < 1, \\ 0, & t > 1, \end{cases}$$
where $u(0) = 1$ and $u'(0) = 0$.

11. Let $f(t)$ be a periodic function with period p. That is, $f(t+p) = f(t)$ for all $t > 0$. Show that the Laplace transform of f is given by
$$F(s) = \frac{1}{1-e^{-ps}} \int_0^p f(r)e^{-rs}dr.$$
(Hint: break up the interval $(-\infty, +\infty)$ into subintervals $(np, (n+1)p)$, calculate the transform on each subinterval, and finally use the geometric series $1 + x + x^2 + \cdots = \frac{1}{1-x}$.)

12. Show that the Laplace transform of the periodic, square-wave function that takes the value 1 on intervals $[0, a)$, $[2a, 3a)$, $[4a, 5a)$,..., and the value -1 on the intervals $[a, 2a)$, $[3a, 4a)$, $[5a, 6a)$,..., is $\frac{1}{s}\tanh\left(\frac{as}{2}\right)$.

13. Write a single line formula for the function that is 2 between $2n$ and $2n+1$, and 1 between $2n-1$ and $2n$, where $n = 0, 1, 2, 3, 4, \ldots$.

4.5 Point Sources

Many physical and biological processes have source terms that act at a single instant of time. For example, we can idealize an injection of medicine (a "shot") into the blood stream as occurring at a single instant; a mechanical system, for example, a damped spring-mass system in a shock absorber on a car, can be given an impulsive force by hitting a bump in the road; an electrical circuit can be closed only for an instant, which leads to an impulsive, applied voltage.

To fix the idea, let us consider an RC circuit with a given emf $e(t)$ and with no initial charge on the capacitor. In terms of the charge $q(t)$ on the capacitor, the governing circuit equation is

$$Rq' + \frac{1}{C}q = e(t), \quad q(0) = 0. \tag{4.8}$$

This is a linear first-order equation, and if the emf is a continuous function, or piecewise continuous function, the problem can be solved by the methods presented in Chapter 2 or by transform methods. We use the latter. Taking Laplace transforms and solving for $Q(s)$, the Laplace transform of $q(t)$, gives

$$Q(s) = \frac{1}{R}\frac{1}{s + 1/RC}E(s),$$

where $E(s)$ is the transform of the emf $e(t)$. Using the convolution property we have the solution

$$q(t) = \frac{1}{R}\int_0^t e^{-(t-\tau)/RC}e(\tau)d\tau. \tag{4.9}$$

But presently we want to consider a special type of electromotive force $e(t)$, one given by an voltage impulse that acts only for a single instant (i.e., a quick surge of voltage). To fix the idea, suppose the source is a 1 volt battery. Imagine that the circuit is open and we just touch the leads together at a single instant at time $t = a$. How does the circuit respond? We denote this unit voltage input by $e(t) = \delta_a(t)$, which is called a **unit impulse** at $t = a$. The question is how to define $\delta_a(t)$, an energy source that acts at a single instant of time. At first it *appears* that we should take $\delta_a(t) = 1$ if $t = a$, and $\delta_a(t) = 0$, otherwise. But this is not correct. To illustrate, we can substitute into (4.9) and write

$$q(t) = \frac{1}{R}\int_0^t e^{-(t-\tau)/RC}\delta_a(\tau)d\tau. \tag{4.10}$$

If $\delta_a(t) = 0$ at all values of t, except $t = a$, the integral must be zero because the integrand is zero except at a single point. Hence, the response of the circuit is $q(t) = 0$, which is incorrect! Something is clearly wrong with this argument and our tentative definition of $\delta_a(t)$.

The difficulty is with the "function" $\delta_a(t)$. We must come to terms with the idea of an impulse. Actually, having the source act at a single instant of time is an idealization. Rather, such a short impulse must occur over a very small interval $[a-\varepsilon/2, a+\varepsilon/2]$, where ε is a small positive number. We do not know the actual form of the applied voltage over this interval, but we want its average value over the interval to be 1 volt. Therefore, let us define an *idealized* applied voltage by

$$\begin{aligned} e_{a,\varepsilon}(t) &= \begin{cases} \frac{1}{\varepsilon}, & a-\varepsilon/2 < t < a+\varepsilon/2 \\ 0, & \text{otherwise}, \end{cases} \\ &= \frac{1}{\varepsilon}(h_{a-\varepsilon/2}(t) - h_{a+\varepsilon/2}(t)). \end{aligned}$$

These idealized impulses are rectangular voltage inputs that get taller and narrower (of height $1/\varepsilon$ and width ε) as ε gets small. But their average value over the small interval $a-\varepsilon/2 < t < a+\varepsilon/2$ is always 1; that is,

$$\int_{a-\varepsilon/2}^{a+\varepsilon/2} e_{a,\varepsilon}(t)dt = 1.$$

This property should hold for all ε, regardless of how small. It seems reasonable therefore to define the unit impulse $\delta_a(t)$ at $t = a$ in a limiting sense, having the property

$$\int_{a-\varepsilon/2}^{a+\varepsilon/2} \delta_a(t)dt = 1, \quad \text{for all } \varepsilon > 0.$$

Engineers and scientists used this condition, along with $\delta_a(t) = 0, \ t \neq a$, for decades to define a unit, point source at time $t = a$, called the **delta function**, and they developed a calculus that was successful in obtaining solutions to equations having point sources. But, actually, the unit impulse is not a function at all, and it was shown in the mid-20th century that the unit impulse belongs to a class of so-called **generalized functions** whose actions are not defined pointwise, but rather by how they act when integrated against other functions. Mathematically, the unit impulse is defined by the **sifting property**

$$\int_0^\infty \delta_a(t)\phi(t)dt = \phi(a).$$

That is, when integrated against any nice function $\phi(t)$, the delta function picks out the value of $\phi(t)$ at $t = a$. We check that this works in our problem. If we use this sifting property back in (4.10), then for $t > a$ we have

$$q(t) = \frac{1}{R}\int_0^t e^{-(t-\tau)/RC}\delta_a(\tau)d\tau = \frac{1}{R}e^{-(t-a)/RC}, \quad t > a,$$

which is the correct solution. Note that $q(t) = 0$ up until $t = a$, because there is no source. Furthermore, $q(a) = 1/R$. Therefore the charge is zero up to time a, at which it jumps to the value $1/R$, and then decays away.

To deal with differential equations involving impulses we can use Laplace transforms in a formal way. Using the sifting property, with $\phi(t) = e^{-st}$, we obtain

$$\mathcal{L}[\delta_a(t)] = \int_0^\infty \delta_a(t)e^{-st}dt = e^{-as},$$

which is a formula for the Laplace transform of the unit impulse function. This gives, of course, the inverse formula

$$\mathcal{L}^{-1}[e^{-as}] = \delta_a(t).$$

The previous discussion is highly intuitive and lacks a careful mathematical base. However, the ideas can be made precise and rigorous. We refer to advanced texts for a thorough treatment of generalized functions. Another common notation for the unit impulse $\delta_a(t)$ is $\delta(t-a)$. If an impulse has magnitude f_0, instead of 1, then we denote it by $f_0\delta_a(t)$. For example, an impulse given a circuit by a 12 volt battery at time $t = a$ is $12\delta_a(t)$.

Example 4.16

Solve the initial value problem

$$u'' + u' = \delta_2(t), \quad u(0) = u'(0) = 0,$$

with a unit impulse applied at time $t = 2$. Taking the transform,

$$s^2U(s) + sU(s) = e^{-2s}.$$

Thus

$$U(s) = \frac{e^{-2s}}{s(s+1)}.$$

Using the table it is simple to find

$$\mathcal{L}^{-1}\left[\frac{1}{s(s+1)}\right] = 1 - e^{-t}.$$

Therefore, by the shift property, the solution is

$$u(t) = \mathcal{L}^{-1}\left[\frac{e^{-2s}}{s(s+1)}\right] = (1 - e^{-(t-2)})h_2(t).$$

The initial conditions are zero, and so the solution is zero up until time $t = 2$, when the impulse occurs. At that time the solution increases with limit 1 as $t \to \infty$. See figure 4.4.

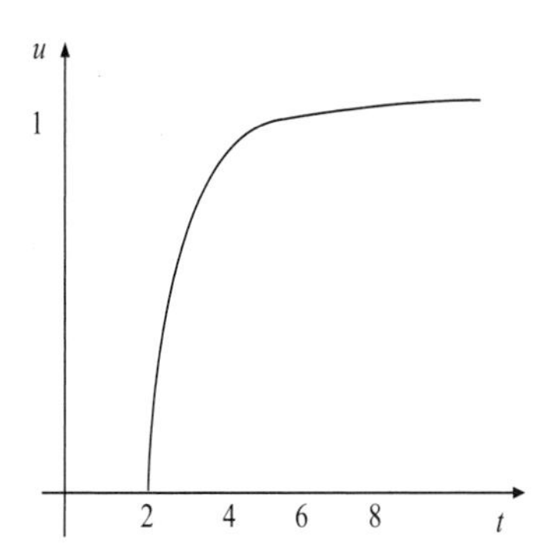

Figure 4.4 Solution in Example 4.16.

EXERCISES

1. Compute $\int_0^\infty e^{-2(t-3)^2}\delta_4(t)dt$.

2. Solve the initial value problem

$$\begin{aligned} u' + 3u &= \delta_1(t) + h_4(t), \\ u(0) &= 1. \end{aligned}$$

   Sketch the solution.

3. Solve the initial value problem

$$\begin{aligned} u'' - u &= \delta_5(t), \\ u(0) &= u'(0) = 0. \end{aligned}$$

   Sketch the solution.

4. Solve the initial value problem

$$\begin{aligned} u'' + u &= \delta_2(t), \\ u(0) &= u'(0) = 0. \end{aligned}$$

   Sketch the solution.

5. Invert the transform $F(s) = \frac{e^{-2s}}{s} + e^{-3s}$.

6. Solve the initial value problem

$$\begin{aligned} u'' + 4u &= \delta_2(t) - \delta_5(t), \\ u(0) &= u'(0) = 0. \end{aligned}$$

7. Consider an LC circuit with $L = C = 1$ and $v(0) = v'(0) = 0$, containing a 1 volt battery, where v is the voltage across the capacitor. At each of the times $t = 0, \pi, 2\pi, 3\pi, ..., n\pi, ...$ the circuit is closed for a single instant. Determine the resulting voltage response $v(t)$ on the capacitor.

8. Compute the Laplace transform of the unit impulse in a different way from that in this section by calculating the transform of $e_{a,\varepsilon}(t)$, and then taking the limit as $\varepsilon \to 0$. Specifically, show

$$\mathcal{L}[e_{a,\varepsilon}(t)] = \mathcal{L}[\frac{1}{\varepsilon}(h_{a-\varepsilon/2}(t) - h_{a+\varepsilon/2}(t))] = \frac{1}{s}e^{-as}\frac{2\sinh\frac{\varepsilon s}{2}}{\varepsilon}.$$

Then use l'Hospital's rule to compute the limit

$$\lim_{\varepsilon\to 0}\frac{2\sinh\frac{\varepsilon s}{2}}{\varepsilon} = s,$$

thereby showing

$$\mathcal{L}[\delta_a(t)] = e^{-as}.$$

4.6 Table of Laplace Transforms

$u(t)$	$U(s)$
e^{at}	$\frac{1}{s-a}$
t^n	$\frac{n!}{s^{n+1}}, \quad n = 0, 1, 2, 3, ...$
t^a	$\frac{\Gamma(a+1)}{s^{a+1}}$
$\sin kt$	$\frac{k}{s^2+k^2}$
$\cos kt$	$\frac{s}{s^2+k^2}$
$\sinh kt$	$\frac{k}{s^2-k^2}$
$\cosh kt$	$\frac{s}{s^2-k^2}$
$e^{at} \sin kt$	$\frac{k}{(s-a)^2+k^2}$
$e^{at} \cos kt$	$\frac{s-a}{(s-a)^2+k^2}$
$\frac{1}{a-b}(e^{at} - e^{bt})$	$\frac{1}{(s-a)(s-b)}$
$t^n e^{at}$	$\frac{n!}{(s-a)^{n+1}}$
$u'(t)$	$sU(s) - u(0)$
$u''(t)$	$s^2U(s) - su(0) - u'(0)$
$u^{(n)}(t)$	$s^nU(s) - s^{n-1}u(0) - \cdots - u^{(n-1)}(0)$
$u(at)$	$\frac{1}{a}U(\frac{s}{a})$
$h_a(t)$	$\frac{1}{s}e^{-as}$
$u(t)e^{at}$	$U(s-a)$
$\delta_a(t)$	e^{-as}
$h_a(t)u(t-a)$	$U(s)e^{-as}$
$\int_0^t u(\tau)v(t-\tau)d\tau$	$U(s)V(s)$
$f(t)h_a(t)$	$e^{-as}\mathcal{L}[f(t+a)]$

5 Linear Systems

Up until now we have focused upon a single differential equation with one unknown state function. Yet, most physical systems require several states for their characterization. Therefore, we are naturally led to study several differential equations for several unknowns. Typically, we expect that if there are n unknown states, then there will be n differential equations, and each DE will contain many of the unknown state functions. Thus the equations are coupled together in the same way as simultaneous systems of algebraic equations. If there are n simultaneous differential equations in n unknowns, we call the set of equations an n-dimensional system.

5.1 Introduction

A two-dimensional, linear, homogeneous system of differential equations has the form

$$x' = ax + by, \tag{5.1}$$

$$y' = cx + dy, \tag{5.2}$$

where a, b, c, and d are constants, and where x and y are the unknown states. A solution consists of a pair of functions $x = x(t)$, $y = y(t)$, that, when substituted into the equations, reduce the equations to identities. We can interpret the solution geometrically in two ways. First, we can plot $x = x(t)$ and $y = y(t)$

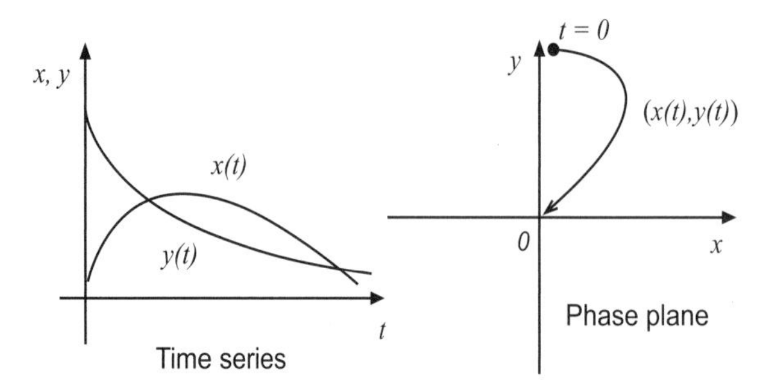

Figure 5.1 Plots showing the two representations of a solution to a system for $t \geq 0$. The plot to the left shows the time series plots $x = x(t)$, $y = y(t)$, and the plot to the right shows the corresponding orbit in the xy-phase plane.

vs. t on the same set of axes as shown in figure 5.1. These are the time series plots and they tell us how the states x and y vary in time. Or, second, we can think of $x = x(t)$, $y = y(t)$ as parametric equations of a curve in an xy plane, with time t as the parameter along the curve. See figure 5.1. In this latter context, the parametric solution representation is called an **orbit**, and the xy plane is called the **phase plane**. Other words used to describe a solution curve in the phase plane, in addition to orbit, are **solution curve**, **path**, and **trajectory**. These words are often used interchangeably. In multi-variable calculus the reader probably used the position vector $\mathbf{x}(t) = x(t)\mathbf{i} + y(t)\mathbf{j}$ to represent this orbit, where $\mathbf{i}$ and $\mathbf{j}$ are the unit vectors, but here we use the column vector notation

$$\mathbf{x}(t) = \begin{pmatrix} x(t) \\ y(t) \end{pmatrix}.$$

To save vertical space in typesetting, we often write this column vector as $(x(t), y(t))^{\mathrm{T}}$, where "T" denotes transpose; transpose means turn the row into a column. Mostly we use the phase plane representation of a solution rather than the time series representation.

The linear system (5.1)–(5.2) has infinitely many orbits, each defined for all times $-\infty < t < \infty$. When we impose **initial conditions**, which take the form

$$x(t_0) = x_0, \quad y(t_0) = y_0,$$

then a single orbit is selected out. That is, the **initial value problem**, consisting of the system (5.1)–(5.2) and the initial conditions, has a unique solution.

Equations (5.1)–(5.2) also give geometrical information about the direction of the solution curves in the phase plane in much the same way as the slope field of a single differential equation gives information about the slopes of a solution curve (see Section 1.1.2). At any point (x, y) in the xy plane, the right sides of (5.1)–(5.2) define a vector

$$\mathbf{v} = \mathbf{x}' = \begin{pmatrix} x' \\ y' \end{pmatrix} = \begin{pmatrix} ax + by \\ cx + dy \end{pmatrix},$$

which is the tangent vector to the solution curve that goes through that point. Recall from multi-variable calculus that a curve $(x(t), y(t))^{\mathrm{T}}$ has tangent vector $\mathbf{v} = (x'(t), y'(t))^{\mathrm{T}}$. We can plot, or have software plot for us, this vector at a large set of points in the plane to obtain a vector field (a field of vectors) that indicates the "flow", or direction, of the solution curves, as shown in figure 5.2. The orbits fit in so that their tangent vectors coincide with the vector

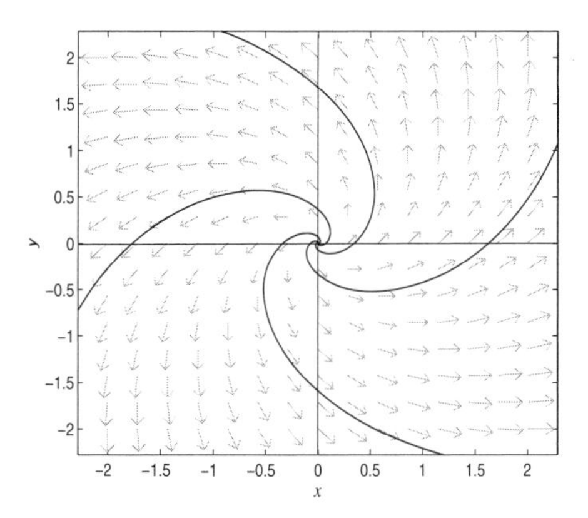

Figure 5.2 In the phase plane, the vector field $\mathbf{v} = (x - y, x + y)^{\mathrm{T}}$ associated with the system $x' = x - y$, $y' = x + y$ and several solution curves $x = x(t)$, $y = y(t)$ which spiral out from the origin. The vector field is tangent to the solution curves. The orbits approach infinity as time goes forward, i.e., $t \to +\infty$, and they approach the origin (but never reach it) as time goes backward, i.e., $t \to -\infty$.

field. A diagram showing several key orbits is called a **phase diagram**, or **phase portrait,** of the system (5.1)–(5.2). The phase portrait may, or may not, include the vector field.

Example 5.1

We observed in Chapter 3 that a second-order differential equation can be reformulated as a system of two first-order equations. For example, the damped, spring-mass equation

$$mx'' = -kx - cx'$$

can be rewritten as

$$\begin{aligned} x' &= y, \\ y' &= -\frac{k}{m}x - \frac{c}{m}y, \end{aligned}$$

where x is position or displacement of the mass from equilibrium and y is its velocity. This system has the form of a two-dimensional linear system. In this manner, mechanical problems can be studied as linear systems. With specific physical parameters $k = m = 1$ and $c = 0.5$, we obtain

$$\begin{aligned} x' &= y, \\ y' &= -x - 0.5y. \end{aligned}$$

The response of this damped spring-mass system is a decaying oscillation. Figure 5.3 shows a phase diagram.

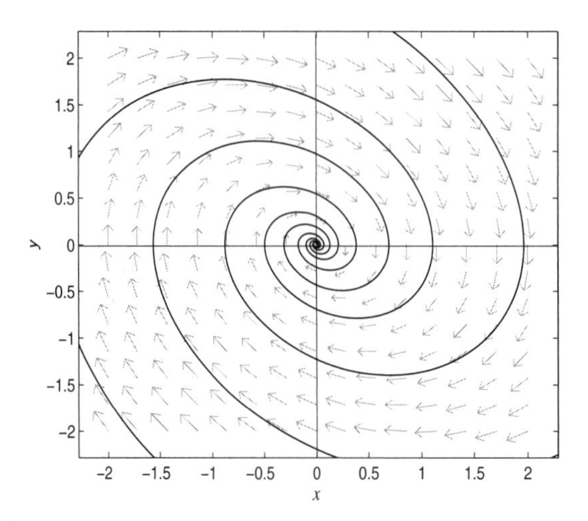

Figure 5.3 Phase plane diagram and vector field for the system $x' = y$, $y' = -x - 0.5y$, showing several orbits, which are spirals approaching the origin. These spirals correspond to time series plots of x and y vs t that oscillate and decay.

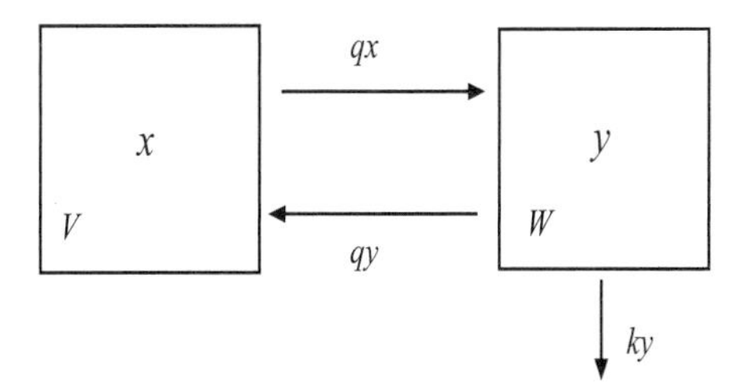

Figure 5.4 Two compartments with arrows indicating the flow rates between them, and the decay rate.

Example 5.2

In this example we generalize ideas presented in Section 1.3.5 on chemical reactors, and the reader should review that material before continuing. The idea here is to consider linked chemical reactors, or several different compartments. Compartmental models play an important role in many areas of science and technology, and they often lead to linear systems. The compartments may be reservoirs, organs, the blood system, industrial chemical reactors, or even classes of individuals. Consider the two compartments in figure 5.4 where a chemical in compartment 1 (having volume V liters) flows into compartment 2 (of volume W liters) at the rate of q liters per minute. In compartment 2 it decays at a rate proportional to its concentration, and it flows back into compartment 1 at the same rate q. At all times both compartments are stirred thoroughly to guarantee perfect mixing. Here we could think of two lakes, or the blood system and an organ. Let x and y denote the concentrations of the chemical in compartments 1 and 2, respectively. We measure concentrations in grams per liter. The technique for finding model equations that govern the concentrations is the same as that noted in Section 1.3.5. Namely, use mass balance. Regardless of the number of compartments, mass must be balanced in each one: the rate of change of mass must equal the rate that mass flows in or is created, minus the rate that mass flows out or is consumed. The mass in compartment 1 is Vx, and the mass in compartment 2 is Wy. A mass flow rate (mass per time) equals the volumetric flow rate (volume per time) times the concentration (mass per volume). So the mass flow rate from compartment 1 to 2 is qx and the mass flow rate from compartment 2 to 1 is qy. Therefore, balancing rates gives

$$\begin{aligned} Vx' &= -qx + qy, \\ Wy' &= qx - qy - Wky, \end{aligned}$$

where k is the decay rate (grams per volume per time) in compartment 2. The volume W must appear as a factor in the decay rate term to make the dimensions correct. We can write this system as

$$\begin{aligned} x' &= -\frac{q}{V}x + \frac{q}{V}y, \\ y' &= \frac{q}{W}x - \left(\frac{q}{W} + k\right)y, \end{aligned}$$

which is the form of a two dimensional linear system. In any compartment model the key idea is to account for all sources and sinks in each compartment. If the volumetric flow rates are not the same, then the problem is complicated by variable volumes, making the problem nonhomogeneous and time dependent. But the technique for obtaining the model equations is the same.

EXERCISES

1. Verify that $\mathbf{x}(t) = (\cos 2t, -2\sin 2t)^{\mathrm{T}}$ is a solution to the system

$$x' = y, \quad y' = -4x.$$

Sketch a time series plot of the solution and the corresponding orbit in the phase plane. Indicate the direction of the orbit as time increases. By hand, plot several vectors in the vector field to show the direction of solution curves.

2. Verify that

$$\mathbf{x}(t) = \begin{pmatrix} 2e^t \\ -3e^t \end{pmatrix}$$

is a solution to the linear system

$$x' = 4x + 2y, \quad y' = -3x - y.$$

Plot this solution in the xy plane for $t \in (-\infty, \infty)$ and find the tangent vectors along the solution curve. Plot the time series on $-\infty < t < +\infty$.

3. Consider the linear system

$$x' = -x + y, \quad y' = 4x - 4y,$$

with initial conditions $x(0) = 10$, $y(0) = 0$. Find formulas for the solution $x(t), y(t)$, and plot their time series. (Hint: multiply the first equation by 4 and add the two equations.)

4. In Example 5.2 let $V = 1$ liter, $W = 0.5$ liters, $q = 0.05$ liters per minute, with no decay in compartment 2. If $x(0) = 10$ grams per liter and $y(0) = 0$, find the concentrations $x(t)$ and $y(t)$ in the two compartments. (Hint: add appropriate multiples of the equations.) Plot the time series and the corresponding orbit in the phase plane. What are the concentrations in the compartments after a long time?

5. In Exercise 4 assume there is decay in compartment 2 with $k = 0.2$ grams per liter per minute. Find the concentrations and plot the time series graphs and the phase plot. (Hint: transform the system into a single second-order equation for x.)

6. In the damped spring-mass system in Example 5.1 take $m = 1$, $k = 2$, and $c = \frac{1}{2}$, with initial conditions $x(0) = 4$ and $y(0) = 0$. Find formulas for the position $x(t)$ and velocity $y(t)$. Show the time series and the orbit in the phase plane.

7. Let q and I be the charge and the current in an RCL circuit with no electromotive force. Write down a linear system of first-order equations that govern the two variables q and I. Take $L = 1$, $R = 0$, and $C = \frac{1}{4}$. If $q(0) = 8$ and $I(0) = 0$, find $q(t)$ and $I(t)$. Show a time series plot of the solution and the corresponding orbit in the qI phase plane.

5.2 Matrices

The study of simultaneous differential equations is greatly facilitated by matrices. Matrix theory provides a convenient language and notation to express many of the ideas concisely. Complicated formulas are simplified considerably in this framework, and matrix notation is more or less independent of dimension. In this extended section we present a brief introduction to square matrices. Some of the definitions and properties are given for general n by n matrices, but our focus is on the two- and three-dimensional cases. This section does not represent a thorough treatment of matrix theory, but rather a limited discussion centered on ideas necessary to discuss solutions of differential equations.

A square array A of numbers having n rows and n columns is called a **square matrix** of size n, or an $n \times n$ matrix (we say, "n by n matrix"). The number in the ith row and jth column is denoted by a_{ij}. General 2×2 and 3×3 matrices have the form

$$A = \begin{pmatrix} a_{11} & a_{12} \\ a_{21} & a_{22} \end{pmatrix}, \quad A = \begin{pmatrix} a_{11} & a_{12} & a_{13} \\ a_{21} & a_{22} & a_{23} \\ a_{31} & a_{32} & a_{33} \end{pmatrix}.$$

The numbers a_{ij} are called the **entries** in the matrix; the first subscript i denotes the row, and the second subscript j denotes the column. The **main diagonal** of a square matrix A is the set of elements $a_{11}, a_{22}, ..., a_{nn}$. We often write matrices using the brief notation $A = (a_{ij})$. An n-**vector** $\mathbf{x}$ is a list of n numbers $x_1, x_2, ..., x_n$, written as a *column*; so "vector" means "column list." The numbers $x_1, x_2, ..., x_n$ in the list are called its **components**. For example,

$$\mathbf{x} = \begin{pmatrix} x_1 \\ x_2 \end{pmatrix}$$

is a 2-vector. Vectors are denoted by lowercase boldface letters like $\mathbf{x}$, $\mathbf{y}$, etc., and matrices are denoted by capital letters like A, B, etc. To minimize space in typesetting, we often write, for example, a 2-vector $\mathbf{x}$ as $(x_1, x_2)^{\mathrm{T}}$, where the T denotes *transpose*, meaning turn the row into a column.

Two square matrices having the same size can be added entry-wise. That is, if $A = (a_{ij})$ and $B = (b_{ij})$ are both $n \times n$ matrices, then the **sum** $A + B$ is an $n \times n$ matrix defined by $A + B = (a_{ij} + b_{ij})$. A square matrix $A = (a_{ij})$ of any size can be multiplied by a constant c by multiplying all the elements of A by the constant; in symbols this **scalar multiplication** is defined by $cA = (ca_{ij})$. Thus $-A = (-a_{ij})$, and it is clear that $A + (-A) = 0$, where 0 is the **zero matrix** having all entries zero. If A and B have the same size, then **subtraction** is defined by $A - B = A + (-B)$. Also, $A + 0 = A$, if 0 has the same size as A. Addition, when defined, is both commutative and associative. Therefore the arithmetic rules of addition for $n \times n$ matrices are the same as the usual rules for addition of numbers.

Similar rules hold for addition of column vectors of the same length and multiplication of column vectors by scalars; these are the definitions you encountered in multi-variable calculus where n-vectors are regarded as elements of $\mathbf{R}^n$. Vectors add component-wise, and multiplication of a vector by a scalar multiplies each component of that vector by that scalar.

Example 5.3

Let

$$A = \begin{pmatrix} 1 & 2 \\ 3 & -4 \end{pmatrix}, \quad B = \begin{pmatrix} 0 & -2 \\ 7 & -4 \end{pmatrix}, \quad \mathbf{x} = \begin{pmatrix} -4 \\ 6 \end{pmatrix}, \quad \mathbf{y} = \begin{pmatrix} 5 \\ 1 \end{pmatrix}.$$

Then

$$\begin{aligned} A + B &= \begin{pmatrix} 1 & 0 \\ 10 & -8 \end{pmatrix}, \quad -3B = \begin{pmatrix} 0 & 6 \\ -21 & 12 \end{pmatrix}, \\ 5\mathbf{x} &= \begin{pmatrix} -20 \\ 30 \end{pmatrix}, \quad \mathbf{x}+2\mathbf{y} = \begin{pmatrix} 6 \\ 8 \end{pmatrix}. \end{aligned}$$

The product of two square matrices of the same size is *not* found by multiplying entry-wise. Rather, **matrix multiplication** is defined as follows. Let A and B be two $n \times n$ matrices. Then the matrix AB is defined to be the $n \times n$ matrix $C = (c_{ij})$ where the ij entry (in the ith row and jth column) of the product C is found by taking the product (dot product, as with vectors) of the ith row of A and the jth column of B. In symbols, $AB = C$, where

$$c_{ij} = \mathbf{a}_i \cdot \mathbf{b}_j = a_{i1}b_{1j} + a_{i2}b_{2j} + \cdots + a_{in}b_{nj},$$

where $\mathbf{a}_i$ denotes the ith row of A, and $\mathbf{b}_j$ denotes the jth column of B. Generally, matrix multiplication is *not* commutative (i.e., $AB \neq BA$), so the order in which matrices are multiplied is important. However, the associative law $AB(C) = (AB)C$ does hold, so you can regroup products as you wish. The distributive law connecting addition and multiplication, $A(B+C) = AB + AC$, also holds. The powers of a square matrix are defined by $A^2 = AA$, $A^3 = AA^2$, and so on.

Example 5.4

Let

$$A = \begin{pmatrix} 2 & 3 \\ -1 & 0 \end{pmatrix}, \quad B = \begin{pmatrix} 1 & 4 \\ 5 & 2 \end{pmatrix}.$$

Then

$$AB = \begin{pmatrix} 2 \cdot 1 + 3 \cdot 5 & 2 \cdot 4 + 3 \cdot 2 \\ -1 \cdot 1 + 0 \cdot 5 & -1 \cdot 4 + 0 \cdot 2 \end{pmatrix} = \begin{pmatrix} 17 & 14 \\ -1 & -4 \end{pmatrix}.$$

Also

$$\begin{aligned} A^2 &= \begin{pmatrix} 2 & 3 \\ -1 & 0 \end{pmatrix}\begin{pmatrix} 2 & 3 \\ -1 & 0 \end{pmatrix} \\ &= \begin{pmatrix} 2 \cdot 2 + 3 \cdot (-1) & 2 \cdot 3 + 3 \cdot 0 \\ -1 \cdot 2 + 0 \cdot (-1) & -1 \cdot 3 + 0 \cdot 0 \end{pmatrix} = \begin{pmatrix} -1 & 6 \\ -2 & -3 \end{pmatrix}. \end{aligned}$$

Next we define multiplication of an $n \times n$ matrix A times an n-vector $\mathbf{x}$. The product $A\mathbf{x}$, with the matrix on the left, is defined to be the n-vector whose ith component is $\mathbf{a}_i \cdot \mathbf{x}$. In other words, the ith element in the list $A\mathbf{x}$ is found by taking the product of the ith row of A and the vector $\mathbf{x}$. The product $\mathbf{x}A$ is not defined.

Example 5.5

When $n = 2$ we have

$$A\mathbf{x} = \begin{pmatrix} a & b \\ c & d \end{pmatrix}\begin{pmatrix} x \\ y \end{pmatrix} = \begin{pmatrix} ax + by \\ cx + dy \end{pmatrix}.$$

For a numerical example take

$$A = \begin{pmatrix} 2 & 3 \\ -1 & 0 \end{pmatrix}, \quad \mathbf{x} = \begin{pmatrix} 5 \\ 7 \end{pmatrix}.$$

Then

$$A\mathbf{x} = \begin{pmatrix} 2 \cdot 5 + 3 \cdot 7 \\ -1 \cdot 5 + 0 \cdot 7 \end{pmatrix} = \begin{pmatrix} 31 \\ -5 \end{pmatrix}.$$

The special square matrix having ones on the main diagonal and zeros elsewhere else is called the **identity matrix** and is denoted by I. For example, the 2×2 and 3×3 identities are

$$I = \begin{pmatrix} 1 & 0 \\ 0 & 1 \end{pmatrix} \quad \text{and} \quad I = \begin{pmatrix} 1 & 0 & 0 \\ 0 & 1 & 0 \\ 0 & 0 & 1 \end{pmatrix}.$$

It is easy to see that if A is any square matrix and I is the identity matrix of the same size, then $AI = IA = A$. Therefore multiplication by the identity matrix does not change the result, a situation similar to multiplying real numbers by the unit number 1. If A is an $n \times n$ matrix and there exists a matrix B for which $AB = BA = I$, then B is called the **inverse** of A and we denote it by $B = A^{-1}$. If A^{-1} exists, we say A is a **nonsingular** matrix; otherwise it is called **singular**. One can show that the inverse of a matrix, if it exists, is unique. We never write $1/A$ for the inverse of A.

A useful number associated with a square matrix A is its determinant. The **determinant** of a square matrix A, denoted by $\det A$ (also by $|A|$) is a number found by combining the elements of the matrix is a special way. The determinant of a 1×1 matrix is just the single number in the matrix. For a 2×2 matrix we define

$$\det A = \det \begin{pmatrix} a & b \\ c & d \end{pmatrix} = ad - cb,$$

and for a 3×3 matrix we define

$$\det \begin{pmatrix} a & b & c \\ d & e & f \\ g & h & i \end{pmatrix} = aei + bfg + cdh - ceg - bdi - ahf. \tag{5.3}$$

Example 5.6

We have

$$\det \begin{pmatrix} 2 & 6 \\ -2 & 0 \end{pmatrix} = 2 \cdot 0 - (-2) \cdot 6 = 12.$$

There is a general inductive formula that defines the determinant of an $n \times n$ matrix as a sum of $(n-1) \times (n-1)$ matrices. Let $A = (a_{ij})$ be an $n \times n$ matrix, and let M_{ij} denote the $(n-1) \times (n-1)$ matrix found by deleting the ith row and jth column of A; the matrix M_{ij} is called the ij **minor** of A. Then $\det A$ is defined by choosing any fixed column J of A and summing the elements a_{iJ} in that column times the determinants of their minors M_{iJ}, with an associated sign ($\pm$), depending upon location in the column. That is, for any fixed J,

$$\det A = \sum_{i=1}^{n} (-1)^{i+J} a_{iJ} \det(M_{iJ}).$$

This is called the **expansion by minors** formula. One can show that you get the same value regardless of which column J you use. In fact, one can expand on any fixed row I instead of a column and still obtain the same value,

$$\det A = \sum_{j=1}^{n} (-1)^{I+j} a_{Ij} \det(M_{Ij}).$$

So, the determinant is well defined by these equations. The reader should check that these formulas give the values for the 2×2 and 3×3 determinants presented above. A few comments are in order. First, the expansion by minors formulas are useful only for small matrices. For an $n \times n$ matrix, it takes roughly $n!$ arithmetic calculations to compute the determinant using expansion by minors, which is enormous when n is large. Efficient computational algorithms to calculate determinants use row reduction methods. Both computer algebra systems and calculators have routines for calculating determinants.

Using the determinant we can give a simple formula for the inverse of a 2×2 matrix A. Let

$$A = \begin{pmatrix} a & b \\ c & d \end{pmatrix}$$

and suppose $\det A \neq 0$. Then

$$A^{-1} = \frac{1}{\det A} \begin{pmatrix} d & -b \\ -c & a \end{pmatrix}. \tag{5.4}$$

So the inverse of a 2×2 matrix is found by interchanging the main diagonal elements, putting minus signs on the off-diagonal elements, and dividing by the determinant. There is a similar formula for the inverse of larger matrices; for completeness we will write the formula down, but for the record we comment that there are more efficient ways to calculate the inverse. With that said, the inverse of an $n \times n$ matrix A is the $n \times n$ matrix whose ij entry is $(-1)^{i+j} \det(M_{ji})$, divided by the determinant of A, which is assumed nonzero. In symbols,

$$A^{-1} = \frac{1}{\det A}((-1)^{i+j} \det(M_{ji})). \tag{5.5}$$

Note that the ij entry of A^{-1} is computed from the ji minor, with indices transposed. In the 3×3 case the formula is

$$A^{-1} = \frac{1}{\det A} \begin{pmatrix} \det M_{11} & -\det M_{21} & \det M_{31} \\ -\det M_{12} & \det M_{22} & -\det M_{32} \\ \det M_{13} & -\det M_{23} & \det M_{33} \end{pmatrix}.$$

Example 5.7

If

$$A = \begin{pmatrix} 1 & 2 \\ 4 & 3 \end{pmatrix},$$

then

$$A^{-1} = \frac{1}{\det A} \begin{pmatrix} 3 & -2 \\ -4 & 1 \end{pmatrix} = \frac{1}{-5} \begin{pmatrix} 3 & -2 \\ -4 & 1 \end{pmatrix} = \begin{pmatrix} -\frac{3}{5} & \frac{2}{5} \\ \frac{4}{5} & -\frac{1}{5} \end{pmatrix}.$$

The reader can easily check that $AA^{-1} = I$.

Equations (5.4) and (5.5) are revealing because they seem to indicate the inverse matrix exists only when the determinant is nonzero (you can't divide by zero). In fact, these two statements are equivalent for any square matrix, regardless of its size: A^{-1} exists if, and only if, $\det A \neq 0$. This is a major theoretical result in matrix theory, and it is a convenient test for invertibility of small matrices. Again, for larger matrices it is more efficient to use row reduction methods to calculate determinants and inverses. The reader should remember the equivalences

$$A^{-1}\text{exists} \Leftrightarrow A \text{ is nonsingular} \Leftrightarrow \det A \neq 0.$$

Matrices were developed to represent and study linear algebraic systems (n linear algebraic equations in n unknowns) in a concise way. For example, consider two equations in two unknowns x_1, x_2 given in standard form by

$$\begin{aligned} a_{11}x_1 + a_{12}x_2 &= b_1 \\ a_{21}x_1 + a_{22}x_2 &= b_2. \end{aligned}$$

Using matrix notation we can write this as

$$\begin{pmatrix} a_{11} & a_{12} \\ a_{21} & a_{22} \end{pmatrix} \begin{pmatrix} x_1 \\ x_2 \end{pmatrix} = \begin{pmatrix} b_1 \\ b_2 \end{pmatrix},$$

or just simply as

$$A\mathbf{x} = \mathbf{b}, \tag{5.6}$$

where

$$A = \begin{pmatrix} a_{11} & a_{12} \\ a_{21} & a_{22} \end{pmatrix}, \quad \mathbf{x} = \begin{pmatrix} x_1 \\ x_2 \end{pmatrix}, \quad \mathbf{b} = \begin{pmatrix} b_1 \\ b_2 \end{pmatrix}.$$

A is the **coefficient matrix**, $\mathbf{x}$ is a column vector containing the unknowns, and $\mathbf{b}$ is a column vector representing the right side. If $\mathbf{b} = \mathbf{0}$, the zero vector, then the system (5.6) is called **homogeneous**. Otherwise it is called **nonhomogeneous**. In a two-dimensional system each equation represents a line in the plane. When $\mathbf{b} = \mathbf{0}$ the two lines pass through the origin. A solution vector $\mathbf{x}$ is represented by a point that lies on both lines. There is a unique solution when both lines intersect at a single point; there are infinitely many solutions when both lines coincide; there is no solution if the lines are parallel and different. In the case of three equations in three unknowns, each equation in the system has the form $\alpha x_1 + \beta x_2 + \gamma x_3 = d$ and represents a plane in space. If $d = 0$ then the plane passes through the origin. The three planes represented by the three equations can intersect in many ways, giving no solution (no common intersection points), a unique solution (when they intersect at a single point), a line of solutions (when they intersect in a common line), and a plane of solutions (when all the equations represent the same plane).

The following theorem tells us when a linear system $A\mathbf{x} = \mathbf{b}$ of n equations in n unknowns is solvable. It is a key result that is applied often in the sequel.

Theorem 5.8

Let A be an $n \times n$ matrix. If A is nonsingular, then the system $A\mathbf{x} = \mathbf{b}$ has a unique solution given by $\mathbf{x} = A^{-1}\mathbf{b}$; in particular, the homogeneous system $A\mathbf{x} = \mathbf{0}$ has only the trivial solution $\mathbf{x} = \mathbf{0}$. If A is singular, then the homogeneous system $A\mathbf{x} = \mathbf{0}$ has infinitely many solutions, and the nonhomogeneous system $A\mathbf{x} = \mathbf{b}$ may have no solution or infinitely many solutions.

It is easy to show the first part of the theorem, when A is nonsingular, using the machinery of matrix notation. If A is nonsingular then A^{-1} exists. Multiplying both sides of $A\mathbf{x} = \mathbf{b}$ on the left by A^{-1} gives

$$\begin{aligned} A^{-1}A\mathbf{x} &= A^{-1}\mathbf{b}, \\ I\mathbf{x} &= A^{-1}\mathbf{b}, \\ \mathbf{x} &= A^{-1}\mathbf{b}, \end{aligned}$$

which is the unique solution. If A is singular one can appeal to a geometric argument in two dimensions. That is, if A is singular, then $\det A = 0$, and the two lines represented by the two equations must be parallel (can you show that?). Therefore they either coincide or they do not, giving either infinitely many solutions or no solution. We remark that the method of finding and multiplying by the inverse of the matrix A, as above, is not the most efficient method for solving linear systems. Row reduction methods, introduced in high school algebra (and reviewed below), provide an efficient computational algorithm for solving large systems.

Example 5.9

Consider the homogeneous linear system

$$\begin{pmatrix} 4 & 1 \\ 8 & 2 \end{pmatrix} \begin{pmatrix} x_1 \\ x_2 \end{pmatrix} = \begin{pmatrix} 0 \\ 0 \end{pmatrix}.$$

The coefficient matrix has determinant zero, so there will be infinitely many solutions. The two equations represented by the system are

$$4x_1 + x_2 = 0, \quad 8x_1 + 2x_2 = 0,$$

which are clearly not independent; one is a multiple of the other. Therefore we need only consider one of the equations, say $4x_1 + x_2 = 0$. With one equation in two unknowns we are free to pick a value for one of the variables and solve for the other. Let $x_1 = 1$; then $x_2 = -4$ and we get a single solution $\mathbf{x} = (1, -4)^{\mathrm{T}}$. More generally, if we choose $x_1 = \alpha$, where α is any real parameter, then $x_2 = -4\alpha$. Therefore all solutions are given by

$$\mathbf{x} = \begin{pmatrix} x_1 \\ x_2 \end{pmatrix} = \begin{pmatrix} \alpha \\ -4\alpha \end{pmatrix} = \alpha \begin{pmatrix} 1 \\ -4 \end{pmatrix}, \quad a \in \mathbf{R}.$$

Thus all solutions are multiples of $(1, -4)^{\mathrm{T}}$, and the solution set lies along the straight line through the origin defined by this vector. Geometrically, the two equations represent two lines in the plane that coincide.

Next we review the **row reduction method** for solving linear systems when $n = 3$. Consider the algebraic system $A\mathbf{x} = \mathbf{b}$, or

$$\begin{aligned} a_{11}x_1 + a_{12}x_2 + a_{13}x_3 &= b_1, \\ a_{21}x_1 + a_{22}x_2 + a_{23}x_3 &= b_2, \\ a_{31}x_1 + a_{32}x_2 + a_{33}x_3 &= b_3. \end{aligned} \tag{5.7}$$

At first we assume the coefficient matrix $A = (a_{ij})$ is nonsingular, so that the system has a unique solution. The basic idea is to transform the system into the simpler *triangular form*

$$\begin{aligned} \widetilde{a}_{11}x_1 + \widetilde{a}_{12}x_2 + \widetilde{a}_{13}x_3 &= \widetilde{b}_1, \\ \widetilde{a}_{22}x_2 + \widetilde{a}_{23}x_3 &= \widetilde{b}_2, \\ \widetilde{a}_{33}x_3 &= \widetilde{b}_3. \end{aligned}$$

This triangular system is easily solved by back substitution. That is, the third equation involves only one unknown and we can instantly find x_3. That value is substituted back into the second equation where we can then find x_2, and those two values are substituted back into the first equation and we can find x_1. The process of transforming (5.7) into triangular form is carried out by three admissible operations that do not affect the solution structure.

1. Any equation may be multiplied by a nonzero constant.
2. Any two equations may be interchanged.
3. Any equation may be replaced by that equation plus (or minus) a multiple of any other equation.

We observe that any equation in the system (5.7) is represented by its coefficients and the right side, so we only need work with the numbers, which saves writing. We organize the numbers in an **augmented array**

$$\begin{pmatrix} a_{11} & a_{12} & a_{13} & b_1 \\ a_{21} & a_{22} & a_{23} & b_2 \\ a_{31} & a_{32} & a_{33} & b_3 \end{pmatrix}.$$

The admissible operations listed above translate into row operations on the augmented array: any row may be multiplied by a nonzero constant, any two rows may be interchanged, and any row may be replaced by itself plus (or minus) any other row. By performing these row operations we transform the augmented array into a triangular array with zeros in the lower left corner below the main diagonal. The process is carried out one column at a time, beginning from the left.

Example 5.10

Consider the system

$$\begin{aligned} x_1 + x_2 + x_3 &= 0, \\ 2x_1 - 2x_3 &= 2, \\ x_1 - x_2 + x_3 &= 6. \end{aligned}$$

The augmented array is

$$\begin{pmatrix} 1 & 1 & 1 & 0 \\ 2 & 0 & -2 & 2 \\ 1 & -1 & 1 & 6 \end{pmatrix}.$$

Begin working on the first column to get zeros in the 2,1 and 3,1 positions by replacing the second and third rows by themselves plus multiples of the first row. So we replace the second row by the second row minus twice the first row and replace the third row by third row minus the first row. This gives

$$\begin{pmatrix} 1 & 1 & 1 & 0 \\ 0 & -2 & -4 & 2 \\ 0 & -2 & 0 & 6 \end{pmatrix}.$$

Next work on the second column to get a zero in the 3,2 position, below the diagonal entry. Specifically, replace the third row by the third row minus the second row:

$$\begin{pmatrix} 1 & 1 & 1 & 0 \\ 0 & -2 & -4 & 2 \\ 0 & 0 & 4 & 4 \end{pmatrix}.$$

This is triangular, as desired. To make the arithmetic easier, multiply the third row by $1/4$ and the second row by $-1/2$ to get

$$\begin{pmatrix} 1 & 1 & 1 & 0 \\ 0 & 1 & 2 & -1 \\ 0 & 0 & 1 & 1 \end{pmatrix},$$

with ones on the diagonal. This triangular, augmented array represents the system

$$\begin{aligned} x_1 + x_2 + x_3 &= 0, \\ x_2 + 2x_3 &= -1, \\ x_3 &= 1. \end{aligned}$$

Using back substitution, $x_3 = 1$, $x_2 = -3$, and $x_1 = 2$, which is the unique solution, representing a point $(2, -3, 1)$ in $\mathbf{R}^3$.

If the coefficient matrix A is singular we can end up with different types of triangular forms, for example,

$$\begin{pmatrix} 1 & * & * & * \\ 0 & 1 & * & * \\ 0 & 0 & 0 & * \end{pmatrix}, \quad \begin{pmatrix} 1 & * & * & * \\ 0 & 0 & * & * \\ 0 & 0 & 0 & * \end{pmatrix}, \text{ or } \begin{pmatrix} 1 & * & * & * \\ 0 & 0 & 0 & * \\ 0 & 0 & 0 & * \end{pmatrix},$$

where the $*$ denotes an entry. These augmented arrays can be translated back into equations. Depending upon the values of those entries, we will get no solution (the equations are inconsistent) or infinitely many solutions. As examples, suppose there are three systems with triangular forms at the end of the process given by

$$\begin{pmatrix} 1 & 1 & 3 & 0 \\ 0 & 1 & 2 & 5 \\ 0 & 0 & 0 & 7 \end{pmatrix}, \quad \begin{pmatrix} 1 & 0 & 3 & 3 \\ 0 & 0 & 1 & 1 \\ 0 & 0 & 0 & 0 \end{pmatrix}, \text{ or } \begin{pmatrix} 1 & 2 & 0 & 1 \\ 0 & 0 & 0 & 0 \\ 0 & 0 & 0 & 0 \end{pmatrix}.$$

There would be no solution for the first system (the last row states $0 = 7$), and infinitely many solutions for the second and third systems. Specifically, the second system would have solution $x_3 = 1$ and $x_1 = 0$, with $x_2 = a$, which is arbitrary. Therefore the solution to the second system could be written

$$\begin{pmatrix} x_1 \\ x_2 \\ x_3 \end{pmatrix} = \begin{pmatrix} 0 \\ a \\ 1 \end{pmatrix} = a \begin{pmatrix} 0 \\ 1 \\ 0 \end{pmatrix} + \begin{pmatrix} 0 \\ 0 \\ 1 \end{pmatrix},$$

with a an arbitrary constant. This represents a line in $\mathbf{R}^3$. A line is a one-dimensional geometrical object described in terms of one parameter. The third system above reduced to $x_1 + 2x_2 = 1$. So we may pick x_3 and x_2 arbitrarily, say $x_2 = a$ and $x_3 = b$, and then $x_1 = 1 - 2a$. The solution to the third system can then be written

$$\begin{pmatrix} x_1 \\ x_2 \\ x_3 \end{pmatrix} = \begin{pmatrix} 1-2a \\ a \\ b \end{pmatrix} = a \begin{pmatrix} -2 \\ 1 \\ 0 \end{pmatrix} + b \begin{pmatrix} 0 \\ 0 \\ 1 \end{pmatrix} + \begin{pmatrix} 1 \\ 0 \\ 0 \end{pmatrix},$$

which is a plane in $\mathbf{R}^3$. A plane is a two-dimensional object in $\mathbf{R}^3$ requiring two parameters for its description.

The set of all solutions to a homogeneous system $A\mathbf{x} = \mathbf{0}$ is called the **nullspace** of A. The nullspace may consist of a single point $\mathbf{x} = \mathbf{0}$ when A is nonsingular, or it may be a line or plane passing through the origin in the case where A is singular.

Finally we introduce the notion of independence of column vectors. A set of vectors is said to be a linearly independent set if any one of them cannot be written as a combination of some of the others. We can express this statement mathematically as follows. A set (p of them) of n-vectors $\mathbf{v}_1, \mathbf{v}_2, ..., \mathbf{v}_p$ is a **linearly independent set** if the equation[1]

$$c_1\mathbf{v}_1 + c_2\mathbf{v}_2 + \cdots + c_p\mathbf{v}_p = \mathbf{0}$$

forces all the constants to be zero; that is, $c_1 = c_2 = \cdots = c_p = 0$. If all the constants are not forced to be zero, then we say the set of vectors is **linearly dependent**. In this case there would be at least one of the constants, say c_r, which is not zero, at which point we could solve for $\mathbf{v}_r$ in terms of the remaining vectors.

Notice that two vectors are independent if one is not a multiple of the other.

[1] A sum of constant multiples of a set of vectors is called a **linear combination** of those vectors.

In the sequel we also need the notion of linear independence for **vector functions**. A vector function in two dimensions has the form of a 2-vector whose entries are functions of time t; for example,

$$\mathbf{r}(t) = \begin{pmatrix} x(t) \\ y(t) \end{pmatrix},$$

where t belongs to some interval I of time. The vector function $\mathbf{r}(t)$ is the position vector, and its arrowhead traces out a curve in the plane given by the parametric equations $x = x(t),\ y = y(t),\quad t \in I$. As observed in Section 5.1, solutions to two-dimensional systems of differential equations are vector functions. Linear independence of a set of n-vector functions $\mathbf{r}_1(t), \mathbf{r}_2(t), ..., \mathbf{r}_p(t)$ on an interval I means that if a linear combination of those vectors is set equal to zero, for all $t \in I$, then the set of constants is forced to be zero. In symbols,

$$c_1\mathbf{r}_1(t)+c_2\mathbf{r}_2(t)+\cdots+c_p\mathbf{r}_p(t) = \mathbf{0},\ \ t \in I, \quad \text{implies} \quad c_1 = 0,\ c_2 = 0, ..., c_p = 0.$$

Finally, if a matrix has entries that are functions of t, i.e., $A = A(t) = (a_{ij}(t))$, then we define the derivative of the matrix as the matrix of derivatives, or $A'(t) = (a'_{ij}(t))$.

Example 5.11

The two vector functions

$$\mathbf{r}_1(t) = \begin{pmatrix} e^{2t} \\ 7 \end{pmatrix}, \quad \mathbf{r}_2(t) = \begin{pmatrix} 5e^{2t} \\ \sin t \end{pmatrix}$$

form a linearly independent set on the real line because one is not a multiple of the other. Looked at differently, if we set a linear combination of them equal to the zero vector (i.e., $c_1\mathbf{r}_1(t) + c_2\mathbf{r}_2(t) = \mathbf{0}$), and take $t = 0$, then

$$c_1 + 5c_2 = 0, \quad 7c_1 = 0,$$

which forces $c_1 = c_2 = 0$. Because the linear combination is zero for all t, we may take $t = 0$.

Example 5.12

The three vector functions

$$\mathbf{r}_1(t) = \begin{pmatrix} e^{2t} \\ 7 \end{pmatrix}, \quad \mathbf{r}_2(t) = \begin{pmatrix} 5e^{2t} \\ \sin t \end{pmatrix}, \quad \mathbf{r}_3(t) = \begin{pmatrix} 1 \\ 3\sin\frac{t}{2} \end{pmatrix},$$

form a linearly independent set on $\mathbf{R}$ because none can be written as a combination of the others. That is, if we take a linear combination and set it equal to zero; that is, $c_1\mathbf{r}_1(t) + c_2\mathbf{r}_1(t) + c_3\mathbf{r}_1(t) = \mathbf{0}$, for all $t \in \mathbf{R}$, then we are forced into $c_1 = c_2 = c_3 = 0$ (see Exercise 15).

EXERCISES

1. Let
$$A = \begin{pmatrix} 1 & 3 \\ 2 & 4 \end{pmatrix}, \quad B = \begin{pmatrix} -1 & 0 \\ 3 & 7 \end{pmatrix}, \quad \mathbf{x} = \begin{pmatrix} 2 \\ -5 \end{pmatrix}.$$
Find $A + B$, $B - 4A$, AB, BA, A^2, $B\mathbf{x}$, $AB\mathbf{x}$, A^{-1}, $\det B$, B^3, AI, and $\det(A - \lambda I\ )$, where λ is a parameter.

2. With A given in Exercise 1 and $b = (2, 1)^T$, solve the system $A\mathbf{x} = \mathbf{b}$ using A^{-1}. Then solve the system by row reduction.

3. Let
$$A = \begin{pmatrix} 0 & 2 & -1 \\ 1 & 6 & -2 \\ 2 & 0 & 3 \end{pmatrix}, \quad B = \begin{pmatrix} 1 & -1 & 0 \\ 2 & 1 & 4 \\ -1 & -1 & 1 \end{pmatrix}, \quad \mathbf{x} = \begin{pmatrix} 2 \\ 0 \\ -1 \end{pmatrix}.$$
Find $A + B$, $B - 4A$, BA, A^2, $B\mathbf{x}$, $\det A$, AI, $A - 3I$, and $\det(B - I)$.

4. Find all values of the parameter λ that satisfy the equation $\det(A - \lambda I) = 0$, where A is given in Exercise 1.

5. Let
$$A = \begin{pmatrix} 2 & -1 \\ -4 & 2 \end{pmatrix}.$$
Compute $\det A$. Does A^{-1} exist? Find all solutions to $A\mathbf{x} = \mathbf{0}$ and plot the solution set in the plane.

6. Use the row reduction method to determine all values m for which the algebraic system
$$2x + 3y = m, \quad -6x - 9y = 5,$$
has no solution, a unique solution, or infinitely many solutions.

7. Use row reduction to determine the value(s) of m for which the following system has infinitely many solutions.
$$\begin{aligned} x + y &= 0, \\ 2x + y &= 0, \\ 3x + 2y + mz &= 0. \end{aligned}$$

8. If a square matrix A has all zeros either below its main diagonal or above its main diagonal, show that $\det A$ equals the product of the elements on the main diagonal.

9. Construct simple homogeneous systems $A\mathbf{x} = \mathbf{0}$ of three equations in three unknowns that have: (a) a unique solution, (b) an infinitude of solutions lying on a line in $\mathbf{R}^3$, and (c) an infinitude of solutions lying on a plane in $\mathbf{R}^3$. Is there a case when there is no solution?

10. Let
$$A = \begin{pmatrix} 0 & 2 & -1 \\ 1 & 6 & -2 \\ 2 & 0 & 3 \end{pmatrix}.$$

    a) Find $\det A$ by the expansion by minors formula using the first column, the second column, and the third row. Is A invertible? Is A singular?

    b) Find the inverse of A and use it to solve $A\mathbf{x} = \mathbf{b}$, where $\mathbf{b} = (1, 0, 4)^{\mathrm{T}}$.

    c) Solve $A\mathbf{x} = \mathbf{b}$ in part (b) using row reduction.

11. Find all solutions to the homogeneous system $A\mathbf{x} = \mathbf{0}$ if
$$A = \begin{pmatrix} -2 & 0 & 2 \\ 2 & -4 & 0 \\ 0 & 4 & -2 \end{pmatrix}.$$

12. Use the definition of linear independence to show that the 2-vectors $(2, -3)^T$ and $(-4, 8)^T$ are linearly independent.

13. Use the definition to show that the 3-vectors $(0, 1, 0)^{\mathrm{T}}$, $(1, 2, 0)^{\mathrm{T}}$, and $(0, 1, 4)^{\mathrm{T}}$ are linearly independent.

14. Use the definition to show that the 3-vectors $(1, 0, 1)^{\mathrm{T}}$, $(5, -1, 0)^{\mathrm{T}}$, and $(-7, 1, 2)^{\mathrm{T}}$ are linearly dependent.

15. Verify the claim in Example 5.12 by taking two special values of t.

16. Plot each of the following vector functions in the xy plane, where $-\infty < t < +\infty$.
$$\mathbf{r}_1(t) = \begin{pmatrix} 3\cos t \\ 2\sin t \end{pmatrix}, \quad \mathbf{r}_2(t) = \begin{pmatrix} 1 \\ 3 \end{pmatrix} t, \quad \mathbf{r}_3(t) = \begin{pmatrix} t \\ t+1 \end{pmatrix} e^{-t}.$$
    Show that these vector functions form a linearly independent set by setting $c_1\mathbf{r}_1(t) + c_2\mathbf{r}_1(t) + c_3\mathbf{r}_1(t) = \mathbf{0}$ and then choosing special values of t to force the constants to be zero.

17. Show that a 3×3 matrix A is invertible if, and only if, its three columns form an independent set of 3-vectors.

18. Find $A'(t)$ if

$$A(t) = \begin{pmatrix} \cos t & t^2 & 0 \\ 2e^{2t} & 1 & \sin 2t \\ 0 & \sqrt{2t} & \frac{-5}{t^2+1} \end{pmatrix}.$$

5.3 Two-Dimensional Systems

5.3.1 Solutions and Linear Orbits

A two-dimensional linear system of differential equations

$$\begin{aligned} x' &= ax + by, \\ y' &= cx + dy, \end{aligned}$$

where a, b, c, and d are constants, can be written compactly using vectors and matrices. Denoting

$$\mathbf{x}(t) = \begin{pmatrix} x(t) \\ y(t) \end{pmatrix}, \quad A = \begin{pmatrix} a & b \\ c & d \end{pmatrix},$$

the system can be written

$$\begin{pmatrix} x'(t) \\ y'(t) \end{pmatrix} = \begin{pmatrix} a & b \\ c & d \end{pmatrix} \begin{pmatrix} x(t) \\ y(t) \end{pmatrix},$$

or

$$\mathbf{x}'(t) = \begin{pmatrix} a & b \\ c & d \end{pmatrix} \mathbf{x}(\mathbf{t}).$$

We often write this simply as

$$\mathbf{x}' = A\mathbf{x}, \tag{5.8}$$

where we have suppressed the understood dependence of $\mathbf{x}$ on t. We briefly reiterate the ideas introduced in the introduction, Section 5.1. A solution to the system (5.8) on an interval is a vector function $\mathbf{x}(t) = (x(t), y(t))^T$, that satisfies the system on the required interval. We can graph $x(t)$ and $y(t)$ vs. t, which gives the state space representation or **time series** plots of the solution. Alternatively, a solution can be graphed as a parametric curve, or vector function, in the xy plane. We call the xy plane the **phase plane**, and we call a solution curve plotted in the xy plane an **orbit**. Observe that a solution is a vector function $\mathbf{x}(t)$ with components $x(t)$ and $y(t)$. In the phase plane, the orbit is represented in parametric form and is traced out as time proceeds. Thus, time is not explicitly displayed in the phase plane representation, but it is a parameter along the orbit. An orbit is traced out in a specific direction as time increases, and we usually denote that direction by an arrow along the curve. Furthermore, time can always be shifted along a solution curve. That is, if $\mathbf{x}(t)$ is a solution, then $\mathbf{x}(t-c)$ is a solution for any real number c and it represents the same solution curve.

Our main objective is to find the phase portrait, or a plot of key orbits of the given system. We are particularly interested in the **equilibrium solutions** of (5.8). These are the *constant* vector solutions $\mathbf{x}^*$ for which $A\mathbf{x}^* = \mathbf{0}$. An equilibrium solution is represented in the phase plane as a point. The vector field vanishes at an equilibrium point. The time series representation of an equilibrium solution is two constant functions. If $\det A \neq 0$ then $\mathbf{x}^* = \mathbf{0}$ is the only equilibrium of (5.8), and it is represented by the origin, $(0, 0)$, in the phase plane. We say in this case that the origin is an **isolated equilibrium**. If $\det A = 0$, then there will be an entire line of equilibrium solutions through the origin; each point on the line represents an equilibrium solution, and the equilibria are not isolated. Equilibrium solutions are important because the interesting behavior of the orbits occurs near these solutions. (Equilibrium solutions are also called critical points by some authors.)

Example 5.13

Consider the system

$$\begin{aligned} x' &= -2x - y, \\ y' &= 2x - 5y, \end{aligned}$$

which we write as

$$\mathbf{x}' = \begin{pmatrix} -2 & -1 \\ 2 & -5 \end{pmatrix} \mathbf{x}.$$

The coefficient determinant is nonzero, so the only equilibrium solution is represented by the origin, $x(t) = 0$, $y(t) = 0$. By substitution, it is straightforward to check that

$$\mathbf{x}_1(t) = \begin{pmatrix} x(t) \\ y(t) \end{pmatrix} = \begin{pmatrix} e^{-3t} \\ e^{-3t} \end{pmatrix} = \begin{pmatrix} 1 \\ 1 \end{pmatrix} e^{-3t}$$

is a solution. Also

$$\mathbf{x}_2(t) = \begin{pmatrix} e^{-4t} \\ 2e^{-4t} \end{pmatrix} = \begin{pmatrix} 1 \\ 2 \end{pmatrix} e^{-4t}$$

is a solution. Each of these solutions has the form of a constant vector times a scalar exponential function of time t. Why should we expect exponential solutions? The two equations involve both x and y and their derivatives; a solution must make everything cancel out, and so each term must basically have the same form. Exponential functions and their derivatives both have the same form, and therefore exponential functions for both x and y are likely candidates for solutions. We graph these two independent solutions $\mathbf{x}_1(t)$ and $\mathbf{x}_2(t)$ in the phase plane. See figure 5.5. Each solution, or orbit, plots as a ray traced from infinity (as time t approaches $-\infty$) into the origin (as t approaches $+\infty$). The slopes of these ray-like solutions are defined by the constant vectors preceding the scalar exponential factor, the latter of which has the effect of stretching or shrinking the vector. Note that these two orbits approach the origin as time gets large, but they never actually reach it. Another way to look

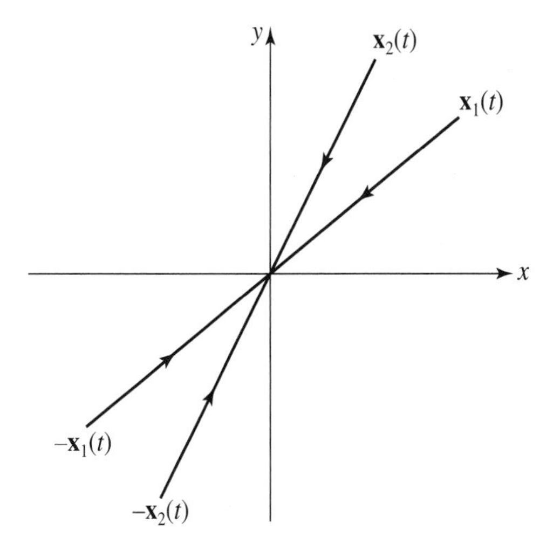

Figure 5.5 $\mathbf{x}_1(t)$ and $\mathbf{x}_2(t)$ are shown as linear orbits (rays) entering the origin in the first quadrant. The reflection of those rays in the third quadrant are the solutions $-\mathbf{x}_1(t)$ and $-\mathbf{x}_2(t)$. Note that all four of these linear orbits approach the origin as $t \to +\infty$ because of the decaying exponential factor in the solution. As $t \to -\infty$ (backward in time) all four of these linear orbits go to infinity.

at it is this. If we eliminate the parameter t in the parametric representation $x = e^{-4t}$, $y = 2e^{-4t}$ of $\mathbf{x}_2(t)$, say, then $y = 2x$, which is a straight line in the xy plane. This orbit is on one ray of this straight line, lying in the first quadrant.

Solutions of (5.8) the form $\mathbf{x}(t) = \mathbf{v}e^{\lambda t}$, where λ is a real constant and $\mathbf{v}$ is a constant, real vector, are called **linear orbits** because they plot as rays in the xy-phase plane.

We are ready to make some observations about the structure of the solution set to the two-dimensional linear system (5.8). All of these properties can be extended to three, or even n, dimensional systems.

1. **(Superposition)** If $\mathbf{x}_1(t)$ and $\mathbf{x}_2(t)$ are any solutions and c_1 and c_2 are any constants, then the **linear combination** $c_1\mathbf{x}_1(t) + c_2\mathbf{x}_2(t)$ is a solution.

2. **(General Solution)** If $\mathbf{x}_1(t)$ and $\mathbf{x}_2(t)$ are two linear independent solutions (i.e., one is not a multiple of the other), then all solutions are given by $\mathbf{x}(t) = c_1\mathbf{x}_1(t) + c_2\mathbf{x}_2(t)$, where c_1 and c_2 are arbitrary constants. This combination is called the **general solution** of (5.8).

3. **(Existence-Uniqueness)** The initial value problem

$$\mathbf{x}' = A\mathbf{x}, \quad \mathbf{x}(t_0) = \mathbf{x}_0,$$

where $\mathbf{x}_0$ is a fixed vector, has a unique solution valid for all $-\infty < t < +\infty$.

The existence-uniqueness property actually guarantees that there are two independent solutions to a two-dimensional system. Let $\mathbf{x}_1$ be the unique solution to the initial value problem $\mathbf{x}_1' = A\mathbf{x}_1, \quad \mathbf{x}_1(0) = (1,0)^{\mathrm{T}}$ and $\mathbf{x}_2$ be the unique solution to the initial value problem $\mathbf{x}_2' = A\mathbf{x}_2, \quad \mathbf{x}(0) = (0,1)^{\mathrm{T}}$. These must be independent. Otherwise they would be proportional and we would have

$$\mathbf{x}_1(t) = k\mathbf{x}_2(t),$$

for all t, where k is a nonzero constant. But if we take $t = 0$, we would have

$$(1,0)^T = k(0,1)^T,$$

which is a contradiction.

The question is how to determine two independent solutions so that we can obtain the general solution. This is a central issue we address in the sequel. One method to solve a two-dimensional linear system is to eliminate one of the variables and reduce the problem to a single second-order equation.

Example 5.14

(*Method of Elimination*) Consider

$$\begin{aligned} x' &= 4x - 3y, \\ y' &= 6x - 7y. \end{aligned}$$

Differentiate the first and then use the second to get

$$\begin{aligned} x'' &= 4x' - 3y' = 4(4x - 3y) - 3(6x - 7y) \\ &= -2x + 9y = -2x + 9(-\frac{1}{3}x' + \frac{4}{3}x) \\ &= -3x' + 10x, \end{aligned}$$

which is a second-order equation. The characteristic equation is $\lambda^2+3\lambda-10=0$ with roots $\lambda = -5, 2$. Thus

$$x(t) = c_1 e^{-5t} + c_2 e^{2t}.$$

Then

$$y(t) = -\frac{1}{3}x' + \frac{4}{3}x = 3c_1 e^{-5t} + \frac{2}{3}c_2 e^{2t}.$$

We can write the solution in vector form as

$$\mathbf{x}(t) = \begin{pmatrix} x(t) \\ y(t) \end{pmatrix} = c_1 \begin{pmatrix} e^{-5t} \\ 3e^{-5t} \end{pmatrix} + c_2 \begin{pmatrix} e^{2t} \\ \frac{2}{3}e^{2t} \end{pmatrix}.$$

In this form we can see that two independent vector solutions are

$$\mathbf{x}_1(t) = \begin{pmatrix} e^{-5t} \\ 3e^{-5t} \end{pmatrix}, \quad \mathbf{x}_2(t) = \begin{pmatrix} e^{2t} \\ \frac{2}{3}e^{2t} \end{pmatrix},$$

and the general solution is a linear combination of these, $\mathbf{x}(t) = c_1\mathbf{x}_1(t) + c_2\mathbf{x}_2(t)$. However simple this strategy appears in two dimensions, it does not work as easily in higher dimensions, nor does it expose methods that are easily adaptable to higher-dimensional systems. Therefore we do not often use the elimination method.

But we point out features of the phase plane. Notice that $\mathbf{x}_1$ graphs as a linear orbit in the first quadrant of the xy phase plane, along the ray defined by of the vector $(1, 3)^T$. It enters the origin as $t \to \infty$ because of the decaying exponential factor. The other solution, $\mathbf{x}_2$, also represents a linear orbit along the direction defined by the vector $(1, 2/3)^T$. This solution, because of the increasing exponential factor e^{2t}, tends to infinity as $t \to +\infty$. Figure 5.6 shows the linear orbits. Figure 5.7 shows several orbits on the phase diagram obtained by taking different values of the arbitrary constants in the general solution. The structure of the orbital system near the origin, where curves veer away and approach the linear orbits as time goes forward and backward, is called a **saddle point** structure. The linear orbits are sometimes called **separatrices** because they separate different types of orbits. All orbits approach the separatrices as time gets large, either negatively or positively.

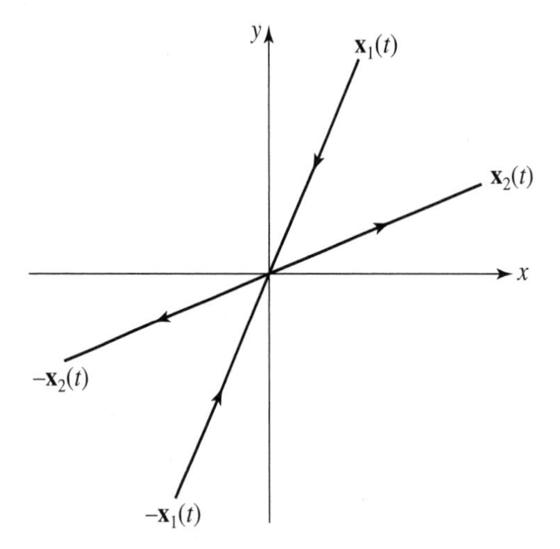

Figure 5.6 Linear orbits in Example 5.14 representing the solutions corresponding to $\mathbf{x}_1(t)$ and $\mathbf{x}_2(t)$, and the companion orbits $-\mathbf{x}_1(t)$ and $-\mathbf{x}_2(t)$. These linear orbits are called separatrices.

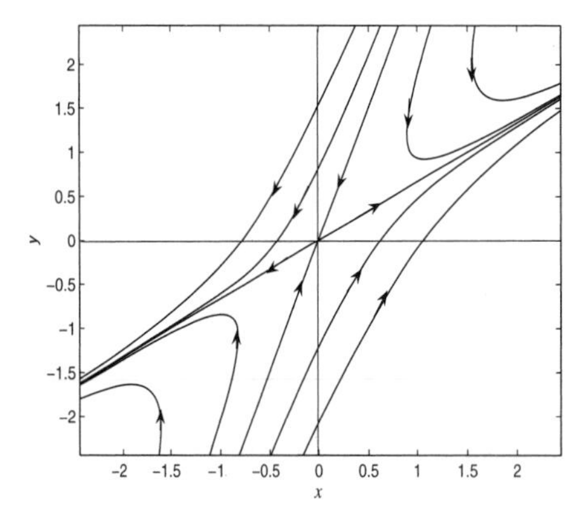

Figure 5.7 Phase portrait for the system showing a saddle point at the origin.

5.3.2 The Eigenvalue Problem

Now we introduce some general methods for the two-dimensional system

$$\mathbf{x}' = A\mathbf{x}. \tag{5.9}$$

We assume that $\det A \neq 0$ so that the only equilibrium solution of the system is at the origin. As examples have shown, we should expect an exponential-type solution. Therefore, we attempt to find a solution of the form

$$\mathbf{x} = \mathbf{v}e^{\lambda t}, \tag{5.10}$$

where λ is a constant and $\mathbf{v}$ is a nonzero constant vector, both to be determined.

Substituting $\mathbf{x} = \mathbf{v}e^{\lambda t}$ and $\mathbf{x}'=\lambda \mathbf{v}e^{\lambda t}$ into the (5.9) gives

$$\lambda \mathbf{v}e^{\lambda t} = A(\mathbf{v}e^{\lambda t}),$$

or

$$A\mathbf{v} = \lambda \mathbf{v}. \tag{5.11}$$

Therefore, if a λ and $\mathbf{v}$ can be found that satisfy (5.11), then we will have determined a solution of the form (5.10). The vector equation (5.11) represents a well-known problem in mathematics called the **algebraic eigenvalue problem**. The eigenvalue problem is to determine values of λ for which (5.11) has a nontrivial solution $\mathbf{v}$. A value of λ for which there is a nontrivial solution $\mathbf{v}$ is called an **eigenvalue**, and a corresponding $\mathbf{v}$ associated with that eigenvalue is called an **eigenvector**. The pair $\lambda, \mathbf{v}$ is called an **eigenpair**. Geometrically we think of the eigenvalue problem like this: A represents a transformation that maps vectors in the plane to vectors in the plane; a vector $\mathbf{x}$ gets transformed to a vector $A\mathbf{x}$. An eigenvector of A is a special vector that is mapped to a multiple (λ) of itself; that is, $Ax = \lambda x$. In summary, we have reduced the problem of finding solutions to a system of differential equations to the problem of finding solutions of an algebra problem—every eigenpair gives a solution.

Geometrically, if λ is real, the linear orbit representing this solution lies along a ray emanating from the origin along the direction defined by the vector $\mathbf{v}$. If $\lambda < 0$ the solution approaches the origin along the ray, and if $\lambda > 0$ the solution goes to infinity along the ray. The situation is similar to that shown in figure 5.6. When there is a solution graphing as a linear orbit, then there is automatically a second, opposite, linear orbit along the ray $-\mathbf{v}$. This is because if $\mathbf{x} = \mathbf{v}e^{\lambda t}$ is a solution, then so is $-\mathbf{x} = -\mathbf{v}e^{\lambda t}$

To solve the eigenvalue problem we rewrite (5.11) as a homogeneous linear system

$$(A-\lambda I)\mathbf{v} = \mathbf{0}. \tag{5.12}$$

By Theorem 5.8 this system will have the desired nontrivial solutions if the determinant of the coefficient matrix is zero, or

$$\det(A-\lambda I) = 0. \tag{5.13}$$

Written out explicitly, this system (5.12) has the form

$$\begin{pmatrix} a-\lambda & b \\ c & d-\lambda \end{pmatrix} \begin{pmatrix} v_1 \\ v_2 \end{pmatrix} = \begin{pmatrix} 0 \\ 0 \end{pmatrix},$$

where the coefficient matrix $A-\lambda I$ is the matrix A with λ subtracted from the diagonal elements. Equation (5.13) is, explicitly,

$$\det \begin{pmatrix} a-\lambda & b \\ c & d-\lambda \end{pmatrix} = (a-\lambda)(d-\lambda) - cb = 0,$$

or equivalently,

$$\lambda^2 - (a+b)\lambda + (ad-bc) = 0.$$

This last equation can be memorized easily if it is written

$$\lambda^2 - (\mathrm{tr} A)\lambda + \det A = 0, \tag{5.14}$$

where $\mathrm{tr} A = a+d$ is called the **trace** of A, defined to be the sum of the diagonal elements of A. Equation (5.14) is called the **characteristic equation** associated with A, and it is a quadratic equation in λ. Its roots, found by factoring or using the quadratic formula, are the two eigenvalues. The eigenvalues may be real and unequal, real and equal, or complex conjugates.

Once the eigenvalues are computed, we can substitute them in turn into the system (5.12) to determine corresponding eigenvectors $\mathbf{v}$. Note that any multiple of an eigenvector is again an eigenvector for that same eigenvalue; this follows from the calculation

$$A(c\mathbf{v}) = cA\mathbf{v} = c(\lambda\mathbf{v}) = \lambda(c\mathbf{v}).$$

Thus, an eigenvector corresponding to a given eigenvalue is not unique; we may multiply them by constants. This is expected from Theorem 5.8. Some calculators display normalized eigenvectors (of length one) found by dividing by their length.

As noted, the eigenvalues may be real and unequal, real and equal, or complex numbers. We now discuss these different cases.

5.3.3 Real Unequal Eigenvalues

If the two eigenvalues are real and unequal, say λ_1 and λ_2, then corresponding eigenvectors $\mathbf{v}_1$ and $\mathbf{v}_2$ are independent and we obtain two independent solutions $\mathbf{v}_1 e^{\lambda_1 t}$ and $\mathbf{v}_2 e^{\lambda_2 t}$. The general solution of the system is then a linear combination of these two independent solutions,

$$\mathbf{x}(t) = c_1 \mathbf{v}_1 e^{\lambda_1 t} + c_2 \mathbf{v}_2 e^{\lambda_2 t},$$

where c_1 and c_2 are arbitrary constants. Each of the independent solutions represents linear orbits in the phase plane, which helps in plotting the phase diagram. All solutions (orbits) $\mathbf{x}(t)$ are linear combinations of the two independent solutions, with each specific solution obtained by fixing values of the arbitrary constants.

Example 5.15

Consider the linear system

$$\mathbf{x}' = \begin{pmatrix} -\frac{3}{2} & \frac{1}{2} \\ 1 & -1 \end{pmatrix} \mathbf{x}. \tag{5.15}$$

The characteristic equation (5.14) is

$$\lambda^2 + \frac{5}{2}\lambda + 1 = 0.$$

By the quadratic formula the eigenvalues are

$$\lambda = -\frac{1}{2}, -2.$$

Now we take each eigenvalue successively and substitute it into (5.12) to obtain corresponding eigenvectors. First, for $\lambda = -\frac{1}{2}$, we get

$$\begin{pmatrix} -1 & \frac{1}{2} \\ 1 & -\frac{1}{2} \end{pmatrix} \begin{pmatrix} v_1 \\ v_2 \end{pmatrix} = \begin{pmatrix} 0 \\ 0 \end{pmatrix},$$

which has a solution $(v_1, v_2)^{\mathrm{T}} = (1, 2)^{\mathrm{T}}$. Notice that any multiple of this eigenvector is again an eigenvector, but all we need is one. Therefore an eigenpair is

$$-\frac{1}{2}, \begin{pmatrix} 1 \\ 2 \end{pmatrix}.$$

Now take $\lambda = -2$. The system (5.12) becomes

$$\begin{pmatrix} \frac{1}{2} & \frac{1}{2} \\ 1 & 1 \end{pmatrix} \begin{pmatrix} v_1 \\ v_2 \end{pmatrix} = \begin{pmatrix} 0 \\ 0 \end{pmatrix},$$

which has solution $(v_1, v_2)^{\mathrm{T}} = (-1, 1)^{\mathrm{T}}$. Thus, another eigenpair is

$$-2, \begin{pmatrix} -1 \\ 1 \end{pmatrix}.$$

The two eigenpairs give two independent solutions

$$\mathbf{x}_1(t) = \begin{pmatrix} 1 \\ 2 \end{pmatrix} e^{-t/2} \quad \text{and} \quad \mathbf{x}_2(t) = \begin{pmatrix} -1 \\ 1 \end{pmatrix} e^{-2t}. \tag{5.16}$$

Each one plots, along with its negative counterparts, as a linear orbit in the phase plane entering the origin as time increases. The general solution of the system (5.15) is

$$\mathbf{x}(t) = c_1 \begin{pmatrix} 1 \\ 2 \end{pmatrix} e^{-t/2} + c_2 \begin{pmatrix} -1 \\ 1 \end{pmatrix} e^{-2t}.$$

This is a two-parameter family of solution curves, and the totality of all these solution curves, or orbits, represents the phase diagram in the xy plane. These orbits are shown in figure 5.8. Because both terms in the general solution decay as time increases, all orbits enter the origin as $t \to +\infty$. And, as t gets large, the term with $e^{-t/2}$ dominates the term with e^{-2t}. Therefore all orbits approach the origin along the direction $(1, 2)^{\mathrm{T}}$. As $t \to -\infty$ the orbits go to infinity; for large negative times the term e^{-2t} dominates the term $e^{-t/2}$, and the orbits become parallel to the direction $(-1, 1)^{\mathrm{T}}$. Each of the two basic solutions 5.16 represents linear orbits along rays in the directions of the eigenvectors. When both eigenvalues are negative, as in this case, all orbits approach the origin in the direction of one of the eigenvectors. When we obtain this type of phase plane structure, we call the origin an **asymptotically stable node**. When both eigenvalues are positive, then the time direction along the orbits is reversed and we call the origin an **unstable node**. The meaning of the term stable is discussed subsequently.

An initial condition picks out one of the many orbits by fixing values for the two arbitrary constants. For example, if $\mathbf{x}(0) = (1, 4)^{\mathrm{T}}$, or we want an orbit passing through the point $(1, 4)$, then

$$c_1 \begin{pmatrix} 1 \\ 2 \end{pmatrix} + c_2 \begin{pmatrix} -1 \\ 1 \end{pmatrix} = \begin{pmatrix} 1 \\ 4 \end{pmatrix},$$

giving $c_1 = 5/3$ and $c_2 = 2/3$. Therefore the unique solution to the initial value problem is

$$\begin{aligned} \mathbf{x}(t) &= \frac{5}{3} \begin{pmatrix} 1 \\ 2 \end{pmatrix} e^{-t/2} + \frac{2}{3} \begin{pmatrix} -1 \\ 1 \end{pmatrix} e^{-2t} \\ &= \begin{pmatrix} \frac{5}{3} e^{-t/2} - \frac{2}{3} e^{-2t} \\ \frac{10}{3} e^{-t/2} + \frac{2}{3} e^{-2t} \end{pmatrix}. \end{aligned}$$

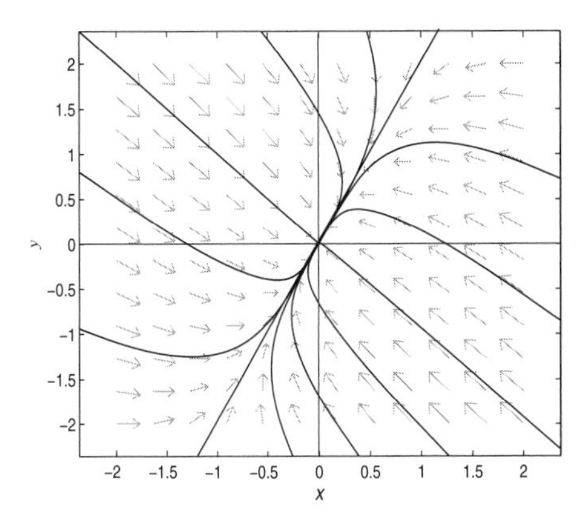

Figure 5.8 A node. All orbits approach the origin, tangent to the direction $(1, 2)$, as $t \to \infty$. Backwards in time, as $t \to -\infty$, the orbits become parallel to the direction $(-1, -1)$. Notice the linear orbits.

Example 5.16

If a system has eigenpairs

$$-2, \begin{pmatrix} 3 \\ 2 \end{pmatrix}, \quad 3, \begin{pmatrix} -1 \\ 5 \end{pmatrix},$$

with real eigenvalues of opposite sign, then the general solution is

$$\mathbf{x}(t) = c_1 \begin{pmatrix} 3 \\ 2 \end{pmatrix} e^{-2t} + c_2 \begin{pmatrix} -1 \\ 5 \end{pmatrix} e^{3t}.$$

In this case one of the eigenvalues is positive and one is negative. Now there are two sets of opposite linear orbits, one pair corresponding to -2 approaching the origin from the directions $\pm(3, 2)^{\mathrm{T}}$, and one pair corresponding to $\lambda = 3$ approaching infinity along the directions $\pm(-1, 5)^{\mathrm{T}}$. The orbital structure is that of a **saddle point** (refer to figure 5.7), and we anticipate saddle point structure when the eigenvalues are real and have opposite sign.

5.3.4 Complex Eigenvalues

If the eigenvalues of the matrix A are complex, they must appear as complex conjugates, or $\lambda = a \pm bi$. The eigenvectors will be $\mathbf{v} = \mathbf{w} \pm i\mathbf{z}$. Therefore, taking the eigenpair $a+bi$, $\mathbf{w}+i\mathbf{z}$, we obtain the *complex* solution

$$(\mathbf{w}+i\mathbf{z})e^{(a+bi)t}.$$

Recalling that the real and imaginary parts of a complex solution are real solutions, we expand this complex solution using Euler's formula to get

$$\begin{aligned}(\mathbf{w}+i\mathbf{z})e^{at}e^{ibt} &= e^{at}(\mathbf{w}+i\mathbf{z})(\cos bt + i\sin bt)\\ &= e^{at}(\mathbf{w}\cos bt - \mathbf{z}\sin bt) + ie^{at}(\mathbf{w}\sin bt + \mathbf{z}\cos bt).\end{aligned}$$

Therefore two *real*, independent solutions are

$$\mathbf{x}_1(t) = e^{at}(\mathbf{w}\cos bt - \mathbf{z}\sin bt), \quad \mathbf{x}_2(t) = e^{at}(\mathbf{w}\sin bt + \mathbf{z}\cos bt),$$

and the general solution is a combination of these,

$$\mathbf{x}(t) = c_1e^{at}(\mathbf{w}\cos bt - \mathbf{z}\sin bt) + c_2e^{at}(\mathbf{w}\sin bt + \mathbf{z}\cos bt). \tag{5.17}$$

In the case of complex eigenvalues we need not consider both eigenpairs; each eigenpair leads to the same two independent solutions. For complex eigenvalues there are no linear orbits. The terms involving the trigonometric functions are periodic functions with period $2\pi/b$, and they define orbits that rotate around the origin. The factor e^{at} acts as an amplitude factor causing the rotating orbits to expand if $a>0$, and we obtain spiral orbits going away from the origin. If $a<0$ the amplitude decays and the spiral orbits go into the origin. In the complex eigenvalue case we say the origin is an **asymptotically stable spiral point** when $a<0$, and an **unstable spiral point** when $a>0$.

If the eigenvalues of A are purely imaginary, $\lambda = \pm bi$, then the amplitude factor e^{at} in (5.17) is absent and the solutions are periodic of period $\frac{2\pi}{b}$, given by

$$\mathbf{x}(t) = c_1(\mathbf{w}\cos bt - \mathbf{z}\sin bt) + c_2(\mathbf{w}\sin bt + \mathbf{z}\cos bt).$$

The orbits are closed cycles and plot as either concentric ellipses or concentric circles. In this case we say the origin is a **(neutrally) stable center**.

Example 5.17

Let

$$\mathbf{x}' = \begin{pmatrix} -2 & -3 \\ 3 & -2 \end{pmatrix}\mathbf{x}.$$

The matrix A has eigenvalues $\lambda = -2 \pm 3i$. An eigenvector corresponding to $\lambda = -2+3i$ is $\mathbf{v}_1 = [-1\ i]^{\mathrm{T}}$. Therefore a complex solution is

$$\begin{aligned}\mathbf{x} &= \begin{pmatrix} -1 \\ i \end{pmatrix} e^{(-2+3i)t} = \left[\begin{pmatrix} -1 \\ 0 \end{pmatrix} + i\begin{pmatrix} 0 \\ 1 \end{pmatrix}\right] e^{-2t}(\cos 3t + i\sin 3t)\\ &= \begin{pmatrix} -e^{-2t}\cos 3t \\ -e^{-2t}\sin 3t \end{pmatrix} + i\begin{pmatrix} -e^{-2t}\sin 3t \\ -e^{-2t}\cos 3t \end{pmatrix}.\end{aligned}$$

Therefore two linearly independent solutions are

$$\mathbf{x}_1(t) = \begin{pmatrix} -e^{-2t}\cos 3t \\ -e^{-2t}\sin 3t \end{pmatrix}, \quad \mathbf{x}_2(t) = \begin{pmatrix} -e^{-2t}\sin 3t \\ -e^{-2t}\cos 3t \end{pmatrix}.$$

The general solution is a linear combination of these two solutions, $\mathbf{x}(t) = c_1\mathbf{x}_1(t) + c_2\mathbf{x}_2(t)$. In the phase plane the orbits are spirals that approach the origin as $t \to +\infty$ because the real part -2 of the eigenvalues is negative. See figure 5.9. At the point $(1,1)$ the tangent vector (direction field) is $(-5,1)$, so the spirals are counterclockwise.

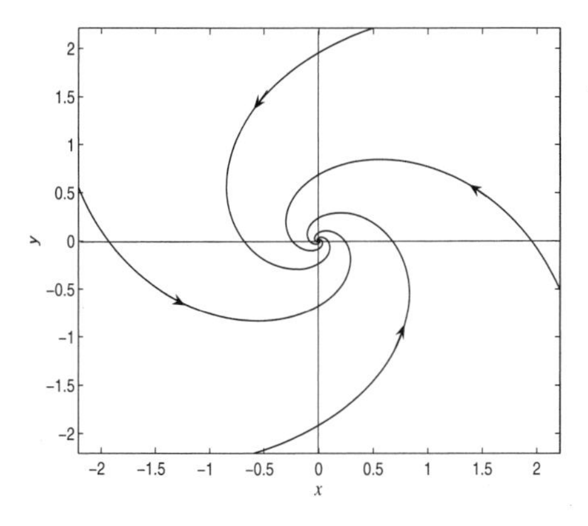

Figure 5.9 A stable spiral from Example 5.17.

5.3.5 Real, Repeated Eigenvalues

One case remains, when A has a repeated real eigenvalue λ with a single eigenvector $\mathbf{v}$. Then $\mathbf{x}_1 = \mathbf{v}e^{\lambda t}$ is one solution (representing a linear orbit), and we need another independent solution. We try a second solution of the form $\mathbf{x}_2 = e^{\lambda t}(t\mathbf{v} + \mathbf{w})$, where $\mathbf{w}$ is to be determined. A more intuitive guess, based on our experience with second-order equations in Chapter 3, would have been $e^{\lambda t}t\mathbf{v}$, but that does not work (try it). Substituting $\mathbf{x}_2$ into the system we get

$$\begin{aligned} \mathbf{x}_2' &= e^{\lambda t}\mathbf{v} + \lambda e^{\lambda t}(t\mathbf{v} + \mathbf{w}), \\ A\mathbf{x}_2 &= e^{\lambda t}A(t\mathbf{v} + \mathbf{w}). \end{aligned}$$

Therefore we obtain an algebraic system for $\mathbf{w}$:

$$(A - \lambda I)\mathbf{w} = \mathbf{v}.$$

This system will always have a solution $\mathbf{w}$, and therefore we will have determined a second linearly independent solution. In fact, this system always has infinitely many solutions, and all we have to do is find one solution. The vector $\mathbf{w}$ is called a **generalized eigenvector**. Therefore, the general solution to the linear system $\mathbf{x}' = A\mathbf{x}$ in the repeated eigenvalue case is

$$\mathbf{x}(t) = c_1 \mathbf{v} e^{\lambda t} + c_2 e^{\lambda t}(t\mathbf{v} + \mathbf{w}).$$

If the eigenvalue is negative the orbits enter the origin as $t \to +\infty$, and they go to infinity as $t \to -\infty$. If the eigenvalue is positive, the orbits reverse direction in time.

In the case where the eigenvalues are equal, the origin has a nodal-like structure, as in Example 5.13. When there is a single eigenvector associated with the repeated eigenvalue, we often call the origin a **degenerate node**. It may occur in a special case that a repeated eigenvalue λ has two independent eigenvectors vectors $\mathbf{v}_1$ and $\mathbf{v}_2$ associated with it. When this occurs, the general solution is just $\mathbf{x}(t) = c_1 \mathbf{v}_1 e^{\lambda t} + c_2 \mathbf{v}_2 e^{\lambda t}$. It happens when the two equations in the system are decoupled, and the matrix is diagonal with equal elements on the diagonal. In this exceptional case all of the orbits are linear orbits entering $(\lambda < 0)$ or leaving $(\lambda > 0)$ the origin; we refer to the origin in this case as a **star-like** node.

Example 5.18

Consider the system

$$\mathbf{x}' = \begin{pmatrix} 2 & 1 \\ -1 & 4 \end{pmatrix} \mathbf{x}.$$

The eigenvalues are $\lambda = 3, 3$ and a corresponding eigenvector is $\mathbf{v} = (1,1)^{\mathrm{T}}$. Therefore one solution is

$$\mathbf{x}_1(t) = \begin{pmatrix} 1 \\ 1 \end{pmatrix} e^{3t}.$$

Notice that this solution plots as a linear orbit coming out of the origin and approaching infinity along the direction $(1,1)^{\mathrm{T}}$. There is automatically an opposite orbit coming out of the origin and approaching infinity along the direction $-(1,1)^{\mathrm{T}}$. A second independent solution will have the form $\mathbf{x}_2 = e^{3t}(t\mathbf{v} + \mathbf{w})$ where $\mathbf{w}$ satisfies

$$(A - 3I)\mathbf{w} = \begin{pmatrix} -1 & 1 \\ -1 & 1 \end{pmatrix} \mathbf{w} = \begin{pmatrix} 1 \\ 1 \end{pmatrix}.$$

This equation has many solutions, and so we choose

$$\mathbf{w} = \begin{pmatrix} 0 \\ 1 \end{pmatrix}.$$

Therefore a second solution has the form

$$\mathbf{x}_2(t) = e^{3t}(t\mathbf{v} + \mathbf{w}) = e^{3t}\left[\begin{pmatrix} 1 \\ 1 \end{pmatrix} t + \begin{pmatrix} 0 \\ 1 \end{pmatrix}\right] = \begin{pmatrix} te^{3t} \\ (t+1)e^{3t} \end{pmatrix}.$$

The general solution of the system is the linear combination

$$\mathbf{x}(t) = c_1\mathbf{x}_1(t) + c_2\mathbf{x}_2(t).$$

If we append an initial condition, for example,

$$\mathbf{x}(0) = \begin{pmatrix} 1 \\ 0 \end{pmatrix},$$

then we can determine the two constants c_1 and c_2. We have

$$\mathbf{x}(0) = c_1\mathbf{x}_1(0) + c_2\mathbf{x}_2(0) = c_1\begin{pmatrix} 1 \\ 1 \end{pmatrix} + c_2\begin{pmatrix} 0 \\ 1 \end{pmatrix} = \begin{pmatrix} 1 \\ 0 \end{pmatrix}.$$

Hence

$$c_1 = 1, \quad c_2 = -1.$$

Therefore the solution to the initial value problem is given by

$$\mathbf{x}(t) = (1)\begin{pmatrix} 1 \\ 1 \end{pmatrix} e^{3t} + (-1)\begin{pmatrix} te^{3t} \\ (t+1)e^{3t} \end{pmatrix} = \begin{pmatrix} (1-t)e^{3t} \\ -te^{3t} \end{pmatrix}.$$

As time goes forward ($t \to \infty$), the orbits go to infinity, and as time goes backward ($t \to -\infty$), the orbits enter the origin. The origin is an unstable node.

How to draw a phase diagram. In general, to draw a rough phase diagram for a liner system all you need to know are the eigenvalues and eigenvectors. If the eigenvalues are real then draw straight lines through the origin in the direction of the associated eigenvectors. Label each ray of the line with an arrow that points inward toward the origin if the eigenvalue is negative and outward if the eigenvalue is positive. Then fill in the regions between these linear orbits with consistent solution curves, paying attention to which "eigendirection" dominates as $t \to \infty$ and $t \to -\infty$. Real eigenvalues with the same sign give nodes, and real eigenvalues of opposite signs give saddles.If the eigenvalues are purely imaginary then the orbits are closed loops around the origin, and if they are complex the orbits are spirals. They spiral in if the eigenvalues have negative real part, and they spiral out if the eigenvalues have positive real part. The direction (clockwise or counterclockwise) of the cycles or spirals can be determined directly from the direction field, often by just plotting one vector in the vector field. Another helpful device to get an improved phase diagram is to plot the set of points where the vector field is vertical (the orbits have a vertical tangent) and where the vector field is horizontal (the orbits have a horizontal tangent). These sets of points are found by setting $x^{prime} = ax + by = 0$ and $y^{prime} = cx + dy = 0$, respectively. These straight lines are called the x- and y-**nullclines**.

Example 5.19

The system

$$\mathbf{x}' = \begin{pmatrix} 2 & 5 \\ -2 & 0 \end{pmatrix} \mathbf{x}$$

has eigenvalues $1 \pm 3i$. The orbits spiral outward (because the real part of the eigenvalues, 1, is positive). They are clockwise because the second equation in the system is $y' = -2x$, and so y decreases ($y' < 0$) when $x > 0$. Observe that the orbits are vertical as they cross the nullcline $2x + 5y = 0$, and they are horizontal as they cross the nullcline $x = 0$. With this information the reader should be able to draw a rough phase diagram.

5.3.6 Stability

We mentioned the word stability in the last section. Now we extend the discussion. For the linear system $\mathbf{x}' = A\mathbf{x}$, an **equilibrium solution** is a constant vector solution $\mathbf{x}(t) = \mathbf{x}^*$ representing a point in the phase plane. The zero-vector $\mathbf{x}^* = \mathbf{0}$ (the origin) is always an equilibrium solution to a linear system. Other equilibria will satisfy $A\mathbf{x}^* = \mathbf{0}$, and thus the only time we get a nontrivial equilibrium solution is when $\det A = 0$; in this case there are infinitely many equilibria. If $\det A \neq 0$, then $\mathbf{x}^* = \mathbf{0}$ is the only equilibrium, and it is called an **isolated equilibrium**. For the discussion in the remainder of this section we assume $\det A \neq 0$.

Suppose the system is in its zero equilibrium state. Intuitively, the equilibrium is stable if a small perturbation, or disturbance, does not cause the system to deviate too far from the equilibrium; the equilibrium is unstable if a small disturbance causes the system to deviate far from its original equilibrium state. We have seen in two-dimensional systems that if the eigenvalues of the matrix A are both negative or have negative real parts, then all orbits approach the origin as $t \to +\infty$. In these cases we say that the origin is **asymptotically**

stable node (including degenerate and star-like nodes) or an **asymptotically stable spiral point**. If the eigenvalues are both positive, have positive real parts, or are real of opposite sign, then some or all orbits that begin near the origin do not stay near the origin as $t \to +\infty$, and we say the origin an **unstable node**indexnode, unstable (including degenerate and star-like nodes), an **unstable spiral point**, and a **saddle**, respectively. If the eigenvalues are purely imaginary we obtain periodic solutions, or closed cycles, and the origin is a center. In this case a small perturbation from the origin puts us on one of the elliptical orbits and we cycle near the origin; we say a center is **neutrally stable**, or just **stable**, but not asymptotically stable. Asymptotically stable equilibria are also called **attractors** or **sinks**, and unstable equilibria are called **repellers** or **sources**. Also, we often refer to asymptotically stable spirals and nodes as just stable spirals and nodes—the word asymptotic is understood.

Summary. We make some important summarizing observations that should be remembered for future use (Chapter 6). For two-dimensional systems it is easy to check stability of the origin, and sometimes this is all we want to do. The eigenvalues are roots of the characteristic equation

$$\lambda^2 - (\mathrm{tr}A)\lambda + \det A = 0.$$

By the quadratic formula,

$$\lambda = \frac{1}{2}(\mathrm{tr}A \pm \sqrt{(\mathrm{tr}A)^2 - 4\det A}).$$

One can easily checks the following facts.

1. If $\det A < 0$, then the eigenvalues are real and have opposite sign, and the origin is a saddle.
2. If $\det A > 0$, then the eigenvalues are real with the same sign (nodes) or complex conjugates (centers and spirals). Nodes have $\mathrm{tr}A)^2 - 4\det A > 0$ and spirals have $\mathrm{tr}A)^2 - 4\det A < 0$. If $\mathrm{tr}A)^2 - 4\det A = 0$ then we obtain degenerate and star-like nodes. If $\mathrm{tr}A < 0$ then the nodes and spirals are stable, and if $\mathrm{tr}A > 0$ they are unstable. If $\mathrm{tr}A = 0$ we obtain centers.
3. If $\det A = 0$, then at least one of the eigenvalues is zero and there is a line of equilibria.

An important result is that *the origin is asymptotically stable if, and only if,* $\mathrm{tr}A < 0$ and $\det A > 0$.

EXERCISES

1. Find the eigenvalues and eigenvectors of the following matrices:

$$A = \begin{pmatrix} -1 & 4 \\ -2 & 5 \end{pmatrix}; \quad B = \begin{pmatrix} 2 & 3 \\ 4 & 6 \end{pmatrix}; \quad C = \begin{pmatrix} 2 & -8 \\ 1 & -2 \end{pmatrix}.$$

2. Write the general solution of the linear system $\mathbf{x}' = A\mathbf{x}$ if A has eigenpairs 2, $(1,5)^{\mathrm{T}}$ and -3, $(2,-4)^{\mathrm{T}}$. Sketch the linear orbits in the phase plane corresponding to these eigenpairs. Find the solution curve that satisfies the initial condition $\mathbf{x}(0) = (0,1)^{\mathrm{T}}$ and plot it in the phase plane. Do the same for the initial condition $\mathbf{x}(0) = (-6,12)^{\mathrm{T}}$.

3. Answer the questions in Exercise 2 for a system whose eigenpairs are -6, $(1,2)^{\mathrm{T}}$ and -1, $(1,-5)^{\mathrm{T}}$.

4. For each system find the general solution and sketch the phase portrait. Indicate the linear orbits (if any) and the direction of the solution curves.

a) $\mathbf{x}' = \begin{pmatrix} 1 & 2 \\ 3 & 2 \end{pmatrix} \mathbf{x}.$

b) $\mathbf{x}' = \begin{pmatrix} -3 & 4 \\ 0 & -3 \end{pmatrix} \mathbf{x}.$

c) $\mathbf{x}' = \begin{pmatrix} 2 & 2 \\ 6 & 3 \end{pmatrix} \mathbf{x}.$

d) $\mathbf{x}' = \begin{pmatrix} -5 & 3 \\ 2 & -10 \end{pmatrix} \mathbf{x}.$

e) $\mathbf{x}' = \begin{pmatrix} 2 & 0 \\ 0 & 2 \end{pmatrix} \mathbf{x}.$

f) $\mathbf{x}' = \begin{pmatrix} 3 & -2 \\ 4 & -1 \end{pmatrix} \mathbf{x}.$

g) $\mathbf{x}' = \begin{pmatrix} 5 & -4 \\ 1 & 1 \end{pmatrix} \mathbf{x}.$

h) $\mathbf{x}' = \begin{pmatrix} 0 & 9 \\ -1 & 0 \end{pmatrix} \mathbf{x}.$

5. Solve the initial value problem

$$\mathbf{x}' = \begin{pmatrix} 2 & 1 \\ -1 & 0 \end{pmatrix} \mathbf{x}, \quad \mathbf{x}(0) = \begin{pmatrix} 1 \\ -1 \end{pmatrix}.$$

6. Consider the system
$$\mathbf{x}' = \begin{pmatrix} 1 & -2 \\ -2 & 4 \end{pmatrix} \mathbf{x}.$$
   a) Find the equilibrium solutions and plot them in the phase plane.
   b) Find the eigenvalues and determine if there are linear orbits.
   c) Find the general solution and plot the phase portrait.

7. Determine the behavior of solutions near the origin for the system
$$\mathbf{x}' = \begin{pmatrix} 3 & a \\ 1 & 1 \end{pmatrix} \mathbf{x}$$
   for different values of the parameter a.

8. For the systems in Exercise 4, characterize the origin as to type (node, degenerate node, star-like node, center, spiral, saddle) and stability (unstable, neutrally stable, asymptotically stable).

9. Consider the system
$$\begin{aligned} x' &= -3x + ay, \\ y' &= bx - 2y. \end{aligned}$$
   Are there values of a and b where the solutions are closed cycles (periodic orbits)?

10. In an individual let x be the excess glucose concentration in the blood and y be the excess insulin concentration (positive x and y values measure the concentrations above normal levels, and negative values measure concentrations below normal levels). These quantities are measured in mg per ml and insulin units per ml, respectively, and time is given in hours. One simple model of glucose–insulin dynamics is
$$\begin{aligned} x' &= -ax - by, \\ y' &= cx - dy, \end{aligned}$$
    where $-ax$ is the rate glucose is absorbed in the liver and $-by$ is the rate it is used in the muscle. The rate cx is the rate insulin is produced by the pancreas and $-dy$ is the rate degraded by the liver. A set of values for the constants is $a = 3$, $b = 4.3$, $c = 0.2$, and $d = 0.8$. If $x(0) = 1$ and $y(0) = 0$ find the glucose and insulin concentrations and graph time series plots over a 4 hour period.

11. Find a two-dimensional linear system whose matrix has eigenvalues $\lambda = -2$ and $\lambda = -3$.

12. Rewrite the damped spring-mass equation $mx'' + cx' + kx = 0$ as a system of two first-order equations for x and $y = x'$. Find the characteristic equation of the matrix for the system and show that it coincides with the characteristic equation associated with the second-order DE.

13. Consider an RCL circuit governed by $LCv'' + RCv' + v = 0$, where v is the voltage on the capacitor. Rewrite the equation as a two-dimensional linear system and determine conditions on the constants R, L, and C for which the origin is an asymptotically stable spiral. To what electrical response $v(t)$ does this case correspond?

14. What are the possible behaviors, depending on γ, of the solutions to the linear system

$$\begin{aligned} x' &= -\gamma x - y, \\ y' &= x - \gamma y. \end{aligned}$$

15. Show that A^{-1} exists if, and only if, zero is not an eigenvalue of A.

16. For a 2×2 matrix show that the product of the two eigenvalues equals its determinant, and the sum of the two eigenvalues equals its trace.

17. For a 2×2 matrix A of a linear system, let p equal its trace and q equal its determinant. Sketch the set of points in the pq-plane where the system has an asymptotically stable spiral at the origin. Sketch the region where it has a saddle points.

5.4 Nonhomogeneous Systems

Corresponding to a two-dimensional, linear homogeneous system $\mathbf{x}' = A\mathbf{x}$, we now examine the **nonhomogeneous system**

$$\mathbf{x}' = A\mathbf{x} + \mathbf{f}(t), \tag{5.18}$$

where

$$\mathbf{f}(t) = \begin{pmatrix} f_1(t) \\ f_2(t) \end{pmatrix}$$

is a given vector function. We think of this function as the driving force in the system.

To ease the notation in writing the solution we define a **fundamental matrix** $\Phi(t)$ as a 2×2 matrix whose columns are two independent solutions to the associated homogeneous system $\mathbf{x}' = A\mathbf{x}$. So, the fundamental matrix is a square array that holds both vector solutions. It is straightforward to show that $\Phi(t)$ satisfies the *matrix* equation $\Phi'(t) = A\Phi(t)$, and that the general solution to the homogeneous equation $\mathbf{x}' = A\mathbf{x}$ can therefore be written in the form

$$\mathbf{x}_h(t) = \Phi(t)\mathbf{c},$$

where $\mathbf{c} = (c_1, c_2)^{\mathrm{T}}$ is an arbitrary constant vector. (The reader should do Exercise 1 presently, which requires verifying these relations.)

The variation of constants method introduced in Chapter 2 is applicable to a first-order linear system. Therefore we assume a solution to (5.18) of the form

$$\mathbf{x}(t) = \Phi(t)\mathbf{c}(t), \tag{5.19}$$

where we have "varied" the constant vector $\mathbf{c}$. Then, using the product rule for differentiation (which works for matrices),

$$\begin{aligned} \mathbf{x}'(t) &= \Phi(t)\mathbf{c}'(t) + \Phi'(t)\mathbf{c}(t) = \Phi(t)\mathbf{c}'(t) + A\Phi(t)\mathbf{c}(t) \\ &= A\mathbf{x} + \mathbf{f}(t) = A\Phi(t)\mathbf{c}(t) + \mathbf{f}(t). \end{aligned}$$

Comparison gives

$$\Phi(t)\mathbf{c}'(t) = \mathbf{f}(t) \quad \text{or} \quad \mathbf{c}'(t) = \Phi(t)^{-1}\mathbf{f}(t).$$

We can invert the fundamental matrix because its determinant is nonzero, a fact that follows from the independence of its columns. Integrating the last equation from 0 to t then gives

$$\mathbf{c}(t) = \int_0^t \Phi(s)^{-1}\mathbf{f}(s)ds + \mathbf{k},$$

where $\mathbf{k}$ is a arbitrary constant vector. Note that the integral of a vector functions is defined to be the vector consisting of the integrals of the components. Substituting into (5.19) shows that the general solution to the nonhomogeneous equation (5.18) is

$$\mathbf{x}(t) = \Phi(t)\mathbf{k} + \Phi(t)\int_0^t \Phi(s)^{-1}\mathbf{f}(s)ds. \tag{5.20}$$

As for a single first-order linear DE, this formula gives the general solution of (5.18) as a sum of the general solution to the homogeneous equation (first term) and a particular solution to the nonhomogeneous equation (second term). Equation (5.20) is called the **variation of parameters formula** for systems. It is equally valid for systems of any dimension, with appropriate size increase in the vectors and matrices.

It is sometimes a formidable task to calculate the solution (5.20), even in the two-dimensional case. It involves finding the two independent solutions to the homogeneous equation, forming the fundamental matrix, inverting the fundamental matrix, and then integrating.

Example 5.20

Consider the nonhomogeneous system

$$\mathbf{x}' = \begin{pmatrix} 4 & 3 \\ -1 & 0 \end{pmatrix} \mathbf{x} + \begin{pmatrix} 0 \\ t \end{pmatrix}.$$

It is a straightforward exercise to find the solution to the homogeneous system

$$\mathbf{x}' = \begin{pmatrix} 4 & 3 \\ -1 & 0 \end{pmatrix} \mathbf{x}.$$

The eigenpairs are 1, $(1,-1)^{\mathrm{T}}$ and 3, $(-3,1)^{\mathrm{T}}$. Therefore two independent solutions are

$$\begin{pmatrix} e^t \\ -e^t \end{pmatrix}, \begin{pmatrix} -3e^{3t} \\ e^{3t} \end{pmatrix}.$$

A fundamental matrix is

$$\Phi(t) = \begin{pmatrix} e^t & -3e^{3t} \\ -e^t & e^{3t} \end{pmatrix},$$

and its inverse is

$$\Phi^{-1}(t) = \frac{1}{\det \Phi} \begin{pmatrix} e^{3t} & 3e^{3t} \\ e^t & e^t \end{pmatrix} = \frac{1}{-2e^{4t}} \begin{pmatrix} e^{3t} & 3e^{3t} \\ e^t & e^t \end{pmatrix} = -\frac{1}{2} \begin{pmatrix} e^{-t} & 3e^{-t} \\ e^{-3t} & e^{-3t} \end{pmatrix}.$$

By the variation of parameters formula (5.20), the general solution is

$$\begin{aligned}
\mathbf{x}(t) &= \Phi(t)\mathbf{k} + \begin{pmatrix} e^t & -3e^{3t} \\ -e^t & e^{3t} \end{pmatrix} \int_0^t -\frac{1}{2} \begin{pmatrix} e^{-s} & 3e^{-s} \\ e^{-3s} & e^{-3s} \end{pmatrix} \begin{pmatrix} 0 \\ s \end{pmatrix} ds \\
&= \Phi(t)\mathbf{k} - \frac{1}{2} \begin{pmatrix} e^t & -3e^{3t} \\ -e^t & e^{3t} \end{pmatrix} \int_0^t \begin{pmatrix} 3se^{-s} \\ se^{-3s} \end{pmatrix} ds \\
&= \Phi(t)\mathbf{k} - \frac{1}{2} \begin{pmatrix} e^t & -3e^{3t} \\ -e^t & e^{3t} \end{pmatrix} \begin{pmatrix} 3\int_0^t se^{-s} ds \\ \int_0^t se^{-3s} ds \end{pmatrix} \\
&= \Phi(t)\mathbf{k} - \frac{1}{2} \begin{pmatrix} e^t & -3e^{3t} \\ -e^t & e^{3t} \end{pmatrix} \begin{pmatrix} 3 - 3(t+1)e^{-t} \\ \frac{1}{9} - (\frac{t}{3} + \frac{1}{9})e^{-3t} \end{pmatrix} \\
&= \begin{pmatrix} k_1 e^t - 3k_2 e^{3t} \\ -k_1 e^t + k_2 e^{3t} \end{pmatrix} + \begin{pmatrix} t + \frac{4}{3} \\ -\frac{4}{3}t - \frac{13}{9} \end{pmatrix}.
\end{aligned}$$

If the nonhomogeneous term $\mathbf{f}(t)$ is relatively simple, we can use the method of undetermined coefficients (judicious guessing) introduced for second-order equations in Chapter 3 to find the particular solution. In this case we guess a particular solution, depending upon the form of $\mathbf{f}(t)$. For example, if both components are polynomials, then we guess a particular solution with both components being polynomials that have the highest degree that appears. If

$$\mathbf{f}(t) = \begin{pmatrix} 1 \\ t^2 + 2 \end{pmatrix},$$

then a guess for the particular solution would be

$$\mathbf{x}_p(t) = \begin{pmatrix} a_1 t^2 + b_1 t + c_1 \\ a_2 t^2 + b_2 t + c_2 \end{pmatrix}.$$

Substitution into the nonhomogeneous system then determines the six constants. Generally, if a term appears in one component of $\mathbf{f}(t)$, then the guess must have that term appear in all its components. The method is successful on forcing terms with sines, cosines, polynomials, exponentials, and products and sums of those. The rules are the same as for single equations. But the calculations are tedious and a computer algebra system is often preferred.

Example 5.21

We use the method of undetermined coefficients to find a particular solution to the equation in Example 5.20. The forcing function is

$$\begin{pmatrix} 0 \\ t \end{pmatrix},$$

and therefore we guess a particular solution of the form

$$\mathbf{x}_p = \begin{pmatrix} at + b \\ ct + d \end{pmatrix}.$$

Substituting into the original system yields

$$\begin{pmatrix} a \\ c \end{pmatrix} = \begin{pmatrix} 4 & 3 \\ -1 & 0 \end{pmatrix} \begin{pmatrix} at + b \\ ct + d \end{pmatrix} + \begin{pmatrix} 0 \\ t \end{pmatrix}.$$

Simplifying leads to the two equations

$$\begin{aligned} a &= (4a + 3c)t + 4b + 3d, \\ c &= -b + (1 - a)t. \end{aligned}$$

Comparing coefficients gives

$$a = 1, \quad b = -c = \frac{4}{3}, \quad d = -\frac{13}{9}.$$

Therefore a particular solution is

$$\mathbf{x}_p = \begin{pmatrix} t + \frac{4}{3} \\ -\frac{4}{3}t - \frac{13}{9} \end{pmatrix}.$$

EXERCISES

1. Let
$$\mathbf{x}_1 = \begin{pmatrix} \phi_1(t) \\ \phi_2(t) \end{pmatrix}, \quad \mathbf{x}_2 = \begin{pmatrix} \psi_1(t) \\ \psi_2(t) \end{pmatrix}$$
be independent solutions to the homogeneous equation $\mathbf{x}' = A\mathbf{x}$, and let
$$\Phi(t) = \begin{pmatrix} \phi_1(t) & \psi_1(t) \\ \phi_2(t) & \psi_2(t) \end{pmatrix}$$
be a fundamental matrix. Show, by direct calculation and comparison of entries, that $\Phi'(t) = A\Phi(t)$. Show that the general solution of the homogeneous system can be written equivalently as
$$c_1\mathbf{x}_1 + c_2\mathbf{x}_2 = \Phi(t)\mathbf{c},$$
where $\mathbf{c} = (c_1, c_2)^{\mathrm{T}}$ is an arbitrary constant vector.

2. Two lakes of volume V_1 and V_2 initially have no contamination. A toxic chemical flows into lake 1 at $q + r$ gallons per minute with a concentration c grams per gallon. From lake 1 the mixed solution flows into lake 2 at q gallons per minute, while it simultaneously flows out into a drainage ditch at r gallons per minute. In lake 2 the the chemical mixture flows out at q gallons per minute. If x and y denote the concentrations of the chemical in lake 1 and lake 2, respectively, set up an initial value problem whose solution would give these two concentrations (draw a compartmental diagram). What are the equilibrium concentrations in the lakes, if any? Find $x(t)$ and $y(t)$. Now change the problem by assuming the initial concentration in lake 1 is x_0 and fresh water flows in. Write down the initial value problem and qualitatively, without solving, describe the dynamics of this problem using eigenvalues.

3. Solve the initial value problem
$$\begin{pmatrix} x' \\ y' \end{pmatrix} = \begin{pmatrix} 3 & -1 \\ 1 & 1 \end{pmatrix}\begin{pmatrix} x \\ y \end{pmatrix} + \begin{pmatrix} 1 \\ 2 \end{pmatrix}, \quad \begin{pmatrix} x(0) \\ y(0) \end{pmatrix} = \begin{pmatrix} 1 \\ 2 \end{pmatrix}$$
if
$$\Phi = \begin{pmatrix} 1+t & -t \\ t & 1-t \end{pmatrix} e^{2t}$$
is a fundamental matrix.

4. Solve the problem in Exercise 3 using undetermined coefficients to find a particular solution.

5. Consider the nonhomogeneous equation

$$\mathbf{x}' = \begin{pmatrix} -5 & 3 \\ 2 & -10 \end{pmatrix} \mathbf{x} + \begin{pmatrix} e^{-t} \\ 0 \end{pmatrix}.$$

Find the fundamental matrix and its inverse. Find a particular solution to the system and the general solution.

6. In pharmaceutical studies it is important to model and track concentrations of chemicals and drugs in the blood and in the body tissues. Let x and y denote the amounts (in milligrams) of a certain drug in the blood and in the tissues, respectively. Assume that the drug in the blood is taken up by the tissues at rate $r_1 x$ and is returned to the blood from the tissues at rate $r_2 y$. At the same time the drug amount in the blood is continuously degraded by the liver at rate $r_3 x$. Argue that the model equations which govern the drug amounts in the blood and tissues are

$$\begin{aligned} x' &= -r_1 x - r_3 x + r_2 y, \\ y' &= r_1 x - r_2 y. \end{aligned}$$

Find the eigenvalues of the matrix and determine the response of the system to an initial dosage of $x(0) = x_0$, given intravenously, with $y(0) = 0$. (Hint: show both eigenvalues are negative.)

7. In the preceding problem assume that the drug is administered intravenously and continuously at a constant rate D. What are the governing equations in this case? What is the amount of the drug in the tissues after a long time?

8. An animal species of population $P = P(t)$ has a *per capita* mortality rate m. The animals lay eggs at a rate of b eggs per day, per animal. The eggs hatch at a rate proportional to the number of eggs $E = E(t)$; each hatched egg gives rise to a single new animal.

   a) Write down model equations that govern P and E, and carefully describe the dynamics of the system in the two cases $b > m$ and $b < m$.

   b) Modify the model equations if, at the same time, an egg–eating predator consumes the eggs at a constant rate of r eggs per day.

   c) Solve the model equations in part (b) when $b > m$, and discuss the dynamics.

   d) How would the model change if each hatched egg were multi-yolked and gave rise to y animals?

5.5 Three-Dimensional Systems

In this section we give some examples of solving three linear differential equations in three unknowns. The method is the same as for two-dimensional systems, but now the matrix A for the system is 3×3, and there are three eigenvalues, and so on. We assume $\det A \neq 0$. Eigenvalues λ are found from the characteristic equation $\det(A-\lambda I)=0$, which, when written out, is a cubic equation in λ. For each eigenvalue λ we solve the homogeneous system $(A-\lambda I)\mathbf{v}=0$ to determine the associated eigenvector(s). We will have to worry about real, complex, and equal eigenvalues, as in the two-dimensional case. Each eigenpair $\lambda, \mathbf{v}$ gives a solution $\mathbf{v}e^{\lambda t}$, which, if λ is real, is a linear orbit lying on a ray in $\mathbf{R}^3$ in the direction defined by the eigenvector $\mathbf{v}$. We need three independent solutions $\mathbf{x}_1(t), \mathbf{x}_2(t), \mathbf{x}_3(t)$ to form the general solution, which is the linear combination $\mathbf{x}(t)=c_1\mathbf{x}_1(t)+c_2\mathbf{x}_2(t)+c_3\mathbf{x}_3(t)$ of those. If all the eigenvalues are real and unequal, then the eigenvectors will be independent and we will have three independent solutions; this is the easy case. Other cases, such as repeated roots and complex roots, are discussed in the examples and in the exercises.

If all the eigenvalues are negative, or have negative real part, then all solution curves approach (0,0,0), and the origin is an asymptotically stable equilibrium. If there is a positive eigenvalue, or complex eigenvalues with positive real part, then the origin is unstable because there is at least one orbit receding from the origin. Three-dimensional orbits can be drawn using computer software, but the plots are often difficult to visualize.

Examples illustrate the key ideas, and we suggest the reader work through the missing details.

Example 5.22

Consider the system

$$\begin{aligned} x_1' &= x_1 + x_2 + x_3 \\ x_2' &= 2x_1 + x_2 - x_3 \\ x_3' &= -8x_1 - 5x_2 - 3x_3 \end{aligned}$$

with matrix

$$A = \begin{pmatrix} 1 & 1 & 1 \\ 2 & 1 & -1 \\ -8 & -5 & -3 \end{pmatrix}.$$

Eigenpairs of A are given by

$$-1, \begin{pmatrix} -3 \\ 4 \\ 2 \end{pmatrix}, \quad -2, \begin{pmatrix} -4 \\ 5 \\ 7 \end{pmatrix}, \quad 2, \begin{pmatrix} 0 \\ 1 \\ -1 \end{pmatrix}.$$

These lead to three independent solutions

$$\mathbf{x}_1 = \begin{pmatrix} -3 \\ 4 \\ 2 \end{pmatrix} e^{-t}, \quad \mathbf{x}_2 = \begin{pmatrix} -4 \\ 5 \\ 7 \end{pmatrix} e^{-2t}, \quad \mathbf{x}_3 = \begin{pmatrix} 0 \\ 1 \\ -1 \end{pmatrix} e^{2t}.$$

Each represents a linear orbit. The general solution is a linear combination of these three; that is, $\mathbf{x}(t) = c_1\mathbf{x}_1(t) + c_2\mathbf{x}_2(t) + c_3\mathbf{x}_3(t)$. The origin is unstable because of the positive eigenvalue.

Example 5.23

Consider

$$\mathbf{x}' = \begin{pmatrix} 1 & 0 & 2 \\ 0 & 3 & 0 \\ 2 & 0 & 1 \end{pmatrix} \mathbf{x}.$$

The eigenvalues, found from $\det(A - \lambda I) = 0$, are $\lambda = -1, 3, 3$. An eigenvector corresponding to $\lambda = -1$ is $(1, 0, -1)^{\mathrm{T}}$, and so

$$\mathbf{x}_1 = \begin{pmatrix} 1 \\ 0 \\ -1 \end{pmatrix} e^{-t}$$

is one solution. To find eigenvector(s) corresponding to the other eigenvalue, a double root, we form $(A - 3I)\mathbf{v} = 0$, or

$$\begin{pmatrix} -2 & 0 & 2 \\ 0 & 0 & 0 \\ 2 & 0 & -2 \end{pmatrix} \begin{pmatrix} v_1 \\ v_2 \\ v_3 \end{pmatrix} = \begin{pmatrix} 0 \\ 0 \\ 0 \end{pmatrix}.$$

This system leads to the single equation

$$v_1 - v_3 = 0,$$

with v_2 arbitrary Letting $v_2 = \beta$ and $v_1 = \alpha$, we can write the solution as as

$$\begin{pmatrix} v_1 \\ v_2 \\ v_3 \end{pmatrix} = \alpha \begin{pmatrix} 1 \\ 0 \\ 1 \end{pmatrix} + \beta \begin{pmatrix} 0 \\ 1 \\ 0 \end{pmatrix},$$

where α and β are arbitrary. Therefore there are two, independent eigenvectors associated with $\lambda = 3$. This gives two independent solutions

$$\mathbf{x}_2 = \begin{pmatrix} 1 \\ 0 \\ 1 \end{pmatrix} e^{3t}, \quad \mathbf{x}_3 = \begin{pmatrix} 0 \\ 1 \\ 0 \end{pmatrix} e^{3t}.$$

Therefore the general solution is a linear combination of the three independent solutions we determined:

$$\mathbf{x}(t) = c_1\mathbf{x}_1 + c_2\mathbf{x}_2 + c_3\mathbf{x}_3.$$

We remark that a given eigenvalue with multiplicity two may not yield two independent eigenvectors, as was the case in the last example. Then we must proceed differently to find another independent solution, such as the method given in Section 5.3.3 (see Exercise 2(c) below).

Example 5.24

If the matrix for a three-dimensional system $\mathbf{x}' = A\mathbf{x}$ has one real eigenvalue λ and two complex conjugate eigenvalues $a \pm ib$, with associated eigenvectors $\mathbf{v}$ and $\mathbf{w} \pm i\mathbf{z}$, respectively, then the general solution is, as is expected from Section 5.3.2,

$$\mathbf{x}(t) = c_1\mathbf{v}e^{\lambda t} + c_2e^{at}(\mathbf{w}\cos bt - \mathbf{z}\sin bt) + c_3e^{at}(\mathbf{w}\sin bt + \mathbf{z}\cos bt).$$

EXERCISES

1. Find the eigenvalues and eigenvectors of the following matrices:

$$A = \begin{pmatrix} 2 & 3 & 0 \\ 0 & 6 & 2 \\ 0 & 0 & -1 \end{pmatrix}; \quad B = \begin{pmatrix} 2 & 3 & 4 \\ 2 & 0 & 2 \\ 4 & 2 & 3 \end{pmatrix}; \quad C = \begin{pmatrix} 1 & 0 & 1 \\ 0 & 1 & 0 \\ 1 & -1 & 1 \end{pmatrix}.$$

2. Find the general solution of the following three-dimensional systems:

a) $\mathbf{x}' = \begin{pmatrix} 3 & 1 & 3 \\ -5 & -3 & -3 \\ 6 & 6 & 4 \end{pmatrix}\mathbf{x}$. (Hint: $\lambda = 4$ is one eigenvalue.)

b) $\mathbf{x}' = \begin{pmatrix} -0.2 & 0 & 0.2 \\ 0.2 & -0.4 & 0 \\ 0 & 0.4 & -0.2 \end{pmatrix}\mathbf{x}$. (Hint: $\lambda = -1$ is one eigenvalue.)

c) $\mathbf{x}' = \begin{pmatrix} 2 & 1 & -2 \\ -1 & 0 & 0 \\ 0 & 2 & -2 \end{pmatrix} \mathbf{x}$. (Hint: see Section 5.3.3.)

d) $\mathbf{x}' = \begin{pmatrix} 1 & 0 & 1 \\ 0 & 1 & 0 \\ 1 & -1 & 1 \end{pmatrix} \mathbf{x}$

3. Find the general solution of the system

$$\begin{aligned} x' &= \rho x - y, \\ y' &= x + \rho y, \\ z' &= -2z, \end{aligned}$$

where ρ is a constant.

4. Consider the system

$$\mathbf{x}' = \begin{pmatrix} 0 & 1 & 2 \\ 1 & 0 & 2 \\ -1 & -2 & -3 \end{pmatrix} \mathbf{x}.$$

a) Show that the eigenvalues are $\lambda = -1, -1, -1$.

b) Find an eigenvector $\mathbf{v}_1$ associated with $\lambda = -1$ and obtain a solution to the system.

c) Show that a second independent solution has the form $(\mathbf{v}_2 + t\mathbf{v}_1)e^{-t}$ and find $\mathbf{v}_2$.

d) Show that a third independent solution has the form $(\mathbf{v}_3 + t\mathbf{v}_2 + \frac{1}{2}t^2\mathbf{v}_1)e^{-t}$ and find $\mathbf{v}_3$.

e) Find the general solution and then solve the initial value problem $\mathbf{x}' = A\mathbf{x}$, $\mathbf{x}(0) = (0, 1, 0)^{\mathrm{T}}$.

6 Nonlinear Systems

Nonlinear dynamics is common in nature. Unlike linear systems, where we can always find explicit formulas for the solution, nonlinear systems can seldom be solved. For some nonlinear systems we even have to give up on existence and uniqueness. So nonlinear dynamics is much more complicated than linear dynamics, and therefore we rely heavily on qualitative methods to determine their dynamical behavior. As for linear systems, equilibrium solutions and their stability play a fundamental role in the analysis.

6.1 Nonlinear Models

6.1.1 Phase Plane Phenomena

A two-dimensional nonlinear autonomous system has the general form

$$x' = f(x, y) \quad (6.1)$$

$$y' = g(x, y), \quad (6.2)$$

where f and g are given functions of x and y that are assumed to have continuous first partial derivatives in some open region in the plane. This regularity assumption on the first partial derivatives guarantees that the initial value problem associated with (6.1)–(6.2) will have a unique solution through any point in the region. Nonlinear systems arise naturally in mechanics, circuit theory, compartmental analysis, reaction kinetics, mathematical biology, economics, and other areas. In fact, in applications, most systems are nonlinear.

Example 6.1

We have repeatedly noted that a second-order equation can be reformulated as a first-order system. As a reminder, consider Newton's second law of motion for a particle of mass m moving in one dimension,

$$mx'' = F(x, x'),$$

where F is a force depending upon the position and the velocity. Introducing the velocity $y = x'$ as another state variable, we obtain the equivalent first-order system

$$\begin{aligned} x' &= y \\ y' &= \frac{1}{m}F(x, y). \end{aligned}$$

Consequently, we can study mechanical systems in an xy-phase space rather than the traditional position–time space.

In this chapter we are less reliant on vector notation than for linear systems, where vectors and matrices provide a natural language. We review some general terminology of Chapter 5. A **solution** $x = x(t)$, $y = y(t)$ to (6.1)–(6.2) can be represented graphically in two different ways (see figure 5.1 in Chapter 5). We can plot x vs t and y vs t to obtain the time series plots showing how the states x and y vary with time t. Or, we can plot the parametric equations $x = x(t)$, $y = y(t)$ in the xy phase plane. A solution curve in the xy plane is called an **orbit**. On a solution curve in the phase plane, time is a parameter and it may be shifted at will; that is, if $x = x(t)$, $y = y(t)$ is a solution, then $x = x(t - c)$, $y = y(t - c)$ represents the same solution and same orbit for any constant c. This follows because the system is autonomous. The **initial value problem** (IVP) consists of the solving the system (6.1)–(6.2) subject to the **initial conditions**

$$x(t_0) = x_0, \quad y(t_0) = y_0.$$

Geometrically, this means finding the orbit that goes through the point (x_0, y_0) at time t_0. If the functions f and g are continuous and have continuous first partial derivatives on $\mathbf{R}^2$, then the IVP has a unique solution. Therefore, two different solution curves cannot cross in the phase plane. We always assume conditions that guarantee existence and uniqueness.

As is true for their linear counterparts, there is an important geometric interpretation for nonlinear systems in terms of vector fields. For a solution curve $x = x(t)$, $y = y(t)$ we have $(x'(t), y'(t)) = (f(x(t), y(t)), g(x(t), y(t)))$. Therefore, at each point (x, y) in the plane the functions f and g define a vector $(f(x, y), g(x, y))$ that is the tangent vector to the orbit which passes through that point. Thus, the system (6.1)–(6.2) generates a **vector field.** A different way to think about it is this. The totality of all orbits form the *flow* of the vector field. Intuitively, we think of the flow as fluid particle paths with the vector field representing the velocity of the particles at the various points. A plot of several representative or key orbits in the xy-plane is called a **phase diagram** of the system. It is important that f and g do not depend explicitly upon time. Otherwise the vector field would not be stationary and would change, giving a different vector field at each instant of time. This would spoil a simple geometric approach to nonlinear systems. Nonautonomous systems are much harder to deal with than autonomous ones.

Among the most important solutions to (6.1)–(6.2) are the constant solutions, or **equilibrium solutions**. These are solutions $x(t) = x_e$, $y(t) = y_e$, where x_e and y_e are constant. Thus, equilibrium solutions are found as solutions of the algebraic, simultaneous system of equations

$$f(x, y) = 0, \quad g(x, y) = 0.$$

The time series plots of an equilibrium solution are just constant solutions (horizontal lines) in time. In the phase plane an equilibrium solution is represented by a single point (x_e, y_e). We often refer to these as equilibrium points. Nonlinear systems may have several equilibrium points. If an equilibrium point in the phase plane has the property that there is a small neighborhood about the point where there are no other equilibria, then we say the equilibrium point is **isolated**.

Example 6.2

If a particle of mass $m = 1$ moves on an x-axis under the influence of a force $F(x) = 3x^2 - 1$, then the equations of motion in the phase plane take the form

$$\begin{aligned} x' &= y, \\ y' &= 3x^2 - 1, \end{aligned}$$

where the position x and the velocity y are functions of time t. Here we can obtain the actual equation for the orbits in the xy-phase plane, in terms of x and y. Dividing the two equations[1] gives

$$\frac{dy/dt}{dx/dt} = \frac{dy}{dx} = \frac{3x^2 - 1}{y}.$$

[1] Along an orbit $x = x(t)$, $y = y(t)$ we also have y as a function of x, or $y = y(x)$. Then the chain rule dictates $\frac{dy}{dt} = \frac{dy}{dx}\frac{dx}{dt}$.

Separating variables and integrating yields

$$\int y dy = \int (3x^2 - 1) dx,$$

or

$$\frac{1}{2}y^2 = x^3 - x + E, \tag{6.3}$$

where we have chosen the letter E to denote the arbitrary constant of integration (as we soon observe, E stands for total energy). This equation represents a family of orbits in the phase plane giving a relationship between position and velocity. By dividing the equations as we did, time dependence is lost on these orbits. Equation (6.3) has an important physical meaning that is worth reviewing. The term $\frac{1}{2}y^2$ represents the kinetic energy (one-half the mass times the velocity-squared). Secondly, we recall that the potential energy $V(x)$ associated with a conservative force $F(x)$ is $V(x) = -\int F(x)dx$, or $F(x) = -dV/dx$. In the present case $V(x) = -x^3 + x$, where we have taken $V = 0$ at $x = 0$. The orbits (6.3) can be written

$$\frac{1}{2}y^2 + (-x^3 + x) = E,$$

which states that the kinetic energy plus the potential energy is constant. Therefore, the orbits (6.3) represent constant energy curves. The total energy E can be written in terms of the initial position and velocity as $E = \frac{1}{2}y^2(0) + (-x(0)^3 + x(0))$. For each value of E we can plot the locus of points defined by equation (6.3). To carry this out practically, we may solve for y and write

$$y = \sqrt{2}\sqrt{x^3 - x + E}, \quad y = -\sqrt{2}\sqrt{x^3 - x + E}.$$

Then we can plot the curves using a calculator or computer algebra system. (For values of x that make the expression under the radical negative, the curve will not be defined.) Figure 6.1 shows several orbits. Let us discuss their features. There are two points, $x = \sqrt{\frac{1}{3}}$, $y = 0$ and $x = -\sqrt{\frac{1}{3}}$, $y = 0$, where $x' = y' = 0$. These are two equilibrium solutions where the velocity is zero and the force is zero (so the particle cannot be in motion). The equilibrium solution $x = -\sqrt{\frac{1}{3}}$, $y = 0$ has the structure of a center, and for initial values close to this equilibrium the system will oscillate. The other equilibrium $x = \sqrt{\frac{1}{3}}$, $y = 0$ has the structure of an unstable saddle point. Because $x' = y$, for $y > 0$ we have $x' > 0$, and the orbits are directed to the right in the upper half-plane. For $y < 0$ we have $x' < 0$, and the orbits are directed to the left in the lower half-plane. For large initial energies the system does not oscillate but rather goes to $x = +\infty$, $y = +\infty$; that is, the mass moves farther and farther to the right with faster speed.

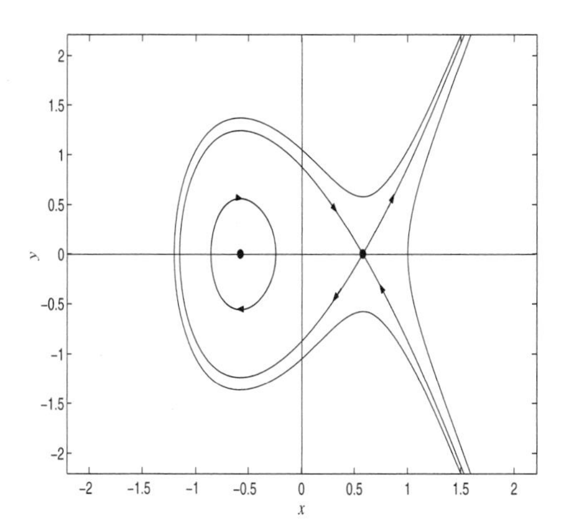

Figure 6.1 Plots of the constant energy curves $\frac{1}{2}y^2 - x^3 + x = E$ in the xy-phase plane. These curves represent the orbits of the system and show how position and velocity relate. Time dependence is lost in this representation of the orbits. Because $x' = y$, the orbits are moving to the right (x is increasing) in the upper half-plane $y > 0$, and to the left (x is decreasing) in the lower half-plane $y < 0$.

Example 6.3

Consider the simple nonlinear system

$$x' = y^2, \tag{6.4}$$

$$y' = -\frac{2}{3}x. \tag{6.5}$$

Clearly, the origin $x = 0$, $y = 0$, is the only equilibrium solution. In this case we can divide the two equations and separate variables to get

$$y^2y' = -\frac{2}{3}xx'.$$

Integrating with respect to t gives

$$\int y^2y'dt = -\int \frac{2}{3}xx'dt + C,$$

where C is an arbitrary constant. Changing variables in each integral, $y = y(t)$ in the left integral and $x = x(t)$ in the right, we obtain

$$\int y^2dy = -\int \frac{2}{3}xdx + C,$$

or

$$y^3 = -x^2 + C.$$

Rearranging,

$$y = (C - x^2)^{1/3}.$$

Consequently, we have obtained the orbits for system (6.4)–(6.5) in terms of x and y. These are easily plotted (e.g., on a calculator, for different values of C), and they are shown in Figure 6.2. This technique illustrates a general

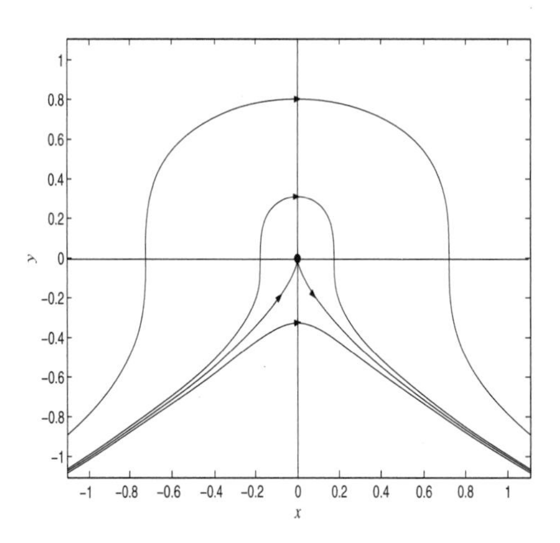

Figure 6.2 Phase diagram for $x' = y^2$, $y' = -\frac{2}{3}x$. Because $x' > 0$, all the orbits are moving to the right as time increases.

method for finding the equation of the orbits for simple equations in terms of the state variables alone: divide the differential equations and integrate, as far as possible. With this technique, however, we lose information about how the states depend on time, or how time varies along the orbits. To find solution curves in terms of time t, we can write (6.4) as

$$x' = y^2 = (C - x^2)^{2/3},$$

which is a single differential equation for $x = x(t)$. We can separate variables, but the result is not very satisfying because we get a complicated integral. This shows that time series solutions are not easily obtained for nonlinear problems. Usually, the qualitative behavior shown in the phase diagram is all we want. If we do need time series plots, we can obtain them using a numerical method, which we discuss later.

We point out an important feature of the phase diagram shown in figure 6.2. The origin does not have the typical type of structure encountered in Chapter 5 for linear systems. There we were able to completely characterize all equilibrium solutions as saddles, spirals, centers, or nodes. The origin for the nonlinear system (6.4)–(6.5) is not one of those. Therefore, nonlinear systems can have an unusual orbital structure near equilibria.

Why are the equilibrium solutions so important? First, much of the "action" in the phase plane takes place near the equilibrium points, so analysis of the flow near those points is insightful. Second, physical systems often seek out and migrate toward equilibria; so equilibrium states can represent persistent states. We think of x and y as representing two competing animal populations. If a system is in an equilibrium state, the two populations coexist. Those populations will remain in the equilibrium states unless the system is perturbed. This means that some event (e.g., a bonanza or catastrophe), would either add or subtract individuals from the populations without changing the underlying processes that govern the population dynamics. If the inflicted population changes are *small*, the populations would be bumped to new values *near* the equilibrium. This brings up the stability issue. Do the populations return to the coexistent state, or do they change to another state? If the populations return to the equilibrium, then it is a persistent state and **asymptotically stable**. If the populations move further away from the equilibrium, then it is not persistent and **unstable**. If the populations remain close to the equilibrium, but do not actually approach it, then the equilibrium is **neutrally stable**. For each model it is important to discover the *locally* stable equilibrium states, or persistent states, in order to understand the dynamics of the model. In Example 6.2 the saddle point is unstable and the center is neutrally stable (figure 6.1), and in Example 6.3 the equilibrium is unstable (figure 6.2). For an unstable equilibrium, orbits that begin near the equilibrium do not remain near. Examples of different types of stability are discussed in the sequel.

The emphasis in the preceding paragraph is on the word *local*. That is, what happens if *small* changes occur near an equilibrium, not large changes. Of course, we really want to know what happens if an equilibrium is disturbed by all possible changes, including an arbitrarily large change. Often the adjectives **local** and **global** are appended to stability statements to indicate what types of perturbations (small or arbitrary) are under investigation. However, we cannot usually solve a nonlinear system, and so we cannot get an explicit resolution of global behavior. Therefore we are content with analyzing local stability properties, and not global stability properties. As it turns out, local stability can be determined because we can approximate the nonlinear system by a tractable linear system near equilibria (Section 6.3).

EXERCISES

1. Consider the uncoupled nonlinear system $x' = x^2$, $y' = -y$.

   a) Find a relation between x and y that describes the orbits. Are all the orbits contained in this relation for different values of the arbitrary constant?

   b) Sketch the vector field at several points near the origin.

   c) Draw the phase diagram. Is the equilibrium stable or unstable?

   d) Find the solutions $x = x(t)$, $y = y(t)$, and plot typical time series. Pick a single time series plot and draw the corresponding orbit in the phase plane.

2. Consider the system $x' = -\frac{1}{y}$, $y' = 2x$.

   a) Are there any equilibrium solutions?

   b) Find a relationship between x and y that must hold on any orbit, and plot several orbits in the phase plane.

   c) From the orbits, sketch the vector field.

   d) Do any orbits touch the x-axis?

3. Consider the nonlinear system $x' = x^2 + y^2 - 4$, $y' = y - 2x$.

   a) Find the two equilibria and plot them in the phase plane.

   b) On the plot in part (a), sketch the set of points where the vector field is vertical (up or down) and the set of points where the vector field is horizontal (left or right).

4. Do parts (a) and (b) of the previous problems for the nonlinear system $x' = y + 1$, $y' = y + x^2$.

5. A nonlinear model of the form

$$\begin{aligned} x' &= y - x \\ y' &= -y + \frac{5x^2}{4 + x^2}, \end{aligned}$$

   has been proposed to describe cell differentiation. Find all equilibrium solutions.

6. Find all equilibria for the system $x' = \sin y$, $y' = 2x$.

7. Consider the nonlinear system $x' = y$, $y' = -x - y^3$. Show that the function $V(x, y) = x^2 + y^2$ decreases along any orbit (i.e., $\frac{d}{dt}V(x(t), y(t)) < 0$), and state why this proves that every orbit approaches the origin as $t \to +\infty$.

8. Consider the nonlinear system $x' = x^2 - y^2$, $y' = x - y$.

   a) Find and plot the equilibrium points in the phase plane. Are they isolated?

   b) Show that, on an orbit, $x + y + 1 = Ce^y$, where C is some constant, and plot several of these curves.

   c) Sketch the vector field.

   d) Describe the fate of the orbit that begins at $(\frac{1}{4}, 0)$ at $t = 0$ as $t \to +\infty$ and as $t \to -\infty$.

   e) Draw a phase plane diagram, being sure to indicate the directions of the orbits.

6.1.2 The Lotka–Volterra Model

Nonlinear equations play an central role in modeling population dynamics in ecology. We formulate and study a model involving predator–prey dynamics. Let $x = x(t)$ be the prey population and $y = y(t)$ be the predator population. We can think of rabbits and foxes, food fish and sharks, or any consumer-resource interaction, including herbivores and plants. If there is no predator we assume the prey dynamics is $x' = rx$, or exponential growth, where r is the *per capita* growth rate. In the absence of prey, we assume that the predator dies via $y' = -my$, where m is the *per capita* mortality rate. When there are interactions, we must include terms that decrease the prey population and increase the predator population. To determine the form of the predation term, we assume that the rate of predation, or the the number of prey consumed per unit of time, per predator, is proportional to the number of prey. That is, the rate of predation is ax. Thus, if there are y predators then the rate that prey is decreased is axy. Note that the interaction term is proportional to xy, the product of the number of predators and the number of prey. For example, if there were 20 prey and 10 predators, there would be 200 possible interactions. Only a fraction of them, a, are assumed to result in a kill. The parameter a depends upon the fraction of encounters and the success of the encounters. The prey consumed cause a rate of increase in predators of εaxy, where ε is the conversion efficiency of the predator population (one prey consumed does not mean one predator born). Therefore, we obtain the simplest model of predator–prey interaction, called the **Lotka–Volterra model**:

$$\begin{aligned} x' &= rx - axy \\ y' &= -my + bxy, \end{aligned}$$

where $b = \varepsilon a$.

To analyze the Lotka–Volterra model we factor the right sides of the equations to obtain

$$x' = x(r - ay), \quad y' = y(-m + bx). \tag{6.6}$$

Now it is simple to locate the equilibria. Setting the right sides equal to zero gives two solutions, $x = 0$, $y = 0$ and $x = m/b$, $y = r/a$. Thus, in the phase plane, the points $(0, 0)$ and $(m/b, r/a)$ represent equilibria. The origin represents extinction of both species, and the nonzero equilibrium represents a coexistent state. To determine properties of the orbits we usually plot curves in the xy plane where the vector field is vertical (where $x' = 0$) and curves where the vector field is horizontal ($y' = 0$). These curves are called the **nullclines**. They are not (usually) orbits, but rather the curves where the orbits cross vertically or horizontally. The x-nullclines for (6.6) , where $x' = 0$, are $x = 0$ and $y = r/a$. Thus the orbits cross these two lines vertically. The y-nullclines, where $y' = 0$, are $y = 0$ and $x = m/b$. The orbits cross these lines horizontally. Notice that the equilibrium solutions are the intersections of the x- and y-nullclines. The nullclines partition the plane into regions where x' and y' have various signs, and therefore we get a picture of the direction of the flow pattern. See figure 6.3. Next, along each nullcline we can find the direction of

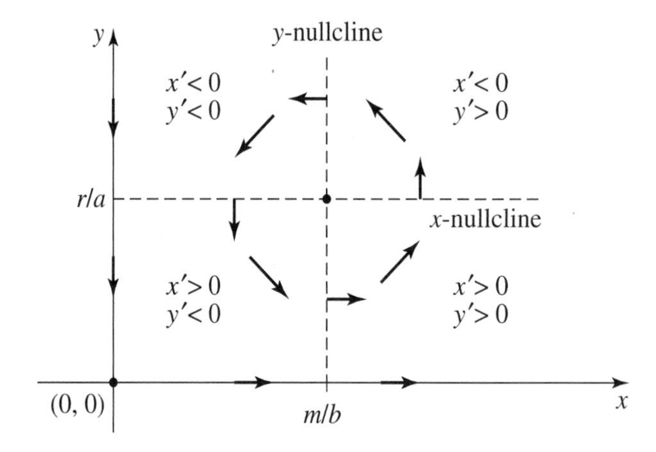

Figure 6.3 Nullclines (dashed) and vector field in regions between nullclines. The x and y axes are nullclines, as well as orbits.

the vector field. For example, on the ray to the right of the equilibrium we have $x > m/b$, $y = r/a$. We know the vector field is vertical so we need only check the sign of y'. We have $y' = y(-m + bx) = (r/a)(-m + bx) > 0$, so the vector field points upward. Similarly we can determine the directions along the other three rays. These are shown in the accompanying figure 6.3. Note that $y = 0$ and $x = 0$, both nullclines, are also orbits. For example, when $x = 0$ we have $y' = -my$, or $y(t) = Ce^{-mt}$; when there are no prey, the foxes die out. Similarly, when $y = 0$ we have $x(t) = Ce^{rt}$, so the rabbits increase in number.

Finally, we can determine the direction of the vector field in the regions between the nullclines either by selecting an arbitrary point in that region and calculating x' and y', or by just noting the sign of x' and y' in that region from information obtained from the system. For example, in the quadrant above and to the right of the nonzero equilibrium, it is easy to see that $x' < 0$ and $y' > 0$; so the vector field points upward and to the left. We can complete this task for each region and obtain the directions shown in figure 6.3. Having the direction of the vector field along the nullclines and in the regions bounded by the nullclines tells us the directions of the solution curves, or orbits. Near $(0, 0)$ the orbits appear to veer away and the equilibrium has a saddle point structure. The equilibrium $(0, 0)$ is unstable. It appears that orbits circle around the nonzero equilibrium in a counterclockwise fashion. But at this time it is not clear if they form closed paths or spirals, so more work is needed.

We attempt to obtain the equation of the orbits by dividing the two equations in (6.6). We get

$$\frac{y'}{x'} = \frac{dy}{dx} = \frac{y(-m + bx)}{x(r - ay)}.$$

Rearranging and integrating gives

$$\int \frac{r - ay}{y} dy = \int \frac{bx - m}{x} dx + C.$$

Carrying out the integration gives

$$r \ln y - ay = bx - m \ln x + C,$$

which is the algebraic equation for the orbits. It is obscure what these curves are because it is not possible to solve for either of the variables. So, cleverness is required. If we exponentiate we get

$$y^r e^{-ay} = e^C e^{bx} x^{-m}.$$

Now consider the y nullcline where x is fixed at a value m/b, and fix a positive C value (i.e., fix an orbit). The right side of the last equation is a positive number A, and so $y^r = Ae^{ay}$. If we plot both sides of this equation (do this!—plot a power function and a growing exponential) we observe that there can be at most two intersections; therefore, this equation can have at most two solutions for y. Hence, along the vertical line $x = m/b$, there can be at most two crossings; this means an orbit cannot spiral into or out from the equilibrium point, because that would mean many values of y would be possible. We conclude

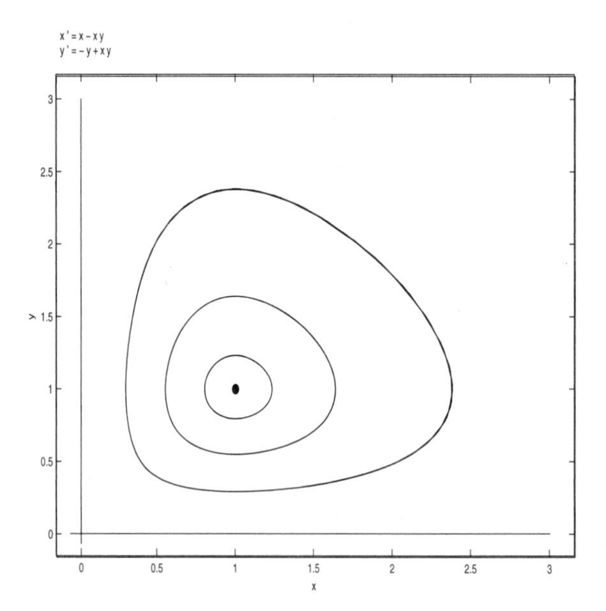

Figure 6.4 Closed, counterclockwise, periodic orbits of the Lotka–Volterra predator–prey model $x' = x - xy$, $y' = -y + xy$. The x-axis is an orbit leaving the origin and the y-axis is an orbit entering the origin.

that the equilibrium is a center with closed, periodic orbits encircling it. A phase diagram is shown in figure 6.4. Time series plots of the prey and predator populations are shown in figure 6.5. When the prey population is high the predators have a high food source and their numbers start to increase, thereby eventually driving down the prey population. Then the prey population gets low, ultimately reducing the number of predators because of lack of food. Then the process repeats, giving cycles.

The Lotka–Volterra model, developed by A. Lotka and V. Volterra in 1925, is the simplest model in ecology showing how populations can cycle, and it was one of the first strategic models to explain qualitative observations in natural systems. Note that the nonzero equilibrium is neutrally stable. A small perturbation from equilibrium puts the populations on a periodic orbit that stays near the equilibrium. But the system does not return to that equilibrium. So the nonzero equilibrium is stable, but not asymptotically stable. The other equilibrium, the origin, which corresponds to extinction of both species, is an unstable saddle point with the two coordinate axes as separatrices.

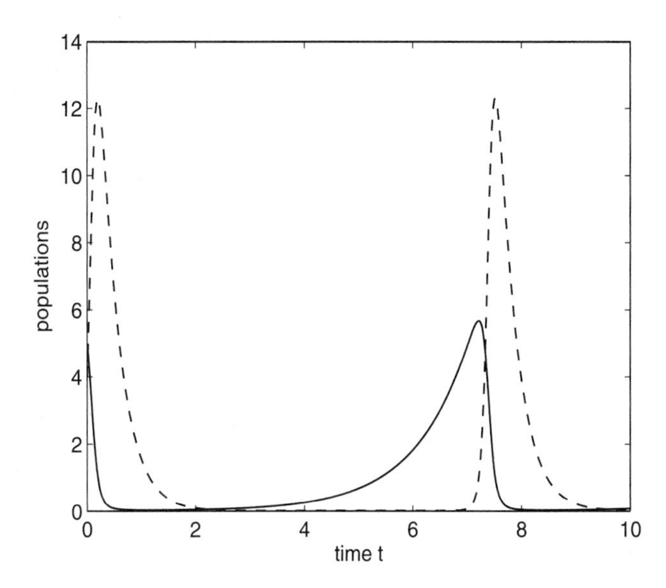

Figure 6.5 Time series solution to the Lotka–Volterra system $x' = x - xy$, $y' = -3y + 3xy$, showing the predator (dashed) and prey (solid) populations.

6.1.3 Holling Functional Responses

Ecology provides a rich source of problems in nonlinear dynamics, and now we take time to introduce another one. In the Lotka–Volterra model the rate of predation (prey per time, per predator) was assumed to be proportional to the number of prey (i.e., ax). Thinking carefully about this leads to concerns. Increasing the prey density indefinitely leads to an extremely high consumption rate, which is clearly impossible for any consumer. It seems more reasonable if the rate of predation would have a limiting value as prey density gets large. In the late 1950s, C. Holling developed a functional form that has this limiting property by partitioning the time budget of the predator. He reasoned that the number N of prey captured by a *single* predator is proportional to the number x of prey and the time T_s allotted for searching.[2] Thus $N = aT_sx$, where the proportionality constant a is the effective encounter rate. But the total time T available to the predator must be partitioned into search time and total handling time T_h, or $T = T_s + T_h$. The total handling time is proportional to the number captured, $T_h = hN$, where h is the time for a predator to handle a

[2] We are thinking of x and y as population numbers, but we can also regard them as *population densities*, or animals per area. There is always an underlying fixed area where the dynamics is occurring.

single prey. Hence $N = a(T - hN)x$. Solving for N/T, which is the predation rate, gives

$$\frac{N}{T} = \frac{ax}{1 + ahx}.$$

This function for the predation rate is called a Holling type II response, or the Holling disk equation. Note that $\lim_{x\to\infty} \frac{ax}{1+ahx} = 1/h$, so the rate of predation approaches a constant value. This quantity N/T is measured in prey per time, per predator, so multiplying by the number of predators y gives the predation rate for y predators.

If the encounter rate a is a function of the prey density (e.g., a linear function $a = bx$), the the predation rate is

$$\frac{N}{T} = \frac{bx^2}{1 + bhx^2},$$

which is called a Holling type III response. Figure 6.6 compares different types of predation rates used by ecologists. For a type III response the predation is turned on once the prey density is high enough; this models, for example, predators that must form a "prey image" before they become aware of the prey, or predators that eat different types of prey. At low densities prey go nearly unnoticed; but once the density reaches an upper threshold the predation rises quickly to its maximum rate.

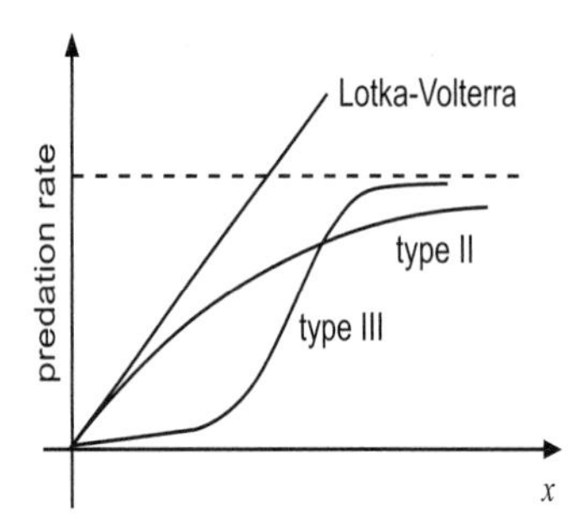

Figure 6.6 Three types of predation rates studied in ecology.

Replacing the linear predation rate ax in the Lotka–Volterra model by the **Holling type II response**, we obtain the model

$$\begin{aligned} x' &= rx - \frac{ax}{1 + ahx}y, \\ y' &= -my + \varepsilon\frac{ax}{1 + ahx}y. \end{aligned}$$

We can even go another step and replace the linear growth rate in the model by a more realistic logistics growth term. Then we obtain the **Rosenzweig–MacArthur** model

$$\begin{aligned} x' &= rx(1 - \frac{x}{K}) - \frac{ax}{1 + ahx}y, \\ y &= -my + \varepsilon\frac{ax}{1 + ahx}y. \end{aligned}$$

Else, a type III response could be used. All of these models have very interesting dynamics. Questions abound. Do they lead to cycles? Are there persistent states where the predator and prey coexist at constant densities? Does the predator or prey population go to extinction? What happens when a parameter, for example, the carrying capacity K, increases? Some aspects of these models are examined in the Exercises.

Other types of ecological models have been developed for interacting species. A model such as

$$\begin{aligned} x' &= f(x) - axy, \\ y' &= g(y) - bxy \end{aligned}$$

is interpreted as a **competition model** because the interaction terms $-axy$ and $-bxy$ are both negative and lead to a decrease in each population. When both interaction terms are positive, then the model is called a **cooperative model**.

6.1.4 An Epidemic Model

We consider a simple epidemic model where, in a fixed population of size N, the function $I = I(t)$ represents the number of individuals that are infected with a contagious illness and $S = S(t)$ represents the number of individuals that are susceptible to the illness, but not yet infected. We also introduce a removed class where $R = R(t)$ is the number who cannot get the illness because they have recovered permanently, are naturally immune, or have died. We assume $N = S(t) + I(t) + R(t)$, and each individual belongs to only one of the three classes. Observe that N includes the number who may have died. The evolution of the illness in the population can be described as follows. Infectives communicate the disease to susceptibles with a known infection rate; the susceptibles become infectives who have the disease a short time, recover (or die), and enter the removed class. Our goal is to set up a model that describes how the disease progresses with time. These models are called **SIR models**.

In this model we make several assumptions. First, we work in a time frame where we can ignore births and immigration. Next, we assume that the population mixes homogeneously, where all members of the population interact with one another to the same degree and each has the same risk of exposure to the disease. Think of measles, the flu, or chicken pox at an elementary school. We assume that individuals get over the disease reasonably fast. So, we are not modeling tuberculosis, AIDS, or other long-lasting or permanent diseases. Of course, more complicated models can be developed to account for all sorts of factors, such as vaccination, the possibility of reinfection, and so on.

The disease spreads when a susceptible comes in contact with an infective. A reasonable measure of the number of contacts between susceptibles and infectives is $S(t)I(t)$. For example, if there are five infectives and twenty susceptibles, then one hundred contacts are possible. However, not every contact results in an infection. We use the letter a to denote the **transmission coefficient**, or the fraction of those contacts that usually result in infection. For example, a could be 0.02, or 2 percent. The parameter a is the product of two effects, the fraction of the total possible number of encounters that occur, and the fraction of those that result in infection. The constant a has dimensions time^{-1} per individual, aN is a measure of the the average rate that a susceptible individual makes infectious contacts, and $1/(aN)$ is the average time one might expect to get the infection. The quantity $aS(t)I(t)$ is the infection rate, or the rate that members of the susceptible class become infected. Observe that this model is the same as the law of mass action in chemistry where the rate of chemical reaction between two reactants is proportional to the product of their concentrations. Therefore, if no other processes are included, we would have

$$S' = -aSI, \quad I' = aSI.$$

But, as individuals get over the disease, they become part of the removed class R. The **recovery rate** r is the fraction of the infected class that ceases to be infected; thus, the rate of removal is $rI(t)$. The parameter r is measured in time^{-1} and $1/r$ can be interpreted as the average time to recover. Therefore, we modify the last set of equations to get

$$S' = -aSI, \tag{6.7}$$

$$I' = aSI - rI. \tag{6.8}$$

These are our working equations. We do not need an equation for R' because R can be determined directly from $R = N - S - I$. At time $t = 0$ we assume there are I_0 infectives and S_0 susceptibles, but no one yet removed. Thus, initial conditions are given by

$$S(0) = S_0, \quad I(0) = I_0, \tag{6.9}$$

and $S_0 + I_0 = N$. SIR models are commonly diagrammed as in figure 6.7 with S, I, and R compartments and with arrows that indicate the rates that individuals progress from one compartment to the other. An arrow entering a compartment represents a positive rate and an arrow leaving a compartment represents a negative rate.

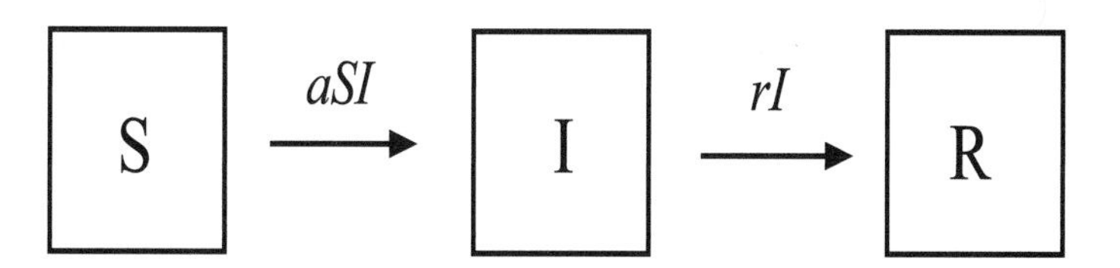

Figure 6.7 Compartments representing the number of susceptibles, the number of infectives, and the number removed, and the flow rates in and out of the compartments.

Qualitative analysis can help us understand how a parametric solution curve $S = S(t)$, $I = I(t)$, or orbit, behaves in the SI-phase plane. First, the initial value must lie on the straight line $I = -S + N$. Where then does the orbit go? Note that S' is always negative so the orbit must always move to the left, decreasing S. Also, because $I' = I(aS - r)$, we see that the number of infectives increases if $S > r/a$, and the number of infectives decreases if $S < r/a$. So, there are two cases to consider: $r/a > N$ and $r/a < N$. That is, it makes a difference if the ratio r/a is greater than the population, or less than the population. The vertical line $S = r/a$ is the I nullcline where the vector field is horizontal. Let us fix the idea and take $r/a < N$. (The other case is requested in the Exercises.) If the initial condition is at point P in figure 6.8, the orbit goes directly down and to the left until it hits $I = 0$ and the disease dies out. If the initial condition is at point Q, then the orbit increases to the left, reaching a maximum at $S = r/a$. Then it decreases to the left and ends on $I = 0$. There are two questions remaining, namely, how steep is the orbit at the initial point, and where on the S axis does the orbit terminate. Figure 6.8 anticipates the answer to the first question. The total number of infectives and susceptibles cannot go above the line $I + S = N$, and therefore the slope of the orbit at $t = 0$ is not as steep as -1, the slope of the line $I + S = N$. To analytically resolve the second issue we can obtain a relationship between S and I along a solution curve as we have done in previous examples. If we divide the equations (6.7)–(6.8) we obtain

$$\frac{I'}{S'} = \frac{dI/dt}{dS/dt} = \frac{dI}{dS} = \frac{aSI - rI}{-aSI} = -1 + \frac{r}{aS}.$$

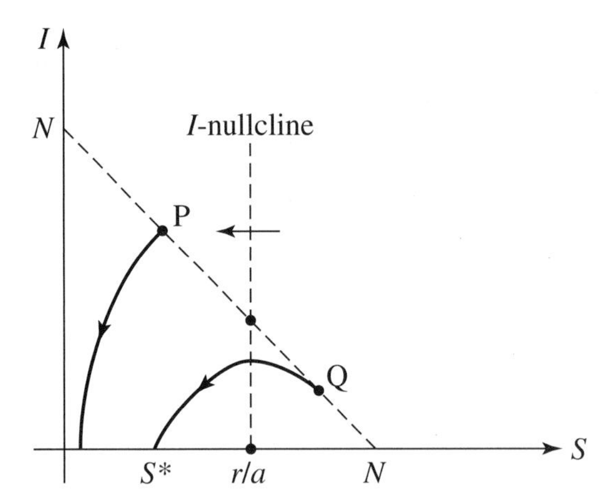

Figure 6.8 The SI phase plane showing two orbits in the case $r/a < N$. One starts at P and and one starts at Q, on the line $I + S = N$. The second shows an epidemic where the number of infectives increases to a maximum value and then decreases to zero; S^* represents the number that does not get the disease.

Thus

$$\frac{dI}{dS} = -1 + \frac{r}{aS}.$$

Integrating both sides with respect to S (or separating variables) yields

$$I = -S + \frac{r}{a}\ln S + C,$$

where C is an arbitrary constant. From the initial conditions, $C = N - (r/a)\ln S_0$. So the solution curve, or orbit, is

$$I = -S + \frac{r}{a}\ln S + N - \frac{r}{a}\ln S_0 = -S + N + \frac{r}{a}\ln\frac{S}{S_0}.$$

This curve can be graphed with a calculator or computer algebra system, once parameter values are specified. Making such plots shows what the general curve looks like, as plotted in figure 6.8. Notice that the solution curve cannot intersect the I axis where $S = 0$, so it must intersect the S axis at $I = 0$, or at the root S^* of the nonlinear equation

$$-S + N + \frac{r}{a}\ln\frac{S}{S_0} = 0.$$

See figure 6.8. This root represents the number of individuals who do not get the disease. Once parameter values are specified, a numerical approximation of S^* can be obtained. In all cases, the disease dies out because of lack of infectives. Observe, again, in this approach we lost time dependence on the orbits. But the qualitative features of the phase plane give good resolution of the disease dynamics. In the next section we show how to obtain accurate time series plots using numerical methods.

Generally, we are interested in the question of whether there will be an epidemic when there are initially a small number of infectives. The number $R_0 = \frac{aS(0)}{r}$ is a threshold quantity called the *reproductive number*, and it determines if there will be an epidemic. If $R_0 > 1$ there will be an epidemic (the number of infectives increase), and if $R_0 < 1$ then the infection dies out.

EXERCISES

1. In the SIR model analyze the case when $r/a > N$. Does an epidemic occur in this case?

2. Referring to figure 6.8, draw the shapes of the times series plots $S(t)$ and $I(t)$ on the same set of axes when the initial point is at point Q.

3. In a population of 200 individuals, 20 were initially infected with an influenza virus. After the flu ran its course, it was found that 100 individuals did not contract the flu. If it took about 3 days to recover, what was the transmission coefficient a? What was the average time that it might have taken for someone to get the flu?

4. In a population of 500 people, 25 have the contagious illness. On the average it takes about 2 days to contract the illness and 4 days to recover. How many in the population will not get the illness? What is the maximum number of infectives at a single time?

5. In a constant population, consider an SIS model (susceptibles become infectives who then become susceptible again after recovery) with infection rate aSI and recovery rate rI. Draw a compartmental diagram as in figure 6.7, and write down the model equations. Reformulate the model as a single DE for the infected class, and describe the dynamics of the disease.

6. If, in the Lotka–Volterra model, we include a constant harvesting rate h of the prey, the model equations become

$$\begin{aligned} x' &= rx - axy - h \\ y' &= -my + bxy. \end{aligned}$$

   Explain how the equilibrium is shifted from that in the Lotka–Volterra model. How does the equilibrium shift if both prey and predator are harvested at the same rate?

7. Modify the Lotka–Volterra model to include *refuge*. That is, assume that the environment always provides a constant number of the hiding places

where the prey can avoid predators. Argue that

$$\begin{aligned} x' &= rx - a(x-k)y \\ y' &= -my + b(x-k)y. \end{aligned}$$

How does refuge affect the equilibrium populations compared to no refuge?

8. Formulate a predator–prey model based on Lotka–Volterra, but where the predator migrates out of the region at a constant rate M. Discuss the dynamics of the system.

9. A simple cooperative model where two species depend upon mutual cooperation for their survival is

$$\begin{aligned} x' &= -kx + axy \\ y' &= -my + bxy. \end{aligned}$$

Find the equilibria and identify, insofar as possible, the region in the phase plane where, if the initial populations lie in that region, then both species become extinct. Can the populations ever coexist in a nonzero equilibrium?

10. Beginning with the SIR model, assume that susceptible individuals are vaccinated at a constant rate ν. Formulate the model equations and describe the progress of the disease if, initially, there are a small number of infectives in a large population.

11. Beginning with the SIR model, assume that recovered individuals can lose their immunity and become susceptible again, with rate μR, where r is the recovery rate. Draw a compartmental diagram and formulate a two-dimensional system of model equations. Find the equilibria. Is there a disease-free equilibrium with $I = 0$? Is there an endemic equilibrium with $I > 0$?

12. Two populations X and Y grow logistically and both compete for the same resource. A competition model is given by

$$\frac{dX}{d\tau} = r_1 X\left(1 - \frac{X}{K_1}\right) - b_1 XY, \quad \frac{dY}{d\tau} = r_2 Y\left(1 - \frac{Y}{K_2}\right) - b_2 XY.$$

The competition terms are b_1XY and b_2XY.

a) Scale time by r_1^{-1} and scale the populations by their respective carrying capacities to derive a dimensionless model

$$x' = x(1-x) - axy, \quad y' = cy(1-y) - bxy,$$

where a, b, and c are appropriately defined dimensionless constants. Give a biological interpretation of the constants.

b) In the case $a > 1$ and $c > b$ determine the equilibria, the nullclines, and the direction of the vector field on and in between the nullclines.

c) Determine the stability of the equilibria by sketching a generic phase diagram. How will an initial state evolve in time?

d) Analyze the population dynamics in the case $a > 1$ and $c < b$.

13. Consider the system

$$x' = \frac{axy}{1+y} - x, \quad y' = -\frac{axy}{1+y} - y + b,$$

where a and b are positive parameters with $a > 1$ and $b > \frac{1}{a-1}$.

a) Find the equilibrium solutions, plot the nullclines, and find the directions of the vector field along the nullclines.

b) Find the direction field in the first quadrant in the regions bounded by the nullclines. Can you determine from this information the stability of any equilibria?

6.2 Numerical Methods

We have formulated a few models that lead to two-dimensional nonlinear systems and have illustrated some elementary methods of analysis. In the next section we advance our technique and show how a more detailed analysis can lead to an overall qualitative picture of the nonlinear dynamics. But first we develop some numerical methods to solve such systems. Unlike two-dimensional linear systems with constant coefficients, nonlinear systems can rarely be resolved analytically by finding solution formulas. So, along with qualitative methods, numerical methods come to the forefront.

We begin with the Euler method, which was formulated in Section 2.4 for a single equation. The idea was to discretize the time interval and replace the derivative in the differential equation by a difference quotient approximation, thereby setting up an iterative method to advance the approximation from time to time. We take the same approach for systems. Consider the nonlinear, autonomous initial value problem

$$\begin{aligned} x' &= f(x,y), \quad y' = g(x,y), \\ x(0) &= x_0, \quad y(0) = y_0, \end{aligned}$$

where a solution is sought on the interval $0 \le t \le T$. First we discretize the time interval by dividing the interval into N equal parts of length $h = T/N$, which is the stepsize; N is the number of steps. The discrete times are $t_n = nh$, $n = 0, 1, 2, ..., N$. We let x_n and y_n denote approximations to the exact solution values $x(t_n)$ and $y(t_n)$ at the discrete points. Then, evaluating the equations at t_n, or $x'(t_n) = f(x(t_n), y(t_n))$, $y'(t_n) = g(x(t_n), y(t_n))$, and then replacing the derivatives by their difference quotient approximations, we obtain, approximately,

$$\begin{aligned} \frac{x(t_{n+1}) - x(t_n)}{h} &= f(x(t_n), x(t_n)), \\ \frac{y(t_{n+1}) - y(t_n)}{h} &= g(x(t_n), x(t_n)). \end{aligned}$$

Therefore, the **Euler method** for computing approximations x_n and y_n is

$$\begin{aligned} x_{n+1} &= x_n + hf(x_n, y_n), \\ y_{n+1} &= y_n + hg(x_n, y_n), \end{aligned}$$

$n = 0, 1, 2, ..., N - 1$. Here, x_0 and y_0 are the prescribed initial conditions that start the recursion process.

The Euler method can be selected on calculators to plot the solution, and it is also available in computer algebra systems. As in Section 2.4, it is easy to write a simple code that calculates the approximate values.

Example 6.4

Consider a mass ($m = 1$) on a nonlinear spring whose oscillations are governed by the second-order equation

$$x'' = -x + 0.1x^3.$$

This is equivalent to the system

$$\begin{aligned} x' &= y, \\ y' &= -x + 0.1x^3. \end{aligned}$$

Euler's formulas are

$$\begin{aligned} x_{n+1} &= x_n + hy_n, \\ y_{n+1} &= y_n + h(-x_n + 0.1x_n^3). \end{aligned}$$

If the initial conditions are $x(0) = 2$ and $y(0) = 0.5$, and if the stepsize is $h = 0.05$, then

$$\begin{aligned} x_1 &= x_0 + hy_0 = 2 + (0.05)(0.5) = 2.025, \\ y_1 &= y_0 + h(-x_0 + 0.1x_0^3) = 0.5 + (0.05)(-2 + (0.1)2^3) = 0.44. \end{aligned}$$

Continuing in this way we can calculate x_2, y_2, and so on, at all the discrete time values. It is clear that calculators and computers are better suited to perform these routine calculations, and Appendix B shows sample computations.

The cumulative error in the Euler method over the interval is proportional to the stepsize h. Just as for a single equation we can increase the order of accuracy with a *modified* Euler method (predictor–corrector), which has a cumulative error of order h^2, or with the classical Runge–Kutta method, which has order h^4. There are other methods of interest, especially those that deal with *stiff* equations where rapid changes in the solution functions occur (such as in chemical reactions or in nerve-firing mechanisms). Runge–Kutta type methods sometimes cannot keep up with rapid changes, so numerical analysts have developed *stiff methods* that adapt to the changes by varying the step size automatically to maintain a small local error. These advanced methods are presented in numerical analysis textbooks. It is clear that the Euler, modified Euler, and Runge–Kutta methods can be extended to three equations in three unknowns, and beyond.

The following exercises require some hand calculation as well as numerical computation. Use a software system or write a program to obtain numerical solutions (see Appendix B for templates).

EXERCISES

1. In Example 6.4 compute x_2, y_2 and x_3, y_3 by hand.

2. Compute, by hand, the first three approximations in Example 6.4 using a modified Euler method.

3. (Trajectory of a baseball) A ball of mass m is hit by a batter. The trajectory is the xy plane. There are two forces on the ball, gravity and air resistance. Gravity acts downward with magnitude mg, and air resistance is directed opposite the velocity vector $\mathbf{v}$ and has magnitude kv^2, where v is the magnitude of $\mathbf{v}$. Use Newton's second law to derive the equations of motion (remember, you have to resolve vertical and horizontal directions). Now take $g = 32$ and $k/m = 0.0025$. Assume the batted ball starts at the origin and the initial velocity is 160 ft per sec at an angle of 30 degrees elevation. Compare a batted ball with air resistance and without air resistance with respect to height, distance, and time to hit the ground.

4. Use a calculator's Runge-Kutta solver, or a computer algebra system, to graph the solution $u = u(t)$ to

$$\begin{aligned} u'' + 9u &= 80\cos 5t, \\ u(0) &= u'(0) = 0, \end{aligned}$$

on the interval $0 \le t \le 6\pi$.

5. Plot several orbits in the phase plane for the system

$$x' = x^2 - 2xy, \quad y' = -y^2 + 2xy.$$

6. Consider a nonlinear mechanical system governed by

$$mx'' = -kx + ax' - b(x')^3,$$

where $m = 2$ and $a = k = b = 1$. Plot the orbit in the phase plane for $t > 0$ and with initial condition $x(0) = 0.01$, $x'(0) = 0$. Plot the time series $x = x(t)$ on the interval $0 \le t \le 60$.

7. The **Van der Pol equation**

$$x'' + a(x^2 - 1)x' + x = 0$$

arises in modeling RCL circuits with nonlinear resistors. For $a = 2$ plot the orbit in the phase plane satisfying $x(0) = 2$, $x'(0) = 0$. Plot the time series graphs, $x = x(t)$ and $y = x'(t)$, on the interval $0 \le t \le 25$. Estimate the period of the oscillation.

8. Consider an influenza outbreak governed by the SIR model (6.7)–(6.8). Let the total population be $N = 500$ and suppose 45 individuals initially have the flu. The data indicate that the likelihood of a healthy individual becoming infected by contact with an individual with the flu is 0.1%. And, once taken ill, an infective is contagious for 5 days. Numerically solve the model equations and draw graphs of S and I vs. time, in days. Draw the orbit in the SI phase plane. How many individuals do not get the flu? What is the maximum number of individuals that have the flu at a single time.

9. Refer to Exercise 8. One way to prevent the spread of a disease is to quarantine some of the infected individuals. Let q be the fraction of infectives that are quarantined. Modify the SIR model to include quarantine, and use the data in Exercise 8 to investigate the behavior of the model for several values of q. Is there a smallest value of q that prevents an epidemic from occurring?

10. The **forced Duffing equation**

$$x'' = x - cx' - x^3 + A\cos t$$

models the damped motion of a mass on a nonlinear spring driven by a periodic forcing function of amplitude A. Take initial conditions $x(0) = 0.5$, $x'(0) = 0$ and plot the phase plane orbit and the time series when $c = 0.25$ and $A = 0.3$. Is the motion periodic? Carry out the same tasks for several other values of the amplitude A and comment on the results.

6.3 Linearization and Stability

For nonlinear systems we have learned how to find equilibrium solutions, nullclines, and the direction of the vector field in regions bounded by the nullclines. What is missing is a detailed analysis of the orbits near the equilibrium points, where much of the action takes place in two-dimensional flows. As mentioned in the last section, we classify equilibrium points as (locally) asymptotically stable, unstable, or neutrally stable, depending upon whether small deviations from equilibrium decay, grow, or remain close. To get an idea of where we are going we consider a simple example.

Example 6.5

Consider

$$x' = x - xy, \quad y' = y - xy. \tag{6.10}$$

This is a simple competition model where two organisms grow with constant *per capita* growth rates, but interaction, represented by the xy terms, has a negative effect on both populations. The origin $(0, 0)$ is an equilibrium point, as is $(1, 1)$. What type are they? Let's try the following strategy. Near the origin both x and y are small. But terms having products of x and y are even smaller, and we suspect we can ignore them. That is, in the first equation x has greater magnitude than xy, and in the second equation y has magnitude greater than xy. Hence, near the origin, the nonlinear system is approximated by

$$x' = x, \quad y' = y.$$

This linearized system has eigenvalues $\lambda = 1, 1$, and therefore $(0, 0)$ is an unstable node. We suspect that the nonlinear system therefore has an unstable node at $(0, 0)$ as well. This turns out to be correct.

Let's apply a similar analysis at the equilibrium $(1, 1)$. We can represent points near $(1, 1)$ as $u = x - 1$, $v = y - 1$ where u and v are small. This is the same as $x = 1 + u$, $y = 1 + v$, so we may regard u and v as small deviations from $x = 1$ and $y = 1$. Rewriting the nonlinear system (6.10) in terms of u and v gives

$$\begin{aligned} u' &= (u+1)(-v) = -v - uv, \\ v' &= (v+1)(-u) = -u - uv, \end{aligned}$$

which is a system of differential equations for the small deviations. Again, because the deviations u and v from equilibrium are small we can ignore the products of u and v in favor of the larger linear terms. Then the system can be approximated by

$$u' = -v, \quad v' = -u.$$

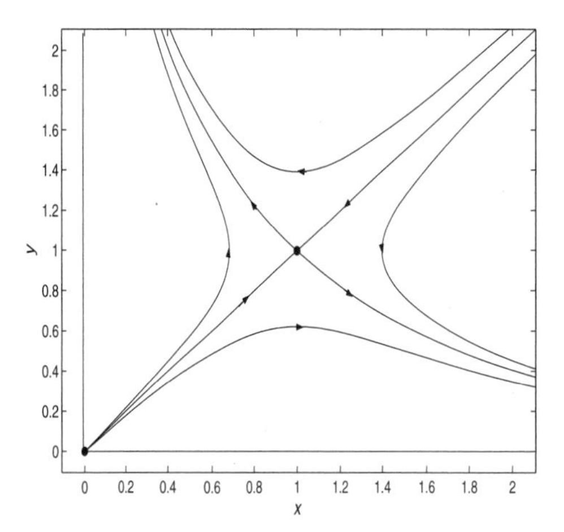

Figure 6.9 Phase portrait for the nonlinear system (6.10) with a saddle at $(1,1)$ and an unstable node at $(0,0)$.

This linear system has eigenvalues $\lambda = -1, 1$, and so $(0,0)$ is a saddle point for the uv-system. This leads us to suspect that $(1,1)$ is a saddle point for the nonlinear system (6.10). We can look at it in this way. If $x = 1+u$ and $y = 1+v$, and changes in u and v have an unstable saddle structure near $(0,0)$, then x and y should have a saddle structure near $(1,1)$. Indeed, the phase portrait for (6.10) is shown in figure 6.9 and it confirms our calculations. Although this is just a toy model of competition with both species having the same dynamics, it leads to an interesting conclusion. Both equilibria are unstable in the sense that small deviations from those equilibria put the populations on orbits that go away from those equilibrium states. There are always perturbations or deviations in a system. So, in this model, there are no persistent states. One of the populations, depending upon where the initial data are, will dominate and the other will approach extinction.

If a nonlinear system has an equilibrium, then the behavior of the orbits near that point is often mirrored by a linear system obtained by discarding the small nonlinear terms. We already know how to analyze linear systems; their behavior is determined by the eigenvalues of the associated matrix for the system. Therefore the general idea is to approximate the nonlinear system by a linear system in a neighborhood of the equilibrium and use the properties of the linear system to deduce the properties of the nonlinear system. This analysis, which is standard fare in differential equations, is called **local stability**

analysis. So, we begin with the system

$$x' = f(x,y) \tag{6.11}$$
$$y' = g(x,y). \tag{6.12}$$

Let $\mathbf{x}^* = (x_e, y_e)$ be an isolated equilibrium and let u and v denote small deviations (often called **small perturbations**) from equilibrium:

$$u = x - x_e, \quad v = y - y_e.$$

To determine if the perturbations grow or decay, we derive differential equations for those perturbations. Substituting into (6.11)–(6.12) we get, in terms of u and v, the system

$$\begin{aligned} u' &= f(x_e + u, y_e + v), \\ v' &= g(x_e + u, y_e + v). \end{aligned}$$

This system of equations for the perturbations has a corresponding equilibrium at $u = v = 0$. Now, in this system, we discard the nonlinear terms in u and v. Formally we can do this by expanding the right sides in Taylor series about point (x_e, y_e) to obtain

$$\begin{aligned} u' &= f(x_e, y_e) + f_x(x_e, y_e)u + f_y(x_e, y_e)v + \text{ higher-order terms in } u \text{ and } v, \\ v' &= g(x_e, y_e) + g_x(x_e, y_e)u + g_y(x_e, y_e)v + \text{ higher-order terms in } u \text{ and } v, \end{aligned}$$

where the higher-order terms are nonlinear terms involving powers of u and v and their products. The first terms on the right sides are zero because (x_e, y_e) is an equilibrium, and the higher-order terms are small in comparison to the linear terms (e.g., if u is small, say 0.1, then u^2 is much smaller, 0.01). Therefore the perturbation equations can be approximated by

$$\begin{aligned} u' &= f_x(x_e, y_e)u + f_y(x_e, y_e)v, \\ v' &= g_x(x_e, y_e)u + g_y(x_e, y_e)v. \end{aligned}$$

This linear system for the small deviations is called the linearized perturbation equations, or simply the **linearization** of (6.11)–(6.12) at the equilibrium (x_e, y_e). It has an equilibrium point at $(0,0)$ corresponding to (x_e, y_e) for the nonlinear system. In matrix form we can write the linearization as

$$\begin{pmatrix} u' \\ v' \end{pmatrix} = \begin{pmatrix} f_x(x_e, y_e) & f_y(x_e, y_e) \\ g_x(x_e, y_e) & g_y(x_e, y_e) \end{pmatrix} \begin{pmatrix} u \\ v \end{pmatrix}. \tag{6.13}$$

The matrix $J = J(x_e, y_e)$ of first partial derivatives of f and g defined by

$$J(x_e, y_e) = \begin{pmatrix} f_x(x_e, y_e) & f_y(x_e, y_e) \\ g_x(x_e, y_e) & g_y(x_e, y_e) \end{pmatrix}$$

is called the **Jacobian matrix** at the equilibrium (x_e, y_e). Note that this matrix is a matrix of numbers because the partial derivatives are evaluated at the equilibrium. We assume that J does not have a zero eigenvalue (i.e., $\det J \neq 0$). If so, we would have to look at the higher-order terms in the Taylor expansions of the right sides of the equations.

We already know that the nature of the equilibrium of (6.13) is determined by the eigenvalues of the matrix J. The question is: does the linearized system for the perturbations u and v near $u = v = 0$ aid in predicting the qualitative behavior in the nonlinear system of the solution curves near an equilibrium point (x_e, y_e)? The answer is yes in all cases except perhaps when the eigenvalues of the Jacobian matrix are purely imaginary (i.e., $\lambda = \pm bi$), or when there are two equal eigenvalues. Stated differently, the phase portrait of a nonlinear system close to an equilibrium point looks essentially the same as that of the linearization provided the eigenvalues have nonzero real part or are equal. Pictorially, near the equilibrium the small nonlinearities in the nonlinear system produce a slightly distorted phase diagram from that of the linearization. We summarize the basic results in the following items.

1. If $(0, 0)$ is asymptotically stable for the linearization (6.13), then the perturbations decay and (x_e, y_e) is asymptotically stable for the nonlinear system (6.11)–(6.12). This will occur when J has negative eigenvalues, or complex eigenvalues with negative real part.

2. If $(0, 0)$ is unstable for the linearization (6.13), then some or all of the perturbations grow and (x_e, y_e) is unstable for the nonlinear system (6.11)–(6.12). This will occur when J has a positive eigenvalue or complex eigenvalues with positive real part.

3. The exceptional case for stability is that of a center. If $(0, 0)$ is a center for the linearization (6.13), then (x_e, y_e) may be asymptotically stable, unstable, or a center for the nonlinear system (6.11)–(6.12). This case occurs when J has purely imaginary eigenvalues.

4. The borderline cases (equal eigenvalues) of degenerate and star-like nodes maintain stability, but the type of equilibria may change. For example, the inclusion of nonlinear terms can change a star-like node into a spiral, but it will not affect stability.

This means if the linearization predicts a regular node, saddle, or spiral at $(0, 0)$, then the nonlinear system will have a regular node, saddle, or spiral at the equilibrium (x_e, y_e). In the case of regular nodes and saddles, the directions of the eigenvectors give the directions of the tangent lines to the special curves that enter or exit the equilibrium point. Such curves are called **separatrices** (singular: **separatrix**). For linear systems the separatrices are the linear orbits entering or leaving the origin in the case of a saddle or node.

Sometimes we are only interested in whether an equilibrium is stable, and not whether it is a node or spiral. Stability can be determined by examining the trace of J and the determinant of J. We recall from Chapter 5:

– The equilibrium (x_e, y_e) is asymptotically stable if and only if

$$\mathrm{tr} J(x_e, y_e) < 0 \quad \text{and} \quad \det J(x_e, y_e) > 0. \tag{6.14}$$

Example 6.6

Consider the decoupled nonlinear system

$$x' = x - x^3, \quad y' = 2y.$$

The equilibria are $(0,0)$ and $(\pm 1, 0)$. The Jacobian matrix at an arbitrary (x,y) for the linearization is

$$J(x,y) = \begin{pmatrix} f_x(x,y) & f_y(x,y) \\ g_x(x,y) & g_y(x,y) \end{pmatrix} = \begin{pmatrix} 1-3x^2 & 0 \\ 0 & 2 \end{pmatrix}.$$

Therefore

$$J(0,0) = \begin{pmatrix} 1 & 0 \\ 0 & 2 \end{pmatrix},$$

which has eigenvalues 1 and 2. Thus $(0,0)$ is an unstable node. Next

$$J(1,0) = \begin{pmatrix} -2 & 0 \\ 0 & 2 \end{pmatrix}, \quad J(-1,0) = \begin{pmatrix} -2 & 0 \\ 0 & 2 \end{pmatrix},$$

and both have eigenvalues -2 and 2. Therefore $(1,0)$ and $(-1,0)$ are saddle points. The phase diagram is easy to draw. The vertical nullclines are $x = 0$, $x = 1$, and $x = -1$, and the horizontal nullcline $y = 0$. Along the x axis we have $x' > 0$ if $-1 < x < 1$, and $x' < 0$ if $|x| > 1$. The phase portrait is shown in figure 6.10.

Example 6.7

Consider the Lotka–Volterra model

$$x' = x(r - ay), \quad y' = y(-m + bx). \tag{6.15}$$

The equilibria are (0, 0) and $(m/b, r/a)$. The Jacobian matrix is

$$J(x,y) = \begin{pmatrix} f_x(x,y) & f_y(x,y) \\ g_x(x,y) & g_y(x,y) \end{pmatrix} = \begin{pmatrix} r - ay & -ax \\ by & -m + bx \end{pmatrix}.$$

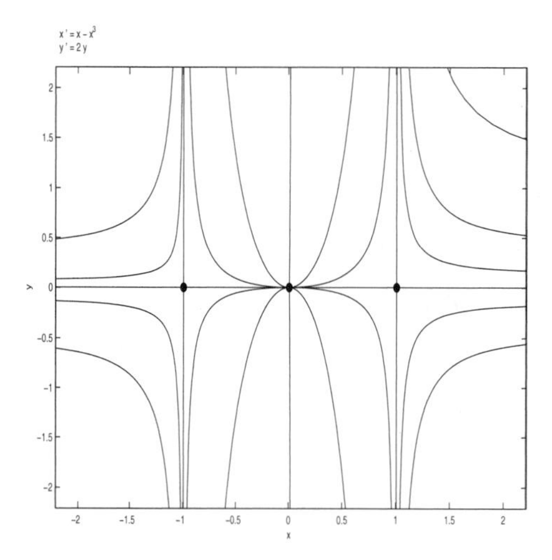

Figure 6.10 Phase diagram for the system $x' = x - x^3, \quad y' = 2y$. In the upper half-plane the orbits are moving upward, and in the lower half-plane they are moving downward.

We have

$$J(0,0) = \begin{pmatrix} r & 0 \\ 0 & -m \end{pmatrix},$$

which has eigenvalues r and $-m$. Thus $(0,0)$ is a saddle. For the other equilibrium,

$$J(m/b, r/a) = \begin{pmatrix} 0 & -am/b \\ rb/a & 0 \end{pmatrix}.$$

The characteristic equation is $\lambda^2 + rm = 0$, and therefore the eigenvalues are purely imaginary: $\lambda = \pm\sqrt{rm}$. This is the exceptional case; we cannot conclude that the equilibrium is a center, and we must work further to determine the nature of the equilibrium. We did this in Section 6.2.1 and found that $(m/b, r/a)$ was indeed a center.

Example 6.8

The nonlinear system

$$\begin{aligned} x' &= \frac{1}{2}x - y - \frac{1}{2}(x^3 + xy^2), \\ y' &= x + \frac{1}{2}y - \frac{1}{2}(y^3 + yx^2), \end{aligned}$$

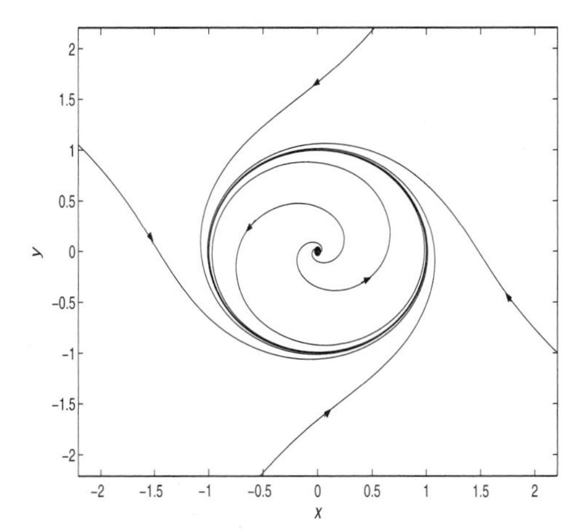

Figure 6.11 Orbits spiral out from the origin and approach the limit cycle $x^2 + y^2 = 1$, which is a closed, periodic orbit. Orbits outside the limit cycle spiral toward it. We say the limit cycle is stable.

has an equilibrium at the origin. The linearized system is

$$\begin{pmatrix} u' \\ v' \end{pmatrix} = \begin{pmatrix} \frac{1}{2} & -1 \\ 1 & \frac{1}{2} \end{pmatrix} \begin{pmatrix} u \\ v \end{pmatrix},$$

with eigenvalues $\frac{1}{2} \pm i$. Therefore the origin is an unstable spiral point. One can check the direction field near the origin to see that the spirals are counterclockwise. Do these spirals go out to infinity? We do not know without further analysis. We have only checked the local behavior, near the equilibrium. What happens beyond that is unknown and is described as the *global behavior* of the system. Using software, in fact, shows that there is cycle at radius one and the spirals coming out of the origin approach that cycle from within. Outside the closed cycle the orbits come in from infinity and approach the cycle. See figure 6.11. A cycle, or periodic solution, that is approached by another orbit as $t \to +\infty$ or as $t \to -\infty$ is called a **limit cycle**.

One can use computer algebra systems, or even a calculator, to draw phase diagrams. With computer algebra systems there are two options. You can write a program to numerically solve and plot the solutions (e.g., a Runge-Kutta routine), or you can use built-in programs that plot solutions automatically. Another option is to use codes developed by others to sketch phase diagrams. One of the best is a MATLAB code, *pplane6,* developed by Professor John Polking at Rice University (see the references for further information).

In summary, we have developed a set of tools to analyze nonlinear systems. We can systematically follow the steps below to obtain a complete phase diagram.

1. Find the equilibrium solutions and check their nature by examining the eigenvalues of the Jacobian J for the linearized system.

2. Draw the nullclines and indicate the direction of the vector field along those lines.

3. Find the direction of the vector field in the regions bounded by the nullclines.

4. Find directions of the separatrices (if any) at equilibria, indicated by the eigenvectors of J.

5. By dividing the equations, find the orbits (this may be impossible in many cases).

6. Use a software package or graphing calculator to get a complete phase diagram.

Example 6.9

A model of vibrations of a nonlinear spring with restoring force $F(x) = -x+x^3$ is

$$x'' = -x + x^3,$$

where the mass is $m = 1$. As a system,

$$x' = y, \quad y' = -x + x^3,$$

where y is the velocity. The equilibria are easily $(0,0)$, $(1,0)$, and $(-1,0)$. Let us check their nature. The Jacobian matrix is

$$J(x,y) = \begin{pmatrix} 0 & 1 \\ -1+3x^2 & 0 \end{pmatrix}.$$

Then

$$J(0,0) = \begin{pmatrix} 0 & 1 \\ -1 & 0 \end{pmatrix}, \quad J(1,0) = J(-1,0) = \begin{pmatrix} 0 & 1 \\ 2 & 0 \end{pmatrix}.$$

The eigenvalues of these two matrices are $\pm i$ and $\pm\sqrt{2}$, respectively. Thus $(-1,0)$ and $(1,0)$ are saddles and are unstable; $(0,0)$ is a center for the linearization, which gives us no information about that point for the nonlinear system. It is easy to see that the x-nullcline (vertical vector field) is $y = 0$, or the x-axis, and the y-nullclines (horizontal vector field) are the three lines $x = 0, 1, -1$. The directions of the separatrices coming in and out of the saddle points are given by the eigenvectors of the Jacobian matrix, which are easily found to be $(1, \pm\sqrt{2})^T$. So we have an accurate picture of the phase plane structure except near the origin. To analyze the behavior near the origin we can find formulas for the orbits. Dividing the two differential equations gives

$$\frac{dy}{dx} = \frac{-x + x^3}{y},$$

which, using separation of variables, integrates to

$$\frac{1}{2}y^2 + \frac{1}{2}x^2 - \frac{1}{4}x^4 = E,$$

where E is a constant of integration. Again, observe that this expression is just the conservation of energy law because the kinetic energy is $\frac{1}{2}y^2$ and the potential energy is $V(x) = -\int F(x)dx = -\int(-x + x^3)dx = \frac{1}{2}x^2 - \frac{1}{4}x^4$. We can solve for y to obtain

$$y = \pm\sqrt{2}\sqrt{E - \frac{1}{2}x^2 + \frac{1}{4}x^4}.$$

These curves can be plotted for different values of E and we find that they are cycles near the origin. So the origin is a center, which is neutrally stable. A phase diagram is shown in figure 6.12. This type of analysis can be carried out for any conservative mechanical system $x'' = F(x)$. The orbits are always given by $y = \pm\sqrt{2}\sqrt{E - V(x)}$, where $V(x) = -\int F(x)dx$ is the potential energy.

In summary, what we described in this section is **local stability analysis**, that is, how small perturbations from equilibrium evolve in time. Local stability analysis turns a nonlinear problem into a linear one, and it is a procedure that answers the question of what happens when we perturb the states x and y a small amount from their equilibrium values. Local analysis does not give any information about global behavior of the orbits far from equilibria, but it usually does give reliable information about perturbations near equilibria. The local behavior is determined by the eigenvalues of the Jacobian matrix, or the matrix of the linearized system. The only exceptional case is that of a center. One big difference between linear and nonlinear systems is that linear systems, as discussed in Chapter 5, can be solved completely and the global behavior of solutions is known. For nonlinear systems we can often obtain only local behavior near equilibria; it is difficult to tie down the global behavior.

One final remark. In Chapter 1 we investigated a single autonomous equation, and we plotted on a bifurcation diagram how equilibria and their stability change as a function of some parameter in the problem. This same type of behavior is also interesting for systems of equations. As a parameter in a given nonlinear system varies, the equilibria vary and stability can change. Some of the Exercises explore bifurcation phenomena in such systems.

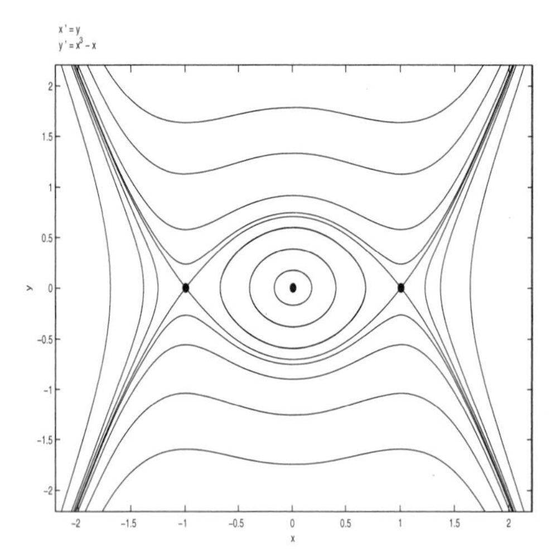

Figure 6.12 Phase portrait of the system $x' = y, \quad y' = -x + x^3$. The orbits are moving to the right in the upper half-plane and to the left in the lower half-plane.

EXERCISES

1. Find the equation of the orbits of the system $x' = e^x - 1, \ y' = ye^x$ and plot the the orbits in phase plane.

2. Write down an equation for the orbits of the system $x' = y, \ y' = 2y + xy$. Sketch the phase diagram.

3. For the following system find the equilibria, sketch the nullclines and the direction of the flow along the nullclines, and sketch the phase diagram:

$$x' = y - x^2, \ y' = 2x - y.$$

What happens to the orbit beginning at $(1, 3/2)$ as $t \to +\infty$?

4. Determine the nature of each equilibrium of the system $x' = 4x^2 - a, \quad y' = -\frac{y}{4}(x^2 + 4)$, and show how the equilibria change as the parameter a varies.

5. Consider the system

$$\begin{aligned} x' &= 2x(1 - \frac{x}{2}) - xy, \\ y' &= y\left(\frac{9}{4} - y^2\right) - x^2 y. \end{aligned}$$

Find the equilibria and sketch the nullclines. Use the Jacobian matrix to determine the type and stability of each equilibrium point and sketch the phase portrait.

6. Completely analyze the nonlinear system

$$x' = y, \quad y' = x^2 - 1 - y.$$

7. In some systems there are snails with two types of symmetry. Let R be the number of right curling snails and L be the number of left curling snails. The population dynamics is given by the competition equations

$$\begin{aligned} R' &= R - (R^2 + aRL) \\ L' &= L - (L^2 + aRL), \end{aligned}$$

where a is a positive constant. Analyze the behavior of the system for different values of a. Which snail dominates?

8. Consider the system

$$\begin{aligned} x' &= xy - 2x^2 \\ y' &= x^2 - y. \end{aligned}$$

Find the equilibria and use the Jacobian matrix to determine their types and stability. Draw the nullclines and indicate on those lines the direction of the vector field. Draw a phase diagram.

9. The dynamics of two competing species is governed by the system

$$\begin{aligned} x' &= x(10 - x - y), \\ y' &= y(30 - 2x - y). \end{aligned}$$

Find the equilibria and sketch the nullclines. Use the Jacobian matrix to determine the type and stability of each equilibrium point and sketch the phase diagram.

10. Show that the origin is asymptotically stable for the system

$$\begin{aligned} x' &= y, \\ y' &= 2y(x^2 - 1) - x. \end{aligned}$$

11. Consider the system

$$\begin{aligned} x' &= y, \\ y' &= -x - y^3. \end{aligned}$$

Show that the origin for the linearized system is a center, yet the nonlinear system itself is asymptotically stable. (Hint: show that $\frac{d}{dt}(x^2 + y^2) < 0$.)

12. A particle of mass 1 moves on the x-axis under the influence of a potential $V(x) = x - \frac{1}{3}x^3$. Formulate the dynamics of the particle in x, y coordinates, where y is velocity, and analyze the system in the phase plane. Specifically, find and classify the equilibria, draw the nullclines, determine the xy equation for the orbits, and plot the phase diagram.

13. A system

$$\begin{aligned} x' &= f(x,y) \\ y' &= g(x,y), \end{aligned}$$

is called a **Hamiltonian system** if there is a function $H(x,y)$ for which $f = H_y$ and $g = -H_x$. The function H is called the **Hamiltonian**. Prove the following facts about Hamiltonian systems.

a) If $f_x + g_y = 0$, then the system is Hamiltonian. (Recall that $f_x + g_y$ is the divergence of the vector field (f,g).)

b) Prove that along any orbit, $H(x,y) =$ constant, and therefore all the orbits are given by $H(x,y) =$ constant.

c) Show that if a Hamiltonian system has an equilibrium, then it is not a source or sink (node or spiral).

d) Show that any conservative dynamical equation $x'' = f(x)$ leads to a Hamiltonian system, and show that the Hamiltonian coincides with the total energy.

e) Find the Hamiltonian for the system $x' = y, \quad y' = x - x^2$, and plot the orbits.

14. In a Hamiltonian system the Hamiltonian given by $H(x,y) = x^2 + 4y^4$. Write down the system and determine the equilibria. Sketch the orbits.

15. A system

$$\begin{aligned} x' &= f(x,y) \\ y' &= g(x,y), \end{aligned}$$

is called a **gradient system** if there is a function $G(x,y)$ for which $f = G_x$ and $g = G_y$.

a) If $f_y - g_x = 0$, prove that the system is a gradient system. (Recall that $f_y - g_x$ is the curl of the two-dimensional vector field (f,g); a zero curl ensures existence of a potential function on nice domains.)

b) Prove that along any orbit, $\frac{d}{dt}G(x,t) \geq 0$. Show that periodic orbits are impossible in gradient systems.

c) Show that if a gradient system has an equilibrium, then it is not a center or spiral.

d) Show that the system $x' = 9x^2 - 10xy^2$, $y' = 2y - 10x^2y$ is a gradient system.

e) Show that the system $x' = \sin y$, $y' = x \cos y$ has no periodic orbits.

16. The populations of two competing species x and y are modeled by the system

$$\begin{aligned} x' &= (K - x)x - xy, \\ y' &= (1 - 2y)y - xy, \end{aligned}$$

where K is a positive constant. In terms of K, find the equilibria. Explain how the equilibria change, as to type and stability, as the parameter K increases through the interval $0 < K \leq 1$, and describe how the phase diagram evolves. Especially describe the nature of the change at $K = 1/2$.

17. Give a thorough description, in terms of equilibria, stability, and phase diagram, of the behavior of the system

$$\begin{aligned} x' &= y + (1 - x)(2 - x), \\ y' &= y - ax^2, \end{aligned}$$

as a function of the parameter $a > 0$.

18. A predator–prey model is given by

$$\begin{aligned} x' &= rx\left(1 - \frac{x}{K}\right) - f(x)y, \\ y' &= -my + cf(x)y, \end{aligned}$$

where r, m, c, and K are positive parameters, and the predation rate $f(x)$ satisfies $f(0) = 0$, $f'(x) > 0$, and $f(x) \to M$ as $x \to \infty$.

a) Show that $(0, 0)$ and $(K, 0)$ are equilibria.

b) Classify the $(0, 0)$ equilibrium. Find conditions that guarantee that $(K, 0)$ is unstable and state what type of unstable point it is.

c) Under what conditions will there be an equilibrium in the first quadrant?

19. Consider the dynamical equation $x'' = f(x)$, with $f(x_0) = 0$. Find a condition that guarantees that $(x_0, 0)$ will be a saddle point in the phase plane representation of the problem.

20. The dynamics of two competing species is given by

$$\begin{aligned} x' &= 4x(1 - x/4) - xy, \\ y' &= 2y(1 - ay/2) - bxy. \end{aligned}$$

For which values of a and b can the two species coexist? Physically, what do the parameters a and b represent?

21. A particle of mass $m = 1$ moves on the x-axis under the influence of a force $F = -x + x^3$ as discussed in Example 6.8.

    a) Determine the values of the total energy for which the motion will be periodic.

    b) Find and plot the equation of the orbit in phase space of the particle if its initial position and velocity are $x(0) = 0.5$ and $y(0) = 0$. Do the same if $x(0) = -2$ and $y(0) = 2$.

6.4 Periodic Solutions

We noted the exceptional case in the linearization procedure: if the associated linearization for the perturbations has a center (purely imaginary eigenvalues) at (0,0), then the behavior of the nonlinear system at the equilibrium is undetermined. This fact suggests that the existence of periodic solutions, or (closed) cycles, for nonlinear systems is not always easily decided. In this section we discuss some special cases when we can be assured that periodic solutions do not exist, and when they do exist. The presence of oscillations in physical and biological systems often represent important phenomena, and that is why such solutions are of great interest.

We first state two negative criteria for the nonlinear system

$$x' = f(x, y) \tag{6.16}$$

$$y' = g(x, y). \tag{6.17}$$

1. (**Equilibrium Criterion**) If the nonlinear system (6.16)–(6.17) has a cycle, then the region inside the cycle must contain an equilibrium. Therefore, if there are no equilibria in a given region, then the region can contain no cycles.

2. (**Dulac's Criterion**) Consider the nonlinear system (6.16)–(6.17). If in a given region of the plane there is a function $\beta(x, y)$ for which

$$\frac{\partial}{\partial x}(\beta f) + \frac{\partial}{\partial y}(\beta g)$$

is of one sign (strictly positive or strictly negative) entirely in the region, then the system cannot have a cycle in that region.

We omit the proof of the equilibrium criterion (it may be found in the references), but we give the proof of Dulac's criterion because it is a simple application of Green's theorem,[3] which was encountered in multi-variable calculus. The proof is by contradiction, and it assumes that there *is* a cycle of period p given by $x = x(t)$, $y = y(t)$, $0 \le t \le p$, lying entirely in the region and represented by a simple closed curve C. Assume it encloses a domain R. Without loss of generality suppose that $\frac{\partial}{\partial x}(\beta f) + \frac{\partial}{\partial y}(\beta g) > 0$. Then, to obtain a contradiction, we make the following calculation.

$$\begin{aligned} 0 &< \int\int_R \left(\frac{\partial}{\partial x}(\beta f) + \frac{\partial}{\partial y}(\beta g)\right) dA = \int_C (-\beta g dx + bf dy) \\ &= \int_0^p (-\beta g x' dt + bf y' dt) = \int_0^p (-\beta g f dt + bf g dt) = 0, \end{aligned}$$

the contradiction being $0 < 0$. Therefore the assumption of a cycle is false, and there can be no periodic solution.

Example 6.10

The system

$$x' = 1 + y^2, \quad y' = x - y + xy$$

does not have any equilibria (note x' can never equal zero), so this system cannot have cycles.

Example 6.11

Consider the system

$$x' = x + x^3 - 2y, \quad y' = -3x + y^3.$$

Then

$$\frac{\partial}{\partial x} f + \frac{\partial}{\partial x} g = \frac{\partial}{\partial x}(x + x^3 - 2y) + \frac{\partial}{\partial x}(-3x + y^3) = 1 + 3x^2 + 3y^2 > 0,$$

which is positive for all x and y. Dulac's criterion implies there are no periodic orbits in the entire plane. Note here that $\beta = 1$.

[3] For a region R enclosed by a simple closed curve C we have $\int_C P dx + Q dy = \int\int_R (Q_x - P_y) dA$, where C is taken counterclockwise.

One must be careful in applying Dulac's criterion. If we find that $\frac{\partial}{\partial x}(\beta f) + \frac{\partial}{\partial y}(\beta g) > 0$ in, say, the first quadrant only, then that means there are no cycles lying entirely in the first quadrant; but there still may be cycles that go out of the first quadrant.

Sometimes cycles can be detected easily in a polar coordinate system. Presence of the expression $x^2 + y^2$ in the system of differential equations often signals that a polar representation might be useful in analyzing the problem.

Example 6.12

Consider the system

$$\begin{aligned} x' &= y + x(1 - x^2 - y^2) \\ y' &= -x + y(1 - x^2 - y^2). \end{aligned}$$

The reader should check, by linearization, that the origin is an unstable spiral point. But what happens beyond that? To transform the problem to polar coordinates $x = r\cos\theta$ and $y = r\sin\theta$, we note that

$$r^2 = x^2 + y^2, \quad \tan\theta = \frac{y}{x}.$$

Taking time derivatives and using the chain rule,

$$rr' = xx' + yy', \quad (\sec^2\theta)\theta' = \frac{xy' - yx'}{x^2}.$$

We can solve for r' and θ' to get

$$r' = x'\cos\theta + y'\sin\theta, \quad \theta' = \frac{y'\cos\theta - x'\sin\theta}{r}.$$

Finally we substitute for x' and y' on the right side from the differential equations to get the polar forms of the equations: $r' = F(r, \theta)$, $\theta' = G(r, \theta)$. Leaving the algebra to the reader, we finally get

$$\begin{aligned} r' &= r(1 - r^2), \\ \theta' &= -1. \end{aligned}$$

By direct integration of the second equation, $\theta = -t + C$, so the angle θ rotates clockwise with constant speed. Notice also that $r = 1$ is a solution to the first equation. Thus we have obtained a periodic solution, a circle of radius one, to the system. For $r < 1$ we have $r' > 0$, so r is increasing on orbits, consistent with our remark that the origin is an unstable spiral. For $r > 1$ we have $r' < 0$, so r is decreasing along orbits. Hence, there is a limit cycle that is approached by orbits from its interior and its exterior. Figure 6.13 shows the phase diagram.

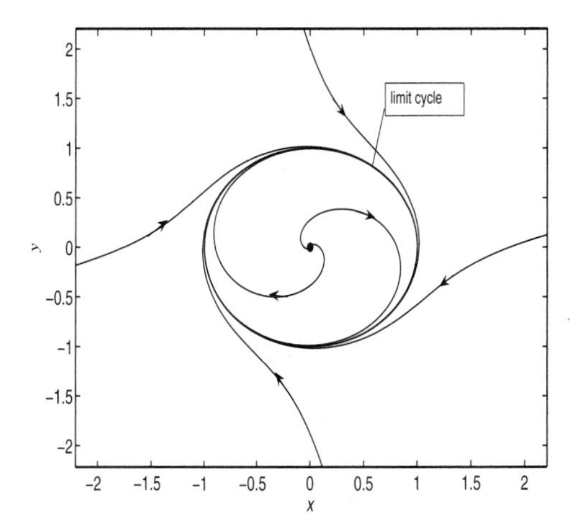

Figure 6.13 Limit cycle. The orbits rotate clockwise.

6.4.1 The Poincaré–Bendixson Theorem

To sum it up, through examples we have observed various nonlinear phenomena in the phase plane, including equilibria, orbits that approach equilibria, orbits that go to infinity, cycles, and orbits that approach cycles. What have we missed? Is there some other complicated orbital structure that is possible? The answer to this question is no; dynamical possibilities in a two-dimensional phase plane are very limited. If an orbit is confined to a closed bounded region in the plane, then as $t \to +\infty$ that orbit must be an equilibrium solution (a point), be a cycle, approach a cycle, or approach an equilibrium. (Recall that a closed region includes its boundary). The same result holds as $t \to -\infty$. This is a famous result called the **Poincaré–Bendixson theorem**, and it is proved in advanced texts. We remark that the theorem is not true in three dimensions or higher where orbits for nonlinear systems can exhibit bizarre behavior, for example, approaching sets of fractal dimension (strange attractors) or showing chaotic behavior. Henri Poincaré (1854–1912) was one of the great contributors to the theory of differential equations and dynamical systems.

Example 6.13

Consider the model

$$x' = \frac{2}{3}x\left(1-\frac{x}{4}\right) - \frac{xy}{1+x},$$
$$y' = ry\left(1-\frac{y}{x}\right), \quad r > 0.$$

In an ecological context, we can think of this system as a predator–prey model. The prey (x) grow logistically and are harvested by the predators (y) with a Holling type II rate. The predator grows logistically, with its carrying capacity depending linearly upon the prey population. The horizontal, y-nullclines, are $y = x$ and $y = 0$, and the vertical, or x-nullcline is the parabola $y = \left(\frac{2}{3} - \frac{1}{6}x\right)(x+1)$. The equilibria are $(1,1)$, and $(4,0)$. The system is not defined when $x = 0$ and we classify the y-axis as a line of *singularities*; no orbits can cross this line. The Jacobian matrix is

$$J(x,y) = \begin{pmatrix} f_x & f_y \\ g_x & g_y \end{pmatrix} = \begin{pmatrix} \frac{2}{3} - \frac{1}{6}x - \frac{y}{(1+x)^2} & \frac{-x}{1+x} \\ \frac{ry^2}{x^2} & r - \frac{2ry}{x} \end{pmatrix}.$$

Evaluating at the equilibria yields

$$J(4,0) = \begin{pmatrix} -\frac{2}{3} & -\frac{4}{5} \\ 0 & r \end{pmatrix}, \quad J(1,1) = \begin{pmatrix} \frac{1}{12} & -\frac{1}{2} \\ r & -r \end{pmatrix}.$$

It is clear that $(4,0)$ is a saddle point with eigenvalues r and $-2/3$. At $(1,1)$ we find $\text{tr}J = \frac{1}{12} - r$ and $\det J = \frac{5}{12}r > 0$. Therefore $(1,1)$ is asymptotically stable if $r > \frac{1}{12}$ and unstable if $r < \frac{1}{12}$. So, there is a bifurcation, or change, at $r = \frac{1}{12}$ because the stability of the equilibrium changes. For a large predator growth rate r there is a nonzero persistent state where predator and prey can coexist. As the growth rate of the predator decreases to a critical value, this persistence goes away. What happens then? Let us imagine that the system is in the stable equilibrium state and other factors, possibly environmental, cause the growth rate of the predator to slowly decrease. How will the populations respond once the critical value of r is reached?

Let us carefully examine the case when $r < \frac{1}{12}$. Consider the direction of the vector field on the boundary of the square with corners $(0,0)$, $(4,0)$, $(4,4)$, $(0,4)$. See figure 6.14. On the left side ($x = 0$) the vector field is undefined, and near that boundary it is nearly vertical; orbits cannot enter or escape along that edge. On the lower side ($y = 0$) the vector field is horizontal ($y' = 0$, $x' > 0$). On the right edge ($x = 4$) we have $x' < 0$ and $y' > 0$, so the vector field points into the square. And, finally, along the upper edge ($y = 4$) we have $x' < 0$ and $y' < 0$, so again the vector field points into the square. The equilibrium at $(1,1)$ is unstable, so orbits go away from equilibrium; but they cannot escape from the

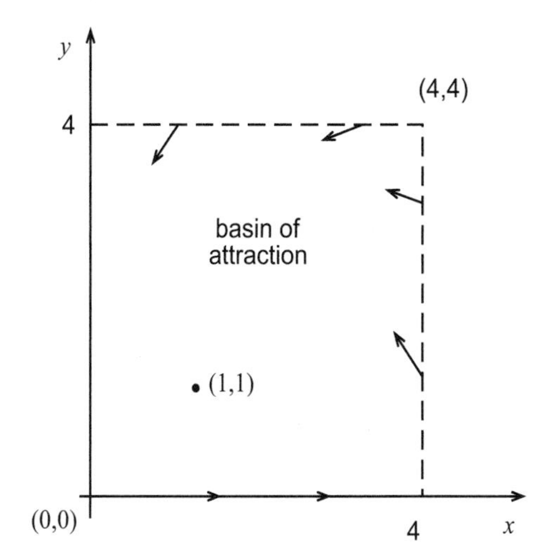

Figure 6.14 A square representing a basin of attraction. Orbits cannot escape the square.

square. On the other hand, orbits along the top and right sides are entering the square. What can happen? They cannot crash into each other! (Uniqueness.) So, there must be a counterclockwise limit cycle in the interior of the square (by the Poincaré–Bendixson theorem). The orbits entering the square approach the cycle from the outside, and the orbits coming out of the unstable equilibrium at $(1,1)$ approach the cycle from the inside. Now we can state what happens as the predator growth rate r decreases through the critical value. The persistent state becomes unstable and a small perturbation, always present, causes the orbit to approach the limit cycle. Thus, we expect the populations to cycle near the limit cycle. A phase diagram is shown in figure 6.15.

In this example we used a common technique of constructing a region, called a **basin of attraction**, that contains an unstable spiral (or node), yet orbits cannot escape the region. In this case there must be a limit cycle in the region. A similar result holds true for annular type regions (doughnut type regions bounded by concentric simple close curves)—if there are no equilibria in an annular region R and the vector field points inward into the region on both the inner and outer concentric boundaries, then there must be a limit cycle in R.

EXERCISES

1. Does the system

$$\begin{aligned} x' &= x - y - x\sqrt{x^2+y^2}, \\ y' &= x + y - y\sqrt{x^2+y^2}, \end{aligned}$$

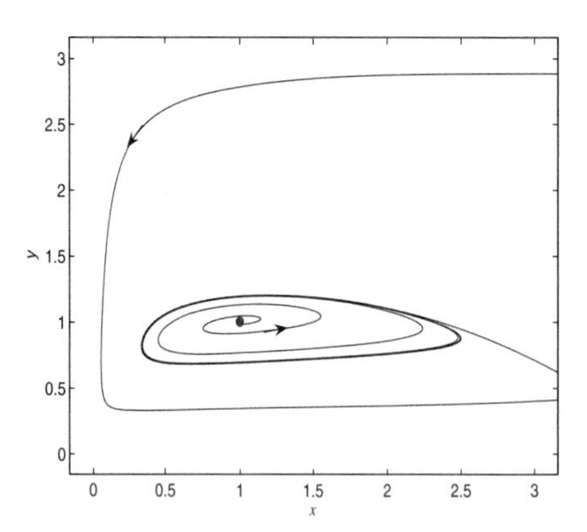

Figure 6.15 Phase diagram showing a counterclockwise limit cycle. Curves approach the limit cycle from the outside and from the inside. The interior equilibrium is an unstable spiral point.

have periodic orbits? Does it have limit cycles?

2. Show that the system

$$x' = 1 + x^2 + y^2,$$
$$y' = (x-1)^2 + 4,$$

has no periodic solutions.

3. Show that the system

$$x' = x + x^3 - 2y,$$
$$y' = y^5 - 3x,$$

has no periodic solutions.

4. Analyze the dynamics of the system

$$x' = y,$$
$$y' = -x(1-x) + cy,$$

for different positive values of c. Draw phase diagrams for each case, illustrating the behavior.

5. An RCL circuit with a nonlinear resistor (the voltage drop across the resistor is a nonlinear function of the current) can be modeled by the Van der Pol equation
$$x'' + a(x^2 - 1)x' + x = 0,$$
where a is a positive constant, and $x = x(t)$ is the current. In the phase plane formulation, show that the origin is unstable. Sketch the nullclines and the vector field. Can you tell if there is a limit cycle? Use a computer algebra system to sketch the phase plane diagram in the case $a = 1$. Draw a time series plot for the current in this case for initial conditions $x(0) = 0.05$, $x'(0) = 0$. Is there a limit cycle?

6. For the system
$$\begin{aligned} x' &= y, \\ y' &= x - y - x^3, \end{aligned}$$
determine the equilibria. Write down the Jacobian matrix at each equilibrium and investigate stability. Sketch the nullclines. Finally, sketch a phase diagram.

7. Let P denote the carbon biomass of plants in an ecosystem and H the carbon biomass of herbivores. Let ϕ denote the constant rate of primary production of carbon in plants due to photosynthesis. Then a model of plant–herbivore dynamics is given by
$$\begin{aligned} P' &= \phi - aP - bHP, \\ H' &= \varepsilon bHP - cH, \end{aligned}$$
where a, b, c, and ε are positive parameters.

   a) Explain the various terms in the model and determine the dimensions of each constant.

   b) Find the equilibrium solutions.

   c) Analyze the dynamics in two cases, that of high primary production ($\phi > ac/\varepsilon b$) and low primary production ($\phi < ac/\varepsilon b$). Determine what happens to the system if the primary production is slowly increased from a low value to a high value.

8. Consider the system
$$x' = ax + y - x(x^2 + y^2), \quad y' = -x + ay - y(x^2 + y^2),$$
where a is a parameter. Discuss the qualitative behavior of the system as a function of the parameter a. In particular, how does the phase plane evolve as a is changed?

9. Show that periodic orbits, or cycles, for the system

$$x' = y, \quad y' = -ky - V'(x)$$

are possible only if $k = 0$.

10. Consider the system

$$x' = x(P - ax + by), \quad y' = y(Q - cy + dx),$$

where a, $c > 0$. Show that there cannot be periodic orbits in the first quadrant of the xy plane. (Hint: take $\beta = (xy)^{-1}$.)

11. Analyze the nonlinear system

$$\begin{aligned} x' &= y - x, \\ y' &= -y + \frac{5x^2}{4 + x^2}. \end{aligned}$$

12. (Project) Consider two competing species where one of the species immigrates or emigrates at constant rate h. The populations are governed by the dynamical equations

$$\begin{aligned} x' &= x(1 - ax) - xy, \\ y' &= y(b - y) - xy + h, \end{aligned}$$

where $a, b > 0$.

a) In the case $h = 0$ (no immigration or emigration) give a complete analysis of the system and indicate in a, b parameter space (i.e., in the ab plane) the different possible behaviors, including where bifurcations occur. Include in your discussion equilibria, stability, and so forth.

b) Repeat part (a) for various fixed values of h, with $h > 0$.

c) Repeat part (a) for various fixed values of h, with $h < 0$.

A
References

1. S. Axler, 1997. *Linear Algebra Done Right*, 2nd ed., Springer-Verlag, New York. (A good second course in linear algebra.)

2. F. Brauer & C. Castillo-Chavez, 2001. *Mathematical Models in Population Biology and Epidemiology*, Springer-Verlag, New York. (Concentrates on populations, disease dynamics, and resource management, for advanced undergraduates.)

3. N. Britton, 2003. *Essential Mathematical Biology*, Springer-Verlag, New York. (An introduction to mathematical biology with broad coverage.)

4. S. C. Chapra, 2005. *Applied Numerical Methods with* MATLAB *for Engineers and Scientists*, McGraw-Hill, Boston. (A blend of elementary numerical analysis with MATLAB instruction.)

5. R. V. Churchill, 1972. *Operational Mathematics*, 3rd ed., McGraw-Hill, New York. (This classic book contains many applications and an extensive table of transforms.)

6. F. Diacu, 2000. *An Introduction to Differential Equations*, W. H. Freeman, New York. (An elementary text with a full discussion of MATLAB, Maple, and Mathematica commands for solving problems in differential equations.)

7. C. H. Edwards & D. E. Penney, 2004. *Applications Manual: Differential Equations and Boundary Value Problems*, 3rd ed., Pearson Education, Upper Saddle River, NJ. (A manual with a detailed discussion and illustration of MATLAB, Maple, and Mathematica techniques.)

8. D. J. Higham & N. J. Higham, 2005. *MATLAB Guide*, 2nd ed., SIAM, Philadelphia. (An excellent source for MATLAB applications.).

9. M. W. Hirsch, S. Smale, & R. L. Devaney, 2004. *Differential Equations, Dynamical Systems, & An Introduction to Chaos*, Elsevier, New York. (A readable, intermediate text with an excellent format.)

10. D. Hughes-Hallet, et al. 2005. *Calculus: Single Variable*, 4th ed, John Wiley, New York. (Chapter 11 of this widely used calculus text is an excellent introduction to simple ideas in differential equations.)

11. W. Kelley & A. Peterson, 2003. *The Theory of Differential Equations*, Pearson Education, Upper Saddle River NJ. (An intermediate level text focuses on the theory of differential equations.)

12. J. H. Kwak & S. Hong, 2004. *Linear Algebra*, 2nd ed., Birkhauser, Boston. (A readable and thorough introduction to matrices and linear algebra.)

13. G. Ledder, 2005. *Differential Equations: A Modeling Approach*, McGraw-Hill, New York. (An introductory text that contains many interesting models and projects in science, biology, and engineering.)

14. J. D. Logan, 1997. *Applied Mathematics*, 2nd ed., Wiley-Interscience, New York. (An introduction to dimensional analysis and scaling, as well as to advanced techniques in differential equations, including regular and singular perturbation methods and bifurcation theory.)

15. J. D. Logan, 2004. *Applied Partial Differential Equations*, 2nd ed., Springer-Verlag, New York. (A very brief treatment of partial differential equations written at an elementary level.)

16. C. Neuhauser 2004. *Calculus for Biology and Medicine*, Pearson Education, Upper Saddle River, NJ. (A biology-motivated calculus text with chapters on differential equations.)

17. J. Polking, 2004. *pplane7* and *dfield7*, http://www.rice.edu/˜polking. (Outstanding, downloadable MATLAB m-files for graphical solutions of DEs.)

18. S. H. Strogatz, 1994. *Nonlinear Dynamics and Chaos*, Addison-Wesley, Reading, MA. (An excellent treatment of nonlinear dynamics.)

19. P. Waltman, 1986. *A Second Course in Elementary Differential Equations*, Academic, New York. (A classic, easily accessible, text on intermediate DEs.)

B
Computer Algebra Systems

There is great diversity in differential equations courses with regard to technology use, and there is equal diversity regarding the choice of technology. MATLAB, Maple, and Mathematica are common computer environments used at many colleges and universities. MATLAB, in particular, has become an important tool in scientific computation; Maple and Mathematica are computer algebra systems that are used for symbolic computation. There is also an add-on symbolic toolbox for the professional version of MATLAB; the student edition comes with the toolbox. In this appendix we present a list of useful commands in Maple and MATLAB. The presentation is only for reference and to present some standard templates for tasks that are commonly faced in differential equations. It is not meant to be an introduction or tutorial to these environments, but only a statement of the syntax of a few basic commands. The reader should realize that these systems are updated regularly, so there is danger that the commands will become obsolete quickly as new versions appear.

Advanced scientific calculators also permit symbolic computation and can perform many of the same tasks. Manuals that accompany these calculators give specific instructions that are not be repeated here.

B.1 Maple

Maple has single, automatic commands that perform most of the calculations and graphics used in differential equations. There are excellent Maple application manuals available, but everything required can be found in the help menu in the program itself. A good strategy is to find what you want in the help menu, copy and paste it into your Maple worksheet, and then modify it to conform to your own problem. Listed below are some useful commands for plotting solutions to differential equations, and for other calculations. The output of these commands is not shown; we suggest the reader type these commands in a worksheet and observe the results. There are packages that must be loaded before making some calculations: `with(plots):` `with(DEtools):` and `with(linalg):` In Maple, a colon suppresses output, and a semicolon presents output.

Define a function $f(t, u) = t^2 - 3u$:

```
f:=(t,u) → t^2-3*u;
```

Draw the slope field for the DE $u' = \sin(t - u)$:

```
DEplot(diff(u(t),t)=sin(t-u(t)),u(t),t=-5..5,u=-5..5);
```

Plot a solution satisfying $u(0) = -0.25$ superimposed upon the slope field:

```
DEplot(diff(u(t),t)=sin(t-u(t)),u(t),t=-5..5,
u=-5..5,[[u(0)=-.25]]);
```

Find the general solution of a differential equation $u' = f(t, u)$ symbolically:

```
dsolve(diff(u(t),t)=f(t,u(t)),u(t));
```

Solve an initial value problem symbolically:

```
dsolve({diff(u(t),t) = f(t,u(t)), u(a)=b}, u(t));
```

Plot solution to: $u'' + \sin u = 0$, $u(0) = 0.5$, $u'(0) = 0.25$.

```
DEplot(diff(u(t),t$2)+sin(u(t)),u(t),t=0..10,
[[u(0)=.5,D(u)(0)=.25]],stepsize=0.05);
```

Euler's method for the IVP $u' = \sin(t - u)$, $u(0) = -0.25$:

```
f:=(t,u) → sin(t-u):
t0:=0: u0:=-0.25: Tfinal:=3:
n:=10: h:=evalf((Tfinal-t0)/n):
t:=t0: u=u0:
for i from 1 to n do
u:=u+h*f(t,u):
t:=t+h:
print(t,u);
od:
```

Set up a matrix and calculate the eigenvalues, eigenvectors, and inverse:

```
with(linalg):
A:=array([[2,2,2],[2,0,-2],[1,-1,1]]);
eigenvectors(A);
eigenvalues(A);
inverse(A);
```

Solve a linear algebraic system:

$A\mathbf{x} = \mathbf{b}$:

```
b:=matrix(3,1,[0,2,3]);
x:=linsolve(A,b);
```

Solve a linear system of DEs with two equations:

```
eq1:=diff(x(t),t)=-y(t):
eq2:=diff(y(t),t)=-x(t)+2*y(t):
dsolve({eq1,eq2},{x(t),y(t)});
dsolve({eq1,eq2,x(0)=2,y(0)=1},{x(t),y(t)});
```

A fundamental matrix associated with the linear system $\mathbf{x}' = A\mathbf{x}$:

```
Phi:=exponential(A,t);
```

Plot a phase diagram in two dimensions:

```
with(DEtools):
eq1:=diff(x(t),t)=y(t):
eq2:=2*diff(y(t),t)=-x(t)+y(t)-y(t)^3:
DEplot([eq1,eq2],[x,y],t=-10..10,x=-5..5,y=-5..5,
{[x(0)=-4,y(0)=-4],[x(0)=-2,y(0)=-2] },
arrows=line, stepsize=0.02);
```

Plot time series:

```
DEplot([eq1,eq2],[x,y],t=0..10,
{[x(0)=1,y(0)=2] },scene=[t,x],arrows=none,stepsize=0.01);
```

Laplace transforms:

```
with(inttrans):
u:=t*sin(t):
U:=laplace(u,t,s):
U:=simplify(expand(U));
u:=invlaplace(U,s,t):
```

Display several plots on same axes:

```
with(plots):
p1:=plot(sin(t), t=0..6): p2:=plot(cos(2*t), t=0..6):
display(p1,p2);
```

Plot a family of curves:

```
eqn:=c*exp(-0.5*t):
curves:={seq(eqn,c=-5..5)}:
plot(curves, t=0..4, y=-6..6);
```

Solve a nonlinear algebraic system: `fsolve({2*x-x*y=0,-y+3*x*y=0},{x,y},{x=0.1..5,y=0..4});`

Find an antiderivative and definite integral:

```
int(1/(t*(2-t)),t); int(1/(t*(2-t)),t=1..1.5);
```

B.2 MATLAB

There are many references on MATLAB applications in science and engineering. Among the best is Higham & Higham (2000). The MATLAB files *dfield7.m* and *pplane7.m*, developed by J. Polking (2004), are two excellent programs for solving and graphing solutions to differential equations. These programs can be downloaded from his Web site (see references). In the table we list several common MATLAB commands. We do not include commands from the symbolic toolbox.

An m-file for Euler's Method. For scientific computation we often write several lines of code to perform a certain task. In MATLAB, such a code, or program, is written and stored in an **m-file**. The m-file below is a program of the Euler method for solving a pair of DEs, namely, the predator–prey system

$$x' = x - 2 * x^2 - xy, \quad y' = -2y + 6xy,$$

subject to initial conditions $x(0) = 1$, $y(0) = 0.1$. The m-file *euler.m* plots the time series solution on the interval $[0, 15]$.

```
function euler
x=1; y=0.1; xhistory=x; yhistory=y; T=15; N=200; h=T/N;
for n=1:N
u=f(x,y); v=g(x,y);
x=x+h*u; y=y+h*v;
xhistory=[xhistory,x]; yhistory=[yhistory,y];
end
t=0:h:T;
plot(t,xhistory,'-',t,yhistory,'--')
xlabel('time'), ylabel('prey (solid),predator (dashed)')
function U=f(x,y)
U=x-2*x.*x-x.*y;
function V=g(x,y)
V=-2*y+6*x.*y;
```

Direction Fields. The `quiver` command plots a vector field in MATLAB. Consider the system

$$x' = x(8 - 4x - y), \quad y' = y(3 - 3x - y).$$

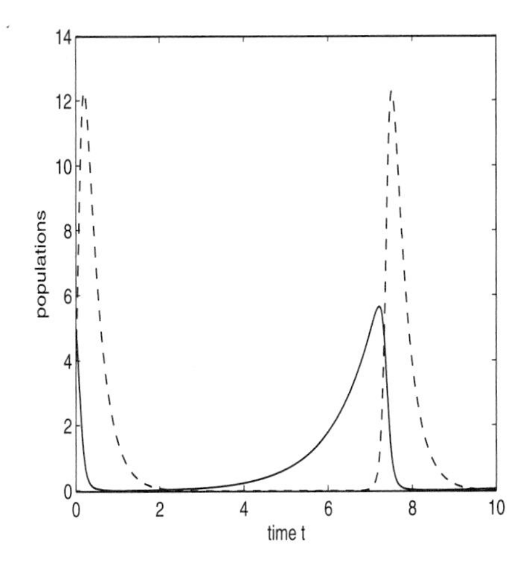

Figure B.1 Predator (dashed) and prey (solid) populations.

To plot the vector field on $0 < x < 3$, $0 < y < 4$ we use:

```
[x,y] = meshgrid(0:0.3:3, 0:0.4:4];
dx = x.*(8-4*x-y);  dy = y.*(3-3*x-y);
quiver(x,y,dx,dy)
```

Using the DE Packages. MATLAB has several differential equations routines that numerically compute the solution to an initial value problem. To use these routines we define the DEs in one m-file and then write a short program in a second m-file that contains the routine and a call to our equations from the first m-file. The files below use the package `ode45`, which is a Runge–Kutta solver with an adaptive stepsize. Consider the initial value problem

$$u' = 2u(1 - 0.3u) + \cos 4t, \quad 0 < t < 3, \quad u(0) = 0.1.$$

We define the differential equation in the m-file:

```
function uprime = f(t,u)
uprime = 2*u.*(1-0.3*u)+cos(4*t);
```

Then we run the m-file:

```
function diffeq
trange = [0  3]; ic=0.1;
[t,u] = ode45(@uprime,trange,ic);
plot(t,u,'*--')
```

Solving a System of DEs. As for a single equation, we first write an m-file that defines the system of DEs. Then we write a second m-file containing a routine that calls the system. Consider the Lotka–Volterra model

$$x' = x - xy, \quad y' = -3y + 3xy,$$

with initial conditions $x(0) = 5$, $y(0) = 4$. Figure B.1 shows the time series plots. The two m-files are:

```
function deriv=lotka(t,z)
deriv=[z(1)-z(1).*z(2); -3*z(2)+3*z(1).*z(2)];

function lotkatimeseries
tspan=[0 10]; ics=[5;4];
[T,X]=ode45(@lotka,tspan,ics);
plot(T,X)
xlabel('time t'), ylabel('populations')
```

Phase Diagrams. To produce phase plane plots we simply plot z(1) versus z(2). We draw two orbits. The main m-file is:

```
function lotkaphase
tspan=[0 10]; ICa=[5;4]; ICb=[4;3];
[ta,ya]=ode45(@lotka,tspan, ICa);
[tb,yb]=ode45(@lotka,tspan, ICb);
plot(ya(:,1),ya(:,2), yb(:,1),yb(:,2))
```

The following table contains several useful MATLAB commands.

MATLAB Command	Instruction
>>	command line prompt
;	seimcolon suppresses output
clc	clear the command screen
Ctrl+C	stop a program
help *topic*	help on MATLAB *topic*
a = 4, A = 5	assigns 4 to a and 5 to A
clear a b	clears the assignments for a and b
clear all	clears all the variable assignments
x=[0, 3,6,9,12,15,18]	vector assignment
x=0:3:18	defines the same vector as above
x=linspace(0,18,7)	defines the same vector as above
+, -, *, /, ∧	operations with numbers
sqrt(a)	square root of a
exp(a), log(a)	e^a and $\ln a$
pi	the number π
.*, ./, .∧	operations on vectors of same length (with dot)
t=0:0.01:5, x=cos(t), plot(t,x)	plots $\cos t$ on $0 \leq t \leq 5$
xlabel('time'), ylabel('state')	labels horizontal and vertical axes
title('Title of Plot')	titles the plot
hold on, hold off	does not plot immediately; releases hold on
for n=1:N,...,end	syntax for a "for-end" loop from 1 to N
bar(x)	plots a bar graph of a vector x
plot(x)	plots a line graph of a vector x
A=[1 2;3 4]	defines a matrix $\begin{pmatrix} 1 & 2 \\ 3 & 4 \end{pmatrix}$
x=A\b	solves Ax=b, where b=[α;β] is a column vector
inv(A)	the inverse matrix
det(A)	determinant of A
[V,D]=eig(A)	computes eigenvalues and eigenvectors of A
q=quad(fun,a,b,tol);	Approximates $\int_a^b \text{fun}(t)dt$, tol = error tolerance
function fun=f(t), fun=t.∧ 2	defines $f(x) = t^2$ in an m-file

C
Sample Examinations

Below are examinations on which students can assess their skills. Solutions are found on the author's Web site (see Preface).

Test 1 (1 hour)

1. Find the general solution to the equation $u'' + 3u' - 10u = 0$.

2. Find the function $u = u(t)$ that solves the initial value problem $u' = \frac{1+t^2}{t}, \quad u(1) = 0$.

3. A mass of 2 kg is hung on a spring with stiffness (spring constant) $k = 3$ N/m. After the system comes to equilibrium, the mass is pulled downward 0.25 m and then given an initial velocity of 1 m/sec. What is the amplitude of the resulting oscillation?

4. A particle of mass 1 moves in one dimension with *acceleration* given by $3 - v(t)$, where $v = v(t)$ is its velocity. If its initial velocity is $v = 1$, when, if ever, is the velocity equal to two?

5. Find $y'(t)$ if
$$y(t) = t^2 \int_1^t \frac{1}{r} e^{-r} dr.$$

6. Consider the initial value problem
$$u' = t^2 - u, \quad u(-2) = 0.$$

Use your calculator to draw the graph of the solution on the interval $-2 \leq t \leq 2$. Reproduce the graph on your answer sheet.

7. For the initial value problem in Problem 6, use the Euler method with stepsize $h = 0.25$ to estimate $u(-1)$.

8. For the differential equation in Problem 6, plot in the tu-plane the locus of points where the slope field has value -1.

9. At noon the forensics expert measured the temperature of a corpse and it was 85 degrees F. Two hours later it was 74 degrees. If the ambient temperature of the air was 68 degrees, use Newton's law of cooling to estimate the time of death. (Set up and solve the problem).

Test 2 (1 hour)

1. Consider the system
$$x' = xy, \quad y' = 2y.$$
Find a relation between x and y that must hold on the orbits in the phase plane.

2. Consider the system
$$x' = 2y - x, \quad y' = xy + 2x^2.$$
Find the equilibrium solutions. Find the nullclines and indicate the nullclines and equilibrium solutions on a phase diagram. Draw several interesting orbits.

3. Using a graphing calculator, sketch the solution $u = u(t)$ of the initial value problem
$$u'' + u' - 3\cos 2t = 0, \quad u(0) = 1, \quad u'(0) = 0$$
on the interval $0 < t < 6$.

4. Solve the initial value problem
$$u' - \frac{3}{t}u = t, \quad u(1) = 0.$$

5. Consider the autonomous equation
$$\frac{du}{dt} = -(u-2)(u-4)^2.$$
Find the equilibrium solutions, sketch the phase line, and indicate the type of stability of the equilibrium solutions.

6. Find the general solution to the linear differential equation

$$u'' - \frac{1}{t}u' + \frac{2}{t^2}u = 0.$$

7. Consider the two-dimensional linear system

$$\mathbf{x}' = \begin{pmatrix} 1 & 12 \\ 3 & 1 \end{pmatrix} \mathbf{x}.$$

   a) Find the eigenvalues and corresponding eigenvectors and identify the type of equilibrium at the origin.

   b) Write down the general solution.

   c) Draw a rough phase plane diagram, being sure to indicate the directions of the orbits.

8. A particle of mass $m = 2$ moves on a u-axis under the influence of a force $F(u) = -au$, where a is a positive constant. Write down the differential equation that governs the motion of the particle and then write down the expression for conservation of energy.

Test 3 (1 hour)

1. Find the equation of the orbits in the xy plane for the system $x' = 4y, \quad y' = 2x - 2$.

2. Consider a population model governed by the autonomous equation

$$p' = \sqrt{2}\,p - \frac{4p^2}{1+p^2}.$$

   a) Sketch a graph of the growth rate p' vs. the population p, and sketch the phase line.

   b) Find the equilibrium populations and determine their stability.

3. For the following system, for which values of the constant b is the origin an unstable spiral?

$$\begin{aligned} x' &= x - (b+1)y \\ y' &= -x + y. \end{aligned}$$

4. Consider the nonlinear system

$$\begin{aligned} x' &= x(1 - xy), \\ y' &= 1 - x^2 + xy. \end{aligned}$$

a) Find all the equilibrium solutions.

b) In the xy plane plot the x and y nullclines.

5. Find a solution representing a linear orbit of the three-dimensional system

$$\mathbf{x}' = \begin{pmatrix} 1 & 2 & 0 \\ 0 & 0 & -1 \\ 0 & 1 & 2 \end{pmatrix} \mathbf{x}.$$

6. Classify the equilibrium as to type and stability for the system

$$x' = x + 13y, \quad y' = -2x - y.$$

7. A two-dimensional system $\mathbf{x}x' = A\mathbf{x}$ has eigenpairs

$$-2, \begin{pmatrix} 1 \\ 2 \end{pmatrix}, \quad 1, \begin{pmatrix} 1 \\ 0 \end{pmatrix}.$$

a) If $\mathbf{x}(0) = \begin{pmatrix} 1 \\ 3 \end{pmatrix}$, find a formula for $y(t)$ (where $\mathbf{x}(t) = \begin{pmatrix} x(t) \\ y(t) \end{pmatrix}$.

b) Sketch a rough, but accurate, phase diagram.

8. Consider the IVP

$$\begin{aligned} x' &= -2x + 2y \\ y' &= 2x - 5y, \\ x(0) &= 3, \quad y(0) = -3. \end{aligned}$$

a) Use your calculator's graphical DE solver to plot the solution for $t > 0$ in the xy-phase plane.

b) Using your plot in (a), sketch $y(t)$ vs. t for $t > 0$.

Final Examination (2 hrs)

1. Find the general solution of the DE $u'' = u' + \frac{1}{2}u$.

2. Find a particular solution to the DE $u'' + 8u' + 16u = t^2$.

3. Find the (implicit) solution of the DE $u' = \frac{1+t}{3tu^2+t}$ that passes through the point $(1, 1)$.

4. Consider the autonomous system $u' = -u(u-2)^2$. Determine all equilibria and their stability. Draw a rough time series plot (u vs. t) of the solution that satisfies the initial condition $x(0) = 1$.

5. Consider the nonlinear system

$$x' = 4x - 2x^2 - xy, \quad y' = y - y^2 - 2xy.$$

Find all the equilibrium points and determine the type and stability of the equilibrium point $(2, 0)$.

6. An RC circuit has $R = 1$, $C = 2$. Initially the voltage drop across the capacitor is 2 volts. For $t > 0$ the applied voltage (emf) in the circuit is $b(t)$ volts. Write down an IVP for the *voltage* across the capacitor and find a formula for it.

7. Solve the IVP

$$u' + 3u = \delta_2(t) + h_4(t), \quad u(0) = 1.$$

8. Use eigenvalue methods to find the general solution of the linear system

$$\mathbf{x}' = \begin{pmatrix} 2 & 0 \\ -1 & 2 \end{pmatrix} \mathbf{x}.$$

9. In a recent TV episode of *Miami: CSI*, Horatio took the temperature of a murder victim at the crime scene at 3:20 A.M. and found that it was 85.7 degrees F. At 3:50 A.M. the victim's temperature dropped to 84.8 degrees. If the temperature during the night was 55 degrees, at what time was the murder committed? Note: Body temperature is 98.6 degrees; work in hours.

10. Consider the model $u' = \lambda^2 u - u^3$, where λ is a parameter. Draw the bifurcation diagram (equilibria solutions vs. the parameter) and determine analytically the stability (stable or unstable) of the branch in the first quadrant.

11. Consider the IVP $u'' = \sqrt{u + t}$, $u(0) = 3$, $u'(0) = 1$. Pick step size $h = 0.1$ and use the modified Euler method to find an approximation to $u(0.1)$.

12. A particle of mass $m = 1$ moves on the x-axis under the influence of a potential $V(x) = x^2(1 - x)$.

    a) Write down Newton's second law, which governs the motion of the particle.

    b) In the phase plane, find the equilibrium solutions. If one of the equilibria is a center, find the type and stability of all the other equilibria.

    c) Draw the phase diagram.

D

Solutions and Hints to Selected Exercises

CHAPTER 1

Section 1.1

2. Try a solution of the form $u = at^m$ and determine a and m.

4. Try a solution of the form $u = at^2 + bt + c$.

6. $u' = u/3 + 2te^{3t}$.

8. (a) linear, nonautonomous; (b) nonlinear, nonautonomous; (c) nonlinear, autonomous; (d) linear autonomous.

9. The derivative of $\sqrt{u}$ is $1/(2\sqrt{u})$, which is not continuous when $u = 0$.

12. The slope field is zero on $u = 0$ and $u = 4$.

13. The nullclines are the horizontal lines $u = \pm 1$. The slope field is -3 on the lines $u = \pm 2$.

15. The nullclines are $u = 0$ and $u = 4\sqrt{t}$.

16. Hint: at a positive maximum $u'' < 0$, and so $u'' - u < 0$, a contradiction.

17. Show the derivative of the expression is zero. In the uu' plane the curves plot as a family of hyperbolas.

18. Use the quotient rule to show the time derivative of u_1/u_2 is zero.

Section 1.2

1. $u = \frac{1}{2}\sin(t^2) + C$. And $u(0) = C = 1$.

2. $u = \frac{2}{3}t^{3/2} + 2\sqrt{t} + C$ is the general solution.

5. $u(t) = \int_1^t e^{-s}\sqrt{s}ds$.

6 . $y = -\frac{1}{4}e^{-4t} + C$.

8. $\frac{d}{dt}(\text{erf}(\sin t)) = \text{erf}'(\sin t)\cos t = \frac{2}{\sqrt{\pi}}e^{-\sin^2 t}\cos t$.

9. (a) If the equation is exact, then $f = h_t$ and $g = h_u$. Then $f_u = h_{tu} = h_{ut} = g_t$. (b)(i) $f_u = 3u^2 = g_t$, and so the equation is exact. Then $h_t = u^3$ implies $h = tu^3 + \phi(u)$. Then $h_u = 3tu^2 + \phi'(u) = 3tu^2$. Hence, $\phi'(u) = 0$, or $\phi(u) = C_1$. Therefore $h = tu^3 + C_1 = C_2$, or $tu^3 = C$.

11. Take the derivative and use the fundamental theorem of calculus to get $u' = -2e^{-t} + tu$, $u(0) = 1$.

Section 1.3.1

1. Use $k = mg/L$.

3. The equation is $mv' = -F$, $v(0) = V$ with solution $v = -(F/m)t + V$. Then $x = \int v dt = -(F/2m)t^2 + Vt$.

6. Mass times acceleration equals force, or $ms'' = -mg\sin\theta$. But $s = l\theta$, so $ml\theta'' = -mg\sin\theta$.

7. (a) $\omega = \sqrt{g/l}$. (b) 2.2 sec.

8. The x and y positions of the cannonball are $x = (v\cos\theta)t$, $y = -\frac{1}{2}gt^2 + (v\sin\theta)t + H$, where θ is the angle of elevation of the cannon.

Section 1.3.2

1. The equilibria are $p = 0$ (stable), $p = a$ (unstable), $p = K$ (stable).

2. Find the equilibria by setting $-\frac{r}{K}p^2 + rp - h = 0$ and use the quadratic formula. We get a positive equilibrium only when $r \geq 4h/K$. If $r = 4h/K$ the single equilibrium is semi-stable, and if $r > 4h/K$ the smaller equilibrium is unstable and the larger one is stable.

3. r has dimensions 1/time, and a has dimensions 1/population. The maximum growth rate occurs at $p = 1/a$. There is no simple formula for the antiderivative $\int \frac{e^{ap}}{p}dp$.

4. Maximum length is a/b.

6. (a) $2\sqrt{u} = t+C$; (b) $u = \ln\sqrt{2t+C}$; (c) $u = \tan(t+C)$; (d) $u = Ce^{3t} + a/3$; (e) $4\ln u + 0.5u^2 = t + C$; (f) $\frac{\sqrt{\pi}}{2}\,\mathrm{erf}(u^2) = t + C$.

9. The amount of carbon-14 is $u = u_0 e^{-kt}$; 13,301 years.

10. 35,595.6 years.

11. $N' = bF_T N(1 - cN/F_T)$ (a logistics type equation with carrying capacity F_T/c).

12. $R = k$ is asymptotically stable. The solution is $R(t) = k\exp(Ce^{-at})$.

13. $p = m$ is unstable. If $p(0) < m$ then population becomes extinct, and if $p(0) > m$ then it blows up.

14. $I' = aI(N - I)$, which is logistics type. The asymptotically stable equilibrium is $I = N$.

Section 1.3.3

1. $p' = 0.2p(1 - p/40) - 1.5$ with equilibria $p = 10$ (unstable) and $p = 30$ (stable). If $p(0) \leq 10$ then the population becomes extinct.The population will likely approach the stable equilibrium $p = 30$.

2. (a) $u = 0$ (unstable), $u = 3$ (stable). (c) $u = 2$ (stable), $u = 4$ (unstable).

3. (a) Equilibria are $u = 0$ and $u = h$. We have $f_u(u) = h - 2u$, and so $f_u(0) = h$ and $f_u(h) = -h$. If $h > 0$ then $u = 0$ is unstable and $u = h$ is stable; if $h < 0$ then $u = 0$ is stable and $u = h$ is unstable. If $h = 0$ there is no information from the derivative condition. A graph shows $u = 0$ is semi-stable.

7. Hint: Plot h vs. u instead of u vs. h.

Section 1.3.4

1. $h = \ln\sqrt{11/4}$. The solid will be 2° at time $2\ln(1/11)/\ln(4/11)$.

2. About 1.5 hours.

5. k is 1/time, q is degrees/time, and θ, T_e, T_0 are in degrees. The dimensionless equation is $\frac{d\psi}{d\tau} = -(\psi - 1) + ae^{-b/\psi}$, with $b = \theta/T_e$ and $a = q/kT_e$.

Section 1.3.5

1. $C(t) = (C_0 - C_{\text{in}})e^{-qt/V} + C_{\text{in}}$.

2. The equation is $100C' = (0.0002)(0.5) - 0.5C$.

5. The equilibrium $C^* = (-q + \sqrt{q^2 + 4kqVC_{\text{in}}})/2kV$ is stable.

7. $C' = -kVC$ gives $C = C_0 e^{-kVt}$. The residence time is $T = -\ln(0.1)/kV$.

8. $C' = -\frac{aC}{b+C} + R$ has a stable equilibrium $C^* = Rb/(a - R)$, where $a > R$. The concentration approaches C^*.

9. $a' = -ka(a - a_0 + b_0)$; $a \to a_0 - b_0$.

10. (b) Set the equations equal to zero and solve for S and P. (c) With values from part (b), maximize aVP_e.

Section 1.3.6

1. $q(t) = 6 - e^{-2t}$, $I(t) = 2e^{-2t}$.

3. The initial condition is $I'(0) = (-RI(0) + E(0))/L$. The DE is $LI'' + RI' + (1/C)I = E'(t)$.

5. Substitute $q = A\cos\omega t$ into $Lq'' + (1/C)q = 0$ to get $\omega = 1/\sqrt{LC}$, A arbitrary.

6. $I'' + \frac{1}{2}(I^2 - 1)I' + I = 0$.

7. $LCV_c'' + RCV_c' + V_c = E(t)$.

CHAPTER 2

Section 2.1

1. (c) $u = \tan(\frac{1}{2}t^2 + t + C)$. (d) $\ln(1 + u^2) = -2t + C$.

2. $u = \ln(\frac{t^3}{3} + 2)$; interval of existence is $(-6^{1/3}, \infty)$.

4. The interval of existence is $(-\sqrt{5/3}, \sqrt{5/3})$.

6. The general solution is $(u-1)^{1/3} = t^2 + C$. If $u = 1$ for all t, then $t^2 + C = 0$ for all t, which is impossible.

8. Integrate both sides of the equation with respect to t. For example, $\int (u_1'/u_1)dt = \ln u_1 + C$.

9. $y' = \frac{F(y)-y}{t}$. Using $y = u/t$ the given DE can be converted to $y' = \frac{4-y^2}{ty}$, which is separable.

10. $u(t) = -e^{-3t} + Ce^{-2t}$.

11. $u(r) = -\frac{p}{4}r^2 + a\ln r + b$.

12. $u(t) = u_0\exp(-\frac{at^2}{2})$. The maximum rate of conversion occurs at time $t = 1/\sqrt{a}$.

13. The IVP is $v' = -32 - v^2/800$, $v(0) = 160$.

14. The governing equation is $u' = -k/u$, with solution $u(x) = \sqrt{C - 2kt}$.

16. $u(t) = u_0 \exp(\int_a^t p(s)ds)$.

17. m is 1/time, b is 1/grasshoppers, and a is 1/(time · spiders). The dimensionless equation is $\frac{dh}{d\tau} = -h - \frac{\lambda h}{1+h}$. The population h approaches zero as $t \to \infty$. Separate variables.

18. (a) $1 - e^{-mt}$; $e^{-ma} - e^{-mb}$. (b) $S(t) = \exp(\int_0^t m(s)ds)$.

Section 2.2

1. $u(t) = \frac{1}{3}t^2 + \frac{C}{t}$.

2. $u(t) = Ce^{-t} + \frac{1}{2}e^t$.

4. $q(t) = 10te^{-5t}$. The maximum occurs at $t = 1/5$.

5. The equation becomes $y' + y = 3t$, which is linear.

6. The equation for y is $y' = 1 + y^2$. Then $y = \tan(t + C)$, $u = \tan(t + C) - t$.

7. $u(t) = e^{t^2}(C + \frac{\sqrt{\pi}}{2}\,\mathrm{erf}(t))$.

8. $u(t) = (u_0 + q/p)e^{pt} - q/p$.

11. (b) Let $y = u^{-1}$. Then $y' = -y - e^{-t}$ and $y = Ce^{-t} - \frac{1}{2}e^t$. Then $u = (Ce^{-t} - \frac{1}{2}e^t)^{-1}$. (c) $y' = -3y/t + 3/t$; $y = 1 + C/t^3$; $u = y^{1/3}$.

12. The vat empties at time $t = 60$. The governing equation is $(60 - t)u' = 2 - 3u$.

14. $S^* = raM/(aM + rA)$.

15. $x' = kx(N - x)$, which is similar to the logistics equation.

16. The IVP is $T' = -3(T - 9 - 10\cos(2\pi t))$, $T(0) = 12$.

18. $P' = N - P$. The stable equilibrium is $P = N$, so everyone hears the rumor.

19. $mv' = mg - \frac{2}{t+1}v$, $v(0) = 0$.

20. (a) $S^* = IP/(I + E)$. (b) The equation is linear: $S' = -\frac{1}{P}(I + E)S + I$. The general solution is $S(t) = S^* + Ce^{-\frac{t}{P}(I+E)}$.(c) Use the formula for S^* for each island and compare.

21 . Ex.2: $u' + u = e^t$ has integrating factor e^t. Multiply by the factor to get $(ue^t)' = e^{2t}$; integrate to get $ue^t = \frac{1}{2}e^{2t} + C$.

Section 2.3

1. The Picard iteration scheme is $u_{n+1}(t) = \int_0^t (1 + u_n(s)^2)ds$, $u_0(t) = 0$. It converges to $\tan t = t + \frac{1}{3}t^3 + \frac{2}{15}t^5 + \cdots$.

2. The Picard iteration scheme is $u_{n+1}(t) = 1 + \int_0^t (s - u_n(s))ds$, $u_0(t) = 1$. We get $u_1(t) = 1 - t + t^2/2 + \cdots$, etc.

Section 2.4

2. $u(t) = e^{\sin t}$.

4. The exact solution is $u_n = u_0(1-hr)^n$. If $h > 1/r$ then the solution oscillates about zero, but the solution to the DE is positive and approaches zero. So we require $h < 1/h$.

7. $u = e^{-t}$. Roundoff error causes the exponentially growing term Ce^{5t} in the general solution to become significant.

CHAPTER 3

Section 3.1

1. $V(x) = x^2/2 - x^4/4$.

2. $mx''x = F(x)x'$. But $\frac{d}{dt}V(x) = \frac{dV}{dx}x' = -F(x)x'$ and $mx''x' = m\frac{d}{dt}(x')^2 = 2mx'x''$. Hence $\frac{1}{2}m\frac{d}{dt}(x')^2 = -\frac{d}{dt}V(x)$. Integrating both sides gives $\frac{1}{2}m(x')^2 = V(x) + C$, which is the conservation of energy law.

5. (a) Make the substitution $v = x'$. The solution is $x(t) = a/t + b$. (b) Make the substitution $v = x'$. The solution is $\int \frac{dx}{a+0.5x^x} = t + b$.

Section 3.2

1. (a) $u = e^{2t}(a + bt)$. (d) $u(t) = a\cos 3t + b\sin 3t$. (e) $u(t) = a + be^{2t}$. (f) $u(t) = a\cosh(\sqrt{12}t) + b\sinh(\sqrt{12}t)$.

4. The solution is periodic if $a = 0$ and $b > 0$; the solution is a decaying oscillation if $a < 0$ and $a^2 < b$; the solution decays without oscillation if $a < 0$ and $a^2 \geq b$.

5. $L = \frac{1}{4}$ (critically damped), $L < \frac{1}{4}$ (over damped), $L > \frac{1}{4}$, (under damped).

6. Critically damped when $9a^2 = 4b$, which plots as a parabola in ab parameter space.

7. $u'' + 2u' - 24u = 0$.

8. $u'' + 6u' + 9u = 0$.

9. $u'' + 16u = 0$.

10. $A = 2$, $B = 0$.

12. $I(t) = \sqrt{10}\sin(t/\sqrt{10})$.

Section 3.3.1

1. (a) $u_p = at3 + bt2 + ct + d$. (b) $u_p = a$. (d) $u_p = a\sin 7t + b\cos 7t$. (f) $u_p = (a + bt)e^{-t}\sin\pi t + (c + dt)e^{-t}\cos\pi t$.

2. (c) $u_p = t^2 - 2t + 2$. (e) $u_p = \frac{9}{2}e^{-t}$.

4. The general solution is $u(t) = c_1 + c_2e^{2t} - 2t$.

5. Write $\sin^2 t = \frac{1}{2}(1 - \cos 2t)$ and then take $u_p = a + b\cos 2t + c\sin 2t$.

7. $u(t) = a\cos 50t + b\sin 50t + 0.02$.

8. The circuit equation is $2q'' + 16q' + 50q = 110$.

Section 3.3.2

2. $u(t) = c_1\cos 4t + c_2\sin 4t + \frac{1}{32}\cos 4t + \frac{1}{8}t\sin 4t$.

3. $L = 1/C\beta^2$.

Section 3.4

1. (b) $u(t) = 2\ln t$.

2. $\beta = 1$.

7. $u(t) = \tan(t + \pi/4)$.

9. The other solution is te^{at}.

11. $\frac{1}{\sqrt{t}}\sin t$.

14. Take the derivative of the Wronskian expression $W = u_1u_2' - u_1'u_2$ and use the fact that u_1 and u_2 are solutions to the differential equation to show $W' = -p(t)W$. Solving gives $W(t) = W(0)\exp(-\int p(t)dt)$, which is always of one sign.

16. The given Riccati equation can be transformed into the Cauchy–Euler equation $u' - \frac{3}{t}u' = 0$.

17. (b) $u_p = -\cos t\ln((1 + \sin t)/\cos t)$. (c) Express the particular solution in terms of integrals. (e) $u_p = t^3/3$.

18. (a) $tp(t) = t \cdot t^{-1} = 1$, and $t^2q(t) = t^2(1 - \frac{k^2}{t^2}) = t^2 - k^2$, which are both power series about $t = 0$.

Section 3.5

2. $u(x) = -\frac{1}{6}x^3 + \frac{1}{240}x^4 + \frac{100}{3}x$. The rate that heat leaves the right end is $-Ku'(20)$ per unit area.

4. There are no nontrivial solutions when $\lambda \leq 0$. There are nontrivial solutions $u_n(x) = \sin n\pi x$ when $\lambda_n = n^2\pi^2$, $n = 1, 2, 3, ...$

5. $u(x) = 2\pi\sqrt{x}$.

6. Integrate the steady-state heat equation from 0 to L and use the fundamental theorem of calculus. This expression states: the rate that heat flows in at $x = 0$ minus the rate it flows out at $x = L$ equals the net rate that heat is generated in the bar.

7 . When ωL is not an integer multiple of π.

8. $\lambda = -1 - n^2\pi^2$, $n = 1, 2, \ldots$.

10. Hint: this is a Cauchy–Euler equation. Consider three cases where the values of λ give characteristic roots that are real and unequal, real and equal, and complex.

Section 3.6

1. (a) $u(t) = c_1 + c_2 \cos t + c_3 \sin t$. (b) $u(t) = c_1 + e^{t/2}(c_2 \cos \frac{\sqrt{3}}{2}t + c_3 \sin \frac{\sqrt{3}}{2}t) + c_4 e^{-t} + t$. (c) $u(t) = c_1 + c_2 t + c_3 \cos t + c_4 \sin t$.

5. $u(t) = e^{3t}(c_1 \cos t + c_2 \sin t) + te^{3t}(c_3 \cos t + c_4 \sin t)$.

Section 3.7

1. (b) $u(t) = Ce^{Rt}$. (c) $u(t) = Ce^{-\sin t} - 1$. (e) $u(t) = \sqrt{t}(c_1 \cos(\frac{\sqrt{7}}{2} \ln t) + c_2 \sin(\frac{\sqrt{7}}{2} \ln t))$. (g) $u(t) = -4t^2 + 6t + C$. (i) A Bernoulli equation. (k) $x(t) = \pm \int \frac{dt}{\sqrt{A-2t}} + B$. (l) Bernoulli equation. (m) $\pm \int_0^u \frac{dw}{\sqrt{A-2w}} = t + B$. (n) Homogeneous equation. (p) Exact equation.

2. $u(t) = (\frac{1}{2} - \sin t)^{-1}$, $-7\pi/6 < t < \pi/6$.

5. If $a \leq 0$ then $u = a$ is the only equilibrium (unstable). If $0 < a < 1$ then there are three equilibria: $u = \pm\sqrt{a}$ (unstable), and $u = a$ (stable). If $a = 1$ then $u = 1$ is unstable. If $a > 1$ then $u = \sqrt{a}$ (stable), $u = a$, $-\sqrt{a}$ (unstable).

6. $r(t) = -kt + r_0$.

7. $p = 0$ (unstable); $p = K$ (stable).

11. $u(t) = \exp(t^2 + C/t^2)$.

12. $u(t) = t - 3t\ln t + 2t^2$.

CHAPTER 4

Section 4.1

1. $U(s) = \frac{1}{s}(e^{-s} - e^{-2s})$.

2. Integrate by parts twice.

3. $L[\sin t] = \frac{1}{1+s^2}$; $L[\sin(t - \pi/2)] = -\frac{s}{1+s^2}$; $L[h_{\pi/2}(t)\sin(t - \pi/2)] = \frac{s}{1+s^2}e^{-\pi s/2}$.

4. $\frac{2!}{(s+3)^3}$.

5. Use $\sinh kt = \frac{1}{2}(e^{kt} - e^{-kt})$.

7. $\frac{1}{s+1}e^{-2(s+1)}$.

9. e^{t^2} is not of exponential order; the improper integral $\int_0^\infty \frac{1}{t}e^{-st}dt$ does not exist at $t = 0$. Neither transform exists.

11. $\frac{1}{s}\frac{1}{1+e^{-s}}$.

12. Hint: Use the definition of the Laplace transform and integrate by parts using $e^{-st} = -\frac{1}{s}\frac{d}{dt}(e^{-st})$.

13. Hint: $\frac{d}{ds}U(s) = \int_0^\infty u(t)\frac{d}{ds}e^{-st}dt$.

14. $\ln(\sqrt{(s+1)/(s-1)})$, $s > 1$.

15. Integrate by parts.

17. (a) Integrate by parts. Change variables in the integral to write $\Gamma(\frac{1}{2}) = 2\int_0^\infty e^{-r^2}dr$.

Section 4.2

2. (c) $\frac{1}{3}t^3e^{5t}$. (d) $7h_4(t)$.

3. (c) $u(t) = \frac{3}{5}e^{3t} + \frac{7}{5}e^{-2t}$. (d) $u(t) = e^t\sin t$. (f) $u(t) = 1$. (i) $\frac{3}{2}\cosh(\sqrt{2}t) - \frac{1}{2}$.

4. Solve for the transforms $X = X(s)$, $Y = Y(s)$ in $sX = X - 2Y - \frac{1}{s^2}$, $sY = 3X + Y$, and then invert.

5. $\frac{d^n}{ds^n}U(s) = \int_0^\infty u(t)\frac{d^n}{ds^n}e^{-st}dt$.

Section 4.3

1. $\frac{1}{2}t\sin t$.

2. $\frac{1}{12}t^4$.

3. $u(t) = u(0)e^{at} + \int_0^t e^{a(t-s)}q(s)ds$.

5. $u(t) = \frac{1}{\omega}\int_0^t \sinh(t-s)f(s)ds$.

7. $u(t) = \int_3^t f(t-s)ds$.

8. $u(t) = \int_0^t (e^{t-s} - 1)f(s)ds$.

9. $U(s) = \frac{F(s)}{1-K(s)}$.

10. (a) $u(t) = \sin(t)$; (b) $u(t) = 0$.

11. The integral is a convolution.

Section 4.4

1. $\frac{2}{s}(e^{-3s} - e^{-4s})$.

2. $e^{-3s}(\frac{2}{s^3} + \frac{6}{s^2} + \frac{9}{s})$.

3. $\frac{1}{6}t^3e^{2t}$.

4. Note $f(t) = 3 - h_2(t) + 4h_\pi(t) - 6h_7(t)$.

5. $t - h_4(t)(t-4)$.

7. $u(t) = e^t$ on $[0,1]$; $u(t) = \frac{e-2}{e}e^t + 2$ on $t > 1$.

8. Solve $q'' + q = t + (9-t)h_9(t)$.

10. $u(t) = 1$, $0 \le t \le 1$; $u(t) = -\cos\pi t$, $t > 1$.

13. $2 - \sum_{n=1}^{\infty}(1)^n h_n(t)$.

Section 4.5

1. $1/e^2$.

3. $u(t) = \sinh(t-5)h_5(t)$.

4. $u(t) = \sin(t-2)h_2(t)$.

5. $u(t) = h_2(t) + \delta_3(t)$.

6. $u(t) = \frac{1}{2}\cos[2(t-2)]h_2(t) - \cos[2(t-5)]h_5(t)$.

7. $v(t) = \sum_{n=0}^{\infty}\sin(t-n\pi)h_{n\pi}(t)$.

CHAPTER 5

Section 5.1

1. The orbit is an ellipse (taken counterclockwise).

2. The tangent vector is $x'(t) = (2, -3)^T e^t$ and it points in the direction $(2, -3)^T$.

3. $x(t) = 8 + 2e^{-5t}$, $y(t) = 8 - 8e^{-5t}$. Over a long time the solution approaches the point (equilibrium) $(8, 8)$.

4. Multiply the first equation by $1/W$, the second by $1/V$, and then add to get $x'/W + y'/V = 0$, or $x/W + y/V = C$. Use $y = V(C - x/W)$ to eliminate y from the first equation to get a single equation in the variable x, namely, $x' = -q(1/V + 1/W)x + qC$. The constant C is determined from the initial condition.

6. Solve the equation $x'' + \frac{1}{2}x' + 2x = 0$ to get a decaying oscillation $x = x(t)$. In the phase plane the solution is a clockwise spiral entering the origin.

7. The system is $q' = I$, $I' = -4q$. We have $q(t) = 8\cos 2t$ and $I(t) = -16\sin 2t$. Both q and I are periodic functions of t with period π, and in the phase plane $q^2/64 + I^2/256 = 1$, which is an ellipse. It is traversed clockwise.

Section 5.2

1. $\det(A - \lambda I) = \lambda^2 - 5\lambda - 2$.

2. $x = 3/2$, $y = 1/6$.

4. $\det(A - \lambda I) = \lambda^2 - 5\lambda - 2 = 0$, so $\lambda = \frac{5}{2} \pm \frac{1}{2}\sqrt{33}$.

5. $\det A = 0$ so A^{-1} does not exist.

6. If $m = -5/3$ then there are infinitely many solutions, and if $m \neq -5/3$, no solution exists.

7. $m = 1$ makes the determinant zero.

8. Use expansion by minors.

10. $\det(A) = -2$, so A is invertible and nonsingular.

11. $\mathbf{x} = a(2, 1, 2)^T$, where a is any real number.

12. Set $c_1(2, -3)^T + c_2(-4, 8)^T = (0, 0)^T$ to get $2c_1 - 4c_2 = 0$ and $-3c_1 + 8c_2 = 0$. This gives $c_1 = c_2 = 0$.

13. Pick $t = 0$ and $t = \pi$.

14. Set a linear combination of the vectors equation to the zero vector and find coefficients c_1, c_2, c_3.

16. $\mathbf{r}_1(t)$ plots as an ellipse; $\mathbf{r}_2(t)$ plots as the straight line $y = 3x$. $\mathbf{r}_2(t)$ plots as a curve approaching the origin along the direction $(1,1)^{\mathrm{T}}$. Choose $t = 0$ to get $c_1 = c_3 = 0$, and then choose $t = 1$ to get $c_2 = 0$.

Section 5.3

1. For A the eigenpairs are 3, $(1,1)^{\mathrm{T}}$ and 1, $(2,1)^{\mathrm{T}}$. For B the eigenpairs are 0, $(3,-2)^{\mathrm{T}}$ and -8, $(1,2)^{\mathrm{T}}$. For C the eigenpairs are $\pm 2i$, $(4, 1 \mp i)^{\mathrm{T}}$.

2. $\mathbf{x} = c_1(1,5)^{\mathrm{T}}e^{2t} + c_2(2,-4)^{\mathrm{T}}e^{-3t}$. The origin has saddle point structure.

3. The origin is a stable node.

4. (a) $\mathbf{x} = c_1(-1,1)^T e^{-t} + c_2(2,3)^T e^{4t}$ (saddle), (c) $\mathbf{x} = c_1(-2,3)^T e^{-t} + c_2(1,2)^T e^{6t}$ (saddle), (d) $\mathbf{x} = c_1(3.1)^T e^{-4t} + c_2(-1,2)^T e^{-11t}$ (stable node), (f) $x(t) = c_1 e^t(\cos 2t - \sin 2t) + c_2 e^t(\cos 2t + \sin 2t)$, $y(t) = 2c_1 e^t \cos 2t + 2c_2 e^t \sin 2t$ (unstable spiral), (h) $x(t) = 3c_1 \cos 3t + 3c_2 \sin 3t$, $y(t) = -c_1 \sin 3t + c_2 \cos 3t$ (center).

6. (a) Equilibria consist of the entire line $x - 2y = 0$. (b) The eigenvalues are 0 and 5; there is a linear orbit associated with 5, but not 0.

7. The eigenvalues are $\lambda = 2 \pm \sqrt{a+1}$; $a = -1$ (unstable node), $a < -1$ (unstable spiral), $a > -1$ (saddle).

9. The eigenvalues are never purely imaginary, so cycles are impossible.

11. There are many matrices. The simplest is a diagonal matrix with -2 and -3 on the diagonal.

13. The system is $v' = w$, $w' = -(1/LC)v - (R/L)w$. The eigenvalues are $\lambda = -R/2L \pm \sqrt{R^2/4L^2 - 1/LC}$. The eigenvalues are complex when $R^2/4L < 1/C$, giving a stable spiral in the phase plane, representing decaying oscillations in the system.

14. The eigenvalues are $-\gamma \pm i$. When $\gamma = 0$ we get a cycle; when $\gamma > 0$ we get a stable spiral; when $\gamma < 0$ we get an unstable spiral.

Section 5.4

2. The equations are $V_1 x' = (q+r)c - qx - rx$, $V_2 y' = qx - qy$. The steady-state is $x = y = c$. When freshwater enters the system, $V_1 x' = -qx - rx$, $V_2 y' = qx - qy$. The eigenvalues are both negative ($-q$ and $-q-r$), and therefore the solution decays to zero. The origin is a stable node.

5. A fundamental matrix is

$$\Phi(t) = \begin{pmatrix} 2e^{-4t} & -e^{-11t} \\ 3e^{-4t} & 2e^{-11t} \end{pmatrix}.$$

The particular solution is $\mathbf{x}_p = -(\frac{9}{42}, \frac{1}{21})^{\mathrm{T}} e^{-t}$.

6. $\det A = r_2 r_3 > 0$ and $\mathrm{tr}(A) = r_1 - r_2 - r_3 < 0$. So the origin is asymptotically stable and both x and y approach zero. The eigenvalues are $\lambda = \frac{1}{2}(\mathrm{tr}(A) \pm \frac{1}{2}\sqrt{\mathrm{tr}(A)2 - 4 \det A}$.

7. In the equations in Problem 6, add D to the right side of the first (x') equation. Over a long time the system will approach the equilibrium solution: $x_e = D/(r_1 + r_2 + r_1 r_3/r_2)$, $y_e = (r_1/r_2)x_e$.

Section 5.5

1. The eigenpairs of A are $2, (1,0,0)^{\mathrm{T}}$; $6, (6,8,0)^{\mathrm{T}}$; $-1, (1,-1,7/2)^{\mathrm{T}}$. The eigenpairs of C are $2, (1,0,1)^{\mathrm{T}}$; $0, (-1,0,1)^{\mathrm{T}}$; $1, (1,1,0)^{\mathrm{T}}$.

2(a). $\mathbf{x} = c_1 \begin{pmatrix} 1 \\ 1 \\ -2 \end{pmatrix} e^{-2t} + c_2 \begin{pmatrix} 3 \\ -3 \\ 2 \end{pmatrix} e^{4t} + c_3 \begin{pmatrix} -1 \\ 1 \\ 0 \end{pmatrix} e^{2t}.$

2(b). $\mathbf{x} = c_1 \begin{pmatrix} 2 \\ 1 \\ 2 \end{pmatrix} + c_2 \begin{pmatrix} \cos 0.2t \\ \sin 0.2t \\ -\cos 0.2t - \sin 0.2t \end{pmatrix} + c_3 \begin{pmatrix} -\sin 0.2t \\ \cos 0.2t \\ -\cos 0.2t + \sin 0.2t \end{pmatrix}.$

2(d). $\mathbf{x} = c_1 \begin{pmatrix} 1 \\ 0 \\ 1 \end{pmatrix} e^{2t} + c_2 \begin{pmatrix} -1 \\ 0 \\ 1 \end{pmatrix} + c_3 \begin{pmatrix} 1 \\ 1 \\ 0 \end{pmatrix} e^{t}.$

4. The eigenvalues are $\lambda = 2, \rho \pm 1$.

CHAPTER 6

Section 6.1.1

1. $y = Ce^{1/x}$, $x(t) = (c_1 - t)^{-1}$, $y(t) = c_2 e^{-t}$.

2. $y = \frac{1}{x^2+C}$. There are no equilibrium solutions. No solutions can touch the x-axis.

3. Two equilibrium points: $(\sqrt{4/5}, 2\sqrt{4/5}), (-\sqrt{4/5}, -2\sqrt{4/5})$. The vector field is vertical on the circle of radius 2: $x^2 + y^2 = 4$. The vector field is horizontal on the straight line $y = 2x$.

4. $(\pm 1, -1)$. The vector field is vertical on the line $y = -1$ and horizontal on the inverted parabola $y = -x^2$.

5. $(0, 0)$, $(1, 1)$, and $(4, 4)$.

6. $(0, n\pi)$, $n = 0, \pm 1, \pm 2, ...$

7. $dV/dt = 2xx' + 2yy' = -2y^4 < 0$.

8. The equilibria are the entire line $y = x$; they are not isolated.

Section 6.1.2

1. There is no epidemic. The number of infectives decreases from its initial value.

2. $I(t)$ increases to a maximum value, then $S(t)$ decreases to the value S^*.

3. $r = 1/3$ and $a = 0.00196$. The average number of days to get the flu is about 2.5 days.

4. $r = 0.25$ and $a = 0.001$, giving $S^* = 93$. Also $I_{max} = 77$.

5. $I' = aI(N - I) - rI$. The equilibrium is $I = aN - r$.

8. $x' = rx - axy, \quad y' = -my + bxy - M$.

9. The equilibria are $(0, 0)$ and $(m/b, k/a)$. The vector field shows that curves veer away from the nonzero equilibrium, so the system could not coexist in that state.

10. $S' = -aSI - \nu, \quad aSI - rI$.

11. $S' = -aSI + \mu(N - S - I), \quad I' = aSI - rI$. The equilibria are $(N, 0)$ and the endemic state $(r/a, I^*)$ where $I^* = \mu(N - r/a)/(r + \mu)$.

Section 6.2

4. Begin by writing the equation as $u' = v, \quad v' = -9u + 80 \cos 5t$.

6. The system is $x' = y, \quad y' = -x/2 + y/2 - y^3/2$.

7. The system is $x' = y, \quad y' = -2(x^2 - 1)y - x$.

8. $S(0) = 465$ with $a = 0.001$ and $r = 0.2$.

9. $S' = -aSI, \quad I' = aSI - rI - qI$.

Section 6.3

1. $y = C(e^x - 1)$.

2. $y^2 - x^2 - 4x = C$.

3. Equilibria are $(0,0)$ (a saddle structure) and $(2,4)$ (stable node) and null-clines: $y = x^2$ and $y = 2x$.

4. $a < 0$ (no equilibria); $a = 0$ (origin is equilibrium); $a > 0$ (the equilibria are $(-\sqrt{a}/2, 0)$ and $(\sqrt{a}/2, 0)$, a stable node and a saddle).

6. $(-1,0)$ (stable spiral); $(1,0)$ (saddle).

8. $(2,4)$ (saddle); $(0,0)$ (stable node). The Jacobian matrix at the origin has a zero eigenvalue.

10. $\mathrm{tr}(A) < 0, \quad \det A > 0$. Thus the equilibrium is asymptotically stable.

11. See Exercise 7, Section 6.1.1.

12. The force is $F = -1 + x^2$, and the system is $x' = y, \quad y' = -1 + x^2$. The equilibrium $(1,0)$ is a saddle and $(-1,0)$ is a center. The latter is determined by noting that the orbits are $\frac{1}{2}y^2 + x - \frac{1}{3}x^3 = E$.

13. (a) $\frac{dH}{dt} = H_x x' + H_y y' = H_x H_y + H_y(-H_x) = 0$. (c) The Jacobian matrix at an equilibrium has zero trace. (e) $H = \frac{1}{2}y^2 - \frac{x^2}{2} + \frac{x^3}{3}$.

14. $(0,0)$ is a center.

15. (c) The eigenvalues of the Jacobian matrix are never complex.

16. $(0,0)$, $(0, \frac{1}{2})$, and $(K, 0)$ are always equilibria. If $K \geq 1$ or $K \leq \frac{1}{2}$ then no other positive equilibria occur. If $\frac{1}{2} < K \leq 1$ then there is an additional positive equilibrium.

17. $a = 1/8$ (one equilibrium); $a > 1/8$, (no equilibria); $0 < a < 1/8$ (two equilibria).

19. The characteristic equation is $\lambda^2 = f'(x_0)$. The equilibrium is a saddle if $f'(x_0) > 0$.

Section 6.4

2. There are no equilibrium, and therefore no cycles.

3. $f_x + g_y > 0$ for all x, y, and therefore there are no cycles (by Dulac's criterion).

4. $(1,0)$ is always a saddle, and $(0,0)$ is unstable node if $c > 2$ and an unstable spiral if $c < 2$.

6. $(0,0)$ is a saddle, $(\pm 1, 0)$ are stable spirals.

7. The equilibria are $H = 0$, $P = \phi/a$ and $H = \frac{\varepsilon\phi}{c} - \frac{a}{b}$, $P = \frac{c}{\varepsilon b}$.

8. In polar coordinates, $r' = r(a - r^2)$, $\theta' = 1$. For $a \leq 0$ the origin is a stable spiral. For $a > 0$ the origin is an unstable spiral with the appearance of a limit cycle at $r = \sqrt{a}$.

9. The characteristic equation is $\lambda^2 + k\lambda + V''(x_0) = 0$ and has roots $\lambda = \frac{1}{2}(-k \pm \sqrt{k^2 - 4V''(x_0)})$. These roots are never purely imaginary unless $k = 0$.

10. Use Dulac's criterion.

11. Equilibria at $(0,0)$, $(1,1,)$, and $(4,4)$.

Index

(continued from page ii)

Frazier: An Introduction to Wavelets Through Linear Algebra
Gamelin: Complex Analysis.
Gordon: Discrete Probability.
Hairer/Wanner: Analysis by Its History. *Readings in Mathematics.*
Halmos: Finite-Dimensional Vector Spaces. Second edition.
Halmos: Naive Set Theory.
Hämmerlin/Hoffmann: Numerical Mathematics. *Readings in Mathematics.*
Harris/Hirst/Mossinghoff: Combinatorics and Graph Theory.
Hartshorne: Geometry: Euclid and Beyond.
Hijab: Introduction to Calculus and Classical Analysis.
Hilton/Holton/Pedersen: Mathematical Reflections: In a Room with Many Mirrors.
Hilton/Holton/Pedersen: Mathematical Vistas: From a Room with Many Windows.
Iooss/Joseph: Elementary Stability and Bifurcation Theory. Second edition.
Irving: Integers, Polynomials, and Rings: A Course in Algebra
Isaac: The Pleasures of Probability. *Readings in Mathematics.*
James: Topological and Uniform Spaces.
Jänich: Linear Algebra.
Jänich: Topology.
Jänich: Vector Analysis.
Kemeny/Snell: Finite Markov Chains.
Kinsey: Topology of Surfaces.
Klambauer: Aspects of Calculus.
Lang: A First Course in Calculus. Fifth edition.
Lang: Calculus of Several Variables. Third edition.
Lang: Introduction to Linear Algebra. Second edition.
Lang: Linear Algebra. Third edition.
Lang: Short Calculus: The Original Edition of "A First Course in Calculus."
Lang: Undergraduate Algebra. Third edition
Lang: Undergraduate Analysis.
Laubenbacher/Pengelley: Mathematical Expeditions.
Lax/Burstein/Lax: Calculus with Applications and Computing. Volume 1.
LeCuyer: College Mathematics with APL.
Lidl/Pilz: Applied Abstract Algebra. Second edition.
Logan: Applied Partial Differential Equations, Second edition.
Logan: A First Course in Differential Equations.
Lovász/Pelikán/Vesztergombi: Discrete Mathematics.
Macki-Strauss: Introduction to Optimal Control Theory.
Malitz: Introduction to Mathematical Logic.
Marsden/Weinstein: Calculus I, II, III. Second edition.
Martin: Counting: The Art of Enumerative Combinatorics.
Martin: The Foundations of Geometry and the Non-Euclidean Plane.
Martin: Geometric Constructions.
Martin: Transformation Geometry: An Introduction to Symmetry.
Millman/Parker: Geometry: A Metric Approach with Models. Second edition.
Moschovakis: Notes on Set Theory.
Owen: A First Course in the Mathematical Foundations of Thermodynamics.
Palka: An Introduction to Complex Function Theory.
Pedrick: A First Course in Analysis.
Peressini/Sullivan/Uhl: The Mathematics of Nonlinear Programming.

Prenowitz/Jantosciak: Join Geometries.
Priestley: Calculus: A Liberal Art. Second edition.
Protter/Morrey: A First Course in Real Analysis. Second edition.
Protter/Morrey: Intermediate Calculus. Second edition.
Pugh: Real Mathematical Analysis.
Roman: An Introduction to Coding and Information Theory.
Roman: Introduction to the Mathematics of Finance: From Risk Management to Options Pricing.
Ross: Differential Equations: An Introduction with Mathematica®. Second edition.
Ross: Elementary Analysis: The Theory of Calculus.
Samuel: Projective Geometry. *Readings in Mathematics.*
Saxe: Beginning Functional Analysis
Scharlau/Opolka: From Fermat to Minkowski.
Schiff: The Laplace Transform: Theory and Applications.
Sethuraman: Rings, Fields, and Vector Spaces: An Approach to Geometric Constructability.
Sigler: Algebra.
Silverman/Tate: Rational Points on Elliptic Curves.
Simmonds: A Brief on Tensor Analysis. Second edition.
Singer: Geometry: Plane and Fancy.
Singer/Thorpe: Lecture Notes on Elementary Topology and Geometry.
Smith: Linear Algebra. Third edition.
Smith: Primer of Modern Analysis. Second edition.
Stanton/White: Constructive Combinatorics.
Stillwell: Elements of Algebra: Geometry, Numbers, Equations.
Stillwell: Elements of Number Theory.
Stillwell: Mathematics and Its History. Second edition.
Stillwell: Numbers and Geometry. *Readings in Mathematics.*
Strayer: Linear Programming and Its Applications.
Toth: Glimpses of Algebra and Geometry. Second Edition. *Readings in Mathematics.*
Troutman: Variational Calculus and Optimal Control. Second edition.
Valenza: Linear Algebra: An Introduction to Abstract Mathematics.
Whyburn/Duda: Dynamic Topology.
Wilson: Much Ado About Calculus.

Chemical Ecology of Vertebrates

Chemical Ecology of Vertebrates is the first book to focus exclusively on the chemically mediated interactions between vertebrates, including fish, amphibians, reptiles, birds, and mammals, and other animals, and plants. Reviewing the latest research in three core areas: pheromones (where the interactions are between members of the same species), interspecific interactions involving allomones (where the sender benefits) and kairomones (where the receiver benefits) This book draws information into a coherent whole from widely varying sources in many different disciplines. Chapters on the environment, properties of odour signals, and the production and release of chemosignals set the stage for discussion of more complex behavioral topics. While the main focus is ecological, dealing with behavior and interactions in the field, it also covers chemoreception, orientation and navigation, the development of behavior, and the practical applications of chemosignals.

Dietland Müller-Schwarze is Professor of Environmental Biology at the State University of New York.

W.H. Freeman Publishing:
Fig. 1.5. From: Moen, A. (1973). *Wildlife Ecology*, p. 71.

The Herpetologists' League, Inc.:
Fig. 7.4. From: Greene, M.J. and Mason, R. T. (2000). *Herpetologica* **56**, 166–175.

Journal of Herpetology:
Fig. 4.8. From: Chelazzi, G. and Delfino, G. (1986). *Journal of Herpetology* **20**, 451–455.

McGraw Hill Education:
Figs. 3.4 and 3.5. From: Duellmann, W. and Trueb, L. (1986). *Biology of Amphibians*.

Nature Publishing:
Fig. 2.3: From: Böcskei, Z. *et al.* (1992). *Nature* **360**, 186–188.

Oxford University Press:
Fig. 5.9b: From: Rasmussen, L.E.L. and Hultgren, B. (1990). *Chemical Signals in Vertebrates*, vol. 5, p. 155.
Fig. 12.4. From: Müller-Schwarze, D. (1990). Chemical Signals in Vertebrates, vol. 5, p. 591.

Sigma Xi, The Scientific Research Society:
Fig. 6.8. From: Peters, R.P. and Mech, L.D. (1975). *American Scientist* **63**, 628–637.

Smithsonian Institution:
Fig. 3.3. From: Weitzman, S.H. and Fink (1985). *Smithsoniam Contributions to Zoology*, **421**, 1–121.

Springer (including Academic Press, Plenum):
Figs. 1.1 and 1.2. From: Regnier, F.E. and Goodwin, M. (1977). *Chemical Signals in Vertebrates*, vol. 1, 115–133.
Fig. 5.9a. From: Stoddart, M. (1983). *The Chemical Ecology of Vertebrate Olfaction*.
Fig. 4.7. From: Benvenuti, S. *et al.* (1992). in: *Chemical Signals in Vertebrates*, vol. 6, 429–434. New York: Plenum.
Fig. 4.13. From: Wiltschko, R. and Wiltschko, W. (1992). *Chemical Signals in Vertebrates*, vol. 6, p. 437.
Fig. 6.14. From: Müller-Schwarze, D. (1992). *Chemical Signals in Vertebrates*, vol. 6, p. 460.
Fig. 7.9. From: Frisch, K. v. (1941). *Zeitschrift für vergleichende Physiologie* **29**, 46–145.
Fig. 8.4. From: Bronson, F. H. and Coquelin, A. (1980). *Chemical Signals in Vertebrates*, vol. 2, p. 256.

Fig. 12.2. *Journal of Chemical Ecology* **21** (1995) p. 1357, Responses of beaver (Castor canadensis Kuhl) to predator chemicals, Engelhart, A. and D. Muller-Schwarze, fig 4. With kind permission of Springer Science and Business Media

Thomson Publishing Services:

Fig. 5.2. From: Bond, Carl E. (1979, 1996). Biology of Fishes.

John Wiley and Sons:

Fig. 3.8. From: Albone, E.S. (1984). Mammalian Semiochemistry, p. 43.

Fig. 5.5. From: Albone, E.S. (1984). Mammalian Semiochemistry, p. 245.

Fig 10.8. With kind permission of the Zoological Society of London.

1

The odorsphere: the environment for transmission of chemical signals

The scent of flowers does not go against the wind, not sandal, rosebay or jasmine, but the scent of the good goes against the wind; a good man is wafted to all quarters.

SUTTAPITAKA ("basket of discourse"), from Pali canon of Theravada Buddhists, ca. 500–250 BC

Land animals exploit the odorsphere, the world of vapors around them. In any given locale, they move in an odorscape, a landscape of volatiles. Even in fish we speak of odors because neurophysiologically the olfactory system is involved, even though water-soluble stimulants are not necessarily volatile. We expect vertebrates to have taken advantage evolutionarily of the physicochemical characteristics of their environment first to select and then to optimize chemical communication. The chemical communication system of a cold-water fish differs vastly from that of a tropical bat. Despite similar biological functions, each system has been shaped by, and is adapted to, a distinct set of environmental circumstances.

In air, temperature, relative humidity, barometric pressure, and air currents not only modulate the movement of molecules from the source but also affect odor reception once the molecules have arrived near the receptors. The evaporation of an odor from a surface such as animal skin, a scent mark, or vegetation is regulated by air temperature, relative humidity, the porosity of the surface, and other compounds present (Regnier and Goodwin, 1977; Figs. 1.1 and 1.2).

The evolution of chemical communication was probably influenced by such additional factors as adsorption of aerial pheromones to vegetation, or waterborne pheromones to suspended clay. The influence of these environmental features has very likely selected for both the choice of chemical constituents of the signals and the appropriate signal-emission behaviors (Gleeson, 1978).

Chemical signals have several advantages over cues in other sensory modalities. They work in darkness, around obstacles, and may last for a long time, ranging from seconds to months. This enables an animal to communicate with others in its absence, or even with

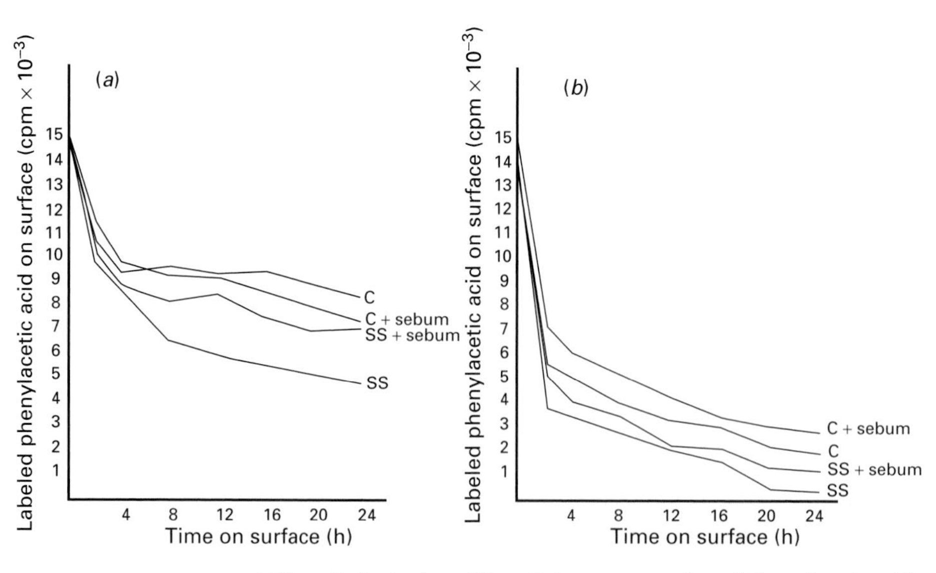

FIGURE 1.1 Effect of relative humidity of air on evaporation of phenylacetic acid from two surfaces in the presence and absence of sebum: stainless steel (SS) and cellulose (C). Relative humidity 0% (*a*) or 100% (*b*), both at 20°C. (Adapted from Regnier and Goodwin, 1977.)

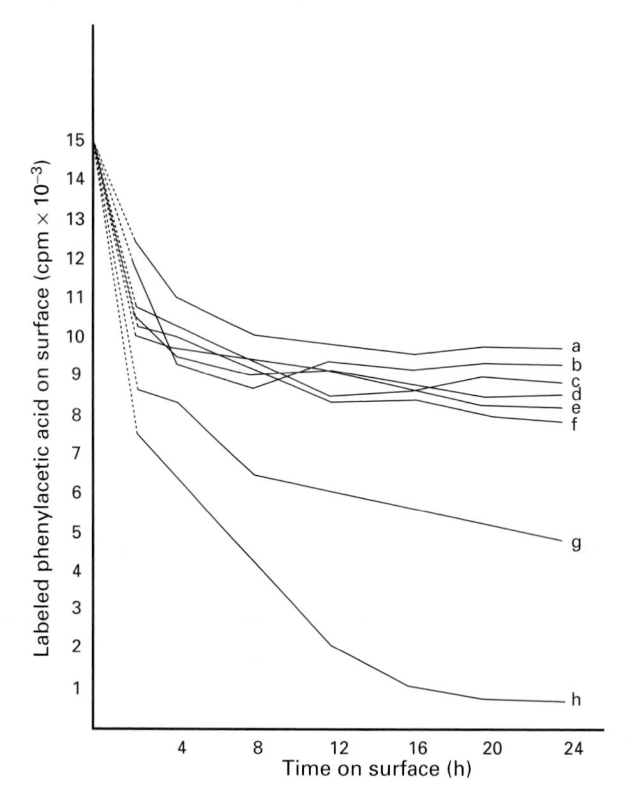

FIGURE 1.2 Effect of surface on evaporation of phenylacetic acid at 0% relative humidity and 20°C. a, montmorillonite; b, mylar plastic; c, glass; d, cellulose; e, kaolinite; f, balsa wood; g, stainless steel; h, platinum. (Adapted from Regnier and Goodwin, 1977.)

itself when returning to a previously scented site. A major disadvantage is a long "fade-out" time that would limit consecutive signals to a slow sequence. In the real world, air or water currents are nearly always present. They transport signals, and the role of diffusion is insignificant because it is too slow.

1.1 Air

For the medium air we can distinguish *close-range* from *long-distance* chemical signals. They differ in the behaviors that animals employ to take advantage of physical processes. Sniffing a scent mark, another animal, or potential food at close range is more than a passive receiving. Exhaling moist air onto the focus of attention moistens the surface and liberates more odorant. Thus, sniffing enables an animal to regulate the release of molecules from the substrate. At zero distance, the odor source can be licked to take up non-volatile compounds. An example of long-distance responses is the approach of seabirds to food odors such as dimethyl sulfide over the ocean. Here air currents and water–air interface transport of molecules are important.

Since virtually all communication by terrestrial vertebrates takes place in the air layer near the ground, the chemical ecologist has to understand the properties of that stratum. Volatiles in higher layers of air, however, may attract vultures to carcasses and could possibly serve as cues for migrating birds. Airborne odors are affected by temperature, relative humidity, barometric pressure, air currents, and vegetation, which, in turn, may influence temperature, humidity, and airflow. Complex interactions of these variables create countless unique environmental conditions for communication patterns of diverse species and for various specific purposes.

1.1.1 Temperature

The vapor pressure of volatile compounds, measured in atmospheres (or millimeters of mercury), varies with temperature. For example, the vapor pressure of acetone increases from 200 to 400 mmHg with a temperature rise from +20 to +40 °C, and that of *n*-heptane from 100 to 400 mmHg with a temperature change from +40 to +80 °C (e.g. Adams *et al.*, 1970). The half-lives of several acetates decreased by two- to fourfold when the temperature was raised from 20 to 30 °C (McDonough *et al.*, 1989). In temperate latitudes, temperatures can vary from about 40 to 0 °C within 24 hours. Therefore, it is important to know the vapor pressure of a given compound for the ambient temperatures under which a particular animal species operates. Diurnal and nocturnal animals may have selected different signal compounds (or mixtures). Do polar and tropical species differ in their choice of compounds for communication? Have cold-climate pheromones higher vapor pressures? Do polar and montane animals use behavioral means such as sniffing at close range to optimize chemoreception? Are there environmental temperatures so low that pheromone communication is impractical, or even not feasible? The animal's ability to escape from ambient conditions by burrowing or seeking other shelter has also to be considered.

The sex pheromone of the male pig is adapted to the ambient temperature. The pheromone, a mixture of 5α-androstenol and 5α-androstenone, is bound to the protein "pheromaxein" in the saliva. At body temperature (+37 °C) most binding is lost after 72 hours. In direct encounters with females, ample pheromone is given off. At +4 °C, however, the binding of the pheromone to the protein is unchanged over 168 hours. Pigs deposit the frothy saliva in the environment during their breeding season in autumn and winter. At the prevailing low temperatures, the pheromone is released only slowly over a longer time (Booth, 1987).

Temperature may significantly affect *chemoreception*. For instance, electrical responses to amylacetate delivered to olfactory receptors of a tortoise, *Gopherus polyphemus*, were little affected by air temperatures between 20 and 30 °C at the nares but changed considerably above and below that range. Up to +35 °C and down to +10 °C, the olfactory response was a "monotonic slowly decreasing function of temperature" (Tucker, 1963; see also Grundvig *et al.*, 1967).

1.1.2 Humidity

The higher the humidity, the more odor molecules evaporate from a surface, because they compete with water molecules for surface sites. Dogs track better on moist ground and/or on humid days, and we are all familiar with the smell of a wet dog. This phenomenon has far-reaching consequences for the choice of compounds and communication patterns in humid versus arid climates. Some field studies have demonstrated behavior differences between wet and dry seasons. For example, the blackbuck, *Antilope cervicapra,* ceases to use dungpiles during the monsoon season (Prasad, 1989). Environments with high temperatures and humidity, such as tropical forests, call for range marks with active compounds that are either large molecules (Alberts, 1992a) or contain effective keeper substances that slowly release volatiles.

An animal can control release of odor molecules from body surfaces, especially skin gland areas, up to a point. But in moist air, volatiles from secretions on skin or hair will evaporate more easily. If only intermittent odor emission is desirable, humidity would interfere with the animal's odor release. Glands that produce such intermittent signals, such as alarm odors, appear to be more developed in species or subspecies in drier climates. An example is the metatarsal gland in the North American deer genus *Odocoileus*. Within one species, the size of the gland, or even its presence or absence, varies with the climate. In populations of white-tailed deer, *Odocoileus virginianus*, the gland becomes smaller (or even absent) from eastern North America to more humid Central America and northern South America. In the western mule deer, *Odocoileus hemionus*, the gland is larger in more arid environments. Finally, comparing the two species, the mule deer of the more arid western North America has a larger metatarsal gland than the white-tailed deer of the more humid east (Müller-Schwarze, 1987). Similarly, of two species of the Indian gerbil genus *Tatera*, the one in a drier climate has a ventral gland, while it is smaller or absent from the species of a more humid environment (Prakash and Idris, 1982; Kumari and Prakash, 1983).

It is important to measure humidity exactly where an animal operates. In a meadow on a summer afternoon, the relative humidity can be over 90% at 5 cm above ground, but only 60% at 60 cm (Geiger, 1965). The implications of this would be different for a deer or a rodent. Furthermore, a species like a deer beds down on the moister ground but applies scent marks on branches in a higher, drier layer of air. The rodent, by comparison, experiences a higher and stable level of humidity in its burrow or cave. Humidity is high in stands of plants, and different plants offer different conditions. For instance, forest edges promote precipitation from fog. East-facing forest edges are in the wind and rain shadow and, therefore, experience less humidity.

Mammalian odor *reception* is modulated by relative humidity. For instance, neotropical bats (*Carollia perspicillata* and *Phyllostomus discolor*) are less able to approach an experimental banana odor correctly if the humidity is lowered from the normal 75% to under 60%. The sensory impairment results from drying out of the nasal mucosa in low humidity (Laska *et al.*, 1986).

It is well known that dogs track better in humid air. Rodents find buried seeds better in wet soil. This is important in arid climates. After rains, yellow pine chipmunks, *Tamias amoenus*, and deer mice, *Peromyscus maniculatus* found experimentally buried seeds of Jeffrey pine, *Pinus jeffreyi*, and antelope bitterbrush, *Purshia tridentata*, better than in dry soil. The recovered number of seeds increased 27- and 15-fold, respectively. In wet soil, seeds take up water rapidly and emanate volatile organic compounds that the rodents exploit. By extension, variations in humidity in arid environments may have profound effects on olfaction-dependent behaviors such as finding food, social interactions, preying, and predator avoidance (Vander Wall 1998).

Rodent species differ in their ability to smell buried seeds: those from arid climates perform better than species from mesic climates. Specifically, Panamint kangaroo rats, *Dipodomys panamintinus*, from arid and semiarid areas of the Great Basin Desert in North America were the only species that found deep caches of seeds under dry conditions. By contrast, chipmunks from more humid eastern North America performed the poorest in finding buried sunflower seeds under dry conditions (Vander Wall *et al.*, 2003). Like the much-studied pesticides, volatiles adsorb to soil particles and desorb from theses particles when the moisture exceeds the thickness of a monomolecular layer, increasing the vapor pressure; this, in turn, facilitates finding buried seeds (Vander Wall, 2003). Furthermore, rodents, use memory as well as odors to find buried seed caches. As moisture favors searching by smell, pilfering occurs more often after rains (Vander Wall, 2000).

1.1.3 Barometric pressure

Hyperbaric pressure may intensify odors or render odoriferous some "odorless" gases such as methane. Professional divers, experimentally exposed to hyperbaric pressures, detected odors of krypton and methane when sniffing these during the decompression phase of a dive. The threshold for krypton was 2 ATA (atmosphere absolute), and 100% positive responses occurred at 6 ATA. For methane, the threshold was 3 ATA (100%: 13 ATA). The thresholds of individuals differed by as much as a factor of three (Laffort and Gortan, 1987).

1.1.4 Air currents

Odors travel in moving air. During their long evolutionary history, animals have adapted to detecting chemical signals from downwind. Many animals integrate chemotaxis with photo-, anemo-, or rheotaxis (Vickers, 2000). Several natural history accounts of large mammals report movements into the wind and presumably toward rain and fresh forage. According to French Camel Corps reports, dromedaries are said to detect water pools and fresh pasture from 40–60 km away. These animals turn into wind blowing from rain clouds and will head into that direction if permitted by the rider (Gauthier-Pilters, 1974). African water buffalo, elephant, and zebra are assumed to "smell rain" and migrate there (Daly, 1988). Arabian oryx, *Oryx leucoryx*, are reported to do the same over distances of about 50 km. One female was documented to have traveled 150 km to an area of fresh rain (Daly, 1988).

Factors affecting currents

Turbulent flow

The air flow in the microclimate of an animal's home range usually is more complicated than basic laminar flow (which does not even exist over large distances and open areas). Fluctuations in air velocity cause turbulence, and stationary objects may create such fluctuations. The air is almost always *turbulent*, and intact parcels of air laden with odor molecules travel in random fashion (Geiger, 1965). The multidirectional turbulence is superimposed on the horizontal wind flow pattern. *Mechanical* turbulence is caused by wind, and *thermal* or *convective* turbulence by heated air rising. The plume from a smokestack loops up and down on a hot day. Such a "looping plume" is shaped by thermal up- and downdrafts, which account for more vertical displacement than the small-scale mechanical turbulence (Thibodeaux, 1979). Clearings in dense forest experience eddies spinning off from the general flow over the treetops, while a sparse stand of trees has a more laminar air flow. Small eddies are typical for the air directly above the air–soil interface (Thibodeaux, 1979). Eddy diffusion varies in space and time: it is less intense and more variable near the ground, and the greatest mixing of air occurs during midday while at night the air is more stable. An exception is the thin layer of air that adheres to the ground, walls, or vegetation. Processes here follow the laws of molecular physics, and not those of eddy diffusion (Geiger, 1965). In most atmospheric odor movements, however, turbulence swamps molecular diffusion. Among insects, turbulent air flow around obstacles such as trees influences the response of a predatory beetle, *Rhizophagus grandis*, to the odor of their prey, bark beetles (Wyatt *et al.*, 1993).

The more stable the air, the higher the concentrations of odor that can be carried over long distances. This, in turn, increases potential communication distances. However, unstable air conditions disperse the odor molecules more "sideways," with a wider "cone;" consequently more individuals can be reached, although the signal is attenuated over a shorter distance. "Parcels" of odor-containing air travel straight, but subsequent parcels travel at different angles, as the wind direction changes over time (David *et al.*, 1982). Fluctuations in speed and direction ("meandering") expose an organism that is fixed in place to changes in odor concentration, including zero levels. This amounts to an on–off effect that, in turn, counteracts habituation of the animal's chemical senses.

Mountains

In the mountains, differential heating during the day causes upslope winds, which move up both sides of a valley, and upvalley winds that blow along the valley. During the night, the directions reverse to downslope and downvalley winds (Geiger, 1965; Fig. 1.3).

Large areas

Pressing problems of air pollution have spawned experiments and models of odor dispersion over larger areas. Strom (1976) and Beaman (1988)

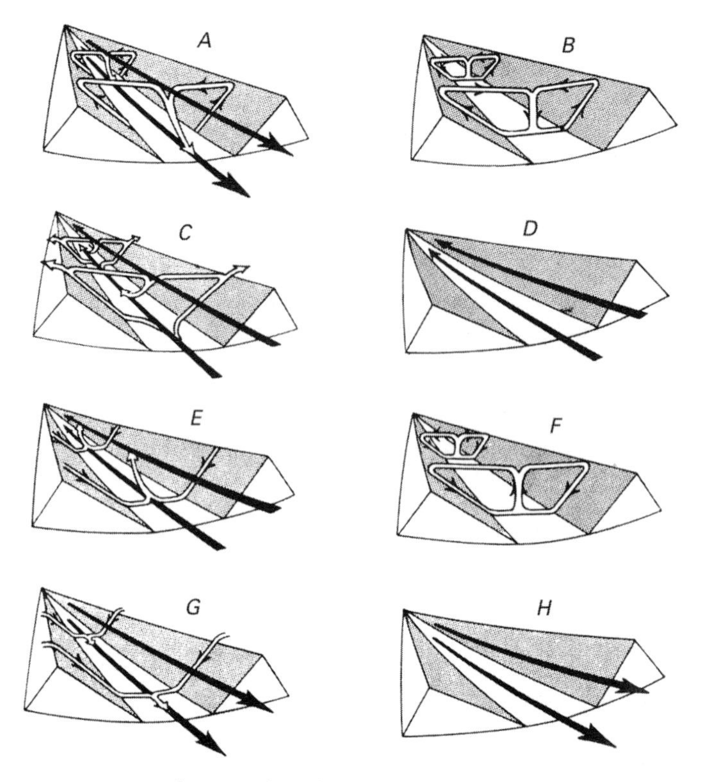

FIGURE 1.3 Changes of wind directions in a valley during 24 hours (from A [night] to H [night again]). Black arrows indicate valley winds and white arrows indicate slope winds. (From Geiger, 1965.)

analyzed transport of stack effluents. Beaman concluded that models of gas dispersion downstream from a source such as a factory predict the actual gas distribution better in rural than in urban areas.

Fluctuations in concentration in eddies

The changes in odor concentration during transport depend on eddy size. Large eddies carry small puffs of odor intact, with little concentration change over time (and distance). Eddies smaller than the puff let the puff grow slowly and concentration decreases, but slowly. Puffs of about the same size as eddies are torn apart, and concentration decreases rapidly. Mixtures of compounds of different molecular sizes will retain their relative concentrations in turbulent flows. Therefore, odor mixtures such as those from plants can contain compounds of high and low molecular weights, even though their separate diffusion rates would be different.

How have animals adapted their communication patterns to prevailing wind patterns? Air flows may be variable, as over slopes, valleys, wood slopes, and forest edges; continuous and strong, as on grasslands, tundra, or the Antarctic;

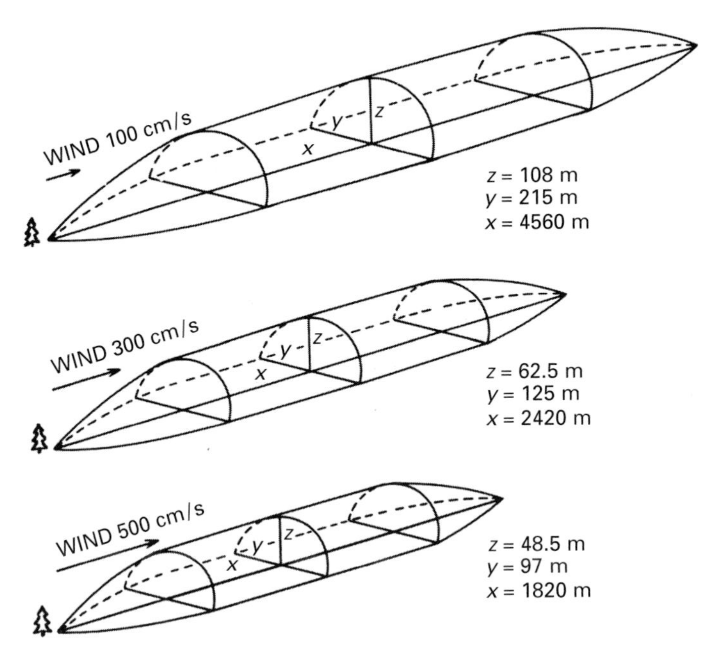

FIGURE 1.4 Active space for different wind speeds. A single female gypsy moth at the tree on the left emits sex attractant. The space where the concentration of pheromone is above threshold (for a searching male) is defined by the dimensions x, y, and z. The active space shrinks with increasing wind velocity. (From Wilson, 1970.)

minimal, as in woods; or absent, as in caves and burrows. In general, animals are expected to rely on wind only for longer-range attraction to food or conspecifics, avoidance, or warning of conspecifics or predators. Finer details, such as the exact outline of a territory, or the identity of an individual, can be assessed by sniffing scent marks or the animals themselves at close range. In some environments, air flow patterns can be very complex, and we have not even asked how vertebrates exploit these for effective communication. For instance, how do nocturnal downvalley winds and the water current along streams affect odor signals used by beaver? Active at night, beaver probably use these for signal propagation between neighbor colonies. They cannot easily exploit the daytime upvalley wind, because they stay in their lodge during the day.

Active space

The biologically significant *active space* where the odor concentration is above threshold is shaped like an overturned boat (Fig. 1.4). (If the molecules were able to spread in all directions, as from an elevated odor source, the active space would assume the shape of a cylinder with pointed ends.)

To calculate the active space, the investigator needs to measure odor emission rate, detection threshold of the animal, and wind speed.

Plume models

Pheromone propagation by wind depends on the release rate of the pheromone (or any other odor) and air movements (turbulent dispersion). In wind, the turbulent diffusivity overwhelms the diffusion properties of a volatile compound or mixture itself. Diffusion properties are now properties of wind structure and boundary surfaces, and preferably termed *dispersion coefficients*. Two models have dominated the discussion of insect pheromone propagation. These are the *time-average model* (Sutton, 1953) and the *Gaussian plume model*.

The time-average model considers the average concentration of airborne materials at sites downwind from a point source. The concentration (or density D) of a pheromone at any one point with the coordinates x (downwind direction), y (horizontal crosswind [transverse] dimension), and z (vertical dimension) can be estimated with the following formula.

$$D_{xyz} = \frac{2Q}{C_y C_z U x^{2-n}} \exp\left[-x^{n-2}\left(\frac{y^2}{C_y^2} + \frac{z^2}{C_z^2}\right)\right]$$

where Q is the release rate; U the mean wind speed; C_y and C_z are horizontal and vertical dispersion coefficients, respectively; and n a parameter ranging between 0 and 1. Wilson and Bossert (1963) have applied this model to pheromones. Dispersion coefficients are functions of atmospheric turbulence, terrain roughness, and vertical wind speed profile. According to Sutton (1953), with light winds of speeds between 100 and 500 cm/s, neutral atmospheric conditions, and level ground: $C_y = 0.4\,\text{cm}^{1/8}$, $C_z = 0.2\,\text{cm}^{1/8}$, and $n = 0.25$.

The release rate Q is doubled to $2Q$ because odor clouds released on or near the ground are "reflected" by this boundary layer. (For elevated odor sources, only Q is used because the molecules can disperse in all directions.)

Most pheromone biologists have used the Sutton formula. For elevated odor sources, a more complicated version of the equation exists. However, compared with the concentration differences at different distances from the source, those between elevated and ground-level sources are minuscule (Elkinton *et al.*, 1984).

Now we introduce the sensitivity (perception threshold) of the receiving animal. According to Bossert and Wilson (1963), the downwind maximum distance $x_{\max}$ (in cm) at which the concentration of an odor remains above threshold is

$$x_{\max} = \left(\frac{2Q}{K\pi C_y C_z U}\right)^{\frac{1}{2}-n}$$

where Q is rate of odorant emission in molecules per second; K is the threshold (molecules/cm^3); C_y and C_z are dispersion constants (0.4 cm$^{1/8}$ and 0.2 cm$^{1/8}$, respectively; see above); U is the mean wind speed; and n is 0.25. For example, the emission rate Q for carcass odors that attract vultures is assumed to be up to 20/day or 7×10^{18} to 1.4×10^{20} molecules/s.

Again, $2Q$ is used instead of Q because the odor is "reflected" from the ground, and twice as many molecules fill the space above ground than would if the odor spread equally in all directions.

The *Gaussian plume model* estimates the average pheromone flux by multiplying the measured odor concentration by mean wind speed, using the following formula (Elkinton *et al.*, 1984). Everything is the same as in the Sutton model, except that σ_y and σ_z, respectively, replace the terms C_y and C_z of the Sutton model. Dispersion coefficients are determined for each experiment separately.

$$D_{xyz} = \frac{Q}{2\pi \sigma_y \sigma_z \, \overline{u}} \exp\left[- \left(\frac{y^2}{2\sigma_y^2} + \frac{z^2}{2\sigma_z^2} \right) \right]$$

Here σ_y and σ_z, the horizontal and vertical diffusion coefficients, are the standard deviations of the cloud dimensions in the horizontal and vertical directions, respectively. They are functions of the downwind distance x and, in this, differ from the constants C_y and C_z of the Sutton model:

$$\sigma_y = \frac{1}{2} C_y \, x^{(2-n)/2} \text{ and } \sigma_z = 2\frac{1}{2} C_z \, x^{(2-n)/2}$$

The rate of dispersion and the values of the dispersion coefficients depend also on terrain and atmospheric conditions. There are "prairie grass" coefficients and values for forests (Fares *et al.*, 1980). Both are discussed by Elkinton *et al.* (1984). (In one more complicated equation, the Gaussian model also considers absorption on the ground surface by introducing a factor α.)

Both the Sutton and the Gaussian models underestimate the width of the active space at each distance, as Elkinton *et al.* (1984) showed. Currently, the Gaussian plume model is preferred over the Sutton model because the dispersion coefficients are measured anew in each experiment. In the Gaussian model, however, the dispersion coefficients are functions of the downwind distances x and correspond to the standard deviation of the vertical and horizontal distribution of the concentration along any axis perpendicular to the mean wind direction. Elkinton *et al.* (1984) pointed out that the Gaussian plume model can be easily applied to a great range of atmospheric stabilities. The dispersion of the plume is affected by temperature changes and resulting turbulence. In the Gaussian model, the size of the active space shrinks as wind speed increases. One has to distinguish concentration (g/cm^3) from flux (g/s per cm^2). The flux increases in direct proportion to wind speed. Flux is more pertinent for stimulation of receptors in animals. Both models consider average concentration over several (3 to 15) minutes. But insects and other animals experience meandering turbulent plumes; consequently most of the time the concentration is below threshold, interrupted by brief exposures to above-threshold concentrations. Therefore, the active space is actually larger than predicted by the models.

For triggering behavior, the concentration at one point in time is more important than the average concentration. Therefore, in the real world, considerable deviation from time-averaging models is observed. In addition to time-averaging models, *peak* concentrations of odors in turbulent systems have to be considered. Aylor (1976) estimated peak concentrations for air currents in forests. Average concentrations, as calculated by the Sutton formula, may be as low as only a few percent of maximum (peak) concentrations. It is often the latter, however, that would trigger an animal's response.

Webster and Weissburg (2001) visualized instantaneous versus time-averaged odor plumes by laser-induced fluorescence. The spatially varying plume at any particular point in time matters more to an animal than an average plume shape. The mean direction and speed of air flow may be relatively constant, but the animal may extract information from concentration differences on very small temporal and spatial scales.

1.1.5 Topography and vegetation

Topography and vegetation create air currents and modify existing wind. Vegetation affects air flow patterns and may adsorb and re-emit odor molecules (Perry and Wall, 1984). The flow over vegetation is "practically exclusively turbulent" (Geiger, 1965). Within vegetation there is little eddy diffusion. The scale of vegetation will affect communication patterns in terrestrial vertebrates: obviously, a vole in a grassy runway, deer in a mature forest, and soaring vultures separated from a carcass by a forest canopy all face different odor propagation problems.

In forests, mechanical turbulence is caused by trees, and temperature inversions by the forest canopy. Ventilation inside a forest is complex and not readily described by existing air flow models (Aylor, 1976).

Forest edges

In the ecologically important *forest edges*, during daytime a "forest breeze" blows from under the tree trunks into the open. However, because the trees act as obstacles, there is no reverse "field breeze" at night, as between land

FIGURE 1.5 "Silent zone" in lee of a hedgerow. Wind-tunnel visualization of air movement over a windbreak, such as a hedgerow, with aid of a bubble generator. Wildlife seek such zones for protection from wind, but information flow through air can be reduced there. (From Moen, 1973.)

and sea. Only in mountainous wooded areas is there a night forest wind that blows into the open country. This downdraft from the radiating crown layer is a night-cooling phenomenon, and not a "forest wind" (Geiger, 1965). Uneven heating inside the forest drives night breezes of about 10 cm/s, but the heterogeneous canopy is more important than this differential heating.

A "quiet zone," about two to three tree heights long, forms in the lee of a forest (Fig. 1.5). It is much shorter behind a forest edge than behind a thin shelter belt. In the open, air currents reach again their maximum speed at a distance of 20 to 70 tree heights from the forest edge (American Meteorological Association, 2000). This will affect the behavior of animals such as deer bedding down or predators chemo-searching for prey; for thermoregulation, an animal should bed down in the "quiet zone," but to receive airborne chemical information on conspecifics, predators, or food it should stay in the wind beyond the "quiet zone."

Area odors

Model experiments have shown the rates of evaporation of volatiles from vegetated areas. The flux of ammonia and related amino compounds from an Australian sheep pasture into the atmosphere was measured by collecting air samples in traps at different heights above ground level. The flux of nitrogen compounds was highest at midday and lowest during the night. Wind and humidity affected flux and gas composition. Pure ammonia was characteristic for cool, humid weather. The pasture lost on average 0.27 kg nitrogen/ha (Denmead *et al.*, 1974). Such "area odors" may serve as important cues for food-seeking herbivores, or migrating mammals or birds.

1.1.6 Chemical environment

In mixtures such as secretions and scent marks, the polarity of the non-pheromonal compounds may greatly affect the release of the active odor molecules (Regnier and Goodwin, 1977). The pH also affects volatility; with higher acidity, a volatile odoriferous base, such as an amine, will form larger amounts of non-volatile salts and, therefore, contribute less to perceived odor. Conversely, odorous acids such as volatile fatty acids will increasingly form salts as the alkalinity increases (Albone, 1984).

1.1.7 Environmental perturbances

More and more artificial changes of the modern environment interfere with communication behavior of animals and humans. Gases, radiation, and gravity all affect responses to odors.

Gases

As an example, low levels of carbon monoxide (CO), together with ingested alcohol, reduce sensitivity to guaiacol, which has a "smokey" or "burnt" odor (Engen, 1986). Smokers take up CO. If they also drink, they could be impaired in their ability to detect smoke from a fire.

Radiation

Patients who were irradiated for 5 weeks with photons of 8 MV of energy at 3 Gy/min, for nasopharyngeal carcinoma or pituitary adenoma, became less sensitive to odors. Their thresholds for amylacetate and eugenol were determined before and several times after radiotherapy. One week after irradiation, the thresholds had increased from 10 to 2 dilution steps. Sensitivity recovered over the subsequent weeks, reaching 6–8 dilution steps 6 months after treatment. The dose received by the olfactory area was estimated as 2 Gy/day (Ophir *et al.*, 1988).

Gravity

The gravitational field affects olfaction as it does vision, audition, or vestibular function. It could impair detection of dangerous fumes or burning electrical equipment in airplanes or space vehicles. Astronauts also report altered perception of food flavors under weightless conditions. Men and women tested with four "scratch-and-sniff" odor samples of the University of Pennsylvania Smell Identification Test identified odors more poorly when in an upside-down position (Mester *et al.*, 1988).

In another experiment, animals were accelerated. Puppies, aged 1–20 days, were rotated in a drum for 5 or 15 minutes and accelerated for a maximum of $3 \times g$. Their response to eucalyptus odor changed from neutral to negative, and that to maternal odor from positive to "uncertain." Odor aversion to eucalyptus lasted longer than that to maternal odor. When visual and auditory stimuli start to control behavior at the age of 12–13 days, the induced odor aversion is more easily suppressed (Kassil and Gulina, 1987).

1.2 Water

Aquatic animals use their chemical senses in all aspects of their lives, from reproductive behavior to feeding, habitat selection, and predator avoidance. The hydrodynamic properties determine the possibilities and limits of chemical communication in water. As a medium, water is as dynamic as air, so that convection and advection are far more important for odor transport than is diffusion. Distribution by currents is even more important in water because compounds of similar molecular weight diffuse four orders of magnitude more slowly than in air (Gleeson, 1978). Diffusion of odorants may be important only in the submillimeter range, while turbulence is typical for water masses above the centimeter range.

1.2.1 Boundary layers

In flowing water, the "boundary layer" is the water mass above the substrate. This boundary layer is defined as that part of the flow where the velocity ranges from 0 to 99% of the velocity of the stream. Most boundary layers of biological interest are turbulent (Webster and Weissburg, 2001).

The behavior of the animal in response to flow is important, not just the flows themselves. An animal searching for the source of an odor moves in the direction of increasing stimulus intensity and stays within the boundaries of the plume. If an animal needs to sample odor in turbulence frequently, it may have to reduce speed, to untenable levels in the case of moths or birds, or to an energetically

Table 1.1 Properties of viscous versus turbulent environments

	Viscous environment	Turbulent environment
Reynolds number	Low	High
Signals	Change slowly	Chaotic structure
Flow	Low	Bulk flow, perceptible to animal
Gradient	Continuous	Intermittent
Animal	Searches in three dimensions for odor boundaries	Relies on flow
Sensory integration	Slower integration time	Rapid integration of brief odor pulses

demanding level in the case of fish. This, in turn, means that it may not arrive in time or expose itself more to predation (Webster and Weissburg, 2001) (Table 1.1).

Depending on the organism, we have to consider water masses and movements ranging from the submillimeter "micropatches" all the way to the oceanographer's gyre on the 1000 km scale. Large bodies of water from freshwater lakes to oceans are stratified. Fish or other animals may sample several of these stable layers and select one that "smells right." Indeed, salmon (*Salmo salar*) with their olfactory sense experimentally obliterated did not sample certain water layers and were unable to select one as fish with intact sense of smell could do (Døving *et al.*, 1985). It has been suggested that the shear between two horizontal layers could be utilized to determine the lower or upper boundary of a particular layer (Westberg, 1984).

1.2.2 Sampling behavior and information currents

Most aquatic animals, notably vertebrates, have to sample *turbulent* water with patchy odor distribution. Two types of stimulus access ("sniffing") have been distinguished in fish: "cyclosmates" such as tuna or lobster, which sample a specific sniff volume in sniffs or flicks, and "isosmates," which sample a steady, ciliary-driven water flow. This latter type of stimulus access is found in slow-moving animals such as catfish, eels, dogfish, or mud snails (Atema, 1988). In turbulent water, patch boundaries will be the sharper the more recently the odor was released. When crossing an odor patch, the animal will learn about the distance to the odor source from the rate of concentration change.

In water, currents may be utilized or even created to communicate chemically. Beaver, river otter, mink, or various fish may simply release their odorants into a stream. Lobster, tail-waving newts, and possibly fanning fish produce *information currents* that propel chemical signals toward recipients. In the lobster, the female sends a current with pheromone into the male's shelter. The male, in turn, draws water toward himself and fans it out into the surroundings, signaling his mating status (Atema, 1986).

Water may be better than air for studying the dynamics of stimulus dispersal, because in this denser medium fluid dynamic processes are attenuated (Atema, 1988). As interindividual distances increase with increasing body size, pheromone communication becomes less and less practical in water, particularly because of signal delays, dilution, and cross-currents.

1.2.3 Water as solvent

The chemical properties of water determine what compounds will leach out of soil and vegetation, and hence what chemicals other plants and animals will be exposed to. For instance, acidic water will extract alkaloids from plant materials.

1.3 Water–air interface

Odor retention, release and propagation on water surfaces are important to many animals, but we know little of how animals exploit or actively manipulate these processes.

Freshwater mammals such as beaver may leave odors on the surface of their ponds and olfactorily sample the water or layer of air immediately above it. Lipids on water may form *micelles*, small blobs of molecules (from Latin *mica*, a grain, crumb, morsel) that enhance evaporation into the air layer by increased chemical potential. Some seabirds hunt by odor (e.g. Hutchison and Wenzel, 1980; Nevitt, 1999). They may respond to prey volatiles (from krill, squid, or fish) that rise to the water surface and evaporate into the air. The air–water equilibrium for dilute solutions can be expressed by using partition coefficients, relative volatility, or Henry's law (Thibodeaux, 1979).

1.4 Influence of setting

The circumstances of an odor experiment are important. For instance, male carp, *Cyprinus carpio*, spermiate in the presence of females that ovulate after having been injected with carp pituitary homogenate (as is already known for goldfish). This effect occurred in earthen ponds (200 m^2) but not in bare tanks

Table 1.2 Different behaviors in different settings

Species	Behavior	Field	Laboratory	Reference
Kin recognition				
Bufo americanus	Associate with kin	Yes	No	Waldman, 1981
Rana cascadae	Associate with kin	Yes	No	O'Hara and Blaustein, 1985
Reproductive behavior				
Carp *Cyprinus carpio*	♂ spermiates in presence of ♀ injected with pituitary homogenate	In earthen ponds: yes	In bare tanks: no	Billard *et al.*, 1989
Domestic cattle	♂ mounts ♀ treated with estrus urine	Pasture: only treated ♀ mounted	Stalls: mount indiscriminately	Sambraus and Waring, 1975
Feeding behavior				
Brown tree snake *Boiga irregularis*	Responds strongly to fish attractants and trapper's lures	No	Yes	D. Chiszar, personal communication
Townsend's vole	Feeding in presence of repellent	Avoids repellent only in open, not in cover	Test without cover may be misleading	Merkens *et al.*, 1991
Mongoose	Aversive conditioning to scented, toxic eggs	Avoid scented, toxic eggs from a distance	Will not avoid at distance, break shell, eat still	Nicolaus and Nellis, 1987
Responses to predator chemicals				
Hedgehog *Erinaceus europaeus*	Avoiding area tainted with badger (predator) feces	Farmland, golf courses: avoid treated area for minutes or hours only	Enclosure: avoid treated area for 2 days	Ward *et al.*, 1997
Chinook salmon	Conditioned to recognize predator (cutthroat trout)	Better survival in creek but only if raised in complex habitat	No survival benefit in hatchery raceway; and none if raised in simple habitat	Berejikian *et al.*, 1999
Gray-tailed voles *Microtus canicaudus*	Mink odor effect on reproductive rate, sexual maturation, juvenile recruitment	No effect	Reproductive behavior suppressed in *M. agrestis* and bank voles, *Clethrionomys glareolus*	Wolff and Davis-Born, 1997

($2\,m^3$) (Billard *et al.*, 1989). In this experiment, the effect may not be clearly pheromonal, as the fish could see each other.

The American toad, *Bufo americanus*, discriminates kin from non-kin (after having been raised in mixed groups of siblings and non-siblings) *only in the field*, but not in the laboratory. The same is true for the cascades frog, *Rana cascadae*. These results have been summarized by Blaustein and Waldman (1992). Conversely, the brown tree snake, *Boiga irregularis*, responds strongly to fish attractants and trappers' lures in the laboratory but hardly at all in the field (D. Chiszar, personal communication, 1994).

Finally, in cattle, bulls respond to female sexual odors on pastures but not in stalls (Sambraus and Waring, 1975). Table 1.2 summarizes these findings.

2

Properties of vertebrate semiochemicals

... the odors of ointments are more durable than those of flowers.
FRANCIS BACON (1561–1626): *Essays, Of Praise.*

On the banks of the Plata I perceived the air tainted with the odour of the male *Cervus campestris*, at half a mile to leeward of a herd; and a silk handkerchief, in which I carried home a skin, though often used and washed, retained, when first unfolded, traces of the odour for one year and seven months.
CHARLES DARWIN: *The Descent of Man*, p. 529.

The structures of vertebrate chemosignals reflect their functions and the environment they are used in. Temporal parameters, spatial range, localizability, intensity, detectability, and information content of the signal depend on both the chemical structure and the operating environment (Alberts, 1992a). The chemical properties involved include functional groups of the molecule; volatility; aromaticity; the number of compounds composing the entire signal; and when, where, and how it is emitted (Alberts, 1992b).

2.1 Functional groups

Functional groups can determine communicative activity of a molecule. For example, hypoxanthine 3N-oxide triggers alarm responses in Ostariophysan fish. The amineoxide (NO) functional group (Fig. 2.1) appears to be essential for the antipredator behavior of fathead minnows, *Pimephales promelas*, and finescale dace, *Phoxinus neogaeus* (Cyprinidae) to occur. Structurally similar molecules lacking the NO group did not release alarm responses in two tetra species (Charicidae). All four species belong to the Ostariophysi, and the response to the NO group is probably widespread in this order, also known as Cypriniformes (Brown *et al.*, 2000).

In another example for functional groups, Gower and Ruparelia (1993) noted that the odoriferous steroids that play a role in communication share certain

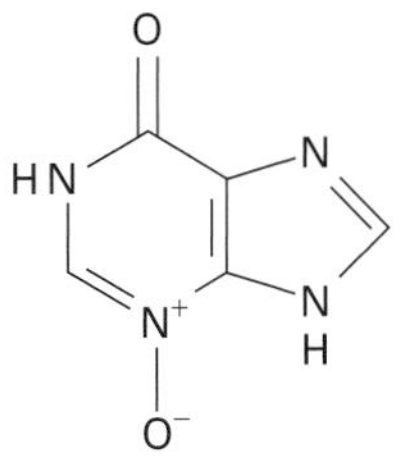

FIGURE 2.1 Hypoxanthine oxide, a fish alarm pheromone.

features. They are relatively volatile, lack any substituent at C-17, are unsaturated at C-16 and C-17, have only one oxygen function at C-3 and hence are very non-polar. The double bond at C-16–17 is not important for the odor to humans, while manipulations at C-3 change the odor quality (Gower *et al.*, 1989).

A third example is the role amino acids such as L-aspartate and L-glutamate have in deterring feeding by the oriental weatherfish, the loach *Misgurnus anguillicaudatus*, presumably signaling a predator. Monoalkyl and dialkyl esters of these amino acids are less repellent. This indicates that the carboxyl group is important for the repellent activity. In the same vein, acetic acid is repellent, while alkyl acetates are less so. Again, the carboxyl group appears to mediate the effect (Harada, 1989).

Further examples of the importance of functional groups for behavior are the responses of sunfish to steroids in beetles, their prey; reactions of birds and mammals to capsaicin-related compounds; and fear behavior of rats when exposed to sulfur compounds from fox urine and feces.

Of several related steroids in the defensive secretion of dytiscid beetles, deoxycorticosterone was most effective. It deterred sunfish from feeding in 94% of the tests. Other steroids (pregnolones) that differed only by lacking a keto group at one carbon atom were either intermediate or not active at all (Gerhart *et al.*, 1991).

Capsaicin analogues that differ in only one functional group affect birds and mammals differently. A change from an acidic phenolic hydroxyl group in vanillyl acetamide to a methoxy group in veratryl acetamide reverses the effect on starlings and rats. The first was aversive to starlings, but attractive to rats, while the opposite was true for veratryl acetamide (Mason *et al.*, 1991).

Finally, fox urine and feces contain sulfur compounds that "stress" rats. In several mercaptoketones, the thiol (mercapto) group was essential for the effect, while the keto group was not (Vernet-Maury *et al.*, 1984).

In addition to their presence, the position of functional groups in a molecule can determine its smell. This is known as "regional selectivity."

Table 2.1 Odor differences of isomers and enantiomers

Compound	Odor perceived by humans
Vanillin	Odor of vanilla
Iso-vanillin	Odorless
(*S*)-4-Carvone	Caraway
(*R*)-4-Carvone	Spearmint
(+)-Androsta-4,16-dien-3-one	Urinous, sweaty, musky, woody (only 10–20% can smell it)
(−)-Androsta-4,16-dien-3-one	Odorless

2.2 Polarity

Most compounds in vertebrates and insects contain polar functional groups (Wheeler, 1977). An intriguing question is whether marking pheromones are less polar than water-soluble ones, such as those in urine. The polarity of non-pheromonal compounds in a mixture greatly affects pheromone release into the environment (Regnier and Goodwin, 1977).

2.3 Solubility

Chemical signals in urine, fresh-, and saltwater are water soluble. In water, many more compounds are pheromones candidates than in air, because molecules of a wide range of sizes are water soluble. Terrestrial scent marks have to survive humidity and precipitation and so here the active components are soluble in lipids.

2.4 Isomers and enantiomers

Humans and insects can discriminate enantiomers of the same compound (Friedman and Miller, 1971; Russell and Hills, 1971) (Table 2.1). Black-tailed deer (*Odocoileus hemionus columbianus*) discriminate between the two geometric isomers and somewhat between the enantiomers of the "deer lactone" (Fig. 2.2) (Brownlee *et al.*, 1969; Müller-Schwarze *et al.*, 1976, 1978a). Young male Asian elephants have more (+) than (−) enantiomer of frontalin in their temporal secretion. Their secretion and (+)-frontalin affect conspecifics little. However, the nearly recemic enantiomer mix from mature bulls attracts ovulating females and repels males (Greenwood *et al.*, 2005)

2.5 Volatiles

Secretions and excretions of terrestrial vertebrates contain compounds with a wide range of volatility. For convenience, the two ends of the continuum

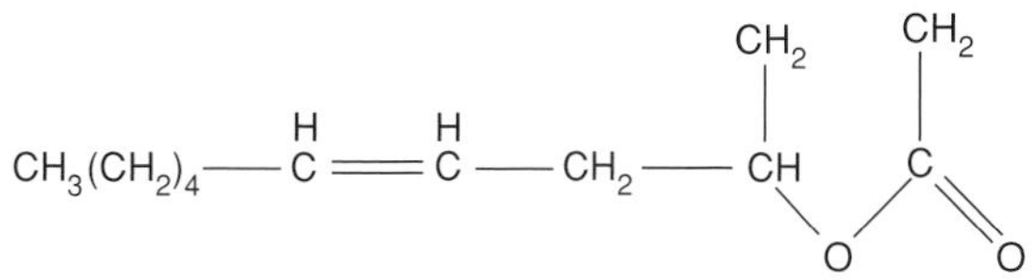

FIGURE 2.2 The "deer lactone: (Z)-4-hydroxydodec-6-enoic acid lactone.

are arbitrarily termed volatiles and non-volatiles. Both can be integral parts of chemical signals between animals. The molecular weight of volatile compounds ranges from about 16 to 300 Dalton (Wilson and Bossert, 1963). Volatility can be increased by elevated temperature.

Long-distance communication via airborne odors requires volatiles, while non-volatiles can effectively be used in short-range communication that requires body contact, as in mutual licking by two animals or the nuzzling and licking of a female by a male in mating behavior. Indeed, mating pheromones are often of low volatility. Many communication systems employ both volatiles and non-volatiles, either simultaneously or in sequences, as when long-range attraction by volatiles is followed by close-range inspection of a conspecific – or of a plant in the case of herbivory – with sniffing, nibbling, biting, or chewing. Acceptance or recognition of young or group members may be the result of exchange of non-volatile signals at close range.

A mammal may emit many volatile compounds. Humans, for instance, give off hundreds of volatiles, many of them chemically identified (Ellin *et al.*, 1974). The volatiles include many classes of compound such as acids (gerbil), ketones, lactones, sulfides (golden hamster), phenolics (beaver, elephant), acetates (mouse), terpenes (elephant), butyrate esters (tamarins), among others. The human samples mentioned before contained hydrocarbons, unsaturated hydrocarbons, alcohols, acids, ketones, aldehydes, esters, nitriles, aromatics, heterocyclics, sulfur compounds, ethers, and halogenated hydrocarbons. Sulfur compounds are found in carnivores, such as foxes, coyotes, or mustelids. The major volatile compound in urine of female coyotes, *Canis latrans,* is methyl 3-methylbut-3-enyl sulfide, which accounts for at least 50% of all urinary volatiles (Schultz *et al.*, 1988).

In the secretions of the anal sacs of dogs (*Canis familiaris*) and coyotes (*C. latrans*) the most abundant volatile compounds are trimethylamine, short-chain (C_2–C_6: acetic, propionic, isobutyric) acids, acetone, and 2-piperidone (Preti *et al.*, 1976).

Free fatty acids up to C_5 are found in urine and other mammalian secretions (Albone, 1984); C_6 and C_8 free fatty acids occur in the anal gland secretion of the aardwolf (Apps *et al.*, 1989), and C_7, C_8, C_9, C_{10}, and C_{12}, "and their isoforms," in the marking fluid of the tiger (Poddar-Sarkar *et al.*, 1991).

The well-known defense secretion from the skunk's (*Mephitis mephitis*) anal glands contains thiols and disulfides (Andersen and Bernstein, 1975) and thioacetates (Wood, 1990). The seven major compounds are (*E*)-2-butene-1-thiol, 3-methyl-1-butanethiol, *S*-(*E*)-2-butenyl thioacetate, *S*-3-methylbutanyl thioacetate, 2-methylquinoline, 2-quinolinemethanethiol, and *S*-2-quinolinemethyl thioacetate (Wood, 1990). Table 13.3 (p. 403) and Fig. 10.9 (p. 403) list some of the sulfur compounds found in Mustelids. Thiols are responsible for the unpleasant smell of the skunk's spray. In the presence of water, they are formed from thioacetates, which also smell rather repulsive. However, the human nose is not necessarily a good judge of what is significant to animals. For instance, a krill extract fraction that humans perceived as fishy did not attract (fish-catching) petrels particularly strongly (Clark and Shah, 1992). At the other extreme, chin gland marks of rabbits appear odorless to humans (Mykytowycz, 1965).

Volatiles from fecal pellets of wild male house mice include ketones, alcohols, and carboxylic acids. Pellets aged over 24 hours had 20 major volatiles whereas fresh pellets had only 15 (Goodrich *et al.*, 1990).

The number of carbon atoms of saturated straight-chain compounds correlates with hydrophobicity, molecular volume, and molecular length. All three could affect the interaction of an odorant ligand with the olfactory receptors (Johnson and Leon, 2001).

The techniques for trapping, concentrating, isolating, and identifying volatiles have been developed by flavor chemists and insect pheromone researchers and are not detailed here. Some of the techniques useful for mammals are summarized in Albone (1984) and Millar and Haynes (1988).

2.6 Non-volatiles

2.6.1 Proteins

The molecular size of non-volatile compounds may be very high. The "mounting pheromone" or "aphrodisin" of the golden hamster, *Mesocricetus auratus*, is a protein with a molecular mass of 17000 daltons (Da) (Singer *et al.*, 1989). It belongs to the alpha-$\alpha_{2\mu}$-globulin superfamily and is very similar to the major urinary proteins (MUPs) in mice (Fig. 2.3), the bovine β-lactoglobulin from cow's milk, and the pyrazine-binding protein in the nasal mucosa.

Mouse urinary proteins

The urine of male mice contains high levels of MUPs, which play an important role in scent communication. Male mice excrete higher levels (up to 20 mg/ml urine) in their urine than females, and secretion increases with sexual

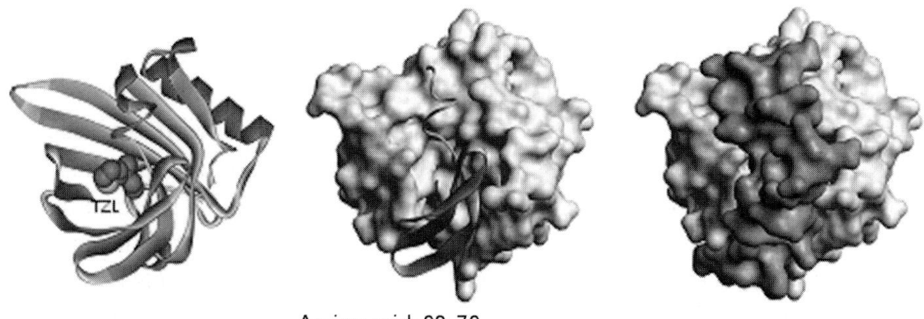

FIGURE 2.3 Major urinary protein (MUP) of mouse, showing region of polymorphic variation. Left: Ribbon model with a volatile ligand (TZL: thiazole) in binding pocket. Center: Ribbon in lower center shows location of variable amino acid sequence. Right: Space-filling model with variable polypeptide chain segment shaded darker, in middle region. (Courtesy Robert J. Beynon.)

maturation. After castration, levels fall to 2.5 to 3.5%, and androgen treatment restores higher levels. This protein type belongs to the lipocalin superfamily and is synthesized in the liver. The liver of males contains three to five times more mRNA for MUPs than that of females. The barrel-shaped (eight-stranded β-barrel), 18 kDa MUP has a hydrophobic cavity that serves as a "container" for volatile ligands (Beynon *et al.*, 1999). This pocket in the molecule binds small hydrophobic molecules and a wide range of odorants can be bound. Mouse urine contains 2-*sec*-butyl-4,5-dihydrothiazole and 2,3-dehydro-*exo*-brevicomin and these volatiles are assumed to signal the presence of a male, while the MUP protein itself seems to trigger puberty in females, acting as a priming pheromone. Volatile ligands attract the attention of male mice, while proteins and non-volatile protein–ligand complexes stimulate counter-marking (Humphries *et al.*, 1999).

The scent mark of the saddle-back tamarin, *Saguinus fuscicollis*, contains water-soluble proteins. The major protein (66 kDa) of the scent mark occurs in the urine, while another major protein (18 kDa) derives from gland secretions. If the proteins are removed by enzymatic degradation, tamarins still discriminate scent marks from different donor types. However, they are able to discriminate scents with intact proteins from those with the proteins digested. Consequently, the proteins are most likely an integral part of the scent image (Belcher *et al.*, 1990).

Proteins also occur in many excretions and glandular secretions, from turtles to mice and humans. Intensive studies are clarifying their role in signaling. In axillary secretion from human males, the main odoriferous acid, (*E*)-3-methyl-2-hexanoic acid, (*E*)-3M2H for short, is bound to two apocrine secretion-binding proteins (ASOB1 and 2) of molecular masses 45 and 26 kDa, respectively. The

ASOB2 is apolipoprotein D (apoD) of the lipocalin or $\alpha_{2\mu}$-microglobulin superfamily of carrier proteins (Preti *et al.*, 1992; Zeng *et al.*, 1996a,b).

The female elephant pheromone (Z)-7-dodecenyl acetate occurs bound to urinary proteins. When it is taken up by a male elephant, the acetate is bound to proteins of the trunk mucus. This might facilitate transport of the pheromone to the vicinity of the sensory epithelium of the vomeronasal organ (Rasmussen and Schulte, 1998).

Frontalin in the temporal gland secretion of Asian elephants is bound to elephant albumin (Schulte *et al.*, 2005). The bulk of the secretion of the chin gland of the rabbit is protein (Goodrich and Mykytowycz, 1972).

2.7 Multicomponent pheromones

2.7.1 Mixtures of volatiles

It was initially thought that, in insects, the major component in a pheromone blend attracted from the longest distance, while the minor components came into play at shorter distances from an odor source such as a "calling" female. However, the male Oriental fruit moth, *Grapholita molesta,* at least, responds from longer distances (100 m) more to the full female pheromone blend of three compounds than to the major component (Linn *et al.*, 1987). Similar tests have not been performed with vertebrates.

For mammals, if not vertebrates in general, multicomponent pheromones appear to be the rule. Such mixtures can comprise compounds of a wide range of volatility. They have been variously termed "odor profile," "pattern," "odor image" (Albone, 1984), "gestalt" (Evans *et al.*, 1978), or "mosaic" (Johnston, 2005). One of the best-investigated odor profiles is that of the scent mark of the saddle-back tamarin, *S. fuscicollis,* (Smith *et al.*, 1985). Here, not even the 16 butyrate esters are sufficient for subspecies recognition. Additional volatiles are also required.

A second example of a mammalian multicomponent pheromone is the puberty-delaying pheromone from female house mice, *Mus musculus*. Two acetate esters and a pyrazine are biologically active in various combinations, but the pyrazine is also active alone (Novotny *et al.*, 1985a).

The chinning response in tree shrews, *Tupaia belangeri*, is triggered by several lipophilic fractions of male urine. The combined fractions are more active than single fractions. Pyrazine compounds and a high concentration of several volatile monocarboxylic acids characterize male urine. Some pyrazine compounds and some monocarboxylic acids release the chinning response (Stralendorff, 1987).

The European badger, *Meles meles*, marks its territory with its subcaudal gland. This gland produces a mixture of many carboxylic acids and other unidentified compounds. In one recent analysis (Buesching *et al.*, 2002a) 21 compounds, mostly fatty acids, were common to all 66 investigated samples, from both sexes. Analyses have indicated that the patterns of these compounds reflected specific characterists such as sex, individuals, group membership, and seasonality. Some or all of these compounds may signal species. The sexes differed in several respects. Females had more measurable peaks (average 36.1) in their chromatograms than males (30.7). (Peaks can represent one or several compounds.) Tetradecanoic acid was more abundant in males than in females. Three not fully identified compounds were more abundant in females than in males in all seasons, except winter. Individuals showed distinct patterns: each badger had a unique combination of 23 to 58 compunds. Group members' profiles were more similar to each other than to those of members of other groups. Consistent seasonal variation occurred in females. Their secretion was less complex in spring than in winter and summer. Certain patterns correlated with reproductive state: dodecanoic acid and two other compounds were more abundant in non-lactating females than in lactating ones. Males with descended testes (i.e. those reproductively active) had more dodecanoic acid than those with non-descended testes. With increasing age, the level of heptadecanoic acid increased (Buesching *et al.*, 2002a). The levels of two monounsaturated C_{16} acids decreased in the course of 2 days, potentially providing information about the age of a scent mark. Individual characteristics remained remarkably consistent from year to year. During the annual cycle, periods of stability of composition alternated with rapid changes (Buesching *et al.*, 2002b). What exact combinations of compounds the badgers themselves use for the various levels of discrimination remains to be seen.

It is important to identify and measure the concentrations of a number of compounds in a mixture simultaneously for several reasons. First, among related compounds there may exist precursors of active ones, and pathways of pheromone synthesis may be elucidated. This is true for steroids in the human axilla. Nixon *et al.* (1988) determined the concentration of five steroids extracted from axillary hair of adult men aged 18 to 40 years. The relationships in concentrations between the two ketones 5α-androst-16-en-3-one and 4,16-androstadien-3-one suggest that axillary bacteria reduce the former to the latter with the aid of the enzyme 4-ene-5α-reductase. Humans have a low olfactory threshold for several 16-androstenes, and the fact that some men have large quantities of 16-androstenes (Nixon *et al.*, 1988) is biologically suggestive.

For several possible functional reasons, pheromones are thought to be mixtures of compounds. In vertebrates, signal specificity is required more at the levels of individual, family, or clan, than the species (in many insects). This specificity arises from "fingerprints" of several compounds mixed in different ratios, and not unique compounds, as in insects. Why are odors so often complex mixtures of compounds? There are several possibilities

- to ensure that an odor is noticed in the first place in case some components vary in concentration over time or between individuals;
- to provide redundancy to counteract environmental obstacles;
- to provide enough variability for distinct specific signals to permit discrimination ("chemical fingerprints," especially in scent marks);
- to increase sensitivity, as a mixture can require fewer molecules of all of its compounds combined to reach an animal's threshold than the sum of the molecules needed to reach threshold for the same compounds tested singly (Laska and Hudson, 1991);
- to render an odor distinct against background odors;
- to be more resistant to sensory adaptation (consequently, mixtures would be better markers for signaling dangers to humans, as in gas leaks) (Cometto-Muñiz and Cain, 1993).

Other complex odors may be extremely variable among individuals and genotypes, and receiving animals have to learn their significance. For example, laboratory mice discriminate the urine odor of different genotypes that vary at the t-locus but no discrete chemosignal compounds have been found in these different genotypes. It is felt that "the variation of the overall pattern of general secondary metabolites" is used for discrimination, instead of specific compounds (Jemiolo *et al.*, 1991). These authors stated that the "lack of discrete chemosignal compounds responsible for the genotype discrimination supports the concept of a non-specific influence of the t-locus (part of the major histocompatibility complex [MHC]) on olfactory cues." This overall pattern is possibly learned, as offspring are exposed to the odor of their parents at an early age. For urine odors determined by the MHC, it was also concluded that they result from multiple and redundant compounds (Tsuchiya *et al.*, 1992). Volatiles alone convey information on MHC genotype, since mice still distinguish corresponding fractions of two odor types even after proteins have been depleted (Singer *et al.*, 1993).

In addition to the simultaneous impact of an odor mixture, differential evaporation may alter the signal over time, indicating the age of a scent. As an example, the major component of the chest gland secretion of the thick-tailed galago, *Galago crassicaudatus*, is benzylcyanide. It evaporates rapidly and is present for only about 1 hour. The other two identified compounds last for several days. In a behavioral corollary, galagos retreat from a scent mark that is less than 1 hour old, while older marks have no such effect (Katsir and Crewe, 1980).

Two different types of scent mark from the same species can evaporate differentially. The brown hyena, *Hyaena brunnea*, pastes two different secretions on the same blade of grass: a brown mark with an odor that dissipates rapidly, and a white mark that – to the human nose – lasts for up to a month (Mills *et al.*, 1980). Similarly, the fade-out time for the anal gland secretion of the African *Helogale undulata rufula*, dwarf mongoose, used for range marking, is about 10 days, while cheek gland secretion, part of the threat display, lasts only 1 to 2 days (Rasa, 1973).

Interspecific chemical cues are also often mixtures. Mixtures of amino acids serve as feeding stimulants in fish. Among mammals, ferrets respond more to mixtures than to pure odors in their foraging responses. The mixtures are thought to contain more information (Apfelbach, 1973).

In some cases, animals do not seem to identify components of complex environmental chemical cues. The pool frog (*Rana lessonae*) provides an example. After being exposed to a mixture of morpholine and β-phenylethanol during development before metamorphosis, the froglets prefer the mixture but are not attracted to either of the two compounds (Ogurtsov and Bastakov, 2001).

2.7.2 Combinations of volatiles and non-volatiles

Volatiles and non-volatiles may be related in at least four ways.

1. Both kinds of compound may combine to form complex signals, sometimes releasing two-step responses. The volatiles attract from a longer distance, while the less-volatile compounds provide more information at close range. For example, volatile lipids from femoral glands are thought to alert iguanas (*Iguana iguana*) to the presence of a conspecific, but a non-volatile protein fraction, picked up from secretion deposits by tongue touches, helps to identify individuals (Alberts and Werner, 1993). Both volatiles and non-volatiles in bovine body fluids from glands at the mucocutaneous junction of the anogenital area stimulate sexual behavior in bulls (Rivard and Klemm, 1990). Cotton-top tamarins, *Saguinus o. oedipus,* typically touch a scent mark to identify conspecifics but are able to rely only on volatiles when forced to do so by a wire screen over the mark (Belcher *et al.*, 1988).
2. The non-volatiles may not possess a signal value themselves but may act as release modulators, or "keeper substances." The MUPs in mice provide an example. Each individual expresses at least 4 to 15 different proteins and the combination differs between individuals in the population. Most of the variation between MUPs occurs on the surface of the molecule and is accessible to chemoreceptors. MUPs are stable over the life of the individual. On urine

FIGURE 2.4 Structure of isopentenyl methyl sulfide.

marks, they are also stable for weeks or months. Volatile metabolites that characterize MHC-different individuals appear to be bound to MHC peptides, to MUPs, or both. Sebum, ubiquitous in mammals, can also serve as a "keeper substance" for volatile signal compounds.

3. A volatile compound from a synthesis pathway such as that leading to the less-volatile steroid sex hormones may signal reproductive status. For example, Jorgenson *et al.* (1978) have suggested that in the red fox, *Vulpes vulpes*, the strong-smelling Δ^3-isopentenylmethyl sulfide (Fig. 2.4) could signal an increase of steroid production, at the start of the mating season. This compound can derive from isopentenyl pyrophosphate, an intermediate in terpene biogenesis, leading to sterols and steroid hormones.
4. The non-volatiles may be precursors of signal compounds, breaking up into volatiles as needed. This has been suggested, for instance, for the esters in the subauricular gland secretion of the pronghorn, *Antilocapra americana* (Müller-Schwarze *et al.*, 1974), and (*E*)-*s*-2-butenyl thioacetate in the anal gland secretion of the hog-nosed skunk, *Conepatus mesoleucus*, which possibly breaks down to the smelly (*E*)-2-butene-1-thiol (Wood *et al.*, 1993). Finally, in brush-tailed possums, oleic acid breaks down into two molecules of nonanal.

2.7.3 Keeper substances

Mammalian secretions typically have large amounts of non-volatile compounds, while the actual chemosignals may represent a rather small part of the total weight. Squalene is one of the more common lipids in glandular secretions such as sebum. Squalene and sebum retard evaporation of volatiles. In laboratory experiments, the volatility of phenylacetic acid, a pheromone compound from the ventral gland of the Mongolian gerbil, *Meriones unguiculatus*, was affected by "keeper substances": being a polar compound, phenylacetic acid is the less volatile the more polar the keeper substance is. Volatility is lowered if a small amount of pheromone interacts with a large amount of solvent (dipole–dipole interaction). For example, polar diethylene glycol tightly retains phenylacetic acid, while non-polar mineral oil retains it only weakly. However, volatility is retarded more by mineral oil than by squalene. Humidity increases volatility as odor molecules compete with water molecules for surface sites

(Regnier and Goodwin, 1977). Less-polar odorants will be less affected by the polarity of their lipid substrate or by relative humidity. For these reasons, the chemical ecologist has to keep in mind that volatility results not only from the nature of the odorant but also from the characteristics of the substrate, and the environmental conditions.

Keeper substances in a scent secretion, and/or the nature of the substrate such as rock, wood or soil, impart longevity to a scent mark. In our own studies, scent marks of captive pronghorn, *A. americana,* were still noticeable to the human nose 4 months after all animals had been removed from a pen. Similarly, scent marks of the aardwolf, *Proteles cristatus,* last for up to 6 months (Apps *et al.*, 1989). 2-Phenoxyethanol in the secretions from the chin gland of the rabbit is typical in dominant animals, serving as a fixative to extend the life of the signal (Hayes *et al.*, 2001).

The urinary lipids in the urine marks of lions and tigers may serve in prolonging pheromone release from the mark into the air (Asa, 1993). Both sexes of tigers, *Panthera t. tigris,* spray a "marking fluid" onto vegetation. The fluid is released through the urinary tract. This lipid-rich fluid contains lipid fixatives at a level of 1–2 mg/ml. These lipids include cholesterol ester, wax esters, triacylglycerols, free fatty acids, diacylglycerols, monoacylglycerols, free sterol, and phospholipid (Poddar-Sarkar, 1996). Wax esters occur in anal gland secretion of beaver, *Castor canadensis* (Grønneberg and Lee, 1984).

2.8 Sex differences

Some male-specific compounds have been identified in mammalian secretions that may have a role in communication (Table 2.2)

The anal gland secretion of beaver, *C. canadensis,* differs between the sexes (Grønneberg, 1978–79). The male grey duiker, *Sylvicapra grimmia,* has more 2-heptanone and 2-nonadecanone in its preorbital secretion than the female. Two thiazoles (2-isobutyl-1,3-thiazole and 2-isobutyl-4,5-dihydro-1,3-thiazole) and an epoxy ketone (3,4-epoxy-2-dodecanone) are also more abundant in the male's secretion. Correlated with these chemical differences is the fact that only males scent mark with the preorbital gland (Burger *et al.*, 1990).

In European moles, *Talpa europaea,* a series of carboxylic acids dominates the anal gland secretion of both adult males and anestrous females. These acids disappear in proestrous and estrous females but are present again in pregnant and lactating females. The acids are also absent in juveniles. The anal scent appears to constitute a "keep out" signal, and its absence in estrous females permits mating (Khazenehdari *et al.*, 1996).

Table 2.2 Some male-specific compounds in mammalian secretions with a role in communication

Species	Compound	Source	Reference
Boar *Sus scrofa*	16-Androstenes	Saliva	Melrose *et al.*, 1971
Goat *Capra hircus*	4-Ethyloctanoic acid	Cornual organ	Sugiyama, 1983
Asian elephant *Elephas maximus*	Frontalin	Temporal gland	Rasmussen and Greenwood, 2003
Humans *Homo sapiens*	(*E*)-3-Methyl-2-hexenoic acid	Axilla	Zeng *et al.*, 1991
Spotted skunk *Spilogale putorius*	(*E*)-2-Butene-1-thiol	Anal glands	Wood *et al.*, 1991
Striped skunk *Mephitis mephitis*	Thiols, thioacetates	Anal glands	Wood, 1990
Egyptian mongoose *Herpestes ichneumon*	2,4,6,10-Tetramethyl 1-undecanoic acid	Anal glands	Hefetz *et al.*, 1984
Wolf *Canis lupus*	Isopentyl sulfide and 3,5-dimethyl-2-octanone	Urine	Raymer *et al.*, 1984

The mixtures of volatiles in the urine of tree shrews, *T. belangeri*, are sex specific but no single sex-specific compounds have been found (Stralendorff, 1987).

Axillary odors of male and female humans contain the same compounds (C_6–C_{11} acids and *E*-3-methyl-2-hexenoic acid), but in different ratios. The characteristic odor resides in acidic compounds (Zeng *et al.*, 1996b).

Other chemical studies did not find sex or seasonal differences in the composition of mammalian scents. No sex differences in the composition of mixtures of volatile compounds from glands have been found in the brushtailed possum, *Trichosurus vulpecula*, for example. The same profiles of low-molecular-weight branched carboxylic acids were found in paracloacal gland secretions of males and females (Woolhouse *et al.*, 1994). Branched carboxylic acids also occurred in the preorbital gland secretion of a female sika deer (*Cervus nippon*) (Wood, 2004). Comparisons between the compositions of secretions in different, related species permit assumptions about functional adaptations and possible evolutionary pathways. Such comparisons are available for five *Mustela* species (Brinck *et al.*, 1983), and three species of hyenas (Buglass *et al.*, 1990).

2.9 Life expectancy of chemical signals

Volatiles evaporate, and signals become weaker over time. In mixtures, evaporation of the more-volatile compounds can change the quality of the odor and the range over which it can be detected. This means that decaying and changing signals potentially provide information about the age of the scent mark and, in turn, the recent behavioral history of the odor donor.

Ferkin and Johnston (1995a) aged anogenital area scent and odor from the posterolateral region in meadow voles (*Microtus pennsylvanicus*) for 15 minutes to 30 days. Males "preferred" female to male anogenital odor if it was 10 days old or less. Females preferred male to female anogenital odor if its age was 25 days or less. Both sexes preferred posterolateral odor of males to that of females if it was up to 1 day of age. In a second experiment, both sexes preferred fresh odors from either source over the same scents that were older. In conclusion, information about sex may get lost with the age of the scent mark.

Pheromones in urine will suffer degradation, hydrolysis, oxidation, and ultraviolet radiation effects. For example, the (Z)-7–12-acetyl derivative in elephant urine will gradually hydrolyze (Rasmussen, 1988). In this case, the lipoprotein carriers of the elephant acetate may also determine the life time of the signal besides serving to filter and select odorants, confer specificity, and play a critical role in the transport and transfer of an active ligand to the vomeronasal organ (Rasmussen and Schulte, 1998).

Keeper substances and commonly used marking substrates such as clay and wood can extend the "fade-out time" of the signal. Licking or exhaling on scent marks or the body surface of another individual liberates "on-demand odors". Alberts (1992a) characterized such signals as having multiple rise times.

2.10 Spatial range of odor signals

Animals boost the range of a signal in various ingenious ways. Multiple scent marks, distributed over territory or home range, reach more receivers and assure reception by one individual as it moves about and misses certain places. Species using multiple marks are commonplace. Examples are hyenas, wolves, antelopes, and beavers. Placing a scent mark as high as possible maximizes its range (active space).

2.11 Interaction of olfactory and visual signals

Rocks, stumps, trees, scraped soil, scarred plant stems, or otherwise disturbed vegetation attract the attention of animals who then examine the spot more closely for olfactory information left by conspecifics. The "urine balls" made by sandrats of the Algerian Sahara provide an elaborate example (Daly and Daly, 1975). Sandrats dig up sand, urinate on it and work it into a ball that other sandrats examine. Urine balls from estrous females strongly attract males.

FIGURE 2.5 Visual signal is combined with odor release when an alarmed pronghorn (*Antilocapra americana*) flares his "rump patch" of long hair covering the ischiadic gland. (Photograph: D. Müller-Schwarze.)

Crepuscular and nocturnal animals can use visual signals to attract attention to their scent glands. Hair tufts or brightly colored hair that stands out against the background of the animal's fur highlight many odor-producing skin glands, especially when hair is erected during a display. Black-tailed and white-tailed deer spread their tarsal hair tuft, located on the tarsal gland. Pronghorn spread white hair on the ischiadic glands ("rump patch;" Fig. 2.5) when alarmed. White, orange, and black hairs draw attention to the dorsal gland of the rock hyrax. Guinea pigs and cuis swivel their rump toward a conspecific and flash the pink inner skin of their perineal pouches as a threat (Rood, 1972).

Visual anomalies in the environment, such as those created by thrashing, pawing, or tearing of bark, render olfactory marks more detectable (Roberts and Gosling, 2001). Finally, the height of marks on tree trunks, saplings, or rocks may signal the size of the marking animal.

3

Odor production and release

The odour emitted must be of considerable importance to the male, inasmuch as large and complex glands, furnished with muscles for everting the sack, and for closing or opening the orifice, have in some cases been developed. The development of these organs is intelligible through sexual selection, if the most odoriferous males are the most successful in winning the females, and in leaving offspring to inherit their gradually-perfected glands and odours.

CHARLES DARWIN: *The Descent of Man*, p. 530

Among mammals, which mostly "think through their noses," it is not surprising that *olfactory* marking of their territory plays a large role. The most diverse methods have been chosen, various scent glands evolved, and the most remarkable displays of depositing urine and feces developed; of these, the leg-lifting of the domestic dog is most familiar to everyone.

KONRAD LORENZ: *On Aggression (1963)*, p. 53 of original German version, translated by author.

Chemosignals in vertebrates come from a great variety of sources. These sources include excretions, secretions, material recycled from other organisms, and even from the environment.

3.1 Signals in excretions

Metabolites in urine or feces provide the energetically least expensive, and evolutionarily probably the original, chemical signals in vertebrates. "Much of history of evolution has concerned the development by living things of responses to metabolites, sometimes their own and sometimes produced by others. Those organisms which developed 'satisfactory' responses succeeded, and those which did not, failed." (Lucas, 1944). Interested parties, such as members of the opposite sex, can then "spy" and read pertinent information about sexual and dominance status, health and body condition, quality of diet, and more. For instance, female goldfish release sex pheromones in their urine that

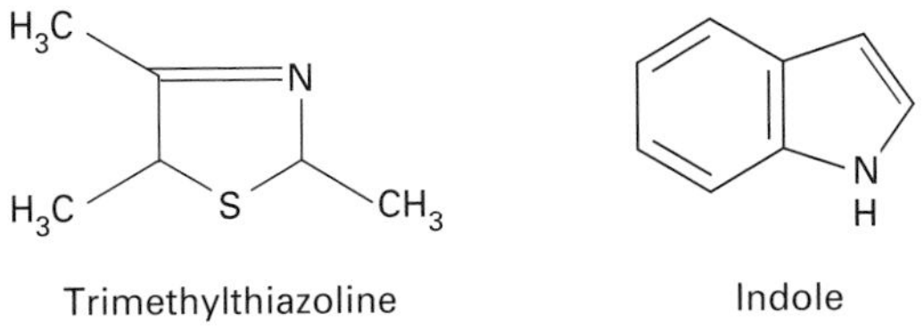

Trimethylthiazoline Indole

FIGURE 3.1 Structures of trimethylthiazoline, found in red fox feces, and indole, found in dog feces (also in coal tar, orange blossoms).

are precursors of androgens and estrogens. In terrestrial animals, potentially many volatile, but also less-volatile, components of urine or feces may serve as chemical cues. For instance, experiments suggested that in the shrew, *Crocidura russula*, males locate females by markers in urine or fecal pellets (Cantoni and Rivier, 1992).

3.1.1 Urine

The volatiles in urine of mice, rats, coyotes, white-tailed deer, and many other species carry chemical cues. Male–male aggression in mice is triggered by a urinary mixture of 2-*sec*-butyl-4,5-dihydrothiazole and 3,4-dehydro-*exo*-brevicomin (Novotny *et al.*, 1985b; Harvey *et al.*, 1989). Dominant males emit (*E*,*E*)-α-farnesene and (*E*)-β-farnesone (Harvey *et al.*, 1989). Adrenal-dependent urinary volatiles, found in female mice, delay puberty in other females. The time of the first vaginal estrus in delayed by appoximately 2 days by *n*-pentyl-acetate, *cis*-2-penten-1-yl acetate, and 2,5-dimethylpyrazine, *Odocoileus hemionus columbianus*, (Novotny *et al.*, 1985a). The "deer lactone" occurs in urine of black-tailed deer and is applied to the hair tuft of the tarsal gland on the deer's hock. There it forms part of a recognition odor (Müller-Schwarze, 1971). Beavers use castoreum in their castor sacs for scent marking. This secretion is, in large part, concentrated and perhaps slightly modified urine.

3.1.2 Feces

Red fox, *Vulpes vulpes*, feces contain trimethylthiazoline (2,5-dihydro-2,4,5-trimethyl thiazole; Fig. 3.1) which alarms a prey species, the Norway rat (Vernet-Maury, 1980).

Indole (Fig. 3.1), in dog feces a smelly matabolite has no effect on sheep (Arnould *et al.*, 1998), although its odor is strong and repulsive to humans. However, mixtures of fatty acids and neutral compounds, as well as sulfur compounds, from dog feces inhibit feeding in sheep (Arnould *et al.*, 1998).

Often animals add glandular secretions to their excretions to produce a complex specific signal. Urine marks of lions and tigers are examples. A whitish material in urine marks, first thought to represent anal gland secretion, tested positive for urinary lipids. When dyed with green food dye, anal gland secretion failed to show up in urine marks. These lipids may both prolong pheromone release from the urine mark and indicate the general nutritional condition of the animal, as urinary lipids correlate with kidney fat (Asa, 1993).

Some pheromones in urine may originate in the reproductive tract, rather than being metabolites. In white-tailed deer, *Odocoileus virginianus*, estrus urine elicits some chases and associating by bucks, but vaginal secretion is more active. It is assumed that the sexual signal originates in glands of the reproductive tract and is transported by urine to the outside (Murphy *et al.*, 1994).

3.2 Glandular secretions

Vertebrates, especially mammals, have evolved a bewildering variety of specialized glands that produce secretions which in turn carry chemical signals, independently of excretions.

3.2.1 Fish

In fish, proven or suggested sources of pheromones are, in addition to urine and feces (including bile), the reproductive organs, skin, and secondary sex structures. Several glands are sources of reproductive pheromones in males: the blenny, *Blennius pavo*, has specialized anal fin appendages that enlarge at maturity and produce a factor that attracts females (Lauman *et al.*, 1974). A second type of gland, the caudal gland in males of the glandulocaudine fishes, is suspected to be its source of sex pheromones. It is located at the base of the tail fin and covered by a large, highly modified scale, or several scales (Fig. 3.2). The glands' secretory cells shrink when males are kept isolated, and enlarge with courtship (Nelson, 1964). Courtship behavior includes tail fanning. It is thought that such movements provide a passive pumping mechanism that releases pheromones from the gland and propels them toward the female (Weitzman and Fink, 1985). In fish, female pheromones typically originate in the ovaries.

3.2.2 Amphibia

Amphibians have mucus glands over their whole body to provide the mucus needed to keep their skin moist, and fields of granular glands that secrete alkaloid toxins. In addition to these general glands, salamanders possess glands

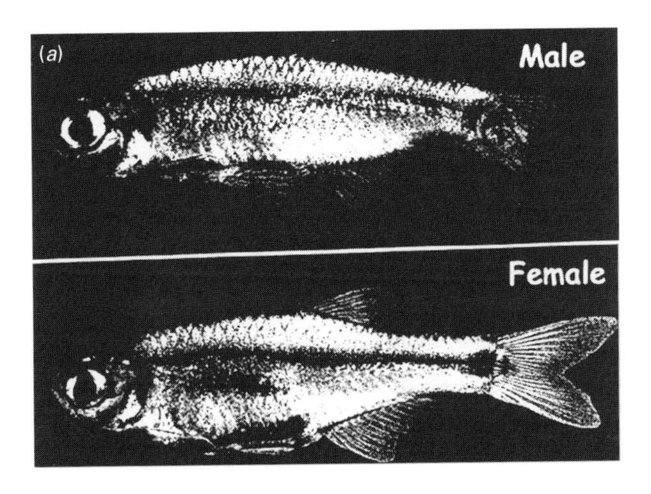

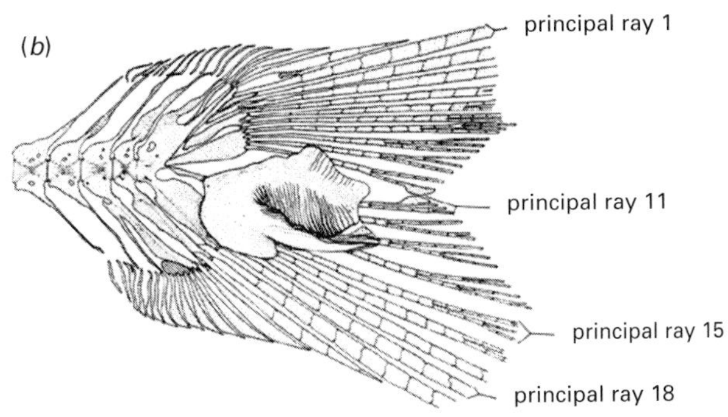

FIGURE 3.2 Glands in fish. (*a*) Caudal gland at the tail base of *Scopeocharax atopodus* (Characidae). Note the difference between male and female. (*b*) The pouch scale is a large, modified scale that covers the caudal gland in a glandulocaudine fish (*Iotabrycon praecox*). (From Weitzman and Fink, 1985.)

of three different types in the cloacal area: cloacal, pelvic, and abdominal glands. These are generally more developed in males. Pelvic and cloacal glands function in spermatophore formation. A communication function of these glands is starting to emerge. Male red-bellied newts, *Cynops pyrrhogaster*, attract females with their abdominal gland secretion. The active ingredient was shown to be a decapeptide, named sodefrin (Kikuyama *et al.*, 1995).

Some salamanders, such as *Plethodon*, have "hedonic" glands on their chin; glands occur also around the eyelids, temporal region, dorsal tail base or entire dorsal region in various species (Fig. 3.3). Finally, the plethodontid salamanders have nasolabial glands on their lips. The glands hypertrophy seasonally in males, and *Plethodon cinereus* show "nose-tapping" and "chin-touching" of the substrate, suggesting a communication function (Tristram, 1997). A "unique cloacal vent gland" has been described for male *Rhyacotriton olympicus* (Dicamptodontidae).

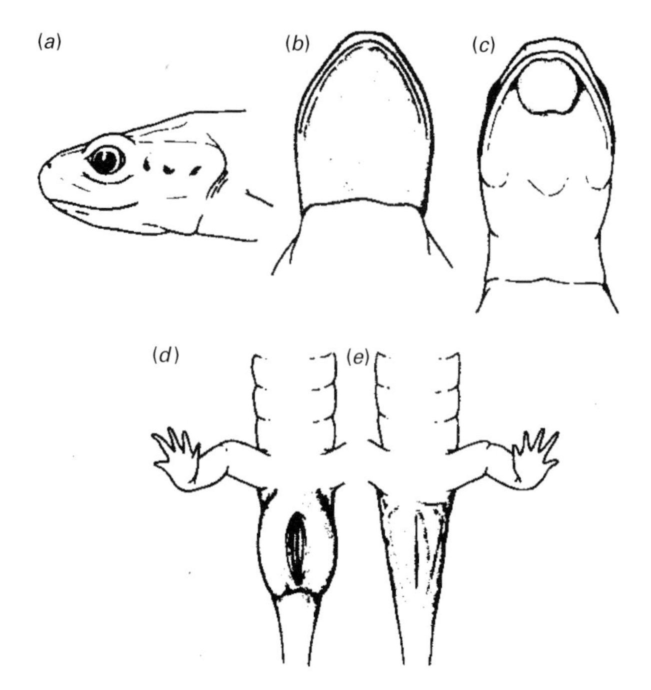

FIGURE 3.3 Glands in salamanders. (*a*) Genial glands on the side of the head of *Notophthalmus viridescens*; (*b*) diffuse submandibular glands of *Taricha torosa*; (*c*) mental gland of *Pseudoeurycea smithii*; (*d, e*) *Ambystoma jeffersonianum*: in the male (*d*) the glandular area around the vent and cloacal papillae is swollen, which is not the case for the cloacal area in the female (*e*). (From Duellman and Trueb, 1986.)

This species may release pheromones during "tail curling" (Sever, 1988). Little is known about glands for communication in anurans, other than the general mucus and granular glands (Fig. 3.4). In the Australian tree frog *Litoria splendida*, the pheromone *splendipherin* is produced in rostral and parotoidal glands on the head of the male (Wabnitz *et al.*, 1999).

3.2.3 Reptiles

In lizards, femoral glands (Fig. 3.5) are arranged in rows on the ventral surface of the hindleg. They are most active in the breeding season and are larger in males than in females. Femoral secretion stimulates tongue flicking and may serve in species recognition and home-range marking. Chemically, more lipids can be extracted from femoral gland secretions during, than outside, the breeding season. Saturated C_{14}–C_{26} fatty acids, unsaturated C_{16}–C_{24} fatty acids, and eight sterols were found, including cholesterol (Alberts *et al.*, 1992).

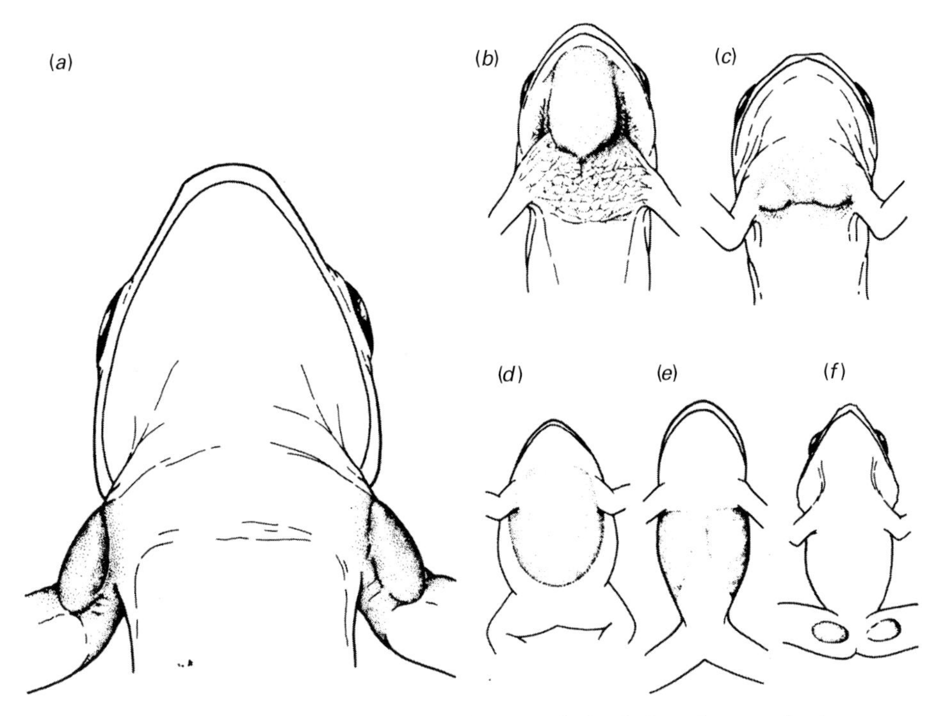

FIGURE 3.4 Glands in male frogs. (*a*) Mental gland of *Kassina senegalensis*; (*b*) pectoral glands of *Leptopelis karissimbensis*; (*c*) humeral glands on the foreleg of *Hylarana albolabris*; (*d*) abdominal gland of *Kaloula verrucosa*; (*e*) ventrolateral glands of *Ptychohyla schmidtorum*; (*f*) femoral glands of *Mantidactylus pseudoasper*. (From Duellman and Trueb, 1986.)

In desert iguanas, *Dipsosaurus dorsalis*, the femoral glands are more developed in males than in females. The percentage of females with active glands varied with the population density of three populations studied. In the one with the highest density, 12.3% of the females had active glands. At the intermediate site, 31.7% of the females had active glands, with 59.2% at the low-density site (Alberts, 1992b).

Snakes have paired "scent glands" that open into the cloaca. Some snakes such as *Natrix* and *Macropisthodon* spp., have "nucho-dorsal" glands under the skin of the dorsal neck area. Their secretions are assumed to serve in defense and/or intraspecific communication (Madison, 1977). Turtles have "mental glands" on their chins, inguinal, and axillary ("Rathke's glands;" Fig. 3.6), lateral, and cloacal glands. Their role in producing chemical signals is little explored. Crocodiles have three main types of "musk glands": mandibular (or throat) musk glands, cloacal, and dorsal glands. The mandibular glands are most active during the breeding season and are discharged during courtship behavior.

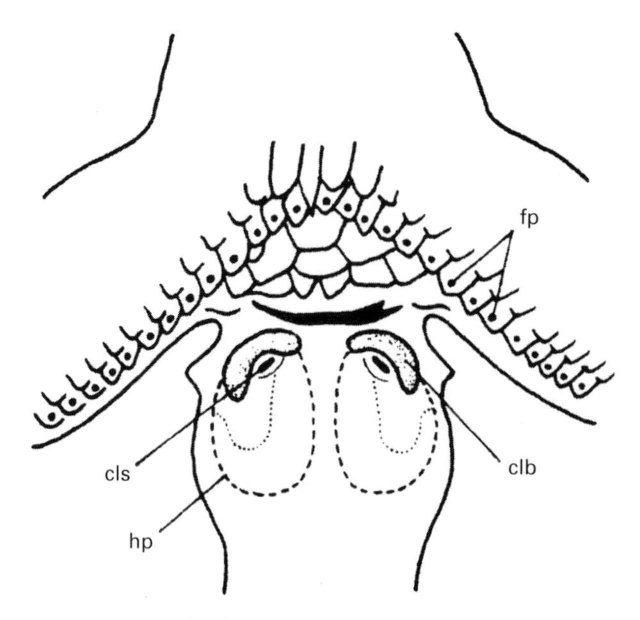

FIGURE 3.5 Glands in lizards are illustrated by the femoral pores and cloacal sacs in a gecko, *Gymnodactylus pulchellus*. clb cloacal bone; cls, cloacal sac opening; fp, femoral pore; hp, hemipenis. (From Bellairs, 1970.)

Skin glands are thought to have evolved from scraped-off corneous material of the skin. Such "uncontrolled semiochemical release" may have given way to "facultative, controlled semiochemical release" of desquamated, loosely joined plugs of keratinous material when the germinal portion of the integumentary gland moved "below" the body surface. Invagination then created a "gland" (Maderson, 1986).

3.2.4 Birds

Birds are not exactly known for scent communication or scent-producing organs. The preen (uropygial) gland is usually considered the only, most developed, and ubiquitous skin gland in birds. It serves to waterproof the plumage and is larger in aquatic than in land birds. However, there are also sex differences in the composition of waxes in the uropygial gland of domestic ducks, possibly for communication (Jacob *et al.*, 1979). In the hoopoe, *Upupa epops*, the uropygial gland is largest at 12 days of age. Surrounded with a ring of specialized feathers, it is said to have a repellent effect on intruders. A predator is squirted with liquid from the large intestine, accompanied by a hissing sound (Sutter, 1946). The roles of cloacal and anal glands in birds are poorly understood (Quay, 1977).

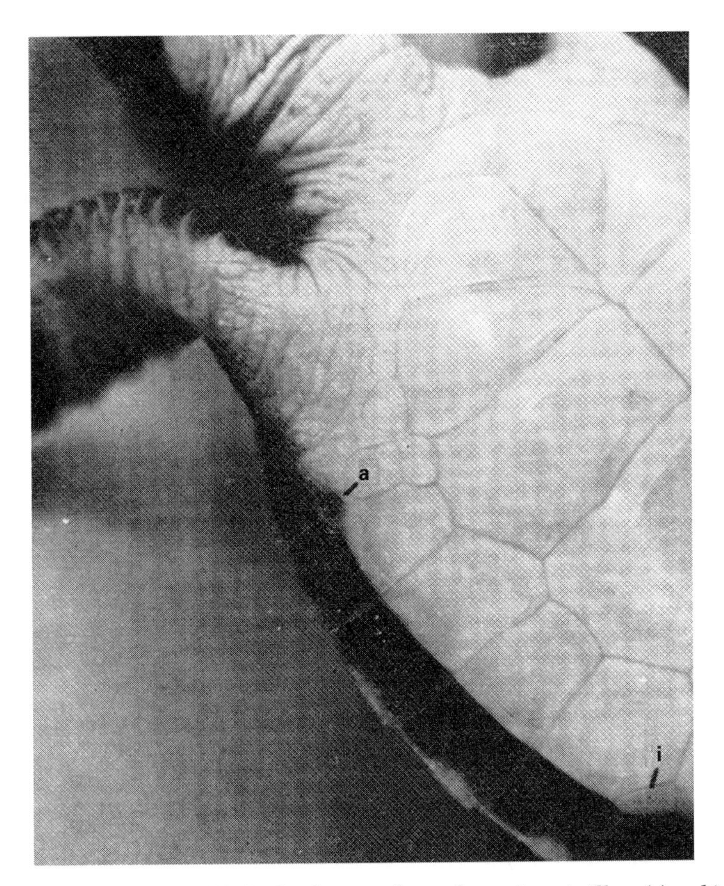

FIGURE 3.6 Rathke's glands in turtles and tortoises. Axillary (a) and inguinal (i) pores (gland openings) are shown in *Chelonia mydas*. (From Ehrenfeld and Ehrenfeld, 1973.)

3.2.5 Mammals

Skin glands in mammals are widespread and diverse. We know so much about their behavioral roles and secretion chemistry that even a brief survey would fill several volumes. Skin glands can occur on all body parts but are found more often on the head, the extremities, and the anogenital region. Animals grow scent glands where vehicles carry lights: front and rear. But there are also neck, dorsal, ventral, and lateral glands. The two basic types of gland component are sebaceous and apocrine (modified sweat) glands (Table 3.1 and Fig. 3.7). Sebaceous gland produce a lipid secretion, while apocrine glands secrete a milk-like aqueous fluid. Both types of gland are associated with hair follicles and secrete into the pilosebaceous canal. By contrast, true sweat (eccrine) glands are not associated with hairs and secrete directly to the skin surface. Sebaceous and apocrine

Table 3.1 Examples of specialized mammalian skin glands

Dominating gland type	Skin gland organ	Species
Holocrine (sebaceous)	Ventral gland	Mongolian gerbil, *Meriones unguiculatus*
	Tarsal gland	Black-tailed deer, *Odocoileus hemionus columbianus*
	Subauricular gland (jaw patch)	Pronghorn, *Antilocapra americana*
Apocrine (modified sweat glands)	Metatarsal gland	Black-tailed deer
	Axillary gland	Humans

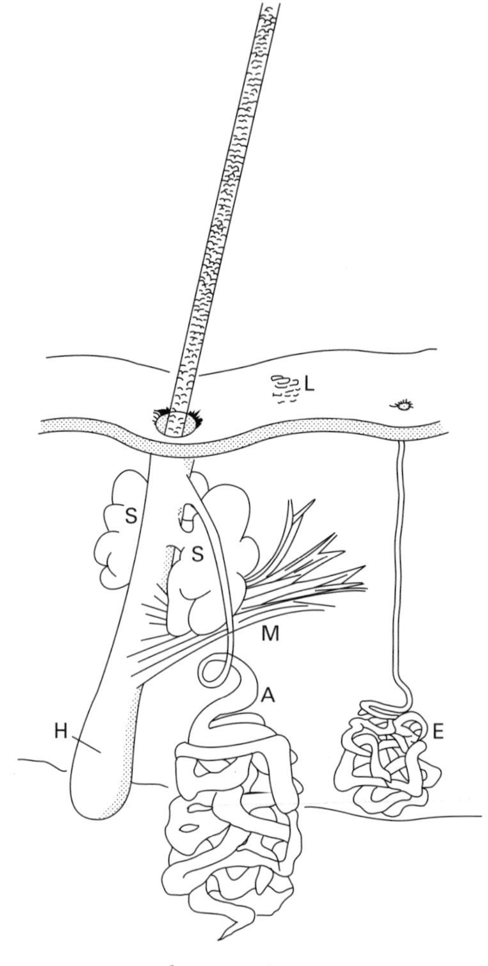

FIGURE 3.7 Elements of mammalian skin glands. S, sebaceous gland; A, apocrine gland; E, eccrine gland; M, errector pili muscle; H, hair follicle. (From Albone, 1984.)

FIGURE 3.8 Subauricular gland (SG: the "jaw patch", a black area below the ear) in male pronghorn, *Antilocapra americana*. (Photograph: D. Müller-Schwarze.)

glands can occur in concentrations of a single type, or combined, depending on the function of a particular gland and its secretion: for scent marking on the substrate, a lipid secretion from the sebaceous glands is called for. An example is the subauricular gland of the pronghorn, *Antilocapra americana* (Fig. 3.8). Another example of a skin gland dominated by sebaceous elements is the tarsal organ (Fig. 3.9) in deer of the genus *Odocoileus* (white-tailed, black-tailed, and mule deer). The odors that are supposed to volatilize fast, as alarm pheromones, typically arise in aqueous secretions from modified sweat glands. Examples are the axillary gland in humans and the metatarsal gland on the hindlegs of deer of the genus *Odocoileus* (Fig. 3.10). In the female brown hyena (*Hyaena brunnea*) two separate glands, one sebaceous, one apocrine, produce a double scent mark (Mills *et al.*, 1980).

In most mammal species the skin glands typically occur only in, or are larger or more numerous in, males. For example, only the male pronghorn has a pair of

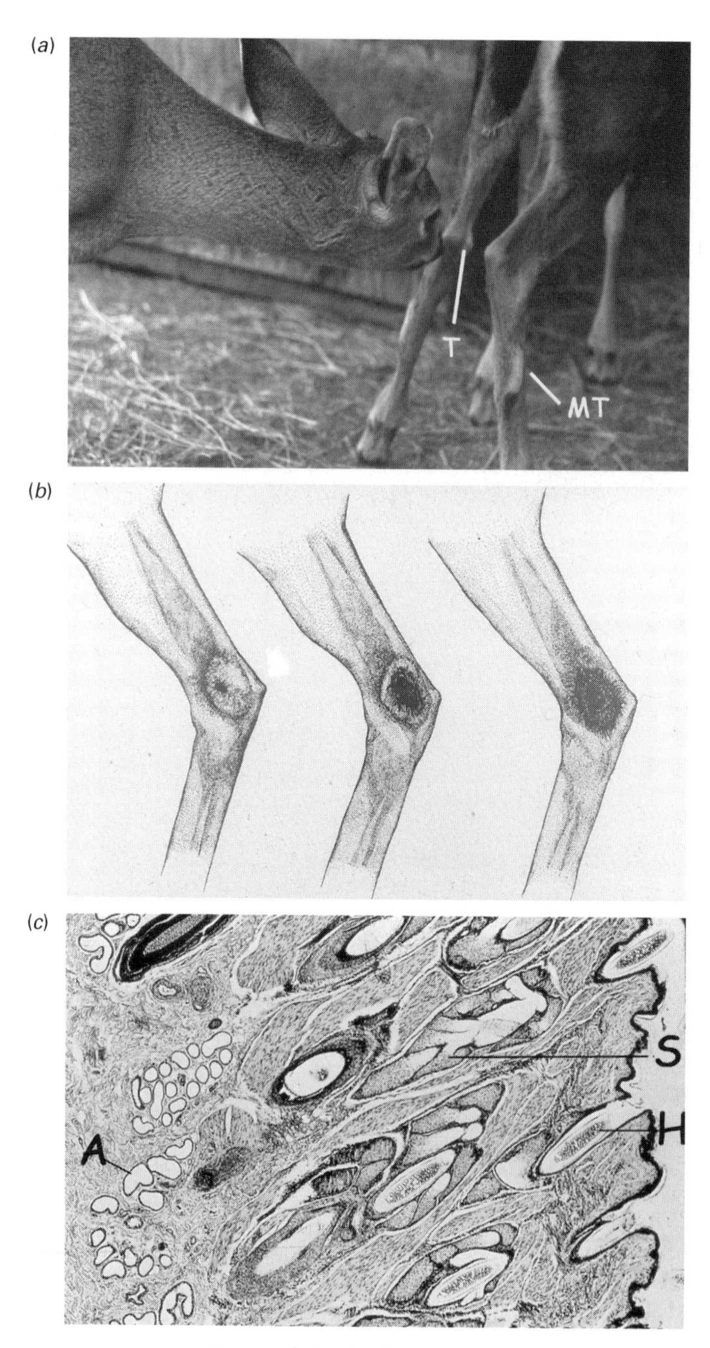

FIGURE 3.9 The tarsal gland of black-tailed deer, dominated by sebaceous glands. (*a*) Location of the gland on the hock (T); a young male sniffs the gland. MT, metatarsal gland. (*b*) Increasing opening of the secretion- and urine-covered tuft of modified hair is seen left to right. (*c*) A section of skin (surface on the right) showing the tarsal gland with its apocrine (A) and sebaceous (S) gland elements and a hair (H). (From Quay and Müller-Schwarze, 1970.)

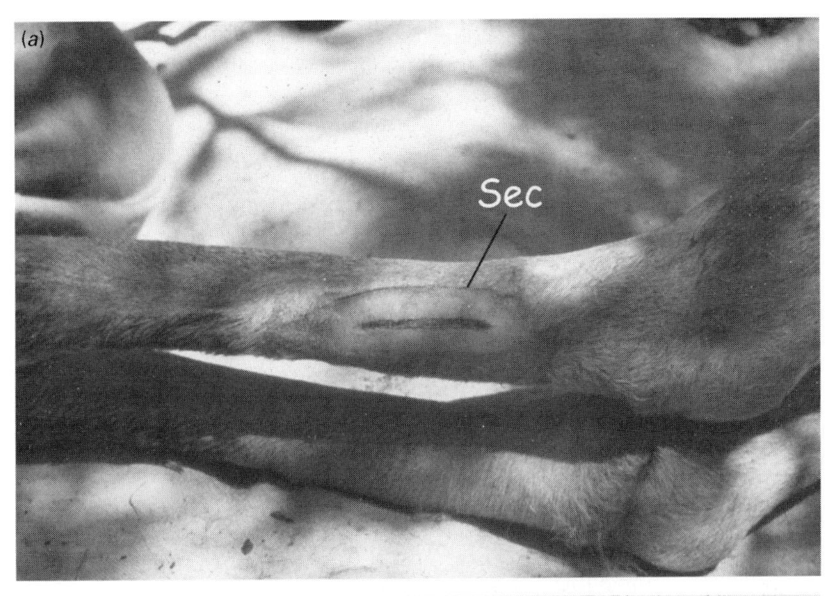

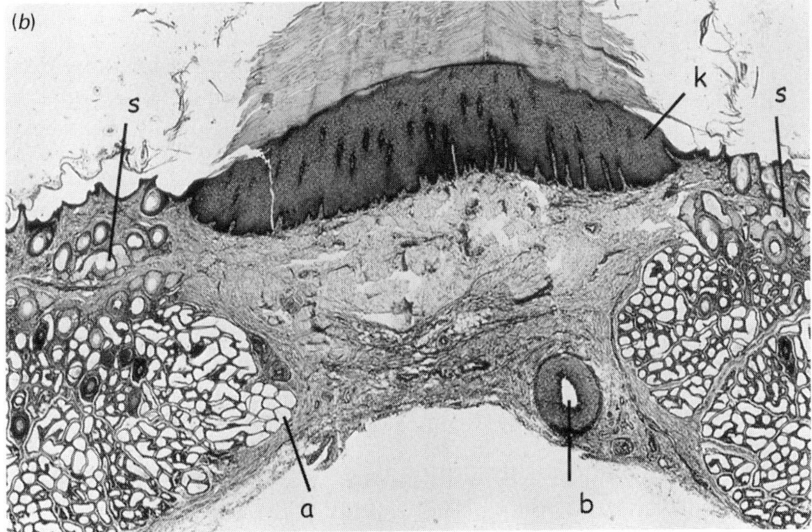

FIGURE 3.10 The metatarsal gland of black-tailed deer, dominated by apocrine glands. (*a*) The location on the outside of the hindleg; there is a dark keratinized ridge in the center with glandular tissue either side. The droplets are secretions (Sec), stimulated by epinephrine injections. (*b*) Histology of a 4-year-old male shows the metatarsal gland. a, apocrine gland portion; b, blood vessel; k, keratinized ridge; s, sebaceous gland. (Photograph (*a*): D. Müller-Schwarze; (*b*) from Quay and Müller-Schwarze, 1970.)

subauricular glands and a single dorsal gland. Male wolves, especially the alpha-male, deposit anal gland secretion on feces more often than do females or juveniles (Asa *et al.*, 1985). However, skin glands can be larger in females, as in the cotton-top tamarin, *Saguinus o. oedipus* (Epple *et al.*, 1988).

The Harderian gland, wrapped behind the eyeball, is especially prominent in rodents. It produces a variable, primarily lipid secretion and drains through the Harderian-lacrymal tract and the external nares. In Mongolian gerbils, the secretion is spread over the body during grooming (Thiessen *et al.*, 1976). A pheromonal function has not been clearly established.

The composition of a scent mark of glandular origin may be quite different from that of the original glandular secretion. For example, the scent marks that rabbits have applied to poles by "chinning" behavior are impoverished compared with the secretion on the chin or the head (Hayes *et al.*, 2002). However many scent marks comprise several different secretions. Two examples are the beaver, which marks the same spot with castoreum and anal gland secretion, and the female giant panda, whose scent mark contains compounds from glands, urine, and the vagina (San Diego Zoo, 2002).

Saliva may also contain specific chemical information for individual recognition (Blass and Teicher, 1980) or kin recognition (Block *et al.*, 1981; Smith and Block, 1990). The best-known example of a salivary pheromone is the mixture of androstenone and androstenol in the submaxillary glands of the boar.

Blood can be a source of pheromones. The chemical cues in secretion from the mucocutaneous junction in domestic cows that stimulate sexual behavior in bulls have been traced back to the blood (Rivard and Klemm, 1989).

3.3 Body odors and body region odors

"Body odor is the sum of all perceivable compounds in excreta and secreta" (Bryant and Atema, 1987). Body odors can change with diet but are nevertheless important in communication: bullhead catfish (*Ictalurus nebulosus*) use body odors in dominance and territorial relationships.

Odors from different body parts of the same individual can differ in their biological effects. In reptiles, dorsal skin odors of colubrid snakes release body bridging in crotaline snakes, but ventral and anal gland secretion do not (Bogert, 1941). Also, the urine of wolf, coyote, and red fox inhibits feeding in herbivorous mammals, while the feces of these species do not (Sullivan *et al.*, 1985a). Hamsters (*Mesocricetus auratus*) did not respond to polecat (*Mustela putorius*) urine but showed an extreme negative reaction to bedding (possibly fur odor) of polecats and other carnivores (Dieterlen, 1959).

Multiple sources can make for complex signals. For example, odors emanating from the anogenital area of the cow can arise from feces, bladder urine, secretions from the upper reproductive tract, vagina, vestibular glands, enlarged sweat and sebaceous glands of the vulval skin, and the activity of microorganisms (Albone *et al.*, 1986). Other examples of complex odor sources are the anal glands of the brown hyena, urine–tarsal gland interactions in black-tailed deer, and the diverse glands in the anal area of the dog (Schaffer, 1940).

Different secretions and body regions can provide redundant signals. Meadow voles, *Microtus pennsylvaticus*, investigate odors from urine, feces, or anogenital area of the opposite sex more than those of the same sex. Males sniffed scents from the mouth of females more than other regions, but females did not discriminate. Finally, males and females investigated odors from the posterolateral region of males more than those of females (Ferkin and Johnston, 1995a).

3.4 Diet influences on odor production and venoms

3.4.1 Fish

Changing the diet of a fish may change the behavior of conspecifics it interacts with subsequently. For instance, if one of a pair of male brown bullhead, *I. nebulosus* (a catfish), is removed from the tank and fed beef liver instead of the usual trout chow and then returned to his partner in their original tank, the resident will behave differently than if the same male is reintroduced without a diet change. The former tank mate is now a "chemical stranger." The behavior changes include loss of territory and more activity by the smaller, manipulated fish and more aggression and activity by the resident fish. These diet-dependent odors are not specialized pheromones, and yet they are probably important social chemical cues in the natural territorial and dominance behavior of bullhead catfish. "Body odor" is the more appropriate term (Bryant and Atema, 1987).

3.4.2 Amphibians

The toxic alkaloids of poison dart frogs appear to depend on diet (Daly *et al.*, 1994, 2000; Caldwell, 1996). Details are described in Section 10.2.2.

3.4.3 Reptiles

Snake venom composition can reflect diet. The Southeast Asian pitviper, *Calloselasma rhodostoma* (Viperidae), preys on different animals in different regions of its range. In some areas, only reptiles are taken; in others endotherms predominate in their diet, and in still others the diet includes a large proportion of amphibians. Accordingly, venom composition varies with diet. Other possible explanations for this variation, such as geographic proximity, or patristic (phylogenetic) distance, were ruled out statistically (Daltry *et al.*, 1996).

3.4.4 Birds

Similar to amphibians, the pitohui, a toxic bird in New Guinea, appears to derive its toxic alkaloids from its invertebrate diet (Dumbacher *et al.*, 2000).

3.4.5 Mammals

Several rodent species produce social odor signals that vary with diet. Lactating laboratory rats produce an odor in their caecal contents that attracts their own young. However, the pups are also attracted to the caecal odor of a different female if she lived on the same diet as their mother. The kind and numbers of caecal bacteria also change with diet. Such a variable odor implies that the young have to learn their mother's cues (Leon, 1975). Male guinea pigs (*Cavia* sp.) investigate female urine for a longer time if the donor females were eating standard guinea pig diet than if they were fed a commercial rat chow. It is possible that chemical signals in the urine provide information on available food sources (Beauchamp, 1976). In addition to fecal and urinary odors, scent gland secretions can also be altered by diet. Young of the Mongolian gerbil, *Meriones unguiculatus*, prefer odors of adults who share their and their parents' diet. The odor of whole animals and that of soiled sawdust or sebum from the ventral scent gland carry the diet-dependent signals (Skeen and Thiessen, 1977). Many compounds that are found in castoreum of beaver, *Castor canadensis*, are known as metabolites of compounds in trees the beaver feeds on (Fig. 3.11).

Diet is only one of the factors responsible for odor variability. For example, genotype and diet contribute additively to the complex odor signatures that permit kin recognition in spiny mice, *Acomys cahirinus* (Porter *et al.*, 1989), and the levels of 5-methyl-2-furoic acid and homogentisic acid in the sternal gland secretion of brushtail possums, *Trichosurus vulpecula*, rise after experimental feeding on *Eucalyptus* for 2 days (Salamon, 1995).

To discriminate diet-dependent odors can be vital in the context of reproductive behavior. Supporting the hypothesis that animals discriminate and prefer potential mates that are in good nutritional condition, Ferkin *et al.* (1997) showed that meadow voles preferred odors of members of their own species that are on a high-protein diet (Table 3.2).

Table 3.2 Some examples of diet-dependent composition or effect of mammalian secretion

Species	Secretion	Dietary factor	Affected compound(s) or response
Brushtail possum *Trichosurus vulpecula*	Sternal gland	Eucalyptus leaves	5-Methyl-2-furoic acid, homogentisic acid
Guinea pig	Urine	Commercial guinea pig food versus rat food	Preference by males
Mongolian gerbil	Ventral gland	Type of laboratory chow	Phenylacetic acid; pups prefer odor of gerbils on same diet as their mother
Meadow vole	Anal gland, urine, feces	Protein, 9, 15 or 25%	Mate choice: voles with high-protein diet preferred

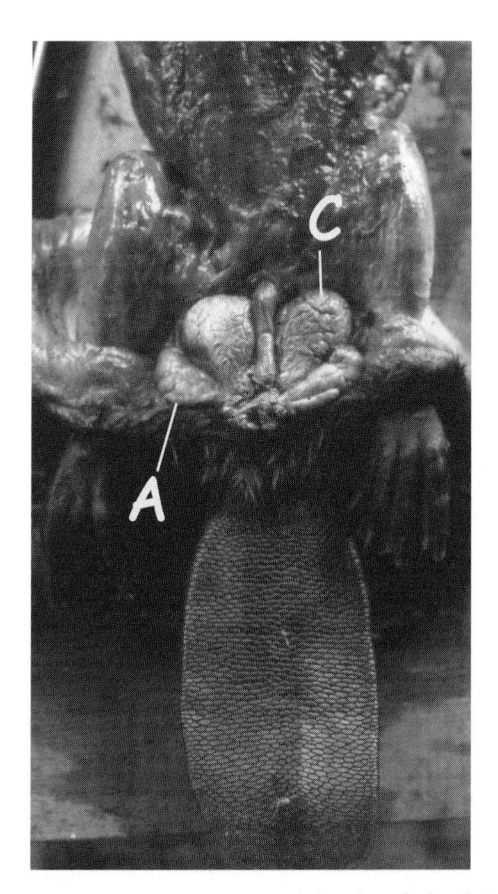

FIGURE 3.11 Castor sacs (C) and anal glands (A) in the beaver. (Photograph, D. Müller-Schwarze; dissection, B. Stagge.)

3.5 Hormonal control of odors in urine and secretions

Numerous studies have shown that reproductive hormones regulate skin gland activity.

3.5.1 Reptiles

The protein-rich secretion of the femoral gland of the agamid lizard, *Amphibolurus ornatus*, depends on hormonal control (Fergusson *et al.*, 1985). In male iguanas, *Iguana iguana*, femoral gland activity and androgen levels are correlated with dominance status (Alberts, 1993). In advanced snakes, the levels of non-volatile integumentary lipids that serve in species identification vary with hormonal state, skin-shed state, and season (Mason, 1992).

3.5.2 Mammals

Testosterone stimulates secretion in sebaceous and apocrine glands in rabbits, rats, guinea pigs, and hamsters, while estrogen in most cases inhibits secretion. Progesterone tends to be neutral (reviewed by Ebling, 1977).

Androgens affect many skin glands. An example is the flank gland of the golden hamster, *M. auratus*. The size and pigmentation, combined in an index, correlate with the relative levels of circulating androgens (Vandenbergh, 1973). The perineal and neck glands of cattle calves are affected by estradiol-17-β, but in different ways. The volume of the sebaceous glands in the perineal region increases, while that of the neck glands decreases. The sweat glands in the perineal region also increase in size while those in the neck do not change (Blasquez *et al.*, 1987). Elephants have high concentrations of testosterone and dihydrotestosterone in their temporal glands (Rasmussen *et al.*, 1984). The size of the snout scent gland (morilla) of the male capybara (*Hydrochaeris hydrochaeris*) is significantly correlated with the testes mass, independently of body size. Age did correlate with morillo and testes size. Capybaras live in stable social groups with a male hierarchy. The dominant male scent-marked most often and had higher reproductive success (Herrera, 1992).

As expected, castration affects gland size, activity, and composition of the secretion profoundly. For example, castration changes the levels of porphyrins, indoles, and proteins in the Harderian gland secretion of the Syrian hamster (Buzzell *et al.*, 1991).

The secretion of the sebaceous glands on the haunch of male and estrous and diestrous female rats varies considerably between individuals. No one single compound is characteristic of sex or reproductive condition. Gas liquid chromatography with principal component analysis of the 22 common peaks showed sex-specific patterns and 79.5% of variation was linked to sexual status (Natynczuk and MacDonald, 1992).

Hormone implants can be used for long-term stimulation or suppression of odor production. Scent marking is stimulated by testosterone in gerbils and wolves (Asa *et al.*, 1990).

Urine composition depends on testosterone, as in male mice (Schwende *et al.*, 1986) and the wolf, *Canis lupus* (Raymer *et al.*, 1986). In the wolf, these volatiles signal both sex and sexual maturity (Raymer *et al.*, 1986). Similarly, composition of female urine changes with the estrus cycle (Schwende *et al.*, 1986).

3.6 Microbial odors

Microbial activity can be critical for odor production in glandular secretions. Frequent body contact in social animals ensures sharing of microorganisms, which, in turn, results in a shared group odor.

In the anal sac contents of the red fox, *V. vulpes*, six genera of bacteria have been identified. The most common is *Clostridium*, with nine species. Also found were *Eubacterium*, *Bacteroides*, *Peptostreptococcus*, *Bifidobacterium*, and *Fusobacterium* (Ware and Gosden, 1980). These bacteria produce fatty acids. Aerobic microorganisms in the anal gland of the red fox include *Streptococcus* spp., *Proteus* spp., coliform bacteria, *Staphylococcus* spp., *Pseudomonas* spp., *Neisseria* spp., and *Bacillus* spp. (Gosden and Ware, 1976).

Svendsen and Jollick (1978) studied the microbes in the castor sacs and anal glands of the beaver, *C. canadensis*. No bacteria cultured from the castor sacs, while the anal sacs contained the aerobe *Escherichia coli* and the anaerobe *Bacteroides fragilis*. Species and numbers of bacteria did not differ between the sexes, age classes, or beaver colonies.

Female laboratory rats seem to depend upon an intact vaginal bacterial flora to be olfactorily attractive to males. When given a choice in a four-arm maze, adult males spend more time with an untreated estrous female than with one whose vaginal bacteria had been killed by daily injections of an antibiotic (Merkx *et al.*, 1988). Generally speaking, individual odors of rats have microbial, genetic, and dietary components (Schellink and Brown, 2000).

Staphylococcus aureus and the yeast *Candida kruzei* dominate the inguinal gland pouches of rabbits, with *Bacillus subtilis*, *E. coli*, and *Streptococcus faecalis* also occurring (Merritt *et al.*, 1982).

Saddleback tamarin monkeys, *Saguinis fuscicollis*, harbor a complex microflora in the secretions of their circumgenital glands. Coagulase-negative staphylococci, Gram-negative bacteria, *Streptococcus* spp., and coryneform bacteria were most abundant, followed by *S. aureus* and *Bacillus* spp. (Nordstrom *et al.*, 1989).

In the human axilla, bacteria produce odoriferous steroids. The concentration of 5α-androst-16-en-one in the axillae of adult men reduced after treating the "superior" (more-producing) axilla with the germicidal agent Povidone-iodine (Bird and Gower, 1982). Faint or acid odors are associated with micrococcaceae, while the more pungent axillary odor that resembles that of androstene and other $C_{19}\Delta^{16}$-androgen steroids correlates with a strong population of lipophilic diphteroids (Leyden *et al.*, 1981). The typical axillary odor is produced by incubation with coryneform bacteria, but not Micrococcaceae, propionibacteria, or Gram-negative organisms. Less-odorous steroids are transformed to the more odorous 5α-androst-16-en-3-one and 5α-androst-16-en-3α-ol (Gower, 1989). The suggested pathway to 3α(β)-androstenol is from 5,16-androstenol via the 4,16- and 5α(β)-derivatives, and finally 3α(β)-androstenol (Gower *et al.*, 1989).

Besides steroids, axillary odor also contains normal, branched, unsaturated aliphatic acids with 6 to 11 carbon members. Most abundant is (*E*)-3-methyl-2-hexenoic acid. Both this compound and its (*Z*)-isomer are found in the aqueous phase hydrolysate and the aqueous phase incubated with bacteria. It is assumed that precursors are water soluble and converted to odoriferous compounds by the axillary flora (Zeng *et al.*, 1992).

Finally, preputial secretion of male mice probably assumes pheromone activity as attractant for females only after having been metabolized by microorganisms (Ninomiya and Kimura, 1988).

3.7 Reservoirs

After a secretion has been produced, it can be stored for later use or even accumulated for massive or repeated signals. Anal sacs of canids, mustelids, and felids, and the castor sacs of beaver are examples. The skunk provides the most dramatic example for such reservoirs with large amounts of often very potent secretions. We do not understand well if and how the various compounds are transformed in these reservoirs.

3.8 Pheromone transport

Many pheromones travel from the tissue of their synthesis via the bloodstream to the surface of the body. They are also transported by larger molecules, notably proteins, both when being emitted by an odor donor ("outgoing") and when being received in the olfactory mucosa of an addressee ("incoming"). In the saliva of the male pig, the pheromone-binding protein pheromaxein binds outgoing 16-androstene steroids. Two compounds serve as pheromone: 5α-androst-16-en-3α-ol (3α-androstenol) and 5α-androst-16-en-3-one (5α-androstenone). Both are produced in the testes, transported by the blood, concentrated in the submaxillary salivary glands, and bound to pheromaxein there. Pheromaxein is formed in the submaxillary salivary glands (reviewed by Booth, 1989). The binding is strongly temperature dependent. At 4 °C the binding of the pheromone to the protein is unchanged over 168 hours. But at 21 and 37 °C (body temperature) most binding is lost after 72 hours. This is ecologically significant: pigs breed in the cold season (i.e. late autumn and winter). The salivary foam is present around the mouth of the boar when courting a sow, but it also deposited in the environment. On the body with its higher temperature, odor release is facilitated, while away from the animal retention of pheromone in saliva is optimized because the degradation of pheromaxein is retarded at lower temperatures (Booth, 1987).

The androst-16-enes in humans are also produced in the testes but spread widely by transport through the body in blood, saliva, seminal fluid, and axillary secretion of males (Gower, 1989).

3.9 Environmental odors for communication

Some chemosignals are not produced by the sending animal itself, but rather appropriated from other individuals, species, or the environment. "Scent rolling" in manure or on carcasses by canids or hyaenids (Drea *et al.*, 2002) and self-anointing by hedgehogs (Poduschka and Firbas, 1968) belong here. Rats use food odors lingering about their snout for communicating information about food sources (Galef and Kaner, 1980; Galef and Stein, 1985).

Mammals often damage plants or the substrate before or during marking, such as when deer thrash a sapling or tree branch with the head or antlers, or scrape the ground before urinating on it. Odors from the plant or soil may become part of the active signal.

Environmental odors can also play a role in interspecific interactions such as preying. The domestic dog uses cues from disturbed soil and crushed plants when tracking and they can be trained to follow tracks that lack scent. When following a scent, such as a human track, they can be diverted by another scent crossing the first pathway. The choice made by the dog is influenced by the odors and cues available at any point (Most and Brückner, 1936; see p. 414).

The environmental odors used for homing in fish, amphibia, and birds, or possibly for long-range navigation, are not yet understood and will be discussed in Chapter 4.

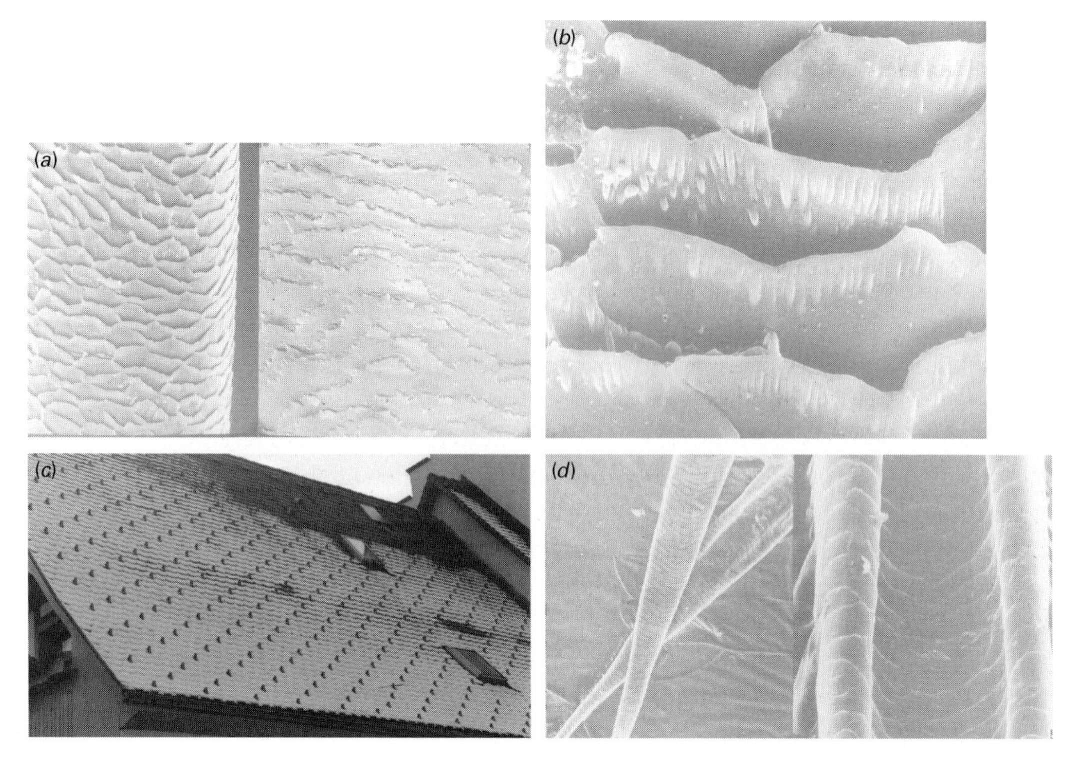

FIGURE 3.12 Osmetrichia (scent hairs). (*a*) Specialized hair from the tarsal gland of the black-tailed deer (left) and human axillary hair (right); (*b*) cuticular scales of tarsal scent hair; (*c*) surface of a snow-retaining roof (Germany); (*d*) modified U-shaped scent hair from the ventral gland of the Mongolian gerbil. (*a, b, d,* from Müller-Schwarze *et al.*, 1977; *c*, photograph by D. Müller-Schwarze.)

3.10 Supporting structures: osmetrichia, muscles

Scent glands often have accessory structures that store secretion, enlarge the surface for scent dispersal, serve as "applicators" during scent marking, or provide opportunities for bacterial action or interaction of compounds that may come from different sources.

Special modified hair in the region of a scent gland can enhance its function. Such "scent hairs" have been termed *osmetrichia* (Müller-Schwarze *et al.*, 1977; Fig. 3.12). They may be stiff bristles with surface chambers formed by their cuticular scales, as in the tarsal tuft of black-tailed deer (Fig. 3.12*a*), spoon or spatula like, as in the ventral gland of the Mongolian gerbil *Meriories unguiculatus* Fig. 3.12*d*), or a "wick" formed by a hollow medulla and vacuolated cortical region, as in the African crested rat, *Lophiomus imhausi* (Stoddart, 1979). Hairs from scent glands in bats of the families Pteropodidae and Molossidae differ from ordinary body hair by being much thicker, and in some species also have a differently sculptured scale pattern on their surface (Hickey and Fenton, 1987). The red fox has heavily sculpted bristles on its supracaudal gland (Brown and MacDonald, 1985).

3.11 Special adaptations for broadcasting chemosignals

The output of signals can increase only within limits because of the cost of producing signal compounds. In addition, as signal strength rises, the effect becomes progressively more inefficient: the perceived odor magnitude increases as a function of signal strength raised to a power of much less than one (i.e. logarithmically). A better way to increase odor detectability is for the receiver to become more sensitive, rather than the sender to emit more material.

Yet there are many ways the producer can improve detectability. Animals increase the spatial range of chemical signals by ingenious means. Many increase the active space, the three-dimensional portion of the animal's home range where the stimulus is above threshold, by broadcasting from an elevated place, as by the mudpiles beavers build for their scent marks. The evaporative surface can be enlarged, as in red deer (wapiti) or goats, which spray urine onto their mane. The saiga antelope (*Saiga tatarica*) wears a large hair tuft below its eye when in its winter coat. It is thought that secretion from the preorbital gland accumulates there on the large surface (Frey and Hofmann, 1997; Fig. 3.13). Male tomb bats saturate their brushlike beard with glandular secretion (Quay, 1970). Hairs on the anal pockets of the male capybara are saturated with secretion that is deposited on vegetation and the ground (MacDonald *et al.*, 1984).

A scent can be propelled into the environment. The most spectacular example is the "manure spreader" of the hippopotamus: the rotating tail sprays about urine and feces (Olivier and Laurie, 1974).

Some mammals have evolved elaborate behaviors to aim odors at conspecifics during critical encounters. Black-tailed deer *O. h. columbianus*, spray urine as a threat. Facing an opponent, a male, female, or fawn rubs the hocks together and urinates over them, soaking the brush-like hair patch on the inner sides of the hocks in the process. Urine odor in the air is noticeable to a human observer farther away than the distance between the interacting animals, and also at a greater distance than during regular urinating. Such rub-urination usually results in spacing of adult deer, while fawns spray urine when suddenly separated from their mother or – in captivity – a caretaker. It is important to know that the hocks

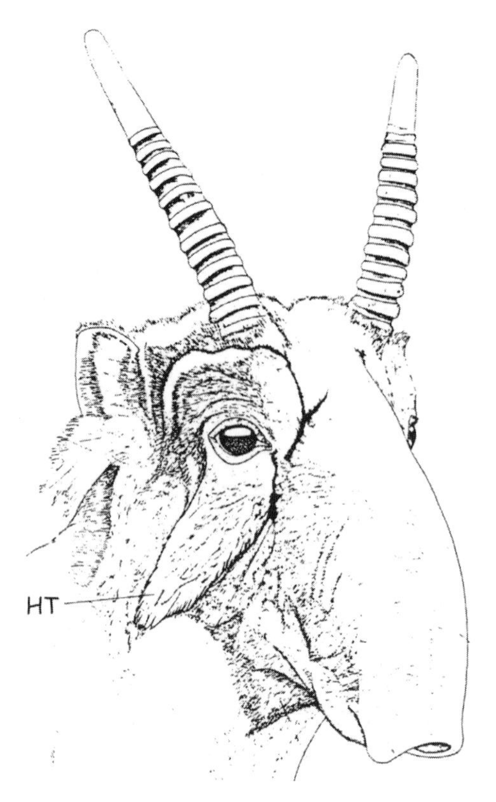

FIGURE 3.13 The hair tuft (HT) of the Saiga antelope is possibly a scent distributor. (From Frey and Hofmann, 1997.)

carry the tarsal gland, dominated by sebaceous glands (Müller-Schwarze, 1971). Gas chromatographic analysis of urine before and after it passes the tarsal tufts showed that during rub-urination material is picked up from the gland and hair surface (D. Müller-Schwarze, unpublished data).

Male bats have pouches on the leading edge of their wings, termed antebrachial sacs. During courtship, males hover in front of, and waft odor toward, roosting females (Bradbury and Vehrencamp, 1977). The sacs themselves do not contain scent glands. Rather, the bat transfers the perfume from other glandular areas on its body. First, a resting male bends his head down toward his genital region and picks up urine with his mouth. Then he licks his wing sac. This is seen as cleaning the pouch. Finally, the male reaches down to the genital region again. This time, he presses his throat with the gular glands to the penis and transfers a droplet of secretion from the penis to the wing sac (Voigt and von Helversen, 1999; Voigt, 2002).

The arrangement of skin glands on the foot of a mouse, for instance, might convey information. As the mouse travels, the glands stamp a pattern on the

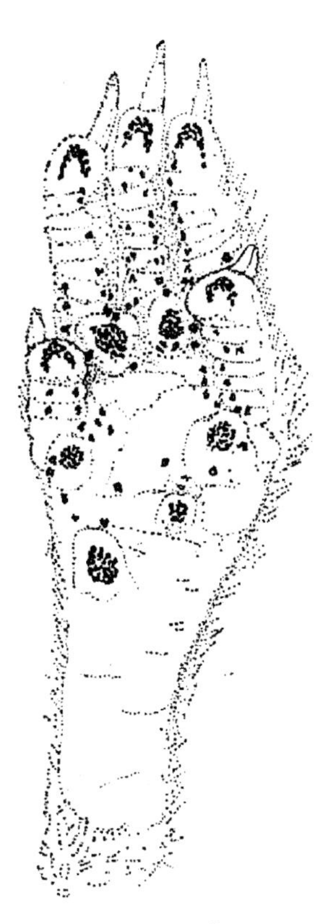

FIGURE 3.14 The pattern of skin glands on the sole of the foot of a mouse may provide information on the direction of travel. (From Ortmann, 1956.)

substrate that may signal size of the mouse and the direction it has been traveling (Ortmann, 1956; Fig. 3.14).

Short-range transfer of chemical factors may require body contact. Male Appalachian woodland salamanders press or slap their mental gland on the nares of the female (Arnold, 1966). The pygmy salamander even pierces the skin of the female's head with his modified teeth and "vaccinates" it with the secretion from his mental gland.

4

Chemical cues in orientation and navigation

Where will I find a patch of fresh grass?
The answer, my friend, is blowing in the wind.
With apologies to Bob Dylan

Nec vident terras, sed in odorem earum natant. (They don't see the land, but swim following its smell.)
(Red deer, *Cervus elaphus*, when supposedly swimming from Cyprus to the mainland of Asia Minor.)

C. PLINIUS SECUNDUS (Pliny the Older): *Naturali Historiae*, Book 8: *Zoology; Land Animals*, 115: *De cervis*, p. 89. Translated into German. Tusculum Library, Heimeran Verlag, 1976.

Humans depending on fishing and hunting have learned at their peril about animal migrations very early on. Today we still marvel at how precisely migrating organisms time their departures and arrivals and almost never fail to find their destinations, even passing on this vital information to their offspring. Young wandering albatrosses circumnavigate the Southern Ocean many times in the course of 8 or more years, before they return to breed at exactly the same place they were born. What cues guide these migrants in their epic journey? We know how insects, fish, and birds use the sun, the stars, and the magnetic field of the earth. But what role do chemical cues play? The physical and biological environments abound with volatile chemicals that vertebrates might use to orient themselves in space. In air, point sources such as a volcano, a small pond, or a tree in bloom are available, as are landscape odors such as those from pinewoods or the ocean. Even ship-borne humans can smell large seabird rookeries on ocean islands from afar. Arriving from sterile Antarctica, an air passenger landing in New Zealand is overwhelmed by its steamy, musty, moldy greenhouse odor. If humans notice such landscape odors, we can expect animals to be much more sensitive and attentive to them. Indeed, early explorers reported striking cases of animals smelling land. Alexander von Humboldt noticed that pigs on board his ship were excited and sniffed toward land when still 30–40 miles from the

Mexican coast with its aromatic vegetation (Faak, 1990). There is every reason to expect that wild animals exploit such landscape odors for their needs.

Vertebrates use many different cues for orientation in space and during navigation when traveling long distances. *Orientation* is defined as moving in relation to an external stimulus, for example light or a feature such as a more complex landmark. A landmark can be a prominent tree, a river, or a coastline. In *navigation*, a goal is approached by means other than the use of landmarks. Navigation is "the self-controlled movement toward an unperceived goal" (Dusenberry, 1992). Navigation is thought to require knowledge of one's geographical position (map component) and a compass. The chemical senses have been implicated in fish migration for a long time, while we are only starting to appreciate this sensory channel for other classes of vertebrate.

4.1 Fish

Among vertebrates, the homing of migrating fish of the anadromous type (undertaking upstream migration) is the classical example of olfactory orientation in space. Such a spawning mechanism was proposed over 100 years ago. Hasler (1954) formulated four testable hypotheses: (1) each stream has a characteristic odor; (2) migrating fish can differentiate among them; (3) the odor of the home stream is learned (imprinted) by young fish; (4) this chemical information is retained in memory and used by adults for orientation during their spawning trips upstream.

4.1.1 Salmon

Numerous experiments with salmon have revealed several facets of their orientation mechanism: Fish can be trained to distinguish odors of different streams; homing behavior is impaired in anosmic experimental fish; the odor is learned early in life, as transplanted presmolt salmon will return to their adopted stream and not their natal stream; and the odor is remembered after only a few days in a particular stream. Atlantic salmon, *Salmo salar*, acquire lasting information about home-stream odor during the period of smoltification (i.e. the transformation from *parr*, which lives in freshwater, to *smolt*, which is salt-water adapted, silvery, and swimming downstream). Conditioning different age groups to L-cysteine and subsequent measuring of their heart rate in response to this compound demonstrated this "imprinting" during a sensitive period (Morin *et al.*, 1987).

In an experiment, hatchery-reared Coho salmon have been imprinted to the artificial odors of morpholine, a heterocyclic amine (C_4H_9NO; 5×10^{-5} mg/l);

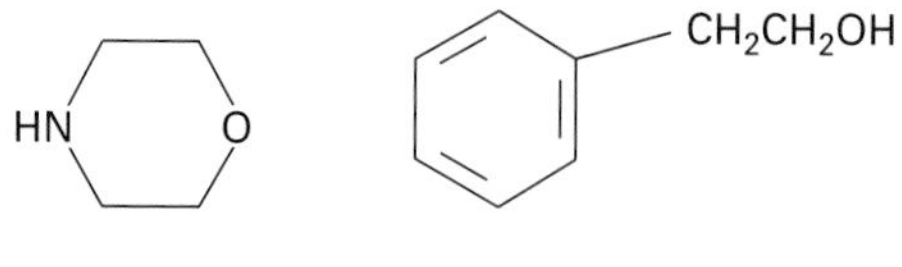

Morpholine β–Phenylethyl alcohol

FIGURE 4.1 Structures of morpholine and phenylethyl alcohol.

and phenylethyl alcohol ($C_8H_{10}O$ at 1×10^{-5} mg/l) (Fig. 4.1) as smolts at the age of about 18 months. (Unconditioned salmon detect morpholine at 1×10^{-6} mg/l water). The fish experienced these odors for 6 weeks between 1 April and 13 May during their presmolt and smolt stage (up to 16 months of age). (Presmolt fish start migrating downstream from their home stream.) In May, the two groups, plus a control group, were tagged and released into Lake Michigan midway between the mouths of two rivers, 4.7 km apart (Fig. 4.2). Eighteen months later, during the upstream spawning migration in autumn, morpholine (final dilution in stream 5×10^{-5} mg/l stream water) was added to one river, and phenylethyl alcohol (5×10^{-3} mg/l) to the other. At 19 monitoring stations along the shore, fish were counted by electrofishing, gill nets, and creel censuses (information from fishermen). The imprinted fish were trapped in both rivers. They had been successfully decoyed by the artificial odors: of the total number of morpholine-exposed salmon that were recovered, 94.1% were found in the morpholine-treated stream in 1 year of the experiment, and 97.6 in the second. Almost 93 % of the alcohol-imprinted fish recovered were caught in the alcohol-scented river (Scholz *et al.*, 1976; Hasler *et al.*, 1978). Hasler and Scholz (1983) have summarized these studies.

4.1.2 Mechanisms

Three mechanisms of home-stream recognition have been proposed: the original *fish stream imprinting* hypothesis by Hasler and Wisby (1951) assumed the memory of one single odor, that of the home stream. The odor is thought to emanate from rocks, soil, and vegetation. The *sequential imprinting* hypothesis by Harden-Jones (1968) postulated successive learning of several water odors along the migration route. Finally, the *pheromone hypothesis* of fish migration sees a role for conspecific odor in the home stream as a critical part of the environmental odor (Nordeng, 1971).

There is evidence for all three hypotheses.

1. The cited experiments by Hasler and coworkers demonstrated fish stream imprinting (Scholz *et al.*, 1976; Hasler *et al.*, 1978).

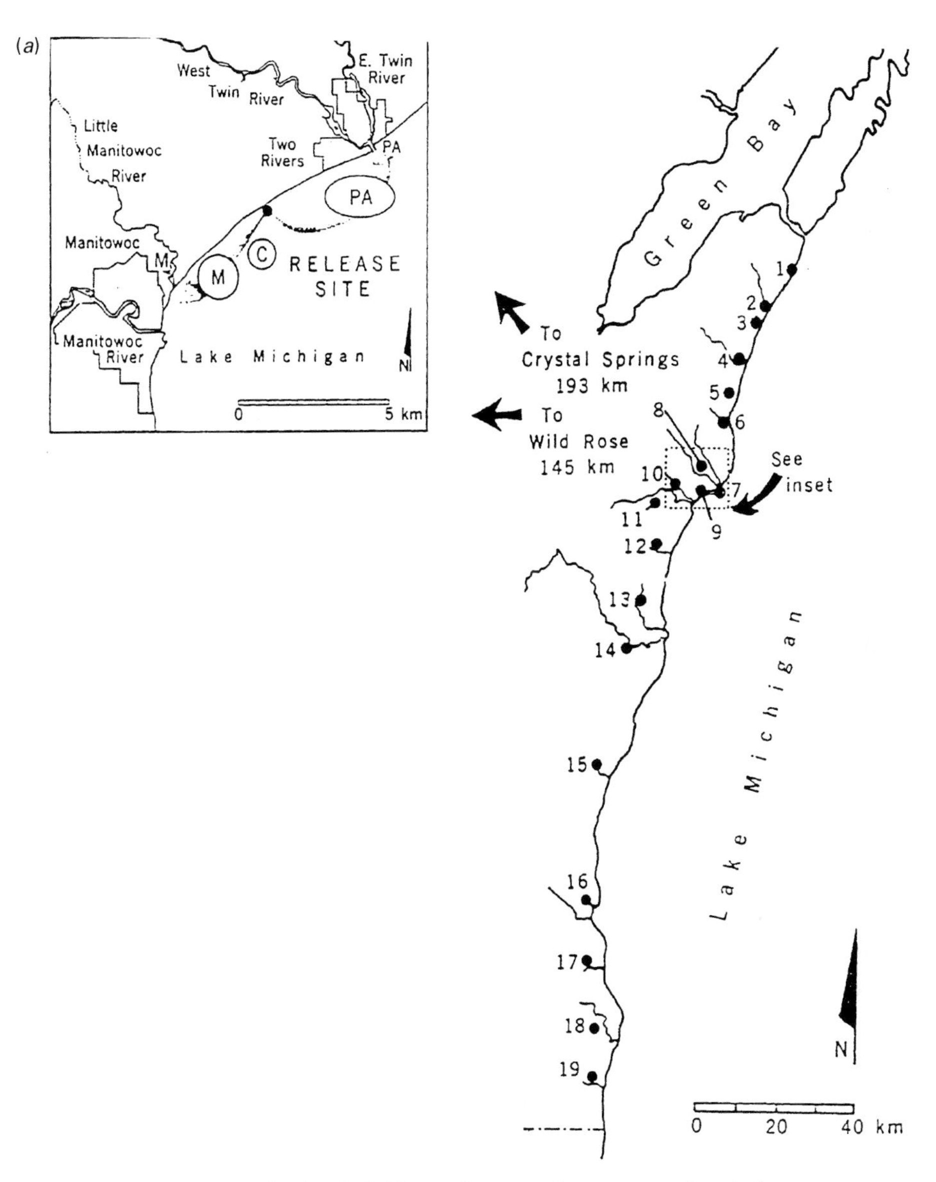

FIGURE 4.2 The classic field experiment to demonstrate chemical stream imprinting in salmon. (*a*) Lake Michigan is on the right in the main picture and the land area of Wisconsin is on the left. Three groups of fish were released between rivers No. 8 (twin Rivers) and 10 (Little Manitowoc River). (This are is shown enlarged in the inset.) One group was imprinted on morpholine (M) and another on phenylethyl alcohol (PA). The third group served as an untreated control (C). To test their chemical orientation, river No. 8 was scented with PA, and river No. 10 with M. Other rivers, numbered 1 to 19, were also monitored for entering fish. They served as control rivers. (*b*) Most M-imprinted salmon showed up in the M-scented river, and most PA-imprinted fish were caught in the PA-scented river. Most untreated control fish swam into unscented control rivers. Almost no imprinted fish appeared in the "wrong" rivers. (From data in Scholz *et al.*, 1976.)

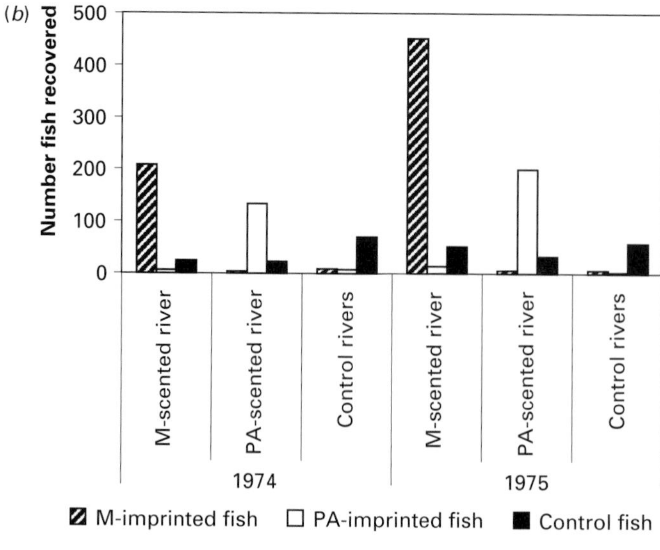

FIGURE 4.2 *(cont.)*

2. When transported 4 or 90 km from their Norwegian home stream into the sea, both intact and anosmic adult Atlantic Salmon, *S. salar*, did not return to their home stream but entered many streams along the coast. These transported fish lacked the opportunity to gather experiences during the outward smolt migration. These experiments support the sequential imprinting hypothesis (Hansen *et al.*, 1987).
3. Support for the pheromone hypothesis has come from preference tests in Atlantic salmon parr and experiments with sea lampreys migrating upstream to spawn.

The bottom-dwelling juveniles (parr) of salmon mark their substrate, gravel, probably with fecal material. The source of the material is the gastrointestinal tract. It originates probably in the liver but is not found in skin mucus. When given a choice, parr prefer the odor of their own strain but are also attracted by that of another strain if only that is available. This suggests a hierarchical order of chemical recognition, ranging from sibling to strain to species. Outside the migrating season, the salmon do not respond to these odors, suggesting a role in orientation (Stabell, 1987). A different experiment also supports Nordeng's pheromone hypothesis. If a basket with smolt of Atlantic salmon is placed in a freshwater pool, salmon will enter the pool (Johannesson, 1987).

Migrating male sea lampreys, *Petromyzon marinus*, are attracted to water from conspecific larvae (Teeter, 1980). This jawless fish originates in the Atlantic Ocean but has invaded the Great Lakes, where it parasitizes fish populations, posing a serious environmental danger. Adults migrate up streams to build nests and spawn in spring. The blind larvae filter feed in stream sediments. After 3 to 18 years, they metamorphose into the parasitic form, migrate downstream, and feed on other fish in lakes or the ocean. After this parasitic life, during which they gain 100-fold in body mass over 15 months, the lampreys stop feeding, start maturing sexually and migrate upstream to spawn. At the end of this 4–8 week migration phase, the fish spawn and die. Since sea lamprey larvae live in less than 10% of 6000 tributaries draining into the Great Lakes, although many more are suited as habitat, these populations may be limited because lampreys attract each other by pheromones (Bjerselius *et al.*, 2000).

In the laboratory as well as in the field, larval waterborne stimuli attracted sexually immature males and females. (The field tests were run from 22:00 to 04:00 hours, the activity time of migrating lampreys.) Certain conditions have to be met: the fish responded only early in the night, and before being fully sexually mature.

Some experiments do not support the pheromone hypothesis. Atlantic salmon grilse (salmon a year older than smolt) do not need a conspecific (smolt) cue for their return from the sea to a saltwater bay where they had been released as smolt. They return to the release site on their own accord but will not proceed from there and enter freshwater pools (Johannesson, 1987). This behavior is not in accordance with the pheromone hypothesis. Similarly, Arctic charr (*Salvelinus alpinus*) could not be lured into a non-natal stream that contained adult conspecifics (Black and Dempson, 1986). Pacific salmon (*Oncorhynchus kisutch*) returned to their natal hatchery, even though on their way from the ocean to their natal area they had to pass a major hatchery where their full-sibling relatives discharged their odor into the water (Brannon and Quinn, 1990).

Barging salmon down the Columbia and Snake rivers impaired their homing. The fish returned but they did not return upstream to spawn, for unclear reasons. It is possible that they failed to imprint on the home stream when barged. To avoid this problem, spillways are recommended (Lorraine Bodo, American Rivers/Columbia River Alliance). This is reminiscent of trucked reindeer in Sweden (see Section 4.5).

Brook trout (*Salvelinus fontinalis*) have both migratory (anadromous) and non-anadromous populations. Fish from each type were tested for their responses to home-stream water. Both preferred home stream over control water but were equally attracted to water from their home-stream and from an unfamiliar stream. Therefore, specific local odors may not be critical cues. Furthermore, anosmic fish made the same choices, indicating that taste may be involved. In the non-migratory population, responses to home-stream water were noted only in summer, not in winter (Keefe and Winn, 1991).

4.1.3 Chemical nature of odor cues

Salmonid fishes such as charr (*S. alpinus*) and grayling (*Thymallus thymallus*) are assumed to use *bile acids* for several functions, most notably homing. Bile acids are prime candidates for this function because they are diverse, specific, potent, and stable. Bile acids elicit electrical responses in the medial part of the olfactory bulb. By contrast, amino acids elicit responses in the lateral part. Taurine-conjugated bile acids are up to 1000 times more potent than methionine. There may be two different types of receptor: one for bile acids and one for amino acids. Because of their "great potency as odorants," their evolutionary history and variability, and their "renowned adherent properties," bile acids are "interesting candidates for specific signals in the aquatic environment" (Døving *et al.*, 1980).

Sea lamprey larvae (*P. marinus*) produce and release petromyzonol sulfate (PS) via their feces, a bile acid not found in any other vertebrate, and allocholic acid (ACA)(Fig. 4.3), which increased swimming activity by adult female sea lampreys in their migratory phase but did not attract them within the confines of the laboratory tanks (Bjerselius *et al.*, 2000). These authors conclude: "our studies lend support to the hypothesis that unique bile acids released by larval sea lamprey attract migratory adults into rivers to spawn." Sea lampreys are very sensitive to P and ACA, and only to these compounds, as revealed by electro-olfactograms, recorded from their olfactory lamellae. The fish detected these two compounds at a concentration of 10^{-12} mol/l (Li *et al.*, 2002). In lamprey streams, the concentrations of PS and ACA are high enough to be detected by adult sea lampreys (Polkinghorne *et al.*, 2001). The migratory pheromone contains at least three compounds (Fine and Sorensen, 2004) and attracts more than one species (Fine *et al.*, 2004).

Eels are catadromous fish that travel downstream to spawn in saltwater. Glass eels are the eels' juvenile stage. They migrate only once from their natal area in the ocean to freshwater bodies, as far as 1000 km inland, growing and developing during that year-long journey. Therefore, eels cannot imprint on a home-stream odor. Yet they are the attracted to bile salts and taurine dissolved in freshwater (Sola and Tosi, 1993). Since taurine occurs in fish mucus, and bile acids in urine and intestinal waste, these stimuli are candidates for pheromones (Fig. 4.3). Geosmin, (*E*)-1,10-dimethyl-(*E*)-9-decalol (Fig. 4.3), produced by actinomycetes, is the most abundant vegetation odor in surface freshwater. Glass eels swim toward this "surface water odor" in the laboratory (Tosi and Sola, 1993) when presented in freshwater. In saltwater, however, the fish avoid the geosmin odor. Salinity affects the response: a very weak geosmin concentration (10^{-13} mg/l) attracted the eels at a salinity of 15°/oo, while a concentration of

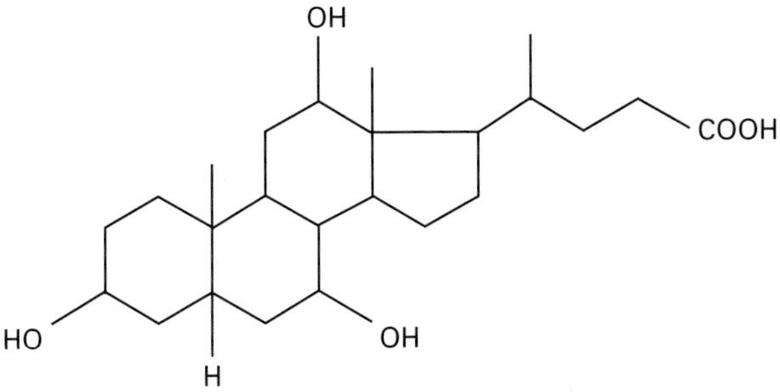

Allocholic acid

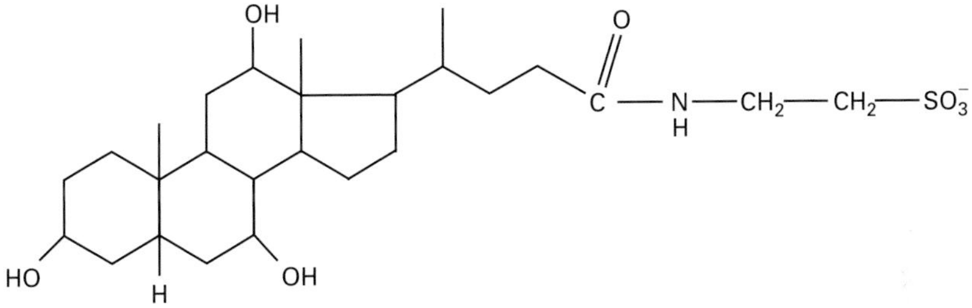

Taurine-conjugated bile acid

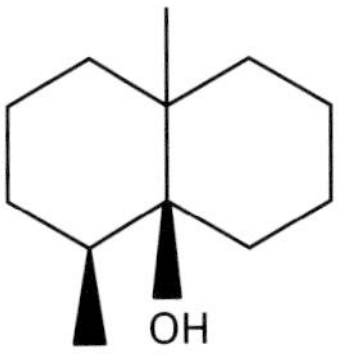

Geosmin

FIGURE 4.3 Structures of allocholic acid (petromyzonol sulfate is identical apart from replacement of the carboxyl group with sulfate), taurine-conjugated bile acid, and geosmin.

10^{-10} mg/l had this effect only at a salinity of 10°/oo (Tosi and Sola, 1993). In summary, it is assumed that glass eels heading for a stream follow decreasing salinity and increasing geosmin odor. Near the coast, geosmin odor may become more important and guide them to a particular stream.

Upstream migrating glass eels also swim toward compounds with earthy and "green" (vegetable) odors. These are pyrazines, thiazoles, and alcohols of the cyclohexanol type (Sola, 1995). One of these compounds, 2-isobutyl-3-methoxypyrazine, is common in rivers and lakes (Hwang *et al.*, 1984).

4.2 Amphibia

Anurans and urodeles are known to home to their ponds. The chemical senses as well as vision and other senses are required for this ability.

4.2.1 Anurans

Among anurans, Fowler's toad, *Bufo woodhousei fowleri*, the Mexican toad, *Bufo valliceps*, and the chorus frogs *Pseudacris clarki*, and *Pseudacris streckeri* prefer the water from their own breeding pond to that of another pond (Grubb, 1973a,b, 1976). Artificial odors can be learned: Mexican toads, tested in a complex maze used odor cues such as anise oil, benzaldehyde, cedarwood oil, citral, and others to orient themselves in a complex environment (Grubb, 1976). The toad *Bufo boreas* uses olfaction for distance orientation, and vision when near the pond (Tracy and Dole, 1969), while the reverse is true for *Salamandra* sp. The green frog (*Rana clamitans*) locates its breeding site with more difficulty when the olfactory tracts are experimentally ablated, but it does not lose the homing ability.

After metamorphosis, first-year individuals of the common European frog *Rana esculenta* stay in their home pond, even if it dries out. Experiments in a two-way choice apparatus showed that these animals prefer to head toward the odor of mud from their home pond as opposed to some other mud sample (Bastakov, 1986). Newly metamorphized (at Gosner stages 43–46) pool frogs (*Rana lessonae*) caught in the wild preferred water from their pond to tap water, and also to water from an unfamiliar pond. When tadpoles were raised in water from their home pond, or water with boiled nettle as food, the froglets later preferred water with these respective stimuli. They had developed a preference during larval development (Ogurtsov and Bastakov, 2001).

The signal for homing appears to be a complex pond odor, originating in vegetation and soil. The frog *Pseudacris triseriata*, tested in a T-maze, chooses the odor of lowland muck and filamentous algae from its breeding habitat over the odor of soil and decaying vegetation from non-breeding upland habitat suspended in distilled water (Martof, 1962).

The dart-poison frog *Dendrobates pumilio* uses odors for homing. This species lives in the understory of lowland tropical forests in Central America. The eggs are laid on land, and the female carries the newly hatched tadpoles on her back to water-filled leaf axils of bromeliads. She feeds the tadpoles with unfertilized eggs, while the male defends the territory. Captive dart-poison frogs tested in a Y-olfactometer chose the odor from their own communal tank over odors from tanks planted with different plants. However, they did not distinguish between the odor of their own tank and that of a tank with bromeliad genetically identical to that in their home aquarium. In addition to plant odors, passive marking possibly takes place, as the frogs had been in their aquarium for 3 weeks before the test (Forester and Wisnieski, 1991).

4.2.2 Urodeles

Among the Salamandridae, displaced Western newts, *Taricha rivularis*, home to the same section of a stream year after year. Blinded newts still homed successfully, and anosmic specimens were reduced in their homing ability but still did not move in random fashion. They return to their home pond from up to 12 km (Twitty, 1966). Another salamander, *Ambystoma maculatum*, tested in the laboratory, preferred paper towels soaked with mud or pond water from their home pond over towels treated with materials from a distant pond (McGregor and Teska, 1989).

4.3 Reptiles

4.3.1 Snakes

Among snakes, over 20 species in five families follow scent trails of conspecifics. The trails are probably used for homing and locating dens and, therefore, serve more purposes than strictly social behavior. Odor trails may be more important for younger snakes and in northern populations. The garter snake *Thamnophis sirtalis* undertakes autumn migrations of up to 17.7 km (Costanzo, 1989). Corn snakes, *Elaphe guttata*, and garter snakes, *Thamnophis radix*, use chemical cues when aggregating and selecting their shelters. The response depends on an intact vomeronasal organ. In corn snakes, a social odor is involved: individual snakes prefer shelters that had been used by groups previously (Eyck and Halpern, 1988). Eastern garter snakes do not depend on their vomeronasal organ for orientation in their home range: severing the vomeronasal nerve does not prevent these snakes from finding previously used sites. However, the distance travelled each day by these avomic snakes (7.7 m) was only about one third of that of controls (24.6 m) (Graves *et al.*, 1993).

4.3.2 Tortoises and turtles

Land tortoises (*Testudo hermanni*), especially males, caught in the wild and displaced 500–1000 m returned to the point of their capture. However, if their olfaction was impaired by washing the nasal cavity with zinc sulfate, their

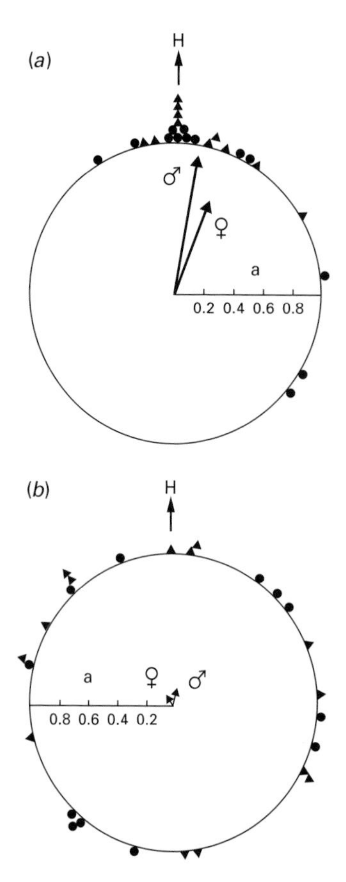

FIGURE 4.4 Homing in intact (*a*) and anosmic (*b*) land tortoises. Triangles indicate males and dots females. While intact tortoises head in the home direction (H), animals rendered anosmic by zinc sulfate treatment scattered in all directions. The inner arrows are resultant vectors for all animals of one sex and represent directedness. a, vector length, with 1.0 being perfect directedness. (From Chelazzi and Delfino, 1986.)

homing was poor (Fig. 4.4). These treated animals did move in many directions and, furthermore, did cover only very short distances during the 14 days of the experiment (Chelazzi and Delfino, 1986).

Freshwater turtles home to their pond when experimentally displaced. Painted turtles, *Chrysemys picta*, homed from 100 m, but not from 1.6 km (Emlen, 1969). They use chemical cues: painted turtles discriminate chemical cues from their home ponds and other ponds. Males and females prefer water from their home pond to that from other ponds (Quinn and Graves, 1998). The Eastern long-necked turtle, *Chelodina longicollis*, of southestern Australia uses solar cues during migration between a permanent lake and an ephemeral swamp, as orientation is impaired on overcast days. Olfactory cues also seem to play a role: in a Y-maze, the turtles headed toward swamp mud and debris in preference to plain water (Graham *et al.*, 1996).

The sea turtles Kemp's ridley, *Lepidochelis kempi*, have been "head-started" by imprinting them to the odor of the water they hatched in. Eggs from the threatened beach at Rancho Nuevo, Mexico were placed in sand and water from Padre Island, Texas and transported to Padre Island. When 4 months old, these turtles were tested in a multiple choice of different kinds of water (Fig. 4.5). They preferred water from Padre Island, indicating that some form of olfactory imprinting had taken place (Grassman *et al.*, 1984). Starting in 1978, 2000 eggs from Rancho Nuevo were hatched annually for 10 years. Later, the hatchlings were released at Padre Island National Seashore (see also Ch.12). In 1996, the first two turtles came ashore at Padre Island and laid a total of 176 eggs. These females were 10 and 13 years old, respectively. The artificial colonization of the Padre Island beaches has been an unqualified success. However, at this time it is not clear what cues the turtles use to return to Padre Island from the ocean. As the young turtles stay in shallow water for a while, they may imprint on any of many factors other than the beach chemistry. These possible other factors include solar cues, star patterns, the Earth's magnetic field, or salinity gradients in the ocean; they may follow other turtles or use still other cues.

4.4 Birds

Homing by olfactory cues has been the most researched topic in bird navigation research for many years. Homing pigeons and petrels use airborne cues for orientation under certain conditions. However, many experimental results do not clearly show that *odors* are the crucial stimuli.

4.4.1 Procellariiforms

Birds' homing to their burrow

Many petrels and shearwaters approach their nest burrow, often located under forest cover, at night. When Leach's petrel (*Oceanodroma leucorrhoa*) return to their nest, they first hover over the spruce-fir canopy near their burrow. Then they plummet to the ground several meters downwind from their nest site and walk upwind to their burrows (Fig. 4.6). In still air, they landed closer to the burrow and followed a more roundabout route than in wind. With external nares plugged or olfactory nerves transected, displaced birds did not return to their burrows for 1 week. In a laboratory two-choice apparatus, breeding petrels

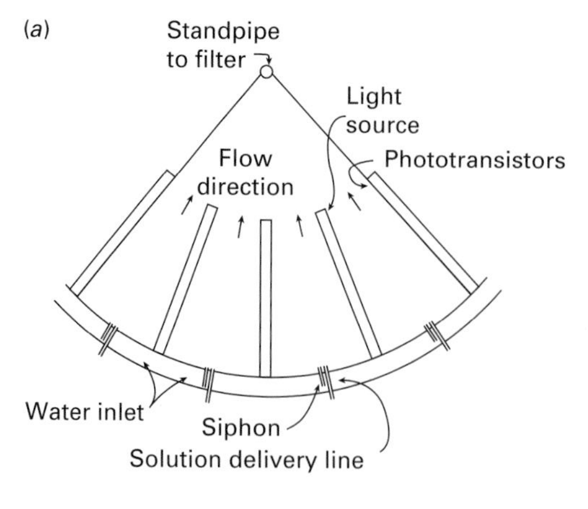

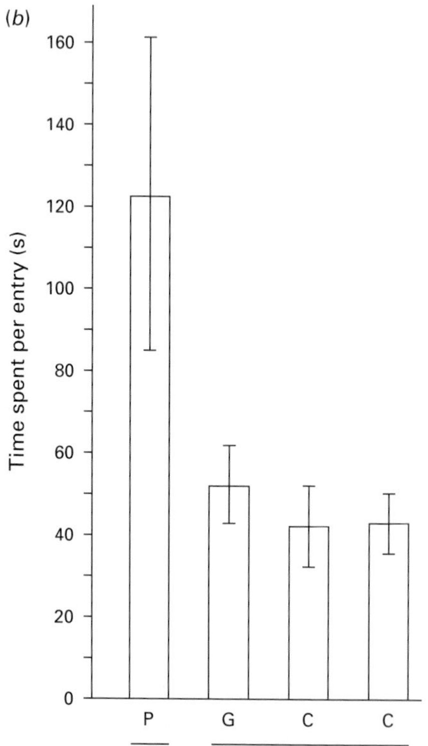

FIGURE 4.5 Sea turtle orientation. (*a*) Bird eye's view of apparatus to test whether sea turtles discriminate sea water samples from different locations. Sea water flows from water inlet toward the standpipe (arrows). The turtle is placed near the standpipe. Phototransistors record when the animals move into different compartments. The water washes of sand come from Padre Island (P) and Galveston (G), and unscented water in two compartments serves as control (C). (*b*) Turtles spent more time in water extract of sand from Padre Island than in that from Galveston. (From Grassman *et al.*, 1984.)

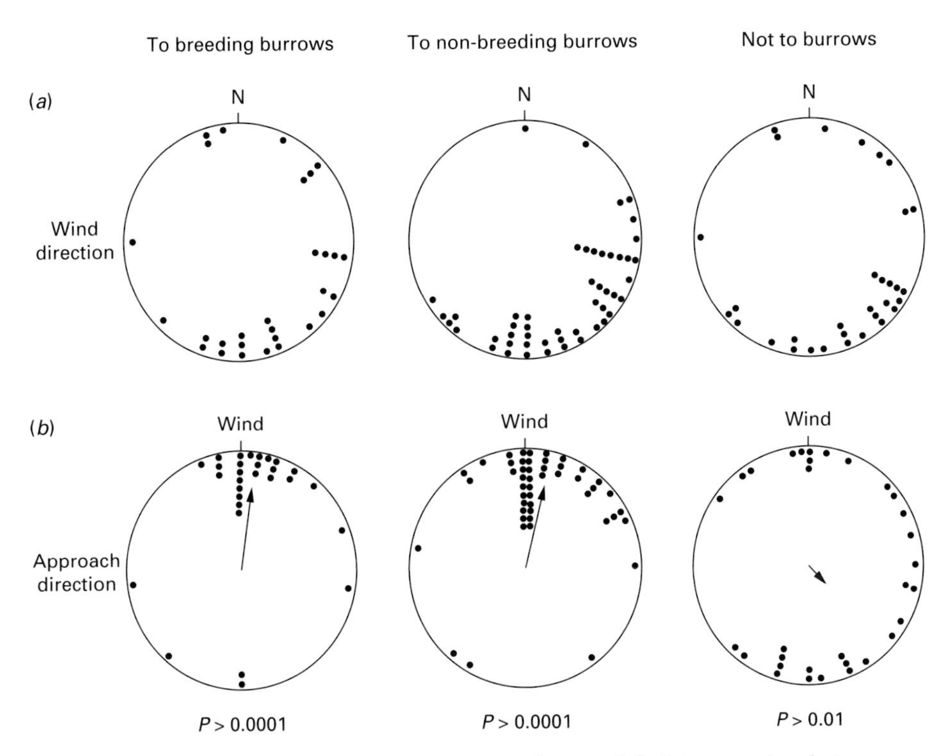

FIGURE 4.6 Orientation of Leach's petrels toward their burrows in relation to wind direction. (*a*) Wind directions measured during observations. (*b*) Approach to burrows on foot on the forest floor related to wind directions. Regardless of its absolute compass direction, wind direction during each observation is depicted as "north" in the diagram. Birds orient into the wind when heading for their burrows, whether breeding or non-breeding, but walk randomly when they are not going to burrows. (From Grubb, 1974.)

preferred an air stream from their own nest material to other forest floor material. These findings suggest that this species uses olfactory cues to locate the nest (Grubb, 1974).

Materials from the birds themselves most likely provide the chemical cues that are implied in this behavior. Petrels use odor in finding their nests early in life. Chicks of the British storm petrel, *Hydrobates pelagicus*, orient toward their nest from 30 cm, and they fail to do so when their nares are blocked. In a T-maze with one arm inserted into a nest burrow, the chicks were able to orient toward a nest crevice (Mínguez, 1997). Other birds may also use odor to orient toward their nest: domestic chicks, *Gallus gallus domesticus*, prefer odor of their own bedding (Jones and Gentle, 1985). Experiments show that possibly only nocturnal burrowing petrels use olfaction to find their nest, while surface nesters and day-active species are visually orientated (Bonadonna and Bretagnolle, 2002).

The snow petrel, *Pagodroma nivea*, of the Antarctic accumulates stomach oil around its nest. These deposits can grow up to 0.5 m thick and have been radiocarbon-dated as 4000 and 9000 years old in the Shackleton range and the Wohlthat Massif, respectively, in Queen Maud Land (Hiller and Wand, 1984). The birds might use odors from these accretions for orientation around their nests. However, a small-scale field experiment in Antarctica's Mühlig- Hofmann mountains suggested that snow petrels do not use odors to find their nest for distances over 1–2 km. The same percentage of birds returned to the breeding colony from those with nasal plugs and those intact (Haftorn *et al.*, 1988). Manx' shearwaters, *Puffinus puffinus*, however, do not seem to use odors for finding their burrows (James, 1986). Blue petrel (*Halobaena caerulea*) find their burrow by smell, as field experiments in the Kerguelen showed (Bonadonna *et al.*, 2004).

Finding food

On the high seas, petrels and albatrosses unfailingly locate patches of krill, their main food. Such patches are as small as a few cubic meters, and the larger ones extend over an area the size of several football fields. On their feeding trips of thousands of kilometers, birds such as black-browed albatrosses or white-chinned petrels find the "needle in a haystack" by celestial cues, the earth magnetic field, and by detecting high odor concentrations in an "olfactory landscape" over the ocean. On such a large scale, they possibly use an odor that emanates from the water when zooplankton such as krill feed on phytoplankton. The relief of the ocean floor is ultimately responsible for this "odor landscape": undersea ridges create upwelling that transports nutrients to the upper water layers, benefiting phytoplankton, the basis of the food chain. We do not know how these seabirds accomplish long-range orientation. Once close to a krill patch, they can use chemical cues blown downwind, or visual cues such as other birds feeding (Nevitt, 1999). The close-range responses of these tube-nosed birds to chemicals released into the air when krill is feeding on phytoplankton are discussed on p. 351. The best known of these food-signaling compounds is dimethyl sulfide.

4.4.2 Homing pigeons

During the last 30 years, several laboratories have examined homing pigeons for their ability to use odors for orientation. Early experiments suggested that olfaction is important in orientation: homing pigeons with their olfactory nerves cut were "generally found to be unable to home from short distances" (Benvenuti *et al.*, 1973). According to Papi's (1976) hypothesis, pigeons learn their loft odor and foreign odors blown their way by the winds. These odors provide the map component, while the sun or magnetic compass is used to deduce flight direction. Papi based his hypothesis on four phenomena. (1) Pigeons prevented from breathing through their nostrils were incapable of correct initial orientation when released at a distance from their loft. (2) If wind deflectors at the loft had shifted air currents, the initial orientation of the birds was also deflected when released at a distance. (3) When an odor was applied to their beaks at the time of release, displaced pigeons flew in the opposite direction of that from which the same experimental odor came on the wind at the loft. (4) Birds transported to the release site by different overland routes differed in their initial flight direction. This effect was abolished if they were transported with their nostrils plugged (reviewed by Papi, 1976).

In a typical early experiment, pigeons were exposed to artificial odorous winds from the time of fledging. One group experienced south wind with olive oil odor, and another group north wind with a mixture of solvent odors (toluene and terpene hydrocarbons). The birds were tested later by applying olive oil or the turpentine mixture to their nostrils while being displaced from their loft. They flew in the direction opposite from that which they had experienced the "training" odor: that is, they "homed". The olfactory information used for navigation seems to be acquired during the first months of life. If the nostrils of young pigeons are covered with masks for 3 days after fledging so that they cannot, or only barely, breathe through their nostrils, they cannot associate odors with wind directions and subsequently home poorly after displacement (Papi *et al.*, 1973). Numerous experiments, however, have raised many questions to which there are no answers at present.

How can a pigeon use olfaction to navigate? The two competing hypotheses are the *mosaic map* (Papi *et al.*, 1972) and the *gradient-map* (Wallraff, 1989) models. In the mosaic map model the pigeon is assumed to qualitatively discriminate different environmental odors that occur in a patchwork over an area of up to 200 km around the home site (loft). The gradient map model postulates quantitative discrimination of long-range environmental odor gradients that extend over thousands of kilometers. So far, neither is convincingly supported (Schmidt-Koenig, 1987).

Airborne stimuli are probably necessary, but not sufficient, for homing. If air samples from future release sites are brought to the loft and pigeons exposed to them, only birds that had been exposed to the odor of their actual release site oriented toward home. In this experiment, the birds were rendered anosmic for the displacement so that they could not pick up odors during the trip to the release site (Kiepenheuer, 1985, 1986).

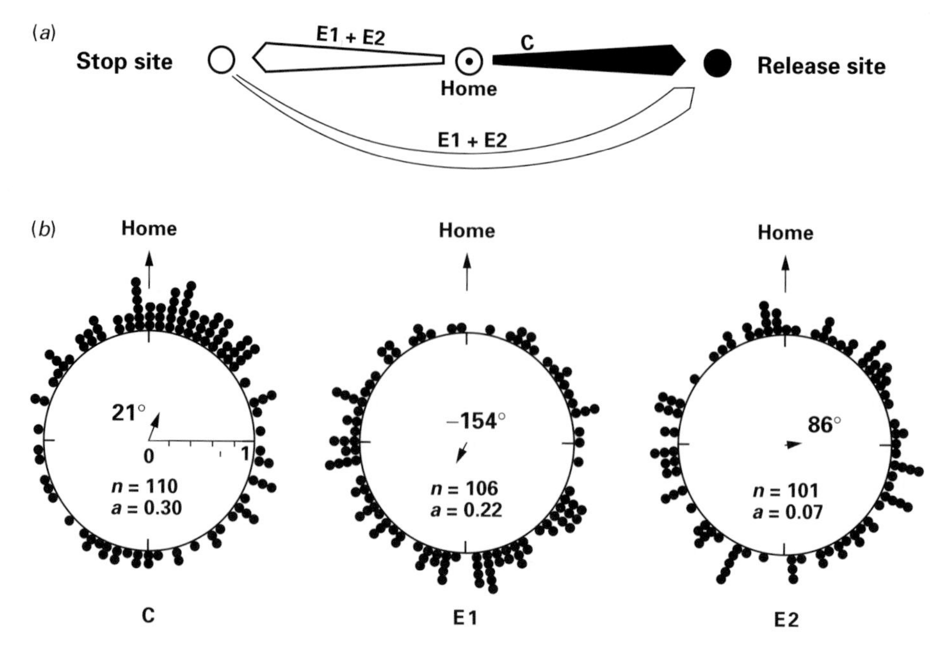

FIGURE 4.7 Pigeons can be "fooled" about where they are in relation to their loft if they are allowed to smell the air at a "distraction site." (*a*) Three groups of pigeons were tested for their homing ability. Experimental groups 1 and 2 (E1 and E2) were brought to a distraction site (stop site) and then transported to the actual release site in the opposite direction from the loft (white arrows). E1 were allowed to smell the local air at the distraction site but not at the release site. E2 were never allowed to smell local air anywhere. Controls (C) were directly transported to the release site (black arrow) and allowed to smell the air there. (*b*) The dots represent individual pigeons' bearings. The controls homed correctly. The E1 group traveled in a direction opposite of home. The E2 group lacked meaningful information and randomly headed in various directions. (From Benvenuti *et al.*, 1992.)

Pigeons can be fooled into using a cue from a "false release site;" before their release, the birds are first brought from their home loft to a site located in the opposite direction from the release site and permitted to smell the air there (Fig. 4.7). In this site simulation test, the released birds flew in the direction opposite from home (Wallraff, 1990; Benvenuti *et al.*, 1992). It is clear that pigeons gain information about the spatial relationship between the strange release site and their home loft by odors they perceive at the unfamiliar site (Benvenuti *et al.*, 1992).

Inexperienced young pigeons, when rendered anosmic before release, have oriented normally in one experiment (Wiltschko and Wiltschko, 1987) but were impaired in another (Gagliardo *et al.*, 1988). In the latter experiment, the birds peculiarly were still directed but flew in the wrong direction (i.e. away from their home loft).

Other experiments, however, have failed to confirm that homing pigeons used odor cues. For instance, pigeons were made anosmic by carrying nasal plugs and being sprayed in their nostrils with the local anesthetic tetracaine. Displaced shortly after treatment and released, they were not impaired in their homing. Their vector length (a measure of "directedness," regardless of where the birds flew), homeward component, and vanishing intervals between time of release and directed departure did not differ consistently or significantly from controls (Wiltschko and Wiltschko, 1987; Wiltschko *et al.*, 1987a). In one experiment, pigeons were transported to the release site unimpaired and allowed to smell the release site for 15 minutes. Then the nostrils of one half of the group were sprayed with tetracaine, and the birds were set free. The control birds were sprayed with propellant only. Birds that were sprayed with tetracaine for the first time were most affected. It is suggested that a non-specific, traumatic effect is responsible for impaired navigation, and not altered olfactory input (Wiltschko *et al.*, 1989).

Strangely, Papi's homing pigeons in Italy appeared to use odors, while the results of experiments in New York were negative (Papi *et al.*, 1978). Later experiments confirmed that the Italian pigeons relied more on odors for homing than pigeons in Germany or the USA. During transfer to the release site, pigeons were rendered anosmic by cotton nasal plugs soaked with citrus oil 30–40 km from their home loft, and the local anaesthetic tetracaine was applied shortly before the release which rendered the birds anosmic for 4–6 hours. The experiments were performed in Pisa, Italy, at Cornell University in Ithaca, New York, and in Frankfurt, Germany (Fig. 4.8). The variables measured were the direction and length of mean vector (i.e. the average travel direction), the deviation of this mean vector from the home direction, the homeward component (derived from mean vector and deviation from home), the median vanishing interval, the median homing speed, and the return rate as a percentage (Wiltschko *et al.*, 1987b). Only the Italian pigeons were impaired, while pigeons in New York and Germany homed normally (Wiltschko and Wiltschko, 1992).

Pigeons appear to use two maps, one olfactory in nature, the other a mosaic of familiar sites. These maps are redundant for intact birds in familiar terrain, but if impaired sensorily, the pigeon can find home without one or the other of these mechanisms. This was shown in experiments with clock-shifted and anosmic pigeons (Luschi and Dall'Antonia, 1993).

Pigeons kept in "deflector lofts" (with panels that change the wind direction that they experienced) were affected in some experiments (Papi, 1990). This may

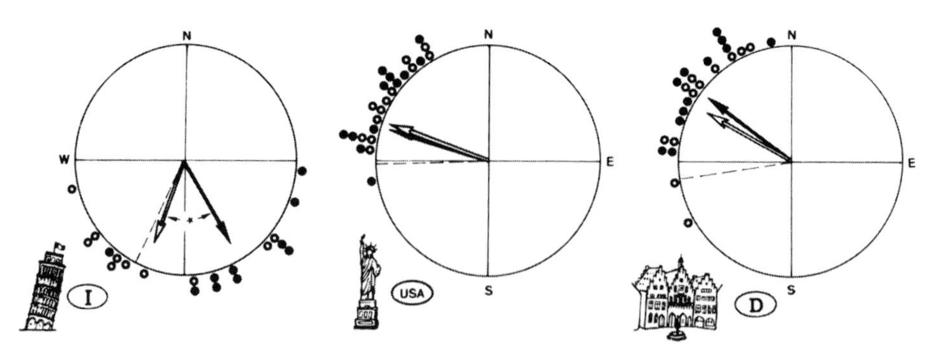

FIGURE 4.8 Pigeons in Italy misorient when deprived of olfaction, while those in New York and Germany were unaffected. In Italy (left), anosmic birds headed into a direction different from home, while no such difference between anosmic and control birds was found in New York and Germany. (Anosmic birds had their nostrils plugged with citrus oil-soaked cotton until release, and an anesthetic was sprayed into their nostrils at the time of release. This renders the birds anosmic for 4–6 hours.) Dashed radius, home direction; solid dots, individual anosmic birds; open circles, control birds; arrows mean directions for anosmic and control birds. (From Wiltschko and Wiltschko, 1992.)

actually have been a consequence of changes in the reflection pattern of polarized light, and not odors (Waldvogel *et al.*, 1988).

The methodology of odor deprivation has been problematical. It is difficult to isolate pigeons from airborne odors. Occlusion of the nostrils reduces the stimulus intensity to 20–30%, but not to zero. Bilateral section of the olfactory nerve irreversibly eliminates responses to weaker stimuli, but responses to higher concentrations remain, probably via trigeminal stimulation. Spraying the nasal cavity with anaesthetic abolishes sensitivity to odors, but the effect is variable and depends on the way the drug is applied (Wallraff, 1988a). Some of the experimental results can possibly be explained by the fact that zinc sulfate produces anosmia that lasts several days, much longer than that induced by xylocaine (Wallraff, 1988a; Benvenuti *et al.*, 1992). Furthermore, using *un*conditioned responses to olfactory stimuli has its disadvantages. (a) The stimulus intensity must be high; no response does not mean that the animal did not receive the stimulus. (b) Without reward or punishment, the response will not be stable over many repetitions. Therefore, more naive birds are needed for testing. (c) This introduces individual-specific amplitudes and response curves to even the same stimulus (Wallraff, 1988a).

What do we know about the olfactory apparatus and performance of homing pigeons? The size of the olfactory bulb lies in the middle range of 151 bird species examined (Bang, 1971). What odors do pigeons use? The home site air, the air that surrounds the pigeon during displacement, and the air at the release site have all been tested. The birds home normally if fiberglass filters remove airborne particles, ranging in size from 0.04 to 2.0 μm from the air around the birds during displacement. However, if charcoal filters are used, they home more poorly over 24–155 km than pigeons in unfiltered air (Wallraff and Foà, 1981). Odor gradients of mixtures in the atmosphere near the ground have been proposed as important for pigeon homing (Wallraff, 1988b).

The concentrations of environmental odors have been calculated to test whether they suffice to serve as odor cues for birds. Each year, there are 17×10^6 metric tons of sulfur released from natural sources between 30° and 50° N alone. Less than 10 kg sulfur and ammonium are emitted daily from 1 ha. Pine-oak forest in the United States gives off 1–5 kg/m^2 α-pinene daily. Air sampling at midday from a scaffold tower above the canopy of a loblolly pine forest in North Carolina yielded α-pinene concentrations as high as 2.5 μg/m^3 air (Arnts *et al.*, 1982). Do these amounts provide concentrations above the olfactory threshold of the pigeons? There are 1–5 parts per billion (ppb) sulfur and ammonium in the atmosphere, and 10–100 ppb α-pinene. These concentrations are below the pigeon's detection thresholds (0.1 to 40 parts per million; Henton *et al.*, 1966). The differences between the performances of the New York and Italian pigeons may be a consequence of atmospheric differences. The Italian experiments were carried out near the coast, which runs in a north–south direction. Indeed, east and west-displaced pigeons oriented better than north and south-displaced birds. Seabreezes from the west may provide odor cues and these will extend for approximately 45 to 85 km inland (Waldvogel, 1987).

Other odor fields possibly available to birds are ammonia from pastures and, over the ocean, dimethyl sulfide released by herbivorous action of plankton on marine plants (Nevitt *et al.*, 1995).

Serious obstacles to olfactory navigation have been pointed out. There are drastic differences between actually observed air currents and *mean* wind directions for a given locality. Also, the odor sources are unstable temporally and spatially. Finally, there is no evidence for long-lived, strong, and stable chemical gradients in the lower atmosphere (Becker and van Raden, 1986). The Cross-Appalachian Tracer Experiment (CAPTEX 1983) showed that an aerosol released in Ohio or Ontario could be found in various concentrations over 300 to 1000 km downwind, with irregular distribution that would provide little directional information to birds for homing (Waldvogel, 1987).

Chemical cues and the earth's magnetic field may interact in pigeon navigation. Pigeons kept in an oscillating magnetic field were affected in their initial orientation only if they could smell atmospheric odors (Wallraff *et al.*, 1986). Olfactory restriction and magnetic deprivation have different effects on homing in pigeons. In one set of experiments, pigeons were exposed to an artificial random magnetic field by one pair of horizontal and two pairs of vertical Helmholtz coils on their heads. Olfactory deprivation was accomplished by plugging the nostrils with cotton and sealing with adhesive tape. Only intact birds (no experimental magnetic field, nostrils not plugged) oriented toward home. Birds prevented from smelling were disoriented; they scattered widely in their departure directions. Magnetically impaired, but olfactorily intact, pigeons were directed, but in their preferred compass direction for that particular year and not towards home (Papi and Ioalè, 1988).

The most important research areas for the future are the central mechanisms in the bird and the nature and distribution of odorous substances in the atmosphere (Benvenuti *et al.*, 1992). Reviewing the current state of olfactory navigation in pigeons, Wallraff (2004) has presented maps of the distribution of different concentrations of six hydrocarbons in a 300 × 300 km area in Germany. Pigeons may well be able to use these atmospheric gas gradients for navigation.

4.5 Mammals

Small mammals are well oriented within their familiar home range but also when exploring beyond that home range, and most likely when dispersing. They home well when experimentally displaced from their home site. Finding one's way within the familiar home range is termed *topographical navigation*, in contrast to operating beyond the home range, where *geographical navigation* takes over. Topographical navigation involves three mechanisms:

- *guided orientation* (also: known as *guidance*): a direct response to objects;
- *path integration*: continuous monitoring and computation of the twists and turns while moving about so that the path back to the starting point ("home") can be found;
- *landmark navigation*: "movement by means of distal visual cues toward a goal not directly perceived" (Alyan and Jander, 1997).

Spatial orientation in the home range has been studied in house mice, domestic rats, golden hamsters, and Mongolian gerbils. All these species use distant (visual) landmarks as reference points. However, most of the laboratory studies focused on very short distances in the 1 m range (e.g. Alyan and Jander, 1994). While most studies in mammals examined how the animals find their way after they have committed their surroundings to memory, we know less on how these animals learn to orient themselves in their home range in the first place. The brain structure instrumental in such spatial learning and memory is the hippocampus.

The "two blind mice" experiment of Alyan and Jander (1994) used the pup-retrieving behavior of female house mice to determine the *visual* cues they use for orientation in an arena of 1 m diameter. A strain of blind mice did not orient as well as intact mice. In different experiments, neither sighted nor blind mice appeared to use olfactory cues such as scent trails in sand, or the odors of wooden blocks in the arena, to find their nest after their arena had been rotated by 90 degrees. Such rotation misled them on their way home, despite landmarks in their arena with odor that should have been familiar to them. We know little on the role of olfactory cues in such short-range orientation.

Given their acute sense of smell, it is reasonable to assume that mammals use environmental odors for orientation, but we lack experimental evidence. Particularly suggestive are observations of desert animals that head into the direction of the wind that blows from bodies of water, areas with rains, or fresh vegetation suited for grazing. Such so-called "smelling of rain" has been reported for camels, antelopes, water buffaloes, and others. Migrating mammals such as caribou, reindeer, or wildebeest may also use conspecific odors on or near trails. These may originate in dung, urine, or interdigital glands. White-footed mice (*Peromyscus leucopus noveboracensis*) homed over distances of about 200 m whether or not they had been made anosmic by treatment with zinc sulfide (Parsons and Terman, 1978).

Even marine mammals, especially baleen whales, may use odors from land or sea for navigation. They may also locate their food, krill, near the surface by sampling the air for "krill odor" (Cave, 1988).

5

Chemoreception

We might fairly gauge the future of biological sciences, centuries ahead, by estimating the time it will take to reach a complete, comprehensive understanding of odor.

LEWIS THOMAS, in: *Late Night Thoughts on Listening to Mahler' s Ninth Symphony*

To appreciate the possibilities and limitations of exchange of chemical information, the chemical ecologist needs to understand how the chemical senses work. There are a number of outstanding reviews and books on chemoreception *per se*. In addition to the general principles of chemoreception, the chemical ecologist is particularly interested in adaptations of anatomy and performance to specific abiotic and biotic conditions that optimize signal detection and maximize information gain. Here odor reception will be treated from an ecological viewpoint and only to the extent that it is an essential step in the behavior of an animal moving about in its environment and attending to vital cues. While sensory studies in the laboratory employ simple and clean stimuli, *noise* and the need for *filtering* are important factors in sensory ecology (Ali, 1978).

5.1 Encounter and exposure: orientation, sniffing, licking, tasting

The ecologist is interested in behaviors that animals employ before odor molecules even reach their chemical sense organs. Animals can manipulate the amount of odorants reaching the chemoreceptors by manipulating airflow toward the receptors, body movements, or locomotion. In its natural environment, an animal may seek or avoid stimulation by odors. Such behaviors cannot be satisfactorily addressed in the confines of laboratory studies. A foraging carnivore or herbivore exposes itself continuously to new chemical signals. Alerted by a sound or visual stimulus, an antelope or deer may turn into the wind and sniff. In many contexts, such orienting movements are the first step in a sequence of behaviors whose later steps are primarily guided by chemical cues.

As any dog owner knows, sniffing, licking, and various body movements in the service of detecting biologically significant chemical signals may be elicited by visual, tactile, auditory, or chemical stimuli, not just by chemical ones. The Russian reflexologists have formally worked out these stimulus–response relationships. Chiszar (1986) has termed such investigatory behavior "dedicated motor patterns" because they are dedicated to bringing sensory receptors in contact with semiochemicals. Such dedicated motor patterns may occur in the total absence of chemical cues. Chiszar has pointed out that many experiments in chemical ecology that measure orienting or investigatory behaviors as dependent variables failed to demonstrate odor effects. A control stimulus may be sniffed as much as the putative semiochemical. As examples, Chiszar discussed the tongue flicking by rattlesnakes after striking a prey animal (strike-induced chemosensory searching), the male guinea pig's investigation of female urine, and the male golden hamster's investigation of vaginal secretion. In all these cases, rates of investigation of various samples do not depend on a particular chemical cue. That is why measuring the intensity of sniffing and/or licking in a bioassay is often a poor indicator of bioactivity of a stimulus. However, complex chemical signals may include volatile "alerting compounds" that attract the attention of conspecifics, while other components in the mixture provide more specific information or trigger typical responses (Müller-Schwarze, 1998).

5.1.1 Fish

Fish manipulate stimulus access to the olfactory organ in different ways. *Isosmates* move water by beating cilia on the olfactory lamellae and thus receive a continuous laminar flow. By contrast, *cyclosmate* species draw in water in pulses by muscles associated with the gills (sniffs or flicks). The commonly observed jaw movements termed "coughing" are a form of "sniffing" in flounders. The pressure in the olfactory sac drops rapidly during such coughing, drawing in water. The jaw movement rotates the lachrymal bone, which in turn, pushes or pulls the olfactory sac. Food odors trigger such coughing (Nevitt, 1991).

5.1.2 Reptiles

Among reptiles, desert iguanas (*Dipsosaurus dorsalis*) may first visually spot conspecific scent marks that absorb long-wave ultraviolet light and then bring their chemoreceptors close. Their femoral gland secretion is non-volatile and active only at close range. In the laboratory, iguanas detected samples of femoral secretion on tiles better in ultraviolet light than under incandescent light. Once close to the scent mark, the iguanas' response is measured as the number of tongue-touches by blindfolded animals. The femoral gland secretion is a pheromone candidate because the gland is larger in males, is activated by testosterone, and there is a cycle of glandular activity that coincides with the breeding season. The secretion contains proteins of 10–30 kDa daltons (Alberts, 1989).

5.1.3 Mammals

Mice are able actively to seek or avoid *priming pheromones* that modulate their ovarian cycle and onset of puberty. Peripubertal female mice avoid the urine odor of adult males, known to accelerate puberty in females, and are more attracted to the odor of grouped adult females. This behavior is particularly effective because the active space of the (almost) non-volatile male pheromone is small, and prolonged exposure is required for the effect to occur (Coppola and O'Connell, 1988). Likewise, prepubertal female mice do not urinate near urine marks of adult males, while grouped, estrous, and diestrous *adult* females do. Such behavior may help young females to avoid exposure to male odors until they reach puberty. This way they would be protected from mating too early, and their eventual reproductive success would be enhanced (Drickamer, 1989a).

Various specialized structures may aid in sampling air or improving directional smelling. The elephant's trunk can be seen as an olfactory periscope; tube-nosed bats (*Murina* and *Harpiocephalus* spp., Vespertilionidae) and hammerhead sharks (Sphyrnidae) represent bizarre examples of specialized olfactory anatomy. Nostrils of lower vertebrates tend to be farther apart than those of higher vertebrates that have more flexible heads and necks (Stoddart, 1983).

5.2 Receiving molecules: chemosensory organs

Apart from taste, vertebrates have five different chemoreceptor systems for airborne chemosignals: the main olfactory system, the vomeronasal organ (VNO), the trigeminal nerve, the septal organ of Masera, and the nervus terminalis. They each will be discussed in turn. All five are fully functional in most mammals (Fig. 5.1).

5.2.1 Main olfactory system

This chapter will only deal with a few facts pertinent to the ecological operations of animals. Excellent reviews and monographs deal with the olfactory system.

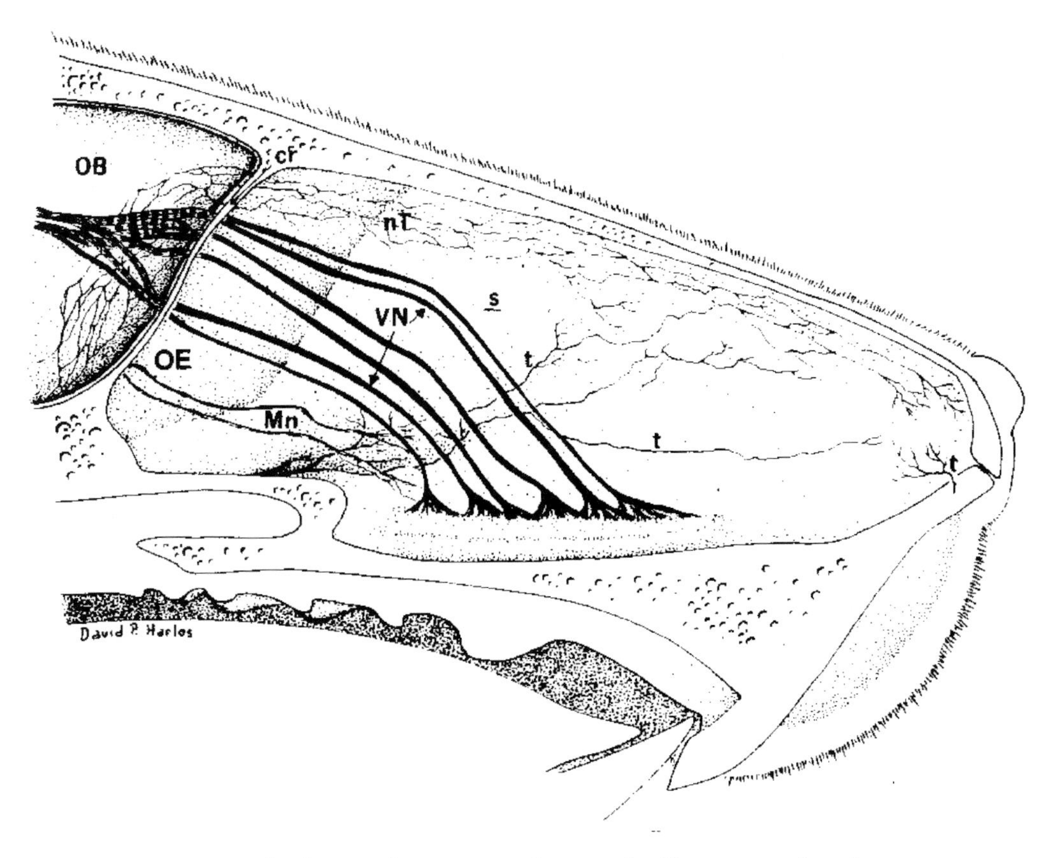

FIGURE 5.1 The five neural components that serve in olfactory reception, shown in an rat; nostrils at upper right, incisors at lower right. The main olfactory bulb (OB) receives axons from the olfactory epithelium (OE) through the cribriform plate (cr). The lines (VN) represent the vomeronasal nerve, conducting signals from the vomeronasal organ above the roof of the mouth. NT, nervus terminalis; t, branches of the trigeminal nerve; Mn, septal organ of Masera; s, septum. (Modified from Graziadei, 1977.).

Overall structure

Fish

Elasmobranchs have their paired olfactory organs on the ventral side near the mouth. As the fish takes the respiratory water current into the mouth, water passes through the olfactory sacs. Thus, elasmobranches use the respiratory water current for supplying the olfactory organ with waterborne stimuli.

By contrast, bony fish have their olfactory organs on the dorsal side of the snout at some distance from the mouth. The olfactory system in fish involves the first (olfactory) cranial nerve, while the ninth (glossopharyngeal) and other nerves serve the sense of taste.

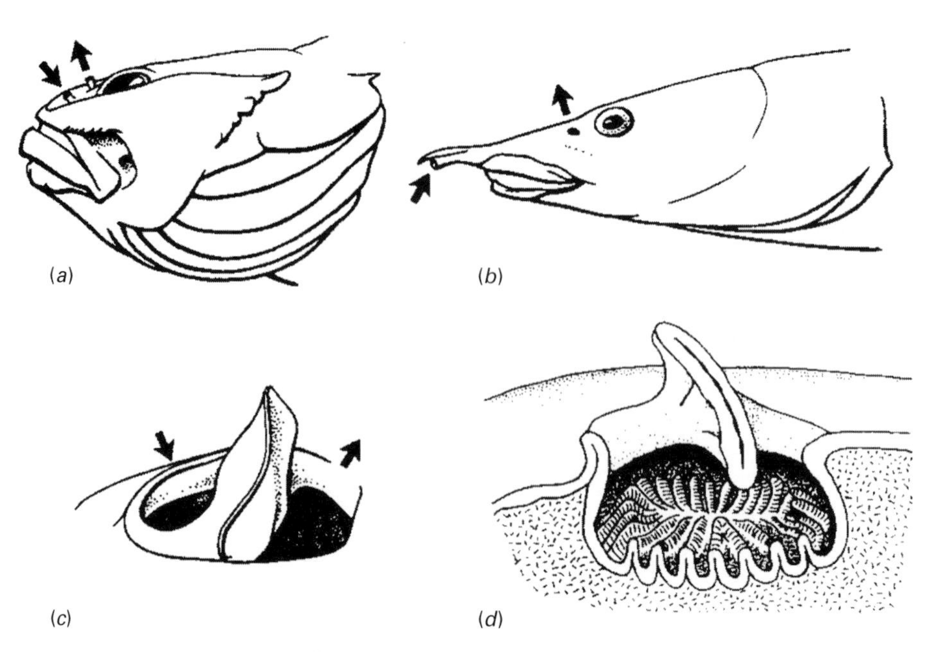

FIGURE 5.2 The olfactory organ in fish. (*a*) The nostril positions in sculpin (Cottidae); (*b*) nostril position in spiny eel (Mastacembelidae); (*c*) a skin flap separates in- and outflow, an arrangement typical for bony fish (here: Catastomidae); (*d*) the olfactory lamellae are located in the floor of the olfactory capsule (here: minnow, *Phoxinus*). (From: C. E. Bond: *Biology of Fishes*.)

The olfactory receptors of fish are located on lamellae, arranged in an oval or round olfactory rosette; this in turn, is housed in the olfactory capsule, a chamber at the front of the head (Fig. 5.2). The olfactory capsule is totally separated from the respiration system. Inlet and outlet openings permit a stream of water to pass through the olfactory capsule. Fish with round rosettes have the fewest chemoreceptors and rely less on olfaction, while elongated rosettes are characteristic for species with a keen sense of smell. Cycloosmatic fish possess accessory sacs, branching off from the main olfactory sacs. The accessory sacs act as pumps drawing water through the olfactory organ, activated by muscles of the jaw and gill regions. With the exception of lungfish, the olfactory sacs in fish are not connected to the mouth. Therefore, the lungfish resemble terrestrial vertebrates in this regard. The olfactory sensory cells are unevenly distributed over the epithelium, with 50000 to 100000 cells/mm^2, on average (Jobling, 1995). The olfactory epithelium sends its information to the olfactory bulb, which is organized into glomeruli, clusters of neurons. The olfactory tract transmits the chemosensory information from the bulb to the brain. The olfactory tract in the Atlantic cod has four nerve bundles that specialize in conveying specific information, leading to four different behaviors: flight, courtship, snapping with open mouth, and food seeking (Jobling, 1995).

In many deep-sea fish, the olfactory capsule and olfactory lobes are much larger in males than in females. The olfactory lobe of the brain may even change in size in the same individual: ripe males of *Gonostoma bathyphilus* (Gonostomatidae) have extremely large olfactory lobes. With sex reversal, their olfactory lobes regress (Badcock, 1986).

In fish, certain odorants activate specific glomeruli in the olfactory bulb. Amino acids stimulate glomeruli in the lateral region of the bulb, while glomeruli in the medial region of the bulb process information on bile acids (Yoshihara *et al.*, 2001).

Reptiles

The forked tongue of squamate reptiles, together with the VNO, has received considerable more attention than the sense of smell *per se*. Recent strong evidence suggests that the forked tongue serves in odor trail location by simultaneously comparing two odor concentrations ("directional smelling"). Ecologically, reptiles that follow prey trails have forked tongues, while ambush hunters do not (Schwenk, 1994). Comparing dinosaur skulls with living reptiles and birds strongly suggests that even in those extinct reptiles the fleshy nostril was located forward (rostral) in the enormous bony nasal aperture – contrary to the traditional assumption of a location farther back on the skull. This would have permitted the dinosaurs to pass the inhaled air over their olfactory epithelium (Witmer, 2001).

Birds

The sense of smell in many birds is better developed than previously thought. The relative size of the olfactory bulb varies from largest in procellariiforms to smallest in songbirds (Bang and Cobb, 1968; Bang, 1971). Olfactory bulb size is correlated with life habits: carnivorous and piscivorous, colonial, burowing and/or sexually monomorph bird species have the largest, seedeaters the smallest (Fig. 5.3). In some species, the numbers of olfactory receptor cells and glomerular mitral cells approach those of mammals with a keen sense of smell, such as the rabbit (Wenzel, 1986). Accordingly, olfactory performance is acute in some bird species. Using Pavlovian techniques and measuring cardiac acceleration, the importance of the olfactory system in the reception of four odorants has been examined in pigeons (Henton *et al.*, 1966). The birds were most sensitive to *n*-butyl acetate (10^{-4} of vapor saturation), followed by butanol ($10^{-4.3}$), *n*-amyl acetate (10^{-4}), and benzaldehyde ($10^{-3.3}$).

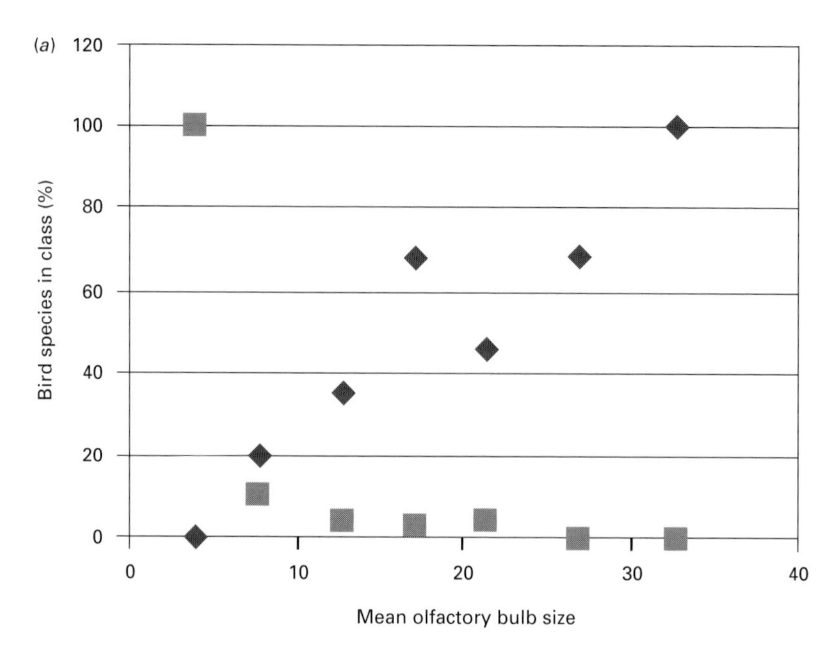

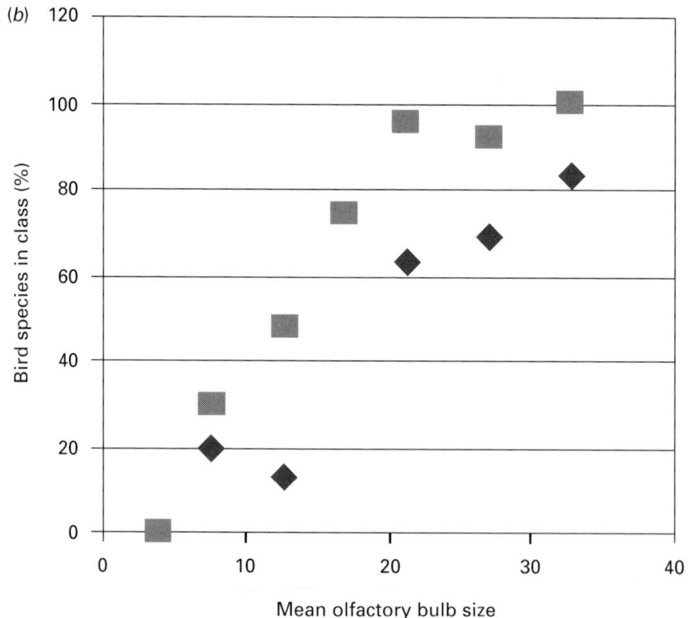

FIGURE 5.3 Food habits and relative size of olfactory bulb in birds. Species were devided into seven classes by mean olfactory bulb size calculated as the largest diameter of the bulb as a percentage of the largest diameter of the whole brain. (*a*) Relationship with food habits; birds with large olfactory bulbs are more likely to be carnivorous or piscivorous (◆), while all bird species in the smallest class are seedeaters (■). (*b*) The larger the olfactory bulb, the more species are colonial breeders (◆) and sexually monomorphic (■). (After data in Bang and Cobb (1968) and Bang (1971.)

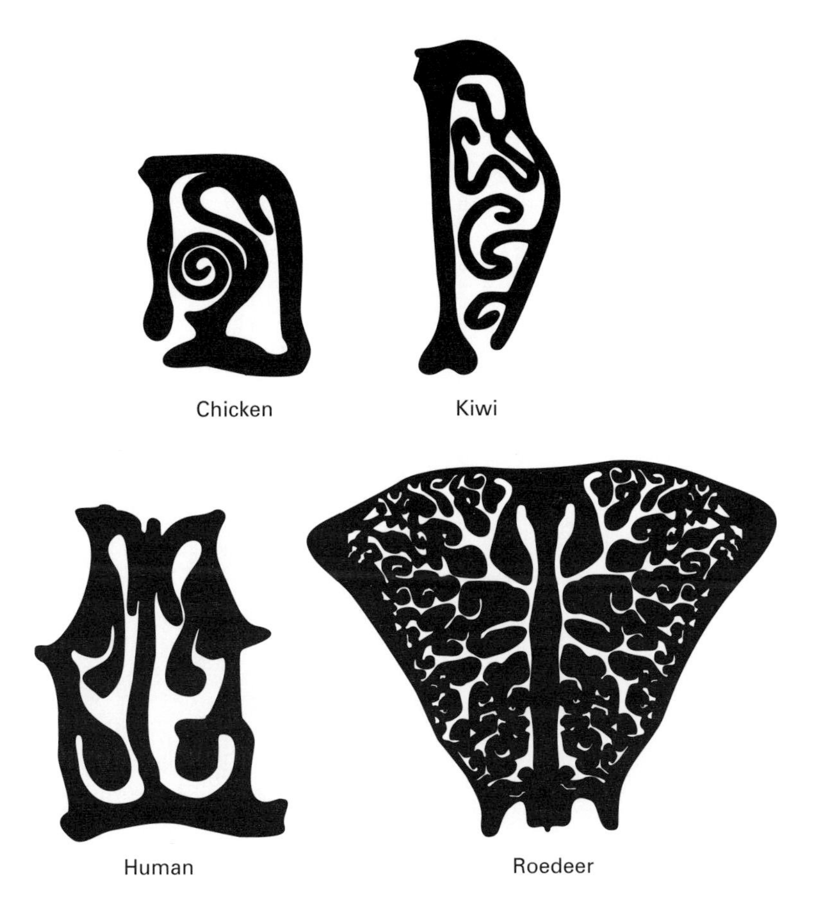

FIGURE 5.4 Different degrees of elaboration of the nasal scrolls (turbinates) in some birds and mammals. Only part of the scrolls contain olfactory epithelium. Note the difference between humans and deer. (Redrawn from various sources.)

Birds smell food not necessarily from a distance. They may do so via the choanae while the food is in the mouth. Waxwings discriminate berries treated with methyl anthranilate from controls only after picking them up with their bills (Avery *et al.*, 1992).

Mammals

In mammals, the main olfactory system is the "work horse" in the perception of odors. Excellent detailed reviews of the mammalian olfactory system are available elsewhere. In brief, the olfactory epithelium is located on a portion of the scroll bones (endoturbinales and posterior part of nasoturbinales; Fig. 5.4), in humans it is located about 1 cm beneath the bridge of the nose. Olfactory reception is affected by several factors, such as the size, shape, or wetness of the nasal passages. In the dog, the olfactory membrane extends over 75–150 cm^2 depending on body size, while in humans it is only 2–4 cm^2. It consists of three

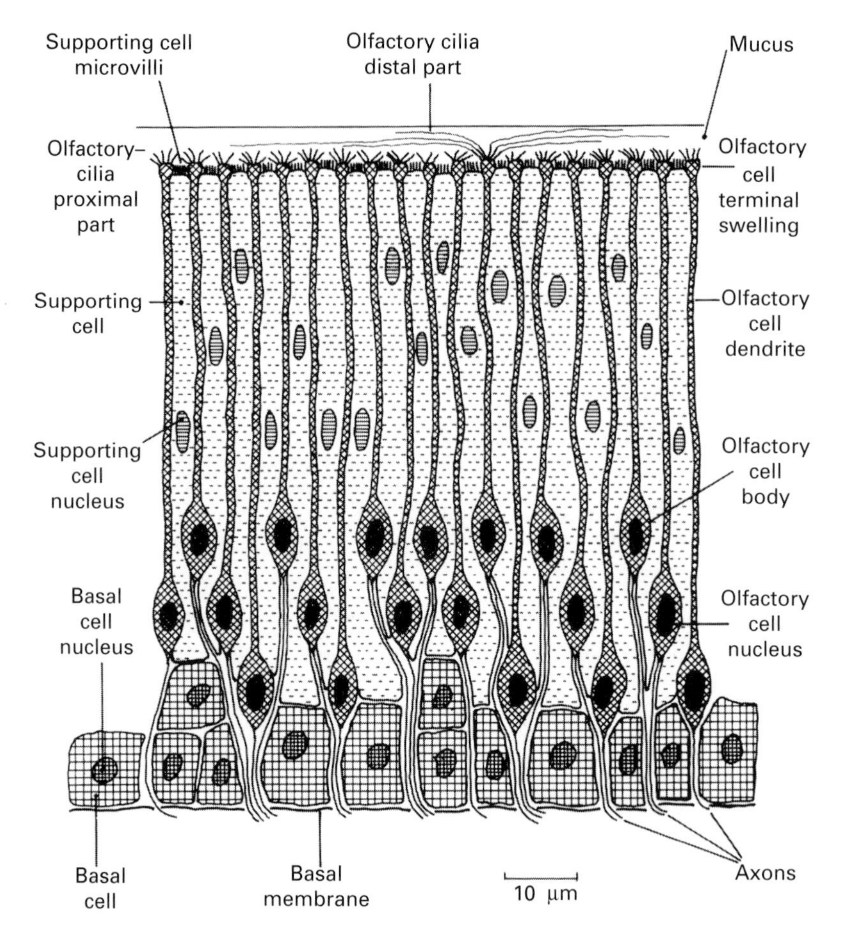

FIGURE 5.5 Structure of olfactory epithelium in mammals, a schematic view. Only one receptor cell is shown with its entire cilia floating in the olfactory mucus layer. (From Albone, 1984.)

cell types: *receptor cells* are bipolar and connected to the olfactory bulb; *supporting cells* space the dendrites out, and *basal cells* are precursors of both receptor and supporting cells (Fig. 5.5). There can be as many as 9.0×10^6 receptors/cm^2. The olfactory epithelium of the mouse contains over 2 million sensory neurons.

The dendrites on the receptor cells have swellings, the olfactory knobs. These knobs have smooth vesicles that may be open to the mucus covering the epithelium. From each knob extend 1 to 150 cilia, floating in the mucus. The precise number of cilia varies with the species. The rabbit, for example, has 10–12 cilia on each knob. The surrounding mucus consists of mucopolysaccharides, lipids, and phosphatides. The axons of the olfactory receptor cells extend through the cribriform plate and terminate in the olfactory bulb (Fig. 5.1).

How olfactory receptors work

The olfactory epithelium contains a large number of different receptors. Buck and Axel (1991) found large families of genes for proteins that appear to be receptors linked to G-proteins and occur only in the olfactory epithelium. These are the largest gene families known to date for any animal genome. Mouse and rat have over 1000 olfactory receptor genes (*OR*), humans 500–750, and zebrafish and catfish around 100 (Mombaerts, 1999). The genes code for a range of receptors that recognize diverse odorants. Does olfaction use "labeled lines" or "diffuse signals" ("broad tuning")? A study of olfactory receptor neurons *in vivo* in anesthetized rats supports the notion of broad tuning. As many as 10 different odors can excite the same olfactory receptor neuron as measured by the electro-olfactogram; 32% responded to six different odors, and only 12% to just one single odor. The study employed odors from the terpene, camphor, aromatic and straight-chain ketone groups, previously used in frogs (Duchamp-Viret *et al.*, 1999). The several compounds that one olfactory receptor type responds to share certain molecular features. The rat neurons responded more broadly but with less sensitivity to odors than did those in the frog. One single olfactory receptor neuron is thought to express only one or only few of the 1000 *OR* genes. Some consider the "one neuron, one *OR*" hypothesis far from proven (Mombaerts, 1999); however, recent work has shown how feedback from an expressed *OR* gene inhibits the activation of other *OR* genes, ensuring the "one receptor – one olfactory neuron rule" (Serizawa *et al.*, 2003).

Molecular events at the olfactory receptors

The mucus of the olfactory epithelium contains odorant-binding proteins that are specific to various odorants, such as anisole, camphor, benzaldehyde (cherry-almond odor), 2-isobutyl-3-methoxypyrazine (green bell pepper odor), and 5α-androst-16-en-3-one (urine odor) (summarized by Leffingwell, 2001).

Odorants are thought to bind to integral membrane receptors on the cilia of the olfactory sensory neurons. The receptors are thought to be specific; different olfactory neuron types recognize different odorants that share certain characteristics (Buck, 1993). The odorant receptors transduce signals via interactions with G-proteins (so-called because guanosine trisphosphate is involved in their activation). These G-protein-coupled exhibit seven hydrophobic domains (Fig. 5.6). Variation in the amino acid sequence of the transmembrane domain may account for specificity and selectivity of odor reception.

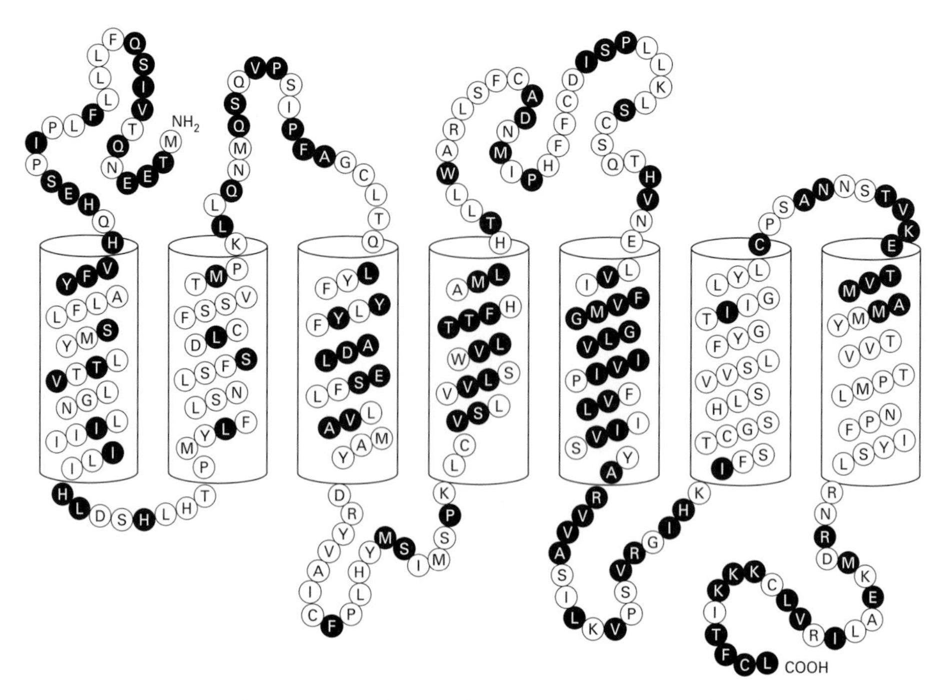

FIGURE 5.6 A transmembrane olfactory receptor. The sections shown as cylinders reside in the cell wall. These are the seven transmembrane hydrophobic domains. Loops on top (and the N-terminus) are outside the cell, those at the bottom (and the C-terminus) are inside the cell. For clarity, the "cylinders" are spread out but, in the cell, they are bundled into a ligand-binding pocket. The residues shown as black dots are especially diverse among known odorant receptors. This diversity enables the receptor family to interact with many different odor molecules. (From Buck, 1996.)

The receptors start a second messenger cascade that is initiated by activation of G-proteins in the cell. These, in turn, interact with membrane-bound adenylyl cyclase, which catalyzes the formation of cyclic adenine monophosphate (cAMP) and opening of cAMP-gated cation channels. Depolarization then brings about an action potential, which travels along the axon of the olfactory sensory neuron. Many of the molecular components of this cascade are olfactory specific.

To be ready for the next odor stimulus, β-adrenoceptor kinase (Bark) 2 inactivates a receptor only a tenth of a second after the first stimulation occurred. This kinase phosphorylates the activated receptor, which allows another protein, B-arrestin, to bind to the receptor and inactivate it. This is a specific example of a group of enzymes that deactivate hormone or neuroreceptors (Dawson *et al.*, 1993).

The first olfactory neuron-specific protein, termed olfactory marker protein was isolated by Frank Margolis in 1972 and was cloned in the 1980s. This 19 kDa cytoplasmatic protein is unique to the mature olfactory system and is found in vertebrates from salamanders to humans (reviewed by Margolis *et al.*, 1993).

In the olfactory epithelium, different *OR* genes are expressed in different regions so olfactory neurons of one type are concentrated in a particular zone. For instance, the K4 and K7 subfamilies are always expressed together in the same (middle) region, while the K18 subfamily is always located in a different region dorsal and medial from K4 and K7 (Ressler *et al.*, 1993). This spatial segregation is preserved in the projections of the olfactory receptor neurons to the olfactory bulb (Ressler *et al.*, 1994). However, in each of the four zones, many different types of olfactory receptor neuron occur in a mosaic. Such zonal arrangements exist in mouse, rat, and zebrafish and may represent an efficient wiring system conducive to easy decoding (Mombaerts, 1999).

Processing in the olfactory bulb

The olfactory bulb contains glomeruli where the dendrites of mitral cells and tufted cells concentrate. The mouse has about 1800 glomeruli in its olfactory bulb. In the rabbit, the input from 5×10^7 receptor cells converges on 1900 glomeruli (Fig. 5.7). In some bats, 900 receptor cells converge on each secondary olfactory neuron.

Associated with the glomerulus are two additional cell types, the periglomerular and granular cells. Both establish lateral connections between tufted and mitral cells. With 1000 *OR* genes estimated in mouse and rat, each olfactory receptor type may correspond to two glomeruli (Mombaerts, 1999). Thus, a glomerulus is the site where the axons of olfactory sensory neurons expressing a specific *OR* type converge, an important stage in integrating olfactory information (Mombaerts, 1999).

With about 1000 *OR* genes, there are about two glomeruli for each. The connection of olfactory neurons to the olfactory bulb follows two principles: zone-to-zone projection and glomerular convergence (Mori *et al.*, 1999). An optical imaging technique, termed intrinsic signal imaging, identifies how specific glomeruli respond to particular odors in living animals. A light illuminates the glomeruli through an opening or thinned spot in the skull above the olfactory bulb. Changes in neural activity affect levels of blood oxygenation and light-scattering properties of neural membranes. (The hemoglobin of the blood absorbs light, and the membranes scatter light in varying ways.) This causes

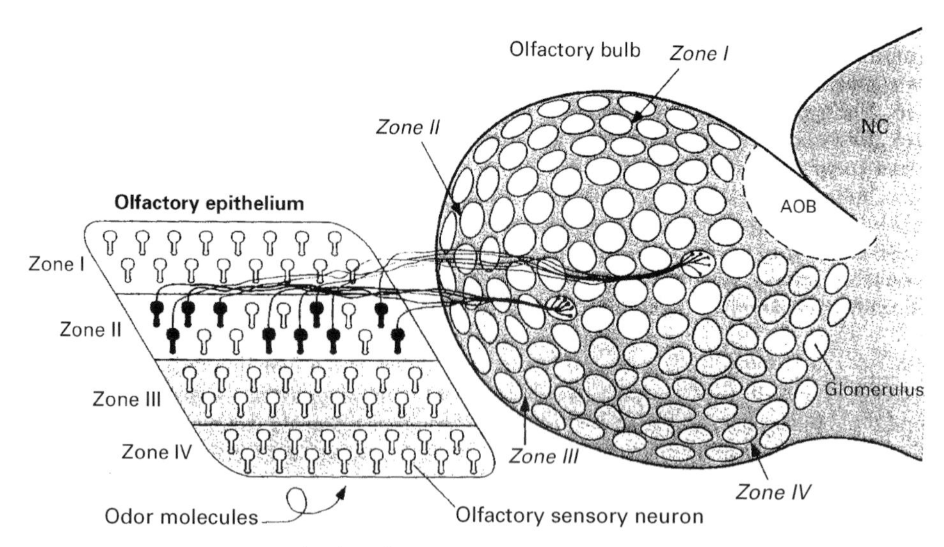

FIGURE 5.7 Projection of receptor input from olfactory epithelium onto glomeruli in the main olfactory bulb in mice. The epithelium is organized into four zones defined by expression of odorant receptors. Olfactory neurons of a particular zone project to a corresponding zone in the bulb. Axons of these olfactory neurons that express the same odorant receptor (such as those shown in black) converge to a small number of glomeruli. AOB, accessory olfactory bulbs, NC, nucleus coeruleus. (From Mori *et al.*, 1999.)

changes in the light bouncing back that can be visualized as differently colored small areas of the olfactory bulb (Rubin and Katz, 1999).

Processing at higher centers

After only one synapse – in the glomerulus – connections with the cortex are made. Approximately 24 output cells emerge from each glomerulus. We are now starting to understand how the olfactory bulb connects functionally to the cortex. Linda Buck and her colleagues have labeled receptor cells by inserting the gene for a marker protein called barley lectin next to the gene for a mouse olfactory receptor protein. The lectin marker could be demonstrated in both the neuron carrying that receptor and in connecting neurons. In this way, the investigators traced the connections from receptor neuron through the olfactory bulb to the olfactory cortex. Each of two *OR* genes thus marked resulted in one or two stained glomeruli on each side of the main olfactory bulb. The olfactory cortex showed several clusters of stained neurons. Each mouse exhibited the same pathways, showing for the first time a clear organization in the projection from the bulb to the cortex. Moreover, projections from different receptor types converge in the cortex, permitting comparison and integration of information. While in each half of the bulb, each receptor is represented in 1 out of 1000 glomeruli (0.1% of the bulb's area), a receptor's projections cover about 5% of the area in the cortex. This suggests that a single cortical neuron receives input from as many as 50 different olfactory receptors. Leading smell researchers see this as a means for the brain to analyze and distinguish the often complex smells.

Beyond the olfactory bulbs, complex signal processing takes place. Connections between the olfactory bulbs permit summation of input from the two olfactory mucosae, thus enhancing sensitivity (Bennett, 1968). The central olfactory apparatus involves many brain areas, including the mediodorsal thalamic nucleus and the prefrontal insular cortex (cited in Ferreira *et al.*, 1987), the hippocampus, fornix, gyrus supracallosus and dentalus, septum lucidum, and uncus.

Other developmental and functional aspects

A number of pertinent observations have been made.

- Odor exposure during early ontogeny (e.g. 2-week-old rats) alters the mitral cells in specific regions of the olfactory bulb (Døving and Pinching, 1973; review by Reasner, 1987).
- Domestication can change the olfactory system. In the domestic pig, most olfactory structures in the forebrain are 30% smaller than in wild boar (Kruska and Rohrs, 1974).
- The olfactory receptors can be stimulated by bloodborne odors. Garlic odor, carried in the blood, can be perceived olfactorily (Maruniak *et al.*, 1983).
- The main olfactory system is designed with considerable redundancy. Rats that have been bulbectomized on one side still have the same absolute threshold and intensity difference threshold as intact rats (Slotnick and Schoonover, 1984).
- Centrifugal neural pathways affect the processing of meaningful biological information in the olfactory bulb; rates of habituation particularly appear to be modulated this way. (Responses to food and sex odors habituate little.) This centrifugal innervation also has long-lasting effects on olfactory learning. In short, the olfactory bulb is part of a neural network (rather than a strict one-way street for incoming stimuli) that supports learning and memory for biologically significant odors (Gervais *et al.*, 1988).

The main olfactory system mediates numerous behaviors that will be discussed in the section on signaling and priming pheromones, and also interspecific responses. For any given mammalian species, the role of this system in particular behaviors cannot be easily predicted. For instance, in prairie voles, *Microtus ochrogaster*, females associate strongly with familiar males. This response is lost after removal of one or both olfactory bulbs. Bulbectomy reduces mating behavior but does not eliminate it, while – unlike in mice – maternal behavior is not affected (Williams *et al.*, 1992a).

5.2.2 Vomeronasal system

The vomeronasal system, also known as the accessory olfactory system, consists of chemoreceptors, organized into the VNO, the vomeronasal nerve, its terminal, the accessory olfactory bulb, and more central pathways. First described by Jacobson in 1811, the VNO has been studied intensely. We now know how stimuli reach it and what behaviors it mediates. The VNO occurs in amphibians, reptiles, and mammals. Among mammals, it is best developed in marsupials and monotremes. In birds it only appears during embryogenesis. The VNO and its function are best known for squamate reptiles, particularly snakes, and rodents and ungulates among the mammals.

Amphibia

The best developed vomeronasal system among amphibia is found in plethodontids, a family of lungless salamanders that originated in the Appalachian mountains of North America, where diverse forms still abound. This family enjoys special chemosensory adaptations and the VNO is situated anteriorly in the nasal cavity (Fig. 5.8). The genus *Plethodon* (woodland salamanders) has been well investigated. In the red-backed salamander, *Plethodon cinereus*, nasolabial grooves lead to the VNO; the behavior of nose tapping draws material up the nasolabial grooves, which probably facilitates stimulus detection (Dawley, 1987, 1992). Nose tapping is the functional equivalent of tongue flicking in snakes. Both behaviors pick up material for chemical evaluation. The red-backed salamander uses its VNO in finding prey such as *Drosophila*, although other senses are also important (Placyk and Graves, 2002).

There are sex differences: in woodland salamanders (*Plethodon*) and brook salamanders (genus *Eurycea*) of the lungless salamander family Plethodontidae, females have smaller VNOs than males (Dawley, 1992). There are also species differences: the VNO in *Eurycea wilderea* is larger than that in *Plethodon cinereus*, and the size of the hypertrophied nasal grooves, called cirri, correlate with size of VNO (Dawley, 1992).

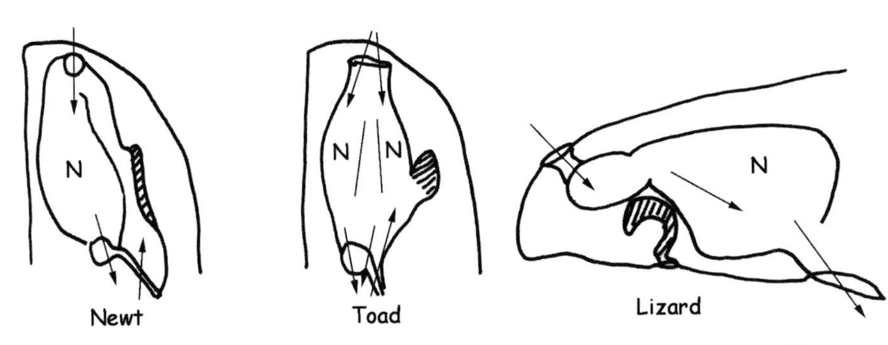

FIGURE 5.8 Location of the vomeronasal organ (VNO, cross-hatched) in amphibians and reptiles. Arrows show air entering through the external nares at top and exiting through the choanae at bottom. From there, VNO receives air with odors. N' nasal cavity. In lizard, VNO receives stimuli from mouth cavity below, with help from tongue. (Redrawn after Romer, 1959.)

In the frog *Rana temporaria*, the VNO shares the nasal compartments with the main olfactory organ. The VNO has three cavities, and water enters from the external nares via two fissures. The VNO is used to sample water while the frog is submerged, while above water air is inhaled and the olfactory system stimulated (Døving *et al.*, 1993).

Reptiles

The reptilian VNO develops from the nasal cavity but is separate from it in the adult stage. The best-known behavior involving the VNO in reptiles is tongue flicking. The forked or simple tongue samples air and is assumed to transport odorants into the openings of the VNO in the roof of the mouth, although this view has been challenged (Schwenk, 1994). The VNO processes volatiles and non-volatiles, while the nasal olfactory system is limited to volatile stimuli. Odorants can be directly moved into the VNO by pressing the back of the tongue against the vomeronasal duct or indirectly after first wiping the tongue over the sublingual plicae (folds) (Schwenk, 1994). The VNO is highly developed in snakes and lizards, though turtles possess some VNO tissue. The VNO of garter snakes projects to the accessory olfactory bulb in a zone-to-zone fashion, as in several mammalian species.

The VNO plays an important role in feeding behavior in garter snakes (Halpern and Frumin, 1979). Lizards also appear to use their VNO for food detection. Examples are the scincid lizard *Chalcides ocellatus* (Graves and Halpern, 1990) and the desert iguana *D. dorsalis* (Cooper and Alberts, 1991). The latter

Table 5.1 Role of vomeronasal organ in behavior of amphibians and reptiles

Class and taxon	Context	Behavior	Reference
Amphibia			
Plethodontids (lungless salamanders): woodland salamanders (*Plethodon*)	Chemical evaluation	Nose tapping	Dawley 1987, 1992
Frog (*Rana temporaria*)	When submerged	Sample water	Døving *et al.*, 1993
Reptiles			
Garter snakes (*Thamnophis* spp.),	Feeding	Response to earthworm wash	Halpern *et al.*, 1985
Lizard *Dipsosaurus dorsalis*	Food finding	Detection of carrot extract	Cooper and Alberts, 1991
Red-eared turtle *Pseudemys scripta*	Chemoreception in water	Accessory bulb activity to salt solutions and soluble vapors	Hatanaka and Hanada, 1987

were unable to detect carrot extract odor or cologne on cotton swabs if their vomeronasal ducts had been sealed experimentally (Cooper and Alberts, 1991).

In garter snakes (*Thamnophis sirtalis*) a VNO-mediated stimulus can reinforce behavior: dried earthworm wash or earthworm bits can be used to reward correct performance in a conditioned response to an arbitrary stimulus, such as dots versus stripes in a Y-maze (Halpern *et al.*, 1985). In red-eared turtles the VNO "is considered to involve aqueous chemoreception in water." Salt solutions and soluble vapor substances generated activity in the accessory olfactory bulb (Hatanaka and Hanada, 1987).

Table 5.1 lists some behavioral roles of the VNO in amphibians and reptiles.

Mammals

Structure of the vomeronasal organ

The VNO of mammals consists of a pair of parallel tubes located above the palate on either side of the nasal septum (Fig. 5.9). The organ communicates with the outside by the nasopalatine duct, which, depending on the taxon, may open (a) to the mouth cavity via the incisive duct and incisive papilla (Fig. 5.10); (b) to the nasal cavity, as in alcelaphine antelopes, such as Topi, *Damaliscus korrigum*, and Coke's hartebeest, *Alcelaphus cokii* (Hart *et al.*, 1988), and the horse (here the VNO does not "get it from the horse's mouth, but rather is paid through the nose"); or (c) both (e.g. the ring-tailed lemur, *Lemur catta*; Bailey,

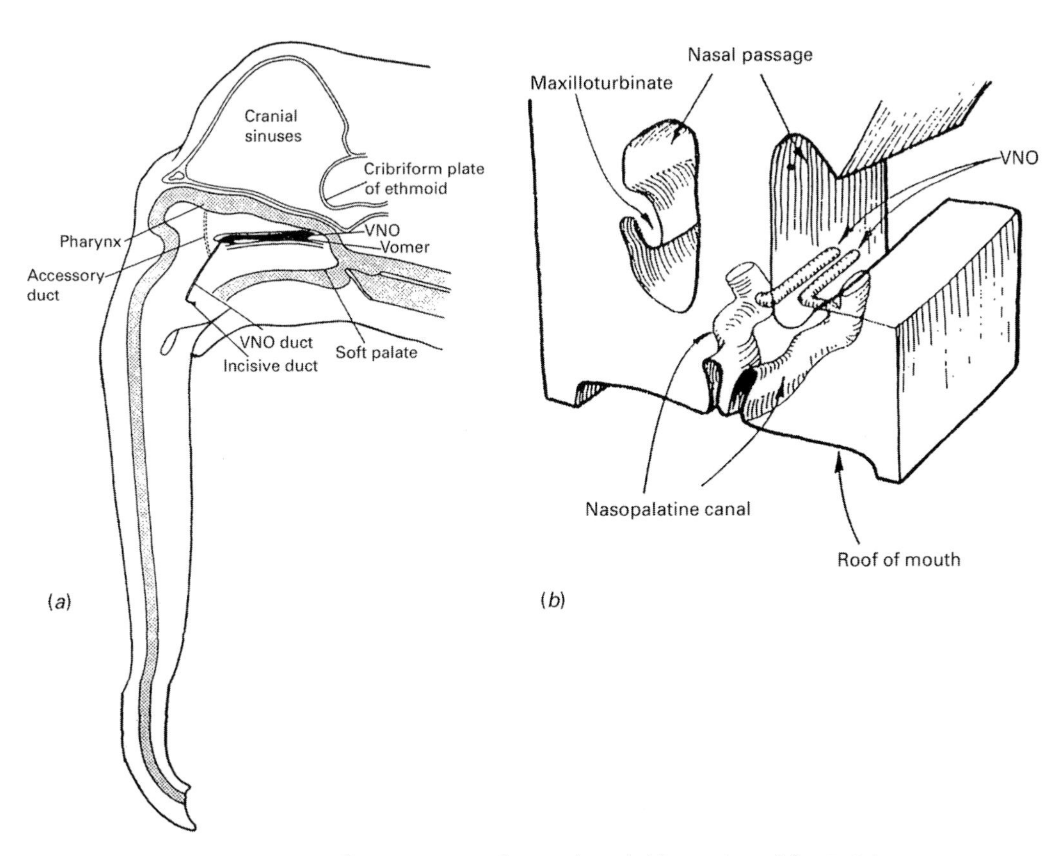

FIGURE 5.9 Mammalian vomeronasal organ (VNO). (*a*) Location of the VNO in a female Asian elephant, *Elephas maximus*. (*b*) Location of the VNO in the European hedgehog, *Erinaceus europaeus*. ([*b*] From Rasmussen and Hultgren, 1990; [*a*] from Stoddart, 1983, after Poduschka and Firbas 1968).

1978). There may be intermediate stages: the common wildebeest (*Connochaetes taurinus*) lacks incisive papillae but has small incisive ducts.

Blood vessels, cavernous tissue, and muscles surround the VNO, and all in turn are enclosed by the vomer bone capsule. This arrangement permits the lumen of the VNO to be compressed by swelling of surrounding tissue and opened by deflation, permitting stimuli to move out or in. This mechanism is known as the vomeronasal pump (Meredith and O'Connell, 1979).

Receptor cells

The VNO is lined with bipolar receptor cells (Fig. 5.11). The receptor cells possess microvilli, in contrast to the cilia on the receptor cells of the main olfactory epithelium. The VNO of male and female mice have sensory cells with receptors that respond to male urine, and others that respond to female

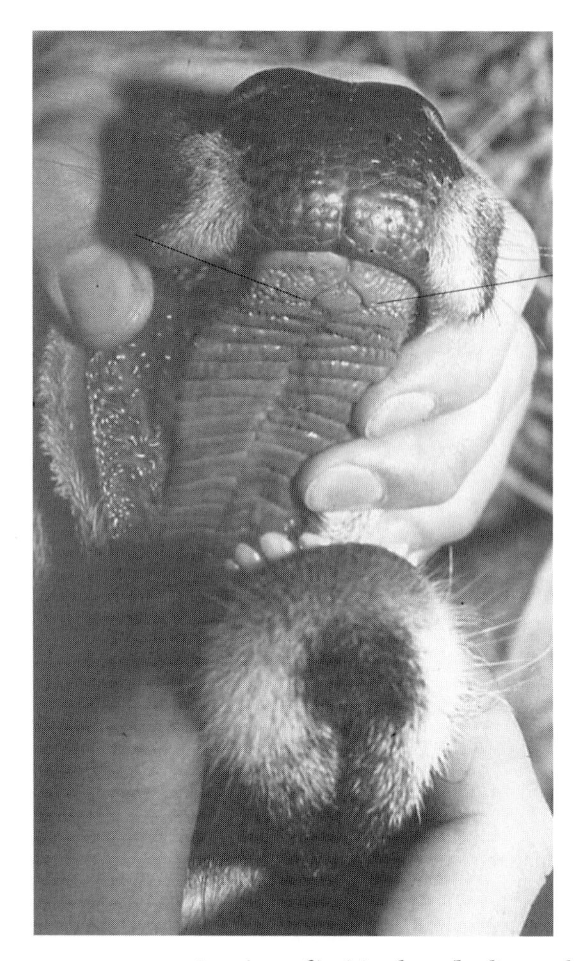

FIGURE 5.10 Openings of incisive ducts (leading to the vomeronasal organ) in the roof of the mouth of a male black-tailed deer, *Odocoileus hemionus columbianus*. The two openings are the small pores in the left and right corner of the rhomboid groove close to the upper lip (indicated by thin lines). (Photograph: D. Müller-Schwarze.)

urine (Holy *et al.*, 2000). The VNO can be very sensitive. VNO neurons, exposed to six pheromonal compounds from mice, responded to concentrations near 10^{-10} mol/l (Leinders-Zufall *et al.*, 2000).

Mechanisms of stimulus access

Stimuli gain access to the organ via the incisive foramen in the roof of the mouth or the nasal cavity. The vomeronasal pump, described for the hamster by Meredith and O'Connell (1979), aids in moving molecules into the organ. This way the animal can control stimulus access and its intensity. Experiments

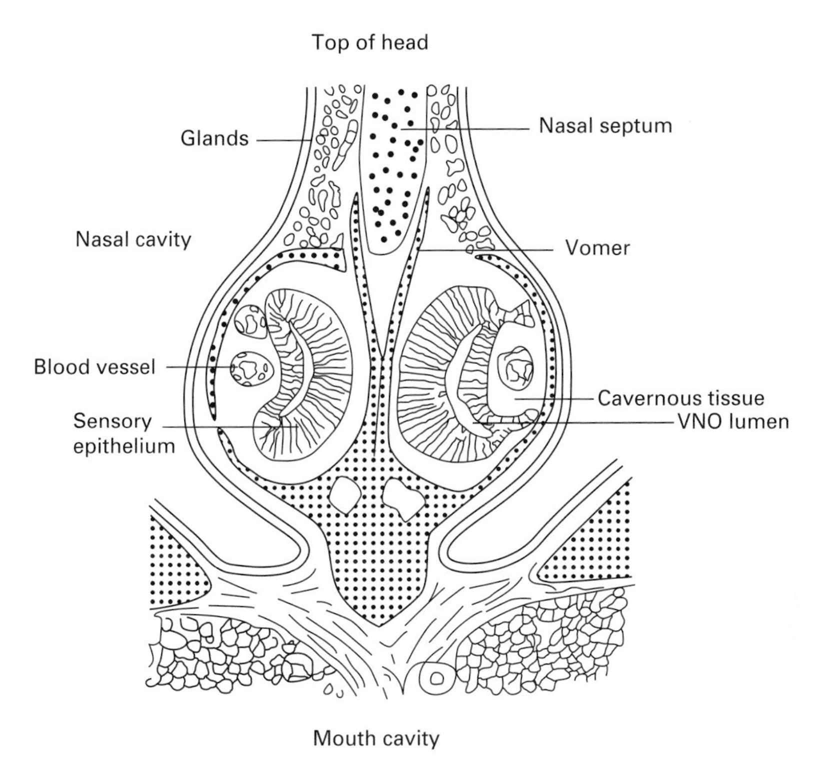

FIGURE 5.11 Structure and location of the vomeronasal organ in the mouse, seen in a transverse cross-section through the middle of the vomeronasal organ. (Redrawn after Døving and Tritier, 1998.)

with dyed materials have shown that non-volatiles reach the VNO in a variety of behavior contexts. For instance, the VNO of male pine voles (*Microtus pinetorum*) and meadow voles (*Microtus pennsylvanicus*) takes up female urine during investigation, and also materials from the fur during grooming, and even dyes applied to food (Wysocki *et al.*, 1980). The VNO plays an important role in mediating reproductive behavior.

Even in primates, the vomeronasal system can be functional. If the *liquid* phase of urine from several species of the same genus is micropipetted into the oral opening of the nasopalatine duct of mouse lemurs (*Microcebus murinus*, Prosimia), they respond with sniffing, licking and "VNO testing." By contrast, urine *volatiles* stimulate only the main olfactory system in this species (Schilling *et al.*, 1989).

Flehmen

Flehmen (or "lipcurl"; Fig. 5.12), a behavior typical for males when examining female urine, occurs in several mammalian groups. After the buck

FIGURE 5.12 Flehmen in a male pronghorn, *Antilocapra americana*. (Photograph: D. Müller-Schwarze.)

has brought urine to his upper palate by licking or mouthing movements, he raises his head, curls up the upper lip, and breathes deeply and rhythmically. Various species differ in details of flehmen. Estes (1972) postulated that the VNO receives material during flehmen. Studies of male goats with dyed urine showed that indeed urine enters the VNO during flehmen (Hart, 1983). In several better-investigated species such as black-tailed deer, *Odocoileus hemionus columbianus*, (Müller-Schwarze, 1979) or domestic sheep (Bland and Jubilan, 1987) Flehmen occurs *least* often during estrus. This indicates the analytical role of flehmen but does not exlude a stimulatory function. Wildebeest males do flehmen to female urine, but lick their nostrils during flehmen. This licking may deliver stimulus material to the VNO via the nasal cavity, perhaps to compensate for the reduced oral access to the VNO in this group of mammals (Hart *et al.*, 1988).

Central projections from the vomeronasal organ

The VNO axons form the vomeronasal nerve, which connects with the accessory olfactory bulb. In contrast to the main olfactory system, axons from neurons expressing a particular vomeronasal receptor project to glomeruli of the accessory olfactory bulb (Mombaerts, 1999). The apical VNO region, with mostly type 1 (V_1) receptors, projects to the anterior (rostral) part of the accessory olfactory bulb, while the basal region (with V_2 receptors) projects to the posterior part. The representation of sensory neurons of the VNO in the glomeruli of the accessory bulb differs from that of the main olfactory system: to make up for the smaller number of receptor types, the VNO achieves odor decoding by a more complicated spatial representation. "Neurons expressing the same receptor gene project to many different glomeruli, while a single glomerulus may receive input from more than one receptor type" (Keverne, 1999). Beyond the accessory bulb, more complexity occurs: one mitral cell can gather information from more than one glomerulus. It is thought that these arrangements ultimately serve as a "difference detector" (Keverne, 1999), most other analysis being done by the main olfactory system.

From the accessory bulb, projections lead to the medial nuclei and posteriomedial portion of the cortical nuclei of the amygdala. The amygdala in turn is connected to higher centers via the stria terminalis, and hypothalamic structures. Thus, the accessory olfactory system represents a neural pathway separate from the main olfactory system. Both project into absolutely different parts of the amygdaloids (Powers and Winans, 1975; Powers *et al.*, 1979).

Functional magnetic resonance imaging permits the central processing of urinary pheromones to be followed. The VNO and accessory olfactory bulb are involved in processing pheromonal stimuli. In an Australian marsupial, the brown antechinus (*Antechinus stuartii*), some projections from the accessory olfactory bulb respond indiscriminately to male and female urine odors while others are activated only by male urine cues. These findings show that urinary pheromones may act on the hypothalamic–pituitary–adrenocortical axis via the paraventricular nucleus of the hypothalamus (Toftegaard *et al.*, 2002).

Role of the vomeronasal organ in behavior

In mice, extirpation of the VNO has many behavioral effects. It reduces aggression in male mice (Bean, 1982; Wysocki *et al.*, 1986), reduces urine marking and aggressive behavior in sexually naive male mice (Maruniak *et al.*, 1986); lowers marking to some extent also in sexually experienced males (Labov and

Table 5.2 Chemosensory systems required for behaviors in golden hamsters

Behavior	Chemosensory system required
Species recognition	
Responses to vaginal secretion marks	Main olfactory system
Responses to flank gland secretion	Main olfactory system
Reproduction	
Androgen responses of male to vaginal secretion	Vomeronasal organ
Ultrasonic calling	Vomeronasal organ and main olfactory system

After Johnston, 1992.

Wysocki, 1989); reduces ultrasonic vocalizations by male mice when a female is present (Wysocki *et al.*, 1982); and abolishes maternal aggression in primiparous female mice (Lepri *et al.*, 1985). Deafferentiation of the VNO in male mice impairs sexual behavior. The effect is greatest if the male was deafferentiated before experience with any adult female (Clancy *et al.*, 1984). Aggressive behavior in male mice is less reduced after deafferentiation of the VNO if they had had fighting experience (Daranzo *et al.*, 1983).

Wysocki *et al.* (1986) have reviewed the relationships between learning and the VNO. The VNO plays a role in responses to infant conspecifics. Intact virgin male rats kill newborn pups. If their VNO is removed surgically, the rate of this infanticide decreases (Mennela and Moltz, 1988). A virgin female rat will accept foster pups if her VNO has been deafferentiated (Fleming *et al.*, 1979). The same is true for male rats. They will show maternal behavior toward pups after removal of the VNO (Saito, 1986a). Lactating rats are impaired in their retrieving of pups if their VNO has been extirpated (Saito, 1986b).

In the golden hamster, the VNO is essential for certain reproductive behaviors, while the main olfactory system mediates responses that involve species recognition (Johnston, 1992; Table 5.2). As in mice, removal of the VNO impairs sexual behavior in male golden hamsters, but only if carried out before the animal had had sexual experience (Meredith, 1986). The same is true for ultrasonic vocalizations.

Virgin male prairie voles (*M. ochrogaster*) are less likely to sire offspring if their VNO is removed (Wekesa and Lepri, 1992). In the guinea pig, stimulation of the VNO is assumed to be "inherently reinforcing." Male guinea pigs cease to head-bob to female urine after their VNO is surgically removed (Beauchamp *et al.*, 1985).

The role of the vomeronasal organ in priming

The VNO is extremely important in mediating endocrine responses to primer pheromones. Puberty acceleration in female rats by male urine odors can be prevented by electrolytic damage to the vomeronasal nerve. Also, effects of male urine odor such as shortening of the estrus cycle (see Ch. 8) can be eliminated by section of the vomeronasal nerve, or bilateral electrocoagulation of the accessory olfactory bulb (Sánchez-Criado, 1982). In rats, the odor of males stimulates ovulation in females, an effect that is lost if the VNO is extirpated (Johns *et al.*, 1978). Female prairie voles, *M. ochrogaster*, respond to odors from males with reproductive activation. Surgical removal of the VNO from adult females impedes this reproductive activation by the stud male. The weights of the uterus and the ovaries of these females were lower than those of normal or sham-operated individuals. However, the females without a VNO were still able to locate food by chemical cues (Lepri and Wysocki, 1987).

At the endocrinological level, the VNO mediates the surge of luteinizing hormone and testosterone in males after exposure to females. This surge does not occur if a male with deafferentiated VNO is exposed to an anesthetized female or her urine (Wysocki *et al.*, 1983). But VNO-deafferentiated males will show a surge in luteinizing hormone in response to *awake* females (Coquelin *et al.*, 1984). In female mice, stimulation of the VNO by male urinary cues activates the limbic system (discussed in Ch. 8). The roles of the VNO in the behavior of some rodents are listed in Table 5.3.

Genetic basis of function in vomeronasal organs

Some sensory neurons of the VNO express two gene superfamilies, termed *V1r* and *V2r*, that encode over 240 proteins of the seven-transmembrane type (Matsunami and Buck, 1997). These G-protein-linked putative pheromone receptors are distantly related to the main olfactory system's receptors. Receptors of the VNO are linked to different G-proteins, and their extracellular N-terminal domains are longer than those of the receptors in the main olfactory system. (V_1 receptors are linked to G_i-proteins and V_2 receptors to G_o-proteins). The intracellular excitation mechanism in VNO sensory neurons also differs from that in the main olfactory systems; instead of linking to adenylyl cyclase, the VNO receptors activate the phosphoinositol second messenger system. This has been demonstrated in several mammalian species. In hamsters, aphrodisin increases inositol 1,4,5-trisphosphate (IP_3) levels in VNO membranes. Boar seminal fluid and urine stimulate increases of IP_3 in the VNO of the female pig. (However, in the pig, the VNO is not necessarily essential for responses to pheromones [Dorries *et al.*, 1997]).

Table 5.3 Role of the vomeronasal organ in rodent behavior

Species	Sex	Context	Behavior	Reference
Mouse *Mus domesticus*	M	Aggression	Aggressive behavior; urine marking	Bean, 1982; Wysocki *et al.*, 1986;
	M	Sexual behavior	Ultrasonic sounds	Wysocki *et al.*, 1982
	F	Maternal behavior	Maternal aggression	Lepri *et al.*, 1985
Rat *Rattus norvegicus*	M	Infanticide	Kills newborn pups	Mennela and Moltz, 1988
	F	Maternal behavior	Neophobia to pups by virgin F	Fleming *et al.*, 1979
	F	Maternal behavior	Retrieving pups	Saito, 1986b
Prairie vole *Microtus ochrogaster*	M	Reproduction	Siring offspring	Wekesa and Lepri, 1992
Golden hamster *Mesocricetus auratus*	M	Mating	Vaginal secretion elicits mounting	Clancy *et al.*, 1984
	M	Sexually native male	Sexual behavior; ultrasonic sounds	Meredith, 1986
Guinea pig *Cavia porcellus*	M	Female urine	Head bobbing	Beauchamp *et al.*, 1985
Meadow vole *Microtus pennsylvanicus*	M, F	Feeding	Food dye found in VNO	Wysocki *et al.*, 1985
Pine vole *Microtus pinetorum*	M, F	Feeding	Food dye found in VNO	Wysocki *et al.*, 1985

M, male; F, Female; VNO, vomeronasal organ.

In rats, both male and female urine activate both subtypes of G-protein. (Lipophilic volatile odorants act on G_i, and a urinary lipocalin protein on G_0). There are specific receptors for urinary lipocalins in the VNO (Krieger *et al.*, 1999). In summary, the two subtypes of receptor may be activated by distinct ligands; V_2 receptors are particularly receptive to non-volatile proteins (summarized in Keverne, 1999).

Deleting a cluster of 16 *V1r* genes (2 of the 12 *V1r* gene families; representing about 12% of the *V1r* repertoire) impairs sexual and maternal behavior. Gene-deficient male mice initiate fewer sexual encounters, and lactating females attack intruders less. Furthermore, both mutant male and female mice showed "specific anosmia" in that VNO did not respond to three of eight tested pheromonal compounds (6-hydroxy-6-methyl-3-heptanone, *n*-pentyl acetate, and isobutylamine) (Del Punta *et al.*, 2002).

Peptides that serve as ligands for major histocompatibility complex (MHC) class I molecules can activate vomeronasal sensory neurons. These peptides contain nine amino acid residues and activate sensory neurons from the V_2 receptor

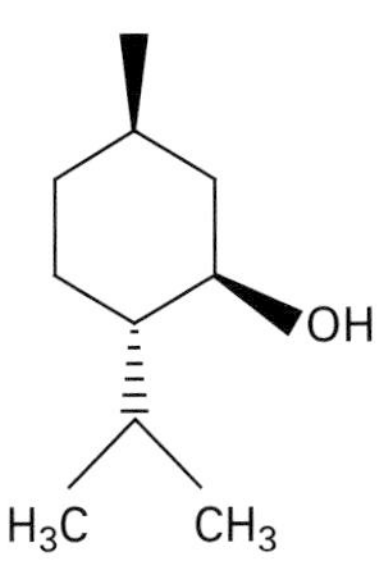

FIGURE 5.13 Structure of menthol, a trigeminal irritant.

family, which are located in the basal (deep) layer of the vomeronasal epithelium. This finding establishes a connection between the MHC molecules that signal individuality (genotypic diversity) and receptor mechanisms that read information about individuality (individual recognition), thus influencing social behavior (Leinders-Zufall *et al.*, 2004).

Why do animals need a VNO? Meredith (1983) suggested three possible reasons: (a) VNO receptors may be more sensitive to certain chemicals, but also more vulnerable to damage; (b) reception of certain chemicals may require special glandular secretions over the receptors, which is only possible in the VNO lumen; and (c) the information reaching the brain via the VNO may be so potent in triggering behavioral or physiological processes that inadvertent triggering should be avoided.

5.2.3 Trigeminal nerve

The trigeminal nerve, or fifth cranial nerve, innervates the face and eyes. It responds not only to touch, painful stimuli, and temperature but also to irritating and noxious chemical stimuli. Pungent, tickling, tingling, stinging, burning, cool, warm, and painful sensations such as those of ammonia, menthol, peppermint, eucalyptol, or carbon dioxide are mediated by the trigeminal nerve. Menthol (Fig. 5.13) is perceived in the nasal cavity as cold at lower concentrations and hot at high concentrations. Mustard oil (allyl isothiocyanate), onion (containing diallyl sulfide), hot Chile powder (capsaicin; Fig. 13.1), and mace spray (capsaicin) also stimulate the trigeminal receptors. In humans, the trigeminal nerves plays an important part in chemoreception.

In birds, the trigeminal nerve appears to play a role in food selection. Starlings more easily accepted commercial feed treated with otherwise avoided coniferyl benzoate after bilateral section of the ophthalmic branch of the trigeminal nerve. Therefore, the trigeminal nerve may help to protect the animal by detecting plant defense compounds. Many of these compounds are astringent or irritating (Jakubas and Mason, 1991).

Blocking trigeminal input by subcutaneous injections of capsaicin does not critically affect perception of olfactory or taste stimuli in starlings (Mason *et al.*, 1987a). Yet, the trigeminal nerve may interact with the olfactory system in perceiving stimuli: patients with unilateral destruction of the trigeminal nerve report a lower intensity of the odors of propanol and butanol for the impaired nostril than for the intact one (Cain, 1978). Trigeminal signals may add a painful or irritating note to food. At low levels, substances stimulating the trigeminal nerve may actually enhance the flavor of foods, as do carbon dioxide in soft drinks, capsaicin in hot peppers and sauces, acetic acid in vinegar, and allyl-isothiocyanate in mustard (Rozin and Schiller, 1980).

5.2.4 Septal organ of Masera

The septal organ is a small patch of sensory epithelium on the wall of the septum, in the anterior part of the nasal cavity, and ventral to the olfactory epithelium. It is found primarily in rodents, has chemical receptors similar to olfactory receptors, and is sensitive to volatile odorants. It projects into the main olfactory bulb, but not into the accessory olfactory bulb (Pedersen and Benson, 1986). Because of its "forward" location, the septal organ may serve as an "early-warning system" that arouses resting or sleeping animals when volatiles are present (Wysocki, 1989).

5.2.5 Nervus terminalis

Lesioning of the terminal nerve system in hamsters disrupts mating behavior (Demski and Northcutt, 1983; Wirsig and Leonard, 1985). The terminal nerve is probably the sole source of all brain neurons containing gonadotropin-releasing hormone (Schwanzel-Fukuda and Pfaff, 1989).

5.2.6 Taste

In fish, both taste and olfactory stimuli are waterborne. However, taste involves the seventh, ninth or tenth cranial nerves, in contrast to the first cranial nerve for smell. Elasmobranchs have their taste buds in the mouth and pharynx, but in bony fish they occur around the gills, on barbels and pectoral fins, and also scattered over the rest of the body surface. They crowd particularly in the roof of the mouth, forming the palatal organ. The taste receptor cells are arranged as a bundle to form a taste bud. Like other vertebrates, fish have receptors for sweet, sour, salty, and bitter. For instance, goldfish reject quinine-treated food pellets (Jobling, 1995). Many fish species are particularly sensitive to acidic taste characteristics. The responses of fish to amino acids will be discussed in Chapter 12.

In addition to taste buds, many fish have solitary chemosensory cells, which resemble taste sensory cells and are distributed over the whole surface of the fish body. Some play a role in food detection, while in other species they appear to help in detecting predator chemicals (Jobling, 1995).

Among the rest of the vertebrates, many social, sexual, or feeding stimuli are non-volatile. Contact with the material is necessary, and animals typically sample by licking. Licking often is one step in a sequence of responses, as in male ungulates' sampling of female urine. Here the animal first smells, then approaches and licks the urine before it exhibits flehmen, thought to serve in moving an odor sample to stimulate yet another chemical sense organ, the VNO. Some authors assume that taste contributes little to the perception of pheromones (Wysocki, 1989).

Structure

Taste receptors on the mammalian tongue occur in three different papillal types: fungiform, foliate, and circumvallate. The pinkish fungiform papillae are located around the edge of the tongue and can be visualized with milk or food color. There are approximately 12 circumvallate papillae arranged in a V shape at the back of the tongue. Finally, the foliate papillae lie in small grooves on the side of the rear of the tongue. One papilla will contain as few as two to five taste buds or as many as 200. Each onion-shaped taste bud in turn contains 50–100 receptor cells. Taste receptors have microvilli, like the vomeronasal receptors. The five types of cell mediate sweet, sour, salty, bitter, and umami (meaty, savory) sensations. (Umami is the taste of glutamate, one of the 20 amino acids making up proteins in meat, fish and legumes. Monosodium glutamate is a food additive.) Each taste bud has an opening, the taste pore, through which the microvilli of the taste cells reach out into the coat of saliva. Food chemicals, called tastants, are dissolved in saliva. The often-published "map" of the tongue with a neat geography of receptors segregated by taste quality is considered outdated. Taste buds in all areas of the tongue can respond to all taste qualities (Smith and Margolskee, 2001).

Function

Molecular events at taste receptors

After the saliva has carried the tastants into the taste bud, they interact with the taste receptors on the surface of the cells, or with ion channels, which are pore-like proteins. Salty and sour tastants act through ion channels, and sweet and bitter sensations are mediated by surface receptors. The different taste submodalities rely on specific mechanisms: Na^+ flux through Na^+

channels accounts for salty taste; sour taste results from H^+ blockade of K^+ or Na^+ channels; and bitter and sweet rely on G-protein mechanisms, involving the G-protein subunit gustducin, which resembles transducin, known to transduce light signals into electrical impulses in the retina (McLaughlin *et al.*, 1993). Genetically engineered mice with defunct gustducin do not distinguish between sweet and bitter (Wong and Gannon, 1996).

The sweet, sour, salty, and bitter primary tastes are thought to serve nutrition, pH levels, ion balance, and protection from toxins, respectively. Bitter receptors deserve a closer look. Far from mediating only a general "bitter" sensation, taste cells on the rat's tongue can distinguish different bitter compounds. The family of bitter receptors consists of 50 to 100 related proteins. Individual taste bud cells express the genes for most of these receptors. This, at first, suggested that a cell would fire when stimulated by any of a great number of bitter compounds. However, 65% of the cells responded to only one of five different bitter compounds tested. About 25% responded to two compounds, and only 7% to three or more. (The cells' activity was monitored visually by fluorescence, which detected the release of Ca^{2+} inside the receptor cell [Caicedo and Roper, 2001]). In short, different taste cells appear to be tuned to different bitter compounds. They may be specialist cells rather than generalists. This specificity, though, seems to be at odds with the fact that one cell has so many different receptor types. As soon as many more than the five bitter compounds have been tested, we will know whether the taste cells are more generalist than appears now.

Central processing of responses

Researchers have oscillated between emphasizing specificity of neurons ("labeled lines") and responses to a spectrum of tastants by one cell. More recently, patterns of activation of a number of sensory cells are favored for coding specific taste sensations (Smith and Margolskee, 2001). Neural distinction of different tastes requires simultaneous activation of different cell types. The brain receives a single channel of information, simply "bitter" for a number of different compounds.

5.3 Structure–activity relationships

5.3.1 Receptor cells: generalists and specialists

Fish have receptor cells specific for certain compounds. Examples are the reception of the fish toxins tetrodotoxin (TTX) and saxitoxin (STX) by rainbow trout, *Salmo gairdneri*, and Arctic charr, *Salvelinus alpinus*. Both toxins are extremely potent taste stimuli. Not only are the receptors extremely sensitive to these compounds (Section 5.5), but specialized receptor cells appear to exist. Cross-adaptation experiments showed that the receptor(s) for TTX are probably distinct from those that detect amino acids and bile salts. Further, TTX and STX do not share the same receptor populations. The extreme sensitivity of these fish probably protects them from ill effects by poisonous prey (Yamamori *et al.*, 1988).

Since prostaglandins serve as fish pheromones, specific receptors for these compounds have been searched for. Of 12 prostaglandins tested with Arctic charr, six gave clear electro-olfactogram responses at a concentration of under 10^{-8} mol/l. The threshold for the two most potent compounds, prostaglandin $F_{2\alpha}$ and its synthetic analogue dimethyl prostaglandin $F_{2\alpha}$, was 10^{-11} mol/l. The concentration–response curve on a semilog graph gave a typical sigmoidal shape, saturating at 5×10^{-8} mol/l. This suggests a single receptor type. The receptors are highly stereospecific. If the stereochemistry of the chemical bond of the hydroxyl group at C-9, C-11, or C-15 is changed, the receptor affinity for the compound is reduced by more than two log units. High species specificity was observed: lake char (*Salvelinus namaycush*) are also sensitive and show specificity, but none of the prostaglandins tested were active in rainbow trout (*O. salvelinus mykiss*) and brook char (*S. fontinalis*) at the concentration of 10^{-8} mol/l (Sveinsson and Hara, 1990).

5.3.2 Stimulus generalization and generalist receptors

How specific are the responses to certain compounds? Will slight changes in the size, shape, or functional groups of the molecule render it unrecognizable for a certain response?

The methods to investigate the specificity of behaviorally active compounds include *spontaneous responses* of untrained animals and *discrimination tests*, where a discriminated stimulus is substituted by another closely related compound to detect the degree of *generalization* from one stimulus to another. In field studies, the first is the method of choice.

An example for stimulus generalization are responses of rats to stress-inducing odors. Laboratory rats of the Wistar strain respond to predator odors, specifically mercapto compounds in fox droppings, with stress reactions, for example avoidance behavior such as "freezing" and increased plasma corticosterone concentrations (Vernet-Maury *et al.*, 1984). The rats were trained to avoid water scented with a mercapto odorant that contained both a keto- and a sulfhydryl group (4-mercapto-4-methyl-2-pentanone). As the animals licked a waterspout, a mild electric shock was applied to their tongue. When different compounds were tested thereafter, the rats avoided compounds with similar

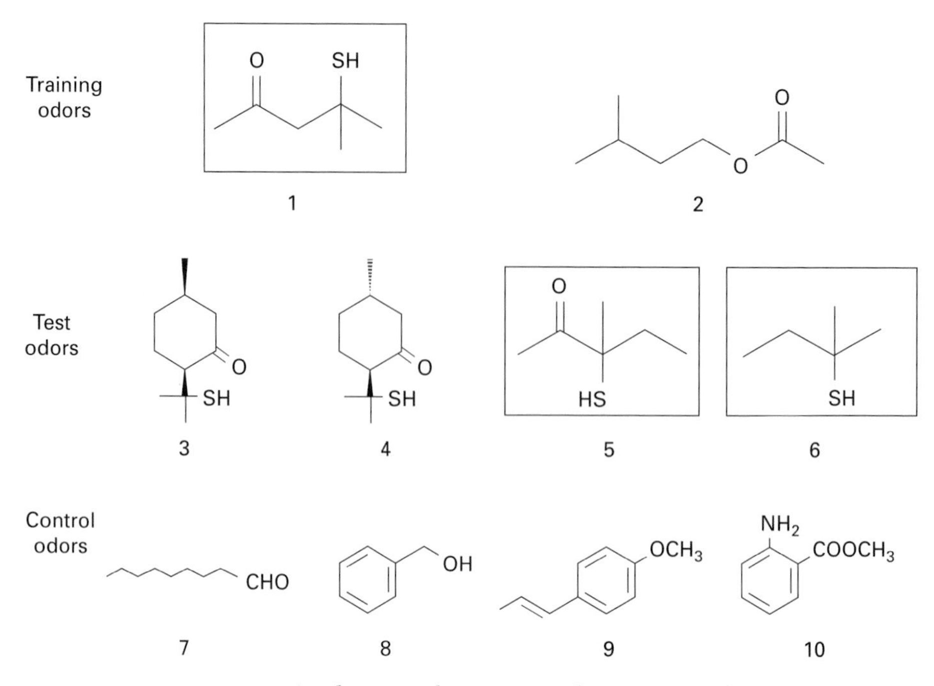

FIGURE 5.14 Stimulus generalization. Rats that experienced electric shocks when drinking water laced with compound 1, later also avoided compounds 5 and 6, which are structurally similar. The keto group was not necessary for the effect. See text. (After Fombon and Polak 1987.)

structure containing the sulfhydryl group, but the keto group was not essential for the effect (Fombon and Polak, 1987; Fig. 5.14).

Other examples of stimulus generalization are the alarm chemicals in salamanders (Mason and Stevens, 1981a) and behaviorally irrelevant pure odors (Braun and Marcus, 1969).

5.4 Neural pathways and decoding

The long-held dichotomy of "macrosmatic" and "microsmatic" vertebrates is no longer tenable. Neuroanatomists had assumed that taxa with relatively few olfactory receptor cells and small olfactory bulbs would also be inferior in olfactory performance (threshold and number of compounds detected) to those with more receptors and larger bulbs. However, we now know from single-cell recordings that a particular receptor cell type can respond to a wide range of odor compounds that share certain features. Keverne (1983) pointed out that the olfactory bulbs act as a filter, while more complex pattern analysis takes place in the neocortex. The more evolved the neocortex, as in primates, the greater the potential for decoding olfactory messages. We now know that birds, cetaceans, and primates, formerly considered "microsmates," command impressive chemoreception.

Many vertebrate odors consist of several to many components. How mammals in particular process and use information coded in such mixtures of odorants is still not clear. One experiment posed the question "do rats discriminate complex odors by recognizing unshared components, or do they treat such odors as gestalten, i.e. in a 'unitary fashion'?" Rats were trained to distinguish three-component odors (ABC versus ABD) by rewarding them at one of the odors with water. They did not recognize compounds C or D when each was presented alone. Also, the rats treated the mixtures AC or AD (two-thirds of original mixtures) as novel odors, with no evidence that they remembered these from the ABC and ABD combinations. This indicates that the animals treated complex odors as units, or "gestalts." However, inclusion of a previously learned one-compound odor into a mixture changed learning of that odor: In a different experiment, rats were preexposed on day 1 to compounds X and Y. X was rewarded (X+), Y was not (Y−). The following day, this group and a not preexposed group were tested with mixtures ABC+ (rewarded) and ABX− (not rewarded). The preexposed animals were slightly slower in learning this reversal from the originally rewarded compound X to one to be avoided (as part of a mixture). This shows that the single compound was remembered in some form (Staubli *et al.*, 1987).

5.5 Odor detection thresholds

The odor sensitivity of an experimental animal needs to be known to appreciate communication distances and performance, but it is not easily measured. The *detection threshold* has to be distinguished from the *recognition threshold*. The former is the concentration at which an odor is noticed, and the latter – typically much higher – the concentration at which behavior tests in animals or verbal responses in humans show that the odor has been recognized as a specific signal, compound, or mixture.

5.5.1 Fish

Eels, *Anguilla anguilla*, are extremely sensitive to odors. Their detection threshold for β-phenylethyl alcohol lies at the unbelievable concentration of 3×10^{-18} mol/l. This corresponds to 1 ml of this alcohol diluted in 58 times the volume of Lake Constance (80 km long) in southern Germany where these experiments were performed. It has been calculated that only three molecules are in the olfactory sac at any one time (Teichmann, 1957, 1959). Coho salmon detect morpholine in water at a concentration of 1×10^{-7} mg/l (Hassler and Kucas, 1988).

Rainbow trout (*S. gairdneri*) and Arctic char (*S. alpinus*) are extremely sensitive to TTX and STX. Electrophysiological recordings from the palatine (seventh cranial) nerve showed that the rainbow trout's threshold concentration of TTX is 2×10^{-7} mol/l. At 10^{-5} mol/l the response is still four times as strong as that to 10^{-3} mol/l of L-proline, the most potent amino acid for this species. The threshold for STX is lower (10^{-8} mol/l), but the response reached its maximum at 10^{-6} mol/l. Arctic char also had a lower threshold for TTX (10^{-8} mol/l) than STX (10^{-7} mol/l). The response magnitude never exceeded that of 10^{-3} mol/l for L-proline. Such a highly sensitive gustatory receptor system is probably effective in protecting predatory fish from the effects of toxins of their prey (Yamamori *et al.*, 1988).

Young (2–3 months old, 80–120 mm long) Russian, Siberian, and stellate sturgeons (*Accipenser gueldenstaedtii, A. baeri*, and *A. stellatus*) had a threshold of 0.001 mmol/l for 19 amino acids (Kasumyan, 1994).

Thresholds of the olfactory receptors of male goldfish are 35 pg/ml water for the prostaglandin $F_{2\alpha}$, and 100 times less for its 15-keto-derivative. The males' receptor threshold for 17,20-progesterone (from females) is "a tiny fraction of 1 pg/ml water." Three grams (one teaspoonful) would provide an above-threshold stimulus when diluted in $500 \times 500 \times 500$ m water (Bjerselius and Olsen, 1993). In lampreys, testosterone from males attracts females at a concentration of 29 pg/ml water but urine with a testosterone concentration of 29×10^{-4} pg/ml is active! (Adams *et al.*, 1987).

The threshold for amino acids can vary with the hunger level. Hungry sablefish, *Anoplopoma fimbria*, have a lower threshold (1.4×10^{-11} mol/l) than satiated fish (4.4×10^{-8} mol/l) (Løkkeborg *et al.*, 1995).

5.5.2 Amphibians and reptiles

Behavioral thresholds for *n*-butyl acetate and *n*-butyl alcohol in the tiger salamander (*Ambystoma tigrinum*) were 8.9×10^{-5} and 6.7×10^{-5} mol/l, respectively (Mason and Stevens, 1981b). The threshold for amyl acetate has been measured as 10^{-4} mol/l in frogs and 10^{-7} mol/l in turtles (Enomoto *et al.*, 1992).

5.5.3 Birds

Since we now know that birds rely on odors in a variety of contexts, we are interested in how sensitive they are to odorants and how their olfactory performance compares with mammals. Birds' sensitivity has been measured in different ways: behavioral thresholds differ from electrophysiological thresholds and can be either spontaneous or experimentally conditioned. The latter pushes an animal more to its limits.

Foraging turkey vultures, *Cathartes aura*, find carcasses by smell. *Behaviorally* measured, turkey vultures are about as sensitive to ecologically relevant odorants of carcasses as pigeons are to heptane, hexane, pentane, and amyl acetate, commonly used odorants (Smith and Paselk, 1986). (According to Henton *et al.*, [1966] the behavioral thresholds of pigeons ranged from 0.1 to 39.7 ppm). The carrion odorants tested were butanoic acid (BA), ethanethiol (ET), and trimethylamine (TMA): BA has the odor of rancid fat and is a byproduct of decomposition of fat, carbohydrates, and protein; thiols, including ethanethiol, are formed from the breakdown of sulfur-containing amino acids (they are also used to give natural gas an odor); and TMA has a fishy odor and is released during decomposition of muscle tissue. Both BA and ET elicited olfactory responses in turkey vultures (Stager, 1964). The thresholds measured by Smith and Paselk (1986) were 1×10^{-6} mol/l for BA and ET, and 1×10^{-5} mol/l for TMA. However, a turkey vulture should smell it only from 0.17 m above a carcass, as estimates of concentrations near carrion have shown. The threshold needed to detect a carcass from 3 to 60 m altitude would lie between 1×10^{-10} and 1×10^{-12} mol/l orders of magnitude lower than actually observed in vultures. The detection rate would be 1 to 20 moles per day. For comparison, humans detect BA and ET at a concentration of 1×10^{-11} mol/l.

Brown-headed cowbirds (*Molothrus ater*) can discriminate ethyl butyrate and *s*-limonene, as evidenced by cardiac conditioning. Furthermore, they discriminate among concentrations of ethyl butyrate. To discriminate the two odorants, vapor saturation must be at least 0.6%. This suggests that for ethyl butyrate the discrimination sensitivity is at least 1.9×10^{13} molecules/ml, or 0.76 ppm (Clark and Mason, 1989).

Electrophysiologically measured thresholds for butanoic acid and ethanethiol in Manx shearwater (*Puffinus puffinus*) and black-footed albatross (*Diomedea nigripes*) are as low as 0.01 ppm (Wenzel and Sieck, 1966). More electrophysiological thresholds for some compounds in tree swallows and cedar waxwings (Clark, 1991), starlings (Clark and Smeraski, 1990), and brown-headed cowbirds (Clark and Mason, 1989) are listed in Table 5.4.

5.5.4 Mammals

The most celebrated mammalian olfactory detector is the dog's nose. As a predator, the dog locates its prey by "air scenting" (following a gradient of airborne odors) and tracking. Practitioners are familiar with the extreme olfactory sensitivity of the dog. The detection threshold for butyric acid has been determined as 9×10^3 molecules/cm^3 air (Neuhaus, 1953). Considering that

Table 5.4 Thresholds for odorants in birds, obtained by behavioral, cardiac, and electrophysiological techniques

Species	Compound	Threshold (ppm)	Reference
Manx shearwater *Puffinus puffinus*	Amyl acetate, trimethylpentane	0.01	Wenzel and Sieck, 1966
Black-footed albatross *Diomedea nigripes*	Amyl acetate, trimethylpentane	0.01	Wenzel and Sieck, 1966
Turkey vulture *Cathartes aura*	Butanoic acid	1.0 (1.0 1 $\times$ 10^{-6} mol/l)	Smith and Paselk, 1986
	Ethanethiol	1.0 (1.0 1 $\times$ 10^{-6} mol/l)	
	Trimethylamine	1.0 (1.0 1 $\times$ 10^{-5} mol/l)	
Domestic fowl *Gallus gallus*	Heptane	0.31–0.57	Stattelman *et al.*, 1975
	Hexane	0.64–1.00	
	Pentane	1.58–2.22	
Pigeon *Columba livia*	Heptane	0.29–0.38	Stattelman *et al.*, 1975
	Hexane	1.53–2.98	Henton, *et al.*, 1966
	Pentane	16.45–20.76	Snyder and Peterson, 1979
	Amyl acetate	0.31–29.8	
	Ethanethiol	10080	
	Butanethiol	13825	
Northern bobwhite *Colinus virginianus*	Heptane	2.14–3.49	Stattelman *et al.*, 1975
	Hexane	3.15–4.02	
	Pentane	7.18–10.92	
Black-billed magpie *Pica pica*	Ethanethiol	8400	Snyder and Peterson, 1979
	Butanethiol	13 416	
Starling *Sturnus vulgaris*	Cyclohexanone	Varies seasonally (0.3% vapor sat. [3.778 $\times$ 10^{14} molecules])	Clark and Smeraski, 1990
Tree swallow *Tachycineta bicolor*	Cyclohexanone	73.4–317.8	Clark, 1991
Cedar waxwing *Bombycilla cedrorum*	Cyclohexanone	6.8–86.46	Clark, 1991
Brown-headed cowbird *Molothrus ater*	Ethyl butyrate	0.76	Clark and Mason, 1989

the number of olfactory receptor cells in dogs ranges from 125 to 225 million, there would be one molecule of odorant for every 82 sensory receptors, on average, if all molecules reached the olfactory epithelium. In reality much fewer molecules reach the epithelium and it was concluded that olfactory receptor cells may respond to only one molecule (Neuhaus and Müller, 1954). Of six fatty acids tested, ranging from acetic acid (C_2) to caprylic acid (C_8), dogs

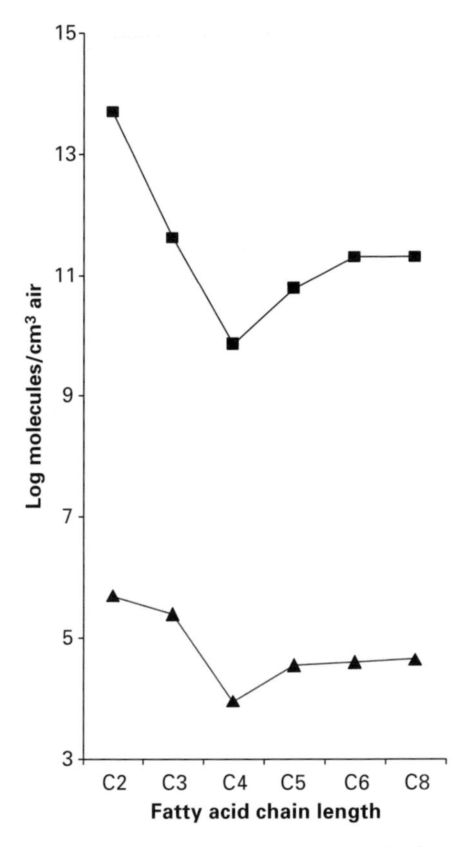

FIGURE 5.15 Detection thresholds for fatty acids in dogs (▲) and humans (■). Both species are most sensitive to butyric acid (C_4). (From data in Neuhaus, 1953.)

were most sensitive to butyric acid (C_4) (Fig. 5.15). Thresholds were higher for fatty acids with lower or higher chain lengths. Humans showed the same V-shaped threshold curve (Neuhaus 1953). However, compared with humans, dogs (beagles) were found to be 300 times more sensitive to amyl acetate (Krestel *et al.*, 1984), and 1000 to 10000 times more sensitive to α-ionone (Marshall and Moulton, 1981). Laboratory rats proved more sensitive than humans to *n*-propanol (8-fold), benzaldehyde (10-fold), cyclohexanone (10-fold), isobutyl *n*-butyrate (30-fold), and *n*-heptanol (50-fold) (Laing, 1975).

Thresholds can vary between and within individuals, with the estrus cycle, and with the chemical background. Some benchmarks will be given here. In laboratory (Wistar) rats, the olfactory detection threshold for ethyl acetate was measured as 7.3×10^{-5} vol% in subadults, while adult rats were more sensitive (1.4×10^{-5} vol%) (Apfelbach *et al.*, 1990). (By comparison, among insects the gypsy moth is 10^{13} times as sensitive. The threshold for bombykol, defined as the concentration at which 50% of the males fan their wings, was approximately 1×10^{-8} g/cm^3 air).

Neotropical bats (*Desmodus rotundus, Artibeus literatus,* and *Phyllostomus discolor*) are very sensitive to butanoic acid: their detection threshold lies between 1.5×10^{-10} and 1.5×10^{-11} mol/l (Schmidt, 1975). The short-tailed fruit bat (*Carollia perspicillata*) has detection thresholds for 18 odorants ranging from 3.6×10^{13} to 2.7×10^{10} molecules/cm^3 air. The animals were most sensitive to fruit-typical compounds such as ethyl butyrate (5.4×10^{10} molecules/cm^3), *n*-pentyl acetate (2.8×10^{10} molecules/cm^3), and linalool (1.8×10^{11} molecules/cm^3), suggesting nutritional specialization of chemoreception (Laska, 1990). The bats increased their respiration rate from a basal rate of 2–4 Hz to as much as 12 Hz when confronted with an odor of high concentration.

A complex biologically relevant odor, such as that of a fruit for fruit bats, may require fewer molecules overall for detection than each of its single components alone (Laska *et al.*, 1990; Laska and Hudson, 1991). Vampire bats (*Plasmodus rotundus*) have a threshold of 1.5×10^{-10} mol/l to 1.5×10^{-11} mol/l for butanoic acid (Schmidt, 1975). Similarly, subthreshold amounts of two or more single compounds may become noticeable when mixed together. Dogs responded to mixtures of subthreshold concentrations of butyric plus isovaleric acids and of butyric plus *n*-caproic acids (Neuhaus 1956a).

Some biologically relevant smells may have two "functional thresholds," the higher one resulting in a qualitatively different odor. For instance, geosmin in high concentrations smells like "musty basement" or "soil," in low concentration like "beets."

In some mammals, the sexes differ in their odor thresholds. The sow detects the boar pheromone androstenone (5α-androsten-16-en-3-one) easily at 3.67×10^{-3} mol/l, while this concentration is near or below the threshold for boars. Boars initially could not detect the odor at all but became more sensitive after repeated exposure (Dorries *et al.*, 1991).

Humans may be programmed to be more sensitive to natural food contaminants. For instance, (−)geosmin (*trans*-1,10-dimethyl-*trans*-9-decalol) occurs in earth, natural surface water, and in foods in contact with soil or water, such as beets, clams, or fish. Geosmin is a microbial, fungal, or algal metabolite. It is a water pollutant and off-flavor compound. The naturally occurring (−)enantiomer has a threshold 11 times lower than the (+) enantiomer (Polak and Provasi, 1992).

Humans are extremely sensitive to hydrogen sulfide, most likely as a danger signal. The recognition threshold lies at 0.00047 ppm (vol) (cited in Cain, 1978). Only 40 molecules of methyl mercaptan, distributed over several receptors, sufficed for detection (cited in Harborne, 1993).

The thresholds of human subjects can vary over minutes, hours, or days as much as between individuals. Therefore, measured interindividual thresholds may reflect nothing more than having been recorded at different levels of these sensitivity fluctuations (Stevens *et al.*, 1988).

Considering the biologically important 16-androstenes, humans are most sensitive to 5α-androst-16-en-3-one (threshold: 0.2 ppb). The detection threshold for 4,16-androstadien-3-one is 1.0 ppb and that for 3α-androstenol is 6.2 ppb (Amoore *et al.*, 1977).

Odor thresholds in humans appear to be more influenced by the environment than genetics. The detection thresholds for acetic acid, isobutyric acid, and cyclohexanone varied as much in monozygotic twins as in fraternal twins. Instead, smoking and diabetes were related to lower olfactory sensitivity, and body fatness and alcohol consumption to greater sensitivity to the latter two compounds (Hubert *et al.*, 1980).

Thresholds are affected by the complexity of an odor. In tests with mixtures of 3, 6, and 12 compounds, human subjects varied less from one another in their thresholds the more complex the odor was. The same trend was observed within individuals (Laska and Hudson, 1991).

In general, perceived odor intensity is a power function of the odor concentration. The exponent is below 1, meaning that boosting the perceived intensity requires increasingly more odor production. As odor production becomes more and more costly, fine-tuning of chemoreception (instead of odor production) becomes the more attractive alternative evolutionarily.

5.6 Hormonal influences on chemoreception

Sensory performance often varies between the sexes and with different stages of the reproductive cycle. In humans, females outperform males at all ages (Doty, 1981; Doty *et al.*, 1984). Even infants show this difference: at 2 days old, girls could discriminate cherry from ginger odor, but boys could not (Balogh and Porter, 1986).

Olfactory thresholds also vary with the ovarian cycle. Women are most sensitive to odors around the time of ovulation, when estrogen levels are highest, and less sensitive to odors during menstruation. This may have sensory–physiological reasons. During menstruation the mucus layer on the olfactory epithelium is thicker and more likely to trap molecules, while the thin mucus layer at the time of ovulation renders the receptors more accessible. The thickness of the mucus layer, in turn, is controlled by testosterone and estrogen (Mair *et al.*, 1978).

Sex differences in olfactory performance have been described for many mammals, implying hormonal differences. In domestic cats, as in other species, males typically perform flehmen in response to conspecific urine. But spayed females can be stimulated to show flehmen by administering testosterone. If paired with estrogen-treated females, they frequently inspect the genital area of the female partner and subsequently exhibit flehmen. Males flehmen in 80% of the time to either female or male urine that is applied to the naso-oral surface. Testosterone-treated females flehmen to 90% of the male urine samples, and to 70% of those from females (Hart and Leedy, 1987).

Males also depend on testosterone for olfactory performance. If in hamsters testosterone is converted to estrogen by subcutaneous silastic implants of the aromatase inhibitor 1,4,6-androstatriene-3,17-dione, their sexual sniffing decreases. They sniff less toward novel females and no longer discriminate between males and females (Steel and Hutchinson, 1987). Castration affects odor detection performance in male rats (Doty and Ferguson-Segall 1989).

Central norepinephrine modulates systems that regulate the attraction to conspecific odors in the rat. A noradrenergic neurotoxin (DPS4), lowers the level of norepinephrine in the olfactory bulb, the olfactory cortex, and frontal cortex. When given to male rats, their response to odors from females is impaired: while sexually experienced, intact males are attracted to pine shavings from the nest of a female and her litter, males treated with DSP4 are not attracted. Furthermore, while normal males enter a chamber with pup-scented shavings more often if odor from anesthetized females is added, males treated with DSP4 do not respond in this way (Cornwell-Jones *et al.*, 1988).

Hormonal events around parturition affect olfactory interactions between mother and offspring. Expulsion of the fetus is followed by liberation of oxytocin in the brain. The oxytocin level rises also in the olfactory bulb. This suggests that oxytocin modulates olfactory processing at the level of the olfactory bulbs (Kendrick *et al.*, 1988). Vaginal stimulation during birth increases noradrenergic activity, especially in the olfactory bulb (Kendrick *et al.*, 1988), possibly resulting in olfactory memory (Rosser and Keverne, 1985). Evidence for a noradrenergic influence on bulbar neural networks during parturition comes from studies with rats. If a female rat's central noradrenergic projection to the olfactory bulb is experimentally lesioned prior to parturition, cannibalism will result, but not general anosmia or gross impairment of maternal behavior (Dickinson and Keverne, 1988).

5.7 Chemoreception and age

Humans perform at their olfactory best during the third to fifth decade of life. A marked decline occurs after the seventh decade. Over 50% of people aged from 65 to 80 years have major impairments. In one study, 75% of subjects 80 years of age or older suffered such loss. The gradual loss of sensitivity to odors is responsible for frequent complaints by the elderly that food lacks flavor. The elderly are also disproportionally often poisoned by household gas (Doty *et al.*, 1984).

Aging does not necessarily affect the process by which overall intensity in different sensory modalities is determined. Young people (18 to 21 years) and older people (61 to 94 years, average 74.5 years) differed in their magnitude estimations of almond odor and almond taste. However, they did not differ when presented with the taste of sucrose or the visual task of estimating the length of a line (Enns and Hornung, 1988). Likewise, young and old did not differ in their magnitude estimations for odor of lime (Stevens and Cain, 1985).

5.8 Interaction between chemical senses

5.8.1 Olfaction-vomeronasal organ interaction

Male mice learn about female cues. In the presence of females, a male emits ultrasonic vocalizations. These vocalizations become less frequent after extirpation of the VNO. The more experience a male has had with females, the smaller the deficit he suffers. The learning is mediated by interaction between the main olfactory system and the accessory olfactory system: after vomeronasal deprivation, olfactory cues maintain the behavior (Wysocki *et al.*, 1986).

5.8.2 Odor–taste interactions

Odorants and tastants in foods interact in complicated ways. For instance, strawberry odor enhances the sweetness of whipped cream, while peanut butter odor does not, and strawberry odor did not enhance the saltiness of sodium chloride (Frank and Byram, 1988).

The best investigated odor-taste interactions occur in conditioned flavor aversions. Tastes that precede a delayed food-related illness are often avoided after only one experience. Odors are not avoided under similar conditions. However, if taste and odor are presented together before the "malaise," animals will avoid odor when encountered later by itself. Taste affects odor, but not vice versa. If only the taste intensity is increased, both taste and odor aversion increase. Conversely, if only the odor stimulus is increased, only the odor aversion increases (Garcia *et al.*, 1986).

5.8.3 Olfaction–trigeminus interactions

Interactions of trigeminal fibers and olfactory chemoreception have been extensively studied. In frogs, trigeminal fibers lead into the olfactory mucosa. The trigeminal system could thus modulate olfaction peripherally as well as centrally. Indeed, antidromic electrical stimulation of the ophthalmic branch of the trigeminal nerve evokes slow potentials in the olfactory mucosa, modifies the activity of receptor cells, and modulates responses to odors. Trigeminal stimulation by odorants might trigger local axon reflexes that induce the sensory neurotransmitter substance P, which, in turn, elicits electrical responses in the mucosa (Bouvet *et al.*, 1989). Blocking trigeminal input by subcutaneous injections of capsaicin does not critically affect perception of olfactory or taste stimuli in starlings (Mason *et al.*, 1987a).

An interaction between "main olfaction" and the trigeminal somatosensory system has been proposed to facilitate "directional smelling." For carbon dioxide and menthol, human subjects were able to tell the odor direction in 96% of cases, while the results for hydrogen sulfide and vanillin were random (Kobal *et al.*, 1989). Rats can discriminate odor direction in one sniff (Rajan *et al.*, 2006).

5.8.4 Chemoreception and other sensory modalities: mammals

Chemical cues often have to be accompanied, preceded, or succeeded by visual, auditory, or tactile stimuli for appropriate behavior to occur. Bats, for example, use both olfactory and acoustic information for individual recognition (Kunz, 1982).

Visual cues can help in the detection of a scent mark in the first place. This may be accomplished in different ways. An animal may use a visually prominent natural landmark for scent marking, such as a rock, stump, or root along a frequently traveled trail. Other species produce their own visually conspicuous support for a scent mark. Carnivores, such as otters (*Lutra* sp.), use feces, called spraint, as carrier for secretion from the anal gland (Chanin, 1985). Beaver, *Castor* sp., have carried this behavior one step further. They build a platform for their scent mark by dredging up mud from their pond and depositing it on the bank. This scent mound is marked with secretion from the castor sacs and anal glands. Finally, the scent mark itself may be visually conspicuous, as that of the brown hyena, *Hyaena brunnea*. During scent marking, the two lobes of its anal glands produce white and black secretion in succession. This results in a bicolored scent mark on a blade of grass that has the black part above the white one and separated from it by 1 cm (Kruuk, 1972; Fig. 6.7).

Relatively non-volatile materials such as the proteins (10–30 kDa) and lipids in the femoral gland secretion of desert iguanas, *D. dorsalis*, may strongly absorb long-wave ultraviolet (300–500 nm). Desert iguanas actually locate scent marks better under ultraviolet light (Alberts, 1989).

Numerous studies have addressed the relative importance of different senses in modulating behavior. Mating behavior of domestic sheep is an example. Rams use mostly vision for sexual activity but smell for seeking the partner. Odors are important for distinguishing estrous from non-estrous ewes. In ewes, blindfolding interfered with seeking rams (Fletcher and Lindsay, 1968).

One sense can be substituted for another in feeding behavior in sheep. Even if smell, taste, and touch are impaired, sheep still show food preferences but undesirable plants become more acceptable. However, the senses contribute specific information: flowering stems are not taken if smell only is impaired. It is assumed that flowers have an attractive smell to sheep. Taste appears to be important in discriminating fine- and broad-leave *Dactylis glomerata* (from different geographic areas). Finally, it is thought that the sense of touch is responsible for the fact that brome grasses are always highly acceptable (Arnold, 1966). Merino weaners learned from experienced adult sheep to eat wheat, a novel food. They learned rapidly, even if one or two senses were impaired. Only weaners with impairment of olfaction, vision, and hearing failed to eat wheat for all 5 days of the experiment; weaners with no "teachers" also did not eat the wheat (Chapple and Wodzicka-Tomaszewska, 1987).

Spiny mice, *Acomys cahirinus*, use all senses interchangeably in locating prey (Langley, 1988). These are cases of "adaptive redundancy" at the level of the sensory modalities. As is well known, deaf–blind humans use smell and taste more.

6

Signaling pheromones I: discrimination and recognition

My dear friend, the last time you were so good as to come and see me – for nobody comes any more to see the wretched invalid I am – I was obliged to take the chair you sat in and keep it out in the courtyard for three days: it was impregnated with scent.

MARCEL PROUST, according to Léon Pierre-Quint, 1925

Optimus odor in corpore est nullus [The best body odor is none].

SENECA, Epistulae ad Lucilium, *Epis* CVIII, 16.

Signaling pheromones are animal-produced, interindividual chemicals that modulate behavior in conspecifics. Like visual and auditory signals, they have comparatively rapid effects: exchange of signals takes seconds or minutes. (Priming pheromones [Ch. 8], by comparison, trigger slower endocrine or developmental processes.) The pheromone concept, originally based on insects (Karlson and Lüscher, 1959), has been debated for vertebrates, notably mammals (e.g. Beauchamp *et al.*, 1976; Johnston, 2001). Often it is better to use the term "body odors" to avoid particular assumptions. Now the term pheromones is widely used for vertebrates, without any particularly narrow definition implied.

A pheromone is functionally defined as a conspecific compound(s) that affects a receiver. Sources such as urine or gland secretions typically contain many compounds of which only some are pheromonally active. So in most cases, a pheromone is more than a single compound and less than a secretion ("scent"), it is rather a group of active compounds in a secretion or excretion that supply information to, or change behavior in, another conspecific.

The following text discusses first the ability of animals to distinguish and recognize other animals by odors without necessarily exhibiting specific behaviors and then the behaviors that are modulated by status signals. Chapter 7 discusses the sexual and evolutionary implications of signaling pheromones.

6.1 Familiarization with environment and objects

Strange as it sounds, the absence of odor often serves as a powerful stimulis; unfamiliar and still unmarked objects or areas prompt vigorous scent marking. This applies to many mammals, including ungulates, such as pronghorn, *Antilocapra americana* (Müller-Schwarze *et al.*, 1972) rodents, such as house mice, *Mus musculus* (Hurst, 1987), and carnivores (Kleiman, 1966).[1] For example, pine martens, *Martes martes*, mark most consistently unmarked objects and do not mark objects that already carry their own odor. It is concluded that marking primarily serves familiarization.

In isolated male mice, own odor regulates the amount of urine deposited in marking. If it is present, they mark less, while clean surfaces and also other males' urine trigger more frequent marking (Daumae and Kimura, 1986). In our laboratory experiments, students are impressed by how a mouse stops at a clean tile in the middle of a soiled open field. A "scent-the-habitat" function for odors from both sexes has been assumed for the gland secretions in the brushtail possum, *Trichosurus vulpecula*, since no sex differences in chemical composition were found (Woolhouse *et al.*, 1994).

The "security blankets" of children belong in this context. Apart from tactile stimuli, they also provide a familiar odor, particularly in strange surroundings. Many children will not easily accept a familiar blanket, piece of cloth, or tattered teddy bear that has been washed. A particularly inventive 3-year old boy held a cloth toy close to his nose when tired or stressed. Asked why he did so, he answered "to get the smell right" (Russell, 1983).

6.2 Familiar versus non-familiar social odors

Discrimination of familiar from unfamiliar social odors is widespread among vertebrates. It serves for recognition and interaction with group (or colony) members. Being familiar with other individuals not only reduces aggression but also may facilitate sexual behavior. Group members can be discriminated on the basis of familiarization with individuals; sharing common environmental odors, such as those of nest or diet; "labeling" of one animal by another; and sharing individual odors that result in a colony odor, as in sugar gliders, *Petaurus breviceps* (Schultze-Westrum, 1965). These can combine into rather complex body odors, determined by diet, maternal labeling, and genetic factors, as for example in spiny mice, *Acomys* sp. (Porter *et al.*, 1989).

[1] Animal names follow the original literature. Thus, the house mouse appears here variously as *Mus musculus*, *Mus musculus domesticus*, or *Mus domesticus*. Retaining the original version may facilitate searching the literature. "Charr" is Canadian for "Char."

6.2.1 Discrimination of own odor versus that of other conspecifics

Reptiles

Lizards have femoral glands that are more developed in males. The secretion is proteinaceous, and its function is largely unknown. Desert iguanas (*Diposaurus dorsalis*) probably use femoral secretion for advertising their home range. Iguanas experimentally exposed to own secretion and that of a strange male tongue flicked oriented toward their own tails in response to own secretion. This may be "*scent matching*" of scent with "scent owners." They marked primarily their own secretion. This may indicate self-recognition (Alberts, 1992c). Blue-tongued skinks (*Tilique scincoides*) also discriminate their own odor from that of other conspecifics (Graves and Halpern, 1991).

Mammals

In male house mice, urine marks from other males trigger intensified marking, while grouped male mice do not distinguish between own odor and that of other males (Daumae and Kimura, 1986). The aboriginal house mouse *Mus spretus* carries or pushes away fresh fecal pellets. Experiments demonstrated that they recognize their own feces. Compared with fecal pellets of other conspecifics of both sexes, these mice are more likely to remove their own and are less likely to investigate them (Hurst and Smith, 1995). Mice (*Mus* sp.) avoid the odor of mice from other groups. Here it is not necessary to be familiar with the odor of the other group. Odors of mice from discontinuous, completely strange demes are also avoided (Cox, 1989). In male mice, volatiles from feces of strange males depress the heart rate more than volatiles from own feces (Goodrich *et al.*, 1990). In pine marten, *M. martes*, own scent marks reduced marking behavior, while all other types of scent mark were marked in turn (deMonte and Roeder, 1990). Pine martens did not respond differentially to familiar/unfamiliar martens, males versus females, or other species such as stone marten or genets (deMonte and Roeder, 1990).

6.2.2 Familiar individuals

Exposure to the odor of *particular* conspecifics can affect behavior toward these individuals when encountered.

Birds

Antarctic prions (*Pachyptila desolata*) have a "musky" smell and appear to use conspecific odors to recognize their burrow. These birds chose the odor of their breeding partner over that of another conspecific in Y-maze experiments on the Kerguelen archipelago. They also preferred a conspecific's odor to their own, and the latter to "no odor" (Bonadonna and Nevitt, 2004).

Rodents

In meadow voles, *Microtus pennsylvanicus*, males are more aggressive toward familiar males than unfamiliar ones, while females behave in the opposite way: encounters between familiar females entail less agonistic behavior and more "amicable acts" than those between unfamiliar females. This is true for voles kept together or apart in the laboratory and also for voles caught in the wild, whether they had been neighbors or lived at greater distances from one another. In contrast to same-sex encounters, familiarity has no effect on male–female interactions (Ferkin, 1988). Spiny mice (*Acomys* sp.) prefer to huddle with familiar cagemates. They do not recognize each other if rendered anosmic by treatment with zinc sulfate (Matochik, 1988).

6.2.3 The "dear enemy" phenomenon

Ecologically, female meadow voles are territorial, know their neighbors, and are more tolerant of each other. They exemplify the "dear enemy" concept: familiar neighbors reduce aggression toward one another because they pose less threat to each other than newcomers without a territory, who might compete for territory, mates, or resources. Males are dispersal prone, and neighbor relations are more ephemeral. Each male's home range overlaps with those of several females (Ferkin, 1988).

Eurasian beavers (*Castor fiber*) show the dear enemy phenomenon: they respond more strongly (by destroying scent mounds and over-marking them) to experimental scent marks from strangers than from neighbors. This is true for marks containing either of the two beaver secretions castoreum or anal glands secretion (Rosell and Bjørkøyli, 2002). The dear enemy phenomenon is thought to be particularly adaptive in species that maintain multipurpose territories containing the breeding site, mate, and food supply (Temeles, 1994).

Males of the Columbian ground squirrel (*Spermophilus columbianus*) sniff scent marks from oral glands of non-residents longer than marks from residents. However, they marked their own, neighbor's, and strangers' scent equally often (Harris and Murie, 1982). Similarly, male woodchucks, *Marmota monax*, spent more time investigating scent marks of strangers than those of neighbors (Meier, 1991). (In woodchucks, adults and juveniles of both sexes mark objects near their burrows with their oral angle gland.)

Preexposing rodents to the odor of a conspecific can alter the response to this individual when it is again encountered later, compared with responses to conspecifics whose odor is unfamiliar. For example, female brown lemmings, *Lemmus trimocrunatus*, were experimentally exposed to the odor of a male. These females engaged in more contact social behavior with that now "familiar" male than females who had experienced odor from a different male or none at all. The males whose odor the females had experienced ejaculated more frequently than males under the other two conditions (Coopersmith and Banks, 1983).

Female golden hamsters can distinguish individual males by the odors of their flank gland secretion, urine, feces, and soiled bedding. Females were less aggressive and showed more sexual behavior toward males whose odor they had been exposed to before than to unfamiliar males. The preexposed females also had larger litter sizes. Thus, familiarity with the odor of a particular male appears to promote mating and enhance mating success (Tang-Martinez *et al.*, 1993).

Carnivores

Urban feral cats of both sexes sniff marks of sprayed urine more if the donor is a strange cat (from a different town) and respond least to the urine of a familiar cat. Since one male roams over the territories of several females, it is assumed that male–male competition has selected males that spray mark more and respond more strongly to urine marks (Natoli, 1985).

Another example from carnivores is the odor of anal sac secretion in the ferret, *Mustela furo*. Ferrets discriminate strange from familiar individuals by this odor. (They also use anal sac secretion to distinguish males from females, a familiar individual's from own odor, and fresh from 1-day old odor, but not between fresh and odor only 2 hours old nor anestrous from estrous females [Clapperton *et al.*, 1988]).

Bats

Female pipistrelle bats (*Pipistellus pipistrellus*) recognize familiar individuals by scent. They preferred the odor of females from the same colony to that of females from a different colony (deFanis and Jones, 1995).

Primates

Lemurs (*Lemur fulvus* and *Lemur macacao*) showed more interest in the scent marks of familiar than unfamiliar individuals (Fornasieri and Roeder, 1992).

Table 6.1 Kin recognition by chemical cues in vertebrates

Species	Behavior	Reference
Fish		
Coho salmon *Oncorhynchus kisutch*	Phenotype matching	Quinn and Busack, 1985
Rainbow trout *Oncorhynchus mykiss*	Staying in odorized water current	Brown *et al.*, 1993
Amphibia		
Polyphenic salamanders	Eating parasitized prey	Pfennig, 1997
Toad tadpoles *Bufo americanus*	Negative response to non-siblings in Y-test	Waldman, 1981
Cascade frog tadpoles *Rana cascadae*	Naive tadpoles group with siblings; waterborne cues	O'Hara and Blaustein, 1985 Blaustein and O'Hara, 1981
Mammals		
White-footed mouse *Peromyscus leucopus*	Possibly by odor	Grau, 1982
Spiny mice *Acomys cahirinus*	Associate with familiar siblings anosmics do not discriminate	Porter, 1986
13-lined ground squirrel *Spermophilus tridecemlineatus*	Reduced social exploration when raised together	Holmes, 1984
Beldings ground squirrel *Spermophilus beldingi*	Phenotype matching	Holmes, 1986
Beaver *Castor canadensis*	Differential response to scent marks of relatives and strangers, phenotype matching	Sun and Müller-Schwarze, 1997

6.2.4 Indirect familiarization

Mammals may become familiarized with one another by body odors, without having direct contact. For instance, Columbian ground squirrels, *S. columbianus*, show more-cohesive and less-agonistic behaviors toward unfamiliar non-colony members if they had experienced the odor of these individuals in traps that had been used in both colonies (Hare, 1994).

6.3 Recognizing kin

Many questions in inclusive fitness theory turn on whether and how animals distinguish genetic relatives from non-kin. Vertebrates, from fish to humans, use odors, in addition to cues in other sensory modalities, to recognize kin (Table 6.1). This recognition, in turn, may enable relatives to cooperate, resulting in kin selection, or to adjust reproduction, as in avoiding incest. Generally, there are four kin recognition mechanisms that involve various sensory modalities, including chemoreception (Fletcher and Michener, 1987).

1. *Spatial cues*: individuals from one common place are considered kin.
2. *Familiarity*: cues are learned during earlier association.
3. *Phenotype matching*: to recognize an individual as kin, it is not necessary to have met it before. Instead, it can be recognized by cues shared with oneself or a known related individual (also known as the "armpit effect").
4. *Recognition alleles*: both expression of the identifying cue and the recognition process do not rely on experience.

Tang-Martinez (2001) suggested that kin recognition possibly relies on only one mechanism: "learning, particular associative learning and habituation" (i.e. point 2 in this list).

6.3.1 Fish

Juvenile coho salmon, *Oncorhynchus kisutch*, recognize kin. This is considered to be advantageous for schooling, predator avoidance, and avoiding inbreeding (Quinn and Busack, 1985). Whether fish recognize kin may depend on how they were reared. In Arctic charr, *Salvelinus alpinus*, only fish that were raised with siblings later preferred water scented by unknown siblings over water from non-siblings. Charr raised in isolation do not discriminate the two odors. Even when the isolated fish, at 15 months of age, are reared with siblings for 50–62 days, they still do not discriminate sibling odor. Therefore, it appears that in Arctic charr social experience during the first 15 months of life is necessary for sibling preference (Winberg and Olsén, 1992; Olsén and Winberg, 1996).

The major histocompatibility complex (MHC) plays a role in kin recognition. Arctic charr preferred water scented by a sibling of the same MHC (class II) type over water from a sibling that differed in MHC type from themselves. The charr did not discriminate in this way if they had been living in isolation since fertilization. However, a sibling of a different MHC type was not distinguished from a non-sibling that shared the MHC type of the responding fish. As the earlier experiments, this confirmed the role of early learning in sibling recognition. Arctic charr probably use MHC plus other odors, possibly based on the MHC class I region, for kin recognition (Olsén *et al.*, 2002a,b).

Striped bass, *Morone saxatilis*, juveniles develop in estuaries, notably Hudson River and Chesapeake Bay. Juveniles school during the day but disperse and feed at night. Schools provide the advantages of improved prey location, predator avoidance, and swimming hydrodynamics. How are the fish attracted to each other? In Y-maze tests in the laboratory, juveniles are attracted to both *familiar siblings* and *unfamiliar unrelated* juveniles but they can distinguish between familiar siblings and unfamiliar, unrelated juveniles, and prefer the former. This recognition is probably important as a simple mechanism to re-form schools. Lost or disoriented individuals probably use a species-specific response (Kelley, 1988).

Juvenile Atlantic salmon (*Salmo salar*) and juvenile rainbow trout (*Oncorhynchus mykiss*) that live in kin groups fight less, thus saving energy and reducing the risk of injury (Brown and Brown, 1993). Rainbow trout discriminate unfamiliar kin from non-kin, but not familiar from unfamiliar kin (Brown *et al.*, 1993). This appears to be an example of phenotype matching: kin have matching odors, while kin and non-kin have not.

6.3.2 Amphibia

In tadpoles of the American toad, *Bufo americanus*, the prehatching environment influences postnatal preferences. Tadpoles reared isolated from any conspecifics, with siblings only, or exposed to both siblings and non-siblings differed in their later association with siblings in a test pool. The first and second group preferred to associate with siblings, while the third did not prefer to associate with siblings, unless they had been reared in sibling groups during early development. The tadpoles reared in isolation from conspecifics discriminated paternal, but not maternal half-siblings from full siblings (Waldmann, 1981).

6.3.3 Reptiles

Hatchling green iguanas, *Iguana iguana,* recognize kin by the odor of their feces, but also by their body odor (Werner *et al.*, 1987).

6.3.4 Non-human mammals

It has been known for a long time that certain rodents avoid mating with siblings. This, for example, is true for European common voles, *Microtus arvalis* (Frank, 1954). After living in isolation, adult common voles sniffed non-siblings of the opposite sex more than siblings (Bolhuis *et al.*, 1988). Similarly, female house mice, *M. musculus*, prefer the smell of non-sibling males to that of siblings or males of other strains. This has been interpreted as a sign of *optimal outbreeding* (Gilder and Slater, 1978). Female prairie voles (*Microtus ochrogaster*) spend more time investigating the anogenital and mouth region odors of non-siblings than siblings. They also discriminate saliva and urine of non-siblings from those of siblings. It is thought that examining one male more than another may help in mate choice (Smale *et al.*, 1990). In the prairie vole (Gavish *et al.*, 1984) and the gray-tailed vole, *Microtus canicaudus* (Boyd and Blaustein, 1985), social experience before weaning is necessary for incest avoidance to occur. Laboratory mice choose their mates based on both relatedness and familiarity (Barnard and Fitzsimmons, 1988). By cross-fostering, it is experimentally possible to "create" "non-genetic" or phenotypic "kin" in laboratory mice (Kareem and Barnard, 1982) and deer mice, *Peromyscus leucopus* (Grau, 1982). Female laboratory mice appear to avoid mating with close relatives by preferring genetically dissimilar mates. Specifically, in one experiment, females chose males that were dissimilar at the MHC (Egid and Brown, 1989). This does not apply to members of the same sex: female house mice prefer to nest with other females with whom they share MHC genes (Manning *et al.*, 1992).

In rats, nulliparous females kill unrelated young. Such infanticide is reduced if these females are exposed to bedding soiled by a pregnant rat (Menella and Moltz, 1989), an odor that also reduces infanticide by male rats (Menella and Moltz, 1988).

How does kin recognition develop? Rats learn some kinship characteristics (Hepper, 1983). Rat pups whose mother and siblings were rubbed with different odors prefer the odor of their siblings when tested in a Y-maze (Hepper, 1987). But genetic factors also play a role. Rats can distinguish chemosignals in the urine of other rats that differ only in the MHC. This may be the basis for kin recognition. In the mouse, the MHC is located on chromosome 17. Proteins encoded by the MHC are found on all cell surfaces, are antigen receptors, but also contribute to urinary odors specific to individuals (Hurst *et al.*, 2001). The binding groove on the protein molecule, a binding site for antigenic peptides, may also serve as binding site for volatile signal compounds (Beynon *et al.*, 2002). There are 50 genes in this complex, and they are highly variable, with about 50 alleles. The gene $H\text{-}2K^b$ is the most mutable. This variability lends itself to signaling individual signatures.

6.3.5 Humans

The odors of T-shirts worn by mothers and their children appear similar even to strangers (Porter *et al.*, 1986). Adults can recognize the odor on shirts worn by their full siblings after several months of separation (Porter *et al.*, 1986).

Mothers recognize their newborn babies by odor (Russell *et al.*, 1983; Schaal *et al.*, 1980) even if there had been little contact after a Cesarean delivery (Porter *et al.*, 1983). Mothers distinguished the odor of their own infant from those of two other infants on a non-soiled undershirt that the infant had worn for at least 13 hours. If the mothers had been exposed to their own infant for less than 10 minutes after birth, only 20% were successful at recognition. After

10–60 minutes, 90% succeeded. All mothers exposed to their infant longer than 1 hour identified their odors. Maternal analgesia or anesthesia had no influence on the discrimination (Kaitz *et al.*, 1987). These results show that the infant's odor is most likely very important for identification during the immediate postpartum period (Eidelman *et al.*, 1987). Even fathers, grandmothers, and aunts, after only 0 to 8 hours of exposure to a newborn, can correctly identify an infant 1–2 days old by the odor on its garment (Porter *et al.*, 1986). Human infants actually mark their mothers with saliva. Sleeping infants aged 6–8 weeks responded to the mother's breast odor or to the odor of their own saliva on gauze pads, but not to cow's milk (Russell *et al.*, 1989).

Breast-fed neonates themselves distinguish breast (Macfarlane, 1975; Russell, 1976; Schaal *et al.*, 1980) and axillary (Cernoch and Porter, 1985) odors of their mother from the same body region odors of other lactating women.

6.4 Individual odors

Every individual of a species can have its own olfactory signature or "fingerprint" owing to the many odor sources that make up the body odor and the many compounds in each secretion or excretion. The relative concentrations of all these compounds may vary, providing for endless variety. These individual signatures can potentially be monitored and exploited by conspecifics. "True individual recognition" is a more selective and complex process than "social recognition" (categorization of conspecifics) (Gheusi *et al.*, 1997). Animals are able to discriminate between the odors of *two individuals other than self*, such as its mate and another individual of the same sex and status. Such individual recognition or, better, discrimination (Halpin, 1986) is thought to increase genetic fitness. Examples are recognition of one's mate in species with strong and/or lasting pair bonds, recognition of own offspring, recognition of relatives where favoring of kin results in increased chances of propagation of one's own genes, and recognition of a potential rival such as a territory owner. True recognition of the odor of an individual, as opposed to discrimination of odors of age, sex, or physiological status, has rarely been demonstrated.

6.4.1 Amphibia

Male red-backed salamanders (*Plethodon cinereus*) in Petri dishes were given a choice between own feces and those of another male. They preferred their own. The same was true for "washes" from the cloacal glands (Simon and Madison, 1984). While this may represent discrimination of "own" and "other," rather than true *individual* recognition, another experiment showed discrimination of individual odors in *Plethodon* spp. In a Y-maze, *Plethodon* salamanders discriminated individual odors in addition to sex and species odors (Dawley, 1984).

6.4.2 Non-human mammals

Among rodents, chemosensory discrimination of individuals occurs in mice (Bowers and Alexander, 1967), rats (Krames, 1970), Mongolian gerbils (Halpin, 1974), eastern chipmunks, *Tamias striatus* (Keevin *et al.*, 1981), cavies, *Cavia aperea* (Martin and Beauchamp, 1982), golden hamsters, *Mesocricetus auratus* (Johnston, 1993; Johnston and Jernigan 1994), and prairie voles, *M. ochrogaster* (Newman and Halpin, 1988). Individual odors are also discriminated by African dwarf mongoose, *Helogale undulata rufula* (Rasa, 1973) and the Indian mongoose, *Herpestes auropunctatus* (Gorman 1976), badgers, *M. meles* (Kruuk *et al.*, 1984), and ferrets (Clapperton *et al.*, 1988).

Rats form an *olfactory image* about individuals that they retain in their *olfactory memory*. Gheusi *et al.* (1997) trained rats by operant conditioning to distinguish two other rats. After successful training, the reward and non-reward stimulus rats were switched (reversal test). The responding rat's error rate increased drastically after this reversal. This test avoids novelty effects that confound habituation–dishabituation experiments. In a second type of experiment, rats were first trained to discriminate pairs of rats, and then given the odor (bedding) of these rats. They were able to transfer the discrimination of whole animals to their odor signature alone (Gheusi *et al.*, 1997).

In primates, lemurs distinguish and remember individual odors. In *L. fulves* the anogenital scent mark of an evicted individual was still recognized (i.e., scent-marked) 10 months after the eviction (Fornasieri and Roeder, 1992).

Individual recognition can be important for selecting and correctly recognizing mates, reducing aggression, incest avoidance, and sexual re-arousal, in the Coolidge effect (named after an anecdote about a US President and refers to increased sexual arousal with a new mating partner) (Dewsbury, 1981).

Mate recognition

Male and female Mongolian gerbils, *Meriones unguiculatus*, respond to the urine of their respective mates by investigating it more and uttering more ultrasonic calls than to unfamiliar gerbils of the same sex. Members of same-sex pairs do not respond differently to the odors of the familiar and an unfamiliar individual of the same sex. To recognize the odor of one's mate may be important for pair maintenance by promoting affiliative behavior and reducing aggression (Brown *et al.*, 1988). Similarly, males and females of the prairie vole,

M. ochrogaster, recognize and prefer the odor of their mate to that of a non-mate. This was established with three techniques. First, after habituation to the odor of soiled bedding or urine from one male or female, material from another individual of the same sex was substituted. The voles discriminated this new odor from the previous one. Second, in a Y-maze, females preferred the odors of their mates over those from other mated or unmated males; males preferred the odors of their mates over those from other mated females but they did not discriminate between their mate and a virgin female. Third, when allowed to stay for over 10 hours in a scented Y-maze and provided with nest material, females stayed and built a nest on the side of their mate's odor. Likewise, males build a nest on the side with their mate's odor, when juxtaposed with the odor of a virgin female. It is not difficult to see that to recognize and prefer the odor of one's mate can play a vital role in monogamy (Newman and Halpin, 1988). Females of the shrew, *Crocidura russula*, distinguish flank gland secretion of their mate from that of another male (Cantoni and Rivier, 1992).

Sex differences

Discrimination of individual odors may depend on the sex of the odor donor. In the Virginia opossum (*D. virginiana*), females distinguish individual female, but not male odors (Holmes, 1992).

Genetic basis and chemical variation of odors

In the rat and other rodents, individual odors probably reflect genetic differences. Laboratory rats can distinguish individuals. They discriminate between two intact males, two castrated males, two estrous/proestrous females, two diestrous/metestrous females, or two ovariectomized females. Urine odors differ individually despite differences in the levels of gonadal hormones. Individual recognition may be independent of reproductive state or social status, even though hormone-influenced body odors may be used for individual recognition (Brown, 1988).

In the rat, the major histocompatibility complex (MHC) is responsible for individual odors (Brown *et al.*, 1987a,b). Genetic differences in the *IA* region actually result in different urine odors (Singh *et al.*, 1987).

The relative concentrations of constituents of the anal sac secretion in male stoats, *Mustela erminea*, are distinct for different individuals, but consistent over time for each individual. This possibly permits individual recognition of territory owners (Erlinge *et al.*, 1982). The chemical composition of the anal gland of the otter, *Lutra lutra* (Gorman *et al.*, 1978) and the subcaudal gland of the badger, *Meles meles* (Kruuk *et al.*, 1984) also differ with the individuals.

Body regions

Odors of different body regions can convey different types of information. In the golden hamster, *M. auratus*, five scents provide individual information. These are the flank glands, ear glands, urine, feces, and vaginal secretion. However, hamsters did not discriminate individuals by the odors of saliva, feet, chest fur, back fur, area behind the ear, and flanks from males with flank glands surgically removed (Johnston *et al.*, 1993). The golden hamster recognizes flank gland and vaginal secretions as coming from the same individual (Johnston and Jernigan, 1994).

6.4.3 Humans

Mothers and siblings can distinguish the odor of young children in their family from those of unrelated children of the same age. Parents could identify T-shirts worn by boys or girls as those of siblings or offspring and could correctly identify the T-shirt worn by each of two of their children (aged 35–58 months and 64–94 months (Porter and Moore, 1981). This shows that the identifying odors were individual signatures and not household odors common to several siblings (Porter and Moore, 1981). The odors of identical twins are so similar that humans cannot easily distinguish these (Wallace, 1977). Even dogs confuse the odors of identical twins, especially if not available simultaneously (Kalmus, 1955).

6.5 Odors in parental behavior

6.5.1 Chemosignals from parents in mammals

Offspring of many different vertebrate groups cue in on chemosignals from the mother, father, or both, and vice versa. Here we focus on mammals. From birth on, odors play a central role in discrimination of individuals. Odors from the mother fall into two categories: general odors, usually signifying a food source, and individual odors that the young use to recognize their mother as a specific individual.

Nipples, milk, and suckling

Nipple pheromone in altricial mammals

Newborn rabbits, *Oryctolagus cuniculus*, nurse only once per day for 5–7 minutes. During this short bout, pups can drink up to 25% of their body weight. They respond first to vibration and tactile cues from the approaching mother and then find the nipples with the aid of chemical cues. If one covers the nipples with

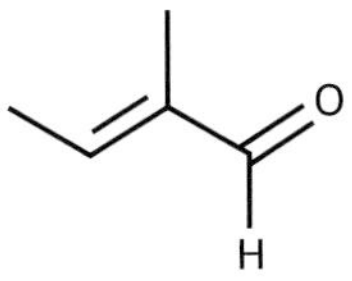

FIGURE 6.1 Structure of 2-methylbut-2-enal, the rabbit mammary pheromone.

plastic, there is little or no sucking. Earlier studies postulated that the "highly stereotyped and reliable response of rabbit pups to the odour presented by the nursing doe seems to qualify for consideration as a true pheromone" (Hudson and Distel, 1983). Newborn rabbits prefer the odor of females in early, rather than late, lactation; in the prenursing rather than postnursing phase; and nipple areas over other abdominal, or back, areas (Coureaud *et al.*, 2001). The mammary pheromone, a compound in rabbit milk that triggers searching by head movements and grasping the nipple, is 2-methylbut-2-enal (Fig. 6.1) (Schaal *et al.*, 2003). The pheromone is specific to rabbits and not produced by, or effective in, hares, rats, or mice. The nipple is the source of the mammary pheromone. Milk before it passes the nipple is not active as pheromone. It remains to be seen whether the pheromone comes from skin glands, symbiotic microorganisms, or compounds complexed to proteins or lipids that are secreted in the mammary glands (Moncomble *et al.*, 2005).

Likewise, odors on the domestic sow's ventrum regulate attachment of piglets to her nipples. Piglets take longer to attach to the nipples after washing the ventrum of the mother with organic solvent or impairing piglet olfaction either by blocking their nares or by flushing their olfactory system with lidocaine, a local anaesthetic. Attachment is still possible, since the response is controlled by tactile and taste stimuli as well as olfaction (Morrow and McGlone, 1988).

Kittens rely little on olfactory stimuli to attach to the nipples of their mother. Instead, tactile cues are important. The mouth area and the trigeminal projection field mediate these tactile sensations (Blass *et al.*, 1988).

Milk

During the days following birth, the odor of milk may activate young mammals. In rats at least, 3- to 9-day-old pups respond to the odor of milk by probing. This behavior wanes at 12 days of age. It is assumed that the odor of milk acquires its activating effect during the first hours and days by the suckling experience. If the dam's diet contains an artificial odor such as eucalyptol, this odor elicits as much activity, mouthing, and probing as milk odor Terry and Johanson, 1987). In rat pups 8 days old or older, suckling is critically dependent on olfaction. If 7-day-old rats are bulbectomized and then tested for nipple attachment 24 hours later, they will not attach to the nipples, lose weight, and become moribund. Cortical lesions have no such effects (Risser and Slotnick, 1987).

Precocial Mammals

Lambs of domestic sheep use different senses for different stages of attachment (bonding, or imprinting) to their mothers. During their first hours, lambs stand up, nose the ewe, and suck. Blindfolded lambs did not stand. When the blindfold was removed, they approached the udder and sucked. This shows the importance of vision for this first step. After spraying the nasal passage with xylocaine to impair olfaction, the lambs did not suck. Finally, lambs with the upper lip anaesthetized with xylocaine needed longer to touch the udder and suck. In summary, visual stimuli guide the initial approach to the mother (Vince *et al.*, 1987). Smell may not be as important for the lamb at that age as it is for the ewe. Initially, "teat seeking" is directed at any ewe, even though the lamb is able to discriminate the odor of its mother from that of other ewes (Vince and Ward, 1984). The smell of a ewe's *inguinal wax* stimulates the lamb's nosing, munching, sucking, and bunting (Vince and Ward, 1984; Vince and Billing, 1986). Calves of free-ranging Labrador Caribou are attracted to the fresh urine of their mothers (Müller-Schwarze and Müller-Schwarze, 1985).

Fecal odors

Feces may also be a source of maternal odors. Domestic piglets prefer maternal feces odor to water, but they do not respond differently to fecal odors of their mother and other lactating or non-lactating females (Morrow and McGlone, 1987). In the laboratory rat, the odor of caecal material from the mother attracts the young once they are mobile. The diet of the mother affects the odor. It is also diet specific: pups will even be attracted to the caecal odor of another lactating female as long as she is on the same diet as their mother. Stimulated by prolactin, lactating females consume more food and water and thus produce more caecal material. This is released to the environment, as lactating rats do not eat caecal material. At weaning, young rats prefer the diet of their mother even if they had been exposed to maternal diet signals only before the time of weaning (Leon, 1975).

Amniotic fluid

Chemical stimulation of a young mammal starts already *in utero*. Amniotic fluid is highly attractive for rat pups. This response is not necessarily acquired during perinatal exposure to this fluid. Rat pups delivered by Cesarean section with no *ex utero* experience did prefer their own mother's amniotic fluid over that of other females (Hepper, 1987). Amniotic fluid is considered important in initial bonding between mother and offspring.

Stress and fear in rodent pups

Laboratory mouse pups show signs of stress, such as higher rates of ultrasonic calls, when the odor of the mother is removed by exchanging the mother's litter with clean bedding (D'Amato and Cabib, 1987).

Some experiments have failed to demonstrate an effect of maternal odors on stressed offspring. The heart rate rose in rat pups removed from their home cage and placed in an unfamiliar environment; this was taken as an index of "fear." Tests with mothers and soiled bedding from mothers or other rat pups showed that tactile and thermal stimuli reduced "fear" in 16-day old pups, but olfactory cues from the mother or odors from the home cage did not (Siegel *et al.*, 1988).

6.5.2 Filial odors: chemosignals from offspring

Olfactory stimuli emanating from the young guide initial bonding, subsequent recognition, and acceptance for nursing in a variety of mammals.

Genital stimulation during parturition stimulates many aspects of maternal behavior, including attraction to amniotic fluid and forming a selective bond with an alien lamb. This is possibly mediated by stimulation of oxytocin release, and/or activation of afferent noradrenergic pathways in the olfactory bulbs (Poindron *et al.*, 1988).

In domestic sheep, the *amniotic fluid* on the coat of the neonate is important for establishing maternal behavior, but only for inexperienced ewes. Washing of the coat of newborn lambs reduces licking and acceptance at the udder by primiparous ewes and increases aggressive behavior by the mother. Multiparous females, however, did not need amniotic fluid for acceptance or for less-aggressive behavior. They merely licked the lamb less (Levy and Poindron, 1987). In dogs (Dunbar *et al.*, 1981) and rats (Kristal *et al.*, 1981), among other species, the amniotic fluid is also important for establishing the first contact between mother and newborn.

The mother's responses to her offspring's odors may have far-reaching consequences for the young later in life. For instance, if a male rat pup is perfumed on his anogenital area, the mother will lick his anogenital area less. Such males later show diminished "masculine" behavior such as mounting and intromission. Licking of the anogenital area of pups by the mother is assumed to produce greater genital sensitivity to peripheral feedback when adult (Moore, 1984; Birke and Sadler, 1987a).

Cross-fostering

Since time immemorial, animal breeders have had to cross-foster motherless lambs or calves, or had to attach newborn mammals to a mother of a different species. They have been aware of odor barriers and developed methods to overcome them. A ewe will accept a non-related lamb if it has been rubbed with the hide or amniotic fluid of her own, perhaps stillborn, lamb. A classical case of successful cross-fostering between species is a technique employed by Peruvian livestock breeders to produce hybrids between alpacas and vicuñas. The cross is called *paco-vicuña* and combines the large quantity of wool of the alpaca with the fine quality of vicuña hair. To breed an alpaca female with a vicuña male, first a male has to be imprinted on alpacas. A newborn male vicuña is covered with the hide of a newborn alpaca and presented to a lactating female alpaca without young. The young vicuña is accepted and nursed on account of his alpaca odor. Successfully raised by his alpaca mother, he will imprint on, and breed with, alpacas when adult.

Recent experiments have confirmed these practitioners' experiences and revealed the precise mechanisms better. A ewe accepts an alien lamb if it wears a stockinette that had been worn before by her own lamb. The salient cue for acceptance is the odor of her own lamb rather than absence of alien odor (Price *et al.*, 1984). A ewe will even accept an additional lamb if it smells like her own lamb. Such "add-on experiments" work better with primiparous ewes than with multiparous ones. To collect the odor, the own lamb wore a stockinette for 20 hours after birth. The stockinette was then pulled over the alien lamb. It should be noted, however, that visual cues such as the color of the lamb's head also influence the ewe's choice (Martin *et al.*, 1987).

In some species, bonding depends solely on the *mother' s* ability to recognize the odor of her pup, as for instance in the Mexican bat, *Tadarida brasiliensis mexicana*. In this species, the pups do not discriminate odors of their (presumed) mother and other lactating females but mothers discriminate well between the pup they nurse and other pups. They also distinguish the odor of their own muzzle glands from that of randomly selected lactating females. It is possible that in this bat species mothers mark their young with secretion from their muzzle glands (Gustin and McCracken, 1987).

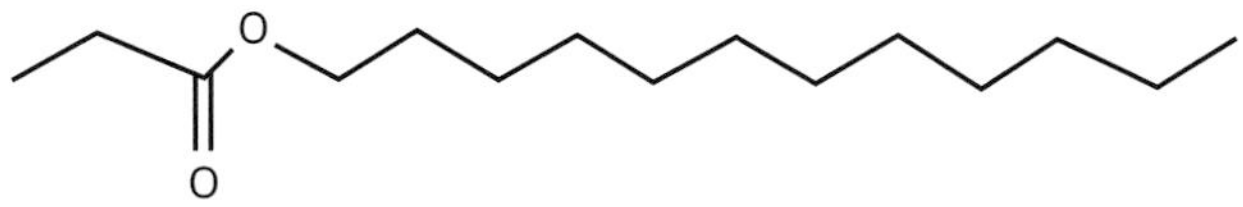

FIGURE 6.2 Structure of dodecyl propionate, found in the preputial gland of rat pups.

Other olfactory effects on maternal behavior

During lactation, female rats eat more, are aggressive toward adult conspecifics, and are less fearful than usual. Lesions of the peripheral and central olfactory system interfere with these behaviors. If the olfactory epithelium is ablated, lactating females eat less, weigh less, and maternal aggression decreases. However, "fear" (i.e. freezing in response to a sound) is not affected by this treatment. Centrally, the olfactory system involves brain structures such as the mediodorsal thalamic nucleus and the prefrontal insular cortex. Thalamic and cortical lesions in these regions lower the frequency of attacking male intruders, but eating behavior and fear responses are left intact (Ferreira *et al.*, 1987).

Behaviorally active chemical compounds

Dodecyl propionate (Fig. 6.2) from the preputial gland of rat pups attracts the mother. She responds by sniffing and licking the secretion, resulting in grooming and stimulating of her young (Brouette-Lahlou *et al.*, 1991).

Paternal behavior

Chemosignals from mate

In the California mouse, *Peromyscus californicus*, both sexes participate in taking care of the young. A chemosignal in the mother's urine maintains parental behavior in the father. This paternal behavior consists of licking the pups and crouching over them in a "nursing position." The active principle resides in the volatile fraction of the female's urine. The paternal behavior was maintained by an experimental dose of 100 μl urine given on the male's nares twice per day (Gubernick, 1990).

In Mongolian gerbils (*M. unguiculatus*; a "biparental" species), males avoid their newborn sons until they are 3 days old. The pups' surge of testosterone around birth is seen as the proximate cause for this behavior. The males appear to respond to an odor from the neonates, because if rendered anosmic by zinc sulfate infusion they do not avoid neonatal pups. The male faces conflicting demands: either attending to the pups or mate-guarding a female during her postpartum estrus. A male engaged in fathering will miss the estrus and mating will be delayed by at least 30 days. It is not clear why only sons have this effect (Clark *et al.*, 2003).

6.6 Species and population discrimination

Discrimination of the own species from other closely related and sympatric species is essential not only for reproductive behavior but also in the contexts of competition for resources and antipredator behavior.

6.6.1 Fish

Juveniles (elvers) of the European eel (*Anguilla anguilla*) migrate upstream in rivers. To test whether they might follow conspecific odors, elvers were taken from the Arno river in Italy and exposed to water in which other juvenile eels had lived for 8 months. They were attracted to this stimulus much more than to uncontaminated water or water from mosquito fish (Pesaro *et al.*, 1981).

Young (i.e. 2-summers old) Arctic charr, *S. alpinus*, are attracted to odors of conspecifics. These odors include those of intestinal contents, fed and starved fish, and urine (Olsén, 1987). In a T-maze fluvarium, fry of Arctic charr of one population was attracted to water from a tank with conspecific fry, while those of another population were not (Olsén, 1990). Juvenile coho salmon, *O. kisutch*, from British Columbia, Canada discriminate between chemical cues from similarly aged conspecifics from their own and a different population. Some experimental results suggest that the odor emanates from the feces. This discrimination of population odors may play a role in imprinting for homing, sibling recognition, or mate choice (Courtnay and Masel, 1997).

Blind cave fish are attracted to water in which conspecifics had lived. This is true for the Congo blind barb, *Caecobarbus geertsi*, and the Somalian cave fish, *Phreatichthys andruzzi* (Berti *et al.*, 1982). The population density can be critical for the strength of a produced chemosignal. The blind cave fish *Astyanax mexicanus* (formerly *Anoptichthys antrobius*; Characidae) is attracted to water in which conspecifics had been. Tested with water from groups of 4, 8, 16, and 32 fish, the response was the stronger the more fish served as stimulus. All other factors, such as familiar versus alien group, did not matter (De Fraipont, 1987). However, test fish may prefer the odor of known individuals, but only for 2 minutes. After that time, the movements of the test fish become random (De Fraipont and Thines, 1986). The Mexican blind cave fish *Astyanax jordani* spends more time in the section of a tank that had been occupied by a conspecific. In the presence of conspecific odor, locomotor activity is reduced in this species (Quinn, 1980).

6.6.2 Amphibians

Courtship pheromones are not necessarily species specific. Pairs of the woodland salamander, *Plethodon shermani*, courted for an equally long time (about 35 to 50 minutes) whether male pheromone from the mental gland of conspecifics or the allopatric species *P. montanus* or *P. yonahlossee* was present, even though the composition of the proteinacous pheromones (plethodontid receptivity factor of these three species differ considerably (Rollmann *et al.*, 2003).

6.6.3 Reptiles

A scincid lizard species, *Eumeces laticeps*, can distinguish the cloacal odors of males of conspecifics from those of *Eumeces fasciatus*. When the odor is presented on cotton-tipped applicators, the lizards flick their tongue more often to their own species' odor. The adaptive advantage of interspecific discrimination may be to recognize sexual rivals or to avoid injury in interspecific fighting (Cooper and Garstka, 1987).

Garter snakes identify their species by tongue flicking at non-volatile, integumentary lipids which they also use in courtship and following conspecific trails. The levels of these lipids fluctuate with hormonal state, skin-shed state, and season (Mason, 1992).

Aggregation pheromones belong in this context. Garter snakes produce an aggregation pheromone in their skin that chemically is a cholesterol ester (Devine, 1977). These snakes can find and follow each other by depositing *trailing* pheromones on vegetation or soil. These have been considered functionally different from the "sexual attractiveness pheromone" (Ford, 1978, 1981, 1982; Garstka and Crews, 1986).

6.6.4 Birds

Social odors have rarely been reported in birds. The respiration rate of wedge-tailed shearwaters (*Puffinus pacificus*) increases in response to the odor of an unfamiliar conspecific (Shallenberger, 1975). Domestic ducks altered social and sexual behavior after bilateral section of the olfactory nerve, or after treatment with amyl acetate or ethyl acrylate (Balthazart and Schoffeniels, 1979). Some bird species are known for their strong characteristic odors. For instance, the hoatzin, *Opisthocomus hoazin*, smells like cow manure (Grimmer, 1962). Oilbirds (*Steatornis* sp.) and hoopoes (*Upupa epops*) are further examples of "smelly" birds. The functions of these odors are still unknown.

6.6.5 Mammals

Mammals easily discriminate the odor of conspecifics from odors of other species.

Investigators have searched for specific compounds that signal the species, while groups of other constituents form the "fingerprint" for individual recognition. The Eurasian otter, *L. lutra*, may provide an example. Its anal sacs contain a secretion with many compounds. One of these, less volatile and hence long lasting, occurred in nearly all individuals, making it a candidate for the signal "otter." The proportions between other compounds varied between individuals and remained constant for 25 days for each individual (Trowbridge, 1983).

Among primates, females of the saddleback tamarin, *Saguinus fuscicollis*, discriminate conspecific scent marks from those of other species or subspecies (Epple *et al.*, 1988), based on a complex mixture of compounds (p. 168). Likewise, *L. fulvus* and *L. macaco* distinguish scent marks by species and "show more interest" in the other species' odor than their own (Fornasieri and Roeder, 1992).

Mole rats of the superspecies *Spalax ehrenbergi* occur in four main chromosome forms: $2n = 52$, 54, 58, and 60. Females of two of these forms (52 and 58) were given choices between soiled bedding (or urine) from males of a homochromosomal or a heterochromosomal form. The females were estrous or diestrous. Only estrous females preferred soiled bedding and urine of homochromosomal males, measured in time spent near the odor samples. Diestrous females showed no preference (Nevo *et al.*, 1976).

6.7 Modulating behavior by status signals

In the second part of this chapter, we discuss communication by pheromones and pheromone-like chemicals in the context of competition, aggression, dominance, and territorial behavior.

6.8 Competition between conspecifics of the same sex

Competition between conspecifics of the same sex may either lead to spacing, often as territorial behavior, or dominance orders among animals that stay together, at least temporarily.

6.8.1 Amphibians

Some salamanders show signs of odor-mediated spacing mechanisms. In Western red-backed (lungless) salamanders, *Plethodon vehiculum*, both sexes avoid the odor of males in a choice test. Ovaska (1988) suggested that the "males use pheromonal markers to space themselves out for mating purposes." Some evidence suggests that male red-spotted newts, *Notophthalmus viridescens*, produce a repellent pheromone that inhibits other males from approaching a cluster of several males courting a female (Park and Propper, 2001).

6.8.2 Mammals

In mammals, dominant individuals typically scent mark more – or in a more effective manner – than subordinate ones.

Marsupials

Dominance status information, coded in whole-body odor, can travel between animals in an air stream. When exposed to the odor of a familiar, dominant male, the sugar glider, *P. breviceps*, increases cardiac and respiration rates within 10 minutes, and levels of glucose and catecholamine in the plasma rise after 30 minutes (Stoddart and Bradley, 1991).

Rodents

In caged laboratory mice, dominant males mark over their entire area, while subordinate neighbors, separated behind a partition, urinate only in the corners (Desjardins *et al.*, 1973; Fig. 6.3). In wild house mice, *M. musculus*, territorial males advertise their "aggressive dominance over other resident and intruder males" by vigorous marking, while subordinate males investigate the dominants' marks. Both use the marks of the dominant males as beacons to orient to their own territory and to avoid areas marked by other dominant males (Hurst, 1990a). Subordinate males also contribute to the communal odor of a group's substrate. A subordinate that is experimentally prevented from contributing his odor, will be attacked more by the dominant and the other subordinate males of the group (Hurst *et al.*, 1993).

One of a pair of fighting mice will be more "confident" if his feces are present: if two male house mice (*M. musculus domesticus*) are placed together in a container along with fecal pellets from one of them, the donor of the fecal pellets shows more and longer aggressive contacts and succeeds more often in

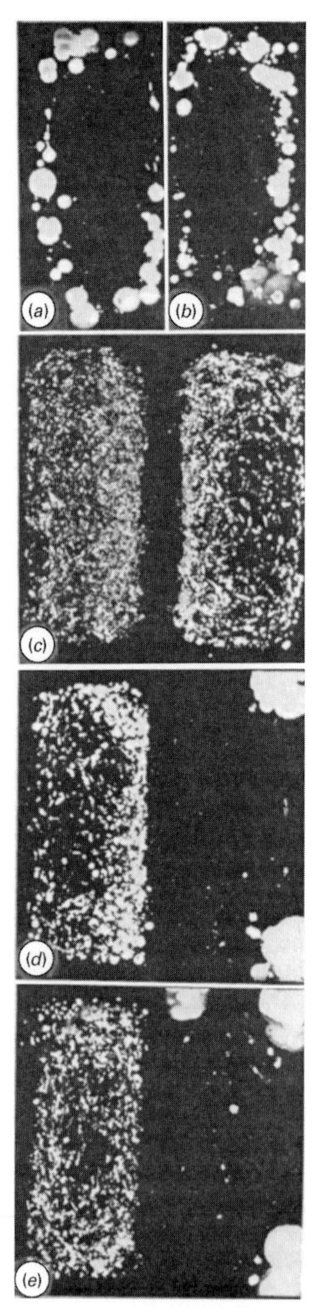

FIGURE 6.3 Ultraviolet light reveals urine marks of dominant and subordinate adult male mice. (*a, b*) Two individually isolated males (without immediate neighbors) urinate along the periphery. (*c*) The same males, but in one cage, only separated by a wire mesh partition. They saturate the entire area with small marks. (*d*) The same animals after an aggressive encounter established the male on the left as dominant. (*e*) The same animals, after five aggressive encounters. The dominant mouse on the left continued to distribute urine over the entire cage floor, while his subordinate neighbor voided urine in corners only. (From Desjardins *et al.*, 1973.)

aggressive encounters (Goodrich *et al.*, 1990). Male mice advertise their dominance by urine marks. In staged encounters between familiar, unfamiliar, dominant, and subordinate males, a urine mark was provided. The dominance status of the urine donor influenced the outcome of the encounter (Hurst, 1993). Male mice also over-mark female urine marks heavily (Hurst, 1989), as is common in ungulates (see below).

Female mice spend more time with the odors of winning male mice than with odors of losers. Even more remarkably, they also spend more time with the odors of the sons of winners than with the odors of sons of losers (Drickamer, 1992).

In other rodents, subordinate males also smell scent marks quite often and so keep informed on the presence, status and activities of higher-ranking group members. For instance, dominant males of the hispid cotton rat, *Sigmodon hispidus*, urine mark more than subordinates. The social status of the male urine donor affects the response of other males to the odor. The response of a reproductive female to feces of either sex depends on her dominance status (Gregory and Cameron, 1989).

Dominant male bank voles (*Clethrionomys glareolus*) over-mark urine and fecal marks from strange dominant males, hierarchically naive males, and sexually unstimulated virgin females. They sniff urine but handle and lick (possibly mask the odor of) feces. Dominant males also scratch their flanks and drag their genitals near conspecific marks, and more so at marks of unknown dominant than unknown naive males. Consequently, in this species, chemical signals provide information on the hierarchical background of an individual in addition to species and sex. Furthermore, bank voles may mask conspecific odors with saliva and specific skin gland secretions (Rozenfeld and Rasmont, 1991).

In the golden hamster, *M. auratus*, both dominant and subordinate males use the flank glands to *communicate* their social status to inhibit overt aggression during encounters (Ferris *et al.*, 1987). However, they do not need their flank glands to *develop* dominant/subordinate relationships.

The dominance status of an individual can be predicted by its scent-marking rate before social interactions take place: Woodchucks (*Marmota monax*) scent mark with their oral glands. When presented with isolated secretion of the oral gland of other woodchucks, future subordinates marked the scent of future dominant animals more often than vice versa (Hébert and Barette, 1989).

Lagomorphs

Rabbits, *O. cuniculus*, "chin-mark" (Fig. 6.4) near their warren entrances and at boundaries to neighboring groups. Only the dominant male marks. This has been demonstrated by comparing gas chromatograms of the chin gland

FIGURE 6.4 A domestic rabbit scent marks by "chinning". (Photograph: R. Mykytowycz.)

secretion of males with those of "environmental" scent marks in the territory of these males. All scent marks contained 2-phenoxyethanol. This compound is typical for dominant males. Furthermore, the scent from the chin gland of the dominant male can be chromatographically found on the forehead of subordinate males and of females, as the dominant buck marks these group members (Hayes *et al.*, 2002). Such "allomarking" is thought to produce a group odor, an olfactory membership badge. However, specific individual relationships between the dominant male and other individuals may also be "recorded" and broadcast this way.

Ungulates

The scent marking of male feral boar, *Sus scrofa*, with metacarpal and preputial glands varies with rank and the social environment. The dominant boar marked more often than the subordinate one when paired, but did not mark at all when alone (Mayer and Brisbin, 1986).

Males of equids urinate and defecate over female urine. In this way, males possibly advertise themselves to other males as a dominance display (Klingel, 1974), while Trumler (1958) proposed that the stallions conceal estrus odors of females from other males. Pronghorn bucks, *A. americana*, cover spots of female urine on the ground in a stereotyped sequence of sniffing, pawing, urination, and defecation, but only when no other bucks are present. This supports the concealment hypothesis (Moodie and Byers, 1989). Males of several other species also cover female urine with their own urine or feces. This is true for blackbuck,

Antilope cervicapra (Dubost and Feer, 1981), dik-dik, *Madoqua kirki* (Hendrichs and Hendrichs, 1971), muntjacs, *Muntiacus reevesi* (Barrette, 1977), and ponies, *Equus caballus* (Feist and McCullough, 1976).

During the rut, a dominant muskox (*Ovibos moschatus*) performs a "superiority display": he tilts his head and walks slowly past his rival in a stiff-legged gait, with the prepuce everted as a 12 cm long pendulous tube. The tube swings about and urine dribbles from the opening. The long belly hair is soaked, emitting a typical strong rutting odor. Washings of the preputial gland contain large amounts of benzoic acid and *p*-cresol. By contrast, the secretion of the muskox infraorbital glands has a "light sweetish, ethereal smell." It contains cholesterol, benzaldehyde, and a homologous series of saturated γ-lactones ranging from 8 to 12 carbons. These lactones smell like the natural secretion (Flood *et al.*, 1989).

6.8.3 Active chemical compounds in mammals

Two compounds in the urine of male mice trigger attacks by other males: 3,4-dehydro-*exo*-brevicomin and 2-*sec*-butyl-4,5-dihydrothiazole (Novotny *et al.*, 1985b). The behavior of male mice depends on the context in which they encounter these compounds. When smeared on castrates or females, they release aggressive behavior, but applied to pups, they inhibit infanticide. Therefore, it appears to be more appropriate to consider the these as signals of "maleness" rather than "aggression pheromones" (Mucignat-Caretta *et al.*, 2004).

The concentrations of 16 constituents of male mouse urine vary with the male's dominance status. Dihydrofurans, ketones, and acetates decreased in subordinates. Two sesquiterpene compounds, α- and β-farnesene, are elevated in dominants' urine 1 week after establishing dominance. The bladder or voided urine of dominants contains more 2-*sec*-butyl-4,5-dihydrothiazole. Four compounds depend on hormones: α- and β-farnesene, dehydro-exo-brevicomin, and 2-*sec*-butyl-4,5-dihydrothiazole. The latter two are absent in urine of immature or castrated males, and testosterone treatment restores their presence. In addition, α- and β-farnesene do not occur in urine of immature males and are merely reduced in urine of castrates. They are not found in bladder urine and originate in the preputial glands (Harvey *et al.*, 1989). While subordinate male mice have reduced levels of farnesenes, levels of their major urinary proteins remain high (Malone *et al.*, 2001).

Dominant males of the European rabbit, *O. cuniculus*, have 2-phenoxyethanol in their chin gland secretion. Behaviorally subordinate males lack this compound. When a subordinate becomes dominant after removal of the originally dominant male, 2-phenoxyethanol starts showing up in his secretion. The perfume industry uses this compound as a fixative. Rabbits perhaps also employ this alcohol as a fixative to keep alive the dominant's odor in the environment (Hayes *et al.*, 2001).

6.8.4 Females

Rodents

Female wild house mice mark frequently to advertise their dominant breeding status to other females. Resident breeding females overmark the urine of other breeding females (Hurst, 1990b).

Female mice "prefer" odors of dominant males over those of submissive males. Such preferences disappear if the dominant male is preputialectomized. A male mouse needs rivals to be aggressive, which, in turn, maintains his preputial gland and hence his attractiveness to females (Hayashi, 1987). Social dominance can be influenced by strange male odor (Hayashi, 1989). In bank voles (*C. glareolus*) the effect of urine odor of a familiar female depends on the status of the receiving female. Before the receiver's pregnancy, the odor inhibits aggression; later in pregnancy and early in lactation it releases aggressive behavior and scent marking. This is thought to aid in spacing (Rozenfeld and Denoël, 1994).

Ungulates

In female cattle, olfactory stimuli are more important than visual cues for dominance relationships. In one experiment, the 10 most dominant of 30 Holstein cows were sprayed with anise oil, or painted. They were kept apart from the herd for 2 hours and then reintroduced. Visual alteration had no effect. However, olfactory alteration resulted in less interaction with other cows in the group. Specifically, other cows investigated the altered individuals *less* and reduced their submissive behavior toward them (Cummins and Myers, 1988).

Here's the rub: "dominance areas" as transition between dominance and territory

White-tailed deer, *Odocoileus virginianus*, bucks compete with one another by three behaviors: they rub the bark off tree saplings with the base of their antlers, scrape the ground at various places in their home range during the rutting season (Fig. 6.5), and rub their muzzle and forehead on overhanging dry twigs while the antorbital gland opens, and its secretion is transferred to the substrate (Fig. 6.6 shows this for a male black-tailed deer). The first two, rubs and scrapes, can occur together or separately. Bucks also defend the area

FIGURE 6.5 A "rub" on a sapling and "scrape" on the ground made by a white-tailed deer buck during the rutting season. Such sites attract females and are thought to constitute a "dominance area." (Photograph: D. Müller-Schwarze.)

around estrous females (Moore and Marchinton, 1974). Only males older than 2.5 years regularly maintain the scrapes and rubs. One male may distribute scrapes and rubs over an area as large as 4 miles long and 1.5 miles wide (Moore and Marchinton, 1974).

6.9 Liquid assets: marking territory and home range with urine and secretions

6.9.1 Fish

Even fish may mark their substrate. Juvenile Atlantic salmon (parr) are attracted to extracts from gravel over which salmon parr had been reared. They

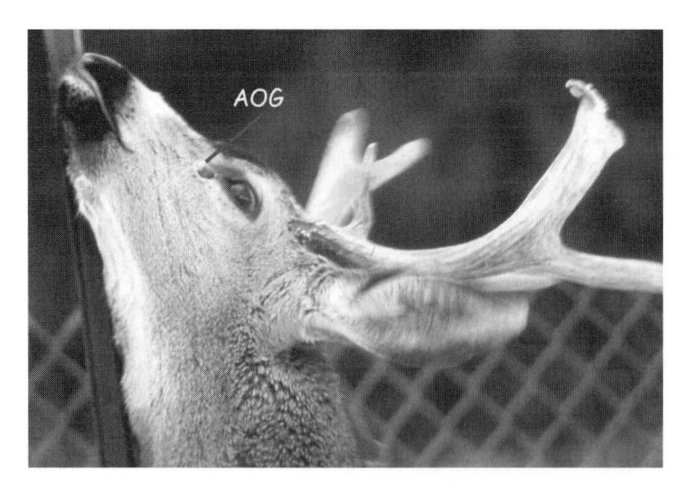

FIGURE 6.6 Secretion marking in the black-tailed deer. A male scent marks by rubbing the side of his head against a post while opening the pouch of the antorbital (AOG) (or preorbital) gland. (Photograph: D. Müller-Schwarze.)

preferred the odor of their own strain. It is assumed that fecal material serves as "scent marks" in this species (Stabell, 1987).

6.9.2 Amphibia

Amphibians appear to use territorial pheromones. The best-investigated species is the red-backed salamander, *P. cinereus*. Females deposit scent on fecal pellets. They are more aggressive to fecal pellets of other females than males are to fecal pellets of other males. These pellets may advertise territories (Horne and Jaeger, 1988). Territory owners of this species use pheromones as the first line of defense, followed by aggressive and submissive signals to intruders, and finally by biting and expelling intruders (Jaeger, 1986).

Salamanders of the ambystomatid family may also use territorial marks. In contrast to red-backed salamanders, which are *repelled* by conspecific odors or fecal pellets (Jaeger and Gergits, 1979; Jaeger, 1986), spotted salamanders, *Ambystoma maculatum*, are *attracted* to paper towels on which conspecifics had lived for 4 days, whether the latter were familiar or unfamiliar (Ducey and Ritsema, 1988).

6.9.3 Reptiles

Reptiles advertise their territories mostly by visual displays, but some olfactory marking may occur. For instance, western fence lizards, *Sceleporis* sp., of North America deposit fecal boli in prime basking sites, which are important for thermoregulation. These fecal boli provide chemical as well as visual cues. Lizards are still attracted to boli that are covered with translucent plastic. The boli possibly signal that the area is occupied, the sex of the marker, breeding or non-breeding status, and the individual's resource-holding power. It is an energetically inexpensive way to maintain a territory (Duvall *et al.*, 1987).

6.9.4 Birds

Birds do not usually rely on scent for territorial marking. Some species, however, such as burrowing owls, apply odoriferous substances to their nest cavities or burrows. The functions are not well understood. Non-avian predators are thought to be the addressees, placing such marking outside the intraspecific behavior of this chapter.

While birds or nests in a number of species are "malodorous," the question is: does any bird "mark" in an area outside its nest, and do conspecifics respond to this chemical cue? Candidates are the "mimuyi," deposits of stomach oil near the nest of Antarctic and snow petrels that have been dated as 9000 and 4000 years old (Hiller and Wand, 1984). Petrels spray stomach oil at intruders as far as 2 m. However, most investigators see bird odors as antipredator, rather than territorial.

6.10 Scent marking in mammals

Scent marking for demarcation of territory or home range of individuals or groups is the hallmark of mammals and this will be discussed in some detail below. Depending on the social organization, scent marking in mammals can take many different forms. Solitary species, such as musk deer, *Moschus* sp., aardvark, *Orycteropus afer*, or aardwolf, *Proteles cristatus*, for example, simply cover their feces with soil or leaves. Pangolins, *Pangolinus arboricoles*, maintain individual distances by means of scent marks (Pages, 1972). Many social species, by contrast, mark in elaborate behaviors with complex or specific spatial and temporal patterns. Examples are the well-studied hyenas and canids such as wolves and coyotes.

6.10.1 Territorial marking in carnivores

Hyenas: a case study in territorial marking

The four species of hyena all scent mark plants with anal gland secretion ("pasting"). Different species concentrate their scent marks in different parts of their territories. In the resource-poor Kalahari, both brown (*Hyaena brunnea*) and spotted (*Crocuta crocuta*) hyenas maintain large group territories of up to 500 and 1000 km^2, respectively. Given their average group sizes of 1 to 9 and 3 to 15 individuals, respectively, each individual would have to mark between 7 and 37 km border on average, were they to concentrate their scent marks at the periphery. This would strain their time and energy budget. Instead, these two species increase their marking from the edge toward the center of their territory where they are active anyway. The striped hyena of the Serengeti, a solitary species, also practices such hinterland marking. With a territory size of 40–70 km^2, the single animal would have to renew scent marks along a 22–30 km circumference. By contrast, border marking is feasible in smaller territories. The spotted hyenas of the Ngorongoro crater live in much larger groups (30–80) than the same species at Kalahari. In addition, their territory is rather small (about 30 km^2). They have only a quarter to two thirds of a kilometer circumference to mark per group member, on average. Indeed, they mark mostly at the border. Finally, the aardwolf, *P. cristatus*, lives alone or in pairs in a small territory (1.5 km^2). This requires only 1.7 to 3.4 km periphery to mark per individual. Consistent with the assumption, the aardwolf follows a border-marking strategy (Gorman and Mills, 1984).

The brown hyena has been studied in detail. Its scent mark, typically on a grass stalk, consists of two parts: an upper black watery portion that loses its odor quickly, and a lower, lipid-rich white part (Fig. 6.7) the odor of which humans still detect after 30 days. The mark is placed at nose level of hyenas, ensuring that newcomers will encounter them easily. Hyenas travel on average 30–40 km per night and paste at a rate of 2.64 marks/km. In the course of a year, an individual may produce 29 000 paste marks, a considerable energetic cost, indicating an important function (Mills *et al.*, 1980).

Hyenas also maintain latrine sites that accumulate 5–50 feces. While the latrine sites are concentrated near, but not on, the border, the paste marks become denser toward the center of the territory, the main area of hyena activity. The driving force for the pasting pattern is not primarily the need to signal at the periphery, which is met by the latrine sites, but the impossibility to mark regularly the long border of a huge territory, as mentioned above (Gorman and Mills, 1984). Would mammals in such situations develop chemical marks that stay fresh longer?

Canids: a second case study in territorial marking

Urine marking by canids has received much attention. Wolves, *Canis lupus*, mark by raised-leg urination (RLU), squat urination (SQU), defecation, and scratching. In Minnesota's Superior National Forest, wolf packs maintain

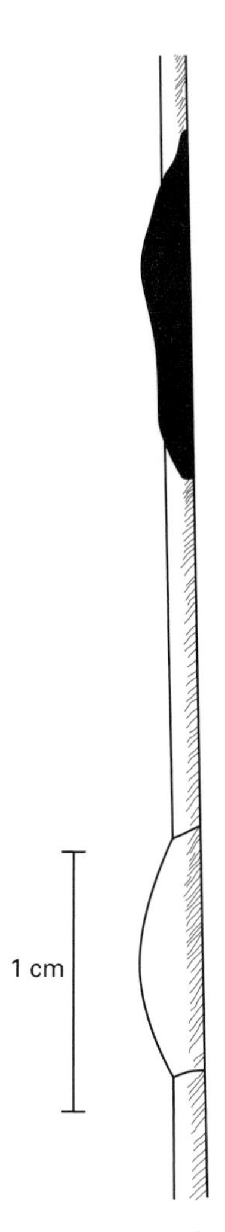

FIGURE 6.7 The double scent mark of the brown hyena. The black mark on top originates from apocrine glands and the white mark from separately located sebaceous glands. (Redrawn from Mills *et al.*, 1980.)

territories that range in size from 125 to 310 km^2. Wolves RLU-mark conspicuous objects in their territory such as blocks of snow, trees, rocks, or snow banks. They re-marked the same spots even though their own odor had not faded completely, nor had other wolves marked there. Fresh marks (2–6 days old) stimulate more marking than those 8 or more days old. Located along trails (Fig. 6.8), scent marks will be encountered often by resident wolves, informing them if they are still in their own territory or reaching the boundary zone with another pack's territory. From the age of a scent mark, wolves can tell how recently they have been in an area. Temporarily single pack members may also read from urination

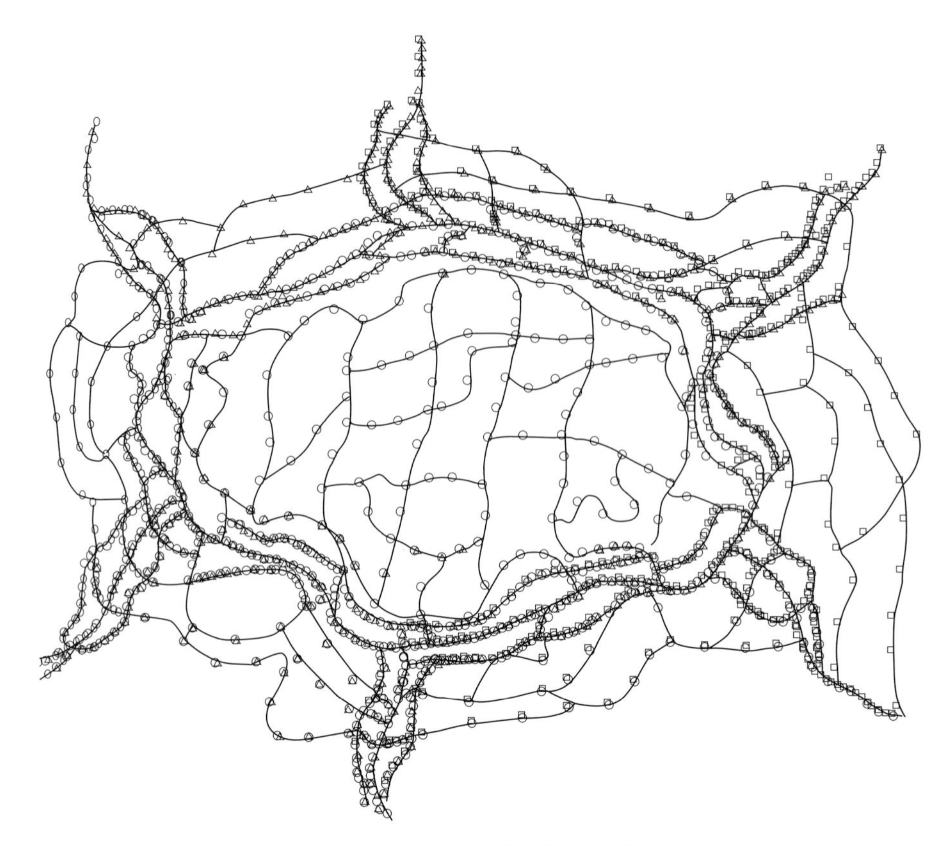

FIGURE 6.8 Schematic sketch of distribution of raised-leg-urination marks in a 20 km wide territory of a wolf pack. Dots are urine marks, lines are trails. Higher marking density along the periphery creates a bowl-shaped pattern. Six neighboring pack territories are indicated by dots in different colors. (From Peters and Mech, 1975.)

and defecation marks whether other pack members have hunted an area recently, if a member is nearby, or which animals are traveling together. Wolves appear to avoid unfamiliar marks as an integral part of unfamiliar terrain (Peters and Mech, 1975).

Urine marking in coyotes, *Canis latrans,* resembles that of wolves. Males also mark with RLU (Fig. 6.9). In a study at Grand Teton National Park, Wyoming, RLU was associated with courtship and mating, traveling, and aggression. Urine marking becomes more frequent during the breeding season (December to February), peaks in March, and reduces in April when the pups are born. Females squat urinate (SQU) year round and this is associated with acquisition and possession of food; it is particularly frequent during the denning season. The marking rate is higher in groups of two or more coyotes. Marking with RLU is higher

FIGURE 6.9 A signature urine mark (left center) on a snow bank made by a coyote. To the right, the animal has defecated, pawed, and partially covered the feces with snow. (Photograph: D. Müller-Schwarze.)

in areas of high rates of intrusion than near the denning area or in areas with less-frequent intrusions. Marking by SQU was most frequent in both denning areas and high-intrusion areas. It was concluded that "scent odours are important in orienting individuals in space but do not represent in and of themselves barriers to movement" (Wells and Bekoff, 1981).

Specialized site creation for marking: beavers

Unlike most terrestrial mammals, beavers maintain territories along a line following watercourses. Both species, the Eurasian beaver, *C. fiber*, and the North American beaver, *C. canadensis*, dredge up mud from the bottom of their home pond, carry it on land, deposit it, and apply a scent mark on top of this mud

FIGURE 6.10 A giant beaver scent mound. (Vertical kitchen knife, center, for scale). (Photograph: D. Müller-Schwarze.)

pile. The mark on these scent mounds consists of castoreum, a urine-derived liquid from the castor sacs, and secretion from the anal glands, also known to trappers as oil sacs (Fig. 6.10).

Scent mounds of North American beavers are concentrated at strategic points in a beaver territory, such as shoreside trailheads of paths leading to feeding grounds. This suggests that beavers might use these scent mounds as "odor beacons" during the night to find their currently active feeding trails. Eurasian beavers make smaller mudpiles or mark directly on rocks or wood. Their scent marks are often concentrated at common boundaries between territories.

It is less likely that beavers will colonize unoccupied beaver sites if these sites are artificially scented with castoreum and anal scent secretion (Welsh and Müller-Schwarze, 1989).

6.10.2 Spatial distribution of scent marks in territory or home range

As we have seen in the carnivore examples, the spatial pattern of scent marking varies between and within species. These comparative studies hint at the function(s) of scent marking, even though we are far from understanding its precise mechanisms. Some species mark throughout their area; some mark more at or near the periphery, while others concentrate their marks around centers of daily activity, or at areas of contact (or conflict) with neighbors or transients, and still others combine some of these patterns (Table 6.2). The scent-marking pattern can vary within a species according to ecological differences, as in the spotted hyena.

Table 6.2 Distribution of scent marks in mammalian territories

Territory area marked by species	Behavior, secretion	Reference
More at periphery		
Wolf *Canis lupus*	Urine marks: RLU	Peters and Mech, 1975
Coyote *Canis latrans*	Urine marks: RLU	Bowen, 1978
Red fox *Vulpes vulpes*	RLU	White *et al.*, 1989
Spotted hyena *Crocuta crocuta*	Feces and AGS	Kruuk, 1972
Brown hyena *Hyaena brunnea*	Pasting	Gorman and Mills, 1984
Aardwolf *Proteles cristatus*	Latrines	Richardson, 1990
Water deer *Hydropotes* sp.	Pasting, feces, urine	Sun *et al.*, 1994
Beaver *Castor fiber*	Feces, urine, castoreum, AGS	Rosell and Nolet, 1997
More at contact zones		
Tiger *Panthera tigris*	Urine, AGS	Smith *et al.*, 1989
Wild European Rabbit *Oryctolagus cuniculus*	Chin gland secretion	Hayes *et al.*, 2002
More in center		
Brown hyena *H. brunnea*	Pasting	Gorman and Mills, 1984
Woodchuck *Marmota monax*	Oral angle glands	Ouellet and Ferron, 1988
Gerenuk *Litocranius walleri*	Antorbital secretion	Gosling, 1981
Honey badger *Mellivora capensis*	Token urination, latrines	Begg *et al.*, 2003
More at "activity centers"		
Beaver *Castor canadensis, C. fiber*	Castoreum, AGS	Müller-Schwarze and Heckman, 1980; Rosell and Nolet, 1997
Wild European rabbit *O. cuniculus*	Chin gland secretion	Hayes *et al.*, 2002
Entire area		
Wolf *C. lupus*	RLU	Paquet and Fuller, 1989
River otter *L. lutra*	Feces, urine, AGS	Trowbridge, 1983,
Thomson's gazelle *Gazella thomsoni*	Antorbital secretion	Walther, 1978
Vicuña *V. vicugna*	Dung piles: feces and urine	Franklin, 1983

RLU, raised leg urination; AGS, anal gland secretion.

Woodchucks, *M. monax*, like other sciurids, possess oral angle glands and rub their muzzle on objects such as fences, woodpiles, shrubs, rocks, or burrow mounds. The vast majority (95%) of the scent marks are within 6 m of the burrow. Males and females mark equally often. While social interactions are not immediately associated with marking, sighting of a conspecific, with or without interaction, "may release scent marking." Woodchucks also mark when moving into a different burrow. These observations suggest that scent marking in woodchucks advertises that a burrow is occupied. It is unclear whether marking is directed at self or conspecifics (Ouellet and Ferron, 1988).

Free-ranging Mongolian gerbils, *M. unguiculatus*, mark more at territorial boundaries, plus at their burrow entrances (Ågren *et al.*, 1989). Similarly, rabbits, *O. cuniculus*, mark throughout their territory but concentrate their scent marks at the center of a group's area, and at boundaries with other groups' territories (Hayes *et al.*, 2002).

Tigers of both sexes mark more at contact zones (Smith *et al.*, 1989), and wolf packs mark more frequently by RLU at the edge of their territory than in the center (Peters and Mech, 1975). Captive red foxes, *Vulpes vulpes*, deposited fecal pellets with anal gland secretion more frequently in the perimeter strip of a 4 ha pen (White *et al.*, 1989).

Thomson gazelles, *Gazella thomsoni*, mark the entire territory. An 8000 m^2 large territory had 18 dung piles and 110 preorbital marks. Dung piles were most frequent in border sections where there were frequent agonistic encounters with neighboring territory owners. In the center of the territory was only one dung pile, near a bedding site. The preorbital marks were arranged in a "broad belt" around the territory. A central area remained unmarked. Walther (1978) concluded that the scent marks were more for the owner's own orientation than for territorial defense.

Another antelope, the gerenuk (*Litocranius walleri*) marks an oval-shaped line in only the core area of its territory, with lines of scent marks extending from it radially like spokes in a wheel. This ensures that intruders will encounter some scent marks, no matter from what angle they approach (Gosling, 1981). Similarly, Eurasian otters, *L. lutra*, in coastal Scotland place feces, urine, and anal gland secretion at so-called spraint stations along trails near the ocean at an average rate of 266 marks/km. This insures that strange otters entering the marked area from the sea will soon encounter a spraint site. Inland trails along streams have only 20 marks/km. Here otters are funneled along trails and will easily encounter scent marks. Otters mark throughout their territories, ensuring that other otters will find a spraint station soon (Trowbridge, 1983).

Linearly arranged territories, as those of beaver, *C. canadensis* and *C. fiber*, along a stream require yet another marking pattern. Trespassing conspecifics are expected to arrive along the watercourse, not usually overland. Therefore, if directed at potential invaders, marking is expected to be particularly intense at the up- and downstream ends of the territory, and less along the banks. Eurasian beavers mark indeed more heavily at territorial borders, and especially at the upstream limit of their territory (Rosell *et al.*, 1998).

Since resources are never completely evenly distributed within a territory, resource-related marking in many species will be concentrated in certain areas. Herbivores with vegetation covering most of their territories will mark the periphery and a ring farther "inland," or a core area ("hinterland marking"), if the territory is very large (Roberts and Gosling, 2001). After all, marking costs the animal in terms of producing chemicals, spending time marking (instead of foraging), and risk of predation. Pronghorn, *A. americana*, males undertake regular, exclusive "marking trips" along the periphery of their territory to refresh their subauricular scent marks (Gilbert, 1973) (Fig. 6.11).

6.10.3 Marking and population density

In some rodents, scent marking intensity reflects population density: the more scent mounds a beaver colony has, the more neighboring colonies exist. This is true for both beaver species (Houlihan, 1989; Rosell and Nolet, 1997; Fig. 6.12). The scent marking syndrome (marking behavior plus gland structure, size, and activity) can vary according to population density within a species in one geographical area. The Indian gerbil, *Tatera indica*, one of the predominant rodents in the Rajasthan desert of northwest India, is solitary in scrub grassland but develops dense urban populations. The urban populations have up to 70 individuals in one burrow. This gerbil makes scent marks by a "perineal drag," depositing urine and sebum from the ventral gland. These glands are larger in males than in females. Low-density populations mark more often than the urban ones. Moreover, the gland is present in more animals of the solitary type (91.4% of males, 38.5% of females) than those of the urban, gregarious type (85.6 and 3.2%). Experimental groups of 1, 3, 6, or 12 wild-caught males or females from solitary grassland populations were kept in one cage. The greater the density, the more they fought and chased each other. However, the frequencies of sebum marking, urine marking, and urination declined with density (Idris and Prakash, 1987).

6.10.4 Marking territory with family odors

In mice, urine is the main source of social odors. Wild house mice (*M. domesticus*) families over time build bizarre small posts of solidified urine by repeated marking (Hurst, 1987; Fig. 6.13). A mouse family is habituated to its own background odor, which permeates its living area. The ubiquitous family odor is dominated by the odor of the dominant male and identifies the home area to residents as well as non-residents. (When the author trapped 26 deer mice over

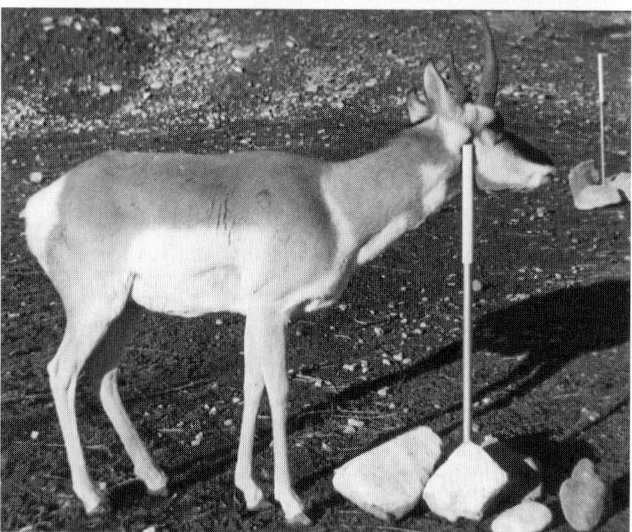

FIGURE 6.11 Marking by a pronghorn buck with his subauricular gland of a thistle in his natural environment (top; Yellowstone National Park) and an experimental Teflon rod (bottom). (Photograph: D. Müller-Schwarze.)

2 weeks in his cabin in the woods, the strong mouse smell disappeared once a big male had been caught.) Any unusual change is detected, and re-marking will restore the familiar background odor. Any novel odor is investigated. However, the urine marks of an unfamiliar family do *not* deter adult males or other mice from invading a new area. Adult males do not investigate, and marked little, the

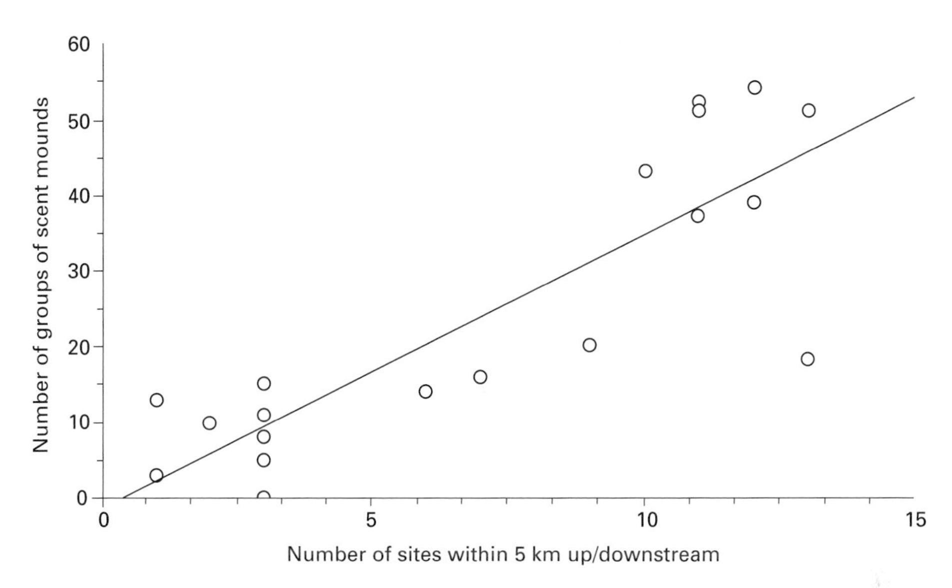

FIGURE 6.12 Correlation between population density and scent marking intensity in the beaver, *Castor canadensis*. (From Houlihan, 1989.)

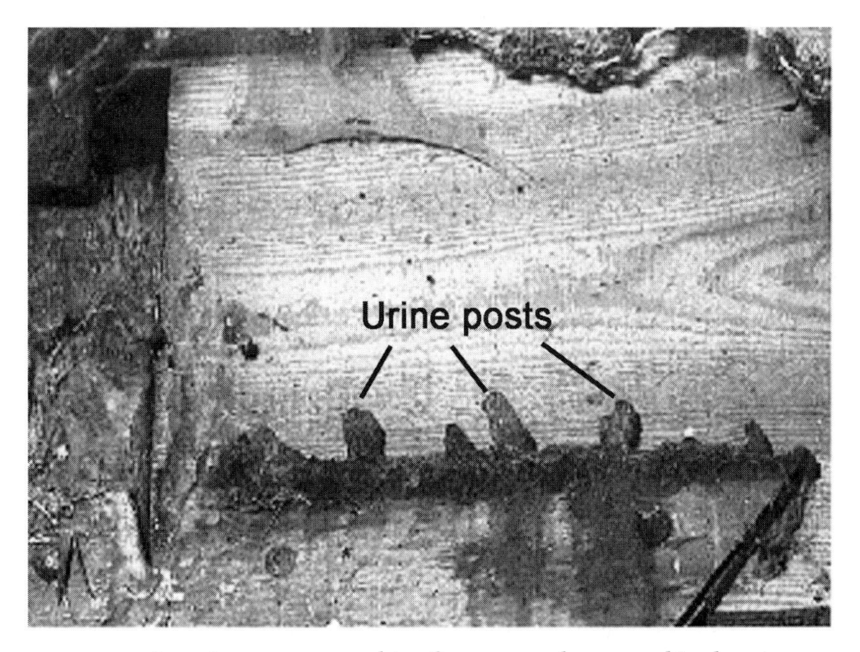

FIGURE 6.13 Scent turrets resulting from repeated scent marking by mice. (Courtesy Jane Hurst.)

urine marks of unfamiliar males. Isolated adult males are attracted to the urine of adult females. Females investigate and mark clean areas and also investigate the marks of adult unfamiliar males. Even a family member's odor can be novel if it is encountered at an unusual concentration, at an unexpected place, or against the background of a habituated odor. A bout of marking by a resident or a familiar neighbor is also treated as "novel" stimulus and can be used to communicate with familiar mice. In this paradox, a *familiar* odor can become a *novel* odor. This happens by virtue of contrast. As signals decay, they become part of the familiar background. Only fresh signals are detected. Immature females maximize their exposure to reproduction-suppressing odor cues: they investigate the urine of resident *females* but *not* the urine of the resident adult *male* (which might accelerate their puberty; see Section 8.4.2). But subadult females investigate urine of *unfamiliar* adult males (Hurst, 1989).

6.10.5 Latrine sites

Many mammals use latrine sites. They often are distributed over the entire home range, but not particularly dense at its periphery. This is true for civets and spotted hyenas (Bearder and Randall, 1978; see also above) and raccoon dogs, *Nyctereutes procyonoides* (Ikeda, 1984). The latrine sites of European badgers, *M. meles*, are large (2–4 m^2), contain feces, urine and secretions from the anal sac and subcaudal gland, and are bigger and more visible at territory borders with neighbor contact (Kruuk, 1978). Badgers locate their latrines under trees, particularly conifers, which are relatively rare in the English study area. Conifers may protect the latrines from rain and render them more conspicuous than they would be near the more common deciduous trees. The badgers also often place the latrines near linear features such as trails, ditches, or fence lines. Different behaviors point to specific functions: all age and sex classes participated in squat marking with the subcaudal gland, indicating a possible general territorial significance. By contrast, digging with forepaws and scratching with hindfeet occurred in the context of mating and may signal breeding condition. The latrines may communicate a variety of types of information (Stewart *et al.*, 2002).

6.10.6 Effects of scent marks

The precise *effects* of territorial scent marks on conspecifics are still a matter of debate. They may warn of a possible agonistic encounter and/or induce a state of increased probability to loose a fight. Gosling (1982) proposed that an intruder matches the scent of a competitor, or of a mark it is seen to have made, with that of other marks nearby. If the odors match, the competitor is probably the territory owner. This mechanism of *scent matching* helps the owner to reduce the costs of territory defense.

Territorial scent marking may greatly decrease the energetic costs of maintaining a resource-defense territory. In fact, a male mammal may only be able to maintain such a territory if he marks it. Otherwise the costs, such as walking, defending, predation risk, staying out in the heat instead of seeking shade, etc., outweigh the reproductive benefits. Indeed, all species with resource territories also scent mark (Gosling, 1986). Contests with other males establish the high social status of a territory owner even before he becomes territorial. It is to the advantage of the territory holder to be recognized as such. This saves energetically expensive and risky fights with intruders. The scent marks are located where encounters with the owner are likely. Odors of marks and the owner can be compared. Such scent matching (Gosling, 1982) is essential for the signal to work; information merely intrinsic to the scent mark, as assumed before, would not suffice. In other herbivores, such as beaver, scent marks may serve in population regulation *before* food becomes a limiting factor (Aleksiuk, 1968).

6.10.7 Countermarking: a visitors' register?

Over-marking is common. The best-known example is urine marking in dogs. How complex a visiting register can dogs read? Does only the top odor count or can they read the complex history of odor layers? Johnston and his group (1995) have experimentally examined the function of over-marking in golden hamsters, *M. auratus*, and Hurst and coworkers (Hurst and Rich, 1999) in wild house mice. In general, male hamsters remember and mark a female's vaginal secretion that is on top of an earlier mark (Johnston, *et al.*, 1995). Female house mice prefer a urine mark of a male mouse that is on top of that of another male. They also prefer a male (or its odor) who had exclusively marked or over-marked other urine marks in a territory that the female had encountered earlier. Male mice do not completely over-mark each other's marks. This provides information about relative competitive ability not otherwise available (Hurst and Rich, 1999).

Neighbors and newcomers, including potential mates, can "read a story": unchallenged scent marks inform of exclusive territory ownership and hence a high-quality individual, thought to be preferred in mate choice. Owners typically will destroy and/or over-mark a challenger's scent mark found in their own territory. This is the basis for the bioassays of beaver pheromones. An experimental scent mound is placed in an occupied territory. Resident beavers obliterate it by pawing and over-marking (e.g. Müller-Schwarze, 1992). In house mice, a single drop of competitor's urine placed in a male's territory not only increases the owner's aggression against competitors but also reduces evasion of this owner by other males and increases challenges against him (Hurst, 1993).

Female mice given a choice between territorial marks of exclusive (one odor) and invaded (mixed odor) territories (in the absence of the owner) "chose" the first odor. Therefore, females use not only the odor of the territory owner but also the absence or presence of marks by competitors (challengers). The absence of any interfering mark identifies the marking male as one of high quality (Hurst and Rich, 1999). In short, individual A can read the story of interactions between individuals B and C and act accordingly to maximize its own fitness.

6.10.8 Factors stimulating scent marking

Scent marking can be stimulated by internal and external factors. Internal factors include most importantly gonadal steroid hormones, as in Mongolian gerbils, *M. unguiculatus* (Yahr, 1977), and the hormone-dependent high dominance status, as in rabbits, *O. cuniculus* (Mykytowycz, 1968). External factors include closeness to neighbors (population density), as in beaver, *C. canadensis* (Müller-Schwarze and Heckman, 1980; Houlihan, 1989), or encountering the mark of another individual, as in canids. Beaver also increased their scent marking drastically on days when transient conspecifics were observed. Even 1 day later the scent marking rate was still elevated (Bollinger, 1980).

Domestic female rabbits mark more when in estrus and prefer to over-mark scent marks by males. They also prefer to mark over chin marks, as opposed to urine and control odors, and prefer to over-mark chin marks from donors living in a long daylight cycle over those from short-day animals (Hudson and Vodermayer, 1992).

6.10.9 Feeding and scent marking

Feeding and scent marking are often connected: gerenuk, *Litocranius walleri*, mark on frequently eaten plants. These would most likely be encountered by an arriving stranger (Gosling, 1981). Tree shrews (*Tupaia glis*) also mark fig trees where they have been feeding (Kawamichi and Kawamichi, 1979). Beavers, *C. canadensis*, frequently place scent marks or scent mounds at feeding sites on the shore of their pond, sometimes where they have felled a tree (author's own observations). Marmosets (*Callithrix* and *Cebuella* spp.) place circumgenital secretion at holes on branches that they have gouged and extracted sap from (Lacher *et al.*, 1981). This marking of favored food trees may communicate information on food resources and avoid aggression, since several groups feed and mark in the same area but avoid each other (Epple *et al.*, 1986; Roberts and Gosling, 2001).

In coyotes, *C. latrans*, SQU by females is associated with acquisition and possession of food, and the denning season (Wells and Bekoff, 1981). Red foxes, *V. vulpes*, urine mark a buried food cache at each visit; the more depleted it is, the more urine odor has accumulated, amounting to "book-keeping" (Henry, 1980). Foxes urine mark inedible food remains on repeated visits. This "no-food-left" signal, in turn, decreases the foxes' interest, and they investigate the site very little (Henry, 1980). Here, an animal may chemo-communicate with *itself* about food. Wolves, *C. lupus*, also urine mark their food caches (Harrington, 1981).

The latrine sites of European badgers, *M. meles*, also relate to food resources. In a 2-year study of a Mediterranean coastal population, the badgers produced more latrines in a year with few fruit because of a drought. This suggests increased defense of a scarce resource (Pigozzi, 1990).

Otters, *L. lutra*, scent mark with feces, called spraints in Britain. One study noted 12 times more frequent marking in winter when food is scarce than in summer. Spraints perhaps advertise depletion of patchy, recurrent resources to other otters and space out the animals. This non-competitive resource partitioning does not require agonistic reinforcement (Kruuk, 1992). Carnivores in general may heavily mark stable resource patches (Macdonald, 1980).

6.10.10 Sex differences in scent marking

Some glands and secretions are unique to one sex, such as preputial glands in males or vaginal secretion in females. Examples of skin glands typical for one sex are the subauricular and dorsal glands in the male pronghorn. Other glands occur in both sexes, such as the castor sacs and anal glands in beavers, and flank glands in golden hamsters. Even when present in both sexes, secretions may differ in composition, visible by differing color or viscosity, as in the anal gland secretion of both species of beavers (Grønneberg, 1978–79; Schulte *et al.*, 1995; Rosell and Sun, 1999). Although urine, feces, and saliva are common to both sexes, sex-specific accessory gland secretions may be added. Also urine marking can be ritualized by one sex, as RLU in canids shows.

Who scent marks in breeding pairs, families or extended kinship groups? In river otter, *L. lutra*, both sexes mark equally often. Tree shrews, *T. glis*, mark by rubbing chin, chest, and anogenital regions against trees. Adult males mark more than adult females and also have more secretion with a stronger odor. Juvenile males and females mark about 10 times less often (Kawamichi and Kawamichi, 1979). In captive red foxes, *V. vulpes*, males produced 84% of fecal pellets (with anal gland secretion) that served as scent marks (White *et al.*, 1989).

As in other sensory modalities, sex differences of anatomical structures such as scent glands, chemical composition of secretions, and behaviors associated with scent communication are more pronounced in species with polygamous mating systems than those with monogamy.

6.10.11 Chemical composition of scent marks

In a few cases we understand the chemical composition of territorial scent marks well enough that synthetic scents have succeeded in triggering typical scent marking in free-ranging mammals. Red foxes, *V. vulpes*, responded during their courtship season (January/February) to an aqueous solution of ethanol, polyethylene glycol, and eight volatile fox urine components on artificial mounds of fresh snow. These artificial scent marks were placed 20–40 m apart along trails traveled by foxes. The foxes marked the "fox urine" mixture more often than a control sample consisting of water, ethanol, and polyethylene glycol (Whitten *et al.*, 1980).

If artificial castoreum scent marks are placed on the banks of a pond, beavers, *C. canadensis*, are more likely to visit, destroy, and re-mark the sites as the complexity of the artificial odor composition increases (Fig. 6.14). While some single phenolics from castoreum such as 4-ethylphenol trigger marking (Müller-Schwarze and Houlihan, 1991), the response increases as the mixture grows to 4, 6, 10, 13, and finally 15 compounds. A mixture of 14 phenolics and 12 neutrals (mostly oxygenated monoterpenes) released responses almost as strong as whole castoreum (Schulte *et al.*, 1995).

The male pronghorn, *A. americana*, scent marks his territory with the subauricular gland. Among the constituents of the subauricular secretion, isovaleric acid (Fig. 6.15) released the strongest marking responses by pronghorn bucks (Müller-Schwarze *et al.*, 1974).

Non-volatile, hydrophilic compounds of high molecular weight in the polypeptide fraction of female (but not male) urine stimulate scent marking in male Mongolian gerbils, *M. unguiculatus*. The major volatile constituents of scent marks (suprapubic/perineal gland secretion) of the saddle back tamarin, *S. fuscicollis*, are butyrate esters with 20 to 28 carbon atoms (Smith *et al.*, 1985).

6.10.12 Time course of scent marking

The seasonal distribution of scent marking often suggests its function. The latrines of European badgers (*M. meles*) are largest in April and October, coinciding more with breeding than foraging, while the smaller, temporary

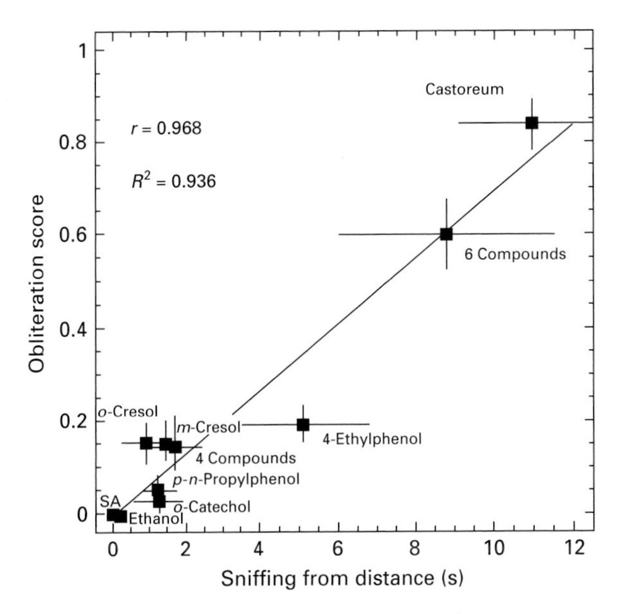

FIGURE 6.14 Free-ranging beavers respond more to mixtures than single compounds from castoreum. Sniffing from a distance was sniffing from water toward experimental scent mark on shore. The obliteration score measured pawing and re-marking of the experimental mark. The control was ethanol. Single castoreum compounds are shown and a four compound mixture groups with these while a six compounds mixture releases a stronger response, and complete castoreum the strongest. Obliteration score-0.014 + 0.07 (sniffing distance). (From Müller-Schwarze, 1992.)

$$\begin{array}{l}\quad CH_3 \\ \quad\;| \\ CH_3CHCH_2COOH\end{array}$$

FIGURE 6.15 Structure of isovaleric acid, which stimulates scent marking by male pronghorn.

defecation sites are most numerous in December and January and correlate with food supply and foraging (Roper *et al.*, 1986). Beaver, *C. canadensis*, scent mark most often from April through June (Svendsen, 1980) when the young disperse from their home colonies and search for new sites to settle. Beavers breed much earlier, in February when in the north their ponds still carry ice. Scent marks most likely signal occupancy to trespassers, who compete for colony space. In Norway, Eurasian beavers, *C. fiber*, also marked most heavily during the time of the subadults' dispersal in May (Rosell *et al.*, 1998).

6.10.13 Age of scent marks

The age of a scent mark may provide important information. Male domestic cats sniff fresh urine marks that are 0.5 to 4 hours old more (and perform flehmen more often) than urine that is over 1 day old. The length of examination may be an attempt to identify the donor and the time of emission (de Boer, 1977).

Urine of estrous females is attractive to males for only 24 hours in mice (Lydell and Doty, 1972) and 48 hours in guinea pigs (Beauchamp and Berüter, 1973). Puberty-delaying cues in urine of grouped female mice (Coppola and Vandenbergh, 1985) and the puberty-accelerating factor in urine of pregnant and lactating female mice are active for 5–7 days. Hamster flank gland and vaginal marks are active for 45 and 100 days, respectively (Johnston and Schmidt, 1979).

The number of measurable volatiles from fecal pellets of house mice, *M. musculus*, increased from 15 to 20 over a period of 24 hours (Goodrich *et al.*, 1990). Urine of female house mice contains a potent but ephemeral pheromone that elicits ultrasonic (70 kHz) mating calls from males, and a longer-lasting, weak pheromone. The first disappears within 15–18 hours, while the latter remains for at least 30 days (Sipos *et al.*, 1995). (For additional information on the life time of chemical signals, see p. 32.)

6.10.14 Complex scent marks

A species may scent mark in two different ways that convey different messages. In the stoat, *M. erminea*, body rubbing is correlated with threat behavior, while the anal drag permeates an area with an individual odor (Erlinge *et al.*, 1982).

7

Signaling pheromones II: sex and alarm pheromones and evolutionary considerations

And all your courtly civet cats can vent, Perfume to you, to me is excrement.

POPE, *Epilogue to the Satires*, Dial. 2,1, 183

Chapter 6 discussed signaling pheromones that allow discrimination, recognition, and broadcasting dominance and territorial status. This chapter explores the role of pheromones and other odors in reproduction, alarm, trail following, and in connection with food. Some evolutionary considerations conclude the discussion of signaling pheromones.

7.1 Sex pheromones: attracting and stimulating

Chemical cues are important to *advertise* one's sex and to *attract* the opposite sex as the first step in sexual behavior. Other functions of sexual signals are to signal current sexual status and to alter the behavior of the potential partner(s) via courtship or scent marking to facilitate mating. Typically, the odor of the opposite sex is attractive, at least in the breeding season. (Priming pheromones are covered separately in Chapter 8.)

7.1.1 Jawless fish (Agnatha)

Traditionally, French fishermen have used the male sex attractant of the sea lamprey, *Petromyzon marinus*. They bait a trap with an adult male and catch numerous females at night. Land-locked sea lamprey such as those in the Great Lakes of North America spawn in streams. Males arrive before the females and build nests. In laboratory choice experiments, ovulating females selected water from spermiating males and increased their searching behavior there. In a natural stream, females responded to male odors from as far as 65 m. The active compound has been identified as a bile acid, specifically 7α,12α,24-trihydroxy-5α-cholan-3-one 24-sulfate. The similar petromyzonol sulfate [Fig. 4.3] has a 3-keto group instead of an a hydroxyl group. It plays a role in migrating of non-reproductive adults. Males release the pheromone at a rate of about 250 μg/h. The pheromone is probably synthesized in the liver and carried in the bloodstream to the gills where, in spermiating males, glandular cells are thought to secrete the pheromone actively. This would make male lampreys "active signalers," instead of females acting as "chemical spies" who cue in on their males' metabolites. Bile acids have advantages: they are more soluble in water and can be produced on a larger scale than steroids or prostaglandins (Li *et al.*, 2002; Yun *et al.*, 2003). To find spawning streams, adult sea lampreys use a migratory pheromone: petromyzonol sulfate plus allocholic acid, which larval sea lampreys release into the water. Larvae of 10 lamprey species produce and release petromyzonol sulfate in large amounts, very little prostaglandin, and only two species had allocholic acid (Fine *et al.* 2004; see also Ch.4).

7.1.2 Bony fish (teleosts)

Sex pheromones are widespread among fish. Fish in murky water or in caves benefit particularly from using chemical cues. For instance, the blind goby *Typhlogobius californiensis* (formerly *Othonops eos*) lives in burrows of the ghost shrimp *Callianassa affinis* and pairs for life. These gobies recognize the sex of conspecifics by odor in water (MacGintie, 1939). Chemically, hormones and their metabolites are common as fish pheromones. Tables 7.1 and 7.2 list some fish pheromones.

Male Pheromones in fish

All male fish pheromones are steroidal (Sorensen and Stacey, 1990). Many male fish release chemicals that attract females, stimulate them to spawn, and inhibit their aggression. As early as 1982, Liley compiled a long list of examples: testes, as in goldfish (Stacey and Hourston, 1982), urogenital fluid, glands on the caudal peduncle (an anal fin appendage), mucus and urine, can be sources of male pheromones.

Waterborne chemical stimuli from males in breeding condition attract female fathead minnows (*Pimephales promelas*). Females distinguish these cues from those of other females or sexually regressed males (Cole and Smith, 1992).

Pacific herring (*Clupeus harengus pallasi*) spawn synchronously by the millions near shore, their suspended milt discoloring the water. Small clusters of males start, then activity spreads to an entire school. Spawning lasts several hours in

Table 7.1 Some pheromones in female fish

Species	Compound(s)	Effect	Reference
Goldfish *Carassius auratus*	17α, 20β-Dihydroxypregnenone (preovulatory pheromone)	Stimulates milt production in male	Sorensen and Stacey, 1990
Goldfish	Prostaglandin $F_{2\alpha}$ and its metabolites (postovulatory pheromone)	Sexual arousal of male	Sorensen and Stacey, 1990
Zebra danio *Brachydanio rerio*	Testosterone glucuronide and estradiol glucuronide	Attracts males	van den Hurk *et al.*, 1987

Table 7.2 Some pheromones in male fish

Species	Compound(s)	Effect	Reference
Sea lamprey	Bile acid	Induces preference and searching in ovulating females	Li *et al.*, 2002
Black goby *Gobius jozo*	Conjugated reduced steroid	Attracts ovulated females	Colombo *et al.*, 1980
African catfish *Glarias gariepinus*	5β-Pregnane-3α, 17α-diol-20-one glucuronide	Attracts and stimulates ovulated females	Resink *et al.*, 1987

small schools, and several days in large ones. The fish move into deeper water, and males and females deposit gametes on submerged vegetation or other substrates, forming trails of sticky eggs or viscous milt. The milt dissipates, rendering the water milky. High milt concentrations may even inhibit spawning and thus regulate density of egg deposition. A male pheromone releases behavior in both males and females. Promiscuous "school spawning" may be the ancestral form of reproduction in teleosts. However, it may be specialized, facilitated by long-lived sperm: maximum fertilization occurs 0.5–2.5 hours after gamete release and sperm is still mobile after 5 days (review by Carolsfeld *et al.*, 1997a,b).

Male brook sticklebacks, *Culaea inconstans*, attract females by chemical cues. Females respond to such cues with courtship and receptivity displays. These consist of raising the head, sinking to the bottom, and remaining stationary. Female brook sticklebacks responded strongly to the male cue during ovulation. Their response waned after that and increased again at the next ovulation, 4–5 days later (McLennan, 2005).

Male bullhead (*Cottocomephorus grewingki*) urine contains a sex attractant for females. It is still active when it had been diluted by 1×10^{-12} (Dmitrieva and Ostroumov, 1986). The male African catfish, *Glarias gariepinus*, produces in his seminal vesicles a sex pheromone that attracts and stimulates ovulated females. One particularly active compound is 5β-pregnane-3α,17α, diol-20-one glucuronide (Resink *et al.*, 1987). Among several steroid glucuronides, this compound released the strongest electrical responses in the olfactory epithelium of females. Holding water from males with their seminal vesicles intact was also active, while absence of the organ or lower levels of glucuronides diminished the response (Resink *et al.*, 1989a). Groups of ovulated catfish are attracted by the steroid conjugate fraction from the seminal vesicle fluid. This fraction contained eight different steroid glucuronides. Without these, the activity was lost. A synthetic mixture of seven steroid glucuronides had a strong dose-dependent effect. A multicomponent pheromone from the seminal vesicle fluid probably attracts the females to the males shortly before spawning (Resink *et al.*, 1989b).

Testosterone, 11β-hydroxytestosterone, and a polyene alcohol, probably farnesol, were found in the urine of male yellowfin Baikal sculpin (*C. grewingki*). Synthesized in the testes, these compounds are excreted with milt (Katsel *et al.*, 1992).

In addition to goldfish and African catfish, chemical analysis of male pheromones has advanced in other teleost fish. Examples are the black goby, *Gobius jozo* (Colombo *et al.*, 1980), the roundgoby (Zielinski *et al.*, 2003), and the zebrafish, *Brachydanio rerio* (van den Hurk *et al.*, 1987). Males of the black goby attract ovulated females to their nests by a C_{19} steroid (etiocholanolone glucuronide). This steroid emanates from a specialized Leydig cell-rich part of the testes, the mesorchial gland (Colombo *et al.*, 1980).

Female pheromones in fish

In various fish species, sexually mature females that are ready to spawn emit chemicals to attract a male and stimulate him to court, release sperm into the water, be more active, build a nest, or assume the courtship coloration. We know more examples than for male pheromones (Liley, 1982). Sources of female pheromones can be ovarian fluid, eggs, mucus, urine, or skin.

Female goldfish (*Carassius auratus*) ovulate in spring, responding to rising temperature, fresh aquatic vegetation, and pheromones. They release eggs within a few hours, requiring close synchronization between the sexes. The female synthesizes the gonadal steroid 17α,20β-dihydroxypregnenone (Fig. 7.1*a*) which induces final oocyte maturation. This hormone, plus

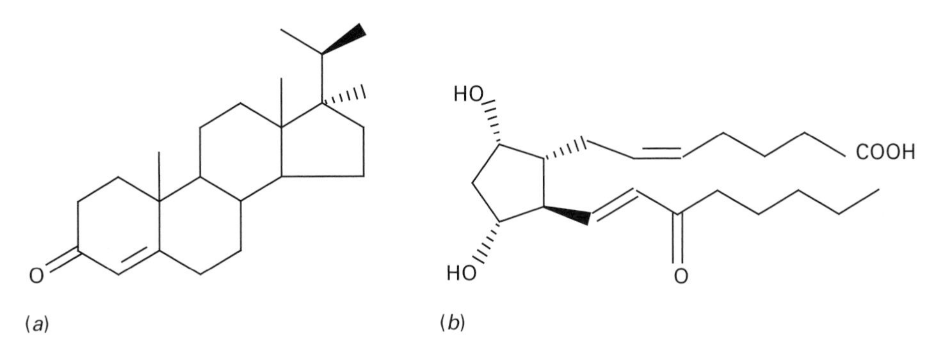

FIGURE 7.1 Female goldfish pheromones. (a) The preovulatory pheromone 17α, 20β-dihydroxypregn-4-en-3 one; (*b*) the postovulatory pheromone 15-keto-prostaglandin $F_{2\alpha}$.

17α,20β-dihydroxypregnenone sulfate and androstenedione are released into the water and affect the male (Sorensen *et al.*, 1990a). The first two compounds stimulate milt production and sexual activity in the male, while androstenedione inhibits these processes (Sorensen *et al.*, 1992). In a comparative study, only goldfish, carp, and hornyhead chub (*Nocomis biguttatus*), all fish species of the minnow family (Cyprinidae), responded to the three steroids, while two members of the Catastomidae did not. Responses to prostaglandins were also weaker in this family (Sorensen *et al.*, 1992).

After ovulation, the female releases prostaglandin metabolites. The prostaglandin $F_{2\alpha}$ and its metabolite 15-keto prostaglandin $F_{2\alpha}$ (Fig. 7.1*b*) are most active in modulating the electro-olfactogram in mature male goldfish. The detection thresholds are 10^{-10} and 10^{-12} mol/l, respectively. These compounds stimulate different receptors, which, in turn, are different from those responding to other olfactory cues (Sorensen *et al.*, 1988). These prostaglandin F derivatives also act as hormones, modulating follicular rupture and stimulating spawning and they arouse the males sexually. Sorensen and Stacey (1990) proposed that most, if not all, postovulatory releasers are prostaglandins. However, although male Atlantic salmon, *Salmo salar*, are attracted to female urine of their own species and that of brown trout, *Salmo trutta*, prostaglandin $F_{2\alpha}$ and one of its metabolites do not appear to be the critical stimuli. These compounds abound in ovarian fluids of both species but occur at low levels in their urine (Olsén *et al.*, 2002a). Sorensen and Stacey (1999) have discussed the evolution of fish hormones into pheromones.

Female fathead minnows, *Pimephalus promelas*, produce a chemical signal that stimulates males to approach and lead them, both courtship patterns. This signal is produced if a female is injected with prostaglandin. (Cole and Smith, 1987).

Table 7.3 Pheromones in amphibians

Species	Sex of sender	Gland	Function	Compound	Reference
Red-bellied newt *Cynops pyrrhogaster*	Male	Abdominal gland (cloaca)	Attracts female	Sodefrin (decapeptide)	Kikuyama *et al.*, 1995
Sword-tailed newt *Cynops ensicauda*	Male	Abdominal gland	Attracts female	Silefrin (decapeptide)	Yamamoto *et al.*, 2000
Magnificent tree frog *Litoria splendida*	Male	Parotoid and rostral glands	Attracts female	Splendipherin (peptide: 25 residues)	Wabnitz *et al.*, 1999

7.1.3 Amphibia

Texas blind salamander *Typhlomolge rathbuni*, a subterranean species, lives in a few caves and wells in the area of San Marcos, Texas. Males and females are attracted by water conditioned by the opposite sex (Bechler, 1986). Males of the salamander *Plethodon cinereus* actually signal their diet quality to the females. Males may feed on termites or ants. Termites provide the better diet because they are rich in lipids, energy, and vitamin B components, and lack the hard cuticle of ants. Female salamanders spend more time near fecal pellets from a male on a termite diet than one that feeds on ants. Also, in a forest, more males with termites in their diet were found associated with females than were males on an ant diet (Walls *et al.*, 1989).

Salamander courtship pheromones

Male red-bellied newts, *Cynops pyrrhogaster* (Salamandridae), attract females with a pheromone that is released into the water from epithelial cells of the abdominal gland of the cloaca. A decapeptide called sodefrin (Ser-Ile-Pro-Ser-Lys-Asp-Ala-Leu-Leu-Lys) is the first amphibian pheromone with female-attracting properties ever chemically identified (Kikuyama *et al.*, 1995). Silefrin in the related sword-tailed newt, *Cynops ensicauda*, is a similar decapeptide and differs from sodefrin in only two amino acid residues (Yamamoto *et al.*, 2000) (Table 7.3).

Another salamander courtship pheromone, identified in 1999, is a protein. During courtship, males of the Appalachian woodland salamander, *Plethodon jordani*, (Plethodontidae), a terrestrial species, actively deliver a pheromone to the female. In the mating season, males develop a "mental gland," located under the chin. During courtship, the female straddles the male's tail, with her chin touching the male's tail base. Together, they walk forward. During this "tail-straddling walk" the male repeatedly turns back and touches the nares of the female with his mental gland, applying pheromone. This pheromone increases the female's receptivity and has been termed plethodontid receptivity factor. Chemically, it is a 22 kDa protein and shows homology with cytokines of the interleukin-6 family (Rollmann *et al.*, 1999). Non-volatile pheromones such as this protein require close contact between the communicating animals, insuring privacy vis-à-vis predators or competitors. While pheromones in the aquatic-breeding salamanders, such as the red-bellied newt mentioned above, attract the female, in terrestrial salamanders, they increase the receptivity of the female. Furthermore, terrestrial salamanders have a prolonged mating season, and insemination can occur several months before oviposition. Therefore, plethodontid courtship pheromones are unique among vertebrates in that they induce changes in receptivity in females that are inseminated long before they lay their eggs (Rollmann *et al.*, 1999). The authors recommend these salamanders as a model system to study changes in female receptivity that occur separately from ovulatory processes. Both the plethodontid receptivity factor, and a second protein (7 kDa, named P_j-7), occur in several isoforms. Salamander populations differ in the composition of their mental gland secretions. Some isoforms may be present or absent, and their ratios often differ between males in one population and also between populations. This divergence of courtship pheromone composition may be the result of sexual selection by conferring different rates of mating success (Rollmann *et al.*, 2000).

Frogs and toads

The first anuran sex pheromone has been chemically identified. Males of the magnificent tree frog (*Litoria splendida*) of Australia produce a sex attractant in the parotoid and rostral glands on their head. One of several peptides secreted by the glands, splendipherin, attracts females. The peptide comprises 25 amino acids residues (Gly-Leu-Val-Ser-Ser-Ile-Gly-Lys-Ala-Leu-Gly-Gly-Leu-Leu-Ala-Asp-Val-Val-Lys-Ser-Lys-Gly-Gln-Pro-Ala-OH) and in the breeding season, the frog produces 10 times as much splendipherin as at other times. In laboratory tests, females approach a pad scented with splendipherin, stay on it, and refuse to budge (Wabnitz *et al.*, 1999).

7.1.4 Reptiles

Odors are important in sexual behavior of snakes, but less so in lizards, which use primarily visual signals.

FIGURE 7.2 A "mating ball" of garter snakes. The large head (center left) belongs to the female, which is surrounded by males. (Photograph courtesy Robert Mason.)

Snakes

Female snakes leave odor trails as they move through vegetation. Their body odor adheres to the anterolateral surfaces of vertical objects. Males then are able to determine the direction of a female's path. The plains garter snake, *Thamnophis radix*, extracts information in this way (Ford and Low, 1984).

Male adders (*Vipera berus*) fight for access to unmated females. Courtship starts in spring only after sexually active males have shed their skins. Males tongue flick at females and court them but attack other shed males. If a recently shed male meets a non-shed male or female, he will ignore both. In summary, the male adder needs a chemical cue from the skin to court or fight (Andren, 1982).

Working with garter snakes, Robert Mason identified the first reptilian sex pheromones. Canadian red-sided garter snake (*Thamnophis sirtalis parietalis*) emerge from their winter dens and form clusters termed "mating balls" (Fig. 7.2). Many males compete for a female. Females attract males with non-volatile saturated and monounsaturated long-chain methyl ketones (Fig. 7.3*a*). Males have squalene (Fig. 7.3*b*) as part of their sex recognition system. Squalene inhibits courtship of females. Males that behave like females to evade male aggression when pursuing females, so-called "she-males", have no or little squalene (Mason *et al.*, 1989a). Thirteen methyl ketones have been identified for females. The major unsaturated ketones are (*Z*)-24-tritriaconten-2-one and

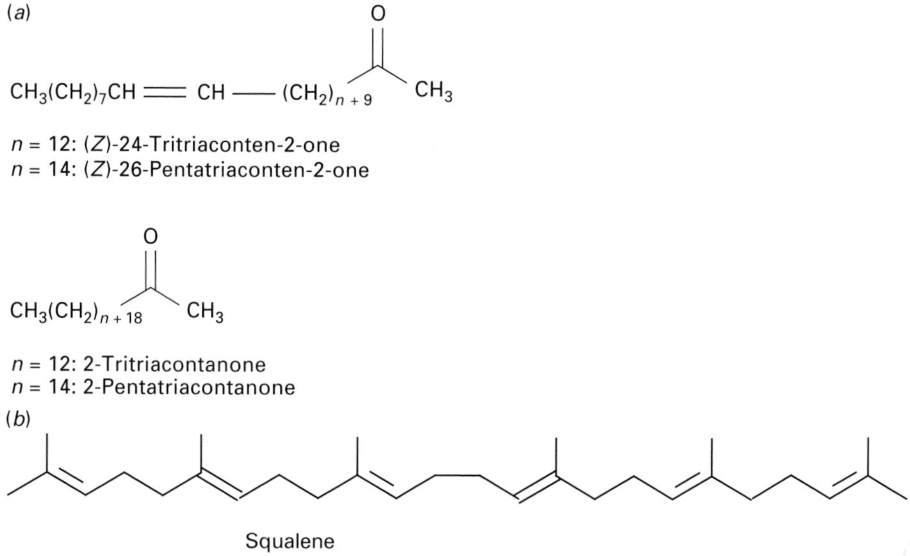

(*a*)

O

$CH_3(CH_2)_7CH{=}CH{-}(CH_2)_{n+9}$ CH_3

n = 12: (*Z*)-24-Tritriaconten-2-one
n = 14: (*Z*)-26-Pentatriaconten-2-one

O

$CH_3(CH_2)_{n+18}$ CH_3

n = 12: 2-Tritriacontanone
n = 14: 2-Pentatriacontanone

(*b*)

Squalene

FIGURE 7.3 Sex pheromones in garter snakes. (*a*) Methyl esters in the attractiveness pheromone of females. (*b*) Squalene, which forms part of the male sex recognition system.

(Z)-26-pentatriaconten-2-one (Fig. 7.3*a*). These and their saturated analogues, 2-tritriacontanone and 2-pentatriacontanone, were synthesized and tested in the field. A blend in natural proportions released courtship behavior and tongue flicking (Mason *et al.*, 1989a). The pheromone originates in the skin, which has seven cell layers. The most basal layer has complete cells with all organelles functioning. As the cells differentiate into the upper (outer) skin layers, they degrade and the organelles decompose. The outermost (surface) layer has flattened cells with no organelles left and the degraded cell constituents oozing out into the interstitial space. As the outer skin is shed, the new skin underneath is coated with the methyl ketones and squalene. Therefore, newly shed female snakes are most attractive to males in terms of their pheromones (R. T. Mason, personal communication, January 2005). The pheromone is a recycled product, stemming from constituents of former cell organelles. Since pheromone is derived from lipids of the epidermis, it qualifies as a protoadapted or exapted (Gould and Vrba, 1982) chemical body constituent.

Male red-sided garter snakes court larger females more than small ones. Even the skin lipid extracts from large females elicit courtship from more males than those from small females. Larger females possess more unsaturated methyl ketones, while small females have more saturated methyl ketones (LeMaster and Mason, 2002).

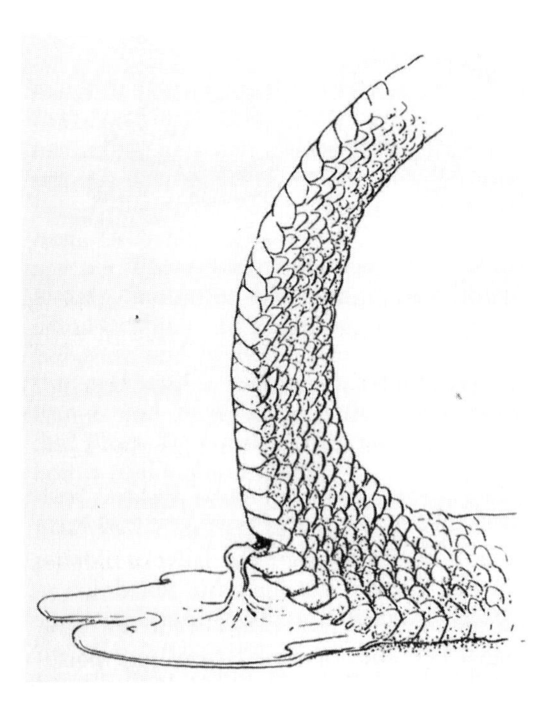

FIGURE 7.4 A female brown tree snake discharges cloacal defense secretion. (From Greene and Mason, 2000.)

Male garter snakes do not pursue already mated females. These females carry a gelatinous mating plug in their cloaca for about 2 days. Copulatory fluids around her cloaca contain a "copulatory pheromone" that inhibits males' advances. Males detect this copulatory pheromone on a female's trail by tongue flicking from a distance (O'Donnell *et al.* 2004).

Female brown tree snakes (*Boiga irregularis*) appear to employ a pheromone to discourage males' courtship. Males start courting a female by tongue flicking her dorsal integument, which is typical for snakes to detect pheromones by the vomeronasal organ. Then the male jerks his head sideways while rapidly tongue flicking, followed by chin rubbing, mounting, and chasing females that move. Aligned side by side, or mounted, the male aligns his cloaca with that of the female, whereupon the male intromits one of his hemipenes. During this process, the female tongue flicks, bobs her head, then mounts the male, and positions her body in front of male's head. This elicits tongue flicking and courtship from the male. The female moves away with head-lifting movements; the male chases and mounts her until their bodies align. If a female rejects a certain male or is not sexually receptive, she may lift her tail and release a clear liquid with a yellow or white precipitate from her cloaca (Fig. 7.4). This secretion, plus the visual signal of the raised tail, stops courtship by males (Greene and Mason, 2000). Female cloacal secretion shortens the duration of a male's courtship and lowers its intensity. This cloacal pheromone specifically inhibits males' courtship of females, as male cloacal secretion has no effect, and ritualized combat between males is not affected by either male or female secretion (Greene and Mason, 2003).

Lizards

Lizards probably rely less on odors for sexual communication. Instead, sex is recognized visually by its typical pigmentation. Males of the viviparous (or common) lizard *Lacerta vivipara* courted males if these were painted with the color pattern of females. However, females were treated the same when painted like males, uniformly black, or left untreated. Males did not change their behavior vis-á-vis females washed with alcohol, especially at their cloacal and femoral regions – which would remove sexual pheromones in garter snakes, *T. radix* (Ross and Crews, 1978). Thus pheromones may not be important in lizards (Bauwens *et al.*, 1987). However, male lizards (*Gerrhosaurus nigrolineatus*) discriminate male from female cloacal secretions (Cooper and Trauth, 1992).

Geckos use chemical cues in mating. In the western banded gecko, *Coleonyx variegatus*, the male cannot determine the gender of another individual from a distance. Rather, he approaches it and grips its tail. Only then will he court a female, but ignore or attack a male. Surgically exchanged tails can guide the response: a female with a male tail is attacked, while the reverse preparation is courted (Greenberg, 1943). Chemical cues in the skin of female leopard geckos (*Eublepharis macularius*) release courting by males before and after the female's skin shedding. During shedding, a female is attacked by the male, presumably because chemical cues are unavailable. The male has the steroids cholestanol, stigmasterol, and stigmastanol, while the female has long-chain saturated and monounsaturated methyl ketones (Mason and Gutzke, 1990).

Tortoises

Tortoises can distinguish sex by chemical cues. A mixture of fatty acids has been extracted from the chin glands of males of several species of *Gopherus*. Applied to the head of a plaster tortoise, these fatty acids trigger head bobbing in females, a component of courtship, while males rammed the models in an aggressive manner (Rose, 1970).

Turtles

Freshwater turtles appear to use waterborne cues to find members of the oposite sex. During the mating season, males of the Iberian peninsula terrapin,

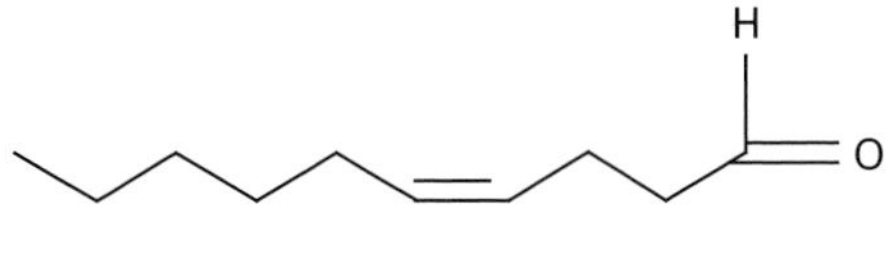

FIGURE 7.5 Both (Z)-4-decenal and octanal are found seasonally elevated in feathers of crested auklets. Auklets are attracted to these compounds.

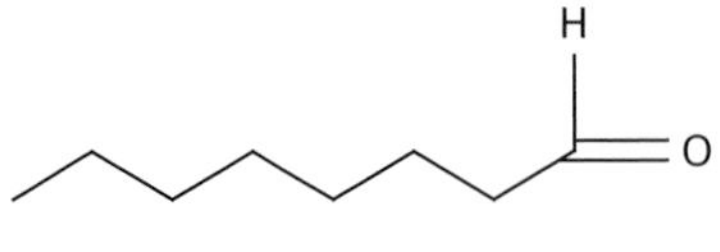

Mauremys leprosa, are attracted to water from females, while females prefer water from other females. Outside the mating season, both females and males avoid water with cues from the opposite sex (Muñoz, 2004).

7.1.5 Birds

Birds are not exactly known for pheromone communication. However, there is some indication that ducks might use sex pheromones. As mentioned in Section 6.6, domestic ducks altered sexual and other behavior after olfactory nerve section and scenting with odorants (Balthazart and Schoffeniels, 1979). The chemical composition of the secretion of the preen (uropygial) gland of mallards, *Anas platyrhynchos*, differs between the sexes during the breeding season, but neither in the quiescent stage nor in ducklings (Jacob *et al.*, 1979).

The crested auklet (*Aethia cristatella*), a social, monogamous seabird of northeast Asia and western Alaska, emits a tangerine-like odor from its feathers. The odor appears strongest on the nape. There the birds smell each other during courtship in what is called "ruff sniff" (Hagelin *et al.*, 2003). The odor is partly caused by (Z)-4-decenal and octanal (Fig. 7.5). These two compounds or odor from conspecific feathers attract crested auklets in a two-way choice apparatus. The compounds are present primarily during the breeding season (Hagelin *et al.*, 2003). At a breeding colony in the Aleutian Islands, Alaska, experimental odor on auklet models attracted males or females only when presented on male models. Odor on female models had no effect on males or females (Jones *et al.*, 2004).

7.1.6 Non-human mammals

Chemical cues attract the sexes and modulate sexual behavior in many or most mammal species. Chemical signals also often reveal the quality and reproductive potential of individuals. In addition to chemical cues, multiple cues in different sensory modalities guide the complex reproductive behavior of mammals.

Male responses to female odors in mammals

Rodents

Mice vary their urination pattern according to urine cues from other mice they find in their environment. Adult male mice urinate most near cues from prepubertal and adult females, while adult females urinate more near cues from males than those from females (Drickamer, 1986). This proximate mechanism is consistent with a communication function.

Male Mongolian gerbils (*Meriones unguiculatus*) scent mark more when a female urine odor is present. The odor cue is hydrophilic, non-volatile, and resides in the non-hydrolyzable high-molecular-weight polypeptide fraction of female gerbil urine. Within a species where body size, form, or coloration is sexually monomorphic, sexual selection may have resulted in odor dimorphism (Blaustein, 1981). Urine of male gerbils and female laboratory mice are not active (Probst and Lorenz, 1987). Experienced males of the golden hamster prefer the odor of a female on the day before maximal receptivity. This is also the day of maximal scent marking by the female. Males least preferred the female odor on the day before diestrus. This is the day when the female chases and attacks the male. Ecologically, this system provides for quick detection and response to impending estrus in a solitary and promiscuous species. The mating advantage for the first male is reflected in pronounced male–male competition (Huck *et al.*, 1989).

Comparative studies have shed light on differences in chemical communication that reflect differences in *social organization*. Examples are the kangaroo rats *Dipodomys merriami* and *Dipodomys spectabilis* of the western United States. In *D. merriami*, one female shares her home range with several males that have separate, smaller home ranges. The reproductive condition of females can be communicated in close contact by anal–nasal circling, and males are not attracted to the urine of estrous females. In *D. spectabilis*, by contrast, the single males and single females defend territories, and females are not in close contact with males. Here, males are attracted to urine of estrous females. The urine seems to be needed as a signal over a greater distance (Randall, 1986).

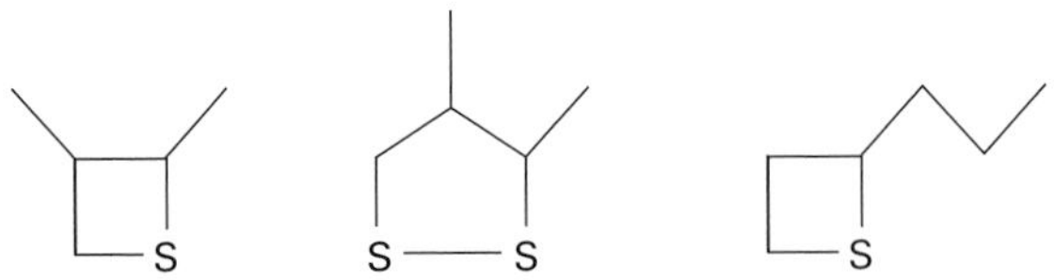

2,3-Dimethyl-thietane | 3,4-Dimethyl-1,2-dithiolane | 2-Propylthietane

FIGURE 7.6 Structures of sulfur compounds in anal gland secretion of ferrets.

Carnivores

Ferrets, *Mustela furo*, distinguish between anal sac secretions of males and females. Males did not discriminate between the anal gland odors of estrous and anestrous females. Chemical sex differences were found, but no seasonal differences. Males had high concentrations of 2,3-dimethylthietane and/or 3,4-dimethyl-1,2-dithiolane (Fig. 7.6). Most individuals had 2-propylthietane. Clapperton *et al.* (1988) concluded that the odor of the anal gland provides sexual and territorial signals and cues to individual identity.

Ungulates

Sex pheromones in large mammals have been difficult to demonstrate. For instance, in white-tailed deer, *Odocoileus virginianus*, females were assumed to leave olfactory signals at the scrape sites that bucks had created on the ground, although females also sometimes scrape the ground with their front hooves (Sawyer *et al.*, 1982). Females rub their forehead at trees near scrapes, urinate near scrapes, walk through and sniff them. The males' responses are not clear. Females possibly also obtain information on the dominance status of bucks. For about 24 hours around conception, the does are found 1.6–2.3 km away from the core area of their own home range. Do they undertake mating excursions to the males' display areas (Sawyer *et al.*, 1989)?

Flehmen (Fig. 5.12) is a behavior shown by male ungulates to primarily female urine. In the stallion, it serves to monitor the estrus cycle, rather than being a step in the actual courtship sequence (Stahlbaum and Houpt, 1989). Table 7.4 lists some examples of the time in the estrus cycle when flehmen is most frequent.

Lek species

In lek species, the male's mating success can depend upon female odors, which attract the females to males. In both Uganda kob (*Kobus kob thomasi*) and Kafue lechwe (*Kobus leche kafuensis*), males attract estrous females to small breeding territories within a lek. Mating success of a male is site specific: it is predicted by the success of the previous occupant of the same territory and not by

Table 7.4 Timing of flehmen during estrus cycle

Species	Time (or peak) of flehmen	Reference
Goat	More during diestrus than estrus	Ladewig *et al.*, 1980
Elephant	More during estrus than diestrus	Rasmussen *et al.*, 1982
Ram	24 h before estrus	Signoret, 1975
Black-tailed deer	Equally often during estrus and diestrus	Müller-Schwarze, 1979

the male's own previous success. Males do not scent mark, but breeding territories are dotted with yellow urine patches from females. Transfer of soil from a successful to an unsuccessful territory increased mating success in the latter more than tenfold. The more successful the soil-donating territory had been, the greater the increase in success at the receiving territory. The mean number of females on the treated territory also increased. On the territories from which soil had been taken, mating success decreased. Successful territories had about 50% more oestrone 3-sulfate in the soil than unsuccessful ones. This is a metabolite of estrogen that occurs in urine of estrous female ungulates. Chance events such as predation by lions, or environmental factors such as predator-harboring cover, may funnel breeding activity to certain territories where females will concentrate and affect males' mating success without enhancing sexual selection (Deutsch and Nefdt, 1992).

Livestock: cattle and sheep

For obvious practical and commercial reasons, the sex attractant(s) of livestock have received considerable attention. In *cattle*, experiments have shown that bulls' responses to sex odors depend on the breeding regimen. Free-ranging bulls with access to cows will prefer to mount a cow that had been scented with urine from an estrous cow to one carrying the urine odor of an anestrous cow. Bulls kept tied up indoors and encountering cows in stalls mount cows indiscriminately (Sambraus and Waring, 1975).

Vulval skin secretion and blood serum contain an active principle, which attracts the bull and releases smelling, salivating, urinating, licking, flehmen, vocalizations, penis protrusion, and mounting. It is not clear whether the blood transports the active principle to or from the vulval skin tissue (Rivard and Klemm, 1989).

The systemic presence of estrus odor in domestic cows was confirmed experimentally: if estrous and diestrous cows are scrubbed with detergent, a bull cannot distinguish between the two. Furthermore, the response of a bull to a diestrous cow was not affected if the cow was marked with vaginal mucus of an

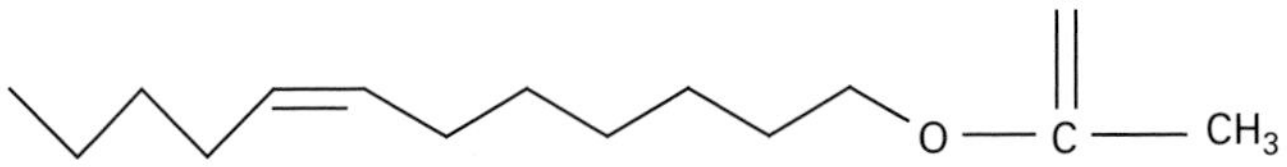

FIGURE 7.7 Structure of the acetate in urine of female elephants.

estrous cow. These results suggest that the estrus odor is not restricted to the vaginal mucus, urine, or feces but emanates from the body surface, particularly from the hindquarters and genital region (Umemura *et al.*, 1988).

Candidate compounds for sex pheromones have been isolated from cervicovaginal mucus of domestic cows. Several diols, ketones, and amines were identified in fractions that released sexual responses in bulls, such as sniffing and licking the sample, flehmen, penile contraction, and preputial secretion (Klemm *et al.*, 1987).

Elephants

Preovulatory urine of female Asian elephant (*Elephas maximus*) contains (Z)-7-dodecen-1-yl acetate (Fig. 7.7). This acetate releases flehmen, and sometimes erections, in males and signals approaching ovulation (Rasmussen *et al.*, 1997). During musth, an annual event, Asian elephants have elevated levels of testosterone and are more aggressive. They secrete from their temporal gland and rub the secretion on trees. They dribble urine, leaving it on their hindlegs and on trails, for others to smell. Before temporal secretion appears, volatiles emanate from the gland. Temporal gland secretion contains frontalin, known as aggregation pheromone in bark beetles. Especially subadult males investigate musth urine, and particularly that of strange bulls. In the current view, musth odors boost males' chances of mating and inform other males of the breeding condition of a bull in the manner of "honest signals" (Schulte *et al.*, 2005; Scott and Rasmussen 2005).

Primates

Some female primates do not advertise estrus by behaviors or physical changes, such as genital swellings. Much sociobiological significance had been attached to this so-called "concealed ovulation." However, chemical cues may still signal ovulation to males. A case in point is the cotton-top tamarin, *Saguinus oedipus*. Scent marks from females during their periovulatory phase were transferred to males' cages. The males showed more penile erections, and mounting of test females, when periovulatory odor was present than in the presence of scent marks from the follicular or luteal phase of the donor females' cycles (Ziegler *et al.*, 1993).

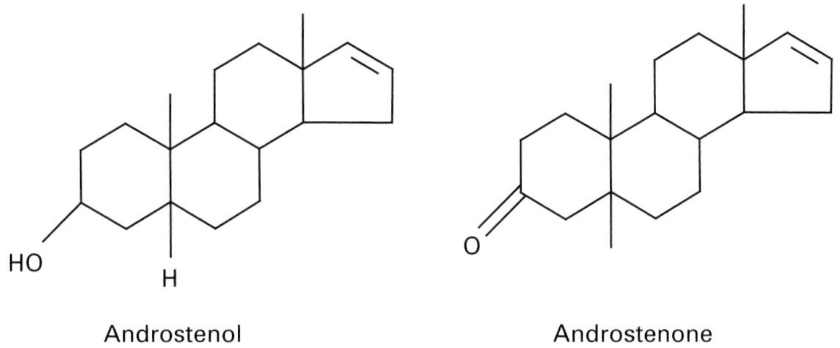

Androstenol Androstenone

FIGURE 7.8 Structures of boar pheromone compounds.

Responses of females to male odors

Odors of male mammals attract females and modulate their behavior in a variety of ways.

Ungulates

The well-known *boar pheromone* was not only one of the first mammalian pheromones identified, but also the first one applied commercially. The saliva from the submaxillary gland contains two steroids. These are 5α-androst-16-en-3-one and 5α-androst-16-en-3α-ol (Fig. 7.8). They are emitted during head-to-head contact in courtship. Both individually stimulate the sow to assume the mating stance, but a mixture of the two is not more active than either compound (Melrose *et al.*, 1971). This may be a case of adaptive redundancy.

The scrapes and rubs produced by bucks of the white-tailed deer, *O. virginianus*, have been interpreted as olfactory attractants for does (Sawyer *et al.*, 1989).

Rodents

In their mate choice, estrous female house mice can detect small genetic differences between males. They prefer to mate with males that are dissimilar in their major histocompatibility complex (MHC). They also spend more time near the odor of dissimilar males than near that of males sharing their own MHC. Females seem to avoid mating with close relatives, and thus maintain genetic diversity (Egid and Brown, 1989). Familial imprinting is an important factor for the development of mating preferences between MHC-dissimilar mice (Yamazaki *et al.*, 1988).

Wild female house mice, *Mus musculus*, prefer the odor of males with the t-complex wild genotype +/+ over males with deleterious mutations (+/t) at the *t* complex. This preference depends on the estrus of the female and her genotype being +/t (Williams and Lenington, 1993). Lethal factors within the t complex play a role in production of the male cue and the expression of preference by females. There may be a second gene (*s*) within the t complex that is involved in expression of female preference (Lenington and Egid, 1985). Preferences by males for +/t females are less dependent on environmental factors and may actually have a strong genetic basis (Egid and Lenington, 1985). No specific chemical compounds characterize mice of different t-locus genotypes, but at certain ages particular volatiles in the urine may be more concentrated. This applies to males and females.

Learning of the parental odor may be important for odor preferences (Jemiolo *et al.*, 1991). Estrous white-footed mice prefer males of intermediate relatedness, or their odors. The levels of reproductive success (i.e. litter size at weaning and offspring weight at weaning) indicated inbreeding depression. Non-estrous females showed no preferences (Keane, 1990).

Female mice are attracted to urine and preputial secretion from male mice. Preputialectomy renders males less attractive. The combination of an androgen-dependent factor in the urine and an androgen-independent factor in preputial secretion is essential for attracting females (Ninomiya and Kimura, 1986, 1988).

The function of the flank gland of the golden hamster, *Mesocricetus auratus*, is not clear but it appears to be involved in signals of sexual and social status and familiarity of the male to the female. Sexually receptive females spend more time near flank scent marks of intact males than castrates, or clean controls. They also stay longer near marks from familiar males than novel males. Finally, these females spend more time near marks of dominant males (compared with subordinate males) (Montgomery-St. Laurent *et al.*, 1988).

Seasonal variation of responses in mammals

Rodents vary seasonally in their responses to odors from the same and opposite sex. The meadow vole, *Microtus pennsylvanicus*, provides an example. Odor preference in a Y-maze and social behavior in dyadic encounters both differed between breeding and non-breeding season. During the breeding season, females are exclusive, and there is more aggression between females than between males or in male–female encounters. Breeding males prefer the odor of females to that of males, and breeding females prefer their own odor and male odor to that of other females. However, non-breeding females prefer each other's odor and are less aggressive. Outside the breeding season, the females' territoriality is relaxed, and they nest in groups. Non-breeding males have no odor preferences. They are solitary, and overwintering groups are, therefore, female biased. Thus, seasonal changes in social organization are correlated with changing odor preferences (Ferkin and Seamon, 1987).

Variation of response with individual reproductive cycles

Male meadow voles, *M. pennsylvanicus*, respond strongly to the anogenital area odor of females only immediately after parturition and for 2 days afterwards, the time of postpartum estrus (Ferkin and Johnston, 1995b). Females, by comparison, responded to male odor throughout their pregnancy and lactation (Ferkin and Johnston, 1995b).

Odor-stimulated flank marking by the female hamster shows a rhythm with a period of 4 days. This may reflect 4-day fluctuations of estrogen and progesterone levels (Albers and Rowland, 1989).

Reproductive status and odor preferences

Young female mice are able to exert some control over their own sexual development and reproductive condition by seeking or avoiding exposure to certain social odors. Prepubertal females avoid odors of adult males, which would hasten their maturation, and prefer the odors from grouped female mice, which slow down sexual development. At puberty, however, their odor preferences are reversed: now they are attracted to male odors and avoid odors of grouped females. Finally, as adults, females strongly prefer male odors and avoid odors of grouped females. Ultimately, the females' changing stimulus preferences during different ontogenetic phases may have important consequences for their reproductive (and inclusive) fitness (Drickamer, 1989b). The specific priming effects will be discussed in Chapter 8.

Chemistry of mammalian sex pheromones

The first two mammalian sex pheromones that were chemically characterized are those of the domestic pig (see p. 54) and of the golden hamster.

The mating behavior of the golden hamster, *M. auratus*, includes two pheromonally guided steps. The female attracts the male with dimethyl disulfide in her vaginal secretion (Singer *et al.*, 1976). Vaginal secretion contains 0.25 ng/mg (0.005 μg/20 mg) dimethyl sulphide and only approximately 200 molecules are needed to attract a male hamster. Dimethyl trisulfide suppresses the response to the former compound (O'Connell *et al.*, 1979). A second, close-range pheromone, *aphrodisin*, triggers mounting by the male. This non-volatile pheromone is a protein of the $\alpha_{2\mu}$-globulin superfamily and has a molecular weight of 17 kDa. Its structure is similar to the retinal binding protein and the pyrazine-binding protein in the nasal mucosa. The pheromone is very specific: three related globulins that share the same overall molecular shape and similar amino acid sequences were not nearly as active: the second most abundant protein in hamster vaginal secretion, the female mouse major urinary protein, and the bovine β-lactoglobulin from cow's milk. Specific structural characteristics, instead of merely the overall molecular shape, appear to be decisive for the high specificity of aphrodisin. Volatiles were experimentally removed from the protein by adsorption chromatography. Therefore, aphrodisin is probably not acting by transporting active ligands to the receptors in the vomeronasal organ (Singer *et al.*, 1989). Aphrodisin produced in *Escherichia coli* by molecular cloning is slightly active, but considerably less active than aphrodisin isolated from the vaginal discharge of the golden hamster. A ligand that is missing in bacterial aphrodisin could be responsible for most of the biological activity (Singer and Macrides, 1990). The gene for aphrodisin is expressed in the vagina, uterus, and Bartholin's gland, even in parotid glands, and can be detected in vaginal discharge before the females reach fertility (Mapert *et al.*, 1999).

In male goats, 4-ethyl-octanoic acid has the most intense goaty odor. This compound also attracts estrous does, and may be a releaser and primer pheromone at the same time (Sugyiama *et al.*, 1981, 1986; Sugiyama, 1983). Another team (Smith *et al.*, 1984) identified 6-transnonenal as typical for the odor of male goats. 4-Ethyl-oct-2-enoic acid gives goat's milk its "goaty" flavor. In the wolf, *Canis lupus*, the volatiles isopentyl sulfide and 3,5-dimethyl-2-octanone occur in male urine, while acetophenone was found in female urine (Raymer, *et al.*, 1984).

The Asian elephant has farnesol, 4-ethylphenol and 4-methylphenol in its temporal gland secretion while the African elephant has only 4-methylphenol. Farnesol levels are inversely related to testosterone levels (Rasmussen and Perrin, 1999). Preovulatory female urine contains (Z)-7-dodecen-1-yl acetate (Rasmussen *et al.*, 1996).

Interspecific detection in mammals

Animals other than conspecifics have been used to detect female sex odors, especially as a bioassay for isolating sex pheromones. Rats distinguish estrous from diestrous urine odor of domestic cows. This discrimination is specific to cows; it is not transferred to urines of goat, pig, or women. Estrous and diestrous urine was discriminated even when diluted 25-fold but not at 1:50 (Dehnhard and Claus, 1988). Dogs have also been used for detecting estrus in cows (Kiddy and Mitchell, 1981). Trained dogs detected estrus odors not only in vaginal mucus or vulva–vestibule samples but also in urine (both naturally voided and catheterized), milk, and blood plasma of cows (Kiddy *et al.*, 1984). The odor can be detected by dogs as early as 3 days before estrus but disappears abruptly on the first day after estrus (Kiddy and Mitchell, 1981).

7.1.7 Humans

Odors are important in the attraction of the sexes, and men and women differ in this. Women (American college students) ranked body odor as more important than any other factor in attraction to men, except "pleasantness." Men, however, ranked "good looks" highest, except for "pleasantness" (Herz and Inzlicht, 2002).

Verbal ratings of strength, pleasantness, and sex attribution have been used as measures of human responses to conspecific sex odors. Several studies have shown that sex, and within the same sex even one's spouse, can be correctly identified by smelling axillary odor on a T-shirt (Russell, 1976; Hold and Schleidt, 1977; Schleidt *et al.*, 1981). Generally, odor intensity and pleasantness are negatively correlated, and strong odors, whether from men or women, are labeled "male." German couples who maintained their regular personal hygiene were less able to distinguish sex by odors on T-shirts. This indicates that cultural personal hygiene patterns can suppress olfactory distinctions between the sexes (Schleidt, 1980).

The apocrine glands in the axilla can secrete enormous amounts of steroids such as dihydrotestosterone and pregnenolone (Brooksbank, 1970). Three single steroid compounds have also been tested on T-shirts. Surprisingly, both sexes attributed androstenol to females. Two other synthetic compounds were attributed to one or the other sex, depending on concentration, and one was perceived as very negative. Androstenol and the two synthetic compounds have very low olfactory thresholds for humans.

The compound responsible for much of the underarm odor is 3-methyl-2-hexenoic acid in both its (*E*) and (*Z*) isomers (Zeng *et al.*, 1991). *Breath* can also communicate gender information. However as with axillary odor, breath from men tends to be classified as stronger and more unpleasant than that of women (Doty *et al.*, 1982).

7.2 Alarm and alert odors

Alarm and alert responses triggered by conspecifics encompass behaviors ranging from increased vigilance to outright fleeing. The cue-sending individuals, in turn, typically are alarmed, disturbed, pursued, caught, or injured by a predator, but also possibly by a conspecific, or perhaps a competitor of another species. Alarm signals exist in all sensory modalities. Alarm calls and visual signals such as raising or flicking the tail in deer and antelope are well known. Chemical alarm signals are more difficult to detect by humans. Therefore, little is known about them in "higher" vertebrates, while fish provide the best-investigated examples.

We have to distinguish alarm signals emitted by a disturbed or stressed, but not captured, animal from "disturbance signals" and chemical cues from injured or captured individuals, termed "damage-released alarm signals" (Chivers and Smith, 1998; Mirza and Chivers, 2001).

7.2.1 Fish

Karl von Frisch (1941) observed minnows, *Phoxinus laevis*, at the edge of an Austrian lake, flee from an injured group member after an attack by a predator such as a pike, *Esox lucius*. Von Frisch induced the minnows to flee by adding macerated minnow skin to a feeding station in the lake (Fig. 7.9). In the laboratory, he introduced skin extracts into a fish tank at the minnows' feeding station. The minnows escaped, swam to the bottom of the tank, and froze. A chemical factor, the "alarm substance" or "Schreckstoff," is released from the skin's club cells (named after their shape) when a fish is injured. Many later experiments have uncovered details, mechanisms, and the distribution of this chemical alarm response among various taxa of fish. Solitary fathead minnows, *P. promelas*, respond to extracts from the skin of injured conspecifics with dashing, freezing and slowing, exploring, or no response (Smith and Lawrence, 1989). Exploring behavior resembles "predator inspection behavior" described for European minnows, *Phoxinus phoxinus*, and is thought to provide information about the predator (Magurran, 1986). Figure 7.10 shows the behavior sequence in the chemical alarm response.

Alarm substances in fish are extremely potent: extract from 1 cm^2 minnow skin creates an active space of over 58 000 liters of water (Smith and Lawrence, 1989).

Guppies avoid a model of a predator such as a pike more when it is paired with guppy extract containing alarm odor. The specific avoidance behaviors of the Trinidadian guppy, *Poecilia reticulata*, include shoaling, dashing, and freezing (Brown and Godin, 1999). Fish can learn to associate the alarm odor with sight of a specific predator, or even a non-predatory fish. Fathead minnows, *P. promelas*, can be conditioned to avoid goldfish or pike by pairing minnow alarm odor with the sight of either species. However, after 2 months, the response to pike, a natural minnow predator, was stronger than that to goldfish, a non-piscivorous exotic. The conditioning is species specific: minnows conditioned to pike did not respond to goldfish and vice versa (Chivers and Smith, 1994). Fathead minnows also learnt a predator odor without seeing the predator at the same time. Fish first exposed to the odor of yellow perch, coupled with alarm substance from other fathead minnows, later survived attacks by yellow perch better than

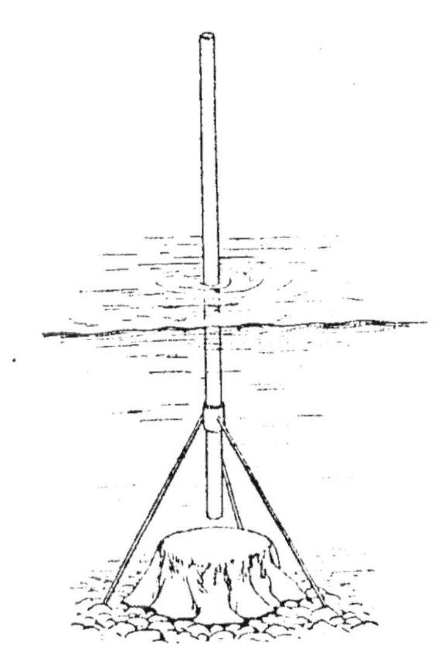

FIGURE 7.9 A field experiment to test alarm pheromone in minnows. The minnows were attracted to a "feeding table" and worms were fed through the tube. When minnows were feeding, macerated minnow skin, containing alarm pheromone, was dropped through the pipe. (From von Frisch, 1941.)

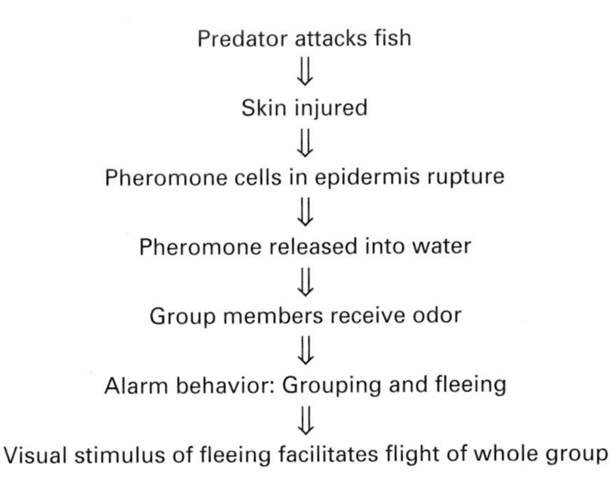

FIGURE 7.10 The sequence of fish responces to an alarm pheromone.

"untrained" fish. This advantage existed only when no cover was available. Simulated aquatic plants in the test tank negated the effect (Gazdewich and Chivers, 2002).

Even a neutral chemical cue can trigger alarm responses in fish if they had experienced it together with a true alarm signal. A coral-reef dwelling goby, *Asterropteryx semipunctatus,* learned to associate a novel chemical cue from a non-predatory fish (water from the tank of the planktivorous damselfish *Acanthochromis polyacanthus*) with danger after this neutral cue was presented together with skin extract of freshly killed conspecifics that contained alarm pheromone. Later the fish responded with alarm (reduced movement and feeding) to the neutral cue alone. Moreover, this response was formed after only one single exposure to the two cues together (Larson and McCormick, 2005).

In addition to behavior changes, exposure to the alarm odor also has physiological effects. For instance, in pearl dace, *Semotilus margarita*, the levels of plasma cortisol and glucose increase 15 minutes after the alarm and are back to normal after 5 hours. The brain concentrations of dopamine, norepinephrine, 5-hydroxytryptamine, or tryptophan did not change (Rehnberg *et al.*, 1987). The fish recovered physiologically much sooner than the behavioral activation; For example, Von Frisch (1941) observed that minnows avoided the site of their encounter with alarm substance for many hours, even days.

Levels of predation risk influence the strength of the alarm response: in the laboratory, female Trinidadian guppies, *Poecilia reticulata*, from a population with much predation, shoaled, dashed, and froze more in response to skin extract from sympatric females than did females from a population that experienced less predation (Brown and Godin, 1999).

The alarm substance (Schreckstoff) has served to test Hamilton's "selfish-herd" theory. Fourteen dace, *Leuciscus leuciscus*, were habituated to minnow schreckstoff, until they no longer responded. They were then joined by a single, naive minnow. Upon adding schreckstoff to the water, the single minnow was alarmed while the school was not. The single minnow moved into the school and became surrounded by other fish on all sides (Krause, 1993). Among alarmed fish, it is "everybody for himself."

The minnow's (*Phoxinus* sp.) alarm substance has been isolated and identified as hypoxanthine 3*N*-oxide (Pfeiffer and Lemke, 1973; Fig. 2.1). The pteridine derivatives isoxanthopterin, 6-acetonylisoxanthopterin, and 2,6-diamino-4-oxodihydropteridine elicit the fright reaction, produce bradycardia, and enhance the dorsal light reaction in giant danios, *Danio malabricus* (Pfeiffer, 1978). The oxide group linked to a nitrogen atom appears to be essential for the alarm response to occur (Brown *et al.*, 2000). Another, structurally different compound with the same group, pyridine *N*-oxide, also stimulates alarm responses such as "fast movement" and remaining "motionless" in juvenile channel catfish (*Ictalurus punctatus*), a species in the superorder Ostariophysae. Non-ostariophysan species such as a cichlid and rainbow trout did not show antipredator behavior vis-à-vis hypoxanthine 3*N*-oxide (Brown *et al.*, 2003).

7.2.2 Amphibia

Tadpoles of the European toad *Bufo bufo* release an alarm pheromone when injured (Eibl-Eibesfeldt, 1949). Alarm pheromones of toad tadpoles may indirectly affect capture efficiency of predators. The giant waterbug, *Lethocerus americanus*, feeds on tadpoles of western toads (*Bufo boreas*) of North America. When a tadpole is damaged, a substance is released that triggers alarm in conspecifics. The tadpoles increase their activity and avoid the feeding site, but only if the injured tadpole is of the same species. The alarm substance released after attack by the giant waterbug also reduces the success rate of the other predator of these tadpoles, the preying naiads of the dragonfly *Aeschna umbrosa*. The success rate is low because the prey is warned, and most likely not as a direct effect on the predator by the released substance since latencies to first contact, first attack, and first capture are not affected by the alarm odor. The tadpoles have giant cells in their skin that may contain bufotoxin (Hews, 1988).

The long-toed salamander, *Ambystoma macrodactylum*, a terrestrial form, avoids areas contaminated with the aqueous extract of ground skin and muscles of conspecifics (Chivers *et al.*, 1996).

7.2.3 Mammals

Alarm odors are very difficult to demonstrate in mammals. Black-tailed deer, *Odocoileus hemionus columbianus*, produce an odor in their metatarsal gland when alarmed, disturbed, or stressed (Fig. 7.11). This odor alerts group members (Müller-Schwarze *et al.*, 1984). Domestic sows stressed by restraining them in a food dispenser without access to food leave an odor that deters other gilts from approaching the feeder. Urine appears to carry this alarm odor (Vieulle-Thomas and Signoret, 1992).

Feral hogs (*Sus scrofa*) in dense scrub vegetation in Florida emitted a typical "pig odor" when fleeing an observer in 36% of 53 episodes, while such an odor was noticeable in only 2 of 18 "control episodes" of feeding or slowly moving ("non-fleeing") hogs. Fleeing pigs often grunted, hissed, or did both, and sometimes did not vocalize: 88.9% of the hissing episodes were accompanied by an odor and 60% of the episodes with both hissing and grunting; 79% of all incidents of hissing, whether by itself or combined with grunting, evidenced a simultaneous odor. When no sound was heard, an odor was noticeable in only 12% (D. Müller-Schwarze, personal observations). Emitting an odor in the face of danger, and then fleeing into dense vegetation, may confuse a potential predator. The predator can be distracted by the lingering odor, while the now invisible prey animal

FIGURE 7.11 Bioassay of the alert odor from the metatarsal gland of black-tailed deer. In foreground compressed air cylinder that creates air current to release MT odor in vessel under tree at feeding station. Left: Deer feeding undisturbed before odor release. Right: Deer become alert when air stream opens valve in odor container literally under their noses. (Photograph: D. Müller-Schwarze.)

can change direction and slip away. Like the severed tail of a lizard, the pig odor might amount to a *pars-pro-toto* diversionary tactic, an antipredator ploy.

7.3 Trail odors

Snakes follow odor trails, presumably to find one another and to migrate to hibernation sites. Male red-sided garter snakes, *Thamnophis sirtalis parietalis*, tested in the field, follow female tracks in spring but not in autumn. Females failed to follow conspecific trails in both spring and autumn (LeMaster *et al.* 2001). In mammals, footprints, especially by animals that also have glands on their feet, can carry information for group members or other conspecifics. Naturalists and hunters have observed deer sniffing tracks of other deer, especially when scattered from a group. Reindeer, *Rangifer tarandus*, have interdigital glands and their secretion has been analyzed (Brundin *et al.*, 1978). In an experimental pen, reindeer sniffed pegs with interdigital secretion that were arranged in a line, resembling a track (Müller-Schwarze *et al.*, 1978b). Details of following non-pheromone trails of other species are discussed in Section 13.4.5.

7.4 Information about food

7.4.1 Rat's breath: exhaled carbon disulfide as food cue

Rats learn from group members about new food sources, and clans may develop "food traditions." How does one rat transmit to another this information about food? This type of intraspecific communication employs both body and foreign odors. Bennet Galef and coworkers (1985) showed that "observer rats" who encounter "demonstrator rats" with food odor on their heads will prefer that food over another when given a choice later. It is important that the food odor is on the head portion of a live rat; if it is applied to the rear end of a rat or to the head of a dead rat, it has no effect (Galef and Stein, 1985).

The food odor conveys information only if associated with breathing. Is breath simply moving air carrying the food odor, or has it a signal odor of its own? Air from the nostrils was sampled and analyzed. It contains carbonyl sulfide and carbon disulfide (Galef *et al.*, 1988). Carbon disulfide proved to be an effective semiochemical to facilitate diet selection in rats: powdered food on cotton was offered to rats, either moistened with a dilute carbon disulfide solution or water only. They were permitted to eat one of these samples and a second type of food in succession. After that, they were injected with lithium chloride. Finally, they were given a choice between the two types of food. They always preferred the food coupled with carbon disulfide during the initial experience (Galef *et al.*, 1988). Adding carbon disulfide and carbonyl sulfide to bait prevents "bait shyness." This procedure has been patented (Patent 4,861,585 to J. R. Mason and B. G. Galef). In field experiments, wild Norway rats consumed three times more bait if carbon disulfide was added. The effect is stronger in females than males. Carbon disulfide looks promising as a rodenticide enhancer (Mason *et al.*, 1988).

House mice, *M. musculus*, also respond to carbon disulfide. Two drops of solutions of different concentrations were added to food pellets. The mice entered an enclosure with treated bait more often, spent more time there, and consumed more food than in an enclosure with water-treated pellets. As in rats, females responded more than males. Higher concentrations of carbon disulfide attracted the mice more than lower ones. Carbon disulfide was more effective than butanol, suggesting that it is a specific cue (Bean *et al.*, 1988).

7.5 Evolutionary considerations

7.5.1 Olfactory isolation mechanisms

Reproductive pheromones are powerful isolating agents between species, as is well known for insects such as moths and drosophilas (Wyatt, 2003).

Fish

Are bony fish (teleosts) reproductively isolated by pheromones? This may depend on the sympatric presence of similar species. Two Southeast Asian species of *Trichogaster* (Belontidae) differed in their discrimination of chemical cues. *T. pectoralis*, the snakeskin gourami, distinguished chemical cues from the two species. It came from stock originally sympatric with three congeners. *T. trichopterus*, the blue gourami, did not discriminate between the two species' odors. It came from a population isolated from congeneric species (McKinnon and Liley, 1987). In general, a given sensory mode should be critical in reproductive isolation if it is (a) an important and efficient mode of communication in the group in question, and (b) involved in the early stages of courtship behavior (Liley, 1982).

Amphibia

Males and females of the North American slimy salamander *Plethodon glutinosus* prefer substrates previously occupied by a conspecific over their own, but they do not discriminate between the sexes. However, in an olfactometer with live salamanders as odor source, male and female *P. glutinosus*, male *Plethodon jordani*, and a related phenotype ("species A") all preferred female over male odors. Male *P. glutinosus* and species A chose conspecific over heterospecific female odors in the olfactometer. This indicates that airborne odors may constitute an important pre-mating isolation mechanism (Dawley, 1984).

Reptiles

Red-sided garter snakes from different regions of Manitoba, Canada show signs of isolation by chemical cues. In choice tests, males from a hibernaculum (overwintering den) in central Manitoba preferred females from their own population to females from western Manitoba. Males from western Manitoba showed no preference. When confronted with experimental trails, males made the same choices. This demonstrated that a chemical factor is involved. Furthermore, the sexual attractiveness pheromone of females, a series of ω-9-cis-unsaturated methyl ketones, varies between the populations. Specifically, the populations differed in the relative concentrations of particular methyl ketones that the females expressed (LeMaster and Mason, 2003).

Mammals

Discrimination of species odors, and by inference, reproductive isolation between species by means of odors, has been demonstrated for many species. Early examples are bank voles, *Clethrionomys glareolus* (Godfrey, 1958), *Peromyscus* spp. (Moore, 1965; Doty, 1972), *Mus* spp. (Bowers and Alexander, 1967), and gerbils (Dagg and Windsor, 1971). Male naked mole rats (superspecies *Spalax ehrenbergi*, Spalacidae) preferred odors from estrous females of their own to a different karyotype in a two-choice apparatus. The odors used were those of soiled bedding and urine from females (Nevo *et al.*, 1976).

Female house mice prefer the odors of males that do not carry *t*-haplotypes. These odors qualify as signal and primer pheromones. (The t complex is a region on chromosome 17.) Males heterozygous for a t-haplotype produce over 90% t-bearing functional gametes; males homozygous for the t-haplotype suffer sterility and homozygosity is generally lethal to embryos of both sexes (Drickamer and Lenington, 1987). The priming issues are discussed on p. 210.

7.5.2 Evolution of pheromone communication

Odors – as any other body features or social signals – are both product and agent of evolutionary processes. Are there extreme results of "runaway evolution"? As such examples are not immediately accessible to our senses, we have yet to discover the olfactory equivalents of the bizarre Irish elk's antlers, the wings of the Argus pheasant, or the antics of the bowerbirds.

We distinguish the evolution of sensory sensitivity, behavioral responses, and of the signals themselves. Without any accompanying changes in signal chemistry, individuals, populations, or species may change their sensitivity or responsiveness to certain compounds or mixtures.

When considering evolution of chemical signaling, the spatial and temporal modulation that the environment provides has to be taken into account. Stimulus propagation characteristics in different environments shape communications patterns. In the acoustical arena, this is true for bird song (Morton, 1975). It also applies to odors. Air currents being most important for odor transport, their presence, force, direction, timing (time of day, for instance), regularity, turbulence, and layering all determine the way an animal transmits its signals coded by volatiles. In addition, temperature, humidity, and other environmental factors, as discussed in Chapter 1, affect the chemical communication system. As a result, animals as disparate as subterranean rodents, small primates of dense tropical forests, or the caribou of the treeless Arctic tundra communicate in vastly different ways purely on environmental considerations. Additional factors such as varying body size or social systems (which, in turn, reflect the environment) leads to enormous potential diversity of signaling systems.

Several comparative studies show that chemical cues may be emphasized in one species, while a closely related species relies more on other modalities. We shall examine why such differences exist.

Fish

In fish, sensory tracking of preexisting cues in evolutionary time permits "spying" by monitoring steroidal pheromones that most likely existed first as hormones and indicate hormonal state. In this way, "high-pass filters" can evolve. These are sensory systems that emphasize certain compounds at low concentrations. Sorensen and Stacey (1999) have discussed the possible evolution of such specific communication channels from early stages of general spying.

The "hormone-to-pheromone hypothesis" proposed for arthropods by Kittredge and Takahashi in 1972 appears to apply to fish. Since fish pheromones are related to steroid hormones, Sorensen and Stacey (1990) favored the "hormone-to-pheromone" hypothesis for three reasons: the preexisting signals are already in synchrony with discrete reproductive events, they are readily excreted into the water, and the evolution of pheromone receptors is simplified by existing internal endocrine receptors. In foraging behavior, fish responses to amino acids as feeding stimuli provide another example of sensory systems becoming attuned to useful stimuli that already exist independently of any other fish "tuning in."

On a larger taxonomic scale, the opposite "pheromone-to-hormone hypothesis" (Haldane, 1955) suggested that the interindividual pheromones of unicellular organisms such as protozoans enabled metazoan development and led to internal chemical communication in and between tissues by hormones in multicellular organisms.

Evolution of a signaling system can also be driven by preexisting sensory biases. The signaler adapts to the receiver's sensory abilities ("sensory drive;" Endler and MacLellan, 1988). In acoustic communication of frogs, such an evolutionary process has been termed "sensory exploitation" (Ryan *et al.*, 1990).

Some taxa may possess pheromone-producing tissues, while others have similar tissues but without pheromones. At the cellular level, extensive comparative studies of fish alarm responses show that cyprinids possess club cells, which release an alarm odor when ruptured by a predator attack. Polypteriformes have very similar club cells. If these cells are homologous, the alarm function in cyprinids is a secondary adaptation (Hugie and Smith, 1987).

Amphibia

European newts of the genus *Triturus* are attracted to the water flowing from a courting pair of the same species. In *T. carnifex*, the crested newt, the response is weak and occurs only in females. This species is highly sexually dimorphic visually. *T. alpestris*, the Alpine newt, is less dimorphic and both sexes respond strongly to the water from the courting pair. Finally, the Italian newt, *T. italicus*, shows the strongest responses. It is not dimorphic at all and exhibits an extended tail-fanning phase during courtship (Belvedere *et al.*, 1988). In summary, each newt species is the more responsive to courtship odors, the less sexually dimorphic it is. This appears to be an example of shifting sensory modalities.

Mammals

In mammals, many chemical signals derive from metabolites in urine or feces or from maintenance compounds such as lipids in skin. In many cases, natural selection appears to alter *not* the signals themselves but rather the *behaviors* of donors and recipients of olfactory cues. This includes depositing, approaching, or avoiding signals. In addition, the *sensitivity* of recipients can evolve (Drickamer, 1986). In other words, evolutionary changes are thought to take place at the level of the central nervous system. Existing chemical cues, such as "body odors" or excretions, can be utilized by the signaler, the recipient, or both for their respective benefits. The response may be genetically anchored, or learned. Chemical signals can derive from excreted metabolites, as in urine of mice and carnivores, or beaver castoreum.

The first step in the evolution of elaborate patterns of odor emission such as scent marking was probably the regular elimination of solid or liquid waste and the inadvertent brushing of the body against vegetation or ground. From this evolved a more deliberate placing of scents at certain places and times with a certain frequency and intensity, resulting in a ritualized dosage of "typical intensity." Finally, supporting bodily structures for the odor signal evolved, such as osmetrichia (scent hairs or bristles). Behavior can also evolve for building supporting structures in the environment, such as the scent mounts that the beaver (*Castor canadensis*) builds from mud.

Intersensory elaboration can improve a signal. Scent release can be combined with visual signals into conspicuous displays. Genets, *Genetta genetta*, rub their flanks on the substrate, a movement accompanied by piloerection. This latter visual display is considered to be *derived* from scent marking. Both are particularly frequent in agonistic encounters (Roeder, 1980). Pronghorn flare the hair of their large white "rump patches" when they release alarm odor from the sciatic glands (Müller-Schwarze, personal observation).

In mammals, males usually have stronger odors, larger and/or more scent glands, scent mark more often, and respond more to alien scent marks. Selection on the basis of odor differences can take place at the level of the individual, deme, population, or subspecies. In polygamous species, intrasexual selection via male–male or female–female competition and sexual selection can be most intense. Most mammals are polygamous, so odor dimorphism is probably widespread (Blaustein, 1981).

Variations of chemical cues

The chemical composition of a glandular secretion may vary considerably between species. Very little is known yet on such differences in vertebrates. Where known, it is not clear whether the differences are genetically controlled, the result of different diets or other ecological or phenotypical variation, or, finally, a combination of different factors.

At the evolutionarily critical level of the subspecies, divergence of the chemical cueing systems of different populations may be not only indicative of but also instrumental in accelerating incipient speciation. Deer mice (*Peromyscus maniculatus*) have lived in North America for many thousands of years, while house mice (*M. musculus*) were introduced only a few hundred years ago. Accordingly, the two species differ in their diversity: House mice have not (yet) developed subspecies, while there are 56 species in the genus *Peromyscus*, and one species of deer mouse has about 20 subspecies. In the laboratory, deer mouse subspecies from locations 3000 km apart cannot effectively communicate by olfactory signals, nor prime each other's reproductive processes chemically. In deer mice, males and female nest together. However, male deer mice of the subspecies *P. m. borealis* from Alberta, brought into the laboratory in Texas, instead killed and cannibalized prepupertal females of the Texas subspecies *P. m. pallescens*, and also *P. m. bairdii* females from Michigan. Priming had also grown disparate: male deer mice from Alberta stimulated uterine growth most in young females from Alberta, and less so in the other two subspecies. House mice from Alberta and Texas, however, were compatible in both signaling and priming (Perrigo and Bronson, 1983).

8

Intraspecific signals: priming pheromones

As the males of most animals search for the females, these odoriferous glands probably serve to excite or charm the female, rather than to guide her to the spot where the male may be found.

CHARLES DARWIN (1896) *Descent of Man and Selection in Relation to Sex*, p. 352

In the rat, olfactory communication results in the synchronization of estrous cycles within a female social group: the majority of females are likely to be at the same phase of their estrous cycles on the same day.

MARTHA McCLINTOCK, (1983) *Chemical Signals in Vertebrates*, vol. 3, p. 159

Priming pheromones are intraspecific chemical stimuli that act on the endocrine system and set in motion slower and often long-lasting processes of growth, maturation, or reproductive state. The difference between signaling and priming pheromones is one of degree. Behavioral responses to pheromones can also be slow and/or are based on physiological changes. Finally, the same compound(s) may release a behavior and also set in motion a physiological response, as occurs in fish and in pigs, for example.

8.1 Fish reproduction

8.1.1 Effects of female pheromones on males

In fish reproduction, the best-investigated pheromone system is that of the goldfish (*Carassius auratus*). Here, sex steroids and prostaglandins play important roles. The female produces two pheromones sequentially: a preovulatory primer pheromone and a postovulatory prostaglandin pheromone that act on the male.

Preovulatory primer pheromone

Female goldfish have an increased level of 17α,20β-dihydroxy-4-pregnen-3-one (17,20βP) in their blood at the time of final maturation of their oocytes (i.e. 10 hours before ovulation and spawning). The pheromone is released across the gill. If 17,20βP is added to aquarium water of male goldfish, they feed less, their level of endogenous gonadotropin rises after 0.5, 1, or 2 hours, and their milt (sperm and seminal fluid) production increases within 4 hours. In females the blood levels of gonadotropin do not increase. Lesions in the anterior telencephalon that disrupt olfactory input and male sexual behavior also block the increase of gonadotropin in response to 17,20βP. Given the presence of 17,20βP in females during oocyte maturation, it is thought that this steroid is a primer pheromone that synchronizes increased male fertility with the time of ovulation (Dulka *et al.*, 1986, 1987, Sorensen and Stacey, 1990). Response to this pheromone is an important factor in reproductive success in male goldfish. Preovulatory 17,20βP has both physiological and behavioral effects. Within 15 minutes, males respond to this stimulus: their blood gonadotropin increases and milt production is stimulated. Thirty minutes after exposure, males are sexually aroused and remain aroused for 10 hours. They have more spawning success, more sperm are produced, and the sperm is more motile (Defraipont and Sorensen, 1993). Males exposed to 17,20βP fertilize a greater percentage of eggs than control males. In competitive spawning, males exposed to this steroid outcompete other males, as shown by microsatellite DNA fingerprinting (Zheng *et al.*, 1997). Sperm quality, rather than spawning activity, appears responsible for this effect: *in vitro* fertilization experiments also showed that sperm from pheromone-exposed males fertilized more eggs than that from non-exposed males (Zheng *et al.*, 1997).

The olfactory system of the male is extremely sensitive to 17,20βP. The fish respond to a concentration of 5×10^{-10} mol/l. This amount (3×10^{14} molecules) is released by a 90 mm female fish into 1 liter water. The females are also very sensitive to 17,20βP. It may stimulate ovulation. Of 47 vitellogenic females 13 ovulated when 17,20βP was added to the water, while only 1 of 43 did so in untreated water (Dulka *et al.*, 1987). Both sexes probably release 17,20βP. Ecologically, this "bisexual" pheromone is thought to synchronize milt production with ovulation and thus coordinate spawning in local populations (Dulka *et al.*, 1987; Sorensen and Stacey, 1990).

Postovulatory prostaglandin pheromone

At the time of ovulation – which is about 12 hours after onset of the gonadotropin II surge in the female – females release less pheromonal steroids than before. They now become sexually active, receptive, and attractive to males.

These phenomena result from the hormonal function of prostaglandin $F_{2\alpha}$, which is produced by the female, and released via urine and across the gill. It stimulates follicular rupture and female spawning. The ovulated eggs are in the oviduct for several hours. During that time they stimulate a 100-fold rise in blood levels of prostaglandin $F_{2\alpha}$.

Released into the water, the prostaglandin also acts as a pheromone. It synchronizes male courtship with female reproductive behavior. Males are very sensitive to this prostaglandin and become sexually aroused (Sorensen and Stacey, 1990). Males detect receptive females by this and a related compound, 15-keto-prostaglandin $F_{2\alpha}$. Females release the latter in urinary pulses that grow more frequent as they enter the vegetation to spawn. The two prostaglandins also act synergistically: both are needed to raise the level of gonadotropin II in males. Sorensen and Stacey (1999) have summarized these studies.

Female and male goldfish also discharge steroids that inhibit milt production in other males. Most potent was androstenedione, having a threshold of 10^{-11} mol/l. These androgens could be functioning as antagonists for 17,20βP receptors (Sorensen *et al.*, 1990b).

Female rainbow trout, *Oncorhynchus mykiss*, also release in their urine 17,20βP. As in goldfish, this pheromone increases the plasma levels of gonadotropin II and testosterone in spermiating males (Scott *et al.*, 1994). Levels of 17,20βP rises within 1 hour of exposure and peak at 3–4 hours. Milt production also increases (Vermeirssen *et al.*, 1997).

Milt production in male carp (*Cyprinus carpio*) can be similarly stimulated by females and possibly a pheromone (Billard *et al.*, 1989). Table 8.1. summarizes some priming pheromones in fish.

The priming effect is not necessarily species specific: ovarian fluid and urine from mature ovulated Atlantic salmon, *Salmo salar*, and brown trout, *Salmo trutta*, affected males of both species. After exposure, maturing males (parr) of both the salmon and the trout developed higher levels of steroid hormones than did controls. Salmon parr had significantly more 17,20βP, 11-ketotestosterone, and testosterone in response to stimuli from both species, while brown trout males only increased their 17,20βP. The amount of strippable milt was elevated only in salmon. Finally, neither species had a changed spermatocrit value (Olsén *et al.*, 2000). Such interspecific effects of priming pheromones bear on the observed hybridization in these and other sympatric species.

8.1.2 Pheromone effects on females

In angelfish, *Pterophyllum scalare*, chemical cues from males stimulate and accelerate spawning in the female. Simultaneous chemical and visual cues stimulate oviposition even more, suggesting an additive effect of the two modalities (Chien, 1973).

Table 8.1 Some priming pheromones in fish

Species	Source and target sex	Pheromone	Effect	Reference
Goldfish *Carassius auratus*	F ⇒ M	Oocyte-maturation-inducing steroid (17,20βP)	⇓ Feeding ⇑ Interaction with females ⇑ Blood gonadotropin II ⇑ milt (within 4 h)	Sorensen and Stacey, 1999
	F ⇒ M	Androstenedione	Inhibits above effects; prevents premature milt release	Sorensen and Stacey, 1999
Rainbow trout *Oncorhynchus mykiss*	F ⇒ M	Urinary pheromone	⇑ gonadotropin II ⇑ 17,20βP ⇑ Testosterone ⇑ Milt production	Vermeirssen *et al.*, 1997
European minnow *phoxinus*	Both sexes	Alarm pheromone	Bradycardia	Pfeiffer and Lamour, 1976
Pearl dace *Semotilus margarita*	Both sexes	Alarm pheromone	⇑ Plasma cortisol, glucose	Rehnberg *et al.*, 1987

17,20βP, 17α,20β-dihydroxy-4-pregnen-3-one; M, male; F, female.

Males of the zebrafish, *Danio rerio*, release a primer pheromone that stimulates ovarian growth and ovulation in females (Chen and Martinich, 1975; van den Hurk *et al.*, 1987).

Under certain conditions, ovulation and plasma gonadotropin levels can be stimulated by ovarian fluid from other ovulated females. The African catfish, *Clarias gariepinus*, is an example. It lives in muddy water and spawns at night, conditions expected to favor pheromone communication. Moreover, this species is a seasonal breeder, spawning after rainfall. The sudden mass spawning requires synchronization. Water from males or ovulated females, and also ovarian fluid of ovulated females, experimentally induced ovulation in 67% of females. Plasma gonadotropin was also increased by these treatments (Resink *et al.*, 1989c).

For fish, it is thought that pheromones derived from hormones. "Hormones are pre-existing signals produced in temporal synchrony with discrete reproductive events" (Sorensen and Stacey, 1990). These hormones are readily excreted into the water. The evolution of olfactory reception mechanisms for pheromones may have been facilitated by already existing internal endocrine receptors. Sorensen and Stacey (1990) argued that "for pheromonal functions to arise mutations are initially required only in the recipient." Specificity of the signal results either from admixing other compounds or from a slight change in the steroid molecule. Such specificity is the result of the selective pressure for reproductive isolation mechanisms (Sorensen and Stacey, 1990).

8.2 Amphibia

Male courtship pheromones can increase the receptivity of the female. In the plethodontid salamander *Desmognathus ochrophaeus* the male courts the female by scraping her dorsum with his specialized premaxillary teeth and swabs the same area with secretion from the "mental gland" on his chin, amounting to an "injection." In an experiment, filter paper with an extract from the excised mental glands of males was placed on the dorsum of females. These treated females mated 28% (59 minutes) sooner than controls (Houck and Reagan, 1990).

The terrestrial salamander *Plethodon jordani* applies courtship pheromone to the female by rubbing or slapping his mental gland directly on the female's nares. This way the pheromone can stimulate the vomeronasal organ and accessory olfactory system. Experimental application of mental gland extracts to the nares of females accelerated the time until active courtship started (Houck *et al.*, 1998).

8.3 Reptiles

In iguanas, *Iguana iguana*, chemical signals from adult males elevate the levels of corticoid steroids in juveniles, indicating stress. Together with visual cues, these cues suppress growth and assertion displays in juveniles (Alberts *et al.*, 1994).

8.4 Mammals

Priming pheromones would be expected to operate foremost in nocturnal and/or subterranean mammals that are precluded from communicating visually, or in species that inhabit enclosed spaces that permit high concentrations of odor to build up. Such a lifestyle would be analogous to those of many social insects, notably termites. The most bizarre social organization found in mammals fits almost exactly that description. Naked mole-rats, *Heterocephalus glaber*, of Kenya, Somalia, and Ethiopia live in large underground colonies of as many as 60 or 80 individuals. Their eyes are very small. Only one female, the "queen," breeds. She is also the colony's largest individual, with prominent teats and a perforate vagina. The remainder of the colony form two or three castes, each with both sexes. The castes are of different body size and perform different tasks. "Frequent workers" dig, forage, and build nests, and are the smallest animals. "Infrequent workers" work less than half as much as frequent workers. Both worker castes carry young about when alarmed. "Non-workers" are the largest animals. The males in this class may mate with the reproductive female and may also assist in the care of the young. The young beg food from all colony members, except the breeding female. Individuals enjoy a long lifespan, and generations overlap. They cooperate in brood care and possibly pass through age-specific roles (polyethism).

All these features parallel those found in eusocial insects, and the naked mole-rat has been recognized as the only eusocial vertebrate (Jarvis, 1981). Naked mole-rats resemble termites more than hymenopterans: they are diploid, male and females form the worker castes, the young contribute to the colony labor, some workers reproduce if the breeding female is removed, and the young obtain food by coprophagy (Jarvis, 1981). A similar social system exists in the Damaraland mole-rat, *Cryptomys damarensis* (Bennett and Jarvis, 1988).

The breeding female needs physical contact to suppress reproduction in colony members. It is possible that her urine carries a regulating pheromone. After urinating at the communal toilet areas, mole-rats scratch their body with their hind feet. This way they may distribute through the colony a pheromone from the breeding female. However, evidence for chemical cues in pheromonal colony regulation of naked mole-rats remains elusive. While soiled bedding and litter do not suppress reproduction in males and females, removal of non-reproducting females from the colony restores ovarian cyclicity. In males, removal from the colony raises urinary testosterone and plasma luteinizing hormone (LH) levels (Faulkes and Abbot, 1993). Even when together with high-ranking non-breeding colony members, these isolated females increased their progesterone levels and their sexual and aggressive behavior. Reproductive suppression probably requires direct contact with the breeding female (Smith *et al.*, 1997). This suggests that primer pheromones may not suppress reproduction, but we do not know the active cues at this time.

8.4.1 Growth and development

Neonatal female mice are *retarded* in their growth if urine from virgin adult female mice is applied to their nostrils, while urine from pseudopregnant females *accelerates* growth (Cowley and Wise, 1972). Both sexes of prairie voles (*Microtus ochrogaster*) arrest their growth when held together as littermates or exposed to air from littermates (Fig. 8.1; Batzli *et al.*, 1977).

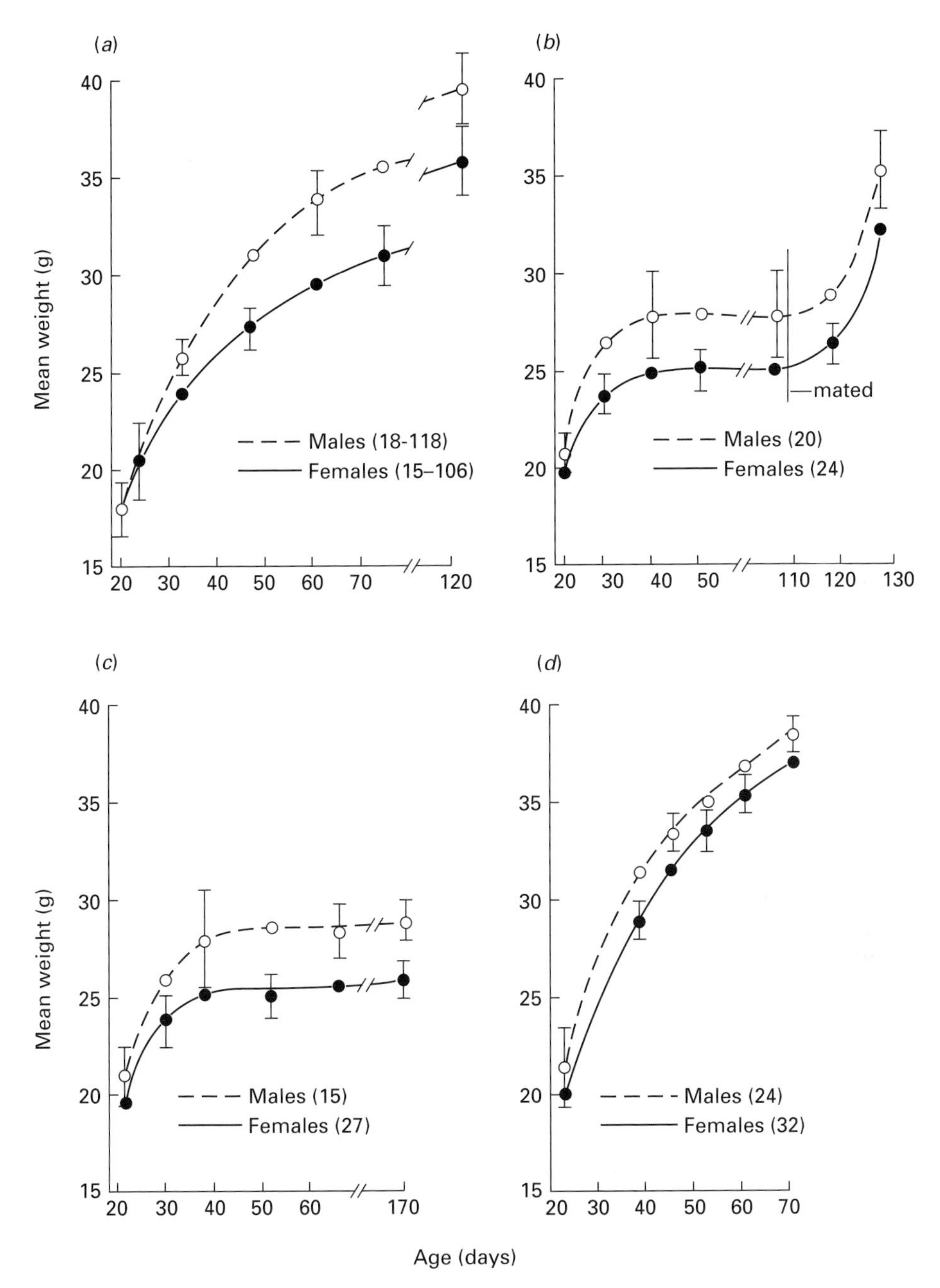

FIGURE 8.1 Priming effect of group members on growth in prairie voles, *Microtus ochrogaster*. (*a*) Normal growth curves for males and females when animals lived in individual cages from age 21 days. (*b*) Growth curves of littermates when kept together, then mated at age 12 weeks. (*c*) Littermates held separately but with air circulation among the cages. (*d*) Growth curves of voles living separately, with no air circulation among cages. (From Batzli *et al.*, 1977.)

OH OH O O

5,5-Dimethyl-2-ethyltetrahydrofuran-2-ol 6-Hydroxy-6-methyl-3-heptanone

FIGURE 8.2 The active compounds in the urine of dominant male mice.

8.4.2 Puberty Acceleration

Males

Adult females or their odor slightly accelerate the puberty of young male mice while the presence of a male inhibits puberty (Vandenbergh, 1971).

Females

A male odor, contained in urine and linked to the presence of testosterone, accelerates puberty in female house mice. Vaginal opening and first estrus take place about 15 days earlier (Vandenbergh, 1969; Lombardi *et al.*, 1976). This pheromone is effective at very low doses (Drickamer, 1982, 1984a) and is detected primarily by the vomeronasal organ (Lomas and Keverne, 1982). Urine from dominant males is effective while that of subordinates is not (Vandenbergh, 1969). Very early on, the pheromone was found to have low volatility (Vandenbergh *et al.*, 1975, 1976). Researchers invoked a protein–pheromone complex that has to be contacted. A male urinary protein had been shown to accelerate puberty in females (Mucignat-Caretta *et al.*, 1995). We now know that the active volatile compound, a lactol, and its hydroxyketo tautomer are complexed to a protein in male mouse urine (Novotny *et al.*, 1999). The lactol (a cyclic vinyl ether) is 5,5-dimethyl-2-ethyltetrahydrofuran-2-ol. It exists in equilibrium with its open form, 6-hydroxy-6-methyl-3-heptanone, a hydroxyketone (Fig. 8.2). Both are implicated in the pheromonal effect of enlarging the uterus in young mice. The lactol has a strong affinity for urinary proteins. Complexing with proteins protects these labile volatiles and provides a slow-release mechanism (Novotny *et al.*, 1999).

The presence of pregnant or lactating females can reduce the effect of the male pheromone (Lombardi and Vandenbergh, 1977). Wild house mice show the same effect (Drickamer, 1979).

Single-locus genetic differences between males can change the urine signal. Maturing female mice attained their first estrus sooner (at 29.9 days) when exposed to urine from males that were +/+ at the *t*-locus than with that from

$+/t$ males (32.6 days). Control mice (treated with water) had their first estrus at 34.7 days (Drickamer and Lenington, 1987).

Puberty acceleration occurs in a number of rodent species. In the Djungarian hamster, *Rhodopus sungorus*, an adult male kept with young females accelerates their uterine and ovarian development. The effect is androgen dependent, as castration eliminates the effect and exogenous testosterone reinstates it. Urine is effective, while ventral sebum is not. There are neural effects of exposure to adult males: an area in the stria terminalis (lateral division of bed nucleus) is smaller in male-exposed females than in females with spontaneous onset of puberty (Reasner and Johnston, 1988). Soon after sexual maturation, at the age of 50 days, females of the Levant vole, *Microtus guentheri*, respond to male urine by increased uterus weight (Benjamini, 1987). Puberty acceleration by a male urine factor in female meadow voles, *Microtus pennsylvaticus*, manifests itself as earlier vaginal perforation, heavier uterus and adrenals, and ovaries with more and larger Graafian follicles. The pheromone responsible for these effects appears to originate at the level of the kidneys rather than the sexual accessory organs (Clulow and Baddaloo, 1987).

Male odor accelerates puberty in the domestic pig. The two steroids 5α-androstenone (5α-androst-16-en-3-one) and 3α-androstenol (5α-androst-16-en-3-ol) play a role in this effect (Booth, 1984).

Puberty acceleration has also been found in a marsupial. The odor left behind in a cage by a male can accelerate the time of first estrus in a female gray, short-tailed opossum, *Monodelphis domestica*, from 161 to 126 days of age (Stonerook and Harder, 1989).

Female odors can also accelerate puberty in young female mice. Urine from singly caged estrous females (Drickamer, 1982) or from pregnant or lactating mice (Drickamer, 1979; Drickamer and Hoover, 1979; Drickamer, 1984b) accelerates puberty in young females and also facilitates estrus in adult mice.

Other factors

Some effects depend on more than the urine stimulus. Reproductively naive females of the prairie vole, *M. ochrogaster*, do not cycle spontaneously. They need a male for stimulation. If male urine is applied to the nares, the uterus weight increases threefold within 24 hours. However, urine alone is not sufficient to induce behavioral estrus as well. This requires pairing with an unfamiliar male. Even though male urine and bedding soiled by males can induce behavioral estrus in some females, most require actual contact with a male (Carter *et al.*, 1980, 1986).

Diet may counteract or substitute for chemical cues that influence sexual maturation. Male California voles, *Microtus californicus*, for instance, are inhibited in their sexual development by odors from their mothers or unrelated females. This effect is eliminated if the regular diet is enriched with fresh lettuce. In young females, uterus growth is stimulated by the odor of males but, in the absence of males, fresh lettuce added to the diet can accomplish the same effect (Rissman and Johnston, 1986). This can be seen as another case of *adaptive redundancy* of cues for an important function.

In primates, effects can be complex. Female cotton-top tamarins, *Saguinus o. oedipus*, require three conditions to start sexual behavior and ovarian cyclicity. These are release from the suppression by their mother, release from suppression by other family members, and direct contact with, or stimulation by, an unfamiliar male (Widowski *et al.*, 1990).

8.4.3 Puberty Delay

Males

Exposing young male prairie deer mice, *Peromyscus maniculatus*, to soiled bedding from adult male conspecifics retards the growth of their testes and seminal vesicles. Male, but not female, urine applied to the nose has the same effect. Removal of the olfactory lobes at the age of 3 weeks blocks this effect (Lawton, 1979). The reproductive development of male California voles, *M. californicus*, is suppressed by chemical cues from the mother (Rissman *et al.*, 1984).

In rabbits, *Oryctolagus cuniculus*, urine of an adult female delays growth and sexual development in unrelated young males, while mother's urine also delays body growth, but it accelerates sexual development in her own sons. This can be seen as females manipulating the reproductive success of nearby conspecifics (Bell, 1986).

Females

Puberty in female mice can be delayed by an odor from adult females that live in groups (Drickamer, 1977). This effect is mediated by the vomeronasal organ (Reynolds and Keverne, 1979). Endocrinologically, the adrenals, and not the ovaries, are required for the pheromone to occur (Drickamer and McIntosh, 1980).

Puberty delay has been documented for feral mice by a clever experiment. Females in populations delimited by a freeway cloverleaf produced urine that delayed puberty in laboratory mice only if their population was very dense (Massey and Vandenbergh, 1980). In another experiment, foreign females were added to a population in a highway cloverleaf. The urine from females in this population delayed puberty in juvenile laboratory females only after the introduction of the additional females. Ecologically, the urine signal in denser populations reduces population growth by extending the generation time. Even though the added females emigrated or died off soon, they stayed long enough to increase social interactions. This, in turn, presumably led to the production of puberty-delaying pheromone in the females' urine. The pheromone was induced rapidly after the population increased, and declined in potency after the population stabilized (Coppola and Vandenbergh, 1987). In addition to timing of puberty, other life-history effects, such as altered litter size or the size of the young, are expected (Coppola, 1986). Crowded female mice do not produce the puberty-delaying pheromone if their vomeronasal organ is experimentally removed (Lepri *et al.*, 1985).

Puberty delay also occurs in other species. Female white-footed mice, *Peromyscus leucopus*, mature later and weigh less if exposed to urine and feces from conspecifics of both sexes (Rogers and Beauchamp, 1976). Bank voles, *Clethrionomys glareolus*, have lower numbers of ovarian follicles and smaller uteri and ovaries if they mature in the presence of several other females. The European pine vole, *Pitymys subterraneus*, lives in sparse populations and urine from even a single female delays puberty in females (Marchlewska-Koj and Kruczek, 1986). Two factors keep subordinate female prairie voles, *M. ochrogaster*, prepubescent: (a) the odor of familiar males, usually father and male siblings; and (b) the inhibitory effect of females such as the mother and female siblings on the young females' response to the odor of unfamiliar males. (The odor of unfamiliar males would help the young female to mature) (Carter and Roberts, 1997).

Female mice have volatile components in their urine that are depressed after adrenalectomy. Six components have been detected, of which three (2-heptanone, *trans*-5-hepten-2-one, and *trans*-4-hepten-2-one) have no apparent effects either as a group or when added to the other three components. The active three components are *n*-pentyl acetate, *cis*-2-penten-1-yl acetate, and 2,5-dimethylpyrazine (Fig. 8.3). If 2,5-dimethylpyrazine is painted daily on the external nares of young female mice from day 21 on, the time of the first vaginal estrus is delayed. These three compounds act in redundant fashion: the two acetate esters delay vaginal estrus by 1.5 days, on average; 2,5-dimethylpyrazine alone delays it by 2.4 days, and a mixture of all three delays it by 1.7 days. (Novotny *et al.*, 1986a).

Stimuli other than chemical cues can play an important role in puberty delay in females. In the California mouse, *Peromyscus californicus*, which is monogamous, physical contact with the mother, and not a urinary chemosignal *per se*, is necessary for delay of puberty of females (Gubernick and Nordby, 1992).

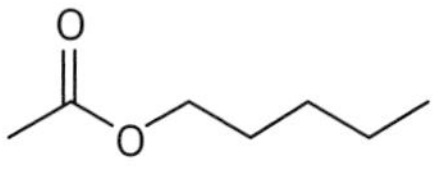

n-Pentyl acetate

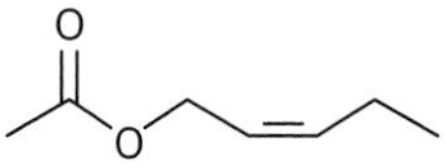

(*Z*)-2-Penten-1-yl acetate

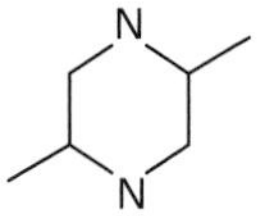

2,5-Dimethylpyrazine

FIGURE 8.3 Puberty-delaying compounds in female mouse urine.

8.4.4 Priming effects in adults: fine-tuning of breeding cycles

Reproductive inhibition in females

In grouped female mice ovulation and estrus are suppressed (Champlin, 1971) and pseudopregnancies occur (van der Lee and Boot, 1955). The odor of crowded female mice (eight per cage) inhibited ovulation in other females. The hormonal control of this effect has been difficult to explain. Early studies indicated that ovariectomized mice were not able to inhibit other females but in other studies, ovariectomized females still inhibited estrus in other females (Marchlewska-Koj, 1990).

Voided or bladder urine from *single* wild house mice also inhibited estrus cycles in females. These two types of urine reduced the normal rate of 2.4 cycles per female in 18 days to only 1.66 and 2.00, respectively. Urine from spayed females was not effective (Pandey and Pandey, 1986).

Synchronization of cycles in females

Urine odor of adult males induces and synchronizes estrus cycles in adult female mice (Whitten, 1958). The cycles grow shorter and more regular when a male is nearby. A male shortens cycles of 7–8 days to 4–5 days. Exposure to male urine for 48 hours has the same effect. This phenomenon is known as the *Whitten effect*. The estrus-inducing activity of male urine is retained even after storage at −4 °C or freeze-drying (Gangrade and Dominic, 1986). This has ecological implications. Marsden and Bronson (1964) synchronized estrus cycles by applying male urine to the nasal region of females. The estrus-stimulating pheromone has multiple sources. It has been found in salivary and preputial glands (Marchlewska-Koj *et al.*, 1990).

Nerve transection experiments have shown that normal estrus cyclicity and behavioral estrus in mice relies on sensory input through the main olfactory bulbs and does not require the accessory olfactory system (Rajendren and Dominic, 1986).

The estrus-synchronizing effect of adult male urine described for female rats has puzzled investigators. Recent experiments and more sophisticated statistical analysis have indicated that apparent synchronization may be actually chance (Schank, 2001).

The wallowing, pawing, and urine-soaking of the body by some deer species is thought to synchronize estrus in the females. If this is so, these behaviors should be more prevalent in alpine and high-latitude species than in tropical species with a long or permanent breeding season.

Induced ovulation

The prairie vole, *M. ochrogaster*, is an induced ovulator. Adult male urine, placed on the nose of females, stimulates weight increase of the uterus. The vomeronasal organ plays a role in this effect (Carter *et al.*, 1980).

Rats ovulate when exposed to soiled bedding from males. The stimulus is nonvolatile, as a wire screen can eliminate this response. Vomeronasal organ occlusion reduces the response, implicating this pathway for priming pheromone reception in this species (Johns *et al.*, 1978). Table 8.2 summarizes the role of the vomeronasal in priming effects in rodents.

In domestic sheep, ewes will ovulate in response to reencountering a ram after 2 or more weeks of separation from adult males. This is true for prepubertal, lactating, and seasonally anovulatory ewes. The primary signal from the ram is chemical in nature, androgen dependent, and resides in "suint," the mixture of secretions from the sebaceous and apocrine glands of the skin and extractable from wool fat (Knight and Lynch, 1980). Urine or preorbital secretions have little effect. The main olfactory system, and not the vomeronasal system, primarily mediates this effect. However, other sensory modalities also modulate the response, as the sexual behavior of the ram intensifies the effect. Neural connections between the main olfactory tract and the anterior hypothalamus play a crucial role. More frequent pulses of luteinizing hormone (LH) are the first measurable event, occurring within minutes of exposure to the ram. This high pulse frequency stimulates follicle growth and estradiol secretion by the ovaries.

Table 8.2 The role of the vomeronasal organ in priming in Mammals

Species	Target sex	Context	Stimulus	Effect	Reference
Rat *Rattus norvegicus*	F	Estrus cycles	M urine	Shortening of cycles	Sanchez-Criado, 1982
	F	Ovulation	M urine	Induced ovulation	Johns *et al.*, 1978
Prairie vole *Microtus ochrogaster*	F	Reproductive activation	M odors	Maintain uterus and ovary weight	Lepri and Wysocki, 1987
	F	Ovulation	M urine	Uterus weight increase	Carter *et al.*, 1980
	F	LHRH release	M urine	Presence of LHRH containing looping fibers in AOB	Reger *et al.*, 1987
Mouse *Mus domesticus*	M	Exposure to F	F urine	Surge of LH and testosterone	Wysocki *et al.*, 1983
	F	Pregnancy block	Strange M urine	Prevention of implantation	Bellringer *et al.*, 1980
	F	Sexual maturation	Dominant M urine	Earlier vaginal opening and first estrus	Lomas and Keverne, 1982

M, male; F, female; LH, luteinizing hormone; LHRH, luteinizing hormone-releasing hormone; AOB, accessory olfactory bulb.

Estradiol subsequently builds up in the blood and first (within 2 to 12 hours) reduces the levels of follicle-stimulating hormone (FSH) and the amplitude of LH pulses, then (within 12 to 48 hours) causes preovulatory surges of LH and FSH. The former promotes ovulation and development of a corpus luteum (reviewed in Martin *et al.*, 1986). Two compounds have been indicated in the effect of the odor of ram's fleece on LH secretion in anestrous ewes. These are 1,2-hexadecanediol and 1,2-octadecanediol. In Merino sheep at least, maximum stimulation of ovulation requires full exposure to a ram, such as "fenceline contact" in pastures. Olfactory cues from the ram's wool, presented in a facemask for the ewe, are ineffective by themselves; visual and tactile stimuli are also important. The Merino breed does not rely as much on olfactory cues as other breeds of sheep (Pearce and Oldham, 1988). The effect is not necessarily species specific: hair extract from male goats stimulates LH release in ewes. For this effect, the accessory olfactory system is not necessary (Signoret *et al.*, 1989).

In domestic cattle, non-estrous cows may become attractive to bulls when in the company of estrous cows. It has been suggested that estrous cows influence their penmates via pheromone(s) (Hradecky, 1989) but other mechanisms may be involved.

An extract of secretion from the sebaceous glands on head and neck of sexually active feral billygoats increases the number of does ovulating. The extract was placed on cotton wool and worn in facemasks. 4-Ethyloctanoic acid and 4-methyloctanoic acid, responsible for the "goaty" odor, were not active but both the free fatty acid and lipid-free non-acid fractions were. The 4-ethyl branched fatty acids are present in the active fraction (Birch *et al.*, 1989).

8.4.5 Stimulation of reproductive activity in males

Male prairie deer mice (*Peromyscus maniculatus bairdii*) in very dense (i.e. asymptotic) populations cease to reproduce. Their testes remain abdominal and are small (20–25% of normal). The testes can assume normal size if such males have contact with reproductively proven females for 30 days. They also recover reproductively if exposed to female urine for 30 days, but contact with females is more effective (Creigh and Terman, 1988).

8.4.6 Pregnancy Block

Inseminated female laboratory mice may fail to implant if they are exposed within 24 hours of coitus to a male (or his odor) that is different from the stud male. The pheromone is contained in urine of intact adult males. It lowers prolactin production by the pituitary of the female, and corpora lutea fail to develop properly, so that pregnancy is not supported endocrinologically. Females with pregnancy thus blocked lose the embryo at the blastula stage and return to estrus within 4–5 days after the original mating (Bruce, 1959). This preimplantation pregnancy block (termed the *Bruce effect*) occurs in at least 12 rodent species such as wild house mice, *Mus musculus* (Chipman and Fox, 1966), deer mice, *P. maniculatus* (Eleftheriou *et al.*, 1962), and seven species of voles, including field voles, *Microtus agrestis* (Clulow and Clarke, 1968), meadow voles, *M. pennsylvaticus* (Clulow and Langford, 1971), and prairie voles, *M. ochrogaster* (Stehn and Richmond, 1975). Pregnancy can also be blocked if the female is exposed to a strange male *after* implantation. This occurs in *P. maniculatus*, *M. pennsylvanicus* and particularly (up to 15 days of gestation) in *M. ochrogaster* (Stehn and Richmond, 1975; Kenney *et al.*, 1977).

In addition to males, females can also induce pregnancy block. An example is the chemical cue from older, probably dominant, female white-footed mice, *P. leucopus* (Haigh, 1987).

Male mouse urine applied directly to the external nares of the female produces pregnancy block (Dominic, 1964). In mice, the vomeronasal mediates this block (Bellringer *et al.*, 1980). The pregnancy-block pheromone of the male may be a product of androgen metabolism, and not one of androgen-dependent tissue. The steroidal anti-androgen cyproterone suppresses the ability of males to induce pregnancy block if given for a long time (42 days). If injected for only 14 days, there is no effect (Rajendren and Dominic, 1988).

The impregnated female mouse retains an *olfactory memory* of the stud male. This memory depends on cervico-vaginal stimulation at mating (Keverne and de la Riva, 1982). For the effect to occur, the female has to be exposed to the stud male's odor for 4–6 hours (Rosser and Keverne, 1985). The memory for the stud male lasts for about 30 days (Kaba *et al.*, 1992). Memory formation is accompanied by synaptic changes in the accessory olfactory bulb (Kaba *et al.*, 1992). The major urinary proteins, currently subjected to intense study, may provide individual information about the male, in addition to the protein's pheromone-binding role.

If the stud male remains near the pregnant female during encounters with a strange male, implantation failure will be prevented. Also, pregnancy block does not occur if the original stud male is reintroduced after 24 hours of separation (Parkes and Bruce, 1961). The active chemical stimulus from the stud male is probably not volatile and may act synergistically with bodily contact between the mates (Thomas and Dominic, 1987a,b). Furthermore, exposure to the original stud male *after* an alien male has caused pregnancy block results in pseudopregnancy. In prairie voles, *M. ochrogaster*, presence of the original mate also prevents strange male odor from terminating the female's pregnancy. In the wild, the pair would share a home range and a nest, but intrusion by strange males could be common, especially when a population is dense. Avoiding pregnancy block is adaptive, considering the short lifespan (45–61 days) of a prairie vole (Hofmann *et al.*, 1987).

The females play an active role in exposing themselves to certain male odors. In wild stock house mice, recently inseminated females avoid the odor of a strange male and prefer that of the stud male, but only up to 8 of the 18 days of gestation. Implantation occurs 4–5 days after insemination. By their behavior, the females protect themselves from pregnancy blockage (Drickamer, 1989b). Non-pregnant female mice do not avoid the odor of a strange male (C. J. Wysocki, unpublished data). The pregnancy block has not been found in rabbits, *O. cuniculus* (Bell, 1986).

8.4.7 Proximal mechanisms: hormonal responses to odors

All the stimulating primer effects in mammals analyzed so far share a common hormonal pathway: the first measurable event is stimulation of LH-releasing hormone (LHRH) release. This, in turn, stimulates LH levels and leads to increases of estrogens in females and testosterone in males. In females, changes in size and function of uterus and ovaries follow, while males respond with the usual testosterone-dependent processes, including production of male pheromone(s).

The LHRH release appears to depend on the vomeronasal organ. Its extirpation in female prairie voles (*M. ochrogaster*) results in the absence of LHRH-containing looping fibers in the accessory olfactory bulb several months later (Reger *et al.*, 1987).

A single drop of male urine, applied on the upper lip of female prairie voles (*M. ochrogaster*) leads to changes of LHRH and norepinephrine concentrations in the olfactory bulb within 1 hour. Rapid increases in serum LH were also observed in these females (Dluzen *et al.*, 1981).

In mice, strange females or their urine increase the levels of plasma testosterone in males (Macrides *et al.*, 1975). Experienced male golden hamsters, on the other hand, do not depend on specific pheromonal odors for this testosterone surge induced by estrous females. Other cues, possibly learned odors from a female, perceived via the main olfactory system, appear to activate the neuroendocrine reflex that results in increased testosterone release (Johnston, 2001).

Anal gland secretion of male short-tailed voles, *M. agrestis*, stimulates plasma testosterone, body weight, and anal gland size in other males. The biological significance of this effect is unclear (Khan and Stoddart, 1986).

Goat pheromones

Most goats in the northern hemisphere breed in fall and winter. Decreasing day length stimulates ovarian cycles. In addition to photoperiod, male pheromones stimulate pulse frequency of LH in the blood plasma of females (Martin *et al.*, 1986). This not only enhances the seasonal onset of ovarian function, but also synchronizes breeding and subsequent lambing.

Male pheromones originate in glands of the parietal region. Scent-saturated hair from billy-goats stimulates LH secretion pulses in females within minutes. For a long time, practitioners have used this method to stimulate female goats that live without access to bucks. Follicular maturation and behavioral estrus require sustained high LH pulse frequency. Follicles in anestrous goats will mature if the animals are exposed to buck hair for 3 days (Claus *et al.*, 2001).

Active compounds have been found in the neutral fraction of buck hair extract (Over *et al.*, 1990). The carbonyl group and a hydroxyl group appear to be important for the biological activity (Claus *et al.*, 2001).

8.4.8 Inhibition of hormones

In a primate, *Microcebus murinus* (Prosimia) male odor depresses the level of plasma testosterone in other males. The lipid fraction (ether extract) of male urine has this effect, while the aqueous fraction does not (Perret and Schilling, 1987). Female tamarins, *Saguinus oedipus*, increase their anogenital scent marking when removed from their natal group or from the company of other cycling females. Their urinary estrone and estradiol also increase. The cyclical pattern of estrone excretion also becomes more pronounced. Males, removed from their natal group or other males, increase sexual behavior, but not scent marking (French *et al.*, 1984).

8.4.9 Adrenocortical activator

Male mice crowded, or "stressed" otherwise, develop enlarged adrenals. For this response to occur no direct contact is needed; odors from crowded male mice suffice (Ropartz, 1968; Wuensch, 1982).

8.4.10 Ecological significance of priming pheromones

The priming pheromones in mice – and presumably other mammals – are crucial signals in feedback mechanisms that synchronize the sexual readiness of the sexes and adjust breeding behavior to ecological conditions and optimize reproduction within the constraints of the environment (Bronson, 1979). Sexual readiness can be adjusted by delaying or accelerating puberty, by modulating estrus cycles in adult females, and by stimulating testosterone production in adult males. Urine odors of adult males accelerate estrus in adult females and accelerate puberty in young females, while urine odors of adult females prime the testosterone in males and delay the puberty and estrus cycles of females, as described above. The Bronson model of priming in mice is shown in Fig. 8.4.

Pregnant and lactating females also produce a pheromone that accelerates puberty in females. Under favorable conditions, this factor is released throughout the year (Drickamer and Hoover, 1979; Drickamer, 1986).

Population dynamics are greatly affected by changes in generation time, especially in fast-breeding small animals. In the female house mouse, the onset of puberty can vary from 4 to 8 weeks of age, depending on whether a young animal is exposed to accelerating male pheromone or inhibitory female pheromone. Field studies have shown that the age of puberty can vary with population density: crowded females inhibit puberty in younger females. The signal seems to be mediated by the vomeronasal organ (Vandenbergh, 1987).

In outdoor enclosures treated with urine and soiled bedding from male mice, mouse populations grew to larger numbers than in water-treated control enclosures. Populations in enclosures treated with urine and soiled bedding from group-caged females, however, grew less than the water controls. Females

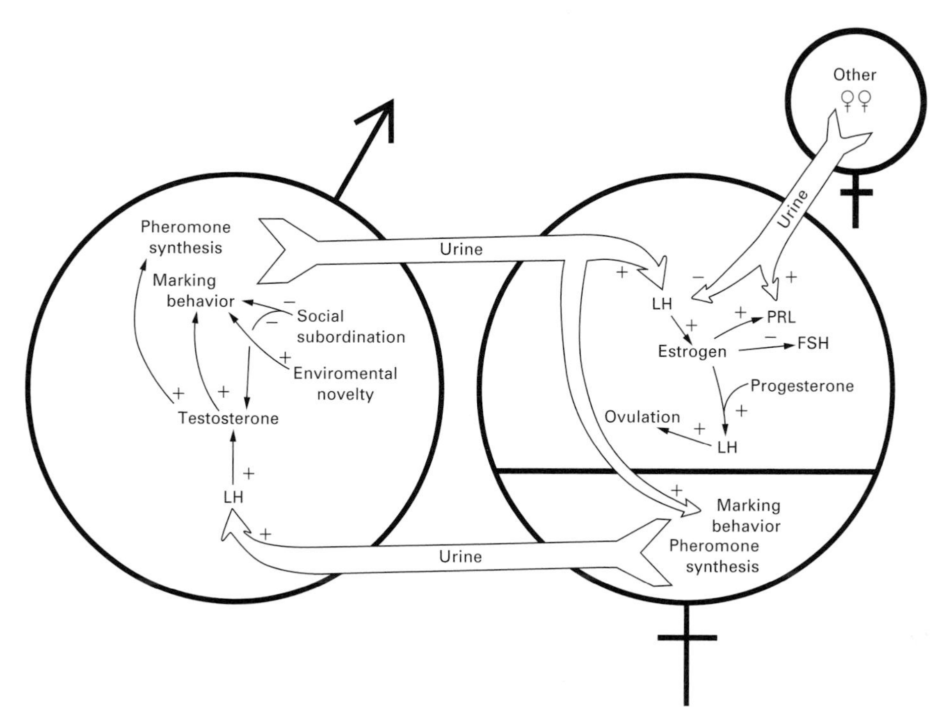

FIGURE 8.4 The Bronson model of priming pheromone actions in rodents. Chemical signaling between males and females constitutes a feedback mechanism that results in accelerated maturation and reproduction. This, in turn, permits the mice to adjust their reproduction and population size quickly to respond to environmental conditions such as sudden food abundance at harvest time. FSH, follicle-stimulating hormone; LH, luteinizing hormone; PRL, prolactin. Stimulation and inhibition are marked by + and −, respectively. (From Bronson and Coquelin, 1980.)

exposed to male cues were in "higher states of reproductive condition," experienced more pregnancies, and matured earlier sexually (Drickamer and Mikesic, 1990).

Small mammals undergo population cycles that each last several years. Mathematical modeling of the influence of several life-history traits showed that, among several factors, changes in age at maturity affected the population cycles most (Oli and Dobson, 1999). These authors suggested that adult females might release puberty-delaying pheromone when the environment is perceived as unfavorable or risky. Signaling poor environmental quality, the pheromone inhibits the maturation of young females. A lower population density ensues, which, in turn, decreases competition for resources (Oli and Dobson, 1999).

Priming pheromones affect the life history of a species. Coppola (1986) pointed out that the recipients benefit from the chemical signals while the signals' producers are little affected. The puberty delay pheromone, in particular, does not seem to be part of a mutually evolved signal system, but rather "unwitting release of metabolic correlates of physiological state." Coppola (1986) proposed to look for pheromonal priming effects on litter size and size of young. Inclusive fitness will depend on such life-history effects of primer pheromones as well as behaviors that regulate exposure to these pheromones, as described by Drickamer (1989c).

Drickamer (1986) suggested that natural selection has had little effect on the chemical *signals* that modulate puberty. Instead, natural selection has shaped the odor sensitivity of the young females and the donors' and recipients' behaviors of depositing, approaching, or avoiding the chemical cues.

8.4.11 Evolutionary significance of pregnancy block

Traditional explanations of the pregnancy block have included insurance of parental care, male competition, protection against infanticide, and promotion of outbreeding. Keverne and Rosser (1986) called such explanations a "blinkered view of evolution."

Instead, the pregnancy block can be seen as one of several possible consequences of the fact that primer pheromones lower the level of prolactin. When the mated female is exposed to a strange male, the male odor is perceived via the vomeronasal organ. Mediated by the accessory olfactory bulb, tuberoinfundibular dopaminergic neurons of the hypothalamus are activated. Dopamine increases and this starts the final common pathway for several pheromonal effects. Release of β-endorphin is inhibited and prolactin levels fall. In the mouse, prolactin is luteotrophic. Lowered prolactin leads to lowered progesterone release from the corpus luteum. This, in turn, releases the hypothalamic–pituitary axis from the inhibitory action of progesterone. LH production is disinhibited and the increase in its levels causes estrogen levels to rise. In an anovulatory female, cyclicity is restored. In short, pregnancy block within 3 days after mating results from lowering of prolactin. Implantation is prevented, the blastocyst is lost, and estrus can occur again (Keverne and Rosser, 1986).

The dopamine agonist bromocriptine produces the male-odor induced effects, while they are prevented by pimozide, which blocks dopaminergic transmission (Kaba *et al.*, 1992).

According to this scheme, the major difference between male and female odor effects is that the hypothalamic dopamine system is inhibited by female odors instead of activated, as described for male odors. Therefore, the converse reproductive effects occur even though they are controlled by the same hormonal system (Keverne and Rosser 1986). All that is needed to produce the various effects is a "maleness" and a "femaleness" pheromone, or simply "odor."

The inverse relationship between dopamine and the β-endorphin neurons is of fundamental evolutionary significance. It is the final common pathway for inhibition of estrus during pregnancy and lactation, "nature's own means of contraception."

How does the female recognize the stud male as familiar? Keverne and Rosser (1986) suggest that the vomeronasal organ is involved, but not the main olfactory system. There are noradrenergic projections from the brain to both the olfactory bulb and the accessory olfactory bulb. If these fibers are lesioned, the accessory bulb is depleted of norepinephrine. The females still suffer the block but fail to recognize the stud male. In this experimental case, the stud male's own pheromones block the pregnancy induced by him. The noradrenergic pathways must be activated by mating. Coitus activates the brain's noradrenergic neurons and those fibers that terminate in the olfactory bulb. The noradrenergic neural system "imprints" recognition. Natural selection has favored recognition of familiar males. Olfactory block to pregnancy circumvents this recognition. This effect is unlikely to occur in the wild, and some consider it a laboratory artifact.

The *memory* that is necessary for the olfactory block to occur involves the synapses between granule cells and mitral cells in the accessory olfactory bulb. During the critical period immediately following copulation, the female is exposed to male pheromone. At the same time, noradrenergic fibers are active and *reduce* the inhibition that granule cells exert on mitral cells. This is considered to improve the signal-to-noise ratio. Such prolonged excitation of the mitral cells, in turn, activates a population of granule cells for the 4 hours when male pheromone and internal norepinephrine impinge on the female's accessory olfactory bulb. This forms an olfactory memory.

In summary, the synapses between the dendrites of the mitral and granule cells of the accessory olfactory bulb appear to be critical for forming the olfactory memory necessary for the pregnancy block. The association of norepinephrine and pheromonally induced activity during the critical period following mating produces a lasting change in the synapse. Once mating and the associated high level of norepinephrine are over, the affected subset of mitral cells experiences more inhibition by their associated granule cells. From then on, when the female encounters the familiar odor of the stud male, the pattern of activated mitral cells will match that of the population of mitral cells whose synapses had been modified and are now more inhibited by their granule cells. Hence the odor will fail to have the effect of blocking pregnancy. Different male odors, however, will not match this subset of mitral cells. They will act on (unmodified) mitral cells that lack the increased feedback inhibition. Because of this, the neuroendocrine mechanisms that lead to pregnancy block will be set in motion (Brennan *et al.*, 1990). There are many different forms of memory. In the pregnancy block process, the changes that constitute memory occur at the most economical place in the nervous system, namely the *first neural relay* in the sensory system (Brennan *et al.*, 1990).

8.5 Priming pheromones in humans?

8.5.1 Puberty

Possible pheromone effects in humans never cease to fascinate us. Are there puberty-accelerating pheromones in humans? Altered pheromonal ambience for girls over the past 150 years has been suggested as an explanation for the decline of the age at menarche in Western industrialized countries (Bierich, 1981). Applying the animal model to humans, close contact with adult females should delay puberty while the presence of an adult male should accelerate it. The socio-economic changes over the last 150 years – so the argument goes – resulted for a girl in less contact with the mother because she worked increasingly more outside the home, and more contact with the father because his work hours declined and he spent increasingly more time at home. Contact with older sisters also declined because families became smaller, and older sisters became less available because they tended to be at school or work. Presence or absence of mother, father, or older siblings of either sex, as well as size of living space (close exposure to more concentrated putative pheromones), should then predict age at menarche in contemporary families (Burger and Gochfeld, 1985). It should be kept in mind that, even if contact with adults indeed shifts puberty in the assumed directions, stimuli in other sensory channels (or combinations) may be responsible. Also, environmental changes, such as nutrition, altered circadian activity rhythms or increased illumination, as in large cities, may affect onset of puberty.

8.5.2 Estrus synchronization

Ever since McClintock (1971) described synchronized estrus cycles in 135 women who roomed together in a women's college dormitory, the search for possible olfactory cues, as occur in other mammals, has continued. The longer the women in McClintock's study spent time together during the academic year (October through March), the closer the times of onset of their menstrual cycles became. In addition, women who had more contact with men had shorter cycles (McClintock, 1971), reminiscent of the Whitten effect.

In a second study at a coeducational college, women who lived without roommates in bedrooms of residence halls or apartments developed menstrual synchrony with "close friends" (Graham and McGrew, 1980). These authors did not find a correlation between cycle length and amount and type of interaction with men.

A third study of 85 20-year-old women found an effect on menstrual synchrony of living together with one, two, or three other women, and of physical activity (Quadagno *et al.*, 1981).

These correlation studies stimulated experiments to identify the active chemical cues. Russell *et al.* (1980) rubbed underarm perspiration from a single woman onto the upper lip of five women aged 19–39 years. After 5 months, the odor-exposed women differed from one another in their onset of the menstrual cycle by 3.4 days, on average, compared with 9.2 days in the control group. Before the experiment, the mean differences had been 9.3 and 8.0 days, respectively. The volunteers were aware of the purpose of the experiment.

The first systematic, double-blind experiments on the influence of male and female odors on women's menstrual cycles were reported by Cutler *et al.*, (1986). and Preti *et al.* (1986), respectively. Neither their subjects nor the administering technician were aware of the true purpose of the study.

Male axillary odor, thought to arise from volatile acids and volatile steroids, was collected on pads worn in the armpit for 6–9 hours. Alcohol extract of these samples was applied to the upper lip of each subject and left there for at least 6 hours. The cycles of women receiving the axillary extract three times per week for an average of 13.5 weeks became less variable and the number of aberrantly long cycles was reduced.

The axillary odor of women, collected over 10 different 3-day sections of the menstrual cycle and applied to the upper lip of recipients, had a different effect. The cycles of the receiving women became more synchronized with the cycle of the donor female (Preti *et al.*, 1986), measured as difference in menses onset in terms of days. Wilson (1987) and Weller and Weller (1993) have criticized these experiments but still they remain pioneering studies in an extremely difficult field.

In a later experiment, two pheromones from axillary odor of young women influenced the estrus cycle of other women (Stern and McClintock, 1998). Armpit secretion that appears odorless to humans was applied daily between the upper lip and nose of healthy young women and had two opposite effects depending on the menstrual phase of the odor donors. Secretion from nine donor women, collected on pads in their armpits during the follicular phase of their cycles (immediately before ovulation), shortened the menstrual cycles in 20 recipients aged 20–35 years. Specifically, the follicular phase of the treated women became shorter. In other words, ovulation was hastened. By contrast, odor from the ovulatory phase of the donors lengthened the follicular phase of the recipients and, therefore, the entire menstrual cycle. Ovulation, and by implication the preovulatory surge of LH, was delayed. In short, interpersonal chemicals can manipulate the timing of ovulation (Stern and McClintock, 1998).

Chemically, androstenol and dehydroepiandrosterone sulfate, found in male axillary secretion (Brooksbank *et al.*, 1974), occur in active samples. Aliphatic acids with 2 to 18 carbon members may also contribute to the axillary odor (Preti *et al.*, 1987).

9

Development of intra- and interspecific chemical communication

The infant, the adult, the aged person, each has is own kind of smell, and . . . it might be possible, within certain limits, to discover the age of a person by his odor. In both sexes puberty, adolescence, early manhood and womanhood are marked by a gradual development of the adult odor of skin and excreta, in general harmony with the secondary sexual developments of hair and pigment.

H. H. ELLIS *Sexual Selection in Man*

The nature–nurture problem revisited: in most vertebrates, early experience of certain odors, interwoven with genetically anchored developmental processes, produces lasting, often irreversible odor recognition, preferences, or avoidance. Such behavioral development often occurs during more or less defined critical windows in time. The development of responses to odors often precedes that of odor production. Neonates already orient towards odors, while many pheromones are not produced until adulthood. Even before hatching or birth, the journey of chemical communication starts in the egg or the uterus. Knowing how chemical communication and chemosensory responses to food or danger develop is essential in areas such as animal husbandry or human behavior.

9.1 Fish

Young fish are attracted to conspecific odors without prior experience. Examples are European eels, *Anguilla anguilla* (Pesaro *et al.*, 1981), Arctic charr, *Salvelinus alpinus* (Olsén, 1987, 1990), and Atlantic salmon (*Salmo salar*). Kin recognition depends on early experience, as in Arctic charr. Fish reared with siblings later preferred water scented by unfamiliar siblings to water scented by non-siblings. By contrast, fish reared singly did not discriminate between water from siblings and water from non-siblings, even though they preferred sibling water to unscented tap water. At 15 months of age, the isolated fish were reared with siblings for 50–62 days. After this, they still made no choice between sibling water and non-sibling water. Thus, sibling preferences appear to be established during the first 15 months of life (Winberg and Olsén, 1992).

9.2 Amphibia

Toad (*Bufo americanus*) tadpoles will recognize their siblings, but this depends on earlier experience. Here, the prehatching environment influences the post-hatching preferences (Waldman, 1981). Tadpoles were reared under three conditions: with siblings only, completely isolated from conspecifics, and sequentially exposed to both siblings and non-siblings. In subsequent choice tests, the first two groups preferred to associate with siblings. The third group did not associate with siblings unless they had been reared with siblings during early development. Tadpoles reared in isolation discriminated paternal, but not maternal, half-siblings from full siblings. In sum, sibling preference can develop without exposure to conspecifics, but early experience is important for normal development of sibling recognition. Table 6.1 (p. 129) contains other examples.

9.3 Reptiles

Numerous studies have shown that neonate or naive reptiles can already show complete species-specific chemical responses to conspecifics or rely on some form of social imprinted responses to food organism or predators, developed by interactions of genetic factors and experience, with each contributing.

9.3.1 Social odors

Neonate garter snakes, *Thamnophis sirtalis*, and brown snakes distinguish conspecific from heterospecific odors (Burghardt 1977, 1983). Newborn timber rattlesnakes, *Crotalus horridus*, are able to follow conspecific odor trails (Brown and MacLean, 1983). Neonate water snakes are attracted to conspecific odor (Scudder *et al.*, 1980) and neonate prairie rattlesnakes, *Crotalus viridis*, to lipoids from the epidermis of adult conspecifics (Graves *et al.*, 1987).

Neonate prairie rattlesnakes, *C. viridis*, show tongue flicking, mouth gaping, and face wiping frequently after birth. These patterns are virtually absent at 2–3 months of age. Face wiping and tongue flicking are directed towards the body of the mother. It has been suggested that some type of olfactory imprinting is served by this behavior.

Odor trails lead from the birthing rookeries of pregnant snakes to their ancestral winter dens. These trails probably help the neonates to find shelter (Graves *et al.*, 1987). Socially naive neonate prairie rattlesnakes were tested for odor preferences in a two-way choice apparatus. The snakes had a choice between clean substrate and substrate marked with exudates from conspecific neonates, conspecific adults, and the predatory bullsnake *Pituophis melanoleucus sayi*. The hatchlings spent more time exploring or resting on areas with conspecific odor and avoided heterospecific odors. They possibly use the conspecific odor to locate the communal hibernacula for overwintering (Scudder *et al.*, 1992).

9.3.2 Antipredator responses

The response to predator odor is little modifiable by experience as two examples illuminate. First, neonate pygmy, *Sistrurus miliarius*, and timber, *C. horridus*, rattlesnakes respond to a predator odor: They "bodybridge" (raise the midsection of their body) when exposed to the odor of ophiophagous colubrid snakes (Marchisin, 1980). Second, king snakes, *Lampropeltis getulus*, prey on pine snakes, *Pituophis melanoleucus*. Pine snake hatchlings kept on soiled bedding from king snakes still prefer pine snake odor over king snake odor despite being familiar with the latter. The predatory king snakes also avoid the king snake odor (Burger *et al.*, 1991).

9.3.3 Feeding behavior

As the feeding behavior of snakes develops, genetic as well as experiential factors come into play. Genetic predispositions have been found at several levels. At their first exposure to food and food odors, neonate snakes respond to odors of foods that are typical for their species. Aquatic forms such as *Thamnophis elegans aquaticus* respond more to fish extracts, and the midland brown snake *Storeria dekayi wrightorum* to extracts from worms and slugs. *Thamnophis r. radix*, which lives on a varied diet, responds to extracts from worms, leech, fish, salamanders, and frogs, while the western smooth green snake, *Opheodrys vernalis blanchardi*, responds to cricket extract, reflecting its diet of insects, spiders, and other arthropods (Burghardt, 1967). Naive snakes from different populations of even the same species can differ behaviorally: in California, neonate coastal *Thamnophis elegans* responds to slugs, while the inland neonates do not (Arnold, 1981a). Arnold also showed that these two populations represent different genotypes. Hybrids between the two populations varied more in their behavior than snakes of either (Arnold 1981b). Within each population, Arnold also found genotypic polymorphism. There is no prenatal food imprinting: the maternal diet from the gastrulation stage of the young onwards had no influence on food preferences in neonates of *Th. sirtalis*.

Table 9.1 The shaping of feeding behavior in garter snakes

Genetic factors

1. Neonate food-naive snakes prefer odors of their "natural" food: eastern plains GS respond to worm, leech, fish, and tadpole odors while western smooth green snakes respond to cricket odor (Burghardt, 1967); coastal neonate *Thamnophis elegans* respond to slugs but the inland form does not (Arnold, 1981a)
2. Population differences: there is geographic variation in genotypes between coast and inland forms of *Th. elegans* (Arnold 1981a)
3. Hybrids between coastal and inland snakes vary more in behavior (Arnold 1981b)
4. There is genotypic polymorphism within populations (Arnold 1981a)

Role of experience

1. No prenatal food imprinting: maternal diet from gastrulation stage on had no effect on preferences by neonate *Th. sirtalis* (Burghardt, 1971)
2. Early postnatal experience modifies feeding in *Th. sirtalis* (Fuchs and Burghardt, 1971; Arnold, 1978)
3. Effect of experience depends on prey type: occurring for fish but not for tadpoles (Arnold, 1978)
4. Injection of LiCl after an earthworm meal induces food aversion in *Th. sirtalis* (Burghardt *et al.*, 1973)

Experience modifies feeding behavior in garter snakes. Early postnatal experience can change feeding behavior in *T. sirtalis*. When two groups of neonates were fed fish (guppy) or redworm, they learnt to prefer the diet they were given (Fuchs and Burghardt 1971). The effect of experience depends on the diet: the snakes learn fish odors, but not those of tadpoles (Arnold 1978). In the same species of snake, injection of lithium chloride after an earthworm meal can produce a food aversion (Burghardt *et al.*, 1973). Table 9.1 summarizes the factors that determine formation of feeding behavior in garter snakes.

Odors can have delayed and unexpected effects. Garter snakes, *T. sirtalis*, were presented with either live earthworm or mosquito fish, *Gambusia affinis*, in a screen-covered bowl for several days. One day after transferring the snakes individually to a box free of prey odors, they were tested with aqueous extracts of fish and worms on cotton swabs. Snakes exposed to fish odor attacked fish extract less, and those exposed to worm attacked worm odor less. This is interpreted as habituation with a possible switch to other prey. This also demonstrates that in any experiment with chemical cues an odor not experienced for 22 hours may still have an effect (Burghardt, 1992).

Experience can further hone behaviors that are already present in naive animals. From 28 to 138 days of age, neonate rattlesnakes (*Crotalus viridis* and *C. horridus*) recognize and attack prey. They also show chemosensory searching and trailing. The initial trailing is "jerky and erratic," but after several feeding experiences becomes more "methodical" (Scudder *et al.*, 1992).

9.3.4 Olfactory "imprinting"

Some reptiles have been successfully *imprinted* to odors early in life. The hallmarks of classical imprinting are: primacy of exposure to a stimulus is more important than recency, a short (or one-time) exposure suffices, there is a critical (or sensitive) period during development (a "window of opportunity") when imprinting is possible and the effect is irreversible during the life of the animal. Green sea turtles, *Chelonia midas,* were exposed to an odor (morpholine or 2-phenylethanol) either during incubation plus 5 days in the nest, or 3 months after hatching in the holding water, or both. They later preferred the familiar odor in a choice experiment only if they had experienced the odor in the nest *and* the post-hatching holding water. Early exposure has no effect if the odor was added only to the nest or only to the holding water. The turtles needed a long period of exposure to the odor (Grassman and Owens, 1987). Therefore, the acquisition of the preference for the odor may be different from classical imprinting, which occurs during a brief period. Similarly, snapping turtles, *Chelydra serpentina,* require 2 weeks of exposure to food odors to be imprinted (Burghardt and Hess, 1966). It is debatable whether we should maintain the term "imprinting" for these "large window" acquisitions of later preferences.

9.4 Birds

Goslings of the greylag goose, *Anser anser* at 10 hours old and food naive will shake their heads when presented with the odors of sage, peppermint, dill, or lavender. All goslings responded at the age of 18 hours. They discriminated different plant odors at 4 to 9 days of age (Würdinger, 1979).

Even bird embryos are sensitive to odors. If domestic fowl embryos were exposed to odors 1 day before hatching (including dichloroethane, formic acid, cineole, and amyl acetate) they responded with increased heart rate, beak clapping, and head shaking to the first three compounds. Amyl acetate had inconsistent results. Blocking the nostrils with wax eliminated the responses (Tolhurst and Vince, 1976).

9.5 Mammals

First, I will discuss the development of *responses* to odors and follow this by the development of *odor production*. Mammals spontaneously respond to odors *in utero* and can also be conditioned to chemical stimuli before birth. Much research has focused on the chemical ecology of the fetus. *Experience* plays an important role in many aspects of sensory development. It will be discussed throughout this section.

9.5.1 Prenatal development

"Spontaneous" preferences

Olfaction

Both *social* and *dietary* odor preferences may be acquired *in utero*. Rat pups at 8 hours of age orient more toward the amniotic fluid of their mother than to that of an unrelated female rat. This preference must be acquired prenatally, as Caesarean-born pups also prefer the mother's amniotic fluid. At birth, rats already know the odor of their kin (Hepper, 1987). The vomeronasal organ probably monitors the chemical quality of the intrauterine environment (Pedersen *et al.*, 1983).

Prenatal olfactory and taste experiences may create standards against which postnatal odors are matched (Schaal, 1988a). Many chemicals, including volatiles, cross the placental barrier and enter the fetal blood. Volatile compounds may diffuse out of the blood and stimulate the olfactory receptors. For instance, the essence of garlic, allyl sulfide, is carried from the mother's blood to that of the fetus. Pregnant rats will readily eat garlic if provided. One clove of garlic consumed daily from gestation day 15 to 21 affects the preferences of a female's offspring. Her pups, tested at 12 days of age, preferred garlic over onion, while pups of mothers without garlic in their diet did not discriminate between garlic and onion. The response is acquired prenatally, since pups of garlic-eating mothers still show that preference even if they have been cross-fostered to a "garlic-free" mother. As to the pathway of the chemical cue, the odor is assumed to diffuse from the capillaries in the nose to the olfactory receptors (Hepper, 1988). Sheep fetuses on gestational day 144 were "externalized" from the uterus and implanted with intranasal catheters and heart rate electrodes. Later, odorants were delivered to the nose through the catheters by means of a syringe while the lambs were in the womb. The odorants were citral (lemon-like odor) and 2-methyl-2-thiazoline, which has a foul odor. The latter significantly reduced the heart rate in both lambs, while citral only slightly accelerated it, and saline (control) had no effect (Schaal *et al.*, 1991). Rabbits also "learn" prenatally odors of aromatic foods eaten by their mother. Postnatally, they prefer such odors (Hudson and Altbäcker, 1994).

Late in gestation (days 19 and 20), rat fetuses respond to intraorally applied lemon solution or citral, a tasteless odor, with behavioral activation. No such response is observed after surgical transection of the olfactory bulb of the fetus

(Smotherman, 1987). To conclude, prenatal olfaction is possibly adaptive by preparing the animal for diet preferences, successful suckling, and socialization (Hepper, 1990).

Taste

The sense of *taste* is also functional in the fetus at the end of the gestation period. In one experiment, rat fetuses received intraoral infusions of milk (a biologically relevant stimulus) or lemon (a novel chemical stimulus) on days 19, 20, or 21 of gestation. They were then observed *in utero*, that is in an "externalized" uterus in saline, removed from the mother's body but still connected to it. Their motor activity showed that they discriminated these stimuli. They responded to milk with delayed fetal movements of low magnitude, while lemon provoked high-magnitude, spiked patterns. Late in gestation, milk elicited stretching, as at the nipple, while lemon caused face wiping, a motor pattern used by rat pups or adults to aversive gustatory stimuli (Smotherman and Robinson, 1987).

A similar study examined the role of *experience*, using the movements of rat fetuses in an "externalized" uterus in response to intraoral infusion of lemon or mint extracts on day 19 of gestation. Prior experience does indeed affect fetal responses. There is *habituation* to repeated stimulation, *central processing* of sensory information, fetal *orienting reflexes* to novel stimuli, and prenatal states exist that are associated with different response patterns (Smotherman and Robinson, 1988).

Aversive conditioning

Rat fetuses exposed to apple juice injected into the amniotic fluid and lithium chloride intraperitoneally on gestational day 20 will avoid the odor of apple juice after birth. They will not attach to nipples that have been treated with apple juice (Stickrod *et al.*, 1982). They will also reduce their wheel-running speed when the odor of apple juice is presented (Smotherman, 1982). This fetal conditioning may help the pups to avoid foods that made their mother sick. Prenatal learning occurs even earlier: a combination of mint odor and lithium chloride given to rat fetuses on day 17 of gestation leads to conditioned aversion to mint 2 days later (Smotherman and Robinson, 1985). The heart rate can be conditioned in response to intrauterine application of apple juice. This olfactory conditioning ontogenetically precedes visual and auditory conditioning of the heart rate, just as somatomotor conditioning does (Sananes *et al.*, 1988; Fig. 9.1).

Particularly alarming are fetal effects of *alcohol* and *drugs* on food-related odor responses in humans. Apart from the severe fetal alcohol syndrome, alcohol can affect the chemosensory behavior of a fetus. Alcohol administered to pregnant female rats impaired odor aversions and preferences in their offspring. A

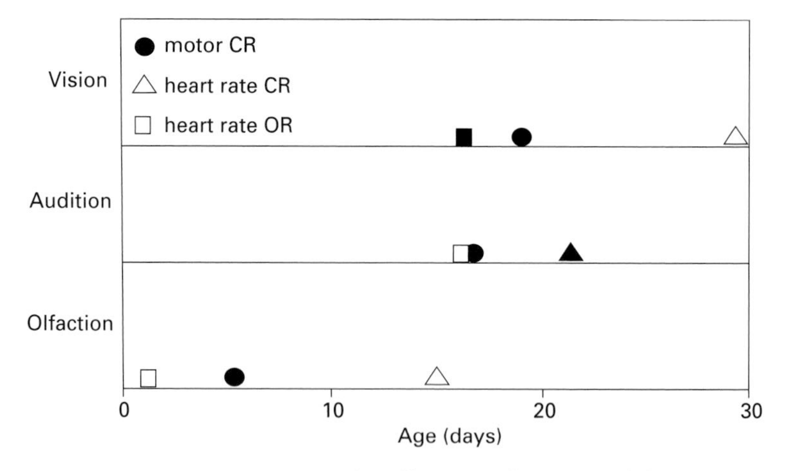

FIGURE 9.1 In rat pups, sense of smell matures first among different sensory modalities. Within the sense of smell, conditioning of somatomotor responses to odors occurs earlier (•) than conditioning of heart rate (△). The earliest occurrence of "orienting response," (i.e. a heart rate deceleration [□]) in response to the first appearance of a conditioned odor. (Redrawn from Sananes *et al.*, 1988.)

liquid diet provided the mothers with either 35 or 0% of the calories in the form of ethanol. The offspring of the group receiving 35% ethanol did *not* develop an odor *aversion* to lemon odor at 10 days of age when this odor was followed by a lithium chloride injection, making the rat sick. Furthermore, if the pups were infused with milk through a cannula, paired with banana odor, the ethanol group failed to develop a preference for banana odor. However, these effects appeared to fade with age. As adults, animals in the ethanol group managed to associate banana odor with drinking water laced with lithium chloride if the odor was present when they became ill (Barron *et al.*, 1988).

Maturation of chemosensory systems

Although the vomeronasal system is specialized to detect stimuli in a liquid environment, it probably is not functional *in utero*, at least in mice. Fluorescent microspheres were not taken up by the vomeronasal organ as the access canal is not open yet *in utero*. In rats, by contrast, the canal is open before birth and the microspheres can be taken up. The olfactory epithelium of the main olfactory system plays a greater role prenatally, as evidenced by the uptake of radiolabeled 2-deoxyglucose (Coppola and Coltrane 1994). Fetal mice respond to amyl acetate and isovaleric acid delivered into the nasal cavity through a tiny cannula (Coppola, 2001). In both rats and mice, the main olfactory system, and not the vomeronasal system, appears to mediate prenatal olfaction (Coppola, 2001).

In utero priming: freemartins

Androgen from the mother or embryonic siblings affects the development of embryos, a form of priming. Males produce testosterone from about day 12 of gestation and male neighbors *in utero* androgenize female siblings. Animal breeders have called such females with male features "freemartins." The effect depends on intrauterine positions. Males on both sides of the female fetus in the uterus (2M condition) have a stronger effect than a single male sibling on only one side (M). Female fetuses with no male neighbor (O condition) provide convenient controls. The androgenizing effects include larger anogenital distance, lower uterine weight, less body fat, delayed puberty, longer and irregular cycles, and an altered sex ratio. In one experiment, control females had 40% male pups, M females 50%, and 2M females 60% while the overall population had 50% males (Vandenbergh and Hotchkiss, 2001). The androgenized females differed in their postnatal behavior from normal females, showing altered odor preferences, urine marking, and olfactory priming of other mice, especially puberty delay (Drickamer, 2001b).

Male mice prefer O or M females or their odor when confronted with soiled traps in the field, while females prefer 2M females and their odor in such contaminated traps. Also, the larger the anogenital distance (i.e. the more masculinized a female), the more urine marking it performed in an arena. The 2M females were less responsive to chemical puberty-delay cues from crowded female mice and also less sensitive to puberty-accelerating male chemosignals (Drickamer, 2001b). They had high levels of serum testosterone and were more infanticidal, while 2M males were less so, instead showing more parental care. Finally, both males and females of the 2M type were more aggressive in encounters with mice of the same sex of the O type (Drickamer, 2001b).

The production of odors in utero can have social and reproductive consequences for the adults: Male mice can smell the fetal MHC odortype as part of the mother's odortype (Beauchamp *et al.*, 2000).

Prenatal humans

The rich chemical ecology of the uterus plays an important role for the human fetus. The chemical senses and metabolism of the fetus are bombarded with its own waste products; the aromas, alcohol, or drugs in the mother's diet; and her inhaled substances such as perfumes or tobacco. Preterm neonates respond reliably to the odor of mint at the gestational age of 7–8 months, and some responses are observed at 6 months of gestational age (reviewed in Schaal, 1988b). Premature newborns, aged 28 to 33 gestational weeks, distinguish the odors of vanilla and butyric acid (Marlier *et al.*, 2001). Prenatal exposure to maternal food flavors via the amniotic fluid is thought to prepare the infant for neonatal preferences (Schaal and Orgeur, 1992). Infants born to mothers who had consumed anise flavor during pregnancy preferred anise odor during their first 4 days after birth. By contrast, offspring of mothers on a diet without anise flavor displayed aversion or neutral responses to anise odor at the same age (Schaal *et al.*, 2000).

Human infants do not discriminate between their mother's amniotic fluid and her milk until 3 days after birth. However, they distinguish well between their mother's amniotic fluid and milk formula. Thus, chemical information acquired *in utero* is used after birth (Schaal *et al.*,2001). Three-day old infants discriminate familiar from unfamiliar amniotic fluid, whether they are being breast-fed or bottle-fed (Schaal *et al.*,1998).

In humans, just as in mice, hormones can act as priming pheromones. Androgens from the mother in the condition known as congenital adrenal hyperplasia masculinize daughters. With increasing use of assisted fertility, heterozygous (fraternal) twins are becoming more frequent. A male sibling will androgenize a female fetus. Both maternal and sibling masculinizing influences have behavioral effects. These include altered spatial ability, sensation seeking, selection of more masculine toys, more masculine-type sexual orientation, and fantasies more related to lesbian or bisexual lifestyles. Moreover, we are just starting to understand how endocrine disruptors in the general environment affect human behavior. So far most of these compounds are estrogen mimics (Vandenbergh and Hotchkiss, 2001).

9.5.2 Postnatal development

Sensory development points to the importance of chemical cues very early in life after birth. From the fourth day of life, mouse pups respond to the odor of their nest (Schmidt *et al.*, 1986). In rats, the main olfactory system processes odor responses during the first days of life.

Newborn mammals use odor cues to find the nipples of their mother and to attach there. Piglets deposit saliva on the teats of their mother that contains 16-androstenes. These androstenes are already present in the saliva of pig fetuses (Gower and Booth, 1986). Piglets learn the odor of their mother within 12 hours after birth. They discriminate between mother and non-mother female odors and are most attracted to odors of feces and skin secretions. Novel odors such as orange or banana are not preferred. The putative maternal pheromone deoxycholic acid was not preferred by piglets. In a Y-maze, piglets were attracted to feces of the mother, not to her colostrum or ventral washings. Ventral washings may be important at close range for nipple finding, however. The critical period for acquiring an attraction to maternal fecal odor lies between birth and 12 hours of age. The preference persists in 14- and 21-day old piglets. Two important reasons for being attracted to maternal feces may be to obtain iron and other nutrients, and to be protected by bile acids against *Escherichia coli* toxins (Morrow-Tesch and McGlone, 1990).

Newborn rabbits (*Oryctolagus cuniculus*) are guided to the nipples of their mother by an odor, the "nipple pheromone" (Hudson and Distel, 1983). The vomeronasal organ is not necessary for this response (Hudson and Distel, 1986). Once mobile, the pups discriminate and are attracted to the anal gland secretion of their mother. They discriminate mother's odor from those of other females or males after the age of 10 days. Behavioral responses are not always obvious; the frequency of nostril movements, indicative of changed respiration rate, is higher in response to mother's anal gland odor and there were fewer "rejection" behaviors in the presence of this secretion. It was concluded that rabbit nestlings acquire the preference for the odor of the mother during the first days of their life (Mykytowycz and Ward, 1971).

The *amniotic fluid*, deposited by a rat on her own teats, can guide her pups to the teats to initiate suckling (Teicher and Blass, 1980; Pedersen and Blass, 1982). In domestic sheep, only inexperienced (primiparous) ewes need amniotic fluid on the lamb's coat for normal development of maternal behavior, such as licking the young and accepting it at the udder. Experienced (multiparous) ewes lick their lambs less if amniotic fluid is removed but otherwise accept them (Levy and Poindron, 1987). The amniotic fluid links mother and offspring chemically, as it stimulates in both directions. In the rat, it serves kin recognition, since pups prefer their mother's amniotic fluid over that of other females, even after caesarean section (Hepper, 1987), as mentioned earlier. The responses by sheep to amniotic fluids of sheep, cows, and goats show some specificity: cow amniotic fluid was always repulsive to sheep, while caprine and ovine amniotic fluid were repulsive only before parturition. Ovine amniotic fluid becomes attractive after parturition while caprine fluid merely loses its repulsiveness (Arnould *et al.*, 1991).

Spiny mice, *Acomys cahirinus*, are the only murid species with precocial young. This permits study of the postnatal development of odor preferences as expressed by active (locomotor) behavior. When given choices of own (familiar), unfamiliar conspecific odor, and "no odor" (clean shavings), infant spiny mice prefer (i.e., stay longer on) odor from another litter (unfamiliar odor) but avoid clean shavings at the age of 3 days. By day 5, they slightly prefer to remain near their parents; by day 7 and 9 they clearly prefer their parents (Birke and Sadler, 1987b).

Odors can affect the survival of newborn rat pups: 100% of pups normally born on gestation day 21 survive, and so do pups delivered by Caesarean section on day 20. If the pups are exposed to odors for 1 hour after birth, the survival rate varies greatly: 9% survive in the presence of the odor of amniotic membranes and placenta; 80% with dimethyl disulfide (which occurs in rat saliva); 75% with no odor; and only 50% with mint odor. In addition to survival, the activity levels of the pups differ. Pups are more active in the presence of odor of amniotic fluid and placenta than they are in the presence of mint odor (Smotherman *et al.*, 1987).

One-day old rat pups show strong *negative* responses to strong taste stimuli such as strong acids or quinine solutions. They gape and flail with their forelimbs. At 12 days of age, they scrape their chin and tread with their paws (Johanson and Shapiro, 1986).

Humans

General odor responses

Newborns respond to odors by rapid breathing and activity changes. They can detect odors, rapidly develop their sensitivity over the first 4 days of life (Lipsitt *et al.*, 1963), discriminate odor qualities and intensities, memorize odors for 1 day (and possibly for life), prefer odors they had experienced earlier, and localize odor sources (e.g. newborns aged 1 to 5.5 day olds turn away from a cotton swab with ammonia; Rieser *et al.*, 1976). Neonates younger than 12 hours show olfactory preferences and aversions even before they experience their first food.

Infants are able to acquire odor preferences on the first day of life. In one experiment, 12 male and 12 female white, healthy, full-term neonates were exposed to the odors of cherry or ginger on a pad taped to the inside of their crib for 24 hours. After this exposure, they were tested for preferences during active sleep (stage II). The behavior was videotaped and the duration of time oriented to each odor measured. Only the female neonates showed a preference for the familiar odor (Balogh and Porter, 1986). Therefore, even on the first day of life, females outperform males, as often described for children and adults (e.g. Yousem *et al.*, 1999).

Development of responses to conspecific odors

In human infants *social odors* are important from a few days after birth, as indicated by head turning in the direction of odors. In 1877, Darwin observed that an infant with its eyes closed would turn its head towards its mother when her breast was brought near. Darwin suggested that the infant might be attracted by the mother's odor or body heat. Hungry sleeping infants turn their heads and show the rooting reflex. At 2 days of age, head turning is random; while some infants orient toward odors with 6–8 days, they all do by 6 weeks. Babies discriminate breast odors: they choose the odor of the mother's breast areola over no odor or odor from another lactating woman (MacFarlane, 1975). Head and arm movements decrease when the infant is exposed to its mother's breast odor (Schaal *et al.*, 1980). Sleeping infants respond with mouth and head movements to bottles with mother's milk but not to empty bottles (Schaal *et al.*, 2001). Infants also discriminate neck and axillary odors. It was concluded that infants are not able to identify the odor of their mothers at 2 days of age but can discriminate her from an unfamiliar mother by the second week after birth. This may be biologically adaptive. Historically, in humans, maternal death rate was high. Therefore, it was adaptive for human babies to form irreversible attachments later than in other mammals. This would permit adoption of alternative parents (MacFarlane, 1975). Breast-fed infants learn to discriminate their mothers' axillary odor, while bottle-fed babies are not able to do so (Cernoch and Porter, 1985). The question arises whether bottle-fed children will subsequently suffer any other, lasting deficiencies in their social behavior.

In summary, postnatal learning, *in utero* experiences, and genetic predispositions play a role in individual olfactory recognition in humans. Learning is demonstrated by the fact that breast-fed, but not bottle-fed, neonates discriminate their mother's odor from that of other lactating women. Artificial odors worn by their mothers are also preferred by infants. Inborn preferences account for the attraction of breast-fed infants to odors of unfamiliar lactating females, although they prefer their own mother's odor. Even bottle-fed infants orient to breast odors from unfamiliar lactating women when presented with the choices of axillary odor, the odor of their own formula, or breast odors from non-lactating females (Porter *et al.*, 1992).

Children 28–36 months of age chose T-shirts worn by their *mothers* over those from other women in over 72% of cases. A group of children aged 45–58 months was over 60% correct (Montagner, 1974). Children of that age group also discriminate between T-shirts worn by siblings and non-siblings (Porter and Moore, 1981).

Preschool children are reputed to differ from adults in their odor preferences. For instance children aged 3–4 years were as likely to "like" the odor of amyl acetate (banana) as synthetic sweat or feces odors. By 6 years of age, their preferences resembled adults, liking banana, and disliking sweaty and fecal odors (Stein *et al.*, 1958). At 4–5 years of age, a shift occurs from positive or neutral to negative characterization of odors of sweat, feces, asa foetida, or butyric acid. However, it is increasingly becoming clear that responses by very young children depend much on how or what they are being asked. For instance, 4–6 year olds respond more positively if asked "tell me if this smells pretty" than when asked "tell me if this smells ugly" (Engen, 1974a). The preferences by 3-year olds in a forced choice played as a game did not differ from those of adults. They were asked to place pleasant odors near a popular puppet ("Big Bird") and unpleasant ones near a doll of a "bad" character ("Oscar the Grouch"). This first study not only found hedonic differences in children under 5 years of age but also established that children are more sensitive than adults to the steroid androstenone. Adults liked androstenone more often than children, and 29% of the children liked the odor of pyridine which adults never did (Schmidt and Beauchamp, 1988). The hedonic hierarchy of some odors is similar for children aged 1, 2, and 3–5 years old as for adults. For instance, the following sequence from most to least liked was the same for all these groups: lavender, amyl acetate, butyric acid, dimethyl disulfide (Bloom, 1975; in Schaal, 1988b).

Odors are not easily forgotten. The recognition of an odor depends on the original coding activity. When trying to recall an odor, life episodes related to that odor and the ability to label and define the odor enhance odor memory. Visual imagery, by comparison, did not enhance odor memory (Lyman and McDaniel, 1986).

Learning responses

Adults continue to associate new odors with pleasant and unpleasant situations in social and sex life, work and recreation, and concerning food and drink. The human patterns of odor recognition and preferences do not merely involve the olfactory nerve and its central projections. Learned associations are formed and stored in memory. To retrieve odor information, we need affective and cognitive components, as well as verbal descriptors. Without the latter, an odor appears familiar but cannot be labeled, the "tip-of-the-nose-phenomenon" (Lawless and Engen, 1977).

In old age, humans experience a precipitous drop in olfactory ability (Doty *et al.*, 1984; Doty, 1986; see Section 5.7). Two excellent recent reviews of the ontogeny of human olfaction and olfactory communication are by Doty (1986) and Schaal (1988b).

I conclude this section with a science fiction idea. If nursing an infant (instead of bottle-feeding) fosters olfactory discrimination (Cernoch and Porter, 1985), would it be desirable and possible to train noses early on? Should babies be breast-fed to give them a chance to become gourmets, sophisticates, wine or tea testers or would we be doing them a favour by bottle-feeding them to render them less sensitive to unpleasant odors such as pollution in our air?

9.6 Learning

The odor preferences of rat pups aged 3–6 days can be reversed by training. If the initially avoided odor of orange extract is presented together with maternal saliva, the pups will later orient toward orange odor. Thus, conditioning can enhance the value of an unfamiliar odor (Sullivan *et al.*, 1986).

The newborn rat is a "natural split-brain preparation" for olfactory learning protocols: it can be trained to associate an odor with a milk reward via just one nostril. If the other nostril is tested, the animal shows no preference. However, at 12 days of age or later, the two sides of the brain are connected and a learned preference occurs with either nostril open. The information is stored *uni*laterally: the animal shows unilateral preference if the commissure is cut *after* training. The maturation of the commissure pathways occurs between 6 and 12 days of age. In summary, unilaterally represented memories remain unilateral, even after bilateral retrieval processes have developed. The mnemonic storage capacity of the brain is increased by confining memory to one side (Kucharski and Hall, 1987).

Home odor can modulate certain behaviors in infant mammals. For instance, rat pups aged 7, 9, and 11 days learned better to escape from a shock if the odor of soiled bedding from their own cage was present than they did in the presence of clean shavings. The familiar odor may reduce a heightened state of arousal in an unfamiliar environment (Smith and Spear, 1981). At the age of 17 days, rats spend less time in the dark half of an apparatus if home nest odor is present, but at 30 days no effect of home nest odor was noted (Richardson and Campbell, 1988). Similarly, 16-day-old rats take 12 trials to learn not to enter the black compartment of a black-and-white shuttle box if trained with clean shavings placed under the apparatus, but only seven trials with soiled litter from their home cage available (Smith and Spear, 1978).

9.6.1 Critical periods

The critical period for developing a preference for the *maternal pheromone* in rat feces occurs at 14 to 16 days of age (Leon and Behse, 1977). Non-rat odors such as pure mint or peppermint can be made as attractive as maternal feces to rat pups under the age of 21 days. At that age, simple exposure suffices. A preference is acquired whether the odor is simply presented or pasted on the mother. But the older the pup, the more important the maternal context becomes, and pups up to 33 days old can acquire peppermint preference only if it is presented on the mother. The waning of the response is also much slower for the odor experienced on the mother. After 20 days of age, the rats acquire preferences less easily (Galef, 1982; Galef and Kaner, 1980). Postnatal conditioning to odors is facilitated not

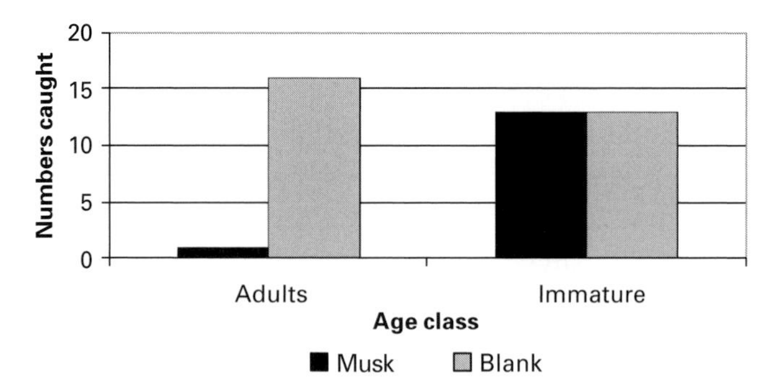

FIGURE 9.2 Immature muskrats do not avoid live traps scented with musk from adult male muskrats, as adults do. (After data from van den Berk and Müller-Schwarze, 1984.)

only by the mother as background but also by *maternal stimulation*. If neonatal rats are exposed to an odor and at the same time receive tactile stimulation that mimics maternal contact, they will prefer that odor later. That odor will also increase uptake of 2-[^{14}C]-deoxyglucose by the olfactory bulb. Various types of moderate stimulation, such as an odor coupled with high humidity, can bring about this preference (Do *et al.*, 1987). This early olfactory learning depends on noradrenergic factors.

Woodmice, *Apodemus sylvaticus*, are able to walk when 11 days old and increase their locomotion between postnatal days 11 and 19. When moving about, they spend more time on home (nest) and male odor than on female odor and clean bedding. This difference is not observed if the woodmice are tested for the first time on day 15, suggesting a critical period for acquiring these odor preferences (Pontet and Schenk, 1988).

Sexually immature mammals often ignore territorial odors that are important signals for adults. For instance, juvenile muskrats, *Ondatra zibethica*, were caught equally often in clean live traps and traps scented with musk (anal gland) secretion from adult male muskrats, while adults avoided the musk (van den Berk and Müller-Schwarze, 1984) (Fig. 9.2).

9.6.2 Long-term effects of early experience: olfactory social imprinting

In mammals, early experience with social odors will have lasting effects on odor preferences. We have to distinguish *filial imprinting* (attachment of young to mother) from *maternal imprinting* (attachment of mother to young) in precocial mammals.

Filial imprinting

In a classical experiment Mainardi (1965) exposed male mouse pups in their nest to an artificial odor, Chanel No. 5. As adults, these mice preferred females treated with Chanel No. 5 over those with regular mouse odor. Mammals differ from birds in that learning plays a critical role in acquiring social preferences; the sensitive period is more drawn out than a narrow "critical period," and effects are not necessarily permanent, and they can be overridden more easily by subsequent relearning.

Even the body odors of mice that genetically differ only at the major histocompatibility complex (MHC) are learnt by their offspring. MHC, called locus *H2* is located on chromosome 17 (out of 20 chromosomes in the mouse). Normally, mice of both sexes prefer to mate with partners of a different MHC type. In one experiment, entire litters of mouse pups were fostered onto parents of the same or a different MHC type. In males, the mating bias appears to be acquired during early development through MHC-controlled signals from the parents: males fostered onto parents of a *different* MHC type mated more often with females of the *same* type. The latter now represented a "strange" odor. The foster rearing produced a reversal of preference; males mated with a female whose odor was different from that of their foster parents. In females, the mating bias was unaffected by the fostering history (Yamazaki *et al.*, 1988).

The MHC odor types are present as early as day 1 after birth. The odor type of a pregnant female is a combination of the fetal and maternal odor types, and a male mouse can smell the fetal odor (Beauchamp *et al.*, 2000).

Long-lasting and specific effects of early odor experience have not yet been reported for many mammal species. In one experiment, rat mothers that raised male pups had their nipples and vagina scented with citral. When tested at 100 days of age, these males ejaculated sooner with sexually receptive, citral-scented females than with controls. Conversely, the latency to ejaculation of males reared with saline-treated mothers was shorter in matings with "normal" females than with citral-scented ones (Fillion and Blass, 1986).

Rabbit pups learnt an odor of their mother in one trial. This odor emanates from the belly of the dam and releases suckling. This rapid learning is considered a form of olfactory imprinting tied to an early sensitive period (Hudson and Altbäcker, 1994).

Maternal imprinting

Maternal imprinting occurs more rapidly than filial imprinting. An ungulate mother such as a goat (Klopfer and Gambale, 1966) or black-tailed deer (Müller-Schwarze and Müller-Schwarze, 1971) will irreversibly and exclusively attach to the odor of her newborn within a few minutes, while the young can still be bonded with their mother up to several days later.

Birth odor of lambs is used by ewes to discriminate their own from alien lambs (Alexander *et al.*, 1987), while in goats the mother appears to "label" her kid with her own odor (Gubernick and Klopfer, 1980).

Imprinting of prey odors

Ferrets, *Mustela furo*, acquire knowledge of prey odors as young animals. This search image is not modifiable at later ages (Apfelbach, 1973).

9.6.3 Odor learning in adults

Even adults can still develop olfactory preferences that contravene those acquired before sexual maturity. Female laboratory mice "imprinted" by the odor of one mouse strain will prefer this odor even more if they are exposed to males of this strain as adults. However, if they are exposed to males of a *different* strain when sexually mature, their original odor preference will be reversed (Albonetti and D'Udine, 1986). Naturally occurring sex or body odors may assume their sexual significance after association with sexual activity: male mice were aroused by a perfume that they had experienced earlier on scented females they had interacted with (Nyby *et al.*, 1978). Practitioners have known that adult mammals can acquire responses after exposure to certain animals. For instance, bulls of the Asian elephant that had been housed near African elephant bulls respond to temporal gland secretion and its three components phenol, 4-methylphenol, and (*E*)-farnesol from the latter species. Asian bulls that had not been associated with African bulls did not respond (Rasmussen, 1988).

Short-term memory for odors

Adult rats investigate a juvenile rat they have encountered 30 minutes before much less than a completely new individual. This "recognition memory" lasts about 80 minutes and is termed a *transient* memory. The anticholinergic drug scopolamine (hyosine) abolishes the decrease of investigation during the second encounter. It was postulated that scopolamine "disturbs chemosensory receptivity in rats" (Soffie and Lamberty, 1988).

Neural mechanisms

The brain sites that are concerned with early olfactory learning of social cues include the olfactory bulb. In *rat pups* blocking the glutamate receptors of the N-methyl-D-aspartate type in the olfactory glomeruli by 2-amino-5-phosphonovaleric acid prevented both the takeup of 2-[^{14}C]-deoxyglucose in the glomeruli and the development of a preference for a *learned maternal odor* (Lincoln *et al.*, 1988).

The organization of the olfactory bulb during the first 2 postnatal weeks may be affected by specific olfactory experiences. Early experience may be incorporated into the firing pattern of olfactory neurons with the result that the mature brain responds in a particular, conditioned way to the now-familiar odor. Such response changes of the mammalian brain to naturally occurring odors as a consequence of specific early experience may be an important component of individual variation (Coopersmith and Leon, 1986).

The processing of biologically relevant odors often changes drastically around the time of giving birth. In sheep, parturition has profound effects on the *ewe's response to lamb odors.* Within 3 hours of parturition, a lasting bond between mother and lamb is formed. Before the lamb is born, ewes reject the smell of amniotic fluid, but they are attracted to it immediately after giving birth. During parturition, the vagina and cervix are stimulated, leading to effects on the brain that render the ewe maternally responsive and capable of bonding with her lamb. Single mitral cells in the olfactory bulbs of ewes respond only to food before parturition but respond also to lamb odors after parturition. Between 3 and 4 days after giving birth, the number of cells responding to lamb odors increases dramatically until 70% of the cells respond to any lamb odor, and the remaining 30% preferentially to the odor of the own lamb. The odor of wool was almost as active as that of the whole lamb. *In vivo* sampling of the external plexiform layer revealed neurotransmitter interaction between granule and mitral cells. Both the excitatory amino acid glutamate, and the inhibitory γ-aminobutyric acid increased during the first 5 minutes of exposure to lamb odor (Kendrick *et al.*, 1992).

Ontogeny of odor production

Odor types distinctive of *H2b* and *Hdh* genotypes already appear in the urine of 1-day-old mouse pups (Yamazaki *et al.*, 1992a). At that age, the normal intestinal bacterial flora is not yet present and hence is not necessary for the odor (Yamazaki *et al.*, 1992b). In addition to MHC, gene(s) on the X and Y chromosomes and other autosomal genes contribute to individual odors (Yamazaki *et al.*, 1986, 1990).

10

Allomones I: chemical defense by animals

The 'bonnacon' [bison] "... emits a fart with the contents of his large intestine which covers three acres, any tree that it reaches catches fire and its pursuers are driven off with the noxious excrement..."

[Clearly, an exaggeration]

ANON (*c.* 1150) *Bestiary* (transl. and ed. T. H. White) New York: Putnam.

Allomones are interspecific semiochemicals that primarily benefit the sender. Animals as well as plants defend themselves chemically against predators. Myriads of ways to deter predation have evolved. Many chemicals have more than one function, being aimed not only at predators but also at parasites, prey, or conspecific competitors. As common denominator of allomones we assume that the inclusive fitness of the sender – rather than the receiver – is enhanced. Chapter 10 deals with defenses by animals, and Chapter 11 with plant defenses against herbivores.

Numerous ingenious chemical defense mechanisms of vertebrates protect against predators, but also against competitors and fierce prey. These range from innocuous odors of prey animals or plants that signal that they are unpalatable to some of the most toxic secretions known in nature. Slower animals tend to rely more on chemical and mechanical defenses. Humans have known animal toxins since time immemorial and medical research has elucidated the mechanisms involved in great detail. However, ecologists have investigated why animals and plants have poisons and venoms in the first place only since the 1950s.

10.1 Fish

Among fish we find some of the most potent toxins in animals. We distinguish *passively toxic fish* from *actively toxic fish*. The former simply have toxins in their tissues, typically taken from some other source such as their diet. The latter produce the poison and have evolved apparatus to discharge, deliver, or inject the toxins.

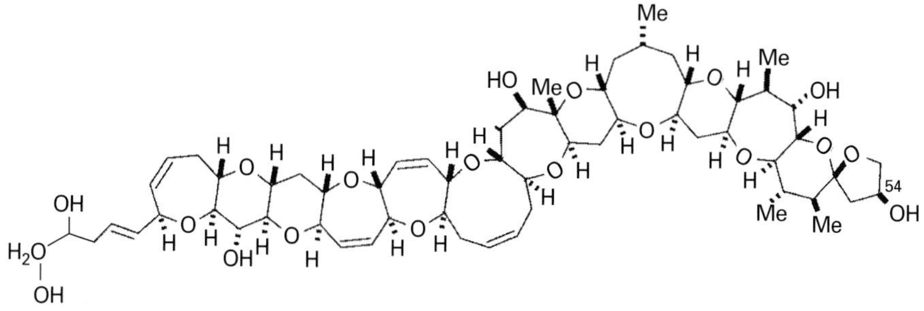

FIGURE 10.1 Ciguatoxin. It passes through food chain from dinoflagellates to herbivorous and carnivorous fish, and humans.

10.1.1 Passively toxic fish

Over 500 species of fish can poison humans (and presumably other consumers) when eaten. The *ciguatoxin* ciguatera moves through the food chain, starting from dinoflagellates (specifically *Gambierdiscus toxicus*). The dinoflagellates reside on macroalgae that herbivorous fish will eat. Carnivorous fish then acquire the toxin from their prey, herbivorous fish, and store it in liver, testes, and intestines, but less in muscles. Humans are poisoned when consuming fish such as mackerel, coral trout, and cod. The toxin is odorless and tasteless, and it survives cooking and freezing. Several hours after the meal, breathing becomes difficult. Poisoning does not lead to immunity. On the contrary, a second poisoning usually proves more serious than the first one. The first report of *ciguatera* poisoning from eating yellowtail tuna appeared in China around AD 700. Ciguatoxins are polyether (polycyclic) compounds that contain a series of rings (Fig. 10.1).

Among the most poisonous of all marine animals are the puffer fish (Tetraodontidae). The ancient Egyptians knew puffer fish to be poisonous and depicted it on the tomb of Pharao Ti in 2500 BC. Puffer fish take up alkaloid toxins such as tetrodotoxin (Fig. 10.2) from algae (*Shewanella* sp.) and bacteria (Yasumoto and Murata, 1993; Yasumoto *et al.*, 1986). The nerve poison resides in liver, gonads, intestines, and skin, but not in the musculature. Therefore, the flesh is generally edible, provided other organs do not contaminate it during processing. The nerve damage starts as a tingling feeling on tongue and lips and spreads down the body. Over 60% of those poisoned die, and death is rapid and violent. Captain James Cook nearly died of (presumably) tetrodotoxin after eating fish in New Caledonia. Eating puffer fish amounts to a game of Russian roulette. In Japan, puffer fish is known as fugu and is prepared by skilled fugu chefs, demanding highest prices. Some toxic species are *Fugu rupripes*, *F. vermicularis*, *F. pardalis*, and the white-spotted ("death") puffer, *Arothron hispidus*.

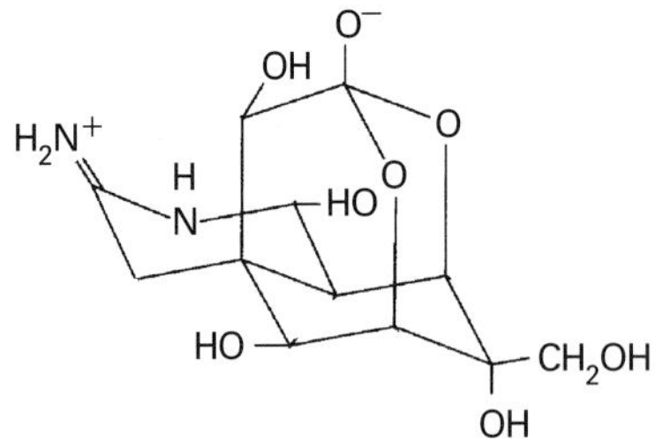

FIGURE 10.2 Tetrodotoxin, a toxin from pufferfish and newts.

Besides the Tetraodontidae, the order Tetraodontiformes has very toxic fish in the families Diodontidae, Canthigasteridae, and possibly Molidae and Triodontidae.

Soles of the genus *Pardachirus* possess one of the better-understood chemical defense systems in fish. The Red Sea Moses sole, *Pardachirus marmoratus*, dubbed "the Red Sea's sharkproof fish," and the related peacock sole, *Pardachirus pavoninus*, of the western Pacific secrete from the mucus glands on their dorsal and anal fins a milky substance that is retained in the mucus coat of the skin. This secretion repels sharks, kills starfish and sea urchins on contact, and is hemolytic. From these secretions, two groups of ichthyotoxic compounds have been isolated. The *pardaxins* are peptides consisting of 33 amino acid residues and physically and pharmacologically resemble melittin, the toxic peptide in the venom of the honeybee, *Apis mellifera*. They are strongly surfactant, owing to hydrophobic and hydrophilic parts of the molecule. Several pardaxins kill killifish, *Oryzias latipes*, within 30 minutes at a concentration of 25 μg/ml. Pardaxins act on the gustatory sense: white-tip reef sharks, *Triaenodon obesus*, escape from pardaxins (100 mg dissolved in 5 ml seawater) that are placed into their mouths. The toxin affects the gills, where osmoregulation is disturbed (Primor *et al.*, 1978). Steroid monoglycosides from *P. pavoninus* have been named *pavoninins* (Fig. 10.3) and those from *P. marmoratus* are *mosesins* (Fig. 10.4). Both together are also known as pavoninins. These pavoninins are both hemolytic and strong repellents for dog sharks, *Mustelus griseus*, and lemon sharks, *Negaprion brevirostris*. They act on the shark's olfactory sense (Tachibana *et al.*, 1984). Needless to say, pardaxins and pavoninins offer fascinating prospects as shark repellents.

Moray eels (Muraenidae) of tropical reefs possess a water-soluble poison. Poisoning occurs from eating the flesh.

10.1.2 Actively toxic fish

About 250 fish species possess various kinds of venom apparatus. The stingray and scorpion fish belong in this category. Stingrays have a

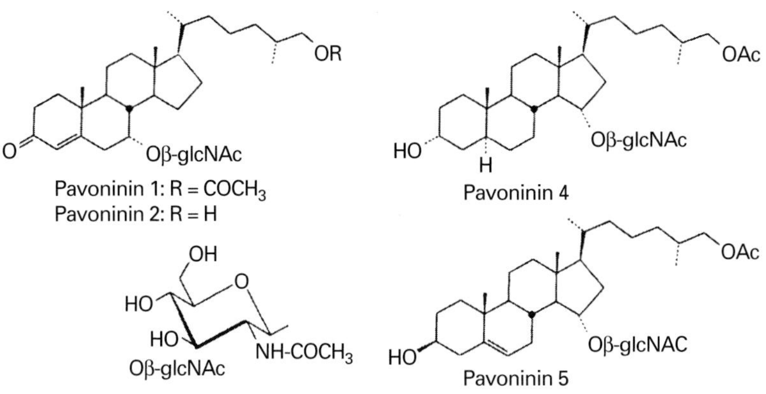

FIGURE 10.3 Pavoninins from the peacock sole.

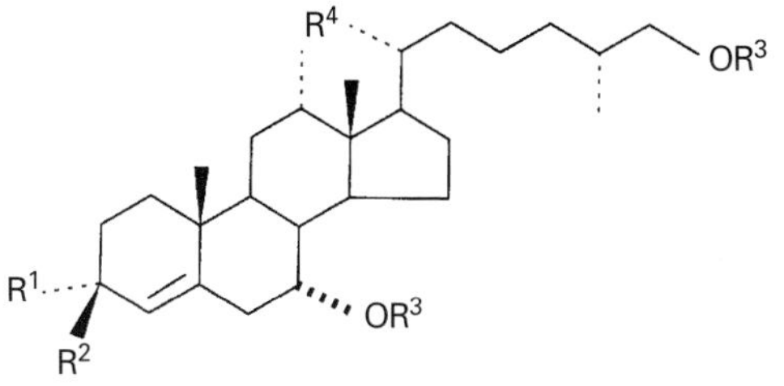

Mosesin 1: R^1 = OH R^2 = H R^3 = 6-Acetyl-β-D-galactose R^4 = OH R^5 = $COCH_3$
Mosesin 5: R^1 = OH R^2 = H R^3 = β-D-galactose R^4 = OH R^5 = $COCH_3$

FIGURE 10.4 Mosesins from the Red Sea Moses sole.

sting apparatus at the end of their tail, with poison glands at the teeth of the spine. The active venom contains proteins with a molecular weight of 100 kDa.

The perfectly camouflaged stonefish (Scorpaenidae) blend into rocks and dead corals and are extremely poisonous. On their dorsal fins, they have 13 needle-like spines with poison glands at their bases, used for defense. As the fish are sluggish, poisoning occurs by stepping on the half-buried fish. It is painful and can be fatal within a few minutes or hours. Divers need to exercise care, because the stings even penetrate rubber. There are about 20 species of stonefish, all occurring in shallow waters and coral reefs of the Indo-Pacific. Two better-known species are *Synanceja trachynis* and *Synaneichthyes verrucosus*. Table 10.1 summarizes some chemical defenses in fish.

Predatory fish may also be affected by alarm pheromones (Section 7.2) of the prey, both directly and indirectly. The alarm odor may act as defense compound that inhibits predator attack or reduces capture rate by inducing predator avoidance in school members of the prey species.

Table 10.1 Defense compounds in fish

Emitting species	Compound	Class of compound	Target species	Effect
Stingrays Dasyatidae	Not a phospholipase	Protein	Defense in general	Pain, death (1%)
Box fishes *Ostraction* spp.	Pahutoxin etc.	Esters of palmitic acid	Predators?	
Red Sea Moses sole *Pardachirus marmoratus*	Pardaxins	Peptides (33 amino acid residues)	Sharks	Affect taste: shark withdraws
	Mosesins	Steroid monoglycosides	Sharks	Affect olfaction, repellent
Peacock sole *Pavonininus pavonininus*	Pavonins	Steroid monoglycosides	Sharks	Shark repellent, affect olfaction
Puffer fish *Fugu* spp.	Tetrodotoxin (found in *Vibrio*, *Pseudomonas* spp.)	Alkaloid	Predators?	Local anesthetic (5–30 minutes); death in 6–24 hours (60% mortality)

10.2 Amphibia

Amphibian skin proliferates with mucus and granular glands. The granular glands produce defense compounds. The chemistry of these secretions is diverse, including biogenic amines, peptides, steroids, and alkaloids. Some of the most toxic compounds known are found in amphibians. Among amphibians, chemical defense is found in sluggish species such as salamanders, larval amphibians, and toads.

10.2.1 Urodeles

Many urodeles produce irritating or even toxic secretions in their skin glands, and some are extremely toxic. The rough-skinned newt (*Taricha granulosa*) is the most toxic of newts from several genera. Its dorsal skin and ovarian eggs contain the very potent tetrodotoxin. Only 0.000 05 ml of newt skin is needed for a lethal peritoneal injection in mice. The skin of one newt can theoretically kill 25 000 mice. When bothered, the newt assumes a characteristic defense posture: it arches, lifts its tail and releases an odor (Brodie *et al.*, 1974).

The level of tetrodotoxin in the skin of a particular female rough-skin newt correlates with that in her eggs. The amounts in the eggs of one clutch vary little, but there is considerable variation between clutches. These findings suggest that the mother passes on the toxin to her eggs. As the amount of toxin is not correlated with egg size, it may not be transferred automatically (passively) with maternal resources to the egg. Rather, the mother may exert some control over the amount allocated to the eggs (Hanfin *et al.*, 2003).

In certain geographical areas, termed "hotspots of snake–newt coevolution," rough-skinned newts are very toxic, and garter snakes (*Thamnophis sirtalis*) have adapted by tolerating high tetrodotoxin levels. The snakes accumulate tetrodotoxin in their livers where it remains for 1 month or longer. After eating one newt, snakes averaged 42 μg tetrodotoxin in their livers. This amount would severely affect avian predators of garter snakes such as harriers, hawks, bittern, or crows, and possibly kill them (Brodie *et al.*, 2002). Mammalian predators would be less sensitive to the toxin (Williams *et al.*, 2004).

Are these toxic compounds of urodeles effective against predators? Short-tailed shrews (*Blarina brevicauda*) prey on these animals; their burrows often contain the remains of salamanders. Laboratory tests confirmed that their chemical defenses are effective: noxious species of salamanders took the shrews longer to kill and had a higher survival rate than non-toxic species. The shrews needed more time to kill the toxic salamanders, and wiped their mouths more often (Brodie *et al.*, 1979).

A non-toxic species of salamander may derive protection from predators by visually resembling a toxic form so closely that predators cannot distinguish between them (*Batesian mimicry*). Free-ranging birds avoid both the toxic red eft (*Notophthalmus viridescens*) and the similar-looking non-toxic red morph of the red-backed salamander (*Plethodon cinereus*). The red-striped morph of *P. cinereus*, which does not resemble the red eft, is eaten (Brodie and Brodie, 1980).

Palatability may change during individual development: the premetamorphic larvae (tadpoles) of seven species of North American frogs, toads, and salamanders are palatable to the predaceous diving beetle *Dytiscus verticalis*, while the metamorphic stage (with limbs) of five of these species is unpalatable. The palatability is inversely correlated with the number, but not the size, of skin glands. Before metamorphosis, the tadpoles can flee quickly, and this primary defense suffices for survival. With legs, the larvae are slower and probably for this reason have to depend more on secondary, chemical defense (Formanowicz and Brodie, 1982).

Predators *learn* to avoid distasteful salamanders. When molested, spotted salamanders (*Ambystoma maculatum*) discharge a white slime, mostly in the tail region. They also raise and wag their tail when a predator is near. In one experiment, four out of five chickens learned to avoid these salamanders by *sight* after

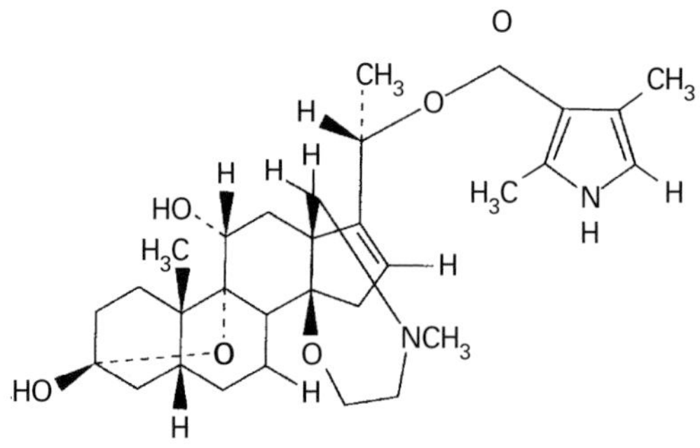

FIGURE 10.5 Batrachotoxin from dart-poison frogs.

14 trials. It is important to know that these salamanders are merely distasteful: the chickens ate 6 of 70 salamanders that were presented without lethal effects (Howard, 1971). The predators probably suffer other, more subtle effects. Mammalian predators also learn to avoid distasteful salamanders: in rats, both primary taste aversion and postingestional cues are important for acquiring a conditioned aversion to tiger salamanders, *Ambystoma tigrinum* (Mason *et al.*, 1982). The gastrointestinal/postingestion cue is the unconditioned stimulus and the food cue the conditional stimulus.

10.2.2 Anurans

The powerful toxins of toads and frogs are well known. The famous dart-poison frogs (Dendrobatids) from South America produce in their rather inconspicuous granular glands some of the most powerful poisons known. Indians have used the toxins on their arrows for many years; the poison of one tiny frog can be used for 50 arrows. The most toxic genus is *Phyllobates*, and the most toxic species *Phyllobates terribilis* of Western Colombia. The hunters merely draw the tip of the dart over the back of *P. terribilis* frogs but impale other species such as *P. bicolor* and *P. aurotaenia* to stimulate a copious flow of skin secretion (Daly and Spande, 1986). Dart-poison frogs are also brightly colored, as warning to potential predators, commonly termed *aposematic coloration*. However, the levels of poison and coloration in some Panamanian poison frogs of the genus *Dendrobates* are not correlated (Daly and Meyers, 1967).

The active principle of dart frog poisons is alkaloids. The study of the dendrobatid poisons led to the discovery of over 200 new alkaloids, including *batrachotoxins* (Fig. 10.5), pumiliotoxins, histrionicotoxins, gephyrotoxins, and decahydroquinolines (Daly *et al.*, 1994). The most common compounds have the basic structure of *piperidine* and include histrionotoxin. In *Phyllobates*, the synthesis of other alkaloids is suppressed in favor of batrachotoxins. These are related to steroids and are some of the most poisonous compounds known. Batrachotoxins affect the Na^+ permeability of mammalian membranes of nerves and motor end plates of muscles irreversibly. This causes arrhythmias, fibrillations, and ultimately failure of the heart. The specific sites of action are the Na^+ channels. Binding of batrachotoxin prevents the normal closing of the channels, resulting in massive influx of Na^+ and permanent depolarization of cells. The nerve cells can no longer transmit signals, and the muscles remain contracted. Histrionotoxin affects the K^+ channel, and pumiliotoxin B the movement of Ca^{2+} (Meyers and Daly, 1983). The poisons do not affect the frogs themselves (Daly *et al.*, 1980) but probably originated from enzymes that regulate water and salt balances in the skin of frogs (Flier *et al.*, 1980).

The level of toxic compounds parallel the diet of the frogs: ants constituted 50 to 73% of the diet of aposematic, toxic species of *Dendrobates*, but only 12–16% of that of non-toxic, cryptic frogs of the genus *Colosthetus*. The lipophilic alkaloids of frogs actually occur in ants (Caldwell, 1996). Some dendrobatid alkaloids appear to be derived from the diet: precoccinelline "most certainly" from small beetles, and pyrrolizidine oximes "most likely from small millipedes" (Daly *et al.*, 1994). In captivity, the toxicity of the frogs declines over time. Panamanian poison dart frogs (*Dendrobates auratus*) produce at least 16 alkaloids but only when feeding on ant-containing leaf litter, while frogs in an aquarium and eating only fruit flies developed no alkaloids. The main alkaloids taken up from ants are pyrrolizidines and indolizidines (Daly *et al.*, 2000). The search is on for prey that might contain other dendrobatid compounds. *D. auratus* from central Panama were introduced into Hawaii and after only about 30 generations, their array of alkaloids had changed, presumably because of the change in diet (Daly *et al.*, 1992).

Frogs of the genus *Dendrobates* are ant specialists. Of the more than 20 structural classes of lipophilic alkaloids found in the frogs, six occur in myrmicine ants. However, many dendrobatid alkaloids such as the batrachotoxins, histrionicotoxins, and pumiliotoxins, have not yet been found in insects and other leaf-litter prey such as beetles and millipedes (Daly *et al.*, 2000). The snake *Liophis epinephelus* feeds on *Dendrobates* and may further bioaccumulate alkaloids.

The African clawed frog, *Xenopus laevis*, well known as a laboratory animal, produces mucus in its granular (poison) glands that affects predatory snakes. The most common frog-eating snakes in the clawed frog's habitat are the African water snakes, *Lycodonomorphus rufulus*, and *Lycodonomorphus laevissimus*. Experiments with snakes from the Cape Town area in South Africa demonstrated the potent effect of the frog's mucus. Live frogs caused the snakes to "yawn" and "gape" more often and to climb away from the prey. Toxin-laden frogs released these behaviors in snakes more often than frogs that had been depleted of toxin

FIGURE 10.6 Skin glands of American toad, *Bufo americanus*. There are large parotoid glands (PG) behind eye. On the back are macroglands ("dorsal warts" [DW]), consisting of clusters of mucus, granules ("poison"), and lipid glands. (Photograph: D. Müller-Schwarze.)

by epinephrine injections. The white, viscous mucus from the granular glands was also active by itself. When given directly to the snakes' mouths, it triggered the same responses as the whole frog, and more so than porcine stomach mucine, which was used as control. The northern water snake of eastern North America, *Nerodia (Natrix) sipedon*, responded even more to the mucus than *Lycodonomorphus* (Zielinski and Barthalmus, 1989).

Aquatic frogs and toads probably need less toxin for defense than terrestrial forms. The compounds in the frogs' mucus are thought to be neuroleptic, blocking dopamine receptors. They also possibly are antibacterial and aid wound repair, and they are known to elevate the level of prolactin, the amphibian juvenile hormone (Barthalmus and Zielinski, 1988).

Toads have skin glands that produce irritating secretions (Fig. 10.6). Mucus glands keep the skin moist. As mucus is an excellent medium for microorganisms, toads and frogs produce toxic compounds with antibiotic properties. Indeed, removal of skin toxins leads to skin infections and death (Habermehl, 1994). Granular glands produce a different kind of secretion. Clusters of both types of gland form macroglands, such as the large *parotoid* glands, located near the ear (*para*, beside or near, and *otid*, ear). (There is no similarity with the mammalian parotid salivary glands.) Numerous "dorsal warts" also contain glands of both types.

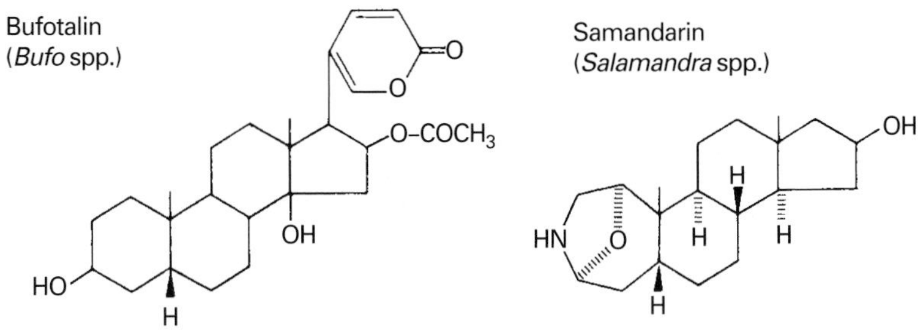

FIGURE 10.7 Bufotalin, found in toad secretion and samandarin, a neurotoxin from salamanders.

The secretions of toad tadpoles affect the behavior of predators. Largemouth bass, *Micropterus salmonides*, even when starved for 1 day, almost totally reject tadpoles of *Bufo americanus* (Fig. 10.6) and *Bufo woodhousei*. However, the hungrier they become, the more tadpoles they will eat. With increasing experience with these tadpoles, the bass take fewer into their mouths and spit more out. In choice experiments, they prefer tadpoles of the spring peeper, *Hyla crucifer*, to those of *Bufo* spp. (Kruse and Stone, 1984).

Predators eat the narrow-mouthed toad, *Gastrophryne carolinensis*, less than other anurans. They contact the toad, bite it, and release it again. Several reptilian and avian predators have been tested, among them the yellow-bellied water snake, *Nerodia erythrogaster*, garter snakes, *T. sirtalis*, snapping turtle, *Chelydra serpentina*, and the black-crowned night heron, *N. nycticorax*. The toad's skin contains a potent toxin: mice injected with extract from 3 mg toad skin per gram body weight of the mouse, died within 15 minutes. Skin from the American bullfrog, *Rana catesbiana* had no effect. Skin secretions also protect the toad from counterattack of ants, its main prey. The granular and mucus glands and their pores appear in tadpoles when the forelimbs emerge. At the same time, the tadpoles become toxic (Garton and Mushinsky, 1979).

The many compounds in skin gland secretions of toads fall into two groups: biogenic amines and steroidal bufogenins. There are two types of biogenic amine: the catecholamines (epinephrine and norepinephrine) and indole alkyl amines. The latter include bufotenin, serotonin, bufotenidin, bufoviridin, bufothionin, and others. One example of steroidal bufogenins is *bufotalin* (Fig. 10.7). If esterized with suberyl arginine, these steroids are called bufotoxins. They act on the heart, similar to digitalis toxins and are potentiated by epinephrine and norepinephrine. Bufotalin occurs in the European toad, *Bufo vulgaris*. *Samandarin* from salamanders (genus *Salamandra*) is structurally similar to bufogenins but acts as a neurotoxin (Fig. 10.7).

Bufo marinus, the giant, or cane toad, of South and Central America has been introduced to control pests, particularly in sugar cane fields, in areas from Florida and Cuba to Hawaii, Fiji, and the Philippines. Mortality of pets poisoned by eating giant toads varies by region. In Hawaii, only 5% of dogs and cats exposed to *B. marinus* die if untreated, while nearly 100% die in Florida. The animal may die within 15 minutes of mouthing or eating the toad (Fowler, 1992).

10.3 Reptiles

10.3.1 Snakes

Naturalists have known peculiar, strong, and persistent body odors of reptiles for some time. For instance, the mamba has a powerful curry-like, not unpleasant, smell that clings to the handlers' skin (Wingate, 1956). Some reptilian secretions possibly serve in defense against predators. The odor of the rat snake *Elaphe climacophora* affects mammalian predators. Brown bears (*Ursus arctos*) retreat from a dead rat snake and drool. They react similarly to a rope treated with the snake's odor (Kano, 1976). There is no indication that canids are repelled by snake odors. Coyotes (*Canis latrans*) are attracted to stations with secretion from the paired glands in the tail base of the western diamondback rattlesnake (*Crotalus atrox*). They rub and roll at these scent stations. Dogs lick, bite, and eat paper treated with the secretion (Weldon and Fagre, 1989). Domestic cats salivate and rub on secretion from a gland on the tail base of the gray rat snake, *Elaphe obsoleta*. The response was stronger than to skin samples of this species and cats ate less of food treated with the secretion (Wright and Weldon, 1990).

The blind snake *Leptotyphlops dulcis* is a specialized burrower that feeds on termites and ant brood. It follows the ant pheromone trails to find its prey (Section 12.1). When attacked by ants, this snake tilts its scales individually so that the skin appears silvery. While it writhes, it covers itself with feces and a clear viscous fluid, discharged from the anus. It may also assume a stationary coiled position. When it resumes searching and feeding, it is no longer attacked by ants. The effect lasts from 3 to 30 minutes (Gehlbach *et al.*, 1968). Here a predatory species protects itself from the defensive actions of its prey.

The cloacal secretion of *L. dulcis* also repels its predators, ophiophagous snakes. All five species of the tested colubrid snakes are repelled, including the ophiophagous species *Diadophis punctatus* and *Lampropeltis triangulum*. However, the cloacal secretion attracts conspecifics. This attraction possibly evolved first as the most important function of the secretion for the blind snakes (Watkins *et al.*, 1969).

Aposematically colored, the yellow-bellied sea snake, *Pelamis platurus* (Hydrophiidae), of the eastern Pacific has venom and is distasteful. It has no known aquatic predators, although remains were found in murray eels and sharks. Predatory fish such as snappers refuse the snake. They reject its meat even when hidden in palatable squid. Predatory fish of the Atlantic ocean, however, ate the sea snake in experiments, and died after 1 of 12 meals (Rubinoff and Kropach, 1970).

Snake venoms

Since predators of snakes (and humans) have to deal with snake venoms as defenses, they are included here, even though they serve in predation. Snake venoms are primarily enzymes (proteins), especially of the phospholipase A_2 type, which breaks down cell membrane phospholipids hydrolytically. Other snake venoms such as *cobrotoxin* contain peptides with 60–70 amino acid residues. Pharmacologically, they have neurotoxic, cytotoxic, anticoagulant, and other effects. The neurotoxins, in turn, can have pre- or postsynaptic effects. Snake venoms with both neurotoxic and hemolytic effects on the heart are known as cardiotoxins. Cytotoxins attach to the cells of blood vessels and cause hemorrhage. Snake venom factors may stimulate or inhibit blood clotting. Finally, platelet-active factors can contribute to hemorrhage.

Snakes themselves are immune to their own venom. Rattlesnakes have antibodies in their serum. Indeed, serum from snakes is more effective in protecting mice from rattlesnake venom than is commercial antivenin. Non-venomous snakes such as the king snake, *Lampropeltis getulus*, which also prey on crotalid rattlesnakes, are resistant to crotalid venom. Mongooses are very skilled in avoiding bites by cobras, but they also are resistant to cobra venom. A large dose, however, will kill even a mongoose. The Virginia opossum is also resistant to crotalid venom and the European hedgehog, *Erinaceus europaeus*, neutralizes some of the hemorrhagic factors in viper poison. Domestic and feral pigs are also impervious to venom by virtue of their thick skin and the subcutaneous fat layer, which retards absorption of venom into the blood. If this protection is circumvented by intramuscular injection, the snake venom can kill a pig. Cats resist snake venom very well. They show local symptoms, vomit, and defecate, but recover after 2 days. Dogs are more sensitive, and horses are more sensitive than cattle. Birds appear not to be protected against snake venom, though raptors take snakes. The feathers are thought to protect against penetration of the skin by the fangs of the snake. Some people such as snake worshippers and individuals who have been repeatedly bitten claim immunity against snake venom. However, there is no evidence for long-term immunity. A herpetologist in Australia had himself hyperimmunized by repeated injections of snake venom over 1 year. At the end of that period, he received eight times the lethal dose of tiger snake, *Notechis scutulatus*, venom and survived. However, just 1 year later, the titer had dropped, and he was probably unprotected. Attempts to vaccinate people in high-risk areas have not been successful. On the contrary, there exists the danger of anaphylaxis (hypersensitivity) from acquired sensitivity to snake venom.

10.3.2 Lizards

Gila monsters (*Heloderma horridum* and *H. suspectum*) bite when bothered. They hold on for 10–15 minutes. Their powerful mouth can be pried open only with a tool such as a screwdriver. Fortunately, we know of no human fatalities. (Cardiac arrest has occurred through excessive fear, alcohol overdose, or infection.) The *gilatoxin* from the oral glands contains phospholipase A_2 and hyaluronidase, which breaks down hyaluronic acid, a bonding agent between cells. This permits more rapid spread of venom between cells. The Mexican bearded lizard, *Heloderma h. horridus*, produces a gilatoxin component, a serine protease with 245 amino acid residues. This compound lowers b lood pressure (Utaisincharoen *et al.*, 1993). Monitor lizards (Varanidae) share nine toxin types with snakes and helodermatids. Oral venoms of the Australian lace monitor, *Varanus varius*, lower blood pressure and slow blood clotting. Prey lose consciousness and bleed to death (Fry *et al.*, 2005).

10.3.3 Turtles

Most aquatic and semiaquatic turtles have two or more pairs of Rathke's glands, which release a secretion through duct openings in the axillary, inguinal, and inframarginal areas. If a probe is applied to the axillary or inguinal regions of loggerhead (*Caretta caretta*) or Kemp's ridley (*Lepidochelys kempi*) turtles, the glands near the stimulation discharge secretion, and those on the opposite side of the body do not (Weldon and Williams, 1988). Disturbed stinkpot turtles (*Sternotherus odoratus*, Kinosternidae) of the eastern USA exude a musky secretion from a pair of glandular openings. The discharge may signal the undesirable taste of the turtle's flesh (olfactory aposematism). Four phenylalcanoic acids in this secretion have been identified. These are phenylacetic, 3-phenylpropionic, 5-phenylpentanoic, and 7-phenylheptanoic acid (Eisner *et al.*, 1977). Table 10.2 lists some chemical defenses in amphibians and reptilians.

Table 10.2 Defense compounds in amphibians and reptiles

Species	Compounds	Class of compound	Effect
California newt *Taricha torosa*	Tetrodotoxin	Alkaloid	Local anesthetic, death
Toad	Epinephrine, norepinephrine,	Biogenic amines	Sympathomimetic
	Bufotalin	Bufogenins (steroids)	Digitalis like (stimulates heart contraction)
	Bufotoxins	Bufogenins esterified with suberylarginine (steroids)	Digitalis like
Salamanders *Salamandra, Triturus* spp.	Samandarin	Steroid	Spasms, inhibits breathing, death
Dart poison frogs, *Phyllobates* spp.	Batrachotoxins	Alkaloids	Na^+ channels
	Pumiliotoxins	Alkaloids	Ca^{2+} channels
	Histrionicotoxins	Alkaloids	K^+ channels
Snakes	Phospholipases A_2	Proteins, peptides (cobrotoxin), amino acids	Neurotoxins
Gila monster *Heloderma horridum, H. suspectum*	Phospholipase A_2	Protein	Neurotoxin
	Hyalurinodase	Protein	Breaks down hyaluronic acid

10.4 Birds

Judging by their variable and often disagreeable taste, birds and their eggs may be chemically defended. In extensive experiments, H. Cott (1954–55) tested birds' eggs and flesh with humans, rats, cats, hedgehogs, ferrets, and hornets. To rats and hedgehogs, the *eggs* of Galliformes and Pelicaniformes were most palatable, least palatable those of Falconiformes and Passeriformes. Humans resembled hedgehogs and rats: all three most preferred eggs of chicken, coot, and kittiwake, while the eggs of the songbirds *Carduelis cannabina, A. schoenobaenus*, and *Sylvia communis* ranked at the bottom of all three species of egg tasters. The feeding habits of the birds played a role: eggs of piscivorous, piscivorous/invertebrate eating, and omnivorous bird species were most palatable. Least palatable were eggs of insectivorous and insectivorous/herbivorous birds. For humans and hedgehogs, no close relationship between the taste of the flesh of a bird species and the palatability of its eggs was found.

Four findings are particularly relevant to the question of chemical defense. First, the *smaller* the eggs the less palatable they are. Smaller eggs have more predators. Second, the eggs of *solitary nesters* are more unpalatable than those of colonial nesters. Third, palatability to the rat varied with nest site *location*: the most-palatable eggs came from birds that nest on cliffs or stacks, or over water. The least-palatable eggs belonged to birds nesting in vegetation less than 90 cm above ground or in holes of trees or masonry. Birds nesting in ground burrows or trees were intermediate. Fourth, *shell coloration* correlated with palatability: cryptic eggs, such as those of ducks and game birds, were considerably more palatable than white/"immaculate" eggs. Combined, these four variables provide a *vulnerability score*. Cott's composite vulnerability index correlated negatively with palatability. Four of the five most palatable species – herring gulls, lesser black-backed gull, great black-backed gull, and gannet – are also the four least-vulnerable species. At the other end of the scale, the six species with the most repugnant eggs, all passerines, fall all in the most vulnerable class. These are the lesser white throat, linnet, great tit, eastern house wren, blue tit, and white throat.

A combination of egg flavor, color, and induced gastrointestinal illness can condition egg predators to avoid eggs on sight. Free-ranging ravens (*Corvus corax*) were presented with surrogate eggs in their breeding territories. The eggs had been injected with a tasteless, illness-producing cholinesterase inhibitor (2,3,5- and 3,4,5-trimethylphenylmethyl carbamate [trade name U.C. 27867]). The treated eggs had one color, the controls another. Of the treated eggs, 70% survived intact, but only 38% of the controls. Such *conditioned taste aversion* most likely plays a major role in the evolution of aposematism and mimicry. Non-toxic mimics may benefit and survive better if they appear *after* a conditioned taste aversion has been established in local predators (Nicolaus, 1987). For application in conservation efforts, it is important that the visual cue (here egg color) suffices for avoidance since "sampling" the eggs would obviously destroy them. Many natural defense compounds have a taste, usually bitter, that plays a role in establishing an aversion. In Nicolaus' study, illness rather than a noxious taste alone was necessary to produce an aversion. Furthermore, predation was lower within raven territories, presumably because they defended their food resources against non-territorial ravens and other predators. Lower predation pressure within breeding territories than outside has been demonstrated for other birds, such as the south polar skua, *Catharacta maccormicki* (Müller-Schwarze and Müller-Schwarze, 1973, 1977).

The palatability of the *flesh* of birds to humans is broadly correlated with their visibility, including plumage color and behavior. Cott (1947) noted in Egypt that hornets (*Vespa orientalis*) fed on a carcass of a dove but ignored that of a kingfisher

FIGURE 10.8 Hornets, *Vespa orientalis*, feed on carcass of a palm dove, *Streptopelia senegalensis aegyptiaca*, left, while ignoring that of a pied kingfisher, *Ceryle rudis rudis*, on the right. (From Cott, 1947.)

(Fig. 10.8). In Zambia, he asked people to rate cooked meat of the pectorals of 200 species of birds from 57 families for their palatability. Humans ranked these birds from "ideal" to "inedible." Hedgehogs, rats, ferrets, and cats followed the same rank order. Birds (and their eggs) that are very vulnerable to predation, such as auks, turacos, kingfishers, starlings, and woodpeckers, tasted the worst. These birds were also more conspicuous than better-tasting species, with blue, red, rufous, white, and black predominating. A general pattern emerged: visibility of the bird was negatively correlated with palatability (Cott, 1947, 1952, 1953, 1954; Cott and Benson, 1969).

Birds may actually defend themselves by toxins. One of the most toxic animal compounds known, a homobatrachotoxin, was found in a New Guinea bird, the hooded pitohui, *Pitohui dichrous* (Dumbacher *et al.*, 1992). These birds are known in New Guinea as "rubbish birds" because the skin tastes bad. Two species of pitohui also emit a "strong sour odor." Three species of the genus *Pitohui* proved to contain homobatrachotoxin, a steroidal alkaloid. The same compound is found in a South American poison-dart frog. The frog and the pitohui are brightly colored (aposematism). People who handle pitohuis suffer watery eyes and runny noses. Five species of *Pitohui* carry batrachotoxins, in some species particularly in the contour feathers of the belly, breast, or legs. The most toxic species are *P. dichrous* and *P. kirhocephalus*. Batrachotoxins also occur in feathers and skin of a bird in a second genus, the blue-capped ifrita, *Ifrita kowaldi*. Populations differed in batrachotoxin levels, suggesting a dietary origin (Dumbacher *et al.*, 2000). A beetle that was found in pitohui stomachs also contains batrachotoxins. The possible targets of the toxins are feather lice, Phthiraptera. They choose a non-toxic over a highly toxic feather. On toxic feathers, the lice die sooner (Dumbacher 2003). The pitohuis' predators are snakes, hawks, such as the brown goshawk, and human hunters, who do not eat them because of the poison (Dumbacher 2003).

The preen gland (uropygial gland) of young hoopoes (*Upupa epops*) is largest at 12 days of age. It possibly has a repellent effect on intruders. A ring of specialized feathers carries and dissipates its odor. Upon approach, a predator is sprayed with a liquid from the large intestine, accompanied by a hissing sound (Sutter, 1946). Petrels (Procellariidae) defend themselves by squirting stomach oil with undigested food at predators and intruders. The aldehydes octanal and decanal, compounds with citrus-like odor in the feathers of crested auklets, appear to repel and kill ectoparasites such as ticks (Douglas *et al.*, 2004). The hoatzin (*Opisthocomus hoazin*) of South America is a sluggish, vulnerable large bird that predators appear to avoid. The bird has a strong, unpleasant smell and Peruvians call it *chancho* (pig). The odor possibly derives from the unique leaf diet of the hoatzin and perhaps protects it from predators.

10.5 Mammals

The duck-billed platypus, *Ornithorhynchus anatinus*, uses a slightly curved poison spur on its ankles (tarsus) in male–male fights over females. It possibly uses it also against predators. In the resting position, the spur is retracted against the leg. Activated, it is raised at a 90 degree angle to the leg. The platypus kicks to insert the spur into the victim. The spur connects to a duct that, in turn, comes from a reservoir downstream from a kidney-shaped crural venom gland located on the medial aspect of the thigh. The venom contains proteins and toxicity varies seasonally, possibly coinciding with the breeding season. Humans struck on their hand will be crippled for months (Vaughan, 1978). Echidnas have a spur but a non-functional gland.

The secretion of the anal glands of skunks epitomizes mammalian defense secretions. Sulfur compounds in the scent of striped skunk (*Mephitis mephitis*) include *trans*-2-butene-1-thiol, 3-methyl-1-butanethiol (Andersen and Bernstein, 1975), *trans*-2-butenyl thioacetate, 3-methylbutanyl thioacetate, and 2-quinolinemethane thiol (Wood, 1990; Fig. 10.9). The thiols, once called mercaptans (because they captured mercury atoms), are like alcohols with a sulfur atom in place of the oxygen. In the striped skunk, the thioacetates break down in water. They yield thiols, which have the more potent, characteristic skunk odor and linger for up to 2 weeks. In the spotted skunk, by contrast, the

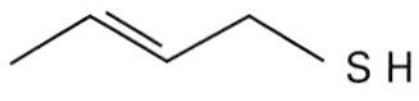

(*E*)-2-Butene-1-thiol (38–44%)

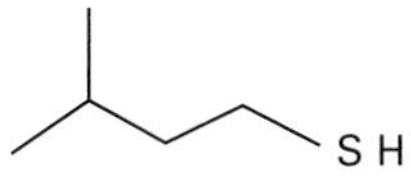

3-Methyl-1-butanethiol (18–26%)

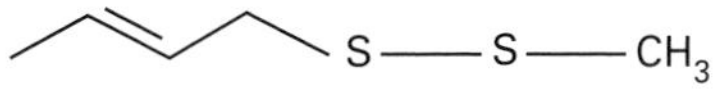

(*E*)-2-Butenyl methyl disulfide (?)

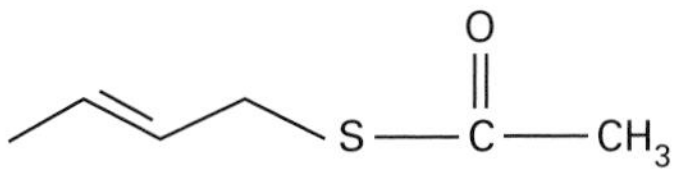

(*S*)-(*E*)-2-Butenyl thioacetate (12–18%)

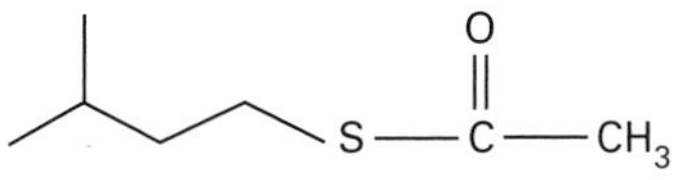

(*S*)-3-Methylbutanyl thioacetate (2–3%)

FIGURE 10.9 Major sulfur compounds in the anal sac of the striped skunk *Mephitis mephitis*. (After Andersen and Bernstein [1975], Wood [1990], and Wood *et al.* [1991].)

thiols are not tied up in slow-release compounds, and the odor dissipates in a few hours (Wood *et al.*, 1991).

The black-and-white contrasts of the skunks are thought to function as a warning color. Intriguingly, a marsupial in New Guinea displays convergent features: the long-fingered triok, *Dactylopsila palpator*, is black and white and also smells like a skunk (Flannery, 1998).

A toxin, blarinatoxin in saliva from the submaxillary and sublingual glands of the North American short-tailed shrew, *Blarina brevicauda*, serves in defense and for paralyzing the prey. Ducts lead to the incisor teeth, which may be modified with channels to direct venom in the wound. Shrew venom is proteolytic and neurotoxic. The poison is not dangerous to humans, who experience a burning sensation and minor swelling that can be painful for several days. To small mammals it can be lethal. It dilates blood vessels and leads to irregular respiration, paralysis, convulsion, and eventually death. In one experiment, one third of bitten mice died. Small mice (body weight under 10 g) died in larger numbers (80%), and often within minutes (average 157 minutes). They were most affected when bitten near the head, and least at tail or feet. The severity of the wound determines mortality. Venom from one shrew can kill 200 mice. The shrews also bit earthworms and cached them comatose. The salivary toxin seems to function in stunning or paralyzing the prey and facilitating storage. It is not important in killing rodents (Tomasi, 1978). Chemically, blarinatoxin is a high-molecular-weight, water-soluble protein with 253 amino acid residues (Kita *et al.*, 2004).

10.6 *Pars pro toto*: decoy odors

Some of the seemingly contradictory observations that a predator is actually *attracted* to a presumed "defense secretion" may be explained by the *pars pro toto* principle. A prey animal in danger, when it is being detected, pursued, or handled by a predator, or simply sneaking away before detection, discharges an odor that attracts the curiosity of the predator. While the odor cloud lingers or is moved by air currents, the prey animal moves off in another direction. This would work especially well when vision is impaired as in the dark, or in heavy vegetation. The emitted odor would serve as *decoy*, the olfactory equivalent of the severed tail of the lizard, and an energetically much cheaper one. Possible examples are the metatarsal odor of the black-tailed deer, *Odocoileus hemionus columbianus*, which is discharged during alarm (Müller-Schwarze *et al.*, 1984), and the secretions from anal sacs in snakes that are sniffed by predators. Similarly, the discharge of stomach oil in procellariid seabirds may also distract predators. Further candidates for chemical *pars pro toto* defense are the odors noticeable to humans when feral pigs flee in dense cover (Section 7.2), and a puff of odor from a fleeing deermouse, *Peromyscus* sp. (Müller-Schwarze, personal observations).

10.7 Invertebrate allomones that deter vertebrate predators

10.7.1 Taste aversion to invertebrate prey

At overwintering sites of the monarch butterfly (*Danaus plexippus*) in Mexico, only one of the three local mouse species, *Peromyscus melanotis*, actually feeds on the butterflies. The monarchs contain cardiac glycosides (CG) and pyrrolizidine alkaloids (PA). All three species of mice have similarly low avoidance thresholds to PA (specifically, monocrotaline). But *P. melanotis* is less sensitive to CG (specifically, digitoxin) than the other two, *Reithrodontomys sumichrasti* and *Peromyscus aztecus*. Laboratory tests indicate that PA is toxic to young mice, but CG is not; even so, PA does not reduce food intake. The mice do not know how to avoid specifically compounds that will damage them so, instead they generally avoid bitter compounds. Since they cannot vomit, mice benefit from avoiding eating many bitter, potentially harmful foods but, by "being on the safe side," they may forgo feeding on potentially useful foods. The three mouse species differ in their ways to overcome the chemical defenses of their prey. Therefore, the effectiveness of the monarch's defenses depends on the species of predator: *P. melanotis* selects monarchs with low CG concentrations, learns to reject the CG-laden cuticle of the butterfly, and is insensitive to the bitter taste and toxic effects of CG. *R. sumichrasti* uses only the first two mechanisms, and *P. aztecus* only the first (Glendinning *et al.*, 1990).

Fish, such as the reef predator *Thalassoma bifasciata*, avoid sponges. Organic extracts of sponges of the sponge genus *Ircinia*, mixed with food, deterred *T. bifasciata* from feeding. Non-volatile furanoseterpene tetronic acids in the extracts are responsible for that effect. By contrast, volatile compounds from the sponges, such as dimethyl sulfide, methyl isocyanide, and methyl isothioyanate, do not help to defend the sponge against fish (Pawlik *et al.*, 2002).

Table 10.3 lists some defense compounds, mostly alkaloid, that invertebrates use to deter vertebrate predators.

10.8 Recycled animal and plant materials

10.8.1 Birds

Ornithologists have known for a long time that certain birds use materials from other animals or plants in or around their nests. The burrowing owl (*Speotyta cunicularia*) of the Western hemisphere lines its burrow with cow dung. It is believed that the dung deters badgers from preying on the owls. Of nests lined with dung, only 8% were lost to badgers, but 54% of the unlined nests (Green, 1988). It is unclear whether the dung masks bird odor or acts as a repellent. Red-breasted nuthatches, *Sitta canadensis*, smear pine resin around their nest cavities. This is assumed to serve a hygienic function, as pine resin is antiseptic, but it is also toxic to some insects (Bent, 1964).

Birds that breed in cavities and reuse their nest sites often incorporate green parts of aromatic plants into their nests. Starlings, *Sturnus vulgaris*, prefer certain species of plants to others. House sparrows, *Passer domesticus*, incorporate neem (margosa) tree, *Azadirachta indica*, leaves into their nests. Extracts from neem leaves repel arthropods and inhibit oviposition (Sengupta, 1981). The aromatic plants are thought to fulfill an important function by keeping down populations of microbes in the birds' nests (Mason and Clark, 1986).

Table 10.3 Allomones: invertebrate compounds that deter vertebrate predators

Vertebrate predator	Invertebrate prey	Compound(s)	Reference
Reef fishes	Orange sponge *Axinella* sp.	Indoles: herbindole A	Herb *et al.*, 1990
Reef fishes *Thalassoma bifasciatum*, *Eupomacentrus partitus*	Caribbean ascidian *Tridemnium sodium*	Macrolides: didemnin B and nordidemnin B	Lindquist *et al.*, 1992
Reef fishes	Sea hares *Stylocheirus longicauda*	Alkaloids: malyngamide A and B, and B acetate, from cyanobacteria	Paul and Pennings, 1991
Blind snake *Leptotyphlops dulcis*	Formicid ant *Neivamyrmex* sp.	Skatole (attracts, but repels other preying snakes and ants)	Watkins *et al.*, 1969
Lizards	Butterflies, months, beetles	Senecionine and lycopsamine type PAs, sequestered from Asteraceae and Boraginaceae	Various authors
Anole lizard *Anolis carolinensis*	Bug *Neacoryphus bicruris* (Heteroptera)	Senecionine type PAs sequestered from *Senecio smallii* (Asteraceae)	MacLain and Shure, 1985
Birds	Taiwan stick insect *Megacrania tsudeii*	Monoterpenoid alkaloid actinidine (3500 × conc. in defense gland)	Chow and Lin, 1986; Ho and Chow, 1993
Gray jay *Perisorius canadensis*	*Eumaeus atala floridana* (Lepidoptera, Lycaenidae)	Cycasin, from *Zamia floridana* (Cycadaceae), stored in larvae	Bowers and Larin, 1989; Bowers and Farley, 1990
Glaucous-winged gull *Larus glaucescens* and sea otter *Enhydra lutris*	Clams	Paralytic shellfish toxins (Guanidins [saxitoxins]), from dinoflagellates	Pawlik, 1993; Kvitek, 1991
Tree sparrow *Passer montanus*	Swallowtail butterfly *Atrophaneura alcinous*	Aristolochic acids (from *Aristolochia debilis*), in test applied to rice grains	Nishida and Fukami, 1989
Mice *Peromyscus* spp., *Reithrodontomys sumichrasti*	Monarch butterfly	Cardiac glycosides and PA	Glendinning, *et al.*, 1990

PA, pyrrolizidine alkaloid.

In a North American study, starlings preferred agrimony (*Agrimonia* sp.), wild carrot (*Daucus carota*), flea bane (*Erigeron* sp.), yarrow (*Achillea millefolium*), red-dead nettle (*Lamium purpureum*), and rough goldenrod (*Solidago rugosa*). In laboratory tests, plants preferred by starlings retarded the hatching of eggs of the louse *Menacanthus* sp. that lives on the feathers but none of the tested plants was effective against the fowl mite *Ornithonyssus sylviarum*. Five of the selected plant species inhibited the growth of at least one bacterial pathogen (*Streptococcus* sp., *Staphylococcus aurealis*, and *Pseudomonas* sp.). They were never effective against *Escherichia coli*.

The total concentration of volatiles was higher in preferred than in other plants. The volatiles in the preferred plants are mostly mono- and sesquiterpenes such as myrcene, α- and β-pinene, and others. One preferred plant, the rough-stemmed goldenrod *S. rugosa* of old fields, contains the sesquiterpenes 2-bornyl acetate and farnesol. These compounds are implicated as potent analogues of insect juvenile hormone, which suppresses molting in arthropods. They possibly impair reproduction of ectoparasites by delaying their development to terminal instars. Reduced populations of parasites would benefit the nestling birds. Volatiles from the plants in question elicit strong responses from the olfactory nerve of the birds. Tests with aromatic plant parts underneath food containers showed that starlings can discriminate different plants by odor (Clark and Mason, 1985; Mason and Clark, 1986). If plant odors were paired with gastrointestinal malaise caused by the bird repellent methiocarb, a conditioned avoidance was produced. If both olfactory nerves were sectioned before the conditioning, no avoidance occurred (Clark and Mason, 1987).

Experiments have shown that aromatic herbs in the nest do not reduce the ectoparasite load in starlings but may stimulate parts of the immune system that mitigate the effects of ectoparasites (Gwinner *et al.*, 2002). Corsica blue tits (*Parus caeruleus ogliastrae*) select 6–10 species of aromatic plants out of the approximately 200 species that occur in their habitat and place them in the nest at the nestling stage. The birds will replenish the aromatic plants when the volatile components have been lost. The chosen plants had proven antibacterial antiviral, antifungal, and insecticidal properties. Interestingly, 40% of the plants involved possess known activity against human immunodeficiency virus 1. (Petit *et al.*, 2002; see p. 379).

10.8.2 Mammals

Both the African hedgehog, *Atelerix pruneri*, and the long-eared hedgehog of Asia, *Hemiechinus auritus*, use defense secretions of toads. The animal takes the secretion up into its mouth and then licks its spines. The spines then become covered with frothy saliva. This behavior does not depend on learning and is released by several species of toads and salamanders, and even by substances such as tobacco juice (Brodie, 1977).

Wedge-capped capuchin monkeys (*Cebus olivaceus*) in the llanos of central Venezuela appropriate the defenses of millipedes to repel mosquitoes. They rub the 10 cm long millipede *Orthoporus dorsovittatus* over their fur, transferring benzoquinones from the millipede. The behavior is most prevalent during the rainy season when mosquitoes are particularly numerous. The mosquitoes also transfer eggs of parasitic bot flies into the skin of the monkeys. We do not yet know whether rubbing milliped benzoquinones into the fur prevents bot fly infestation. Rubbing is a social event: up to four monkeys may share a millipede (Valderrama *et al.*, 2000; see also p. 377 and p. 382).

10.9 The question of coevolution between predator and prey

Coevolution between predator and prey has often been invoked. In one well-investigated case, support for coevolution is weak. In Mexico, the black-headed grosbeak, *Pheucticus melanocephalus*, and the black-backed oriole, *Icterus galbula abeilli*, feed on monarch butterflies, *D. plexippus*, which defend themselves with high levels of cardenolides. The cardenolides are particularly concentrated in the cuticle of the butterfly. The grosbeaks are insensitive to the cardenolides. The oriole bypasses the cuticle and thus avoids emetic dosages of the toxin. It pries open the cuticle and feeds on the contents of the abdomen, which contain less cardenolides than the cuticle. Has the oriole's feeding technique coevolved to breach the defenses of the butterfly? Is the grosbeak insensitive to cardenolides because it responded to these compounds in a coevolutionary process? The answer is no in both cases. These birds are exapted (formerly "preadapted") (Gould and Vrba 1982; Brower *et al.*, 1988): Other orioles that do not feed on butterflies also use the prying movement in feeding. The northern oriole, *Icterus galbula*, pierces fruit and then gapes and eats the pulp without the skin. The black-throated oriole uses the same movements and then laps up the juice with its bifurcated and brushy tongue. The grosbeak is also exapted: early in evolution it raised the amount of toxins it can tolerate (Brower *et al.*, 1988).

10.9.1 Competition

One animal species can suppress a competing species by means of chemical signals. An example of such *competitive exclusion* (*sensu* Gause, 1934), or "interactive segregation," are two species of spiny mice of the genus *Acomys* in southern Israel. They are sympatric and occupy the same rocky habitat. The common spiny mouse, *A. cahirinus*, is nocturnal and forces the golden spiny mouse, *A. russatus*, to be active in the daytime heat of the Negev desert. If *A. cahirinus* is removed from the field, *A. russatus* becomes nocturnal (Shkolnik, 1971). Kept by itself in the laboratory, *A. russatus* is also nocturnal. Odor of feces and urine of the competing species *A. cahirinus*, added to the air supply shifts the activity rhythm of *A. russatus* if the mice are kept in 12 hours darkness and 12 hours light. Under "free-running conditions" (i.e. in constant dim red light), *A. russatus* lower their activity level, as measured by oxygen consumption, in response to these chemical cues from *A. cahirinus* (Haim and Fluxman, 1996).

11

Allomones II: plant chemical defenses against herbivores

The plant world is not colored green; it is colored morphine, caffeine, tannin, phenol, terpene, canavanine, latex, phytohaem–agglutinin, oxalic acid, saponin, L-dopa, etc.
JANZEN (1978)

Herbivores select certain plant species or parts and reject others. Plant defenses determine food choices as much as nutritional value does. Plants can defend themselves mechanically as with thorns, hairs, waxes, or structural fibers, and chemically with secondary plant compounds. Mammals have had to cope with plant defenses since they adapted to an herbivorous lifestyle approximately 85–100 millions years ago (Archibald, 1996).

In 1888, Stahl suggested that plants use toxic chemicals as defense against herbivores based on his feeding experiment with snails. Fraenkel (1959) postulated that "secondary compounds in plants exist solely for the purpose of repelling and attracting insects." We now know that these compounds are aimed at vertebrates, other invertebrates, and microbes as well, and in many cases their roles are still being debated. Here I apply the term "allomone" in its widest sense: compounds that benefit the "sender," even though many are not "signals" in the strict sense. Therefore, "donor" or "originator" organism is a better term. In the metabolism of the herbivore, the receiving organism, such foreign compounds are termed xenobiotics, whether natural or synthetic.

The interactions of plants and vertebrate herbivores can be broken down into several questions.

- What defense compounds do plants use?
- What are their effects on mammalian and avian herbivores?
- What defense strategies do plants use? This includes species differences based on ecological needs; variations of plant defenses between plant parts (leaf, bud, internode, terminal shoot, bark, root) or growth stages that correlate with vulnerability; and seasonal variation of plant defenses.

Table 11.1 Classes of plant compound with antiherbivore defense functions

Class	Examples
Phenolics	
Simple phenolics	Phloroglucinol, gentisic acid, vanillin, caffeic acid, ferulic acid
Complex phenolics	Coumarins, phenolic quinones, lignins, flavonoids, stilbenes, hydrolyzable tannins, condensed (or catechin) tannins, phenolic lipids
Terpenoids	Mono-, sesqui-, di-, and triterpenes
Alkaloids	Nicotine, guinine, morphine, colchicine, strychnine
Saponins	Steroidal saponins, triterpene saponins
Mustard oils	
Cyanogens	Dhurrin, taxiphyllin
Cardiac glycosides	Cardenolides, butadienolides
Proteins	Trypsin inhibitors
Non-protein amino acids	Selenocystathionin, L-mimosine
Phytohemagglutinins (lectins)	

- What are the mechanisms of effective defense? Is the plant chemistry systemic and genetically "hardwired," or is it programmed to respond to need, termed "induced defense"?
- Is intensity of herbivory on specific plants correlated with levels of secondary plant metabolites?
- How do herbivores avoid or deal with ill effects of plant defense compounds?
- Is there evidence of coevolution between herbivorous vertebrates and their food plants?

This first section presents the "cast of characters," the compounds involved in defense.

11.1 Classes of plant defense compound

Many thousands of secondary plant compounds have been identified; and 400000 are suspected to exist. These numbers provide a great incentive for "chemical prospecting" (Eisner, 1989). The most prevalent, broad classes of plant secondary compound are phenolics, alkaloids, and terpenoids (Table 11.1).

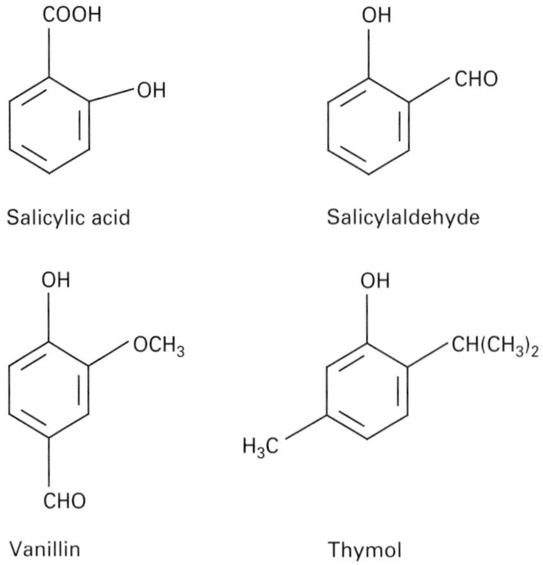

Salicylic acid

Salicylaldehyde

Vanillin

Thymol

FIGURE 11.1 Structures of simple phenolics.

11.1.1 Phenolics

In abundance second only to carbohydrates (Levin, 1971), phenolics occur in plants as complex mixtures. Their name derives from "phen-ol" (benzene), from Greek *phen-* (phainein): "to show." They range from simple molecules such as phloroglucinol to very complex ones such as lignins. Many phenolics are harmful to herbivores, e.g. by binding to proteins, while others are beneficial as antioxidants. Compounds with more than one phenol unit are designated polyphenols. Gallic acid is most readily oxidized and may be the best antioxidant in the context of human nutrition. Salicylic acid is also an antioxidant, but less so than gallic acid. Both may contribute to the "French paradox" – good health despite unhealthy components in diet (de Lorgeril *et al.*, 2002).

Simple phenolics

Simple phenolics are based on a six-carbon ring. Examples are salicylic acid and salicylaldehyde, vanillin, and thymol (Fig. 11.1).

Complex phenolics

An example of more complex phenolics is coniferyl benzoate (Fig. 11.2), which inhibits feeding in birds (Jakubas *et al.*, 1989), as discussed below. It binds to skin proteins and dietary proteins.

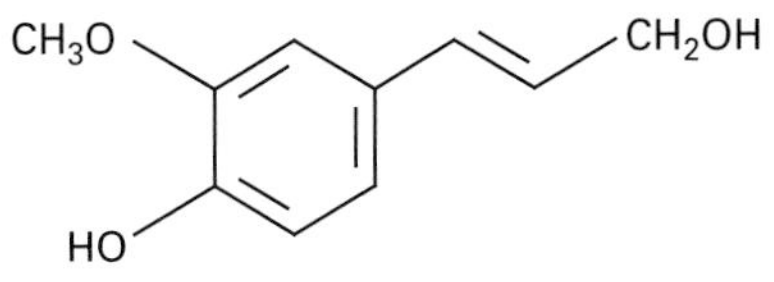

Coniferyl alcohol

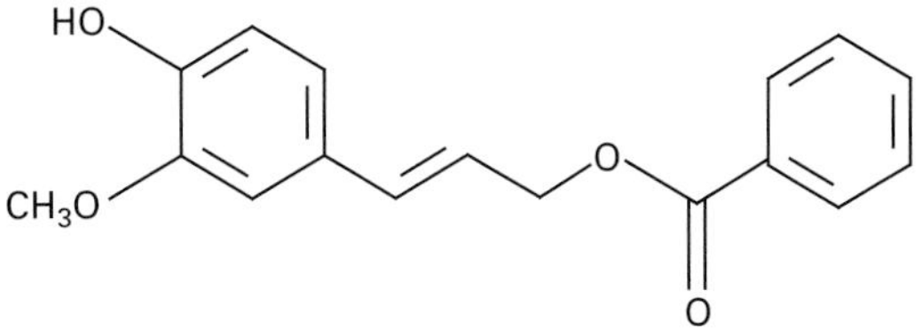

Coniferyl benzoate

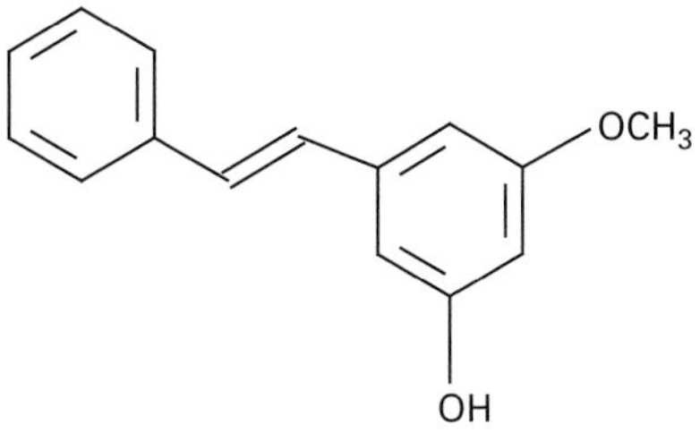

Pinosylvin methyl ether

FIGURE 11.2 Structures of complex phenolics.

Flavonoids

We know of thousands of flavonoids. They have a C6–C3–C6 skeleton and are responsible for the bright colors in flowers. The bitter taste in citrus fruits is caused by flavonone glycosides.

Isoflavonoids have a substituent at C-3. Examples are genistein and formononetin (see Fig. 11.11, below). Isoflavones exert estrogenic activity in mammals, interfere with reproduction in birds, and may render farm animals infertile. For example, sheep feeding on isoflavonoid-containing *Trifolium subterraneum* suffer reduced lambing (Harborne, 1993).

Stilbenes (from the Greek *stilbos*, glistening, after shiny crystals), are two-ring structures of the C6–C2–C6 type; examples are pinosylvin and pinosylvin methyl ether (Fig. 11.2). They play a role as antifeedants for mammals.

Tannins

Tannins have molecular weights of 0.5–20 kDa, with no upper limit. They are responsible for bitter or astringent taste of plants. The dry, astringent sensation brought about by complexation with mucoproteins of the mouth is thought to be repellent to herbivores (Harborne, 1993) while a moderate level of astringency can be pleasing to the human palate. Unripe fruits such as bananas, apples, or persimmon contain high levels of tannins (Mehansho *et al.*, 1987). This is thought to delay consumption by frugivores until the seeds have matured and can benefit from dispersal by animals and possibly from improved germination after passing through their digestive tracts (Section 12.2).

Tannins derive their name from their ability to tan (i.e. they combine with protein). They render plants less palatable and impair digestion by binding with the buccal mucosa, dietary proteins, and digestive enzymes of the animal. Tannins are thought to bind to proteins upon destruction of plant tissue by herbivores. This reduces the nutritive value of the plant to the herbivore. Some tannins such as oak gallotannins, are even toxic to livestock and rabbits (Meyer and Karasov, 1991). However, Martin and Martin (1983) have questioned the role of tannins as plant defense against herbivores.

Two different types of tannin, condensed and hydrolysable tannins, play important roles.

1. **Condensed or catechin tannins** (catechol tannins, Fig. 11.3) are the most common tannins in vascular plants, occurring in three quarters of gymnosperms and over half of the angiosperms. They are unbranched, linear polymers of flavonoid compounds (flavan-3-ols), linked through acid-labile carbon–carbon bonds. Condensed tannins may protect plant cell walls against microbial attack and so may affect microbial fermentation of plant cell walls in herbivores.
2. **Hydrolyzable tannins,** also known as tannic acids, are restricted to the dicotyledons. They have an astringent taste and the ability to tan leather. Their basic building block is gallic acid or ellagic acid (Fig. 11.4). These acids are esterified to the hydroxyl groups of glucose. Since one glucose unit can accommodate esterification with five molecules of gallic acid, the simplest hydrolysable tannin is pentagalloylglucose (Fig. 11.4). Tannins usually occur as mixtures. For instance, "tannic acid" of commerce is a mixture of free gallic acid and various galloyl esters of glucose.

Differences between hydrolyzable and condensed tannins are listed in Table 11.2.

Phenolics are responsible for the coloration of "black-water rivers," particularly in the tropics. An example is the Rio Negro, which remains distinct in its black color for many miles as it flows into the milky brown Amazon. In and near such "black" rivers, fauna and flora are considered impoverished (Swain, 1979).

Lignins are high polymers of phenolic alcohols such as coniferyl alcohol (Fig. 11.2). They increase the toughness of the plant and form wood, obviously useful for defense against herbivores.

11.1.2 Terpenoids

Terpenoids are the most diverse class of organic compounds in plants and the largest group of allelochemicals. The word is derived from "terpentin"

Table 11.2 Differences between hydrolyzable and condensed tannins

	Hydrolyzable tannins	Condensed tannins
Building blocks	Gallic acid, galloyl ester, ester-linked with glucose	Flavanoid units (Procyanidins), catechin
Occurrence in plant	Cell interior	Structural; tightly bound to cellulose wall of plant cell
Possible functions	Anti-herbivore: inhibition of digestion	Protect cell wall against microbes Retard decomposition of leaf litter: slow down release of nutrients (assumed to benefit mother plant) Mostly aimed at microbes, not herbivores
Mechanism	Bind to digestive enzymes of herbivores; variable molecules, specific for proteins	Inhibit pathogen's exo-enzyme system
When abundant	In earliest leaves but low levels in late summer and autumn	Late summer and autumn leaves
Metabolic cost	High: ties up phenylalanine for phenolic portion, plus glucose	Lower

(German for turpentine) and there are approximately 15 000 terpenes. Terpenes are lipophilic, and the building blocks are five-carbon units with the branched carbon skeleton of isopentane. The basic units are sometimes called isoprene (Fig. 11.5*a*), because heat decomposes terpenoids to isoprene. Depending on the number of C_5 units fused, we distinguish mono- (C_{10}), sesqui- (C_{15}), di- (C_{20}), tri- (C_{30}), tetra- (C_{40}) and polyterpenoids [$(C_5)_n$, with $n > 8$]. Alpha-Pinene and borneol (Fig. 11.5*b*) are examples of monoterpenes.

Nepetalactone (Fig. 11.5*c*), a monoterpene lactone (or iridoid) from the volatile oil of catmint, *Nepeta cataria*, excites domestic cats and other felids (Todd, 1962; Palen and Goddard, 1966; Hill *et al.*, 1976).

Three terpenoids (the monoterpenes carvacrol and β-thujaplicine, and the sesquiterpene thujopsene) from the japanese tree *Thujopsis dolabrata* inhibit gnawing by mice. A 1200-year old pagoda in Nara, Japan, made from wood of *Thujopsis*, shows no damage by rodents, insects, or microorganisms. Carvacrol is recommended as a long-term repellent against rodents (Ahn *et al.*, 1995).

The great diversity of terpenes helps to counteract tolerance by herbivores. In all, terpenes are not very toxic to vertebrates. Many mammals ingest a significant amount of terpenoids with their diet. Monoterpenes from pine oil added to the diet reduces food intake in red deer, *Cervus elaphus*, calves (Elliot and Loudon, 1987). The brush-tailed possum, *Trichosurus vulpecula*, detoxifies (+)-α-pinene to alcohol and carboxylic acid derivatives.

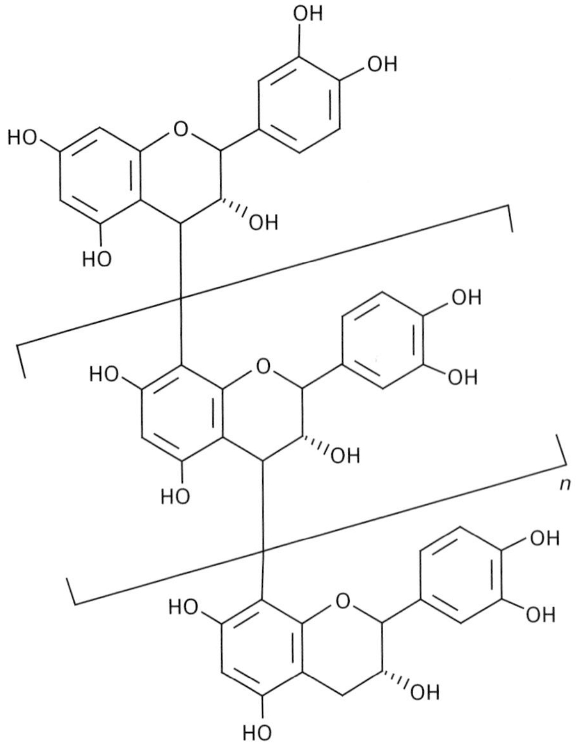

Basic structure of a condensed tannin

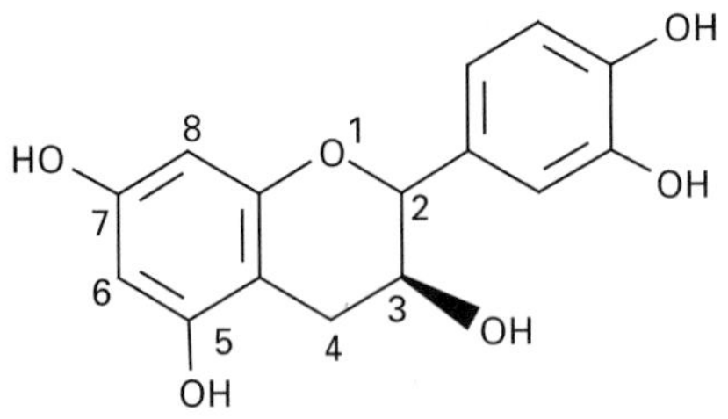

Catechin

FIGURE 11.3 Condensed tannin. Catechin, a flavan-3-ol, is one of the most common building blocks.

Sesquiterpene lactones (e.g. glaucolide A) are "bitter principles" in plants such as the genus *Vernonia* (p. 382). Cottontail rabbits, *Sylvilagus floridanus*, and white-tailed deer, *Odocoileus virginianus*, eat the one species of this genus that does not contain glaucolide A (Mabry and Gill, 1979). Examples of sesquiterpenes are germacrone (Fig. 11.5*d*) in Labrador tea, *Ledum groenlandicum*, and isovelleral in mushrooms.

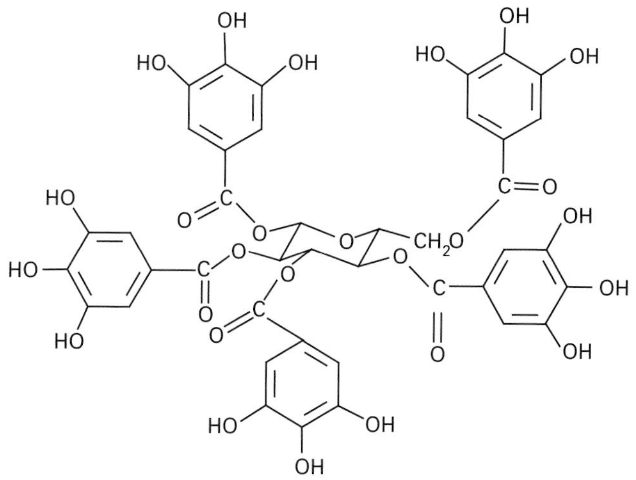

Pentagalloylglucose

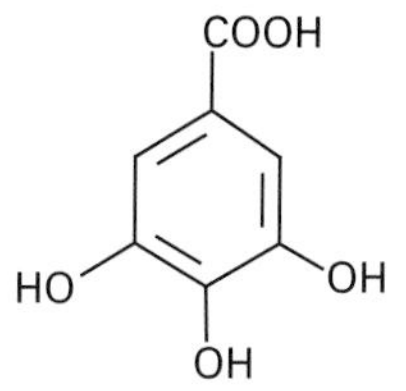

Gallic acid

FIGURE 11.4 Hydrolyzable tannin. In the structure shown, five gallic acid units surround a sugar (hexose).

Diterpenes from Euphorbiaceae are toxic to livestock, and so are daphnane and daphnetoxin, diterpenes from *Daphne mezereum* (Thymelaeaceae). Toxins from *Rhododendron* are diterpenes.

Triterpenes (C30) are common in birches, especially in bark. Papyriferic acid from paper birch, *Betula resinifera*, is a feeding deferent for snowshoe hares, *Lepus americanus* (Fig. 11.5) (Reichardt *et al.*, 1985).

Glycosides of compounds with a triterpenoid origin include saponins, cardiac glycosides, and glycoalkaloids such as solanine (in potatoes).

Saponins

Saponins occur in Liliaceae such as asparagus, in legumes, spinach, and yams. They are triterpenoid glycosides with soap-like properties. Many are glycosides of steroid alcohols, and all have a bitter taste. Two types are

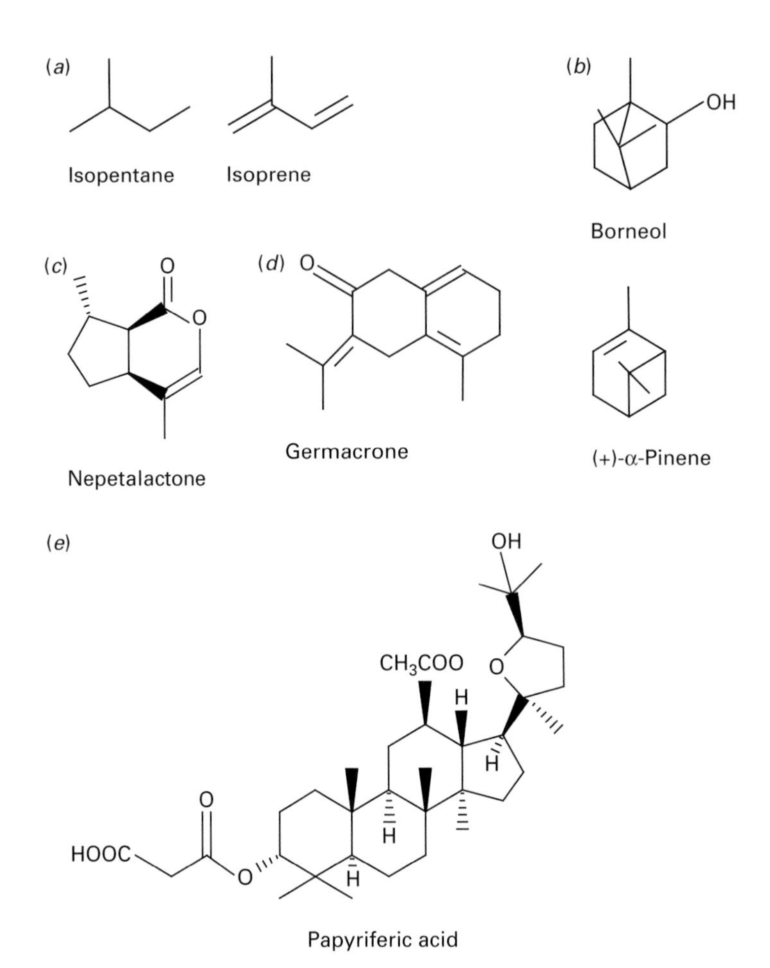

FIGURE 11.5 Terpenoids. (*a*) The basic five-carbon units isopentane and isoprene. (*b*) The monoterpenes borneol and α-pinene. (*c*) A monoterpene lactone, nepetalactone. (*d*) A sesquiterpene, germacrone. (*e*) A triterpene, papyriferic acid, from paper birch.

distinguished: *steroidal saponins* (C_{27} skeleton) and *triterpene saponins* (C_{30}). An example of the former is medicagenic acid (Fig. 11.6) in alfalfa (luzerne), *Medicago sativa*.

Saponins are foaming agents that hemolyze red blood cells. Injected into the bloodstream, they cause rapid lysis of erythrocytes. They also cause bloat in cattle by altering the surface tension of rumen contents and by trapping gas. Poultry are the most sensitive monogastric animals; on a saponin-containing alfalfa diet, they reduce feeding and grow more slowly (Applebaum and Birk, 1979). Mammals avoid saponin-containing plants, although livestock poisoning has been reported. Elephants avoid trees were the foliage contains steroidal saponins

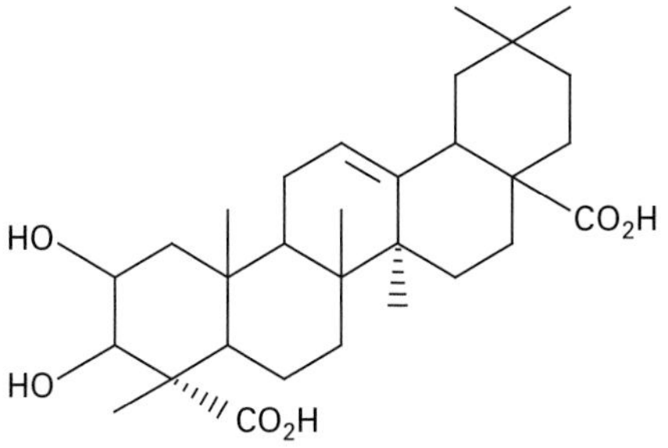

FIGURE 11.6 Medicagenic acid, a steroidal saponin.

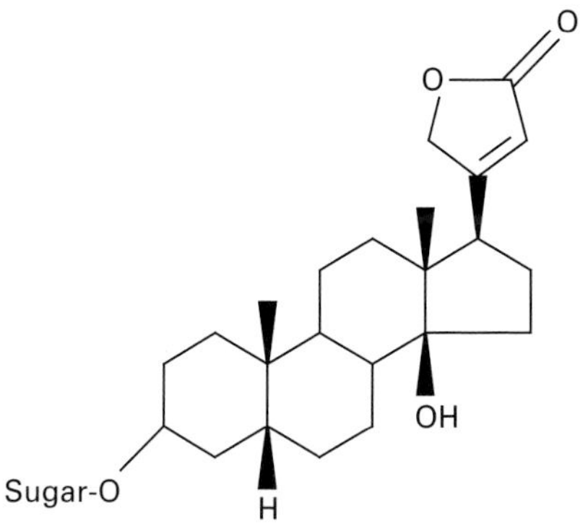

FIGURE 11.7 Cardiac glycosides consist of a sugar, a steroid skeleton, and a lactone ring. A five-membered lactone ring as here occurs in the cardenolide type.

(Jachmann, 1989). Saponins are used as fish poisons and also used as starting materials for steroid hormones.

Cardiac glycosides

Cardiac glycosides are bitter-tasting compounds that consist of three parts: steroid, lactone, and sugar. There are two groups, the cardenolides (Fig. 11.7) and the butadienolides. They are characterized by a 5- or 6-membered lactone ring, and a C_{23} or C_{24} steroid, respectively. Both are common in the Asclepiadaceae (milkweeds). Over 70 species of milkweeds in this family have cardenolide-laden latex. Examples are labriformin and labriformidin in the milkweed species *Asclepias eriocarpa* and *Asclepias labriformis*. These two compounds are toxic to sheep and other livestock (Benson *et al.*, 1979; Nelson *et al.*, 1981). Animals benefit from the warning signal afforded by the bitter taste. If ingested, vomiting removes the toxic agent. Livestock in Africa and North America leave milkweeds alone. Nerium and oleandrin are cardiac glycosides from *Nerium oleander* (Apocynaceae), digitoxin from *Digitalis* sp. (Scrophulariaceae).

Butterflies such as the monarch (*Danaus plexippus*) recycle cardenolides from milkweed (*Asclepias curassavica*), taken up by the caterpillar, as defense against bird predators. Blue jays, *Cyanocitta cristata*, learn to avoid monarchs after becoming sick from eating one. Within 12 minutes the birds become "violently ill." They vomit as often as nine times in 30 minutes but recover fully within half an hour (Brower and Brower, 1964). The birds will also avoid the similarly colored viceroy butterfly, although it is not toxic. Similarly, a grasshopper, *Poekilocerus bufonius*, ingests cardenolides from milkweed and then incorporates them into its noxious foam, which it sprays from a dorsal poison gland against vertebrate predators, such as mice. African hunters used to coat their arrows with extracts of cardenoloid-rich plants.

Cardiac glycosides are medicinally important steroids. Cardenolides are used in heart therapy, as emetics, diuretics, and purgatives. They affect the Na^+/K^+-ATPase and are very toxic at high doses.

Alkaloids

Alkaloids (compounds with basic properties, like ashes of saltwort, *al-qili*) are amino acid-derived compounds that contain nitrogen as part of a heterocyclic ring. They are water soluble, bitter tasting, and occur in about 20% of species of flowering plants. "An alkaloid is a cyclic compound containing nitrogen in a negative oxidation state which is of limited distribution among living organisms" (Pelletier, 1983). Over 10000 compounds belong to this very diverse group. Well-known examples of alkaloids are nicotine, quinine, morphine, colchicine, and strychnine. Coniine in poison hemlock, *Conium maculatum*, is believed to have put to death the philosopher Socrates. Alkaloids are concentrated in vulnerable plant parts such as growing tissues, flowers, seeds, and peripheric tissues such as root bark. In contrast to most other secondary plant compounds, alkaloids may be primarily directed against mammals (Swain, 1977). Alkaloid-rich plants important for livestock are lupines and reed canary grass. Alkaloid-caused livestock losses are high in autumn when alkaloid-rich seeds such as those of lupines mature. *Lupinine* (Fig. 11.8) is one of the active compounds. In the North American west, tall larkspur (*Delphinium barbei*) is toxic to cattle and sheep. The main toxic compounds are the diterpene alkaloids methyllycaconitine and 14-deacetylnudicauline. They block acetylcholine receptors in the central and peripheral nervous systems. This results in muscular paralysis and eventually death (Pfister *et al.*, 1996).

Solanine (Fig. 11.8) occurs in potatoes, particularly in peel (to 1.5 mm depth), eyes, and repaired tissue, and in small and immature, and light-exposed

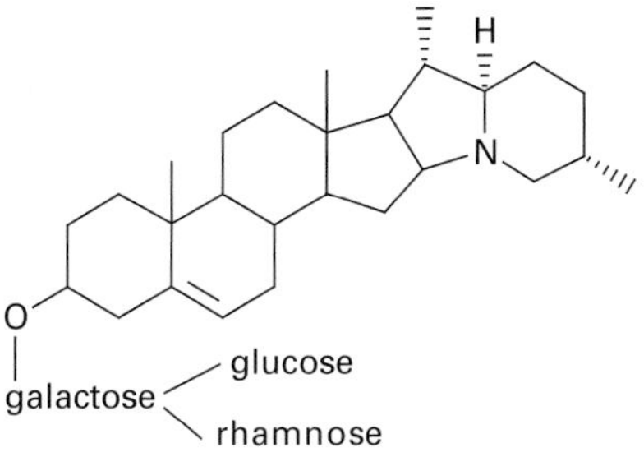

FIGURE 11.8 The alkaloids lupinine and solanine.

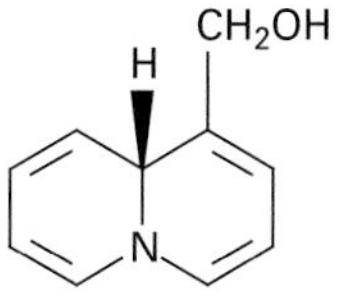

green tubers. Tomatine is found in tomatoes, particularly in green and store-ripened specimens. *Coprine*, from the mushroom *Coprinus atramentarius*, the "alcohol inky," is an alkaloid that inhibits aldehyde dehydrogenase, one of two enzymes that metabolize alcohol. By itself not toxic, the mushroom causes poisoning through accumulation of aldehyde when taken together with alcohol.

Cyanogenic glycosides

Cyanogenic glycosides are water soluble and occur in the vacuoles of plants cells. They consist of glucose attached to an aglycone, are intermediately polar, and are derived from amino acids. Upon damage to the plant tissue by an animal, enzymes release hydrogen cyanide from cyanogenic glycosides. For example, in manioc (see below), linamarin and lotaustralin come in contact with the enzyme linamarase after cells of the outer layer of the tuber are injured.

Hydrogen cyanide is toxic because the cyanide ion has high affinity for metal ions and so binds to the metal-containing cellular respiratory enzymes. Heme proteins such as cytochrome oxidase are complexed, resulting in asphyxiation at the cell level.

Dozens of cyanogenic glycosides occur in higher plants (in 2700 species from 130 families). Examples are dhurrin and taxiphyllin in grasses, and prunasin and amygdalin in the family Rosaceae. The latter two hydrolyze to form benzaldehyde, the "bitter almond" odor. The odor is probably a signal that vertebrate herbivores pay attention to. Proportionally more cultivated plants (perhaps two thirds) are cyanogenic than flowering plants at large (11%; Jones

1998). Seeds of cherries, plums, apples, pears, and other members of the rose family contain cyanogenic glycosides. Deermice consume cherry pits regularly, presumably with no adverse consequences. Intact glucosides are not toxic unless they are ingested along with β-glucosidase. Animals detoxify cyanide by converting it to thiocyanate. While this process protects the body from lethal levels of cyanide, the resulting thiocyanate competes with iodine in the thyroid and can lead to thyroid insufficiency. The body trades an acute, severe danger for chronic poisoning.

Common bracken, *Pteridium aquilinum*, and other ferns are rarely eaten by wildlife. Bracken is cyanogenic, but polymorphic for that trait. Sheep and fallow deer feed only on the acyanogenic form (Cooper-Driver and Swain, 1976). The African savanna grass, *Cynodon plectostachus* (Poaceae), after defoliation by army worms (*Spodophora exempta*), becomes toxic, even deadly, to cattle, because of its cyanide content. The golden bamboo lemur, *Hapalemur aureus*, of Madagascar eats growing tips of cyanogenic bamboo, *Cephalostachyum* ef. *uiguieri*. So far, it is not known how this animal avoids cyanide poisoning. The entire pathway for synthesis of the cyanogenic glucoside *dhurrin* has been transferred from *Sorghum bicolor* to *Arabidopsis thaliana.* The genetically engineered *Arabidopsis* sp. proved resistant to feeding by the flea beetle *Phyllotreta nemorum* of the Chrysomelidae family (Tattersall *et al.*, 2001).

Cyanide content of manioc (*Manihot esculenta*) clones varies with herbivore pressure and soil quality. The more pests and diseases, the more bitter the manioc grows, and "stressed" plants on infertile soil and/or lacking water, also are more bitter. (They do not have the option of outgrowing damage to their tissues.) However, only a few of the about 100 species of *Manihot* have been analyzed. Along the major rivers in Amazonia, cyanide-containing, bitter varieties of manioc dominate, while sweet forms are grown at the Amazon rim and in Central America. Riverine areas have more herbivorous mammals, and larger, semipermanent fields of more sedentary communities attract these herbivores. Frequently shifted small fields are not as vulnerable to herbivory (McKey and Beckerman 1993). Bitterness also protects against human theft in hit-and-run warfare: intruders on the run cannot easily provision themselves from local fields (in McKey and Beckerman, 1993). Current efforts to remove the cyanogenic glycosides, mainly linamarin and lotaustralin, by genetic engineering rely on the fact that the leaves retain cyanogenic compounds (6–40% linamarin) although these compounds are removed from the roots. As little as 1% linamarin remains (Pickrell 2003). It remains to be seen how resistant against pests and pathogens the genetically engineered manioc will be in the long run.

Azetidine 2-carboxylic acid   Proline

FIGURE 11.9 A Non-protein amino acid that interferes with proline metabolism.

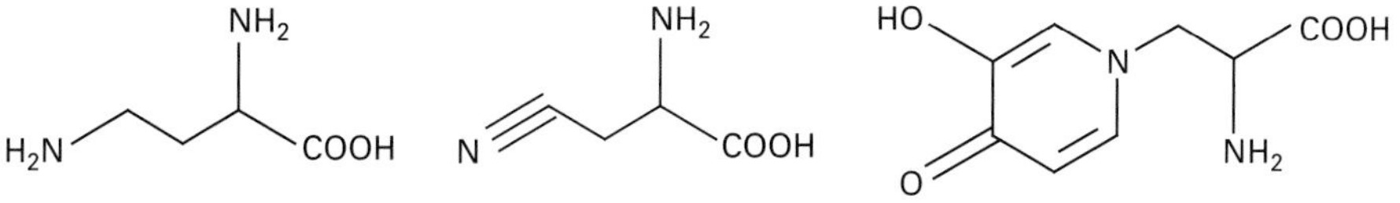

2,4-Diaminobutyric acid   3-Cyanoalanine   L-Mimosine

FIGURE 11.10 Non-protein amino acids.

Non-protein amino acids

About 300 different non-protein amino acids occur in plants. They may be incorporated into proteins in place of the "correct" amino acids. If they are incorporated into enzymes, they can prevent them from functioning. This often leads to death of the animal. For example, azetidine 2-carboxylic acid in lily-of-the-valley, *Convallaria majalis*, and several legumes interferes with synthesis or utilization of structurally similar proline (Fig. 11.9).

Selenocystathionin occurs in seeds of the Central and South American *Lecythis ollaria*. It causes abdominal pain, nausea, vomiting, and diarrhea. *Lathyrus odoratus* contains aminopropionitrile (Rosenthal and Bell, 1979). 2,4-Diaminobutyric acid (Fig. 11.10) in *Lathyrus latifolius* and *Lathyrus sylvestris* causes muscular weakness, paralysis, and death in humans and other mammals. In India, 25 000 cases of this neurolathyrism were reported in a population of 634 000 people (Rosenthal, 1991). 3-Cyanoalanine (Fig. 11.10) from *Vivia sativa* and 15 other *Vivia* species causes convulsion, rigidity, and death.

Heterocyclic compounds such as L-mimosine (Fig. 11.10) cause hair loss in mules, horses, and sheep. Mimosine interferes with the enzyme that forms cysteine from methionine. Cysteine, in turn, is needed in growing hair follicles. Mimosine has been tried as a defleecing agent in sheep. After 2 days of intravenous infusion of mimosine, it takes 15 days for a sheep to drop its entire fleece (Reis *et al.*, 1975). However, it has not yet proven practical in sheep ranching.

L-Canavanine, another non-protein amino acid, occurs in high concentrations in the tropical vine *Dioclea megacarpa* (sea purse, a legume) and is very toxic to mice. Its structure is $NH_2C(NH_2)\text{=}NO(CH_2)_2CH(NH_2)CO_2H$. This is very similar to arginine, with which it interferes.

Oxalic acid

Oxalic acid occurs in sour-tasting plants such as soursob, *Oxalis pes-caprae*, or the better-known *Oxalis acetosella*. Sheep that have ingested too much oxalate suffer labored breathing, "depression," weakness, and death. Rumen microbes degrade oxalate, but excessive doses result in insoluble calcium oxalate, which damages organs such as the rumen or kidneys. Oxalate may reduce calcium levels, and also inhibit respiratory enzymes (Burrit and Provenza, 2000).

11.2 Physiological effects of secondary plant metabolites

At the cellular level, plant secondary metabolites have five major effects on herbivores: (a) alteration of DNA replication, RNA transcription, and protein synthesis; (b) alteration of membrane transport processes; (c) enzyme inhibition and activation; (d) blocking of receptor sites for endogenous chemical transmitters; and (e) affecting the conformation of other macromolecules (Robinson, 1979).

The following will discuss effects at the organ level because this is what the biologist, farmer, zoo keeper, pet owner, or veterinarian deals with. From the viewpoint of the animal, specific life functions, such as digestion, muscle contraction, are affected. While harmful plant compounds mostly act in the digestive tract and the circulation beyond, they can enter cells and interfere with feeding as soon as they enter the mouth. For example, flavonoids in the Mediterranean rockrose *Cistus ladanifer* can prevent skeletal muscles of the mouth in rabbits from relaxing by disturbing the Ca^{2+} transport into the sarcoplasmatic reticulum of the muscle cells, preventing chewing. An avoidance response will result. The active flavonoids are apigenin, quercetin and 3,7-di-O-methyl kaempferol (Sosa *et al.*, 2004).

11.2.1 Inhibition of digestion

Secondary plant compounds can interfere with any stage of digestion. Here we focus on effects in the digestive tract rather than postabsorptive processes such as those taking place in the liver. Phenolics produced the following measurable responses in the voles M. *Microtus ochrogaster* and *Microtus pennsylvanicus*: reduced food digestibility, increased daily food requirements (15–20% with a 6% phenol diet), decreased growth rates, altered organ masses, caused organ lesions, increased excretion of uronic acid, reduced fat deposition, and increased basal metabolism (Lindroth and Batzli, 1983, 1984).

The same diet affects species differently. For instance, mountain hares, *Lepus timidus*, consume much birch in their winter diet, while European hares, *Lepus europaeus*, do not. A high concentration of birch phenolics in the diet causes massive sodium loss via the urine in European hares, but not in mountain hares (Iason and Palo, 1991). Further, among laboratory rodents, hamsters are extremely sensitive to tannins while rats and mice easily adjust within 3 days to doses of condensed tannins that are lethal to hamsters (Mehansho *et al.*, 1987).

Tannins reduce digestion in at least three ways. They reduce the available protein of the plant, as shown for moose (Robbins *et al.*, 1987); inhibit the activity of digestive enzymes (but Blytt *et al.*, suggested in 1988 that condensed tannins may not bind appreciably to digestive enzymes *in vivo*); and inhibit gut microorganisms. A very drastic example is the effect of the condensed tannin quebracho on weanling prairie voles (*M. ochrogaster*). Quebracho inhibits feeding and lowers the digestibility of proteins. Most weanling voles died when fed quebracho (Lindroth and Batzli, 1984).

Ruminants are particularly vulnerable to digestion inhibitors because the short pathway to the rumen provides much less opportunity for detoxification than the longer passage in hindgut fermenters. A phenol from a European birch (*Betula pendula*), platophylloside (5-hydroxy-1,7-bis-(4-hydroxy-phenyl)-3-heptan-one-3-O-β-D-glucopyranoside), inhibits rumen digestion. A significant effect was observed even with amounts below those found in birch (0.8% of dry matter) (Sunnerheim *et al.*, 1988). In black-tailed deer (*Odocoileus hemionus columbianus*), 1.2% of essential oils, containing monoterpenoids and sequiterpenoids, inhibit rumen microbes (Oh *et al.*, 1967).

North American prairie grasses contain polypeptides that act as inhibitors of the proteinase trypsin in mammalian herbivores. The grasses with these trypsin inhibitors are western wheatgrass (*Agropyron smithii*), big bluestem (*Andropogon gerardii*), little bluestem (*Andropogon scoparius*), and blue grama (*Bouteloua gracilis*) (Ross and Detling, 1983).

11.2.2 Growth suppression

The flavonoid quercetin suppresses growth somewhat in prairie voles, *M. ochrogaster*, while "tannic acid," a hydrolyzable tannin, has an even stronger effect (Lindroth and Batzli, 1984). Quercetin is toxic, but does not bind with protein and, therefore, does not affect protein digestibility. Tannic acid has some effect on protein digestibility but is also toxic.

Hamsters, unlike mice and rats, are not able to adjust to high levels of tannins in their diet by producing more proline-rich proteins in their saliva. A diet high in tannin has no effect on the salivary glands, and proline-rich proteins are not induced. Hamsters on a diet containing 2% tannin do not grow. On this diet, at 60 days of age they still weigh the same as at 3 days. After switching diets, the former control animals, now on a 2% tannin diet, lost 20% of their body weight, while the former tannin-fed group now gained weight at almost the normal rate for young hamsters. A 4% quebracho tannin diet killed 75% of hamsters within 3 days, while rats and mice were not adversely affected (Mehansho *et al.*, 1987).

11.2.3 Reproductive inhibition

Dry salt grass, *Distichlis stricta*, has high levels of phenolics after fruiting. Cinnamic acids and their vinyl phenols lower uterus weight, inhibit follicular development, and lead to cessation of reproductive activity in the montane vole *Microtus montanus*. Specific active compounds are 4-vinylguaiacol and 4-vinylphenol, and also *p*-coumaric acid (4-hydroxycinnamic acid) (Berger *et al.*, 1977).

Phytoestrogens

Many plants contain estrogenic compounds. Estrone (Fig. 11.11) is found in seeds of date palms, pomegranates, and apples, and estriol in willow. These may be merely by-products of sterol metabolism, or serve a particular function. Harborne (1993) proposed that plants synthesize steroid hormones to deter feeding by mammals. Estrogenic compounds in plants are thought to upset the delicate hormone balance of mammals.

In Australia, lambing in sheep has dropped to as little as 30% after grazing on subterranean clover, *Trifolium subterraneum*. The clover contains the isoflavonoids formononetin and genistein (Fig. 11.11). These compounds mimic the steroidal nucleus of the natural female hormone estrone. Alfalfa, *M. sativa*, and ladino clover, *Trifolium repens*, contain the even more potent isoflavonoid coumestrol. It is 30 times more active than genistein or formononetin, but occurs in lower concentrations in the plant (Shutt, 1976).

In general, isoflavonoids in clovers and other legume pasture plants pose a danger to reproduction in farm animals because they lead to difficult labor, infertility, and lactation in unbred ewes. In Australia, an estimated one million ewes fail to lamb each year because of "clover disease" (Harborne, 1993).

During the Second World War, women in the Netherlands had to resort to eat tulip bulbs. They blamed their frequent menstrual upsets and ovulation failures on this diet. Harborne (1993) listed garlic, oats, barley, rye grass, coffee, sunflower, parsley, and potato tubers as having effects on estrus in women, but also cows. The active principle may not be a hormone but rather compounds that

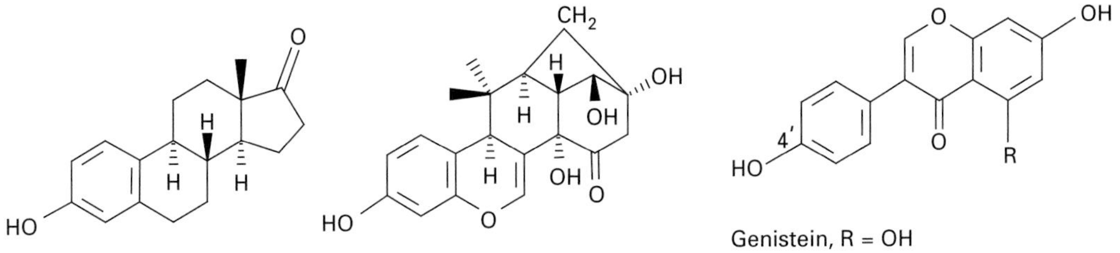

Estrone
(the natural hormone)

Miroestrol
(a plant mimic of hormone)

Genistein, R = OH
Daidzein, R = H
Formononetin, R = H (Me at 4′-OH)

FIGURE 11.11 Phytoestrogens.

mimic hormones. One such compound is *miroestrol* (Fig. 11.11). It was isolated from the roots of the legume *Pueraria mirifica* (Bounds and Pope, 1960). Pregnant women in Burma and Thailand use a root extract of this tree for abortions. Miroestrol is as potent as 17β-estradiol, and more than three times as active as the synthetic drug diethylstilboestrol.

Humans have used a range of plant species to manipulate their own fertility. A test of 10 plant species that bedouins of the Negev desert use for birth control showed that herbivorous rodents are also affected; 6 of the 10 species reduce fertility in laboratory rats. The fruits and shoots of one of these, the tree *Ziziphus spina-christi* (Christ's thorn; Rhamnaceae), delayed puberty and reduced the number of surviving offspring in its natural herbivore, the gerbil *Meriones tristrami*. This effect is thought to be part of a regulatory mechanism: the seeds eaten by the herbivore are dispersed, thus benefiting the plant. The ingested plant material reduces the population of the herbivore, keeping food supply and herbivore population in balance (Shappira *et al.*, 1990). Specifically, the effects of the isoflavonoids formononetin and genistein (Fig. 11.11) may aid in adjusting animal populations to the available food supply. California quail (*Lophortyx californicus*) feed on legumes that vary in growth with rainfall. In good years, the plants are large, with low relative isoflavone concentrations, and the quail breed vigorously. The seeds carry the young through the winter. In dry years, the plants are stunted but rich in isoflavone on a fresh weight basis. The female quail suffer an estrogenic effect and lay fewer eggs (Leopold *et al.*, 1976).

It has been suggested that metabolites resulting from detoxification of plant compounds such as ferulic acid, a detoxification by-product of coniferyl benzoate and analogous compounds, may interfere with reproduction. However, experiments have shown that coniferyl benzoate in the diet of Japanese quail (*Coturnix coturnix*) had no hormonal effects. Rather, costs of detoxication and reduced nutrient utilization deter wild birds such as ruffed grouse, *Bonasa umbellus*, from feeding (Jakubas *et al.*, 1993).

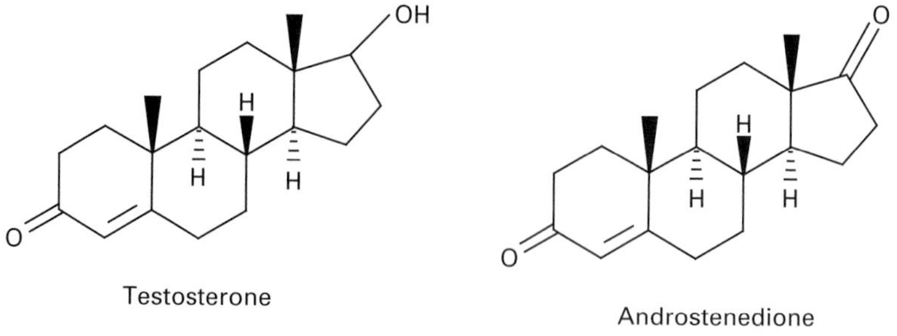

Testosterone

Androstenedione

FIGURE 11.12 Androgens found in plants.

Androgens

Androgenic compounds occur in various plants. Celery stalks and parsnip roots contain trace amounts of androstenedione (Fig. 11.12), the same compound found in boar odor (Claus and Hoppen, 1979). Testosterone is also found in pollen of Scots pine (*Pinus sylvestris*).

11.2.4 Embryonic malformations by Teratogens

Steroidal alkaloids can cause malformations of fetuses during development. Domestic sheep feeding on *Veratrum californicum* often give birth to lambs with grotesquely shaped one-eyed heads. This *cyclopia* is caused by the steroidal alkaloids jervine, cyclopamine, and cycloposine (reviewed by Keeler, 1986).

11.2.5 Carcinogens

For obvious reasons, plant carcinogens have attracted extensive research. Here we review some examples that shed light on the effect of certain plant secondary metabolites on vertebrates, including humans. Tea is one of the best-investigated plant commodities. It contains up to 30% (of dry weight) *tannins* (polyphenols). These may cause esophageal cancer. Esophageal cancer used to be more prevalent in Japan than in Britain. By contrast to the Japanese, the British use milk in tea, which binds tannins and thus protects the esophagus. The Dutch also experienced much esophageal cancer during the nineteenth century when they used to drink tea without milk. The cancer rate dropped when they switched to coffee after they had established coffee plantations in Java and coffee became widely affordable (Morton, 1989). Esophageal cancer is more prevalent in northern Iran, southern Russia, and Turkey, where much tea is consumed (Morton, 1970). Brown and green tea are produced in different ways: brown tea, after withering for 1 day, is rolled which creates flavor and astringency. Enzymes, cell contents and oxygen are mixed during the following *fermentation* at 27 °C for 1–3 hours. Polyphenoloxidase transforms colorless phenolics into astringent tannins. Now the flavor and astringency is fully developed. Finally, the tea is fired at 93 °C, resulting in a 5% moisture content. By contrast, green tea is not fermented. The enzymes are destroyed by steam, and the tea is then rolled and fired.

Maté (*Ilex paraguensis*), widely used as tea in Brazil, Paraguay, Uruguay, and Argentina, contains 9–12% tannin and up to 1.5% caffeine. Esophageal cancer rates are high in the Brazilian state of Rio Grande do Sul (Prudente, 1963) and Uruguay (Kaiser and Bartone, 1966), areas where consumption of maté is particularly high. Khat (*Catha edulis*) leaves, commonly chewed in Aden, Yemen and eastern Africa from Ethiopia to Madagascar, contain from 7 to 15% condensed tannins. Khat is chewed for several hours a day; an Ethiopian can chew as much as 2.5 kg per day. In some areas of heavy khat use, the rate of esophageal cancer is high (Morton, 1989).

Esophageal cancer rates were found to be correlated with consumption levels of sorghum in the Transkei, red wine in Western Europe generally, and high tannin apple cider in northern France (Morton, 1970). In Japan, esophageal cancer occurs particularly often in people who are eating tea-rice gruel, accompanied by tea drinking (Segi, 1975).

Some secondary plant compounds have proven carcinogenic in laboratory rodents. These include D-limonene in orange juice; 5-8-methoxypsoralen in parsley and parsnips; and caffeic acid in coffee, but also in smaller amounts in apples, apricots, broccoli, Brussels sprouts, cabbage, cherries, kale, peaches, pears, and plums (Ames and Gold, 1990; Ames *et al.*, 1990). How dangerous these compounds can be to human health, is still being debated.

11.2.6 Neurotoxicity

Some plants regularly eaten by humans contain neurotoxins that pose serious health problems. On Guam, for example, the seeds of *Cycas circinalis* used to be an important source of carbohydrates. Seeds of *Cycas rumphii* were ground into flour for tortillas. However, the seeds contain β-*N*-methylamino-L-alanine, a suspected excitotoxin that overstimulates and destroys nerve cells. This compound causes a parkinsonism-like disease in macaques (Spencer *et al.*, 1987). Other toxins have been proposed to be responsible for the disease, among them cycasin, another cycad toxin (Stone, 1993).

Among the Guam population, the disease known locally as *lytico-bodig* has been common for hundreds of years (Kurland, 1972). The disease is one of paralysis, tremors, and dementia and is an amyotrophic lateral sclerosis–parkinsonism dementia complex (Lou Gehrig's disease); few cases are now seen (Stone, 1993). The Chamorros of Guam soak the cycad seeds for days and leach out up to 99% of the toxin by repeated rinsing (Duncan, 1991). The disease occurred most often during the time after the Second World War when the Chamorros ate flying foxes, typically prepared in coconut cream. Flying foxes subsist on cycad seeds and accumulate cycasin in their bodies. With depletion of the flying foxes, fewer humans suffered the disease. It has been suggested that the flying foxes were an important link (Cox and Sacks, 2002), but cycasin is water soluble and unlikely to accumulate in the bat body, if indeed it is the cause of the disease. Another, fat-soluble neurotoxin also remains under discussion (Stone, 2002).

11.2.7 Hallucinogens

Many plants contain compounds that cause hallucinations in people. In the book *One River*, Wade Davis describes the work he and his mentor Richard Evans Schulte carried out on the many Amazon plants with hallucinogenic effects in humans. They affect the central nervous system and may be designed to derail the behavior of herbivores, although the precise ecological role of these compounds remains uncertain. The effects of the active compound in cannabis (marijuana), Δ^9-tetrahydrocannabinol (THC), on the mammalian brain promise new insights into its ecological function. Many parts of the brain (but not the brainstem) and the immune and reproductive systems carry cannabinoid receptors. These receptors mediate the action of endogenous cannabinoids, leading to pain relief, loss of short-term memory, sedation, and mild cognitive impairment. External cannabinoids act through this pathway. Although THC may protect a plant from detrimental ultraviolet radiation (cannabis plants contain more THC at higher altitudes), it may induce short-term memory loss in herbivores and thus reduce repeated attacks.

Rodriguez *et al.* (1982) noted that psychoactive compounds also often kill gastrointestinal parasites. They suggested that humans first used the plants to control parasites and discovered the effects on the central nervous system in the process.

The better-known and widely used hallucinogenic plants are San Isidro mushroom (*Psilocybe cubensis*), ergot (*Claviceps*), soma (*Amanita muscaria*), peyote (*Cactus lophophora*), yagé (or ayahuasca), the "vision vine" of the Amazon (*Banisteriopsis caapi*), cannabis (*Cannabis sativa* and *indica*) and perhaps coca (*Erythroxylum novogratense*),

An example of a hallucinogen affecting mammalian herbivores is the isoxazole muscimol (Fig. 11.13), a mushroom toxin. It is a γ-aminobutyric acid agonist in the central nervous system of vertebrates. Muscimol's role in nature

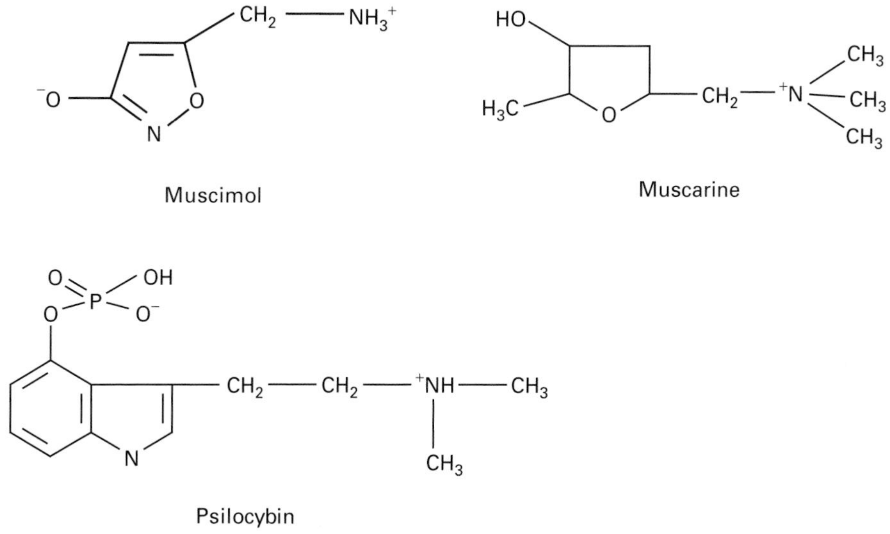

Muscimol

Muscarine

Psilocybin

FIGURE 11.13 Hallocinogens.

is defense against fungivores. Virginia opposums, *Didelphis virginiana*, are fungivorous but avoid muscimol-containing mushrooms such as *A. muscaria.* They develop an aversion after having eaten the mushroom and falling ill (Camazine, 1983).

Another well-known hallucinogen is psilocybin from the mushroom genus *Psilocybe* (Fig. 11.13). Muscarine is an isoxazole derivative (Fig. 11.13). Table 11.3 lists some of the hallucinogens in plants.

11.2.8 Electron transport inhibitors

Inhibition of cytochromes of electron transport system can be caused by cyanogenic glycosides, such as amygdalin (Fig. 11.14) in bitter almonds, *Prunus amygdalus*, linamarin and lotaustralin in clover and birdsfoot trefoil, or dhurrin (Fig. 11.14) in *Sorghum vulgare*. The potent effect of cyanide on cell respiration has given rise to a recent serious conservation problem. In Southeast Asia, divers stun fish on coral reefs with a blast of cyanide to collect them for the aquarium trade. In the process, many fish are killed and the corals bleached, because their symbionts die (e.g. Payne, 2001).

Under special conditions, such as hot weather followed by a freeze and drought, cyanogenic compounds can accumulate in cherry leaves. Caterpillars such as the Eastern tent caterpillar, *Malacosoma americanum*, concentrate the toxic substances further in their bodies. Finally livestock inadvertently ingest their feces and carcasses on pastures or from water tanks. In 2001, this resulted in over

Table 11.3 Hallucinogens from Plants

Plant	Part	Compound	Use	Effect	Users	Place
Yopo tree *Anadenanthera peregrina*	Beans	Cohoba	Snuff	Hallucinogenic	Taino Indians	Hispaniola
Snake plant *Turbina corymbosa*	Seeds	Ololiuqui: LSD-like compounds	Drink	Hallucinogenic	Aztecs	Mexico
Coco plant *Erythroxylon coca*	Leaves	Cocaine	Chewed	Stimulant, analgesic	Incas, South Americans	Peru etc.
Cactus *Lophophora williamsii*	Top: mescal button	Mescaline	Swallowed	Hallucinogenic	Native American Church	Mexico, Texas
Teonancatl *Psilocybe mexicana*	Mushroom	Psilocybin	Eaten	Hallucinogenic	Aztecs, modern Native Americans	Mexico, Central America
Hemp *Cannabis sativa*	Flowers, leaves	Marihuana: tetrahydrocannabinol	Smoked	Sedative	Worldwide	Origin: China
Khat *Catha edulis*	Leaves	Alkaloids: cathinone, cathine	Chewed, tea	Stimulant	Yemenis, Ethiopians, Somalis' etc.	East Arabia, East Africa, Madagascar
Ayahuasca (yagé) *Banisteriopsis caapi*	Bark of vine	Harmine, harmaline	Drink	Hallucinogenic	Shamans, various indigenous groups	Amazon basin
Ebéna *Virola theiodora*	Inner bark	Several tryptamines, including the most potent hallucinogen known from nature: 5-methoxy-*N*,*N*-dimethyltryptamine	Snuff	Hallucinogenic	Yamomami	Amazon basin

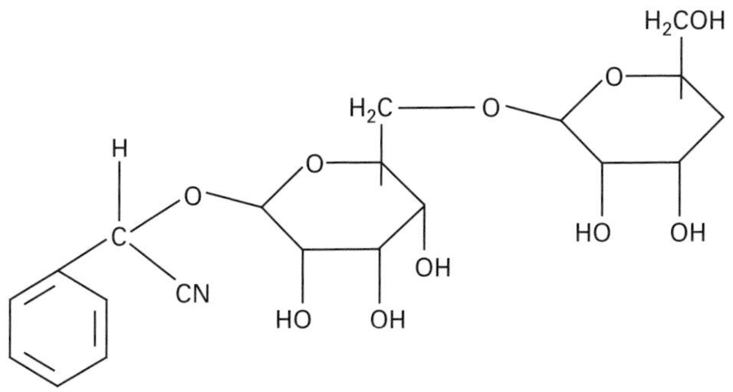

Amydalin

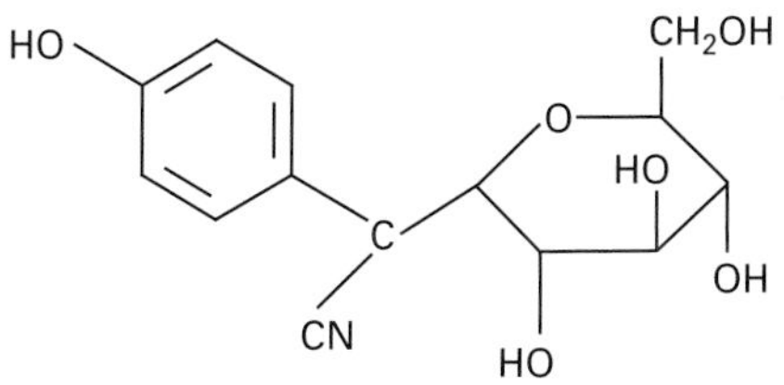

Dhurrin

FIGURE 11.14 Cyanogenic glycosides inhibit the electron transport chain.

500 stillbirths among thoroughbred mares in Kentucky, caused by the blocking of oxygen delivery to the cells by cyanide (Holden, 2001).

11.2.9 Renal effects of chronic poisoning

Voles (*M. pennsylvanicus*) suffer renal lesions (interstitial nephritis) when fed extracts of white clover, *T. repens*. Milder lesions were observed after feeding on reed phalaris (*Phalaris arundinacea*) and timothy (*Phleum pratense*). Many varieties of reed phalaris contain the toxic compounds gramine and tryptamine (Fig.11.15). In summer and autumn, protein levels in the leaves decrease, fiber content goes up, and secondary compounds increase in concentration. Therefore, second growth plants have more toxic effects on voles than the spring plants that grow fast and have lower levels of secondary compounds (Bergeron *et al.*, 1987).

11.2.10 Lethal light: toxicity linked to ultraviolet radiation, or why mammals are nocturnal

Alarmingly, the ozone hole has turned mammalian food into food for thought. The ozone hole permits an increase of ultraviolet radiation in the range 290 to 320 nm, the same wavelengths that induce furocoumarin (also known as

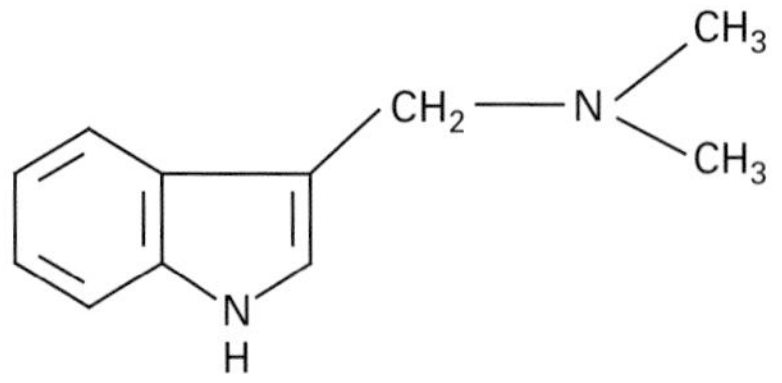

Gramine

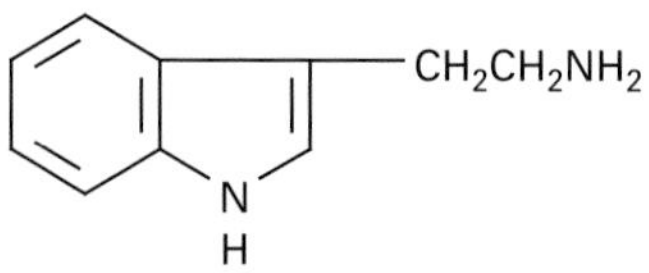

Tryptamine

FIGURE 11.15 Compounds causing renal lesions.

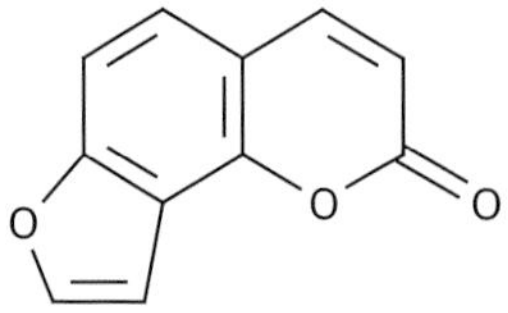

Angelicin

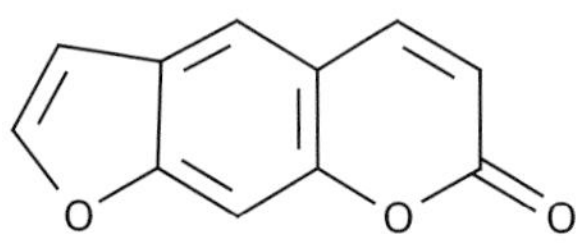

Psoralen

FIGURE 11.16 The furocoumarins.

furanocoumarin) production in certain plants. Many plants contain compounds that are toxic to herbivores after ingestion *only if* the animals are exposed to sunlight. These include the quinone hypericin from St. Johnswort, *Hypericum perforatum.*

A well-investigated group of potent phototoxic compounds are the *furanocoumarins* (psoralens) (Fig. 11.16), compounds that are also used as fish poisons. Furocoumarins absorb ultraviolet at around 330 nm and are converted to triplet stages; these, in turn, produce singlet oxygen that damages amino acids (Fig. 11.17).

Photosensitization by plants containing furocoumarins has been described for geese, chickens, turkeys, ducklings, sheep, dairy cattle, horses, hyrax, and humans, among others. Herbivorous insects and mammals can avoid the toxic effects if they stay in vegetation cover, underground, or lead a crepuscular or nocturnal life. Leaf-rolling behavior in insects has been seen as protection against phototoxic effects of constituents of their diet. How do reptiles that need daylight or direct solar radiation to be active avoid damage by ultraviolet radiation? Does their skin armor screen out the radiation sufficiently?

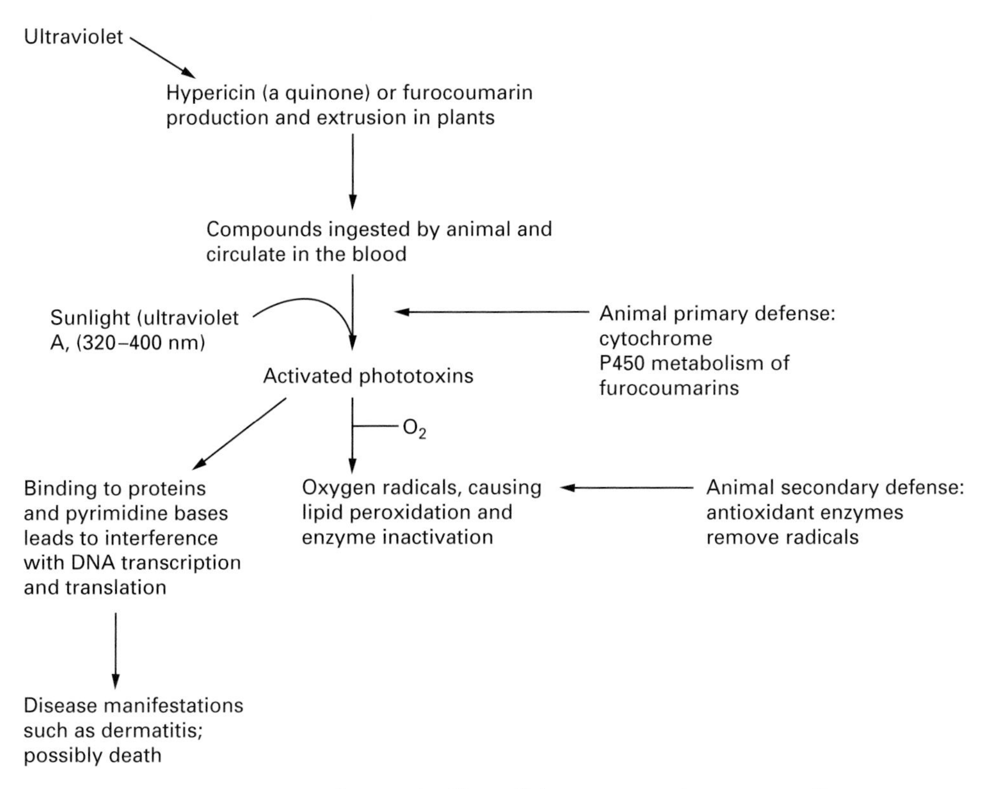

FIGURE 11.17 Phototoxic effects of plant compounds on mammalian herbivores.

Furocoumarins occur particularly frequently in the Leguminosae, Moraceae, Rutaceae, and Umbelliferae (carrot family) (Larson and Berenbaum, 1988). One better-investigated species of the latter family is *Pithuranthos triradiatus*, a desert shrub in the Middle East. Hyrax, *Procavia capensis syriaca*, feed on this plant. After eating branches and staying in the sun, hyraxes develop symptoms of photosensitization: they become apathic, photophobic, and develop injuries around the eyes, and on the back. The animals ate less of older branches, but ingested more furocoumarins, because, although older shoots contain more of the phototoxic compounds imperatorin and isoimperatorin than young shoots, the young shoots are "loaded with higher concentrations of secondary metabolites." Plants seem to avoid autotoxic hazards; there are less furocoumarins in shoots that are still growing and so DNA strands are not damaged in mitotically active tissue (Ashkenazy *et al.*, 1985).

Plants that can cause photodermatitis include carrots, celery, parsnip, dill, fennel, caraway, parsley, lovage, anise, and chervil among the Apiaceae (Umbelliferae, or carrot family), figs (Moraceae), and citrus fruits (grapefruit, lemon, lime, and orange) among the Rutaceae (Berenbaum, 1991). In arid and semiarid areas of North America such photosensitizers are widespread in plants and occur particularly in the Asteraceae (Downum *et al.*, 1989).

Any increase in ultraviolet radiation will exacerbate the danger to herbivores. Ozone depletion by 25% leads to an increase of 50% in ultraviolet B over ambient levels (El-Sayed, 1988). Will livestock be drastically affected if levels rise more in the future? Will animals be able to shift to nocturnal activity? If not, will large portions of the vegetation be rendered useless for domestic stock and wild species or will animals have to be fed indoors? Will humans be affected?

Furocoumarins in the diet can cause photophytodermatitis in humans. Carrots, for instance, contain furanocoumarins. Humans touching leaf surfaces of a variety of plants can pick up the coumarins and produce handlers have suffered from photophytodermatitis. In addition to diet, perfumes can be a source of these compounds. Oil of bergamot contains a furocoumarin and used to be an ingredient of some perfumes. Such perfumes can cause skin rashes upon exposure to sunlight. Eight species of Umbelliferae (carrot family) and Rutaceae (citrus family) were shown to contain several photoactive furanocoumarins. These include xanthotoxin, psoralen, and bergapten. They are more concentrated on younger, spring leaves and least on old, autumn leaves. Zobel and Brown (1990) have suggested that a leaf has an "atmosphere" of sublimating compounds that is toxic to approaching bacteria or fungal spores; furocoumarins are actually antibacterial.

11.2.11 Metabolic costs of detoxification

Secondary plant compounds can be a substantial burden for a herbivore. Nutrients are diverted for conjugation (p. 330) and this can have severe consequences for the organism at low nutrient levels. Malnourished people in developing countries often treat their ailments with herbal cures that, unfortunately, aggravate their problems because precious carbohydrates or amino acids are diverted for detoxification and excretion of plant compounds. Oxidation and conjugation, the two phases of biotransformation (detoxication), require the input of energy. The basal metabolic rate in the meadow vole, *M. pennsylvanicus*, increased by 13.6 to 22.6% when 6% (by dry mass) gallic acid, a phenolic, was added to its diet. This is thought to be the metabolic cost of detoxification (Thomas *et al.*, 1988).

Sheep that are given orcinol intravenously increase their energy expenditure by 5%. (Orcinol is a non-tannin phenolic that occurs in Ericaceae, including heather, *Calluna vulgaris*.) If orcinol and quinol are infused into the rumen, they have no effect on energy expenditure but the intake of digestible energy is reduced. This shows that the effectiveness of the detoxication system is crucial for utilization of toxin-containing plants (Iason and Murray, 1996). It is assumed that microbes in the rumen modify toxins, reducing their systemic effects.

Koalas (*Phascolarctos cinereus*) increase their glucose intake by 20% if they eat *Eucalyptus* spp. foliage, which requires conjugation reactions with glucuronic acid for excretion (Eberhard *et al.*, 1975). Goats and sheep eat more of a toxic diet if given surplus food, which helps detoxication (Provenza, 2004).

11.2.12 Interactions of secondary compounds

Complex interactions may occur if a herbivore ingests a variety of defense compounds from one or several plant species as the various compounds may counteract each other. For instance, tannins can help to reduce the toxic effects of cyanogenic plant compounds. In the papaya, *Carica papaya*, the cyanogenic glycosides tetraphyllin B and prunasin (also present in the passion fruit, *Passiflora edulis*) produce hydrogen cyanide upon enzymatic hydrolysis. The plant also contains protein-precipitating tannins that inhibit enzymes and thereby the enzymatic production of cyanide from tetraphyllin B, prunasin, and also amygdalin. If low levels of hydrogen cyanide are measured in a plant, this may be because tannins are present to interact with the enzymes (Goldstein and Spencer, 1985).

Secondary plant compounds can have negative and positive effects on a herbivore, depending on the form in which they are ingested. In an *in vitro* experiment, rumen microbes from sheep were inhibited by the simple phenols orcinol and quinol: production of gas and volatile fatty acids were reduced. Both compounds reduced digestion of the substrate. By contrast, arbutin, a phenol glycoside, stimulated production of gas and volatile fatty acids. This was probably because of the availability of the fermentable sugar moiety of the glycoside (Murray *et al.*, 1996). Many simple phenolics exist in plants in the form of glycosides. The sugar moiety is liberated in the gastrointestinal tract on hydrolysis by glucosidase enzymes. Therefore, beneficial effects of the sugar moiety may counteract the inhibitory effects of phenolics (Murray *et al.*, 1996).

11.2.13 Effects of toxic plants on herbivore populations

Vole cycles are thought to be influenced, at least in part, by the ratio of toxic plants in the food (Freeland, 1974). According to this hypothesis, increasing numbers of voles deplete the more palatable plant species. At high population levels, their diet would then contain a high percentage of "toxic" plants. This would affect health and growth, especially of young animals, and the population would decline. Data presented by Batzli and Pitelka (1971) in response to this hypothesis, however, did not substantiate the notion that California voles (*M. californicus*) resort to more toxic plants when population levels are high.

Endangered herbivorous mammals that compete with large numbers of livestock may be forced to feed on less-palatable or outright toxic plants, especially if fenced in and poisoning could result in population declines. This has been suggested for the endangered pampas deer, *Ozotoceros bezoarticus*, which shares its – often severely grazed – range with large numbers of cattle and sheep (D. Müller-Schwarze and D. E. Moore, unpublished field observations). High population levels, consumption of tannin-rich browse, and high mortality were correlated in kudu, *Tragelaphus strepsiceros*, in fenced areas (van Hoven, 1991).

Researchers currently debate whether terrestrial ecosystems are regulated "bottom up" by secondary plant metabolites that limit food available to herbivores or "top down" by predators that keep herbivores in check, resulting in lush plant growth. One recent study ruled in favor of the second. Islands of varying sizes, created by the hydroelectric impoundment of Lake Guri in Venezuela, permitted these hypotheses to be tested. On predator-free islands 10 to 100 times more herbivores were living, such as rodents, howler monkeys, iguanas, and leaf-cutter ants, than on the nearby mainland. The densities of seedlings and saplings of canopy trees on these islands were severely reduced, owing to these abundant herbivores (Terborgh *et al.*, 2001).

11.2.14 Potential beneficial effects of secondary plant compounds

Plant secondary metabolites in the diet may actually benefit consumers of plant materials. Many phenolics are antioxidants that mop up free radicals. Gallic acid is particularly active, and resveratrol in grapes and red wine has been in the news for that reason. Long-term ingestion of undesirable plant compounds may boost the immune system. In Liberia, a diet of cassava exposed adults chronically to medium to high levels of cyanogenic glycosides. They had elevated levels of immunoglobulins, especially IgM. Dietary cyanide bound to proteins, and bound forms of cyanide in the diet, may act as antigens. IgM is strongly antibacterial and may play a beneficial role in resistance to tropical bacterial diseases such as Hanson's disease (leprosy), louse-born relapsing fever, yaws, and cholera (Jackson 1994).

Table 11.4 Plant parts differ in chemical defense against vertebrate herbivores

Plant species	Plant part	Herbivore species	Compound	Reference
Green alder *Alnus crispa*	Buds, catkins	Snowshoe hare *Lepus americanus*	Pinosylvin, pinosylvin methyl ether, papyrific acid	Bryant *et al.*, 1983; Clausen *et al.*, 1986
Green alder	J versus M internodes	Snowshoe hare	J has more pinosylvin and pinosylvin methyl ether	Clausen *et al.*, 1986
Paper birch *Betula resinifera*	J growth versus winter-dormant internodes	Snowshoe hare	J has 25 × concentration of papyriferic acid	Reichardt *et al.*, 1985
Quaking aspen *Populus tremuloides*	Scales of flower buds	Ruffed grouse *Bonasa* sp.	Coniferyl benzoate	Jakubas *et al.*, 1989
Balsam poplar *Populus balsamifera*	Buds	Snowshoe hare	Mono- and sesquiterpenes	Reichardt *et al.*, 1990b
White spruce *Picea glauca*	J stages	Snowshoe hare	Camphor	Sinclair *et al.*, 1988
Balsam poplar	J twigs	Snowshoe hare	2,4,6-Trihydroxy-dihydrochalcone	Jogia *et al.*, 1989
Willow *Salix* spp.	J stages	Mountain hare *Lepus timidus*	Phenolic glycosides, catechin	Tahvanainen *et al.*, 1985

J, juvenile; M, mature.

11.3 Chemical defense strategies by plants

11.3.1 To the bitter end: distribution of defense compounds within plants

Plant *parts* often differ in their levels of secondary metabolites (Table 11.4), some containing extremely high levels. For example, the creosote bush, *Larrea tridentata*, of the western United States has phenolic resins concentrated in leaves, amounting to as much as 18% of dry weight. In experiments, desert woodrats, *Neotoma lepida*, selected plant parts of creosote bush with low levels of resins (Meyer and Karazov 1989).

Mountain hares, *L. timidus*, selected birch twigs of intermediate diameter (1.5 to 3.0mm) and attained the best body weight on such a diet. Smaller twigs are high in phenolic glycosides, and larger twigs are least digestible (Palo *et al.*, 1983). In the genus *Populus*, twig bark usually has higher concentrations of phenolics than trunk bark. Also, the levels of these compounds in twigs vary more, and the spectrum of compounds is narrower than in trunk bark (Pearl and Darling, 1968; Palo, 1984).

Particularly valuable and/or vulnerable parts are protected best. Because of their size and seasonal changes, trees provide instructive examples. In green alder, *Alnus crispa*, defense compounds, such as pinosylvin, pinosylvin methyl ether and papyriferic acid, are concentrated in buds (2.6% of dry weight) and catkins (1.7%), and least abundant in the internodes (0.05%). Snowshoe hares, *L. americanus*, reject buds and catkins, while they eat second-year internodes (Bryant *et al.*, 1983; Clausen *et al.*, 1986).

In quaking aspen, *Populus tremuloides*, the scales of the flower buds have the highest concentrations of coniferyl benzoate (Fig. 11.2). Ruffed grouse avoid trees particularly rich in this compound (Jakubas *et al.*, 1989). Experiments have shown that coniferyl benzoate inhibits food consumption (Jakubas and Gullion, 1990). Balsam poplar, *Populus balsamifera*, has redundant compounds, practicing "defensive overkill" against snowshoe hares, who do not eat its buds. The buds contain resins, mainly made up of mono and sesquiterpenes based on cineol, benzyl alcohol, and (+)-α-bisabolol. Juvenile internodes of shoots from stumps are protected by salicaldehyde, while the current annual growth of mature plants is protected by 6-hydroxycyclohexenone (Reichardt *et al.*, 1990b). The buds are defended by 2,4,6-trihydroxydihydrochalcone, but the internodes of juvenile trees by three volatiles: salicaldehyde, 6-hydroxycyclohexenone, and 1,2-cyclohexadione. These volatiles may signal that other noxious compounds are present, constituting an example of chemical aposematism (Eisner and Grant 1981).

The *age* of plant tissues influences palatability in other species as well: Younger plant parts generally have higher concentrations of defense compounds. An example is green alder, *A. crispa*. Juvenile internodes have three times as much pinosylvin and pinosylvin methyl ether as mature internodes. Snowshoe hares, *L. americana*, prefer to feed on mature internodes (Clausen *et al.*, 1986). Similarly, snowshoe hares prefer internodes of the mature growth phase of winter-dormant paper birch, *B. resinifera*, to juvenile growth phase internodes. In perhaps the most dramatic example, the internodes of the juvenile phase contain as much as 30% of the dry weight (25 times as much) as the triterpene papyriferic acid (Fig. 11.5) than mature internodes and this acts as a feeding deterrent (Reichardt *et al.*, 1985). Snowshoe hares also rarely browse juvenile stages of white spruce, *Picea glauca*, a slow-growing late succession tree, but utilize mature stages heavily. Juvenile spruce contained four times as much camphor as mature stages. In experiments, camphor deterred feeding (Sinclair *et al.*, 1988).

Strong chemical defenses of reproductive units such as seeds, bulbs, or tubers, as opposed to the vegetative parts, are commonplace. For instance, potatoes, which have to survive in the ground from one growing season to the next, contain glycoalkaloids such as solanine and chaconine, saponins, a phytohemagglutinin, proteinase inhibitors, sesquiterpenes phytoalexins, and phenols (Johns, 1990). These effectively deter fungi, invertebrates, and rodents.

Female flowers of dioecious trees are better defended than males. Ruffed grouse, *Bonasa umbellus*, prefer male flowers of quaking aspen, *Populus tremuloides*, in winter, and leaves of male specimens in summer (Svoboda and Gullion 1972). Ruffed grouse also feed more on male trees of balsam poplar (Bryant *et al.*, 1991).

11.3.2 Seasonal variation of levels of chemical defenses

Are plants best defended during the season when they are most vulnerable to herbivory? Specifically, do bark and twigs contain their highest level of plant secondary metabolites during the northern winter when they are browsed by mammals small and large, and cannot respond by regrowth? Do they contain leaves with most plant secondary metabolites during the months when insect attacks are most intense? The answers to these questions are complex. It has been suggested that woody boreal plants allocate the largest amount of "defense" compounds to tissues that are essential to photosynthesis and growth and, therefore, contribute significantly to plant fitness (Palo, 1984).

In the European birch, *B. pendula*, the relative concentration of protein, and with it the *in vitro* organic matter digestibility, increases in spring and early summer, while the proportions of cell walls (neutral detergent fiber) and water-soluble phenols, especially catechin, decrease during that time. Such seasonally changing allocation within the plant results in increased palatability to ruminants. The plant can afford to lose biomass to herbivores during the time of rapid growth (Palo *et al.*, 1985). Bark of the birch, *B. pendula* has low tannin levels (4.1–4.8%) in winter, but a maximum (5.8–6.2%) in September (Chernayeva *et al.*, 1982). In the same species, total phenolics reach their maximum (3%) in winter twigs, and are lowest (0.9%) in June (Palo *et al.*, 1983). Leaves of *Betula tortuosa* contain only 7% total phenolics when they emerge, but up to 15% in autumn (Palo, 1984).

Levels of phenolic glycosides in bark of *Salix* and *Populus* spp. generally increase towards the winter. Concentrations are highest during winter dormancy and lowest in September/October (Thieme, 1965). Buds, by comparison, contain the highest levels in May immediately before leaf burst. The leaves contain the least phenolic glycosides in September (Thieme, 1965).

The levels of phenolic glycosides can also vary diurnally. In *Salix fragilis* and *Salix pupurea*, their concentrations decrease by 20% and 40%, respectively, from early morning to late in the day (Thieme, 1965).

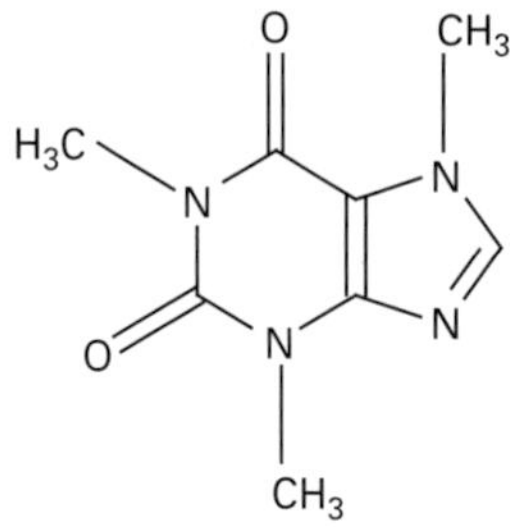

FIGURE 11.18 Caffeine.

11.3.3 Plant secondary metabolites and nutrients in fruits and seeds

Fruits typically are more chemically defended when still unripe, thought to deter frugivores before the seeds have matured. However, in some cases high levels of secondary plant metabolites persist during the ripe stage. For example, among North American nightshade species, ripe fruits of *Solanum carolinense* have high levels of glycoalkaloids, while in *Solanum americanum* they are negligible (Cipollini and Levey, 1997). The specific compounds are α-solasonine and α-solamargine. Among birds and mammals eating the seeds, American robin and opossum, *D. virginiana*, pass almost all seeds intact and are possibly major seed dispersers, while bobwhite quail and deer mice, *Peromyscus maniculatus*, destroyed most of the seeds. All four species, however, preferred the low-glycoalkaloid species and were deterred by the high-glycoalkaloid species. This supports the general toxicity hypothesis as opposed to the directed toxicity hypothesis, which posits that toxins are directed against specific consumers. Highly preferred, nutritious fruits of *S. americanum* have lower levels of glycoalkaloids. This supports the removal rate hypothesis: highly nutritious fruits that consumers remove quickly and predictably will show lower levels of secondary plant metabolites. (The opposite is the nutrient–toxin–titration–hypothesis: the higher the nutritional quality of a fruit, the more secondary metabolites it will have.) Cipollini and Levey (1997) concluded that the primary function of the glycoalkaloids in these *Solanum* fruits is defense against pests and pathogens. This, in turn, will influence the interactions between fruits and seed dispersers. In general, the plants appear to strike a compromise: the fruits contain enough alkaloid to avoid rotting but not so much that animals avoid eating them.

As these examples show, different defenses among plant parts and seasonal differences have to be considered together. Coffee (*Coffea arabica*) is a well-investigated example of increased chemical defenses when the plant is most vulnerable to herbivory. The concentration of the alkaloid caffeine (Fig. 11.18) varies between plant parts and with the growth cycle. In the germinating seed, the developing shoot has little alkaloid as long as it is protected by the endocarp. The caffeine concentration in the seedling is more than twice as high as in the seed (Baumann and Gabriel, 1984). In the young still-soft leaf, the concentration of caffeine increases to the high level of 4% of dry weight. The daily formation of this alkaloid decreases rapidly from 17 to 0.016 mg/g tissue as the leaf matures. The fully expanded leaf remains at that low level (Frischknecht *et al.*, 1986). Leaves are essentially alkaloid free before they die off. The caffeine is probably recycled within the plant, and its nitrogen used for protein synthesis. In the developing fruit, the alkaloid content is high (2% dry weight) early on when the pericarp is still soft but it decreases as the endocarp matures and hardens and is only 0.24% when the fruit is fully ripe. Harborne (1993) emphasized the advantage to the plant of reclaiming nitrogen from the alkaloid molecule as soon as the latter is no longer needed for defense.

The effect of caffeine has been tested with insect herbivores. Tobacco hornworms, *Manduca sexta*, die from a diet with 0.3% caffeine, and the beetle *Callosobruchus chinensis* becomes sterile at a concentration of 1.5% (Nathanson, 1984). However, experiments showed that caffeine does not protect the coffee plant against the leaf miner *Perileucoptera coffeella*, one of the major pests of coffee (Filho and Mazzafera, 2000).

Humans, the world's greatest caffeine-consuming vertebrates, have tested their tolerance limits of toxic effects in often extreme ways. Along the lower Putumayo in southwest Colombia, the Indians cut meter-long lengths from the vine *yoco* (*Paullinia yoco*) and keep it at their homes for up to 1 month. As needed for drinks in the early morning, they rasp the bark and squeeze the milky sap into cold water. They drink the extract from 100 g bark in one serving. The bark contains three times as much caffeine per weight as coffee beans. Even with the incomplete extraction in cold water, one dose delivers as much caffeine as more than 20 cups of coffee! The strong potion suppresses hunger for up to 5 hours, and energizes both men and women for marches of 20 miles or more (Davis, 1996). Too much of a good thing can prove fatal: The French novelist Honoré de Balzac died in 1850 of coffee poisoning. He was 51 years old. To keep writing day and night, he downed four dozens cups of coffee every day (Nathan, 1998).

11.4 Feeding or avoiding? herbivores vis-à-vis plant defenses

11.4.1 Secondary plant compounds, food preferences, and consumption

Herbivores and their food plants live in a delicate balance. We distinguish generalist herbivores from specialists. It is thought that generalism

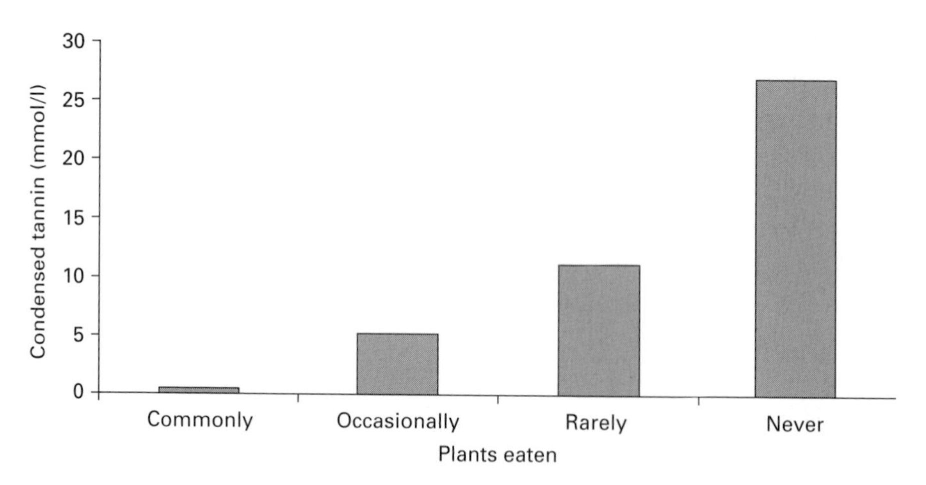

FIGURE 11.19 Condensed tannin concentrations in plants eaten to various degrees by giant tortoises on Adabra Island. (After Swain, 1979.)

evolved to avoid overloading of the herbivore's system with toxic compounds from only one or two plant species (Lindroth, 1988). In temperate and arctic regions, herbivores enjoy a larger number of food plants during summer than in winter. This also reduces the herbivore pressure on any one plant species (Palo, 1984). During the growing season, a plant may be able to outgrow herbivore damage more or less rapidly. This is particularly true for tropical plants. Depending on season and latitude, plants are exposed to different predation pressures: boreal plant species, such as deciduous and coniferous trees, suffer vertebrate herbivory for most or all of the year, while insect and pathogen damage is limited to the short growing season. The following covers feeding patterns of reptiles, birds, and mammals in relation to plant defenses.

Reptiles

Giant tortoises (*Geochelone gigantea*) on Aldabra Island in the Seychelles archipelago are more likely to feed on certain plant species, the less tannins they contain (Swain 1979; Fig. 11.19).

Birds

Ruffed grouse, *B. umbellus*, feed on staminate flower buds and extended catkins of trembling aspen, *P. tremuloides*. In winter these birds select specific trees or clones. Analysis for alkaloids, tannins, and other phenolics showed that feeding preferences were not related to the levels of tannin or total phenolics in flower buds or catkins. However, trees with no history of feeding had buds with higher concentrations of coniferyl benzoate and lower levels of proteins. Catkins have less coniferyl benzoate than buds, but they do not differ in protein content. Extended catkins are probably eaten in spring because they have little bud scale material. Bud scales contain almost all of the non-tannin phenolics found in catkins and dormant buds. Avoiding bud scales appears to be the feeding strategy of the grouse and possibly other birds (Jakubas *et al.*, 1989).

Canada geese, *Branta canadensis*, feed on coastal marsh plants that are low in soluble phenolics. The less phenolics (in percentage dry weight) the leaves of a species contained, the more often geese were observed to feed on it. Nutrient content of the plants was not related to consumption. Tannic acid or quebracho from the tropical American tree *Schinopsis lorentzii* (from *quebrar*, to break, and *hacha*, ax, because of its hard wood) added to *P. pratensis* rations inhibited feeding, and phenolic acids (1% ferulic acid and 1.5% *p*-coumaric acid) were slightly inhibitory. Isolated extracts or pure secondary compounds were never as active as the plant itself. The fractionation may have abolished synergistic effects in addition to changing other factors such as texture or water content (Buchsbaum *et al.*, 1984). Subarctic ptarmigans and grouse also avoid plants with a surface cover of phenolic resins. The resins are antimicrobial and interfere with microbial digestion in the caecum of these birds (Bryant and Kuropat, 1980).

Adult greylag geese, *Anser anser*, respond to non-food plant odors such as sage, violet, lavender, onion, or peppermint with head shaking before they contact the material with their bill. This indicates the role of olfaction in food selection (Würdinger, 1979). The response is present on the first day of life (Section 9.4).

The *ratio* of nutrients to plant secondary metabolites may be more important for food choice by birds and other herbivores than the level of the metabolites per se. Leaves of Arctic willow, *Salix arctica*, develop each spring from closed buds to open buds, rolled leaves, and finally open leaves. During that progression, phenol levels rise, fiber content decreases, and protein content peaks during the open-bud phase. The plant is most vulnerable during the open-bud and rolled-leaves stages when the combination of these three factors favors the herbivore. Greater snow geese, *Chen caerulescens atlantica*, feed on the open-bud and rolled-leaf stages, but not the closed-bud and open-leaf stages. The geese feed during these stages because the high protein content, declining fiber levels, and not-yet-high phenols level provide an optimal compromise. A high phenol:protein ratio lowered feeding more than large concentrations of phenolics alone (Gauthier and Hughes, 1995).

Bullfinches, *Pyrrhula pyrrhula*, feed on ash tree seeds with a high fat content and low tannin levels (Greig-Smith, 1988). Similarly, red-winged blackbirds, *Agelaius phoeniceus*, prefer sunflower seeds with higher oil concentrations, but avoid all levels of anthocyanin. Cultivars with high levels of anthocyanin and low levels of oil may be resistant to these birds (Mason *et al.*, 1989a). Zebra finches, *Taeniopygia guttata*, feed less on lettuce leaves or seed stands of *Setaria italica* that have been scented with oils of peppermint, lavender, sage, eucalyptus, anis, caraway, or crystalline thymol than control leaves (Würdinger, 1990). Finally, house sparrows, *Passer domesticus*, prey less on seeds of those hybrids of the cereal crop sorghum that contain high levels of tannins (Tipton *et al.*, 1970). Table 11.5 summarizes some of these studies.

Thrushes show distinct preferences and aversions for flavors of fruits. European blackbirds, *Turdus merula*, and songthrushes, *Turdus philomelos*, learned to associate the flavor and color of an artificial "fruit" made from flour and lard after only one experience (Sorensen, 1983). Among 11 fruit extracts tested, the birds preferred ivy (*Hedera helix*) most and almost always rejected buckthorn (*Rhamnus cathartica*). Preference for dogrose (*Rosa canina*), elder (*Sambucus nigra*) and hawthorn (*Crataegus monogyna*) was intermediate. The birds were wild caught and, therefore, had experience with wild berries.

Mammals

How much a mammal eats of a given plant often depends on the levels of different classes of chemical constituent, notably nutrients and plant secondary metabolites. As in birds, it is not the plant defense compounds alone, but rather complex balances between nitrogen and carbohydrate contents, levels of defense compounds, and fiber that determine palatability.

Marsupials

The western grey kangaroo (*Macropus fuliginosus*), the major indigenous herbivore in southwestern Australia, avoids plants of the family Myrtaceae (which include *Eucalyptus* spp. cloves, and guava), which includes 3500 species on that continent. Essential oils characterize the Myrtaceae; many species contain two to nine of these. In one experiment, seven Myrtaceae species were not browsed, while comparable species of other families were. All seven species contained the monoterpene 2,5-dimethyl-3-methylene 1,5-heptadine. The kangaroos also browsed one non-Myrtaceae species (*Sollya heterophylla*, Pittosporaceae) very little. This species contained the same monoterpene (Jones *et al.*, 2003).

The brushtail possum, *T. vulpecula*, feeds on eucalyptus leaves. Among the terpenes these contain, 1,8-cineole is the most abundant. Feeding experiments with increasing levels of cineole showed that this compound limits food intake. In the brushtail possum, multiple pathways oxidize cineole, and this total system, rather than any one enzyme, limits the amount of cineole that the animal

Table 11.5 Plant secondary metabolites inhibiting feeding by birds

Bird species	Plant species	Compound(s)	Reference
Red-winged blackbird *Agelaius phoeniceus*	Sunflower seeds	Anthocyanin	Mason *et al.*, 1989b
Bullfinch *P. pyrrhula*	Seeds of ash, *Fraxinus* sp.	Tannins	Greig-Smith, 1988
House sparrow *Passer domesticus*	Sorghum: unripe seeds	Tannins, "astringents"	Tipton *et al.*, 1970
Ruffed grouse *Bonasa umbellus*	Aspen *Populus tremuloides*	Coniferyl benzoate	Jakubas *et al.*, 1989
Canada goose *Branta canadensis*	Coastal marsh plants	Phenolics	Buchsbaum *et al.*, 1984
Greater snow geese *Chen caerulescens atlantica*	Arctic willow *Salix arctica*	Phenols	Gauthier and Hughes, 1995

can process (Boyle and McLean, 2004). Levels of the terpenes 1,8-cineole and limonene correlate significantly with the those of the formylated phloroglucinols, the active antifeedants. This suggests that, as a result of aversive conditioning, the volatile terpenes signal to the mammalian herbivore how unpalatable a particular plant is (Moore *et al.*, 2004).

Rodents

The voles *M. pennsylvanicus* and *M. ochrogaster* did not feed at all on some of 15 tested plant species in the laboratory. Others were eaten to different degrees, and 73% of the variance was explained by the nitrogen contents. The selectivity in the field corresponded to the palatability ratings in the laboratory: *Taraxacum* (dandelion), *Medicago* (alfalfa), and *Trifolium* (clover) spp. were the most palatable. *Andropogon* (salsify) and *Pastinaca* (parsnip) spp. were unpalatable in field and laboratory (Marquis and Batzli, 1989). The unpalatable species contain toxins, specifically alkaloids in *Penstemon digitalis* (Lindroth *et al.*, 1986), glucosinolates in *Barbarea vulgaris* (van Etten and Tookey, 1979), and furocoumarins in *Pastinaca sativa* (Berenbaum, 1981).

Turning to conifers, the meadow vole *M. pennsylvanicus* attacks seedlings of the introduced Norway spruce, *Picea abies*, and the North American red (Norway) pine, *Pinus resinosa*, but not white spruce, *Picea glauca*, and white pine, *Pinus strobus*. The last two have in common β-myrcene and bornyl acetate (Fig. 11.20) which are absent in the former two tree species. These two compounds are also toxic to insects and may be the agents protecting the trees (Roy and Bergeron, 1990a). Total levels of phenolics also tended to be higher in the two species

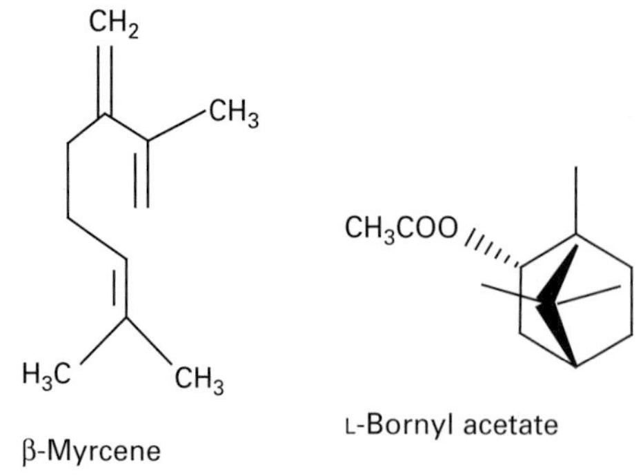

FIGURE 11.20 Secondary metabolites in some conifers.

of conifers that are avoided, but monoterpenes are seen as more important in defense against voles than phenolics (Bucyanayandi *et al.*, 1990).

In winter, Abert's (tassel-eared) squirrels (*Sciurus aberti*) of the southwestern United States feed on the inner bark of ponderosa pine (*Pinus ponderosa*) twigs, besides seeds of pines and piñons. The squirrel selects specific, so-called "feed" or "target" trees that are low in α-pinene while ignoring other nearby trees (Farentinos *et al.*, 1981). Snyder (1992) found low levels of β-pinene and β-phellandrene in the xylem of such "feed trees." Trees defoliated by squirrels suffered up to 90% loss in fitness indicators such as growth, cone production, and seed quality (Snyder 1993). Monoterpene levels in xylem are under strong genetic control. Therefore, Abert's squirrels may play an important role in shaping monoterpene variation in pines (Snyder 1998).

Gray squirrels, S. *carolinensis*, possibly use the high levels of tannins (6–10%) in acorns of red oaks (subgenus *Erythrobalanus*) as cue for burying them as winter cache. Acorns of white oak (subgenus *Lepidobalanus*) are low in tannins (0.5–2.5%) and germinate in autumn, making them unsuitable for caching. Concentrations of tannins and lipids covary in these acorns. Experimental doughballs, prepared from acorn flower, were avoided when tannin was added, but additional fat did not increase feeding unless tannin had been added. Addition of fat attenuated the negative effect of tannin. By avoiding perishable acorns and burying more durable ones, the squirrels optimize their food supply for the entire (winter) season, rather than for shorter periods. The tree may signal to small-mammal scatter hoarders by means of tannin, with the result of improved dispersal (Smallwood and Peters, 1986). In addition, the moist soil may enhance palatability by reducing the tannin level. When acorns of the red oak are abundant, gray squirrels eat only the basal part and discard the tip of the acorn, which contains the better defended radicle (Fig. 11.21)

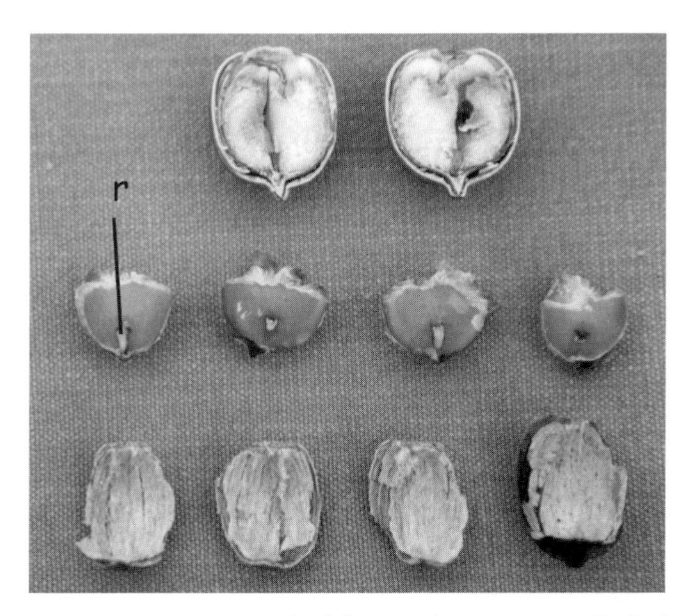

FIGURE 11.21 Gray squirrels have partly eaten acrons of red oak, *Quercus rubrum*. Top shows cross-section through untouched acorn. Two bottom show the rows remains of acorns after feeding. In the middle, there are discarded bottom parts of acorn, which is thought to have a high tannin concentration. The radicle (r) remains intact and such pieces often still germinate. (Photograph: D. Müller-Schwarze.)

Meadow voles, *M. pennsylvanicus*, and sheep eat reed canarygrass according to the alkaloid content of the particular sample (Kendall and Sherwood, 1975; Robinson, 1979). Meadow voles also consume the more alfalfa the less saponins it contains (Kendall and Leath, 1976). Beaver, *Castor canadensis*, feed on the usually avoided red and Scots pines only in late winter (D. Müller-Schwarze, personal observations). It is assumed that the ratio between carbohydrates and terpenes is optimal at that time.

Lagomorphs

Phenols deter Eurasian mountain hares, *Lepus timidus*, from debarking Scots pine, *Pinus sylvestris* (Rousi and Häggman, 1984). In Finland, these hares prefer some willow species (*Salix* sp.) over others in winter, and mature shoots to juvenile shoots. Feeding differences are correlated with phenolic glycoside (Fig. 11.22) content. In feeding experiments in the field, hares ate less oat grains treated with phenylglycoside or catechin (Tahvanainen *et al.*, 1985). The phenolic compound pinosylvin and its monomethyl ether in green alder are highly repellent to snowshoe hare, *L. americanus* (Bryant *et al.*, 1983).

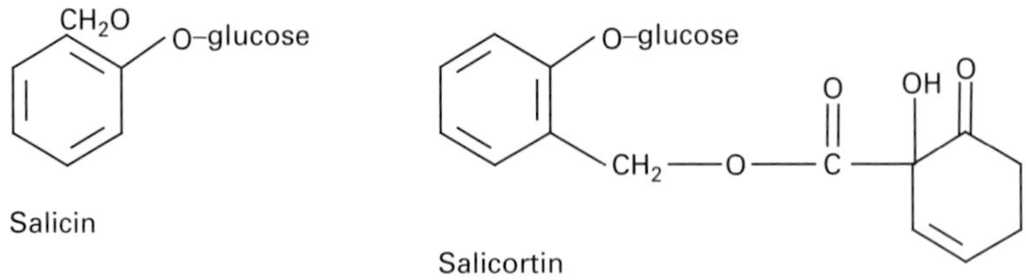

FIGURE 11.22 Phenyl glycosides.

Ruminants

Mule deer, *Odocoileus hemionus hemionus*, prefer juniper (*Juniperus* spp.) that contains lower amounts of volatile oils (i.e. oxygenated monoterpenes). Of three species tested, alligator juniper (*J. deppeana*) had the lowest level of volatile oils and was preferred, while Utah juniper (*J. osteosperma*) and Rocky Mountain juniper (*J. scopulorum*) ranked higher in volatile oil content and lower in preference by deer (Schwartz *et al.*, 1980).

Sagebrush, *Artemisia sp.*, covers 422000 square miles in 11 western states of the USA and Canada. It is important as winter diet of mule deer, *O. h. hemionus*, pronghorn, *Antilocapra americana*, and sage grouse, *Centrocercus urophasianus*. Mule deer prefer sagebrush taxa with low levels of the oxygenated compounds methacrolein and arthole. However, only one fifth or less of utilization of eight taxa of sagebrush leaves is attributable to the levels of the volatiles methacrolein, α-pinene, camphene, β-pinene, arthole, 1,8-cineole, *p*-cymene, and camphor. In October, the value is 15.9%, in winter (March) 20.7%. Therefore, the volatile components may have their greatest effect during intense winter browsing but the sagebrush volatiles actually decrease during winter (Scholl *et al.*, 1977). When chopped alfalfa hay was treated with various sagebrush terpenoids, methacrolein deterred mule deer most. It may guide preferences among different subspecies of big sagebrush, *Artemisia tridentata*, and *p*-cymene between big sagebrush and black sagebrush, *Artemisia nova*. The crude terpenoid fraction with its sesquiterpenes closely correlated with feeding in all four tested taxa of *Artemisia* (Bray *et al.*, 1991).

Moose (*Alces alces*) in northern Europe discriminate between two birch species. They prefer the faster-growing *B. pendula* over *B. pubescens*, most likely because the former has less defense compounds (Danell *et al.*, 1985). This is consistent with the hypothesis of Bryant and Kuropat (1980) that faster growing tree species are less defended chemically. Phenols may defend birch in winter against vertebrate browsers by reducing digestibility and being toxic. Moose also feed less on Scots pine, *P. sylvestris*, if it contains higher levels of pinifolic acid, a diterpene (Danell *et al.*, 1990). Moose choose to utilize particular Scots pine trees

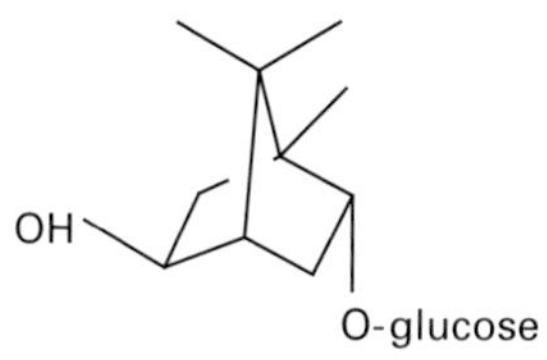

FIGURE 11.23 (1*S*,2*R*,4*S*,5*S*)-Angelicoidenol-2*O*-glucopyranoside (also known as (-)-5-exohydroxyborneol.

according to phenol content, with the concentrations of most phenols internally correlated (Sunnerheim-Sjöberg and Hämäläinen, 1992). A glycoside of the terpene (1*S*,2*R*,4*S*,5*S*)-angelicoidenol, also known as (−)-5-exohydroxyborneol (Fig. 11.23), is a moose feeding deterrent found in Scots pine (Sunnerheim-Sjöberg, 1992). However, no correlation between moose browsing and the composition of ether extracts of Scots pine was found in several studies (Löyttiniemi and Hiltonen, 1978; Löyttiniemi, 1981; and Njemelä and Danell, 1988).

On the South African savanna, large generalist browsers among the ruminants feed on many species. The greater kudu, *T. strepsiceros*, was observed to feed on 60 species of plants in 12 hours. The kudu and two grazer–browsers, the impala (*Aepyceros melampus*) and the Boer goat rejected all plants with more than 5% of condensed tannins in their leaves during the wet season. Otherwise there was no correlation with plant contents (Cooper and Owen-Smith, 1985). The condensed tannins are attached to the plant's cell walls and are thought to protect against microbial and fungal attack (Zucker, 1983). Cooper and Owen-Smith (1985) postulated that the plant defense is really aimed at the microbes, originally in the environment and now in the rumen. Obviously, ungulates need the rumen flora for digestion but become "third parties to the chemical warfare between higher plants and pathogenic microorganisms." Domestic cattle also prefer food with less tannin (Marks *et al.*, 1988), and sheep avoid grazing on lupines with high alkaloid content (Arnold and Hill, 1972).

Proboscids

African elephants avoid plants with high contents of total phenols, steroidal saponins, and lignin (Jachmann, 1989).

Primates

Free-living primates respond to plant chemistry: black-and-white colobus monkeys, *Colobus polykomos*, avoid plant leaves with alkaloids, biflavonoids, and milky latex and prefer to eat leaves with a better ratio of nutrients to digestion inhibitors (McKey *et al.*, 1981). The howler monkey, *Alouatta palliata*, a generalist herbivore, depends on more than one factor in its choice

Table 11.6 Some plants and their secondary compounds that inhibit feeding in mammals

Mammal species	Plant species	Compound(s)
Meadow vole, *Microtus pennsylvanicus*, and *Microtus ochrogaster*	*Penstemon digitalis* *Barbarea vulgaris* *Pastinaca sativa*	Alkaloids Glucosinolates Furocoumarins
Meadow vole	White spruce *Picea glauca*, White pine *Pinus strobus*	β-Myrcene and bornyl acetate
Mountain hare *Lepus timidus*	Scotch pine *Pinus sylvestris*	Phenols
Snowshoe hare *Lepus americanus*	Green alder *Alnus crispa*	Pinosylvin, pinosylvin ether
Mule deer *Odocoileus hemionus*	Juniper *Juniperus* sp.	Oxygenated monoterpenes
Moose *Alces alces*	Scotch pine	Pinifolic acid (diterpene), exohydroxyborneol (terpene)
Greater kudu, *Tragelaphus strepsiceros*, and impala *Aepyceros melampus*	60 species	Condensed tannins
Sheep, meadow vole	Reed canary grass	Alkaloids
Meadow vole	Alfalfa	Saponins
Snowshoe hare	Labrador tea *Ledum groenlandicum*	Germacrone (sesquiterpene)
Opossum *Didelphis* sp.	*Salix* sp.	Salicin derivatives (phenolic glycosides)

of leaves. Young leaves of the monkeys' food plants had, on average, 11% more total phenolics than mature ones. The ratio of protein to fiber is most important, with secondary plant compounds indeed playing a "secondary" role in this case (Milton, 1979). Vervet monkeys, *Cercopithecus aethiops*, select plants (*Acacia tortilis* and *A. xanthophloea*) with low level of tannins for feeding (Wrangham and Waterman, 1981). Rhesus monkeys (*Macaca mulatta*) in Pakistan select food with low tannin content (Marks *et al.*, 1988). Table 11.6 summarizes some of these studies.

Response differences between mammal species

Some defenses are so effective that herbivores do not consume any amount of a plant. Snowshoe hares are deterred from feeding on Labrador tea, *Ledum groenlandicum*, by germacrone, a sesquiterpene, the major compound of its essential oil. Leaves and fresh growth internodes contain large amounts of germacrone (Reichardt *et al.*, 1990a). Yet beaver, *C. canadensis*, in Saskatchewan,

Canada, have been observed to feed on Labrador tea and store it in their food caches (Gunson, 1970). Storing in water may make Labrador tea more palatable.

Neighbor effect

Mammals often avoid palatable plants growing or placed next to effectively defended species. Western grey kangaroos, *M. fuliginosus*, usually browse *Hardenbergia comptoniana* (Fabaceae) heavily. But when grown next to a less-palatable plant from the Myrtaceae family – characterized by essential oils, notably volatile monoterpenes – *Hardenbergia* sp. suffered less browsing by these animals. By comparison, a palatable non-Myrtaceae plant as "nurse plant" next to *H. comptoniana* resulted in heavy browsing of the latter (Jones *et al.*, 2003).

Similarly, Arctic ground squirrels, *Spermophilus parryi plesius*, browse plants with palatable neighbors more than those without (Frid and Turkington, 2001). Field experiments with plants or plant parts treated in different ways have to take into account these neighbor effects. While statistical designs call for random distribution of treatments in experimental plots, the behavior of animals may dictate otherwise. Since the taste or odor of one specimen can overshadow its neighbors, field bioassays work better when same-treatment specimens are clustered. Such "pseudobushes" were used in food choice experiments with beavers (Müller-Schwarze *et al.*, 2001) and field bioassays of plant extracts with Eastern cottontails (Müller-Schwarze and Giner, 2005). We can also recruit these neighbor effects for reducing herbivory of crops since treatment of only certain plants with repellent may also protect untreated neighbours.

Mushrooms

Mushroom poisoning in wild animals is little understood. Many mammals, including squirrels, other rodents, and deer, readily eat mushrooms, even toxic species. Gray squirrels, *S. carolinensis*, feed on *Amanita phalloides*, the death cap. A large portion of the diet of flying squirrels, *Glaucomys sabrinus*, and the California red-backed voles, *Clethrionomys occidentalis*, are hypogeous fungi (Fogel and Trappe, 1978). Red squirrels, *Tamiasciurus hudsonicus*, are famous for consuming toxic mushrooms, among them *Amanita* spp. In laboratory experiments, opossum, *D. virginiana*, readily ate the toxic fly agaric *A. muscaria*, which contains the hallucinogenic muscimol. Opossums vomited after ingesting muscimol-treated, otherwise palatable mushroom species such as *Calvatia gigantea* and *Panellus serotinus*. After one day of eating poisoned mushrooms, the opossum avoided the same and a second, non-poisoned, mushroom species for several days (Camazine, 1983).

Tracking changes in toxicity

A generalist herbivore can base the amount of food it eats on flavor concentrations, if these are correlated with toxicity. Sheep are able to adjust their food intake when the toxicity of food changes. Experimental animals identified plant toxins by associating food flavor added to oats, with post-ingestion consequences. First, if the concentration of a tastable toxin such as lithium chloride (LiCl) is increased, lambs decrease consumption; conversely, they increase consumption if the LiCl concentration is decreased. Second, lambs that had experienced two concentrations each of a bitter and a sweet compound in their barley chow, choose the *lower* concentration after receiving a low dose (125 mg/kg body weight) of LiCl. Finally, lambs were fed oregano-flavored ground barley. Following that, they received LiCl in gelatine capsules orally, in medium, high, or variable doses. The lambs with the medium dosage consumed the most barley, while those receiving high or variable dosages consumed less and by the same degree (Launchbaugh *et al.*, 1993). When toxicity cannot be detected by flavor, their intake depends on the maximum dose of toxin they had experienced.

Herbivores possibly regulate the amount eaten from toxic plants and thus avoid ill effects. Cattle cycle between consuming large and small amounts of tall larkspur, *D. barbeyi*. The principal toxic alkaloid in tall larkspur is methyllycaconitine (Pfister *et al.*, 1997).

Effects of nutritional status on consumption of toxic plants

Toxic plants may kill or harm herbivores if few other plants are available, because the animals are forced to be less discriminating, and because they are more susceptible to the toxins. In an experiment by Wang and Provenza (1996), three groups of lambs were maintained on three different levels of protein. In addition, they received every morning rations that varied in energy (barley) and a toxin (LiCl). The proportions of barley changed every 5 days, for five such periods. In the absence of the toxic LiCl, the lambs, consumption correlated with barley content. However, as LiCl concentrations increased, the lambs ate less of foods high in energy. Remarkably, with increasing food deprivation the lambs ate less and less of the LiCl-containing food, even if it had a high level of energy. This study shows how important interactions between nutritional status and toxicosis can be in herbivores.

11.4.2 When predators cannot cope with toxins of their animal prey

The bearded dragon, *Pogona vitticeps*, from Australia enjoys increasing popularity as a pet in the United States. The species comes from a geographical area lacking very potent chemical defenses in insects. Consequently, the bearded dragon is vulnerable to toxins from prey in its new homeland, fireflies (*Photinus* sp.), particularly, can be deadly as they contain lucibufagin, a steroid resembling digitalis steroids and with the same effect on the heart. The reptile will eat fireflies, fails to neutralize the toxin, and dies as a result. Other pets such as African chameleons and lizards from Eastern Europe suffer similar poisoning by fireflies (Knight *et al.*, 1999).

Absence of plant secondary metabolite effects

In some studies, feeding intensity did not vary with the concentration of the monitored plant defense compounds. Mule deer, *O. hemionus*, use from 0 to 83% of the current year's growth of sagebrush, *Artemisia* sp. The monoterpenoid content varies from 0.75 to 3.62% dry matter. However, no clear correlation between these two has been established in a study of 21 accessions from five taxa (Welch *et al.*, 1983). Penned black-tailed deer, *O. h. columbianus*, consumed Douglas fir of different genetic origins regardless of essential oil contents, even though their preference was correlated with the chlorogenic acid content (Radwan and Crouch, 1978). Similarly, pygmy rabbits, *Brachylagus idahoensis*, did not feed differentially on 15 populations from two subspecies of sagebrush, *A. tridentata* (White *et al.*, 1982). Nevertheless, Welch and McArthur (1981) suggested that sagebrush could be selectively bred for reduced monoterpenes to allow better utilization by mule deer.

In Sweden, phenols did not protect introduced lodgepole pines, *Pinus contorta*, particularly well against debarking by the vole *M. agrestis*. Being an alien tree species may be an important factor (Hansson *et al.*, 1986). The authors pointed out that, in general, any natural defense will work only at low browsing pressure by voles.

11.4.3 Coping: how herbivores overcome plant defenses

> They put arsenic in his meat
> And stared aghast to watch him eat;
> They poured strychnine in his cup
> And shook to see him drink it up:
> They shook, they stared as white's their shirt:
> Them it was their poison hurt.
> –I tell the tale that I heard told.
> Mithridates, he died old. A. E. HOUSMAN

Herbivores have evolved many *offensive adaptations* that counteract the *defensive adaptations* of plants, including antifeeding compounds (Rhoades, 1985).

Herbivores deal with potentially toxic plant secondary metabolites in three different ways: they can avoid particular plants altogether; process the food to improve its palatability; or ingest it and deal with the plant secondary metabolites by biotransforming them. This is where King Mithridates of Pontus (today's Turkey) comes in. Fearing poisoning by his enemies, he ate toxins in increasing amounts and built up his resistance. He survived an assassination attempt and has given his name to a procedure to build up immunity to toxins: "to mithridate."

Food avoidance

Preferences and avoidance such as seeking sweet foods and rejecting bitter-tasting plants can be acquired in three ways. Avoidance acquired in evolutionary time may not require individual learning and is termed *primary aversion*. It occurs to unpleasant taste or odor. A conditioned aversion can be acquired by an individual after a negative experience such as illness by food poisoning. Finally, an individual can learn from relatives or group members to avoid a food, without ever trying out the food itself.

Primary aversion linked to taste

Alkaloids and cyanogenic glycosides taste bitter to humans and have been thought to be universally repellent to higher animals (Bate-Smith, 1972). However, guinea pigs, *Cavia porcellus*, tested with nine bitter substances, reduced their feeding to only two (quinine and sucrose octaacetate), and that only slightly (Nolte *et al.*, 1994a).

Primary avoidance of toxic mushrooms by the fungivorous oppossum, *D. virginiana*, is a good example. Many mushrooms have toxins, including emetics, cathartics, hallucinogens, and liver poisons. The pungent-tasting mushrooms of the family Russulaceae contain *isovelleral*, a sesquiterpenoid dialdehyde. Opossums feed little or not at all on isovelleral-containing species. If isovelleral is added to morsels of the palatable species *Agaricus*, consumption declines by 67 to 22%, depending on concentration. The opossums rapidly eject the morsel, salivate profusely, froth at the mouth, and clean their muzzle. Fatty acids may enhance the peppery sensation of isovelleral and may help to solubilize isovelleral in the saliva (Camazine *et al.*, 1983). However, red squirrels actually prefer peppery mushrooms such as *Russula emetica* or *Lactarius* sp. Similarly, the opossums ate the peppery *Boletus piperatus*. Preference for pepper taste by humans and other mammals may moderately stimulate the induction of detoxication mechanisms to keep them ready for dietary contingencies.

Primary aversion linked to odor

The avoidance of the odor signal may be primary. An odor can be a negative signal acquired during evolution, with no conditioning necessary. However, even anosmic animals can show food aversions. Therefore, taste alone can be sufficient for the aversion. So far, there is no evidence for an intrinsic repellent effect of odor alone. In all studied cases, odor had to be associated with ill effects, including nociception (damage to receptors) or toxicosis (gastrointestinal malaise) (e.g. Provenza *et al.*, 2000).

In *neophobia*, defined as "an individual's avoidance of a novel object" (Barrows, 1995), an animal may totally reject a new food, or take minute amounts to await the consequence of this ingestion. Neophobia, therefore, forms the transition between primary and conditioned aversions.

Conditioned Taste Aversion

Conditioned taste aversions occur after food produced an illness. Both the chemical senses and gastrointestinal feedback are involved.

Postingestive consequences of eating toxic plants may lead to an aversion. For instance, tall larkspur (*Delphinium* sp.) is toxic to cattle, even though it is palatable. Cattle that had been fed larkspur consumed less in subsequent feeding trials. However, animals with esophageal fistulae that had tasted larkspur but not experienced it in their rumen did not reduce their consumption. This shows that chemosensory input alone is not sufficient for producing an aversion (Pfister *et al.*, 1990a). Goats, however, were shown to discriminate between two novel foods with different postingestive consequences. This discrimination depended on both the relative amounts of the two foods eaten and the salience of the flavors of these foods (Provenza, and Lynch, 1994).

An intriguing question is how many plant odors and tastes a herbivorous mammal is able to memorize at any one time, and whether it increases feeding efficiency by grouping several plant species under one odor or taste. In one experiment, lambs were given a meal of five foods, one of which was novel. After 1 to 6 hours, they experienced toxicosis. Subsequently, they decreased their intake of the novel but not the familiar food (Provenza, 1995).

Conditioned aversion to taste plus odor

In a two-step response, an additional, volatile (olfactory) stimulus may signal over a distance that the plant is unpalatable and/or toxic.

Olfactory aposematism (Eisner and Grant, 1981) means associating an odor by conditioning (experience) with an odorless toxin such as nicotine, morphine, or strychnine. This is probably widespread among mammals. First the animal tastes the plant and finds it either unpalatable or suffers ill effects. After that, a different, more volatile compound becomes the signal for immediate avoidance. Opossums learn to associate the typical mushroom odor of the innocuous volatile 1-octen-3-ol with delayed illness from toxic mushrooms (or edible mushrooms experimentally poisoned with the odor- and tasteless toxin muscimol) (Camazine, 1985). Plant odors deter lambs from feeding only when associated with, or followed by, gastrointestinal malaise induced by LiCl. A novel plant odor, such as that of *Astralagus bisulcatus*, an unpalatable, toxic, and sulfur-smelling herb of the North American west, reduces feeding by lambs briefly, but for more than 5 days only if the toxicosis continues (Provenza *et al.*, 2000).

Two speculative possibilities relate to olfactory aposematism. The first is whether there are non-toxic plants that smell or taste like toxic ones. In other words, do plants practice Batesian mimicry? Such mimicry is unlikely, as mammalian herbivores constantly sample plants and thereby test for favorable and adverse effects of eating a particular species. Furthermore, given the keen sense of smell of mammals, two plant species would have to exactly smell alike for mimicry to work. Second, do two distasteful or toxic plant species smell or taste alike so that herbivores can more easily classify dangerous plants and avoid them (Müllerian mimicry) (Eisner and Grant, 1981; Lindroth, 1988; Augner and Bernays, 1998).

Potentiation describes the additive effects of taste and odor that allow a conditioned responce to both that would not occur to either alone. Odor is not necessarily an effective conditioned stimulus, and it has been said that "the nose learns from the mouth". After illness from a poisoned meal, taste is a more potent conditioned stimulus than odor, but taste can potentiate odor. An animal will first smell, then lick a food, and only then reject it. During foraging, odor is perceived before taste. The opposite sequence is true in the intrinsic equilibrium responses of the body: a taste experience leads to an odor becoming a signal (Garcia and Rusiniak, 1980). Learning is extremely quick and lasting: one exposure to a toxic food (single-trial learning) often suffices, and the effect may be lifelong (long-delay learning).

Several sensory modalities can be involved in complex interactions involving at least three cues: rats learnt better to associate the bitter taste of quinine in water with a "context" such as a black or white box if a pyrazine was also present (the specific compound used was 2-methoxy-3-isobutyl pyrazine) (Kaye *et al.*, 1989). It is said the odor "potentiates" learning the connection between taste cue and "context."

A case of taste potentiating a visual cue is that of hawks feeding on mice. If the hawks were routinely fed white mice, and only occasionally a black mouse followed by LiCl injection, the hawks would not eat either black or white mice. However, when a distinctive taste was added to the black mouse, hawks learnt to avoid black mice on sight after experiencing only a single meal of a black mouse that resulted in toxicosis (Brett *et al.*, 1976).

Observational learning

Young animals can learn food choices by observing a parent or other group member without ever eating the plant itself. Many young mammals, such as deer fawns (D. Müller-Schwarze, personal observations), forage close to experienced conspecifics – usually the mother – for long periods of time so they have ample opportunity to learn food preferences. Lambs with their mothers consume more palatable food and become poisoned less often than lambs without mothers (Provenza, 1995).

Stephen's woodrat, *Neotoma stephensi*, specializes on juniper (*Juniperus* sp.) foliage that contains high levels of tannins and terpenoids. Compared with other woodrat species, the young are weaned late, grow slowly, and associate long and closely with their mother. Mothers and young are "sitting nose to nose and eating the same food item" (Vaughan and Czaplewski, 1985). The authors have suggested that during the long preweaning period the young learn to select leaves from particular trees that are low in defensive compounds.

Processing food before ingestion

Some animals "process" plants to reduce defense compounds even before eating. The meadow vole *M. pennsylvanicus* cuts winter branches of white spruce, Norway spruce, white pine, and Norway pine and leaves them on the snow for 2–3 days before eating them. This reduces the levels of condensed tannins and other phenolics by one half, to their summer levels. A high level of protein (12%) and reduced phenolics (1.5% of dry matter) now render the food acceptable. It is not clear how the phenolics are being lost, possibly by polymerization or oxidation (Roy and Bergeron, 1990b).

Pikas (*Ochotona princeps*) of the North American west also practice food conditioning. They store winter food in "hay piles." These food caches consist primarily of Alpine avens, *Acomastylis (Geum) rossii*, which the pikas do not eat during the summer. The Alpine avens contain three to six times more phenolics than the summer diet. These compounds preserve the food. After weeks or months of storage from October to January, the toxins in the cached plants decrease to summer diet levels. As the stored plants lose phenolics, the pikas consume increasingly larger amounts. In short, the pikas practice food conditioning in addition to using natural food preservatives (Dearing, 1997).

Animals that bury acorns and other seeds as winter cache may also practice food processing, tannins being lost during the weeks in the moist soil. This

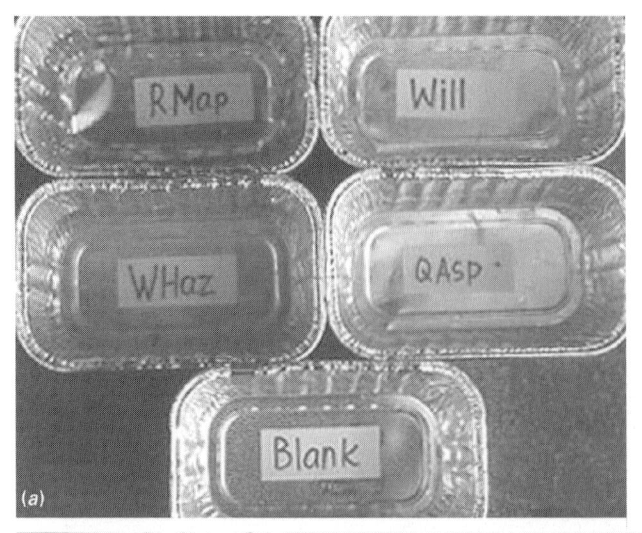

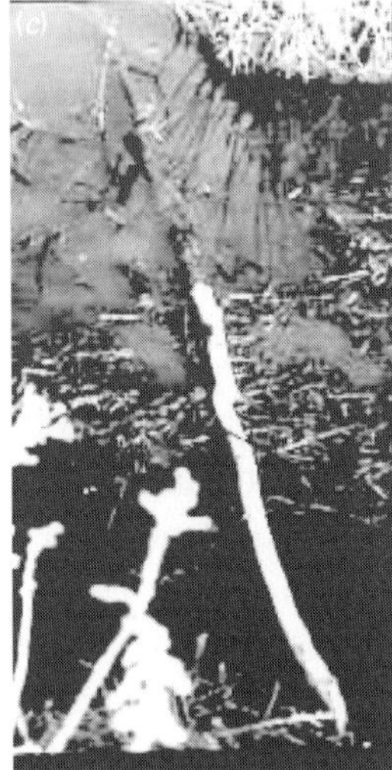

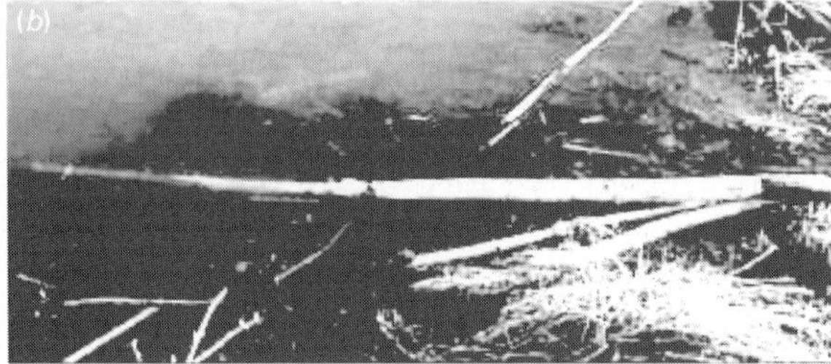

FIGURE 11.24 Leaching: (a): Standard pieces of different trees leach out different amounts of phenolics and other compounds. During 24 hours, red maple and witch hazel (left) stain water a darker brown than willow and quaking aspen (right). (b) and (c): Beavers leave pieces of branches in the water for 1–3 days with some or all bark intact before eating the bark. This presumably leaches out water-soluble plant secondary compounds.

possibly applies to gray squirrels (Smallwood and Peters, 1986) and blue jays (Dixon *et al.*, 1997).

Beavers, *C. canadensis* store branches for weeks and months in the water as winter food caches. They also soak logs and sticks of less-palatable trees in their pond, often for 2–3 days, before they eat them (Fig. 11.24). Field and laboratory experiments suggest that in this way undesirable compounds leach out from the bark (Müller-Schwarze *et al.*, 2001).

FIGURE 11.25 Leaching by humans. (*a*) A cassava plant. (*b*) Ground manioc (cassava) is mixed with water and pressed through tube woven from palm fibers to remove toxic cyanogenic compounds. Segments are collected in a tub. The liquid with toxins is in a pot in the foreground. (Photograph: D. Müller-Schwarze; Paraitepuy Village, Gran Sabana, Venezuela, 2000.)

Humans obviously have devised effective methods of leaching toxins from staples such as bitter manioc, also known as cassava or yucca, *Manihot esculenta* (Euphorbiaceae) (Fig. 11.25). Manoic is native to tropical South America and was introduced to Africa by the Portuguese. The cyanogenic compounds discourage herbivores and protect the crop against microorganisms. Cooking destroys enzymes, but not linamarin, and manioc remains toxic. Therefore, the bitter manioc must be detoxified by first shredding or grating it to bring enzyme and substrate into contact, and then removing the generated hydrogen cyanide as gas or dissolved in water. Many Amazonian societies press a mixture of water and grated manioc through a hose of woven plant fiber (Fig. 11.25*b*) and discard the expressed whitish, toxic liquid. A woman needs 4 to 6 hours daily or every other day for this time-consuming labor (McKey and Beckerman, 1993). Even so, improper processing results in sometimes tragic poisonings. On March 9, 2005, schoolchildren in a village on Bohol Island in the Phillipines were poisoned by deep-fried caramelized cassava and 27 of the children died immediately (*USA Today* March 10, 2005, p. 13A). As sweeter manioc varieties are being introduced, loss to pests will most likely increase. Anecdotal evidence suggests that pigs, baboons, and rodents attack sweet manioc more than bitter varieties. Dried chips of bitter manioc roots store better than those made from sweet varieties (in McKey and Beckerman 1993). Bitter varieties are also said to be more productive. People living along rivers rich in fish use bitter varieties of manioc. One reason may be that fish provides sulfur-containing amino acids, needed for detoxifying hydrogen cyanide with thiosulfate. More nomadic peoples, by comparison, grow sweet manioc and hunt agoutis and other bothersome rodents in their gardens. In general, sweet manioc is grown as a vegetable, while groups who produce manioc flour select for more bitter varieties (McKey and Beckerman 1993).

Biotransformation (detoxication)

Herbivores biotransforms xenobiotics from natural or artificial sources to render them harmless. Disarming potentially toxic compounds is termed detoxication, while the term detoxification refers to correcting a state of toxicosis (Johns, 1990). Herbivores degrade secondary plant compounds in a variety of ways, starting in the mouth, and leading to excretion.

Birds

The diet of some birds contains high levels of monoterpenoids. Sage grouse feed on sagebrush, *A. tridentata,* which contains the monoterpenoids α-pinene, camphene, 1,8-cineole, β-thujone, camphor, and α-terpineol. These compounds constitute 3.23% of the dry matter in *Artemisia* leaves. During passage from the beak to the gizzard, the levels of these compounds are reduced to 64% of that in the leaves, but this is still high (Welch *et al.*, 1989). Further reduction of monoterpenes takes place in the gizzard where the leaves are ground into fragments. Normal body heat volatilizes monoterpenoids. Experiments showed α-pinene to be very volatile under these conditions while camphor is not. Drastic reduction of monoterpenoids occurs between gizzard and duodenum. No monoterpenoids were found in the ceca; therefore the ceca microbes are not endangered. In summary, the bird's body is exposed to far less monoterpenes than the plant content would suggest (Welch *et al.*, 1989).

Detoxification of plant secondary compounds has been suggested as one of the nutritional benefits responsible for the evolution of the unique foregut microbial fermentation system in the hoatzin, *Opisthocomus hoazin,* a South American obligate leaf-eating bird (Fig. 11.26). In this ruminant-like digestive system, crop and lower esophagus produce volatile fatty acids whose concentrations and ratios resemble those of the sheep rumen (Grajal *et al.*, 1989; Grajal, 1995).

FIGURE 11.26 The hoatzin is a leaf-eater that is known in Peru as "chancho" (pig) because of its smell. This bird has few predators, presumably because it sequesters plant compounds that render it unpalatable. (Photograph: D. Müller-Schwarze, Rio Napo, Ecuador, 2003.)

Mammals

Mammals neutralize toxic plant compounds at all stages of ingestion and digestion, aided by both microbes and specialized enzymes. Smaller species can detoxify xenobiotics faster because of their high metabolic rate. Therefore, they are more likely to evolve specialized food habits (Freeland, 1991). As in birds, the fight against ingested plant compounds starts in the mouth. Monoterpenes such as α-pinene are very volatile at body temperature and so their concentration can be reduced by mastication, rumination, and eructation, in addition to the traditionally considered absorption and excretion (Welch *et al.*, 1989).

Browsers such as mule deer, *O. hemionus*, have *proline-rich salivary proteins* that bind tannins; this enables them to feed on tanniniferous plants (Robbins *et al.*, 1991). By contrast, grazers such as sheep and cattle do not possess tannin-binding proteins (Austin *et al.*, 1989; Robbins *et al.*, 1991), and neither do mixed feeders such as goats (Distel and Provenza, 1991). Rat and mouse saliva also contain proline-rich proteins that bind tannins. In human saliva, about 70% of total protein is rich in proline (Fig. 11.9) (Mehansho *et al.*, 1987).

Proline-rich proteins have an affinity for tannins at least three orders of magnitude greater than other proteins, protecting dietary proteins and digestive enzymes (McArthur *et al.*, 1995). The salivary proteins form hydrogen bonds with tannins. They also increase the accessability of the peptide backbone for hydrogen bonding. The interaction is very specific for the protein but not for the tannin. Condensed tannins are slightly more effective as protein precipitants than hydrolyzable tannins. The parotid glands of rats can be induced to produce more proline-rich proteins in response to a tannin-rich sorghum diet. This requires about 3 days (Mehansho *et al.*, 1983). However, meadow voles, *M. pennsylvanicus*, do not appear to produce tannin-binding proteins in response to dietary quebracho tannin (Dietz *et al.*, 1994).

Various mammal species differ in their processing of tannins. Laboratory rats are more resistant than hamsters or prairie voles, and goats more than sheep or cattle. In one experiment, tannin from the quebracho tree (*Schinopsis* sp.) was fed to mule deer, domestic sheep, and black bear (*Ursus americanus*). Mule deer, black bear and laboratory rats have salivary tannin-binding proteins that are very effective. These species excrete up to 98.3% of the tannin in their feces, while in sheep 25% are not found in the feces and so are presumed metabolized with all the physiological consequences. Salivary proteins minimize fecal nitrogen loss by maximizing amounts of tannin bound per unit of protein, minimize tannin absorption and toxic effects by forming stable tannin-protein complexes, and prevent or minimize reduction in fiber digestion (Robbins *et al.*, 1991). High levels of protein in the diet also protect against ill effects of tannins as binding of tannins to protein protects the esophagus (Warner and Azen, 1988).

The good earth: geophagy

Ingestion of mineral substances, termed geophagy, occurs in a wide range of herbivorous birds and mammals. Parrots, cracids, pigeons, doves, deer, elk, elephants, giraffes, zebra, sheep, bears, raccoons, monkeys, tapirs, and peccaris regularly eat soil. Geophagy is still little understood. In addition to uptake of minerals or microorganisms, clay can be helpful in detoxification of secondary plant compounds.

Neotropic parrots make daily morning trips to riverside cliffs to eat clay. The clay-licks of Manu National Park in Peru are especially well known (Fig. 11.27). Macaws and parrots visit the clay-lick more often in the dry season when fresh fruit are rare and they have to rely more on seeds with toxic compounds (Brightsmith, 2002). The value of soil was demonstrated using extracts from a variety of ripe and unripe toxic Peruvian seeds that made up the macaw diet. These extracts were less toxic to brine shrimp after mixing them with soil from

FIGURE 11.27 Use of a day-lick by birds. Top shows mealy parrots. Bottom shows macaws in Manu National Park, Peru. (Photograph: D. Müller-Schwarze, 2004.)

Manu. In another experiment, two groups of captive orange-winged macaws, *Amazona amazonica*, were fed quinidine sulfate with or without clay, respectively. The birds treated with clay had less quinidine circulating in their blood (Gilardi *et al.*, 1999). The layers of clay preferred by the macaws contain high levels of sodium; consequently, the birds may enjoy both detoxication and mineral supplements. Macaws even pass on clay to their chicks when feeding them (Brightsmith, 2002).

Chimpanzees daily eat termite mound clay. The habit may have started by ingesting dirt along with termites. Rats, after rotation that causes motion sickness (stomach upset and diarrhea), eat more clay than normal (Mitchell *et al.*, 1977). Geophagy may be instrumental in the success of rats in habitats with unpredictable diets and the resistance of the species to poisoning by humans. Clay can be eaten prophylactically but also in response to toxic food already in the gastrointestinal tract.

Red-handed howler monkeys, *Alouatta belzebul*, studied in eastern Brazilian Amazonia, ate mostly fruits at the peak of the wet season and switched to leaves during the transition from wet to dry season. The percentages of fruits and leaves, respectively, were 53.5% and 40.8% in the wet season and 18.9 and 77.9% in the dry period. When the animals changed their diet, the leaves were mature and relatively rich in secondary compounds. The monkeys then ate soil from termitaria on 26 occasions, all during the drier season. Such soil was richer in calcium, sodium, and organic carbon than soil from the forest floor. The soil probably adsorbed plant secondary metabolites from the leaves in the digestive tract and helped to excrete these rapidly (De Souza *et al.*, 2002).

Humans are also geophageous. Clay or sand is eaten in many countries (Fig. 11.28) and is prevalent in Asia, the Middle East, Africa, Latin America, and the rural south of the United States. Aristotle described this habit, often practiced by pregnant women. The indigenous people of Mexico and Peru eat clay together with frost-resistant potatoes. The Indians in the southwestern United States used to eat clay with wild tubers after crops had failed. Markets in Guatemala sell clay tablets, known as *esquipolas*, particularly to pregnant women (Fig. 11.28). Hunter–gatherers eat clay and acorns together. Eating dirt during famines in China and medieval Europe may have been connected with the eating of toxic wild tubers. Historically, emperors, kings, and popes have eaten clay as precaution against conspiratorial poisoning.

In the Andes, nearly all of the about 160 varieties of wild potatoes, and two of the eight cultivated species are toxic. Some frost-resistant species that grow above 3600 m have high levels of alkaloids, which are bitter and potentially toxic. These are hybrids between *Solanum stenotonum* and wild potatoes such as *Solanum megistacrolobum* and *Solanum acaule*. Other secondary plant compounds in potatoes are saponins, phytohemagglutinin, proteinase inhibitors, sesquiterpene phytoalexins, and phenols.

The Indians of the Peruvian Altiplano eat potatoes with a dip of clay and a mustard-like herb. They say the clay removes the bitter taste and prevents stomach pains or vomiting after eating large amounts of potatoes. The people who eat clay intend detoxication. This may explain how Indians started to utilize and domesticate wild potatoes. Indeed, experiments have shown that four different types of edible clay adsorb the glycoalkaloid tomatine under simulated

FIGURE 11.28 Commercial products for geophagy in humans. Right, Heilerde (literally healing earth) sold in German health food stores. Left, clay tablets (*esquipolas* or *pan de señor*) eaten by pregnant women against "morning sickness," purchased by the author in markets in Antigua and Chichicastenango, Guatemala.

gastrointestinal conditions (Johns, 1986). Clay also effectively binds the potato's bitter and toxic glycoalkaloid solanine. The amounts of alkaloids in potatoes can be staggering: 100 g fresh wild potatoes typically contain 100 mg tomatine, five times the safe level of 20 mg/100 g. A glycolalkaloid content above 14 mg/100 g fresh weight is distasteful to humans. Cultivated potatoes contain less than 10 mg/100 g. To effectively reduce the level from 100 to 20 mg/100 g potato, only 50–60 mg clay are needed, a miniscule amount compared with what Peruvians actually eat per meal. For detoxication, adsorption to clay is most important in the stomach. In the higher pH of the intestine, less glycoalkaloids are adsorbed as they are less soluble there and, therefore, not taken up (Johns, 1990).

Clay eating probably was extremely important in human evolution: it enabled hominids who did not use fire to eat plants with toxic antifeedants. However, the glycoalkaloids of the potato are heat stable and insoluble in water. Domestication of tomatoes and potatoes probably went hand in hand with clay eating. Johns (1986) suggested that "geophagy is the most basic human detoxification technique with behavior antecedents that are prehominid."

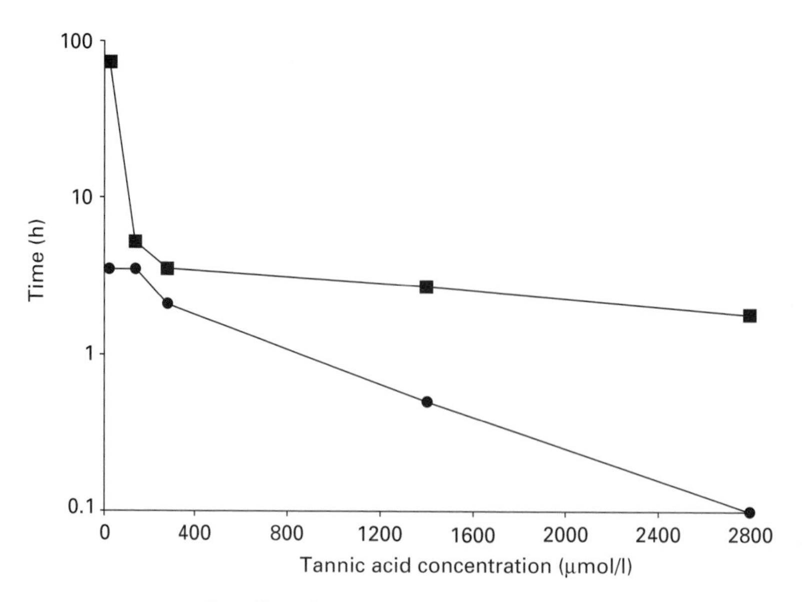

FIGURE 11.29 The effect of tannic acid on time to distress (•) and time to death (■) in fish. (Data taken from Swain, 1979).

Clay may even benefit fish. Swain (1979) found that raising the levels of tannic acid in the water caused increasing discomfort and death to guppies (Fig. 11.29). The addition of clay to the water (52 mg/ml water for a tannic acid concentration of 35 μmol/l) extended the period the guppies survived over six times. Suspended clay particles in "white rivers" may adsorb phenolics and protect fish that cannot easily live in phenolic-rich "black rivers" (Swain, 1979).

Biotransformation

The continuous shedding of surface cells of the digestive tract protects the digestive and absorptive tissues from toxic effects. Once absorbed by the digestive tract, secondary plant compounds of low molecular weight can be excreted via urine, while larger molecules such as monoterpenoids and sesquiterpenoids are excreted via bile and feces. Perspiration, vomitus, milk, hair, exhaled air, and saliva provide other avenues for excretion. For easy excretion, the compounds have to be *water soluble*, contain *polar groups* and be *ionic* at physiological pH. Highly *lipophilic* compounds move across membranes and accumulate in tissues, exerting their toxic effects. Although they have also polar groups, the non-polar effects prevail, and they are not ionic at physiological pH levels. Therefore, the main task of biotransformation is to render the originally lipophilic, hence membrane-threatening, xenobiotics water soluble for excretion.

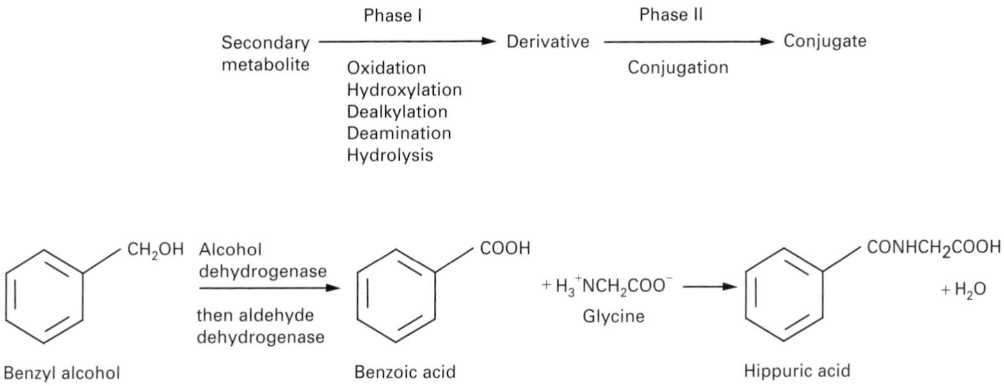

Phase I
Secondary metabolite → Derivative
Oxidation
Hydroxylation
Dealkylation
Deamination
Hydrolysis
Phase II
Derivative → Conjugate
Conjugation

CH_2OH Alcohol dehydrogenase then aldehyde dehydrogenase → COOH $+ H_3^+NCH_2COO^-$ → $CONHCH_2COOH$ $+ H_2O$
Benzyl alcohol
Benzoic acid
Glycine
Hippuric acid

FIGURE 11.30 The two phases of biotransformation of xenobiotics. In this example, an amino acid is used for conjugation. A sugar (glucuronyl) or sulfate can also be used.

Metabolism of ingested plant compounds occurs by two kinds of reactions. *Phase I* reactions are catabolic (enzymatic oxidation, reduction, or hydrolysis) and the products are often more chemically reactive and more polar, making them paradoxically sometimes more toxic. *Phase II* reactions are anabolic (synthetic) and involve conjugation, which usually results in inactive products that can be excreted (Fig. 11.30). Functional groups introduced or uncovered in phase I provide reactive sites for conjugation reactions in phase II. Particularly, phase II produces hydrophilic molecules, making the xenobiotics water soluble. The principal enzymes for metabolizing plant secondary metabolites are located in the liver and gut walls, but also in kidneys, intestines and lungs.

Phase I reactions

Phase I renders xenobiotics more polar. *Oxidation* is the most important process of phase I. It is carried out in the endoplasmic reticulum in many tissues by monooxygenases that contain cytochrome P450 (P450-dependent mixed function oxidases) as electron carrier. These enzymes have evolved in the past billion years in response to plant secondary metabolites. There are a number of P450 gene families.

Alcohols are oxidized to aldehydes by the liver enzyme alcohol dehydrogenase, and aldehydes to carboxylic acids by aldehyde dehydrogenase. In mammals, monooxygenases can be induced by plant secondary metabolites such as α-pinene, caffeine, or isobornyl acetate. *Reduction* is less common and plays a role with ketones that cannot be further oxidized. *Hydrolysis*, the degradation of a compound with addition of water, is also less common than oxidation.

Phase II reactions

Conjugation during Phase II is a synthetic reaction such as combining with carbohydrates (Fig. 11.30) or amino acids, resulting in metabolites of reduced biological activity. Conjugation also produces a highly hydrophilic (water-soluble), less toxic, and more readily excretable ionized product (Freeland and Janzen, 1974). The most common conjugate is glucuronic acid. An important enzyme for conjugation is the UDP-dependent glucuronosyltransferase, responsible for glucuronide formation. Other conjugates are glycosides, sulfates, methyl groups (methylation), amides, glutathione, acetyl groups, and thiocyanates. Species can differ in this regard. For instance, common brushtail possums, *T. vulpecula*, conjugate phenol with glucuronic acid, but carnivorous marsupials use sulfate (Baudinette *et al.*, 1980).

Some mammals depend more on oxygenation (phase I), and others more on conjugation. The common ring-tailed possum, *Pseudocheirus peregrinus*, polyoxygenates terpenes from its eucalyptus diet into highly polar, acidic metabolites that are easily excreted. These metabolites include dicarboxylic acids, hydroxyacids, and lactones. Very little conjugation takes place, possibly to conserve carbohydrate and amino acids (McLean *et al.*, 1993). Generalists appear to conjugate, while specialists oxygenate. Among the generalists, the brushtail possum, *T. vulpecula*, and the laboratory rat excreted more conjugates and less oxygenated metabolites of *p*-cymene, a constituent of eucalyptus leaves. Eucalyptus specialists such as the greater glider, *Petauroides volans*, and the ringtail possum *P. peregrinus*, by comparison, only excreted metabolites with three or four oxygen atoms (Boyle, 1999). These polyoxygenated compounds are more polar and soluble and, therefore, excreted more easily in urine. In this way, the specialists preserve valuable carbohydrates and/or amino acids, which is less critical in the generalist species with their varied diet.

The nutritional value of the diet influences the ability of herbivores to detoxify plant secondary metabolites (Schwass and Finley, 1985). For instance, voles in winter have to consume the bark of birch (*Betula* sp.). This causes stress and leads to poor growth and high mortality among young animals. Birch bark contains phenolics and terpenoids. Both are metabolized by cytochrome P450 monooxidases in phase I and conjugated with glucuronic acid in phase II.

Induced enzymes

In root voles (*Microtus oeconomus*) hepatic phase I detoxication by monooxygenases, specifically ethoxyresorufin-O-dealkylase, can be induced by large doses of birch bark powder in the diet. Conjugation (phase II) is not affected (Harju, 1996).

Degradation speed

The degradation speed differs greatly between species. For instance, the half-life of lysergic acid diethylamide (LSD) is 7 minutes in the mouse, 100 minutes in *Macaca*, 130 minutes in the cat, and 175 minutes in humans. To use another example, ethanol is removed from the blood at a rate of 7 g/h on average, depending on body weight and gender. A quart of beer or 4 oz of whiskey is biotransformed in 5–6 hours. This means that more than 1 gallon of beer or 16 oz of whiskey per day exceeds the enzymatic capacity of the liver. Alcohol would stay in the blood all the time, with destructive consequences.

Acidemia

Many plant secondary metabolites require conjugation, often with glucuronic acid. The conjugation products are strongly acidic and so biotransformation increases the acid load. Accordingly, acidemia is a major consequence of eating diets rich in plant secondary metabolites (Foley *et al.*, 1995). This acid load must be buffered and excreted. An adequate acid balance can be maintained by providing bicarbonate to neutralize excess protons since, at physiological pH, acid metabolites will be almost completely ionized. The hydrogen ion and corresponding anion may be disposed of in different ways. The hydrogen ion can be neutralized in the body by bicarbonate, buffered and retained, or it can be excreted in urine as titratable acid. The anion typically is excreted in urine as ammonium or sodium salts. The bicarbonate needed for buffering the acid load is generated from α-ketoglutarate, which in turn, comes from amino acids (Foley *et al.*, 1995). This requires catabolism of amino acids and results in loss of body protein and depletion of glucose (Illius and Jessop, 1995). To cope with a continuing acid load, the mammal eventually has to break down skeletal muscle to provide glutamine for ammoniogenesis as well. The ammonium ions are needed for excretion of anions as ammonium salts. Any excess ammonium ions are excreted in the urine or combined with bicarbonate to produce urea in the liver (Foley *et al.*, 1995).

Sodium loss

Rabbits form bicarbonate in the gut and absorb it. They do not have to form new bicarbonate in the kidneys and need not excrete ammonium ions in the urine, but they still need to excrete organic anions. These organic anions are accompanied in the urine by sodium or potassium ions, which can generate a severe negative sodium balance for the period that the rabbits are on a browse diet (Iason and Palo, 1991). Therefore, lagomorphs excrete biotransformational organic anions as sodium salts, while most other mammals excrete them as ammonium salts.

Differences in digestive ability can reflect different "ecological lifestyles" of related species. The mountain hare feeds on birch in winter and grass in summer. When fed birch that is high in phenolics, the hare suffers no sodium loss but detoxifies instead. The European hare does not eat birch in winter but eats grass year round. When forced to eat a high-phenolic diet, it loses much sodium through its urine (Iason and Palo, 1991).

Biotransformation by gut flora

The gastrointestinal flora may modify foreign compounds and either render them toxic to their host or assist in detoxication. An example of the former is *cycasin*, a compound in food made from cycad nuts that by itself is not toxic to rats. Hydrolysis by the rat's microflora generates metabolites, one of which is toxic. Germ-free rats suffer no toxic effects from ingested cycasin.

In ruminants, the microbes of the rumen anaerobically detoxify phytochemicals by hydrolysis, reduction, degradation (dealkylation, deamination, decarboxylation, dehalogenation, and ring fission) (Allison, 1978). It has even been suggested that detoxification may be a more important function of the rumen than unlocking nutrients (Morris and García-Rivera, 1955). Ruminants are particularly vulnerable to toxic compounds in their diet because the digestive action by microbes occurs anteriorly in the digestive tract, before many of the compounds that may harm the microbes can be removed. In hindgut fermenters, the essential oils (mono- and sesquiterpenes) may be absorbed and detoxified before they reach the site of microbial activity in the digestive tract, protecting the herbivores microorganisms from ill effects.

11.5 Plant responses to herbivory

11.5.1 Induced defense

Plants fight back against herbivores in a number of ways. Snowshoe hares, *L. americanus*, browse northern deciduous trees or shrubs often severely during peaks of their 10-year population cycle. After browsing, paper birch, *B. papyrifera*, quaking aspen, *P. tremuloides*, balsam poplar, *P. balsamifera*, and green alder, *A. crispa*, grow adventitious shoots. These new shoots contain high levels of terpene and phenolic resins and are very unpalatable to the hares. They are not eaten, even though they are rich in nitrogen and phosphorus. This new generation of unpalatable adventitious shoots may play an important role in the 10-year cycle of the hare populations (Bryant, 1981).

There is some evidence for induction in pasture plants. A furanocoumarin, isopimpinellin, increases in floral stands of wild parsnip, *Pastinaca sativa*, after clipping that simulates herbivory (Nitao, 1988).

In some cases, browsing does not appear to lead to induced defenses. Two species of northern European birch (*B. pendula* and *B. pubescens*) do not seem to respond to moose browsing with increased chemical defenses (Danell *et al.*, 1985). After experimental defoliation, the Emory oak of southwestern North America, *Quercus emoryi*, produces in its regrown leaves 2.5 times as much hydrolyzable tannins than in mature leaves, but lower amounts of condensed tannins. Refoliated branches suffered more, not less herbivory. New leaves have more protein, more water, and are tender. These properties may override any effects of induced chemical defenses, at least for the current growing season (Faeth, 1992), but the induced compounds may still protect a plant later on, especially in winter.

The evidence for induced defenses has grown more complex. For example, *Acacia drepanolobium* in Kenya responded to herbivory by large mammals in unexpected ways. Trees in an area with antelope only were compared with trees exposed to antelope plus elephants and giraffe, and an exclosure without any of these herbivores. The tannin concentration in the leaves increased with increasing pressure by herbivores. However, the trees were richer in tannin in the upper levels (2 m high) than at about 1 m, where more browsing takes place. Also, unexpectedly, the more tannin a tree had, the longer its spines. There was no evidence of a trade-off between chemical and thorn defense. Nor was there a trade-off between chemical defense and amount of growth. Furthermore, different species of ants, some more aggressive than others, may have an effect on the herbivores, although no clear effect of the ants was detected (Ward and Young, 2002).

11.5.2 Chemical defense versus growth rate of plants

A plant, especially a tree or shrub, can escape extensive browsing by either defending itself, chemically or otherwise, or by growing so fast that it more than replaces any biomass losses to herbivores. It may also grow rapidly beyond the reach of its herbivores. Slow-growing boreal and arctic trees are more likely to defend themselves chemically, while this would be less essential for rapidly growing tropical trees. Slowly growing plants on poor soil are thought to invest more into plant defense compounds (Coley *et al.*, 1985). One hypothesis is that fast growing trees are more palatable (Bryant and Kuropat, 1980). In the same boreal forest, two related species may differ along this scale: in Sweden, moose, *A. alces*, prefer the faster-growing birch *B. pendula* over *B. pubescens*. After moderate browsing, *B. pendula* regrows more and larger leaves and larger shoots with more nitrogen than when slightly browsed. This is less true for *B. pubescens*. Experimental clipping also resulted in more browsing during the winter following regrowth. In all, moose find three times more biomass available after browsing (Danell *et al.*, 1985). In *B. pubescens*, phenol levels increase after browsing, while they decrease in *B. pendula* (Palo, 1984).

11.6 The question of coevolution of plants and herbivorous mammals

Coevolution is defined as *reciprocal* stepwise adaptations between at least two species (Ehrlich and Raven, 1964). "Coevolution without the criterion of reciprocity is indistinguishable from evolution and hence a useless concept" (Lindroth, 1988). Consider the following scenario. A plant develops effective antiherbivore defenses. In response, a herbivore counteradapts to circumvent these defenses and is at a competitive advantage over other herbivores. The plant, in turn, responds to this breach of its defenses. In insects, such pairwise reciprocal evolution can take the form of a "chemical arms race" (Dawkins and Krebs, 1979). Coevolution differs from evolution by being narrower, with fewer participants, perhaps even only two species or two populations. In reality, in most ecosystems, many species prey on many other species. Therefore, we can at best speak of *diffuse coevolution*, with a number of participants that exert diluted selection pressures.

Is there coevolution between mammals and their food plants? Herbivorous mammals usually consume many different plant species, and a plant typically is food to more than one herbivore species. Any patterns of coevolution would be diluted, resulting in diffuse rather than narrow coevolution. Even the diets of the koala (eucalyptus) and giant panda (bamboo) are more varied than commonly assumed. The koala feeds on several *Eucalyptus* species and did not thrive in zoos when fed only one species. The giant panda even includes animals in its diet. Nevertheless, Lindroth (1988) saw coevolution as an "attractive hypothesis" for some mammals.

Lindroth (1988) listed five specific criteria for coevolution in mammals:

- long history of close association
- strong mutual selection pressures to produce morphological, physiological, biochemical, and/or behavioral adaptations toward each other
- if the ranges of plant and mammal are not congruent, adaptations should be most pronounced in areas of sympatry

- the *structure of the ecological community* may provide evidence for coevolution
- for mammals, the standard of reciprocal genetic change has to be upheld, but "diffuse coevolution" (groups of populations evolve in response to each other) can be allowed.

Based on these criteria, coevolution between plants and mammals has not been demonstrated satisfactorily. Coevolution is more likely in simpler plant–animal communities. Lindroth (1988) rated coevolution between herbivorous mammals and their food plants an "unsatisfactory 'maybe'."

11.6.1 Possible examples of diffuse coevolution

Three systems possibly result from at least diffuse coevolution: rat kangaroos and legumes in Australia, snowshoe hares and Alaska paper birch in Alaska, and brown and collared lemmings in Arctic North America.

Kangaroos and legumes

The relationship between marsupials and legumes in Australia suggests coevolution: while most mammals die from fluoroacetate concentrations as low as 100 mg/kg body weight, some kangaroos tolerate much higher amounts. Different species of herbivorous marsupials tolerate varying levels of fluoroacetate depending on whether or not they are, or have been, sympatric with the two most toxic plant genera. About 34 species of *Gastrolobium* and *Oxylobium* in Western Australia produce fluoroacetate at levels as high as 0.25% of fresh weight. These two genera occur within the range of the rat kangaroo *Bettongia penicillata*, which tolerates levels of fluoroacetate that would be lethal to other herbivores. By contrast, *Bettongia lesueur* lives in areas that have been devoid of such plants for at least 7000 years. It is moderately resistant to fluoroacetate. Finally, *Bettongia gaimardi*, the most primitive living rat kangaroo and limited to southeast Australia in the past, is not resistant. It is suggested that the genus *Bettongia* radiated westward from its origins in eastern Australia and developed fluoroacetate resistance in its western species (Mead *et al.*, 1985). All three species of gray kangaroos (*Macropus* spp.), by comparison, tolerate fluoroacetate well. This is consistent with the assumption that they first colonized western areas and spread eastward. The eastern gray kangaroo is still able to detoxify fluoroacetate even though it is not now exposed to plants with this compound. The western gray kangaroo regularly feeds on *Gastrolobium* spp. but minimizes FA intake by picking species with the lowest concentrations (Mead *et al.*, 1985). While the tolerance of kangaroos, compared with other mammals, is an impressive adaptation, it is not clear whether the fluoroacetate in the plants is produced in response to herbivorous mammals.

Snowshoe hare and boreal trees

The relationship between the snowshoe hare, *L. americanus*, and birches and willows in the boreal forests of Alaska and Siberia may be an example of moderately diffuse coevolution. In areas of high feeding pressure by hares the levels of chemical defenses in plants are higher than in areas with less pressure. In the high-pressure areas, the hares tolerate higher levels of plant defense compounds. Twigs of birches are covered with phenolic resins such as papyriferic acid, most likely aimed at winter browsers. Birch and willows from Alaska and eastern Asia tend to be particularly resistant to browsing by mammals. These areas were not glaciated during the Pleistocene, allowing ample time for coexisting trees and herbivores to evolve defenses and feeding strategies, respectively (Bryant *et al.*, 1989). On an applied level, trees from these "resistance centers" are a good choice for transplanting to areas with high herbivore pressure.

The 10-year population cycle of hares may be driven, at least in part, by chemical plant defenses (Bryant *et al.*, 1991). During a winter with peak numbers of hares, browsing is severe. Food is reduced so much that the hares must feed on fewer species, which are also of lower quality and heavily defended, such as spruce. The hares start to starve, the population crashes. The growing season after the population crash produces more heavily defended (induced) regrowth of deciduous trees and shrubs. The high levels of defense compounds will persist for two or three winters after the hare population crash. Models show that this can drive the 10-year cycle (May, 1972). The crash of hare populations in response to a diet high in plant secondary metabolites suggests that the animals do not possess effective counteradaptations, weakening the case for coevolution (Lindroth, 1988).

Lemmings

The case of lemmings, according to Lindroth (1988), is an "unsatisfactory 'maybe'" in terms of coevolution. Brown lemming, *Lemmus sibiricus*, and collared lemming, *Dicrostonyx torquatus*, dominate the herbivore community in certain areas of Alaska. Brown lemmings live in wet areas and feed on monocotyledons such as sedges and grasses. Collared lemmings have inhabited the tundra for longer than brown lemmings, live in drier areas, and feed on dicotyledons such as forbs and willow. Plant chemistry appears to explain the diet differences. In the laboratory, brown lemmings thrived poorly on willow and willow-extract-treated lab chow, while collared lemmings ate sedges but grew poorly on rations treated with sedge extract (Jung and Batzli, 1981). The two species of lemming possibly evolved detoxication systems that permit them to specialize on certain food plants but prevent them from dealing with compounds from other species (Lindroth, 1988). In this system, it is not clear what chemicals are involved, nor whether the chemical defenses evolved in response to rodent herbivores (Lindroth, 1988).

11.6.2 Plant defense and wildfires

Wildfires create forest gaps and start new successions of forest plants. Mammals such as snowshoe hares, deer, or beavers take advantage of early successional species. The more hectares burned in an area, the more abundant snowshoe hares become. (In northern mythology, fire "creates hares"). The hares browse birch in winter and the plants respond by induced chemical and other defenses. Thus "selection for anti-browsing defense is greatest where fire has burned the greatest area over evolutionary time" (Clausen *et al.*, 2004). A survey of Alaskan and Canadian sites of lightning-caused wildfires between 1956 and 2002 showed that Alaska paper birch, *Betula neoalaskana*, is more abundant there than is canoe (paper) birch, *B. papyrifera*. The former is better defended chemically than canoe birch and also had more resin glands and higher levels of papyriferic acid in fire areas than the same species from other areas, even when grown together in a garden. So, "selection by an herbivore (snowshoe hare) has resulted in a continental-scale biogeographic pattern in antiherbivore defense" (Clausen *et al.*, 2004).

12

Kairomones and synomones

Kairomones (from the Greek *kairos,* opportune moment, by stealth) are chemical cues from one species that another uses ("spying"). Primarily the receiver of the signal benefits, as in finding prey by odor, or detecting and avoiding predators by chemical cues. Since the cues are available to another species, they are considered "public" signals, in contrast to "private" signals of restricted pheromone systems, intraspecific by definition. Synomones are chemicals that regulate interspecific relationships where both partners benefit.

Section 12.1–12.7 discuss kairomones and Section 12.8 synomones.

12.1 Predator–prey interactions

12.1.1 Prey odors used by predators

Fish

Not surprisingly, much research in sharks, skates and rays has focused on the responses of sharks to human body odors. Human blood attracts sharks, while sweat does not, and urine was even slightly repellent (Tester, 1963). Practitioners use whale meat and mixtures of fish meal and fish oils as shark attractants. In both carnivorous and herbivorous bony fish (Osteichthyes) smell deals with prey odors, social odors, and chemical stimuli in homing, and it is mediated by the first cranial nerve, the olfactory nerve. By contrast, taste serves in detection and selection of food and avoidance of toxic food, and it employs the facial, glossopharyngeal, vagal, and hypoglossal nerves.

Numerous experiments with prey extracts have elucidated the stimuli that guide fish in their feeding behavior. These studies showed:

- rinses of prey organisms attract predatory fish and release food searching behavior
- predators can distinguish rinses of different prey species
- fractions of prey rinses also release feeding responses

- mixtures of amino acids are also active
- single amino acids may trigger feeding responses
- biological activity decreases from complete rinses to amino acid mixtures to single amino acids
- each predator species responds to a different mixture of compounds
- typically, amino acids with three to six carbon atoms and their derivatives serve as chemical cues for predatory fish.

Marine fish

Among the better known commercial fish, the yellowfin tuna, *Thunnus albacares,* responds to the odors of several species of anchovies, such as northern anchovies, *Engraulis mordax,* nehu (inshore anchovy), *Stolephorus purpureus,* and the surf smelt, *Hypomesus pretiosus.* Tuna respond to surf smelt odor only when hungry, and tuna rendered anosmic by nose plugs lose their response. Whole natural rinses are more effective than fractions of these rinses or synthetic mixtures of amino acids. Amino acid constituents are detected at a concentration of 10^{-11} mol/l. The most active amino acid was tryptophan (Williams *et al.*, 1992b). Tuna possibly form "chemical search images" for the most abundant prey at any given moment (Atema *et al.*, 1980). Yellowfin tuna detect prey by amino acid odors complexed with lipid vesicles, which have an onion-skin structure of liposomes. This was the first demonstration of an extracorporeal biological function for liposomes. Tuna detect prey that is beyond visual range. Complexing with lipids may considerably delay the dilution of amino acids from prey to subthreshold levels and hence extend the range of chemical prey detection (Williams *et al.*, 1992b).

Cod, *Gadus morhua,* also respond to amino acids. Among the cod's prey is the lugworm, *Arenicola marina.* The attractive fraction of lugworm rinse elicits strong searching behavior. It is a blend of threonine, serine, glycine, alanine, valine, leucine, and glutamic acid. Individually, only glycine and alanine released significant responses (Pawson, 1977). Extracts of the shrimp *Pandalus borealis* elicit bottom food search in cod. The most abundant amino acids, glycine, alanine, proline, and arginine, are also the most active ones. Singly, glycine was most potent, followed by alanine, as in Pawson's (1977) study. The four amino acids glycine, alanine, proline, and arginine act together synergistically. These four together are more active than the total amino acid pool in the shrimp extract, suggesting that the pool may contain amino acids with antagonistic effects (Ellingsen and Døving, 1986).

Smell receptors in the marine carnivorous Hawaiian goatfish, *Parupeneus porphyreus* (Mullidae), located on the chin barbels, mediate both arousal and food searching in response to prey homogenate and rinse of intact live prey. The barbel nerve (seventh cranial nerve) responds, as in the cod, to those amino acids that are most abundant in its prey: proline, glycine, arginine, glutamic acid, and alanine, but not to those that are present only in trace amounts (Holland, 1978).

The pinfish, *Lagodon rhomboides*, responds three times more strongly to extracts from pink shrimp (*Penaeus duoarum*) than to extract of blue crab (*Callinectus sapidus*). Responses to clam (*Mercenaria campechiensis*), whelk (*Busycon contrarium*), oyster (*Crassostrea virginica*), sea urchin (*Arbacia punctulata*), and striped mullet (*Mugil cephalus*) were 10 times weaker. Mixtures of various amino acids plus betaine were most active, and betaine was so among the single compounds (Carr *et al.*, 1976). When added to a mixture of five amino acids (aspartic acid, glutamine, glycine, isoleucine, and phenylalanine), however, betaine accounted only for 10% of the activity (Carr and Chaney, 1976). Similarly, a mixture of 19 amino acids, plus betaine stimulates feeding behavior in the pigfish, *Orthopristis chrysopterus*. Betaine is the most active single compound, accounting for 39% of the potency of the prey extract (Carr, 1976).

Prey fish may mask their own odors. Some marine fish avoid predation by covering their body odors. Some parrot fish (Scaridae) sleep in a mucus cocoon. It is believed that this covers up its scent and protects it from predation. Table 12.1 summarizes some chemical predator–prey relationships in marine fish.

Freshwater fish

One of the earliest experiments on fish responses to prey odors was that by Parker (1911). He showed that catfish (*Ictalurus* sp.) and killi fish (*Fundulus* sp.) rely on their olfactory nerves for responding to earthworm extract. Bullheads (*Ictalurus* spp.) are attracted to cysteine and respond with feeding (Bardach *et al.*, 1967). The single amino acids L-alanine, L-arginine, and L-proline release the entire feeding sequence of turning, biting, snapping, and masticating in channel catfish, *Ictalurus punctatus* (Valentincic and Caprio, 1992).

Carp, *Cyprinus carpio*, are attracted to cysteine, asparagine, glutamic acid, threonine, and alanine. Extracts from *Tubifex* worms contain at least 17 amino acids. Of these, binary mixtures of one non-polar amino acid and one polar uncharged amino acid attracted carp most and led them to explore the area. Alanine, valine, and glycine proved to be the simplest combination to release significant attraction and exploration (Saglio *et al.*, 1990).

Rainbow trout, *Salmo gairdneri*, are primarily visual feeders, but also use their chemical senses for foraging. They prefer a diet flavored with squid extract to non-treated food. A synthetic mixture of 18 amino acids, two amines, and lactic acid was very active. Only L-forms triggered responses; D-forms were even repellent. Only two combinations of amino acids were active: tyrosine, phenylalanine, and lysine; and tyrosine phenylalanine, and histidine (Adron and Mackie, 1978).

Table 12.1 Prey odors used by marine fish

Predator species	Prey species	Chemical cue[a]	Reference
Yellowfin tuna *Thunnus albacares*	Northern anchovy (*Engraulis mordax*), nehu (*Stolephorus purpureus*), surf smelt (*Hypomesus pretiosus*)	Amino acid mixture: *tryptophan*, on liposomes	Atema *et al.*, 1980 Williams *et al.*, 1992[b]
Cod *Gadus morhua*	Lugworm, *Arenicola marina* Shrimp, *Pandalus borealis*	Mixture: threonine, serine, *glycine*, *alanine*, valine, leucine, glutamic acid, proline, arginine	Pawson, 1977 Ellingsen and Døving, 1986
Pinfish *Lagodon rhomboides*	Pink shrimp (*Penaeus duoarum*)	Mixture: *betaine*, aspartic acid, glutamine, glycine, isoleucine, phenylalanine	Carr *et al.*, 1976
Pigfish *Orthopristis chrysopterus*		*Betaine* and 18 amino acids	Carr, 1976

[a] Italics indicate a particularly active compound.

Arctic charr, *Salvelinus alpinus*, swim into streams that contain mixtures of amino acids. A whole extract of commercial fish food was active at a concentration of 5×10^{-9} mol/l while mixtures of 7, 11, or 18 amino acids were active only at higher concentrations (2×10^{-6} to 5×10^{-6} mol/l) (Olsén *et al.*, 1986).

Fish discriminate between palatable and toxic prey. For instance, largemouth bass, *Micropterus salmonides,* reject the toxic tadpoles of the toad *Bufo americanus* but eat those of the spring peeper, *Hyla crucifer* (Kruse and Stone, 1984).

Herbivorous fish depend on similar chemical stimuli. The redbelly tilapia, *Tilapia zillii,* of Africa responds to amino acids in romaine lettuce (Adams and Johnsen, 1986). Ten amino acids stimulate the electro-olfactogram of the herbivorous grass carp, *Ctenopharyngodon idella.* The electrical response increases exponentially with logarithmic increases of the stimulus concentration from the threshold to 1×10^{-3} mol/l. The detection thresholds were calculated to be $1 \times 10^{-7.15}$ to $1 \times 10^{-8.7}$ mol/l, similar to known thresholds of carnivorous fish. The relative stimulatory efficacy of the ten amino acids at 1×10^{-4} mol/l was used to distinguish five groups of stimuli: cysteine was the most stimulatory, proline the least. Except for arginine, the relative stimulation efficacy of the amino acids

Table 12.2 Prey odors used by freshwater fish

Predator species	Prey species	Chemical cue[a]	Reference
Bullhead *Ictalurus* sp.	Earthworm	Cysteine	Bardach *et al.*, 1967
Channel catfish *Ictalurus punctatus*	Invertebrates, fish aquatic plants	L-Alanine, L-arginine, or L-proline	Valentinicic and Caprio, 1992
Rainbow trout *Salmo gairdneri*	Insects, plankton, crustaceans, fish eggs, smaller fish	Tyrosine, phenylalanine and lysine; or tyrosine, phenylalanine, and histidine	Adron and Mackie, 1978
Atlantic salmon, Salmo salar, fry	Injured crustaceans	Free amino acids	Holm and Walther, 1988
Carp *Cyprinus carpio*	Plants, small animals	Cysteine, asparagine, glutamic acid, threonine, alanine	Saglio *et al.*, 1990
Grass carp *Ctenopharyngodon idella*	Herbivorous	*Cysteine*, arginine	Johnsen *et al.*, 1988

[a] Italics indicate a particularly active compound.

did not differ between carnivorous and herbivorous fish. Arginine, however, is very stimulatory for grass carp but had little effect on carnivorous fish. It was concluded that "feeding niche segregation probably is not facilitated by differential olfactory sensitivities to feeding stimuli" (Johnsen *et al.*, 1988).

Fry of the Atlantic salmon, *Salmo salar*, probably rely on olfactory and gustatory stimuli for their first meal. Injured prey such as small crustaceans will leak free amino acids, which can serve as a feeding signal to the fish fry. Such "handicapped" prey will be easier to catch for the fry. If the prey is dead, and/or its free amino acids are depleted, the fry show no interest in them. In this way, the salmon can optimize its capturing efforts as well as its prey digestion. In laboratory experiments, frozen daphnids leaked as much as 35% of its methionine upon thawing. On their first 3 days of feeding, salmon fry typically chose undepleted daphnids first and virtually all spit-out prey were depleted daphnids (Holm and Walther, 1988). Table 12.2 lists some chemical predator–prey relationships in freshwater fish.

Amphibia

Amphibians usually prey visually and detection is better detected if the prey is moving. However, aquatic salamanders and toads (*Bufo calamita*) use olfaction to locate and catch prey. Vis-à-vis motionless prey, tiger salamanders, *Ambystoma tigrinum*, rely more on olfaction for foraging in darkness or when experimentally deprived of sight (Lindquist and Bachman, 1980). Other species that find their prey by chemical cues alone include the toads *Bufo boreas* and *Bufo* marinus the frog *Rana pipiens*, the salamander *Plethodon cinereus*, and two species of *Triturus*. The newt *Notophthalmus viridescens* locates pill clams, *Musculium rosaceum*, by smell alone. Fossorial anurans (*Rhinophrynus dorsalis* and *Myobatrachus gouldii*) almost certainly smell out their prey (reviewed by Duellman and Trueb, 1986).

Reptiles

Snakes

Two phenomena of reptilian prey searching are well investigated: responses of various snakes to the odors of invertebrates, and rattlesnakes' trailing of envenomated small mammals.

Responses of snakes to food odors illuminate the interaction of genetically anchored preferences with the modifying role of individual feeding experience (see also p. 229). Newborn, food-naive snakes tongue flick and prey attack cotton swabs soaked with water extracts of skin of small invertebrate and vertebrate prey. Species differ according to their natural feeding habits. For example, ecologically different populations of water snakes, *Natrix s. sipedon*, respond to chemicals from different prey species but this response can be modified by experience. Previous exposure to prey odors influences chemosensory responses to the odor. Snakes raised in the laboratory on a diet of goldfish later preferred goldfish extracts. Likewise, snakes caught at a fish hatchery and fed goldfish preferred goldfish odor to all others (Gove and Burghardt, 1975). Wild-caught snakes from a mountain stream preferred extracts from stream fishes, including darter (*Etheostoma ruflineatum*), sculpin (*Cottus carolinae*), minnow (*Notropis coccogenis*), and stone roller (*Campostoma anomalum*). Laboratory-born *offspring* of such mountain-stream snakes preferred extracts of amphibians and fish from the mountain streams, even though they had not experienced them before. Neonates of garter snakes and a number of other species show such preferences (Burghardt, 1966; Burghardt, 1975). Newly hatched food-naive fox snakes, *Elaphe v. vulpine*, preferred extract of baby mouse skin to extracts of many other potential prey species. The corn snake, *E. guttata*, also responded to baby mouse extract, even though it was tested only in comparison with water (Burghardt and Abeshaheen, 1971).

Timber rattlesnakes, *Crotalus horridus*, are ambush hunters. They assume the ambush posture after smelling prey odors. In the laboratory, these snakes recoil the front part of their body into the ambush posture after flicking their tongues in response to water extracts of the surface of white-footed mice and chipmunks, their natural prey. They tongue flicked and showed the ambush posture in response to control odors, such as water rinses of dog, frog, or skink, no more often than to unscented tap water (Clark, 2004).

Hatchling pine snakes, *Pituophis melanoleucus*, of the Pine Barrens in New Jersey prefer mouse odor in a Y-maze more if they have been incubated at a higher temperature (33 °C versus 28 °C). Furthermore, experience plays a role in this species, too. Snakes that had eaten a mouse detect and follow a mouse odor trail, while naive snakes show no such response (Burger, 1991).

Garter snakes

Garter snakes respond to non-volatiles from prey, as tongue contact with prey samples is necessary. The Eastern plains garter snake, *Thamnophis r. radix*, prefers the odor of worms, leeches, fish, tadpoles, and frogs, while the western smooth green snake, *Opheodrys vernalis blanchardi*, prefers only cricket odor (Burghardt, 1967). This reflects the different habitats of the two species: the first forages in wet areas, and the latter in drier upland habitat. Newborn garter snakes, *Thamnophis sirtalis*, but from mothers of different regions such as Wisconsin, Iowa, and Illinois, differed in their food odor preferences (Burghardt, 1970). However, the diet of the mother does not necessarily influence the chemical-cue preferences of newborn snakes: these newborn always preferred worm extract, whether the mother's diet had been worm or fish (Burghardt, 1971).

Sectioning of the olfactory nerve does not eliminate accurate trailing: garter snakes use their vomeronasal organ for earthworm trailing. When the vomeronasal organ is transected or sutured closed, garter snakes are unable to discriminate prey odor from water (Kubie and Halpern, 1978). For trailing of earthworms, the active components are non-volatile, water-soluble, stable macromolecules (Sheffield *et al.*, 1968). The feeding cue is contained in lyophilized collagen of the cuticle of earthworm, *Lumbricus, terrestris*, but not in sandworm (*Nereis*) and is possibly a glycoprotein. Periodate treatment destroys the chemoattractant function, possibly because the carbohydrate residue of the glycoprotein is destroyed (Halpern *et al.*, 1986). The chemical cue is a protein with molecular mass 20 kDa (Wang *et al.*, 1987). It contains glycine, serine, and threonine and has a high hydroxyproline: proline ratio (Kirschbaun *et al.*, 1986). Garter snakes respond to high- and low-molecular-weight fractions of chloroform–methanol extracts from earthworm and fathead minnow, *Pimephales, promelas*, with a similar pattern of tongue flicking and attacking (Burghardt *et al.*, 1988).

Different populations of the same species may differ genetically in their response to diet and diet odors, reflecting an area's predominant diet. The western garter snake, *Thamnophis elegans,* feeds on slugs in the coastal part of its range, and on fish and frogs in the drier inland areas at higher elevations. Food-naive young coastal snakes readily eat slugs, while food-naive inland snakes refuse them. Inland populations face the problem that a snake might attack dangerous leeches along with desirable slugs, while coastal snakes do not experience leeches at all, hence no such confusion. Moreover, isolated newborn snakes from both coastal and inland populations respond strongly to the *odor* of toad tadpoles, but only coastal snakes responded to slug odor (Arnold 1981a).

Rattlesnakes: prey search after striking and envenomation

Preying rattlesnakes first strike and *envenomate* their prey and then release it and let it wander freely for up to several meters. The snake then searches for the envenomated prey, using its chemical senses. This behavior is termed "strike-induced chemosensory searching" (Chiszar and Scudder, 1980). The sequence of preying is striking, high rate of tongue flicking, locating of prey trails, trail following, finding carcass, and finally ingesting (Chiszar *et al.*, 1988a). In the prairie rattlesnake, *Crotalus viridis viridis,* striking intensifies chemosensory trailing (Golan *et al.*, 1982). The banded rock rattlesnake, *Crotalus lepidus,* discriminates between envenomated and non-envenomated mice that are presented to them in mesh bags (Chiszar *et al.*, 1983). Northern Pacific rattlesnakes, *C. viridis oreganus,* that followed a trail of an envenomated mouse were unable to determine the direction the mouse had moved (Smith and Kardong, 2000). The snake distinguished the odor of the nasal–oral area from that of the anogenital area of the mouse, but the mouse has to be envenomated. This discrimination aids in orienting the prey for swallowing it head first (Duvall *et al.*, 1980).

Comparative studies show that rattlesnakes differ in preying and strike-induced chemosensory searching according to their natural feeding habits. *C. viridis* flicks its tongue for a long time after a strike, even if no prey trail odor is present. The twin spotted rattlesnake, *Crotalus pricei*, specializes as a lizard predator and needs the presence of an odor trail for sustained tongue flicking. The requirement of chemical feedback for a high rate of tongue flicking is considered a primitive trait in rattlesnakes (Cruz *et al.*, 1987).

The rattlesnake *C. viridis* searches for the particular odor it had experienced when striking the prey. In one experiment, snakes were induced to strike perfume-treated mice. Then they were exposed to perfumed, but non-envenomated, carcasses. The snakes preferred a carcass with the same odor as the originally struck mouse. In a second experiment, snakes preferred the carcasses of mice on the same diet as the ones they had struck. Thus rattlesnakes form a specific chemical search image (Melcer and Chiszar, 1989). They learn a mouse's chemosensory signature, which is independent of envenomation, and follow its scent trail (Chiszar *et al.*, 1990).

The northern Pacific rattlesnake, *Crotalus viridis oregonus*, seems to need a more complex odor to pursue a mouse after striking and envenomating it. They follow neither venom trail, nor mouse odor, if the two odor trails diverge. This was examined using two stimuli: a trail of an unstruck mouse dragged along the surface (but the snake was permitted to strike this mouse later) and an artificially envenomated, different unstruck mouse. The snake uses chemical cues from two possible sources: cues picked up when the snake is striking and cues in, or produced by, the venom. In terms of relative effectiveness, the cues ranked venom > mouse odor > fang puncture (Lavin-Murcio *et al.*, 1993).

In the wild, rattlesnakes migrate from their hibernacula to the feeding grounds, locate rodent colonies, approach their prey, envenomate by striking, release the wounded rodent, tongue flick at a high rate, locate the trail of the prey, follow the trail, locate the carcass, and finally ingest the rodent. Whether or not (and how soon) the carcass is found, depends half and half on striking and tongue-flicking rate. Whatever influences the tongue-flick rate also affects the finding of the carcass (Chiszar *et al.*, 1988a). Migrating rattlesnakes found deer mice, *Peromyscus maniculatus*, placed in their path by odor, and the odor trail emanated from integument rather than urine (Duvall and Chiszar, 1990). Rattlesnakes, *C. v. viridis*, may assess the population density of their main prey, the deermouse, by taking up mouse odor by tongue flicking and mouth gaping (Duvall, 1986). This permits them to find and hunt in more lucrative areas of high-prey density.

Other snakes also use non-volatiles on prey trails as cues for pursuing their quarry. Prey odors from a large variety of taxa elicit tongue flicking in the predatory king snake, *Lampropeltis getulus* (Brock and Meyers, 1979). King snakes can even distinguish the dorsal skin odor of crotaline snakes from that of colubrid snakes when presented on identically looking cotton swabs (Weldon and Schell, 1984).

The brown tree snake

In the 1940s, the brown tree snake, *Boiga irregularis*, of northern Australia and New Guinea was accidentally introduced to Guam and other islands. It poses a serious danger to the survival of other vertebrate species, especially since it now occurs in densities as high as 16 to 50 snakes per hectare. The snake grows to up to 2.4 m in length and has already done much damage to native birds and bats. Of the 18 more common native species of birds, nine are extinct now, six almost gone, and three very low in numbers. The snakes have reduced the populations of the Mariana fruit bat by preying on their young, and they now feed on chickens and caged birds, their eggs, and other pets.

Efforts to control brown tree snake populations by trapping require knowledge of the cues this species uses for preying. In laboratory experiments, the brown tree snake was unable to locate prey by airborne odors but did follow a trail made by rubbing a rat pup along a tree branch (Chiszar *et al.*, 1988b). Brown tree snakes responded to human blood with an increased rate of tongue flicking (Chiszar *et al.*, 1993a) and 1 in 10 experimental snakes even swallowed a blood-soaked tampon used as a stimulus (Chiszar *et al.*, 1993b). Response to human blood may have serious implications in populated areas on Guam where the brown tree snake is superabundant. Bites by brown tree snakes account for 1 in 1200 emergency room visits here; 80% of these people are bitten while sleeping, and 52% were under 5 years of age. The snakes bite mostly on fingers and hands, chew and try to ingest these body parts while coiling around necks and bodies of their victims. Therefore, these attacks represent attempts at preying, and not defense. Rubbings of human skin are as attractive to brown tree snakes as rubbings from mice. Presumably skin lipids provide a feeding stimulus (Greene *et al.*, 2002).

Dead mice are as attractive as live mice, and dead mouse odor is as active as a dead mouse (Shivik and Clark, 1997). Elucidation of the feeding stimuli used by this species will aid in effective trapping for snake control. Fractions of dead mouse odor are being bioassayed. The best known constituents of rotting carcass (e.g. fish) odor, putrescine (1,4-diaminobutane: $NH_2CH_2CH_2NH_2$) and cadaverine ($NH_2CH_2CH_2CH_2NH_2$) were not active.

Multisensory control of preying in snakes

Visual and chemical cues interact in foraging by natricine snakes. Even visual cues alone can elicit prey attack, especially in aquatic foraging (Drummond, 1985). Aposematic color patterns of prey enhance the learning of prey that induces illness. Garter snakes, *Thamnophis radix haydeni*, were exposed to fish and earthworms presented on black-and-yellow forceps, and then injected with lithium chloride (LiCl). Control prey was offered on green forceps. Later, the snakes avoided food from either forceps, but the aversion to prey paired with black-and yellow was stronger (Terrick *et al.*, 1995).

Learning in snakes

Snakes can learn to discriminate profitable from less-manageable prey. Naive garter snakes, *Thamnophis melanogaster*, attacked both the carrion-eating leech *Erpobdella punctata* and the blood-sucking leech *Haementeria officinalis* even though naive snakes respond less to the odor of *H. officinalis*. The latter thwarted attack or killed the snake. Over 4 weeks, the snakes progressively learned to avoid live *H. officinalis* (Drummond and Garcia, 1995).

Lizards

Lizards also increase tongue flicking after striking prey. The savanna monitor *Varanus exanthematicus* increases its tongue-flicking rate from 25 every 2 minutes to 90 every 2 minutes upon presentation of a mouse if the mouse is removed from its mouth (Cooper, 1989a,b). Among lizards, actively foraging species rely on prey odors, while ambush foragers do not (Cooper, 1994). The vomeronasal organ is important for tongue flicking during foraging in the ocellated skink, *Chalcides ocellatus*. It is assumed that vision is important for attacking fast-moving prey, while chemical identification is required during the subsequent predatory sequence, including consumption of prey (Graves and Halpern, 1990). Indeed, adult males of the broad-headed skink, *Eumeces laticeps*, increased their tongue flicking after they had bitten a neonatal mouse. Cotton swabs with mouse odor also stimulated tongue flicking (Cooper, 1992). Some scincid lizards such as *Eumeces inexpectatus* can discriminate prey odors from control odors from birth (Loop and Scoville, 1972; Burghardt, 1973). One omnivorous lizard, *Gerrhosaurus validus* (Gerrhosauridae, Cordyliformes), increased its tongue-flicking rate when exposed to cricket odor and romaine lettuce odor (their regular food), but not to odor of yellow squash (which they had consistently rejected earlier) (Cooper 2000).

Olfaction is not necessary in other lizard species. For instance, the groundskink, *Scincella lateralis*, which feeds on insects in the ground litter, does not require chemical cues to attack prey (Nicoletto, 1985). Presumably, vision alone suffices for this purpose.

Amphisbaenids

The mostly ant- or termite-eating amphisbaenids use odors for foraging. The amphisbaenan *Blanus cinereus* increased its tongue flicking to the odor of ants presented on a cotton swab but responded little to cologne or water, and not at all to the odor of beetles (López and Salvadore, 1992).

Turtles

Turtles also use chemical cues for food, as shown for the marine loggerhead turtle, *Caretta caretta* (Grassman *et al.*, 1984). Among freshwater turtles, the European pond turtle, *Emys orbicularis*, locates food by chemoreception. This species moves chemical stimuli from the pond water to the chemoreceptors of the nasal and oral cavities by slow "jaw testing movements". The frequency of these movements increases in response to the amino acids alanine, arginine, and glutamine in a dose-dependent fashion (Manteifel and Goncharova, 1992; Manteifel *et al.*, 1992).

Crocodiles

Even crocodilians select prey by odor. If beef, nutria, or rattlesnake meat in perforated paper bags were presented on a pulley system over the water to American alligators, *Alligator mississippiensis*, the alligators contacted more bags with food than empty control bags, and contacted them sooner. They also removed more bags of those that contained meat. They did not distinguish between different types of meat, even though they had been fed nutria meat for years (Scott and Weldon, 1990). In a separate experiment, chloroform extracts of beef elicited more gular pumping behavior in American alligators than did chloroform alone. Gular pumping also increased in response to odors of crayfish, chicken, nutria, catfish, and alligator (Weldon *et al.*, 1990).

Birds

We now know that some bird taxa use their sense of smell in foraging. These include kiwis, vultures, seabirds, and honey guides. Others, such as seed-burying birds, and oilbirds, most likely use olfaction for finding food.

Kiwis

The kiwi (*Apteryx* sp.) has nostrils at the tip of its beak and feeds on earthworms, grubs, and small insects. When given a choice of containers covered with nylon screening and soil that contained either food or only soil, kiwis broke consistently into those with food, indicating that they use chemoreception to locate prey (Wenzel, 1968).

Raptors

Turkey vultures, *Cathartes aura*, find buried carcasses on the floor of tropical forests and will correctly detect meat experimentally hidden under leaves (Owre and Northington, 1961). Ethyl mercaptan released in the path of migrating turkey vultures attracted them to the general area. Once in the area, they seemed to search visually for the exact location of food. This is adaptive in forested and vegetated areas (Stager, 1967). The turkey vulture has the eighth largest olfactory bulb among the 108 bird species investigated by Bang and Cobb (1968). In an experiment at the Smithsonian Tropical Research Institute on Barro Colorado Island in Panama, chicken carcasses were placed 200 or 400 m apart on the floor of the tropical forest and covered with dry leaves. The turkey vultures efficiently located 1-day-old carcasses. However, they did not find animals that had died very recently as easily and they rejected rotten meat. After the turkey vultures, black vultures, *Coragyps atratus*, arrived at the food. This species does not smell the carcasses but visually follows the turkey vultures. By itself, the black vulture forages in open country, and nowadays frequents garbage dumps. Both species eventually removed 90% of the provided food (Houston, 1987).

The large king vulture, *Sarcorhamphus papa*, of South America soars at 300 to 500 m, where presumably little or no odor can be detected. Like the black vulture, the king vulture probably first observes the turkey vultures and two species of yellow-headed vultures (*Cathartes melambrotus* and *Cathartes burrovianus*), which also can smell carcasses. The king vulture appears to use the sense of smell of the other three species for its benefit. In return, the king vulture aids the other vultures by tearing and feeding on the skin of larger animals, making them available to these smaller species. This highlights the interdependence of the members of the community of American forest vultures (Houston, 1984). By comparison, Old World vultures do not forage by smell, stay in more open areas, and do not scavenge in the forest. Using the sense of smell has opened up a new niche to New World vultures.

The turkey vulture's sense of smell has even been recruited to detect leaks in natural gas pipelines. In 1938, Union Oil Co. injected ethanethiol into gas lines and watched for turkey vultures to appear over leaks (Stager, 1964). The vultures were attracted from up to 61 m altitude and 183 m downwind. To achieve this attraction, a detection threshold of 1×10^{-12} to 1×10^{-13} mol/l was necessary. Later, the vulture's detection thresholds were measured and found to be much higher even than that (Smith and Paselk, 1986; see p. 115). It still is not clear what compounds turkey vultures use to detect carcasses and at what concentrations.

Seabirds

Among the Procellariiformes (shearwaters [*Puffinus* spp.] and petrels), Wilson's storm-petrels (*Oceanites oceanicus*), Leach's storm-petrels (*Oceanodroma leucorhoa*), and greater shearwater (*Puffinus gravis*) approached cod liver oil-soaked sponges over water, while sooty shearwaters (*Puffinus griseus*) did not (Grubb, 1972). A variety of other Procellariiformes respond to food odors on the ocean surface. These include black-footed albatrosses (*Diomedea nigripes*), shearwaters (*Puffinus creatopus, P. puffinus, P. bulleri, and P. tenuirostris*) and northern fulmars (*Fulmarus glacialis*). Tuna oil, a tuna fraction, and squid homogenate attracted these birds, while bacon fat, vegetable oil, mineral oil, and petroleum oil did not. They were attracted to slicks contained in plastic pools and to odor-saturated wicks attached to vertical poles on floating inner tubes. Other seabirds such as gulls, terns, cormorants, or pelicans were not attracted to the odors (Hutchison and Wenzel, 1980). In field tests off the California coast, northern fulmars,

F. glacialis, and sooty shearwaters, *P. griseus*, arrived sooner and more reliably at an experimental fraction of cod liver oil than at the whole oil. This may be because of the higher volatility of the fraction. Shearwaters also approached squid and krill from downwind. Seawater or heptane controls did not attract these birds. Non-Procellariids did not differ in their behaviors with respect to stimuli or wind directions (Hutchison *et al.*, 1984). The "fishy odor" of krill is mainly a result of pyrazines and *N,N*-dimethyl-2-phenylethyl amine (Kubota and Kobayashi, 1988). Leach's storm petrel, *Oceanodroma leucorhoa*, was more sensitive to amines than to carboxylic acids from krill. These amines are more volatile than carboxylic acids. However, the carboxylic acid fraction attracted petrels more than did a "fishy" fraction. The more-volatile amines may be a longer-distance attractant while less-volatile carboxylic acids could be responsible for more concentrated searching in a restricted active space. The detection threshold for volatile carboxylic acids is also higher than that of amines (Clark and Shah, 1992).

The anatomy of the northern fulmar's olfactory bulbs is better suited for powerful odor *detection* than for keen odor discrimination. The periglomerular and external tufted cells are relatively sparse. These cells are important for superior odor *discrimination* in macrosmatic mammals. The interior granule cells are also loosely organized (Meisami and Wenzel, 1987).

Several Antarctic petrels distinguished floating sponges soaked with cod liver oil from those soaked with mineral oil. These included Wilson's storm petrel (*O. oceanicus*), snow petrel (*Pagodroma nivea*), and the pintado petrel (*Daption capense*) (Jouventin and Robin, 1984). The snow petrel has the most highly developed olfactory bulbs (Bang, 1965). Non-breeding snow petrels found herring hidden under perforated cups and even pecked at the hand that had handled the fish and not at the other hand. They also found pieces of fish buried in snow. By contrast, South Polar skuas, *Catharacta maccormicki*, never found a piece of fish unless they had seen it buried by the experimenter (Jouventin, 1977).

Sub-Antarctic seabirds were examined in the Crozet archipelago. With 26 species, it boasts the world's highest concentration of Procellariiforms. A sponge treated with cod liver oil was placed in a box on a raft and presented to free-ranging seabirds. A seawater sponge served as control. Wilson's storm petrel, *O. oceanicus*, and the black-bellied storm petrel (*Fregetta tropica*) responded most strongly to the odor. Cape petrels, *D. capense*, white-chinned petrel, *Procellaria aequinoctalis*, and giant petrels, *Macronectes* sp., were also significantly attracted. Albatrosses, prions, and diving petrels showed no interest. In general, only surface feeders, not diving species, responded to the food odor (Lequette *et al.*, 1989).

Marine microalgae such as *Phaeocystis pouchetti* produce dimethyl sulfide and release it into the air, where it can persist for hours or days. Dimethyl Sulfide is formed enzymically from dimethylsulfoniopropionate, particularly upon grazing by zooplankton such as Antarctic krill (*Euphausia superba*). Krill, in turn, is the prey of many seabirds. Experiments have shown that procellariiforms, especially Wilson's and black-bellied storm petrels, white-chinned petrels, and prions are attracted to dimethyl sulfide if it is in the nanomoles per cubic meter range on an oil slick. Dimethyl sulfide is as attractive as cod liver oil to these birds but did not attract albatrosses (black-browed, gray-headed, and wandering) and cape petrels (Nevitt *et al.*, 1995). In the presence of a DMS aerosol, white-chinned petrels also turned more often in the odor plume than they did to a water control, while black-browed albatrosses did not discriminate in this way (Nevitt, 1999).

Another experiment used cod liver oil, crude krill extract, and the krill odor components trimethylamine and pyrazine, with phenylethanol (rose odor) as control. Several procellariiform species were attracted to krill extract. The krill odor components attracted giant petrels, cape petrels, blue petrels, Antarctic petrels, Kerguelen petrels, and black-browed albatrosses. Cape petrels were more attracted to trimethylamine than to pyrazine and cod liver oil but blue petrels responded most to cod liver oil (Nevitt, 1999).

Shorebirds use their sense of taste when probing sand for food. The purple sandpiper, *Calidris maritima*, and the knot, *Calidris canutus*, forage much longer in jars that contain food buried in sand, or sand with an extract of food, than in jars with plain sand (Gerritsen *et al.*, 1983). Table 12.3 lists the responses of various seabirds to prey odors.

Honey guides: of bees and birds, and the badger

Some species of honey guides (Indicatoridae), sparrow-like birds in Africa and Southeast Asia, guide wild mammals to wild honeybee colonies. They are attracted to wax, and not the honey. The tropical African species *Indicator indicator* interacts with the honey badger or ratel, *Mellivora capensis*. The bird smells the wax that contains honey and gives characteristic calls in the presence of the ratel. The badger follows the bird to the wild bees' nest and opens up the hive with its powerful claws. The badger eats the honey, the honey guide the wax. A Portuguese missionary in East Africa reported in 1569 that honey guides flew into his mission, were attracted to the burning altar candles, and ate the beeswax. Honey guides could be kept alive for 32 days on a pure beeswax diet. A 100% beeswax candle put in a tree attracted honey guides only when lit: within 35 minutes, six honey guides appeared (Friedmann, 1955). Honey guides have a large olfactory apparatus (Stager, 1967). Honey guides (*Indicator variegatus* and *Indicator exilis*) have been attracted to mist nets by burning honey comb (Archer and Glen, 1969). The odor of fresh beeswax plays a pheromonal role in the beehive: it stimulates food hoarding in honeybees (Blum *et al.*, 1988). Honey

Table 12.3 Responses of seabirds to food odors

Seabird species	Cod liver oil	Tuna oil, tuna fraction, squid homogenate	Krill extract	Buried fish	Dimethyl sulfide (from phytoplankton)	Reference
Shearwaters (*Puffinus* spp.)						
P. griseus	–	+				1–5
P. gravis	+					1
P. creatopus		+				3
P. puffinus		+				3
P. bulleri		+				3
P. tenuirostris		+				3
Storm petrels						
Wilson's	+				+	1,4,6–9
Leach's	+		+[a]			1–7
Black-bellied	+				+	4–9
Petrels						
Snow	+			+		2, 4
Cape (pintado)	+		+		–	4, 6, 8, 9
White-chinned	+				+	6–9
Giant	+		+			6–8
Blue		+	+			8
Antarctic			+			8
Kerguelen			+			8
Albatrosses	–					6
Black-footed		+				3
Black-browed			+		–	8, 9
Grayheaded					–	9
Wandering					–	9
Prions	–				+	6–9
Diving petrels	–					6
Fulmars northern		+				5
Gulls, terns, cormorants, pelicans						3,5
South Polar skua				–		

[a] Amines, carboxylic acids.

1. Grubb, 1972; 2. Jouventin, 1977; 3. Hutchison and Wenzel, 1980; 4. Jouventin and Robin, 1984; 5. Hutchison *et al.*, 1984; 6. Lequette *et al.*, 1989; 7. Clark and Shah, 1992; 8. Nevitt 1995, 1999; 9. Nevitt *et al.*, 1995.

guides probably have secondarily cued in to the pheromone, which, therefore, also qualifies as a kairomone. Archer and Glen (1969) caught all their honey guide specimens within 50 yards of wild bee colonies, even as long as 2 weeks after bees had been absent from two of the tree hives. The wax of bees plays a vital role in the social organization of honey guides: males of the orange-rumped honeyguide of Asia defend a colony of the bee *Apis dorsata* year round. The wax is a critical resource: females are attracted to these polygynous males on the basis of the resource they defend (Cronin and Sherman, 1976).

Finding food caches

Birds such as nuthatches, nutcrackers, and jays store food by burying seeds such as acoms, beechnuts, and pine seeds in the ground and find it later very well. A single Clark's nutcracker, *Nucifraga columbiana*, buries as many as 33 000 piñon seeds in up to 3750 different caches (Vander Wall, 1982). Experiments have shown that these nutcrackers use memory (Tomback, 1980), visual cues (Vander Wall, 1982), and probably little olfaction to locate buried food caches. Other birds use olfaction. Black-billed magpies, *Pica pica*, discovered buried suet or raisins better when the cache was scented with cod liver oil (Buitron and Nuechterlein, 1985).

The oilbird

The oilbird or guácharo, *Steatornis caripensis,* of South America and Trinidad – a relative of the nighthawks – has large olfactory bulbs (Stager, 1967) and probably uses olfaction for finding the aromatic, spicy fruits of the oil palm during its nocturnal foraging. These birds fly up to 75 km from their cave to feed. Oilbirds were first made famous by Alexander von Humboldt's visit to the Guácharo Cave in Venezuela in 1799. and locals have traditionally turned the fat of young birds into "guácharo butter" for cooking and lighting.

Other senses in birds

Birds do not necessarily distinguish palatable and unpalatable prey by odor. Free-ranging European birds such as chaffinch, house sparrow, robin, starling, blackbird, and song thrush recognized bread pieces treated with quinine and mustard powder only by their size, but not other visual cues or smell. This has a bearing on model–mimic relationships. Batesian mimics are often smaller than their unpalatable models (Marples, 1993).

In ultraviolet light (320–400 nm) birds can see, rather than smell, urine and feces on vole trails. In the laboratory, kestrels, *Falco tinnunculus*, spent more time near ultraviolet-illuminated, artificial urine- and feces-soaked trails of voles, *Microtus agrestis*, than near such trails in visible light, and clean trails under both types of illumination. Likewise, in the field, kestrels hunted, perched, and used nestboxes more often near urine- and feces-soaked artificial vole trails than near artificial trails without urine or feces, and areas with no vole trails (Viitala *et al.*, 1995). Whether and under what conditions kestrels and possibly other birds use the fluorescence caused by the ultraviolet portion of sunlight while hunting remains unclear.

Mammals

Mammals use all senses for foraging. A species' natural history will determine which sensory modality predominates. Nocturnally active species and those living in dense vegetation or underground are expected to be more olfactorily oriented. Predators can be attracted to one odor of a prey species more than to others of the same species. For instance, foxes, ferrets, and cats seek out blood odors of *stressed* Mongolian gerbils, *Meriones unguiculatus*. Cats are attracted to food pellets that are contaminated with a few pellets soaked in stress odor-containing blood of gerbils (Thiessen and Cocke, 1986). Hunting by smell is the hallmark of carnivores. In the following, insectivores, canids, felids, mustelids, and ursids will be discussed.

Insectivores

Shrews (*Sorex cinerus cinereus* and *Blarina brevicauda talpoides*) use olfactory cues to find buried prepupae of the European pine sawfly, *Neodiprion sertifer*. Specifically, these cues play a role in digging, removing, and opening this prey, while taste stimuli guide the eating behavior (Holling, 1958).

Canids

Wolves, *Canis lupus*, detect prey from 300 (27m) yards or even up to 1.5 miles (2.5 km) downwind by "direct scenting," i.e. smelling the air. They may also track initially, followed by direct scenting at close range (Mech, 1970). The same individual may be guided by different senses according to circumstances. For instance, a coyote, *Canis latrans*, uses mostly vision when finding a rabbit in an enclosed room, and olfaction is least important. Outdoors, however, olfaction becomes more important and the coyote approaches from downwind approximately 84% of the time. In addition, wind may interfere with the sounds of the prey outdoors (Bekoff and Wells, 1980).

Felids

Predators of open country, such as Serengeti lions, *Panthera leo*, do not necessarily approach their quarry from downwind. In 300 hunts, Schaller (1972) observed lions approach prey equally often (28% each) from up- and downwind, and in the remainder of cases with wind from the side. This, random approach direction with regard to wind (and odor) suggests that lions primarily use vision in hunting.

Mustelids

Small predators such as mustelids extensively use their sense of smell to locate prey. Moreover, they respond more to familiar prey and *learn* certain odor preferences early in life during a *sensitive period*. In one experiment, pole-cats, *Putorius putorius,* first lived on a diet of domestic chicks and, starting at different ages, later received dead mice and rats in addition. Subsequently, up to 1 year of age, they were tested for prey odor preferences. The later they had been exposed to mouse and rat carcasses, the less they responded to mouse odor. It was concluded that prey odors are learned around the second or third month of life (Apfelbach, 1973). This can be important in captive breeding of rare animals. For instance, captive black-footed ferrets need prairie dog meat while raising their young. They should be exposed to this type of meat early in life. Otherwise, the reintroduction of these captive-raised young might fail.

Ursids

Bears have a keen sense of smell, even though we lack formal and thorough studies. According to popular television programs, polar bears can smell a ringed seal pup buried in snow from distances exceeding 1 mile (1.6 km). The exact distances for the bears prey detection is not yet clear (I. Stirling, personal communication) and only 1 in 20 hunts is successful.

Rodents

Many birds and mammals bury seeds in soil for storage. Mammals that cache food subterraneously include tree and ground squirrels, and deer mice, *P. maniculatus*. Underground storage provides many benefits: competing surface foragers such as birds, deer, and turkeys cannot reach the food and the seeds may loose undesirable secondary plant compounds such as tannins in the moist soil, amounting to food processing. Small mammals use olfaction, spatial memory, and visual landmarks to locate buried seeds. Spatial memory favors the owner of the buried cache, while olfaction permits other individuals to pilfer stored food. For underground hoarding to remain a viable behavior, the animal that caches the food must enjoy a greater chance of finding it again than do competitors.

Deer mice, *P. maniculatus*, readily find buried seeds. In forest reseeding projects, deer mice often remove 70 to 100% of the planted conifer seeds. In the laboratory, deer mice found the aromatic seeds of sugar, Jeffrey, and Ponderosa pine buried in 5 cm peat moss more easily than seeds of wheat, barley, or oats. If these cereal seeds were treated with safflower oil or lecithin-mineral oil, the mice found them more easily. This suggests that deer mice use olfactory cues to find buried seeds (Howard and Cole, 1967; Howard *et al.*, 1968). Deer mice also use olfactory stimuli to find buried prepupae of the European pine sawfly, *N. sertifer*. They distinguish between parasitized and healthy larvae, and between males and females (Holling, 1958).

Yellow pine chipmunks, *Tamias amoenus*, and deer mice of the Sierra Nevada scatter-hoard seeds of Jeffrey pine, *Pinus jeffreyi*, and other trees and shrubs. Knowledgeable foragers found the buried seeds whether the soil was dry or wet. Naive foragers found caches only when the soil was wet and found seeds buried by either species equally well. Few seeds were discovered when the soil was dry regardless of the species of the caching individual or the forager. Wet seeds release organic molecules, which the rodents spy on; these odors are now functioning as kairomones. Under dry conditions, spatial memory gives the owner of the cache the edge over competitors. As the soil moisture rises, owners and competitors vie increasingly more for the seeds and pilfering increases (Vander Wall, 2000).

Primates

Lower primates hunt by smell. To the nocturnal African prosimians *Galagoides demidovii* and *Galago alleni*, the odor of a concealed insect is more important than a visible, but odorless animal. *G. alleni* even distinguishes the odors of the front and rear end of an insect. Aqueous extracts applied to the "wrong" end of the body can trick the galago (Molez-Verriere, 1988).

Marine mammals

Even marine mammals may use airborne odors for food detection. Baleen whales (Mysticeti) feed on krill near the ocean's surface. Oldtime Antarctic whalers noted a "krill odor" near large schools of krill. Cruising at the surface and inhaling periodically, baleen whales may detect krill odor. Anatomically, they have a well-developed olfactory organ (Cave, 1988), in contrast to toothed whales (Kusnetzov, 1988).

12.1.2 Predator odors used by prey

Potential prey species can chemically assess predation risk from a distance and/or from the safety of their refuge by evaluating predator odors in the area (Kats and Dill, 1998). Such odors emanate from the predator itself or its

Table 12.4 Respones of prey fish to predator chemicals

Prey species	Predator species	Stimulus	Response	Reference
Coho salmon *Oncorhynchus kisutch*	Human, bear, dog	L-Serine, L-alanine	Retreat to lower water	Idler *et al.*, 1956; Rehnberg and Schreck, 1986
	Northern squawfish *Ptychocheilus oregonensis*	Body rinse, broken skin extract	Avoidance, plasma cortisone and glucose rise	Rehnberg and Schreck, 1987
Arctic charr *Salvelinus alpinus*	Human	L-serine	Avoidance	Jones and Hara, 1982
Mosquito fish *Gambusia patruelis*	Chain pickerel *Esox niger*; redfin pickerel *Esox americanus*	Mucus coat of pickerel	Swim to upper water levels	George, 1960
Shiners *Notropis texanus and Notropis venustusi* and chubs *Hybopsis aestivalis*	Pikes, *Esox* sp.; sunfish *Lepomis macrochirus*; bass *Micropterus punctulatus*	Rinse from piscivores of North and South America	Swim to bottom of tank; no response to human hand rinse	Reed, 1969
Threadfin shad *Dorosoma petenense*	Large mouth bass *Micropterus salmonides*	Bass rinse	No response (fish move faster than chemicals)	McMahon and Tasch, 1979

"sign" such as tracks, droppings, urine or other scent marks, food remains, or disturbed soil or plants.

Fish

Several fish species, notably salmon, respond to the odors of their predators in an adaptive manner (Table 12.4). Coho and spring salmon retreated to the lower parts of a fish ladder when rinses of human hand, bear paw, dog meat, or sea lion meat were added to the water upstream. Even the odor of deer feet had the same effect (Brett and MacKinnon, 1954). L-Serine was identified as the active compound in mammalian skin (Idler *et al.*, 1956). Arctic char, *Salvelinus alpinus*, also responds to L-serine from human hand wash, while D-serine and several other amino acids were not active. L-Serine is less active than the whole hand wash, suggesting that two or more compounds are required for the full response (Jones and Hara, 1982).

Coho salmon, *Oncorhynchus kisutch*, tested in a two-choice Y-trough, avoided whole body rinses and broken-skin extracts of the predatory northern squawfish, *Ptychocheilus oregonensis*, but not those of the non-predatory largescale sucker,

Catostomus macrocheilus. However, both rinses induced elevated levels of plasma cortisol and glucose, commonly referred to as a stress response. This experiment demonstrates that the behavioral and physiological responses are not necessarily coupled (Rehnberg and Schreck, 1987).

The mosquitofish, *Gambusia patruelis*, avoids its predators, the chain pickerel, *Esox niger*, and redfin pickerel, *Esox americanus*, by swimming to the upper water levels. It also responds to odor from the pickerel mucus coat. This odor survives passage through filter paper and hours of bubbling but loses activity if heated or passed over charcoal (George, 1960).

Shiners (*Notropis texanus* and *N. venustus*) and chubs (*Hybopsis aestivalis*) avoid the water from piscivorous fish, such as pikes (*Esox* sp.), sunfish (*Lepomis macrochirus*), bass (*Micropterus punctulatus*), and two South American cichlids (*Astronotus ocellatus* and *Cichlasoma severum*). They respond by swimming to the bottom of the experimental tank and remaining motionless in clusters. Plain water or human hand rinse do not trigger this response. In this case, predator and prey do not have to occur sympatrically for the avoidance response to occur (Reed, 1969).

Fish avoid more vigorously the odor of predators that have fed on members of their species than that of those on different diets. For example, young Arctic charr avoid water from brown trout fed on Arctic charr and are less wary of that from pellet-fed trout (Hirvonen *et al.*, 2000). Prey fish also reduce their "predator inspection" behavior vis-à-vis predators that have eaten members of their own species. For instance, finescale dace, *Phoxinus neogaeus*, dash toward predators such as yellow perch, *Perca flavescens*, and withdraw. Dace inspect perch models less often if the model is accompanied by water from perch that had eaten dace than if accompanied by water from perch on a swordtail, *Xiphophorus helleri*, diet. Dace produce alarm pheromone, while swordtails do not. The Central American swordtails do not cooccur with finescale dace (Brown *et al.*, 2001).

A chemical cue from piscivorous fish may be responsible for the size and shape change in crucian carp, *Carassius carassius*, in response to the presence of the carp-eating pike, *Esox lucius*. The carp increases its bulk and becomes more difficult to catch by the pike. The chemical cue could emanate from the predator, or from injured or frightened carps (Bronmark and Miner, 1992). Crucian carp (or bronze carp) form schools. Crucian carp show stronger alarm responses to unfamiliar predatory fish such as pike or perch. Where the carp coexist with pike, their alarm behavior is attenuated. Larger predator individuals triggered stronger responses than smaller ones. The diet of the predatory fish affected the responses of crucian carp: after eating prey fish that produce alarm substance, large pike induced more alarm behavior in their prey than small pike. Crucian carp use cues from the predators as well as the alarm substance of their prey. They also distinguish large and small predators and habituate to sympatric predators.

In summary, growing a deep body in response to sympatric predators appears to reduce the need for behavioral alarm behavior (Pettersson *et al.*, 2000).

Other fish species do not respond to predator odors. The threadfin shad, *Dorosoma petenense*, is strongly attracted to odors of its prey such as brine shrimp (*Artemia*) or *Daphnia* spp. but does not respond to those of its predator, the largemouth bass, *M. salmonides*, or conspecifics. Both shad and bass swim faster than chemicals travel in water, which may explain this behavior difference (McMahon and Tash, 1979).

In minnows, taste is not sufficient for predator recognition. Anosmic fathead minnows, *P. promelas*, did not show the flight reaction to the odor of northern pike, *Esox lucius* (Chivers and Smith, 1993). Naive European minnows, *Phoxinus phoxinus*, do not exhibit a fright reaction when first exposed to a predator odor, such as that of pike, *E. lucius*. They develop a conditioned fright response only after experiencing the predator odor in dangerous circumstances, such as when accompanied by schreckstoff (alarm pheromone) of conspecifics. Responses to the odor of non-piscivorous fishes such as tilapia, *Tilapia mariae*, can also be conditioned in this fashion but the responses are much weaker (Magurran, 1989).

Predator density, and hence probability of attack, and the cost to the prey species alter responses to predator cues, including chemical ones, by the same prey species. For instance, in a laboratory tank, Trinidadian guppies, *Peocilia reticulata*, from predator-dense downstream sections stay at greater distances from a hungry predatory largemouth bass, *M. salmonides*, than do guppies from headwater streams that have few predators. Both populations discriminated a hungry bass from a sated one (Licht 1989).

Amphibia

Salamanders

Salamanders and their larvae detect predators by odor cues (Table 12.5). They take refuge and are also distasteful to predators. The salamander *Eurycea bislineata* and tadpoles of the tree frog, *Hyla chrysoscelis*, avoid the odor of the predatory green sunfish, *Lepomis cyanellus*. They do not respond to the odors of the green frog, *Rana clamitans*, and brine shrimp (*Artemia* sp.) used as controls (Petranka *et al.*, 1987). Tadpoles of the small-mouthed salamander, *Ambystoma texanum*, hide in refuges when the odor of their predator, the sunfish *L. cyanellus*, is added to the water. They do not respond to the odors of snapping turtle (*Chelydra serpentina*), water snakes (*Nerodia sipedon*), crayfish (*Oronectus rusticus*), or odorless water. Only larvae from water bodies that have fish do actually respond to fish cues (Kats, 1988). Why do these larvae not respond to the odors of sympatric predaceous crayfish, turtles, and snakes? If further studies confirm these

Table 12.5 Responses of amphibian prey to predator chemicals

Prey species	Predator species	Chemical cue	Response	Reference
Small-mouthed salamander *Ambystoma texanum*: tadpoles	Green sunfish *Lepomys cyanellus*	Rinse	Hide in refuges	Kats, 1988
Salamander *Eurycea bislineata* and tree frog *Hyla chrysocelis*: tadpoles	Green sunfish	Rinse	Avoidance	Petranka *et al.*, 1987
Salamander *Plethodon richmondi*	Ringneck snake *Diadophis punctata*	Traces on substrate	Avoidance	Cupp, 1988
Tadpoles of frogs *Rana lessonae* and *Rana esculenta*	Pike *Esox luteus*	Rinse	Swim, rest, edge use	Stauffer and Semlitch, 1993

laboratory results, one might expect these predators to have a more recent association with small-mouthed salamanders (Kats, 1988).

On land, snakes prey on salamanders and their eggs. The ringneck snake (*Diadophis punctata*) and its prey, the salamanders *Plethodon richmondi* and *Plethodon dorsalis* provide an example. Salamanders were given a choice between a paper towel that had had a ringneck snake on it for 48 hours, and one with clean water. The salamanders of both species avoided the substrate with the odor of the ringneck snake (Cupp, 1988).

In their predator avoidance, salamanders use complex odors that combine chemicals from both predator and prey. In the laboratory, red-backed salamanders, *P. cinereus*, avoid filter papers soaked with water extracts from garter snakes that had been preying on salamanders, while earthworm-fed snakes lacked this effect. Exudations from unfed snakes and extracts from homogenized salamanders had no such alarming effect (Madison *et al.*, 2002).

Frogs and toads

Tadpoles of the two closely related frog species *Rana lessonae* and *Rana esculenta* respond more to chemical cues of their predator, the pike *E. lucius*, than to visual and tactile ones. The strongest swimming, resting, and edge-use behaviors – all considered antipredator responses – occurred to a combination of chemical and tactile cues (Stauffer and Semlitch, 1993). *Rana temporaria* tadpoles became less active when exposed to chemical cues from their predators, perch, *Perca fluviatilis*, and dragonfly larvae, *Aeschna juncea*. Tadpoles of different parentages differed in their responses, suggesting genetic factors. Overall, their behavior appeared to be rather plastic (Laurila, 2000). In one experiment in Sweden, tadpoles of the frog *R. temporaria* avoided water-borne cues from their predator the rainbow trout, *O. mykiss*, by hiding in refuges, while toad tadpoles, *Bufo bufo*, remained unresponsive. The behavior of the frog tadpoles affected lower trophic levels: their reduced grazing behavior resulted in more surviving plant (periphyton) mass (Nyström and Abjörnsson, 2000).

After metamorphosis, juvenile toads generally avoid predators by visual cues, whereas juvenile Great Plains toads, *Bufo cognatus*, and southwestern toads, *Bufo microscaphus*, detect and avoid odors of their respective predators, the eastern plains garter snake, *T. radix*, and the wandering garter snake, *T. elegans*. This was demonstrated by presenting the odors on paper towels on which the snakes had been living for 24 hours. Lizard odor had no effect (Flowers and Graves, 1997).

Ecological and evolutionary aspects

Ephemeral ponds and streams harbor different species of larval amphibians to those in permanent bodies of water. Several factors may be responsible for this, although predation appears to play an important role. In permanent aquatic habitats, predator densities are higher, and fish, including predators, are usually restricted to such permanent bodies of water. Chemical recognition of predators and chemical defense compounds are two ways in which amphibian larvae counteract predation. In one experiment, predator-naive larvae of different amphibian species were exposed to water from tanks with a predatory green fish, *Lepomis cyanellus*. Amphibians from temporary ponds and streams who rarely encounter fish did not avoid the fish odor by hiding. These species are also palatable to predators. By contrast, larvae from permanent ponds often encounter fish and so would usually avoid the fish odor regardless of whether they were palatable or unpalatable. Therefore, the response of amphibian larvae to the odor of a predatory fish correlated the with the probability of fish encounters (i.e. predation risk). Closely related taxa differed in palatability and their responses to predator chemicals. For instance, green frog, *R. clamitans*, and bullfrog, *Rana catesbeiana*, were unpalatable and encounter fish often, while wood frog, *Rana sylvatica*, was palatable and rarely encountered fish. Variation also occurs within one species: small-mouth salamander, *Ambystoma texanum*, from streams with fish avoided the sunfish odor, while salamanders from fishless ponds did not. Thus, natural selection rather than phylogeny appears best to explain the observed differences in antipredator defenses (Kats *et al.*, 1988). Investigators have found that amphibian larvae from permanent bodies of water had at least one of the two chemosensory defenses, while those from ephemeral habitats consistently lacked such defenses. Chemical detection of predatory fish appears to be the major defense of palatable amphibian larvae that coexist with fish predators. These studies imply interaction by chemical cues as a proximate mechanism operating where predators organize animal communities.

Life history shifts

Chemical cues from predators can change the rate of development in amphibians. For example, red-legged frogs, *Rana aurora*, of western North America were raised either in the presence of chemicals from one of their predators, the rough-skinned newt, *Taricha granulosa* (which had eaten red-legged frogs) or in the presence of chemicals from injured conspecifics. In both cases, the tadpoles metamorphosed earlier and at a smaller size than usual (Kiesecker *et al.*, 2002). Similarly, tadpoles of the western toad, *Bufo borealis*, metamorphosed faster when exposed to visual and chemical cues from a predator such as backswimmers (*Notonecta* sp.), or chemical alarm cues from injured tadpoles of their own species (Chivers *et al.*, 1999; Fig. 12.1)

Conservation implications

The decline of amphibian populations has been linked to habitat destruction, increased pathogens, global warming, and ultraviolet radiation, but it may also be linked to failed chemical predator recognition. For example, the California newt, *Taricha torosa*, has decreased or disappeared from streams inhabited by the introduced crayfish, *Procambarus clarki*, and mosquitofish, *Gambusia affinis*. In laboratory and field experiments, the newts, their eggs, and their larvae were successfully attacked and/or eaten by the introduced predators. The newt's chemical predator recognition and its defense by tetrodotoxin do not appear to cope with the introduced species (Gamradt and Kats, 1996).

Reptiles

In the terrestrial environment, olfactory cues have several advantages for detecting and avoiding predators: long-distance propagation by air currents; detection of signals in the dark; reception despite obstacles, such as vegetation; slow fade-out of signals; and possible deception by spatial separation of animal and its scent.

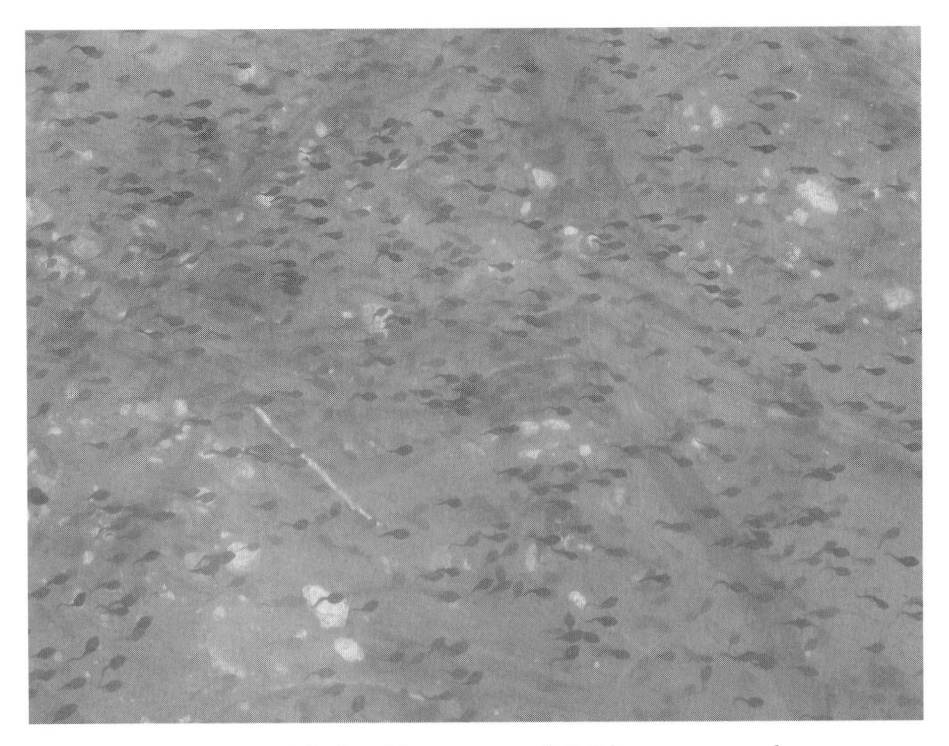

FIGURE 12.1 Tadpoles of the western toad, *Bufo boreas*, metamorphose sooner when exposed to chemical and visual predator cues or alarm substances of conspecific tadpoles (Chivers *et al.*, 1999). (Photograph: D. Müller-Schwarze.)

Snakes

Snakes respond to odors of their predators (Table 12.6). Best known are their reactions to ophiophageous snakes. The odor of the California king snake, *Lampropeltis getulus californiae*, induces *body bridging* in rattlesnakes: the midbody is raised and undulated. Blindfolding does not change this reaction, while a stick rubbed on a king snake elicits the response. Dorsal skin material is most effective. North American rattlesnakes also responded to skin rubbings of the puffing snake *Pseustes sulphureus* from South America, which feeds on birds and lizards. Rubbings from non-ophiophagous snakes of the families of boids, crotalines, and colubrids were ineffective (Bogert, 1941). More than 20 species of crotaline snakes from North, Central and South America show body-bridging when exposed to methanol-soaked cotton balls that had been rubbed on the dorsal skin of ophiophagous snakes (Weldon and Burghardt, 1979).

In a number of species, the active predator odors originate on the dorsal skin. Neonate pygmy rattlesnakes, *Sistrurus miliarius*, and timber rattlesnakes, *C. horridus*, respond to dorsal skin chemicals of the ophiophagous king snakes and indigo snakes, *Drymarchon corais*, but not to those from ventral skin or skin

Table 12.6 Responses of reptiles to predator chemicals

Prey species	Predator species	Cue	Response	Reference
Rattlesnakes	California king snake *Lampropeltis getulus californiae*	Skin rubbings	Body bridging	Bogert, 1941
Rattlesnakes	King snake, human	Air or skin extract, human breath	Increased heart rate	Cowles and Phelan, 1958
Neonate pygmy (*Sistrurus miliarius*) and timber (*Crotalis horridus*) rattlesnakes	King and indigo (*Drymarchon corais*) snakes	Dorsal skin odor	Body-bridging	Marchisin, 1980
Garter snakes, *Thamnophis* sp.	Racer, *Coluber constrictor*, and king snake	Swabs or air	Increased tongue flicking	Weldon, 1982
Pine snake *Pituophis melanoleucus*: hatchlings	King snake	Body odor in Y-maze	Avoidance	Burger, 1989
Lizard *Lacerta vivipara*	*Vipera berus*, smooth snake *Coronella austriaca*	Soiled cage	Increased tongue flicking	Thoen *et al.*, 1986

glands. Furthermore, the rattlesnake *Crotalis atricaudatus* body-bridged to odor of the water snake, *Nerodia fasciata*, which feeds on amphibia and fish (Marchisin, 1980).

Garter snakes (*Thamnophis* sp.) increase their rate of tongue flicking when exposed to swabs or air from snake-eating snakes such as racer, *Coluber constrictor*, and king snake, *L. getulus*, to air that has passed over a king snake, or to skin extract from a king snake. Human breath is also active (Cowles and Phelan, 1958). Odor from the non-predatory hognose snake, *Heterodon nasicus*, does not affect tongue flicking (Weldon, 1982). Hatchling pine snakes, *P. melanoleucus*, also avoided the odor of king snakes in a Y-maze, while they were attracted to conspecific odor (Burger, 1989).

Lizards

Lizards respond to odors of predatory snakes with increased tongue flicking. The common lizard, *Lacerta vivipara*, tongue flicks more when placed in a cage that had held the predator *Vipera berus* or the smooth snake, *Coronella austriaca*, but only slightly more to the odor of the grass snake, *Natrix natrix*, which does not feed on snakes. The behavior also changed to a slow locomotion when in a cage with predaceous snake odor (Thoen *et al.*, 1986). Gekkonid lizards also responded to odors of predatory snakes (Dial *et al.*, 1989).

Lizards that prey on snakes but are also eaten by larger snakes discriminate skin chemicals of snakes very well. Monitor lizards, *Varanus albigularis*, fall in this group. Hatchling monitors attacked harmless snakes but avoided venomous species. However, they accepted meat of all snakes if carefully skinned. The hatchlings tongue flicked to invertebrate prey covered with skin from venomous snakes and rejected these samples (Phillips and Alberts, 1992).

Mammals

Insectivores

Hedgehogs, *Erinaceus europaeus*, avoided areas in an enclosure that were tainted with feces of badger, *Meles meles*, which prey on hedgehogs. The response lasted about 2 days, but free-ranging hedgehogs on golf courses and farmland reduced their feeding near badger odor only for minutes or hours (Ward *et al.*, 1997).

Rodents

Many decades ago, Griffith (1919, 1920) described the white (laboratory) rat's response to cats. From the age of 3 weeks, rats huddle in corners, freeze, and crouch when presented with a cat. They tremble, twitch their muscles, whine, and cease to feed and nurse. A cloth with cat scent or an arena with cat scent sufficed to trigger these responses. Anosmic rats or those confronted with a cat in a glass jar remained unaffected. Hence, the critical cue proved to be odor (Griffith, 1919). Cat feces, urine, heart, or other tissues did not elicit these "fright reactions" (Griffith, 1920).

In settings that permit the recipient to move more freely, many different avoidance responses to predator odors have been observed (Table 12.7). Meadow voles, *Microtus pennsylvanicus*, avoid areas of enclosures that previously had been occupied by short-tailed shrews, *Blarina brevicauda* (Fulk, 1971). This shrew is a voracious predator with venomous saliva that feeds mostly on invertebrates but occasionally takes small mice, voles, or shrews. Free-ranging European field voles, *M. agrestis*, and field mice, *Apodemus sylvaticus*, avoid traps that are scented with anal secretion of weasel, *Mustela nivalis*, or urine from tiger, *Panthera tigris*, and jaguar, *Panthera onca* (Stoddart, 1976). Bank voles avoid areas with weasel odor as much as areas that contain live weasels (Jedrzejewska and Jedrzejewski, 1990).

In Australia, house mice living on islands without predators did not avoid traps treated with predator odors. In areas where the introduced red fox or house cat occur, or the native western quoll, *Dasyurus geoffroyii*, the mice avoided

Table 12.7 Examples of responses of mammals of predator chemicals

Prey species	Predator species	Cue	Response	Reference
Rat *Rattus norvegicus*	Red fox *Vulpes vulpes*	Trimethylthiazoline	"Stress," raised corticosterone	Vernet-Maury *et al.*, 1984
Meadow vole *Microtus pennsylvanicus*	Short-tailed shrew *Blarina brevicauda*	Soiled enclosure	Avoidance	Fulk, 1971
Field voles *Microtus agrestis*; field mice *Apodemus sylvaticus*	Weasel *Mustela nivalis*	Soiled trap	Avoidance of traps	Stoddart, 1976
Field mice and *Clethrionomys glareolus*	Mink *Mustela vison*	2, 2-Dimethylthietane from anal gland	Avoidance of traps	Robinson, 1990
Black-tailed deer *Odocoileus hemionus columbianus*	Coyote *Canis latrans*; puma, *Felis concolor*	Fecal odor	Inhibition of feeding	Müller-Schwarze, 1972

traps with predator odors. They also used dense vegetation on moonlit nights, while mice on the predator-free islands showed no such preference for denser vegetation. Finally, predator-naive and predator-experienced mice, transferred to areas with cats and foxes, differed in their mortality: over twice the number of predator-experienced mice survived compared with naive ones (Dickman, 1992).

Free-ranging North American beaver, *Castor canadensis*, feed less on experimental aspen sticks that have been treated with extracts from predator excrement or urine. Odors from the sympatric coyote and river otter, and extirpated lynx, were most effective, while those from allopatric lion and extirpated wolf odor were less active. However, these response differences between species were small (Fig. 12.2; Engelhart and Müller-Schwarze 1995).

Urine and feces odors of mink applied to soil and vegetation in outdoor enclosures had no effect on gray-tailed voles, *Microtus canicaudus*, in Oregon. The voles did not seek taller vegetation for cover, and their reproduction was not affected. Specifically, reproductive rates, time to sexual maturation, juvenile recruitment, and activity did not change after exposure to mink odors (Wolff and Davis-Born, 1997).

The odor of ferret, *Mustela putorius furo*, urine causes male outbred laboratory *Mus musculus*, to reduce their overmarking of rival urine marks. These mice reduce predation risk at the price of tolerating more intrasexual competition (Roberts *et al.*, 2001).

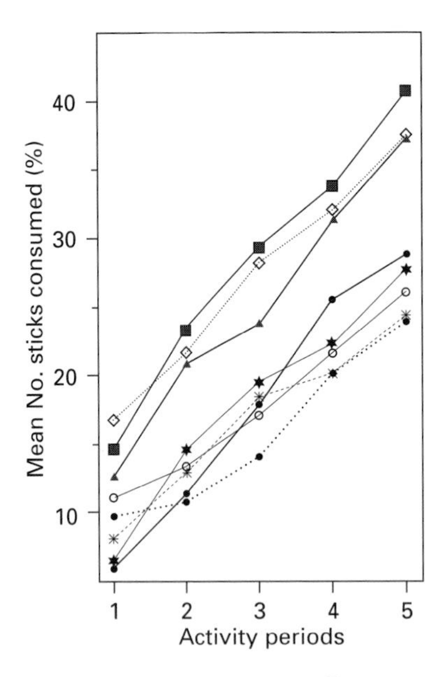

FIGURE 12.2 Responses of beavers to predator chemicals applied to aspen sticks. Activity periods were consecutive 5 days of experiment in two areas in New York and the percentages of sticks consumed is the mean of several replications of the experiment. All treated sticks were punctured to improve the uptake of chemicals; intact sticks were untreated (△), punctured but not treated (□), punctured and treated with the solvent methanol (◇) (the three controls), or treated with extracts from other animals. Treatments that inhibited consumption most were excrement extracts from lynx (∗) and coyote (• . . . •), both sympatric predators (lynx now extirpated). Beavers accepted most readily the three control sets. Other chemicals were from otter (○), wolf (•–•), lion (★). (From Engelhart and Müller-Schwarze, 1995.)

Ungulates

In black-tailed deer, *Odocoileus hemionus columbianus*, fecal odors of sympatric predators (coyote, *C. latrans*, and mountain lion, *Felis concolor*) in vials next to food pellets inhibited feeding, while those of allopatric predators (lion, *Felis leo*, snow leopard, *Uncia uncia*) do not, or very little (Müller-Schwarze, 1972; Fig. 12.3). Note that mammals discriminate between the odors of sym- and allopatric predators, while fish and rattlesnakes do not (pp. 359 and 364). Free-ranging adult female wapiti, *Cervus elaphus canadensis*, respond to the odors of dog urine, and cougar and wolf feces (presented as water slurry) with increased heart rates. It was concluded that the main effect of predator odors may be for assessing the risk of predation (Chabot *et al.*, 1996).

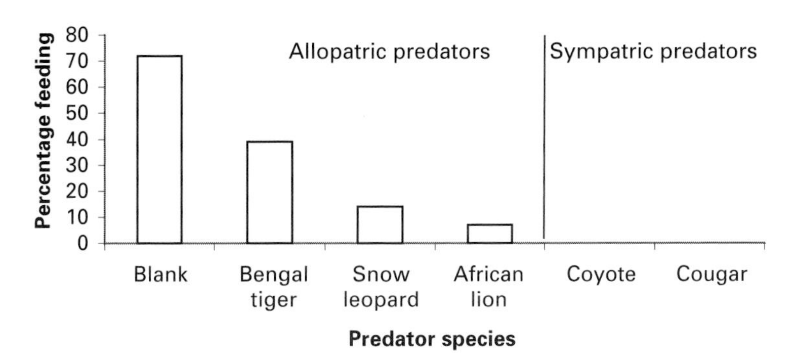

FIGURE 12.3 Responses of black-tailed deer fawns to predator odors.

Primates

Among primates, red-bellied tamarins (*Saguinus labiatus*) sniff and avoid fecal extracts of jaguar (*P. onca*), jaguarundi (*Herpailurus yagouaroundi*), and margay (*Felis wiedi*) more than those from non-predatory mammals. In this case, it is not clear whether these predators actually prey on tamarins (Caine and Weldon, 1989).

Mammals also respond to odors of predatory snakes. Female laboratory mice deposited more fecal boli in a section of an arena that was lined with paper from the cage of rat snakes, *Elaphe obsoleta*, which preys on rodents. Paper from the cage of the rough earth snakes, *Virginia striatulata*, a worm eater, had no effect. The females also ate less from snake-scented food pellets. Males did not respond to these odors (Weldon *et al.*, 1987). Kangaroo rats, *Dipodomys merriami*, avoided sidewinder rattlesnakes, but not after olfactory bulbectomy. Blinded or deafened individuals still avoided the snakes (Webster, 1973). Snake odors alarm even bears: brown bears retreat drooling from the odor of dead or live rat snakes. A rope treated with snake odor still proved active (Kano, 1976).

Chemistry of predator odors

Fox feces contain trimethylthiazoline (Fig. 3.1 p. 37), which by itself induces in rats even stronger "fear" responses than whole feces (Vernet-Maury, 1980). Of 12 compounds from fox feces tested, seven elicited "fear" responses, confirmed by postmortem corticosterone measurements. Some structure–activity relationships emerged. The thiol group is essential, while the keto group is not; molecular size can vary; mercaptoketones can be aliphatic or alicyclic; the thiol can be secondary or tertiary; and the thiol can be alpha or beta to the keto group (Vernet-Maury *et al.*, 1984).

Dickman and Doncaster (1984) suggested that feces and urine of different carnivore species may share chemicals that elicit avoidance responses in small prey mammals. This might explain why prey species avoid chemicals from allopatric or extirpated sympatric predators. Nolte *et al.* (1994b) have suggested that herbivores are repelled by odors from the excreta of various predators because these have in common sulfur compounds resulting from protein digestion. Sulfur compounds from anal gland secretions of several mustelids have been shown to be deterrents for herbivorous prey mammals such as gophers, voles, snowshoe hares, and black-tailed deer (Sullivan and Crump, 1984, 1986a; Sullivan *et al.*, 1988). These effects and the active compounds are further discussed in Chapter 13.

The European rodents *A. sylvaticus* and *Clethrionomys glareolus* tend to avoid traps scented with 2,2-dimethylthietane from the anal sacs of the mink, *Mustela vison,* while it suppresses feeding in European wild rabbits, *Oryctolagus cuniculus* (Robinson, 1990).

Effects of predator diet

When given a choice, mountain beaver, guinea pigs, house mice, and deer mice consumed less food from containers near urine from coyotes on a meat diet than from containers near urine from coyotes on a fruit diet (Nolte *et al.*, 1994b).

Multisensory control of responses

Visual *and* olfactory stimuli contribute to anti-snake responses in mammals: California ground squirrels, *Spermophilus beecheyi,* flag their tail and kick sand at a rattlesnake, *C. viridis,* more often than at a gopher snake, *P. melanoleucus*. The squirrels kicked sand at and approached a snake in a perforated transparent bag more frequently than one in an intact bag. Visual and chemical cues are important, but the latter seem to be the primary releasers (Henessy and Owings, 1979).

Tenrecs from Madagascar respond to urine, feces, or scent gland material from viverrids by showing distinct behavior patterns. The long-tailed tenrec, *Microgale dobsoni,* displays an open-mouth threat to scents of the Madagascar ring-tailed mongoose, *Galidia elegans*. The streaked tenrec, *Hemicentetes semispinosus,* erected its quills and bucked while running toward a cloth impregnated with odor from *G. elegans* or the fanaloka, *Fossa fossa*. The tenrec, *Tenrec ecaudatus,* hissed and erected its hair when exposed to viverrid odors. These tenrecs respond similarly to dogs or humans (Eisenberg and Gould, 1970).

Responses to predator odors raise several questions:

Is the odor detected?
Is it recognized as predator odor?
Is it important information that is used in adaptive behavior?
Is it important energetically (i.e. does it save the animal locomotory energy)?
Is it useful for application as a repellent?
If so, is it an area repellent, feeding inhibitor, or activity depressor?
Can land and wildlife managers use it in assessing predator pressure?

The last three questions will be addressed in Ch. 13.

12.2 Host odors used by parasites

12.2.1 Vertebrates as parasites

Adult sea lampreys, *Petromyzon marinus*, parasitize other fish species. They find brook trout, *Salvelinus fario fontinalis*, and brown trout, *Salvelinus fario trutta*, by odor and can detect 0.014 mg trout extract in water. Trout odor activates sea lampreys even during the rest phase of their 24 hour cycle. When rendered anosmic by nasal plugs, the response disappeared. Even sea lampreys reared in the laboratory in absence of any other potential host fish responded to trout odor, indicating the genetically programmed nature of the response. Of two active amines, isoleucine methyl ester proved to be the most active (Kleerekoper and Mogensen, 1963).

Turning to a *mammalian parasite*, the vampire bat, *Desmodus rotundus*, is very sensitive to butyric acid, a common compound given off by mammals. It detects 0.0039–0.00784 % (vol.) butyric acid. This is lower than the human threshold (Schmidt and Greenhall, 1971).

12.2.2 Vertebrates as hosts

Salmon lice

Salmon lice, *Lepeophtheirus salmonis* (Copepoda) feed on mucus, epithelium, and blood of salmon. The resulting wounds lead to anemia, osmoregulation problems, infections, and disfigured heads. Salmon lice also carry pathogenic viruses and bacteria. Such damage harms the salmon farming industry. Chemical baiting of traps could be a way to control salmon lice. To this end, these copepods were exposed to salmon-conditioned water and extracts from salmon skin, mucus, and flesh, and a number of control substances. Water conditioned by Atlantic salmon, *S. salar*, and turbot, *Scophthalmus maximus*, excited the salmon lice, but they swam in a directed way only in response to salmon stimuli (Hull, 1997; Devine *et al.*, 2000). 6-Methyl-5-hepten-2-one attracts the larvae of salmon lice to their host (Genna *et al.*, 2004). Furthermore, male salmon lice find females on the salmon host by small lipophilic organic molecules (Mordue-Luntz *et al.*, 2004).

Cattle and the tsetse fly

Many arthropod parasites use butyric acid and/or carbon dioxide *emanating* from mammals as kairomones. The well-studied tsetse fly, *Glossina* spp., may serve as an example. Tsetse flies transmit *Trypanosoma* spp., the flagellates that cause sleeping sickness in people, and Nagana sickness in cattle. The flies are attracted to oxen even when these are invisible in underground pits, indicating the chemical nature of the decisive cue (Vale, 1974). The larger the ox, the more flies it attracts (Hargrove and Vale, 1978). Ox odor contains 1-octen-3-ol, acetone, and carbon dioxide. A mixture of these three compounds in their natural ratios is almost as attractive as the total odor from an ox (Vale and Hall, 1985). The effects of carbon dioxide and acetone are additive (Vale, 1980). The three compounds together attract tsetse flies from a long distance, while carbon dioxide alone attracts them into a trap at short range (Vale and Hall, 1985). In a wind tunnel, 1-octen-3-ol triggers upwind flight (Bursell, 1984; Hall *et al.*, 1984). Traps already baited with acetone and 1-octen-3-ol catch more flies if cattle urine is added (Vale *et al.*, 1986). Ox urine also attracts tsetse flies by itself and is used by Kenya tribesmen as bait (Owaga, 1984; Brightwell *et al.*, 1987). The phenolic fraction of cattle urine is particularly active. It contains phenol, 3-methyl phenol, 4-methyl phenol, 3-ethyl phenol, 4-ethyl phenol, 3-propyl phenol, 4-propyl phenol, and 2-methoxy phenol (Bursell *et al.*, 1988). A mixture of 4-methyl phenol and 3-methyl phenol is as attractive to tsetse flies as the total mixture of all eight phenols (Owaga *et al.*, 1988). The diet of livestock affects the active odor: starved oxen are less attractive to tsetse flies but become more so when a fattening diet resumes (Vale, 1981).

A recent study in Ethiopia found the most effective attractant for tsetse flies was acetone, octenol, and cow urine (Belete *et al.*, 2004). An American company markets Mosquito Magnet, a device that purportedly attracts mosquitoes with 1-octen-3-ol, heat, carbon dioxide, and water vapor (Enserink, 2002). Simpler traps are being used in East Africa. A black-and-blue cloth, impregnated with insecticide, attracts mosquitoes with acetone and octenol, or alternatively, buffalo urine (Enserink, 2002).

Urine of a wild host bovid, the African buffalo, *Syncerus caffer*, also attracts tsetse flies. Up to eight phenols are active (Hassanali *et al.*, 1986). The phenols are the same as those in cattle urine described above (Bursell *et al.*, 1988). The phenolics in urine of cattle and buffalo are formed by microbes over several days. Assuming that the hosts move on, tsetse cannot find their hosts by means of (fresh) urine. Urine aged in vessels without soil attracts more tsetse than if soil is present. It is thought that glandular secretions, "processed" by microbes on the body of the host mammals, attract tsetse flies (Madubunyi *et al.*, 1996).

Rodents and fleas

Fleas (Siphonaptera) typically ectoparasitize small to medium-sized burrowing mammals. They find their hosts by odor. The fleas employ behavior that is adapted to the ecology and social behavior of the particular rodent species. For example, in the Negev highlands of Israel, the gerbil *Gerbillus dasyurus* is almost exclusively parasitized by the flea *Xenopsylla dipodilli*, while the spiny mouse, *Acomys cahirinus*, harbors almost exclusively the flea species *Parapulex chephrenis*. In laboratory experiments, odors of whole animals attracted each flea species to its specific host. Moreover, *X. dipodilli* started to approach its gerbil host sooner, and all animals responded, while *P. chephrenis* took longer to start moving, and only 55% of the animals chose an *Acomys* host. This reflects differences in host ecology: the gerbil is solitary and as many as 50% of the animals do not have their own home range and burrow. Therefore, *X. dipodilli* must search for and approach its host more actively. By contrast, the spiny mouse largely nests communally and most animals reside in their burrow for a long time. Accordingly, *P. chephrenis*, its parasite, follows a sit-and-wait strategy (Krasnov *et al.*, 2002).

Humans and mosquitoes

Heat, moisture, and carbon dioxide attract mosquitoes to warm-blooded animals. Mosquitoes that single out humans, such as *Anopheles gambiae* and *Aedes aegypti*, need more specific compounds for host specificity. Incidentally, *A. gambiae* is also attracted by Limburger cheese. Fatty acids are common to cheese and human foot odors, and bacteria of the genus *Brevibacterium* produce fatty acids in both cheese and between human toes. Mosquitoes vary in their preferences: ammonia attracts *A. gambiae*, but *A. aegypti* relies more on lactic acid (Enserink, 2002).

Humans and blood flukes (Trematoda)

Cercariae of the blood fluke (or schistosome) *Schistosoma mansoni* are attracted to humans first by temperature and light. The first signal that facilitates brief attachment to human skin is arginine. Arginine is present in trace amounts on human skin. At the same time, the cercaria's tail is rich in arginine. The cercaria discards its tail and leaves it on the skin surface. So this amino acid may originate from human skin or other cercariae, being kairomone and/or pheromone. The second and third signals are two different mixtures of skin lipids. The third signal is required for penetrating the skin (perceived as "swimmer's itch"), shedding the tail, and transformation from a free-living freshwater form to an endoparasite. Cercariae are even attracted to a dab of human skin lipids applied to the wall of a glass dish. They try to penetrate the glass and actually shed their tails. They even perform this "suicidal" behavior to the lipid cue dispersed in the water (e.g. Feiler and Haas, 1988). This has led Haas to propose that cercaria-infested waters should be sprayed with lipids. The cercariae would then self-destroy.

Humans and leeches

The leech *Hirudo medicinalis* responds to human blood or plasma by a sequence of probing, attachment to skin, biting, and ingestion of blood. Spring water at 38°C releases probing and transient attachment, but not ingestion. Sodium chloride plus arginine is an effective feeding stimulus, sodium chloride alone triggers probing only, and no feeding takes place. Likewise, arginine alone does not lead to feeding. The ions in the salt can be substituted. Lithium ions are as affective as sodium ions but only bromide was as effective as chloride ions. The amino acid is highly specific: D-arginine does not lead to ingestion even in concentrations up to 1000 times the threshold for L-arginine. Other common L-amino acids such as histidine or lysine are not active. Finally, of the arginine analogues, only homoarginine and canavanine were effective feeding stimuli. These two compounds have all three functional groups of arginine intact (Elliot, 1986).

12.3 Eavesdropping

12.3.1 Pheromones as kairomones

Vertebrates using vertebrate signals

In laboratory tests, least weasels, *Mustela nivalis,* were more attracted to estrous than diestrous urine of prairie deer mice, *Peromyscus maniculatus bairdii.* This may increase the weasel's preying chances through the presence of pups in the case of postpartum estrus or estrous females may indicate a higher population density (Cushing, 1984).

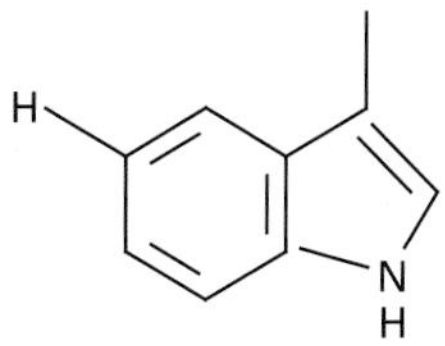

FIGURE 12.4 Skatole, a prey odor used by snakes.

Canids such as wolves or dogs who track their prey most likely eavesdrop on intraspecific signals, such as deposits from interdigital glands in cervids and bovids, or urine marks of rodents.

Finally, the responses of grizzly bear, *Ursus arctos* and polar bear, *Ursus maritimus*, to human menstrual odors during attacks on campers (Cushing, 1983) appears to be a form of eavesdropping, even though the intraspecific function of the odor in humans is not yet understood.

Vertebrates using invertebrate signals

Blind snakes find their prey by using the prey's pheromones. The wormlike Texas blind snake, *Leptotyphlops dulcis*, of the southwestern United States leads a subterranean life and feeds on termites and ant brood. It finds ants by following their pheromone trails (Gehlbach *et al.*, 1968). Other blind snakes such as the American blindsnake, *Typhlops pusillus* (Gehlbach *et al.*, 1971), and the Australian blindsnake, *Ramphotyphlops nigrescens* (Webb and Shine, 1992), also follow odor trails of ants, their prey. Texas blind snakes are attracted to the simple alkaloid *skatole* (methyl indole; Fig. 12.4), an amine with an unpleasant odor from the ant *Neivamyrmex* sp. (Watkins *et al.*, 1969).
The wax pheromones of honeybees that honey guides use as feeding stimuli have been discussed above (p. 352).

Invertebrates using vertebrate signals

The klipspringer, *Oreotragus oreotragus*, is an antelope of the rocky hills (*kopjes*) of eastern and southern Africa, where it is the ecological equivalent of the ibex or chamois. This animal marks its home range by applying secretion from the preorbital glands to blades of grass. Rain washes scent material down the stem, creating a gradient along an odor track. The tick *Ixodes mentzii* (or *I. matopi*) crawls up the stem, following this odor gradient, and then waits at the tip of a grass blade. Now the parasite is ready to transfer to another klipspringer that is attracted to the scent mark and touches the stem while marking in turn (Spickett *et al.*, 1981).

12.3.2 Hormones as kairomones

Host hormones may act as kairomones and attract or otherwise benefit parasites. The female rabbit flea, *Spilopsyllus cuniculi*, has to eat corticoid-laden blood from a pregnant rabbit before its own eggs can mature (Rothschild, 1965).

12.4 The self-anointed: chemical mimicking

Various reptiles, birds, and mammals treat their *own bodies* with odoriferous or toxic secretions and excretions from *other species* (Table 12.8). The benefits of these acquired odors are not well understood. In small prey animals, such as rodents or insectivores, an antipredator function seems plausible. Predators may self-anoint to cover up their own odor to improve their ability to approach prey. A second function for carnivores may be to transfer information about food to members of their own group. For this to be likely, the behavior should be more prevalent in social species. Self-anointing is also thought to deter ectoparasites and/or microbial pathogens. Passive anointing is mere exposure to chemicals of other species, while in active anointing an animal actively applies scents to its own body. Using such acquired odors represents an "extended phenotype" (Weldon 2004). No direct interaction between the odor donor and the target animal, such as a predator or parasite, is necessary. Indeed, as Weldon (2004) pointed out, the donor and the target may be allopatric, not even occurring in the same area.

Young Komodo monitors, *Varanus komodoensis*, roll themselves in the feces of prey animals. This may protect them from predation by larger conspecifics, as these avoid gut contents of their prey by shaking them out before eating the prey (Auffenberg, 1981; Ciofi, 1999).

Among the canids, dogs, wolves, coyotes, and foxes roll in dung and carcasses and other materials such as scent glands of rattlesnakes (Weldon and Fagre, 1989).

The chipmunk, *Eutamias sibiricus asiaticus*, approaches a dead snake, gnaws its skin, chews the gnawed bits, and finally applies them to its own fur. Snake urine, feces, the rectum end section, cloaca, and cloacal sacs release the same behavior. Snakes eat dead mice treated externally with urine from the Japanese rat snake, *Elaphe climacophora*, less readily than untreated controls. This suggests that chipmunks conceal their own odor by self-anointing, reducing predation (Kobayashi and Watanabe, 1986). Chipmunks may rub off the acquired odor on their burrow

Table 12.8 Self-anointing in mammals

Species	Material	From species	Context/function	Reference
Hedgehogs *Erinaceus europaeus*, *Atelerix pruneri*, *Hemiechinus auritus*	Venom	Toad	Self-defense	Brodie, 1977
Siberian chipmunk *Eutamias sibiricus asiaticus*	Carcass	Snake	Self-defense	Kobayashi and Watanabe, 1986
Rice rat *Rattus rattoides*	Anal gland secretion	Weasel *Mustela sibirica*	Self-defense	Xu *et al.*, 1995
Dog	Dung, carcass	Various: cattle	Unclear	Common observation
Capuchin monkey *cebus* sp.	Benzoquinones	Millipede	Mosquito repellent	Valderrama *et al.*, 2000; Weldon *et al.*, 2003

entrance. Thus the material passes from snake to chipmunk to burrow to predators – snake or polecat – who may come along.

The rice rat, *Rattus rattoides*, from southern China also self-anoints, at least in the laboratory. Presented with filter paper carrying anal gland secretion from the weasel *Mustela sibirica*, rice rats chewed the paper and rubbed the pieces against both of their flanks. Even laboratory-born naive rice rats anointed themselves in this fashion. Weasel or fox urine did not release anointing behavior (Xu *et al.*, 1995).

Among insectivores, at least five genera of hedgehogs anoint themselves with toad toxins. Although the exact adaptive function of this behavior is not known, it has been observed that the anointed spines hurt humans more than clean ones (Brodie, 1977).

Capuchin monkeys (*Cebus* spp.) in Venezuela self-anoint with a benzoquinone-secreting millipede. Two of the compounds by themselves also release self-anointing. These are 2-methyl-1,4-benzoquinone and 2-methoxy-3-methyl-1,4-benzoquinone. These compounds repel mosquitoes (Valderrama *et al.*, 2000; Weldon *et al.*, 2003).

12.5 Evolutionary considerations

12.5.1 Ghosts from predation past: history of association of prey and predator

Responses of prey to predator cues depend on predator–prey association in evolutionary time. Nevertheless, some rodents avoid the odor of a predator

Table 12.9 Ghosts of predation past: atavistic responses to chemicals of former predators

Prey species	Predator species	Location	Reference
Orkney vole *Microtus arvalis orcadensis*	Red fox *Vulpes vulpes*	Orkney Islands	Calder and Gorman, 1991
	Stoat *Mustela erminea*	Orkney Islands	Gorman, 1984
Lizard *Podarcis hispanica atrata*	Viper *Vipera latastei*	Spain: Isla Columbreta Grande	van Damme and Castilla, 1996

even if the two species have not been sympatric for many generations. For instance, the lizard *Podarcis hispanica atrata* from the Spanish island Columbreta Grande has not experienced its predator, the viper *Vipera latastei* since the snake's eradication in the 1880s. The lizard responds still to odors of the viper as much as another subspecies, *Podarcis hispanica*, from the mainland where vipers occur today (van Damme and Castilla, 1996).

Deer mice, *P. maniculatus*, on an island (Moresby Island, Gulf Islands, British Columbia) that now lacks their usual predator, the short-tailed weasel, *M. erminea*, still respond to its odor. However, they only show the more delayed and prolonged stress-type, opioid-sensitive behavior. By contrast, this island population has lost its fear and flight responses, which are benzodiazepine sensitive and more immediate. Mainland deer mice that are sympatric with weasels show both types of response (Kavaliers, 1990).

Similarly, on an island in the Orkneys, the vole *Microtus arvalis orcadensis*, introduced there by neolithic settlers between 3700 and 3400 BC, still avoids the anal gland secretion of the stoat *M. erminea*, even though the island has lacked predators for 5000 years (Gorman, 1984). Likewise, the Orkney vole avoids fecal odors of red fox, *Vulpes vulpes*, in live traps in the field and the laboratoy. Moreover, the vole almost never eats bark from Scots pine (*Pinus sylvestris*) saplings treated with fox feces extract (Calder and Gorman, 1991). Avoidance of red fox odor by the Orkney vole is an example of a "ghost from predation past" (Peckarsky and Penton, 1988) (Table 12.9).

12.6 Plant chemicals used by vertebrates

12.6.1 Aromatic plants used by birds

Even bird species with smaller olfactory bulbs respond to odors more than previously assumed. Starlings, *Sturnus vulgaris*, for instance, olfactorily discriminate species of aromatic plants they place in their nests. Three lines of evidence support this First, six plant species evoked strong multiunit electrophysiological responses in the olfactory nerves of adult birds. Second, if plant odors were paired with gastrointestinal malaise caused by the bird repellent methiocarb, a conditioned avoidance was produced. Third, birds with sectioned olfactory nerves did not develop the avoidance (Clark and Mason, 1987). Experiments showed that aromatic herbs in the nest do not reduce the ectoparasite load in starlings but may stimulate those parts of the immune system that mitigate effects of ectoparasites (Gwinner *et al.*, 2002).

Corsica blue tits, *Parus caeruleus ogliastrae*, select 6–10 species of aromatic plants out of about 200 species that occur in their habitat. They add them to their nests during the nestling stage. An ingenious experiment demonstrated that the blue tits maintain an aromatic environment around their nest. Containers filled with invisible aromatic herbs and attached to the nest box reduced the collecting of aromatic plants. Gas chromatographic analysis of volatiles from one plant species (*Achillea ligustica*) showed a rapid loss within 24 hours. Accordingly, the birds started to replenish their aromatic plants about 1 day after the experimenters had placed fresh aromatic plants near the nest. The birds selected aromatic plants that contain proven antibacterial, antiviral, fungicidal, insecticidal, and insect-repellent compounds. These include linalool, camphor, limonene, eucalyptol, myrcene, terpin-4-ol, pulegone, and piperotenone. Moreover, 40% of the plants involved possess activity against the human immunodeficiency virus (Petit *et al.*, 2002).

12.6.2 Regulators of mammal reproduction

Rodents in unpredictable environments such as deserts or alpine and polar habitats cannot easily rely on photoperiod for starting their annual breeding. Instead, many species rely on fresh vegetation to start reproducing. This, in turn, depends on rainfall in deserts and the spring thaw in the mountains and tundra. Reproduction in the montane vole, *Microtus montanus*, of western North America is triggered by (6-MBOA; 6-methoxybenzoxazolinone Fig. 12.5), a cyclic carbamate found in seedlings and growing tissues of grasses and some dicotyledonous plants (Berger *et al.*, 1981; Sanders *et al.*, 1981). The two separate effects of 6-MBOA on reproduction are cueing to time the onset of reproduction and affecting the quality of the reproductive effort. The latter is basically uterotropic in that it stimulates an increase in uterine mass. When given 6-MBOA, montane voles have larger litter sizes and more litters. If both sexes received 6-MBOA, 25% more females than males are produced (Berger *et al.*, 1987). Experimental administration of 6-MBOA to non-breeding winter populations stimulates breeding in montane voles (Berger *et al.*, 1981) and Townsend's

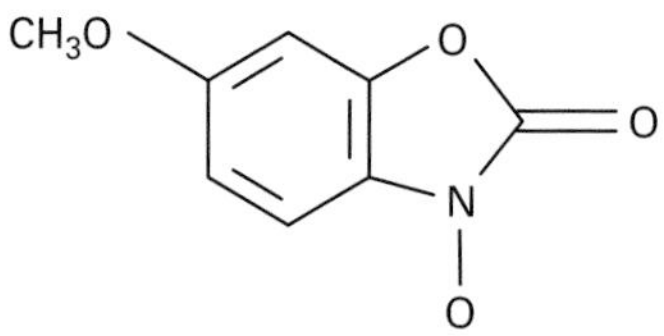

FIGURE 12.5 6-Methoxybenzoxazolinone. This compound occurs in young plants and stimulates reproduction in mammals.

vole, *Microtus townsendii* (Korn and Taitt, 1987). In the woodland vole, *Microtus pinetorum*, 6-MBOA increases uterine and ovarian mass and gonadotropin activity (Schadler *et al.*, 1988).

Ovarian weight is also increased in the laboratory mouse by 6-MBOA (Sanders *et al.*, 1981), and the laboratory rat shows similar responses to those of the woodland vole (Butterstein *et al.*, 1985). Even in predictable environments, such as salt marshes, 6-MBOA may regulate reproduction. The breeding cycles of North American microtine voles, which feed almost exclusively on salt grass, *Distichlis stricta*, parallel the growing season of that grass. It has high levels of 6-MBOA when growing, and little when senescing (Negus *et al.*, 1986).

Ord's kangaroo rat, *Dipodomys ordii* (Heteromyidae), also responds to 6-MBOA. This usually granivorous species of western North America can also consume up to 40–60% of green plants. It becomes reproductively active after rainfall, which nourishes desert annuals. Peritoneal injections of 6-MBOA increased uterus mass in this kangaroo rat. In the field, providing 6-MBOA-treated rolled barley near burrows resulted in more pregnant females. Mere excess food (untreated barley) did not affect reproduction. Litter size was not affected. This compound is but one factor of a complex of stimuli that regulate reproduction in this kangaroo rat (Rowsemitt and O'Connor, 1989). Uterus and ovary weights of Merriam's kangaroo rat, *Dipodomys merriami*, did not change after peritoneal injection of 6-MBOA. This desert animal of western North America may not rely on 6-MBOA as a reproduction trigger; the compound may not even be present in the desert vegetation (McClenaghan, 1987). Table 12.10 summarizes the effects of 6-MBOA in mammals.

Steroids or steroid-mimicking plant compounds affect mammalian reproduction. After giving birth, female muriqui monkeys, *Brachyteles arachnoides*, eat leaves that contain isoflavonoids, which structurally resemble estrogen (see Ch. 10). This might reduce fertility. When the females are ready to reproduce, they consume more "monkey ear," a legume containing a steroid that might boost fertility (Strier and Ziegler, 1994).

Table 12.10 Reproductive effects of 6-methoxybenzoxazolinone in rodents

Mammal species	Diet	Effect	Reference
Montane vole *Microtus montanus*	Grasses, some dicotylons	Timing reproduction; inceased uterine mass: larger and more litters	Berger *et al.*, 1981; Sanders *et al.*, 1981
Townsend's vole, *Microtus townsendi* and montane vole	6-MBOA on rations	Winter breeding	Berger *et al.*, 1981; Korn and Taitt, 1987
Woodland vole *Microtus pinetorum*	6-MBOA	Increased uterine and ovarian weight; more gonadotropin	Schadler *et al.*, 1988
Laboratory mouse	6-MBOA	Increased ovarian weight	Nelson and Shiber, 1990
Laboratory rat	6-MBOA	Increased uterine and ovarian mass; more gonadotropin	Butterstein *et al.*, 1985
Ord's kangaroo rat *Dipodomys ordii*	Green plants	Reproduces after rainfall	Rowsemitt and O'Connor, 1989
	Injected 6-MBOA	Increased uterus mass	
	6-MBOA on rolled barley in field	More pregnant females	

6-MBOA, 6- methoxybenzoxazolinone.

12.6.3 Self-medication

Chimpanzees, *Pan troglodytes*, take up leaves of certain plants and massage them in their mouth between the tongue and the buccal surface. The leaves are swallowed whole without chewing. Whole leaves have been found in chimpanzee feces. The plants include several species of *Aspilia* (Compositae: sunflower family), *Lippia plicata*, *Commelina* spp., and *Ficus exasperata*. They are eaten early in the morning, while most of the regular feeding occurs in the afternoon. The selected plants are widely used in medicine (reviewed by Newton and Nishida, 1990). *Aspilia mossambicensis* contains thiarubrine A, an antifungal, antibacterial, and antihelminthic compound with the appearance of a red oil (Wrangham and Nishida, 1983; Rodriguez *et al.*, 1985). The two diterpenes kaurenoic acid and grandiflorenic acid from the same plant species are strongly uterotonic, and also antibacterial and antihepatotoxic (Page *et al.*, 1992). Chimpanzees carry a heavy load of nematodes, and this plant may control these. An anorexic adult female chimpanzee with a presumed gastrointestinal disorder sucked out the bitter juice from the piths of shoots of the composite *Vernonia amygdalina*, a plant usually not eaten by chimpanzees. This is used in tropical medicine to treat parasitoses and gastrointestinal disorders in humans and domestic animals (Huffman and Seifu, 1989). Chimps in the Mahale Mountains National Park in Tanzania use this plant. Four known sesquiterpene lactones have been isolated, plus seven new steroid glucosides, and two aglycones of these glucosides. *In vitro*, the sesquiterpene lactones are antischistosomal, plasmodicidal, and leishmanicidal. The major steroid glucoside, vernonioside B_1, was also antischistosomal. The glycosides inhibit schistosome movement and egg laying most when the sugar moiety is eliminated. Vernodalin was the most potent antiparasitic compound *in vitro*, but proved toxic *in vivo* to cercariae-infected mice. The pith, chewed by chimps, contains steroid-related compounds, but vernodalin was found only in bark and leaves. Vernodalin and other sesquiterpene lactones may be too toxic for chimps. During the rainy season (mid-October to December) parasite diseases are more prevalent, and levels of vernonioside B increase in stems and leaves yet chimps avoid the leaves (Ohigashi *et al.*, 1994). Chimps possibly practice preventive medicine: they eat more medicinal plants during the rainy season when pneumonia threatens. The chimps may use the forest as a medicine cabinet. After these observations, a new subdiscipline, *zoopharmacognosy* was born. It held its first symposium in 1992.

Capuchin monkeys rub millipedes on their fur and skin to repel insects. Benzoquinones are the active principle. These compounds deter mosquitoes from feeding on human blood in the laboratory. Even captive capuchin monkeys seek benzoquinone-soaked tissues when these are offered by experimenters (Weldon *et al.*, 2003; see also p. 268 and p. 377).

12.6.4 Food preservatives

Pikas, *Ochotona princeps*, of the North American Rocky Mountains store winter food in "hay piles." An individual collects about 28 kg of material. One plant species, *Geum (Acomastylis) rossi*, constitutes up to 75% in some piles. It is almost never eaten in summer and contains considerable levels of phenolics. *Geum* extracts are bacteristatic. Experimental hay piles of this plant retained their biomass better than controls. The pika perhaps uses the phenolics to preserve its food for the mountain winter, lasting several months. Over time, the food loses most of the phenolics and becomes palatable (Dearing, 1997).

12.7 Animal chemicals benefiting plants

12.7.1 Saliva

Saliva of herbivorous mammals and insects (specifically grasshoppers) can stimulate growth of clipped vegetation. Grazers influence pastures not only mechanically but also biochemically via thiamine in saliva. Thiamine in saliva stimulates growth in higher plants when temperatures are low and ascorbic acid is present (Vittoria and Rendina, 1960). Jameson (1964) proposed that thiamine should be applied to partially defoliated plants. In later greenhouse experiments, clipped plants of the grass sideoats grama, *Bouteloua curtipendula*, were treated with bovine saliva, collected from the mouth. This treatment increased forage yields, root yields, and basal area. Thiamine applied to the soil increased the forage yield by as much as 232%, and root yields by 398%. In the field, the increases were less pronounced. These results suggested that the native rangelands may require grazers for vigor and production (Reardon *et al.*, 1972). Sideoats grama grazed by cattle, sheep, or goats regrew significantly more (72.5 cm) than clipped plants (55 cm). Thiamine added to the soil increased forage yield by as much as 42%, but bovine saliva did not affect growth (Reardon *et al.*, 1974).

Saliva and especially urine contain epidermal growth factor, a small protein, which may be very important to stimulate plant growth (Dyer, 1980). Epidermal growth factor from mouse urine stimulates growth in sorghum seedlings and the effect is proportional to the amount used.

Other studies used saliva after it had reached the rumen. This saliva had no effect on plant growth (Johnson and Bailey, 1972; Detling *et al.*, 1980). The effect of thiamine depends on the extent of "grazing": thiamine stimulated growth in clipped grasses as long as they retained 25% of their height (McNaughton, 1985).

Whole saliva from the North American bison, *Bison bison*, stimulated elongation of *Avena* coleoptiles at high pH. Sodium bicarbonate from saliva enhanced growth in a dose-dependent fashion and, therefore, inorganic ions were thought to be the main growth-promoting factor in bison saliva rather than organic compounds (Detling *et al.*, 1981).

12.8 Synomones

Chemicals mediating relationships where both partners benefit are termed synomones.

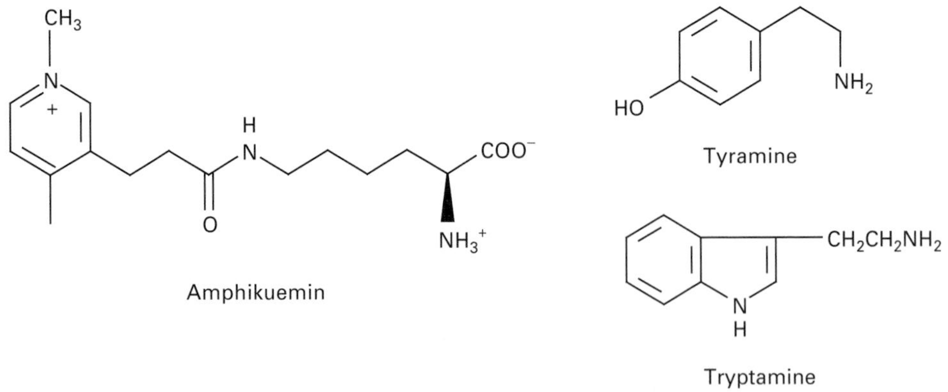

Amphikuemin

Tyramine

Tryptamine

FIGURE 12.6 Synomones attracting anemone fish to its sea anemone partner. Amphikuemin attracts the anemone fish *Amphiprion perideraion* to its sea anemone partner *Radianthus kuekenthali*. Tyramine and tryptamine attract the anemone fish *Amphiprion ocellaris* to its anemone partner *Stoichactis kenti*.

12.8.1 Fish

Anemone fish

The relationship between anemone fish, *Amphiprion* sp., and their sea anemone partners are regulated by alkaloids from the sea anemone. *Amphiprion perideraion* is attracted to the sea anemone *Radianthus kuekenthali* by the simple lysine-derived alkaloid amphikuemin (Fig. 12.6; Murata *et al.*, 1986). Other simple alkaloids, aplysinopsin and dihydroaplysinopsin, also attract *A. perideraion,* regulate its swimming rate and induce their species-specific partnership. A related anemone fish, *Amphiprion ocellaris,* is attracted to the sea anemone *Stoichactis kenti* by tyramine (Fig. 12.6) and tryptamine induces searching by the fish (Murata *et al.*, 1986).

12.8.2 Reptiles, birds and mammals

Fruit flavor, consumption and seed dispersal by birds

Many fruits attract birds by colors, odors, and taste. After consuming the fruit, the birds will disperse the seeds to new habitat. Preference orders exist, and some fruits are rarely eaten by birds (see p. 306). For instance, the flavor of buckthorn, *Rhamnus cathartica,* ranked lowest of 11 fruit extracts tested in blackbirds, *Turdus merula,* and a song thrush, *Turdus philomelus* (Sorensen, 1983). Least preferred fruits contain toxins that deter birds, perhaps to avoid seeds being dispersed to unfavorable habitat. For instance, it is disadvantageous for forest plants to be dispersed by open-country birds to unsuited open habitat (Janzen, 1978; Sorensen, 1983).

Passage of seeds through animals

Fruit- and seed-eating birds and mammals not only disperse seeds but can also mechanically and chemically weaken or break the seed coat. The chemical alteration of seed coats by animal digestive tracts can affect germination rates and patterns, often benefiting the plant in the process. Since this chemical plant–animal interaction takes place in the digestive system, the chemical senses are not directly involved, although the first step, ingestion, is largely guided by olfaction and taste.

Earlier it was thought that plants and animals might have coevolved, the plant depending on seed conditioning by seed-eating birds or mammals. A case in point was the idea that the Tambalacoque tree, *Calvaria major*, on Mauritius, which might have depended on the now extinct dodo for germination. The last dodo died in 1681 and no trees seemed to have germinated from seed during the intervening 300 years (Temple, 1977). To test whether a large bird can enhance germination, Temple force-fed turkeys with 17 of the 5 cm diameter seeds of *Calvaria*. Of the 10 seeds that survived the digestive tract, three germinated. Temple claims that "these may well have been the first *Calvaria* seeds to germinate in more than 300 years." However, more trees have been found since, and the dodo's elevational range was probably not identical with that of *Calvaria* (Owadally, 1979).

Perhaps not by coincidence, another island provides a striking example. On Galapagos, less than 1% of the seeds of the tomato *Lycopersicon esculentum* var. *minor* germinate by themselves. By contrast, the germination rate of tomato seeds fed to two Galapagos tortoises from Indefatigable Island rose to 60–81%, after a 14 to 18 day passage through their digestive tract (Rick and Bowman, 1961).

Even Fish affect the germination rate of seeds that they eat; for example, only 0.5% of untreated seeds of beaked tasselweed, *Ruppia maritima*, and Spiny naiad, *Najas marina*, germinate. Passage through grass carp, *Ctenopharyngodon idella*, increased germination to 5.7% and 10.5%, respectively, and passage through tilapia, *Oreochromis* sp., to 12% and 16% respectively (Agami and Waisel, 1988).

Passage through birds or mammals conditions seeds in several ways. Passage time varies from 30 minutes in some songbirds to up to 60 days in horses (Table 12.11). Seeds may not necessarily be affected, especially during short passage times, and at the other extreme, they may be completely destroyed, as by the

Table 12.11 Time of seed passage through vertebrate digestive tracts

Class	Species	Passage time	Reference
Fish	Tilapia; grass carp	30–60 hours	Agami and Waisel, 1988
Reptiles	Galapagos tortoise	11–12, up to 20 days	Rick and Bowman, 1961
Birds	Turkey	Up to 6 days	Temple, 1977
	Hermit thrush	30 minutes	Krefting and Roe, 1949
	Phainopepla	12–45 minutes	Walsberg, 1975
Mammals	Cattle	2–3, up to 10 days	Peinetti *et al.*, 1993
	Horse	14 (75%), up to 60 days	Janzen, 1981

gizzards of birds, notably gallinaceous birds. Between these extremes lies a continuum. For some plants, only a few seeds remain viable after passing through digestive tracts. For instance, only 7 of 40025 ingested seeds from various plants germinated after passage through the California linnet (Roessler, 1936). About 5 to 10% of the ingested seeds remained viable in several studies.

Birds

Seeds of Tatarian honeysuckle, *Lonicera tatarica*, germinate in 33 days after passing through robins, *Turdus migratorius*, while untreated seeds take 43 days and those passed through the catbird, *Dumetella carolinensis*, take 47 days (Krefting and Roe, 1949). Table 12.12 lists how seven eastern Mediterranean scrubland birds (Izhaki and Safiel, 1990) and pheasants (Swank, 1944) conditioned seeds of various Mediterranean plants. Germination probability rose, dormancy was shortened, and germination time pattern of the seeds became variable.

Mammals

In mammals, gazelles, cattle, and grizzly bear increased germination probability of some seeds (Table 12.13). Elephants enhance germination and seedling survival of the marula plum tree, *Sclerocarya caffra* (Anacardiaceae) (Lewis, 1987). Certain tropical seeds germinate only if passed through the digestive system of monkeys (Hladick and Hladik, 1969).

Many investigators agree that mammalian, and bird, digestive tracts condition seeds to different degrees, especially if several animal species prey on the seeds. In unpredictable environments, such as deserts with little rain, there will always be some seeds ready to germinate. To conditioning must be added the requirement for dispersal, common to all surviving seeds, regardless of any mechanical and chemical abrasion of the seeds. In mainland ecosystems, many plant and animal species interact, so that coevolution between plants and

Table 12.12 How avian frugivores affect seed germination

Bird species	Plant species	Probability of germination	Length of dormancy	Temporal pattern of germination	Process	Reference
Seven eastern Mediterranean scrubland frugivores[a]	Five Mediterranean Species	Increased in 4 of 5 spp.	Decreased in all 5 spp. to 5–90 days	Variable, bird-dependent in 4 spp.	Abrasion and chemical decomposition of seed coat	Izhaki and Safiel, 1990
Pheasant *Phasianus colchicus*	Poison sumac, nightshade, clover, black locust	70% (44–100%) (control: 20%)	Decreased: 10% germinated in 14–35 days (control: 38 days)	not tested	Abrasion, "digestive juices"	Swank, 1944
	Millet, alfalfa, flax	ND	ND	ND		

ND, No differences.

[a] Blackbird (*Turdus merula*), song thrush (*Turdus philomelos*), bulbul (*Pycnonotus barbatus*), blackcap (*Sylvia atricapilla*), orphean warbler (*Sylvia hortensis*), lesser whitethroat (*Sylvia curruca*) and redstart (*Phoenicurus phoenicurus*).

Table 12.13 How mammalian frugivores affect seed germination

Mammal species	Plant species	Germination probability	Length of dormancy	Process	Function	Reference
Bat *Rousettus aegyptiacus*	Fig *Ficus carica*	ND	ND	Inhibitors stay in feces, chemical and physical gut processes	↑ Asynchronous G spreads risk (adaptation to desert rain)	Izhaki *et al.*, 1995
	Mulberry *Morus nigra*	ND	↑G delay			
	Arbutus *Arbutus andrachna*	ND	↑G delay			
Dorcas gazelle	*Acacia* (3 spp.)	↑ to 13% (C: 1%); ↑ to 21% (C: 4%)	↓ Acceleration: 7% / 5days; 13% / 10 days	Bruchid beetle opens seed	Dispersal; adaptation to desert rain pattern	Or and Ward, 2003
Cattle	Caldén *Prosopis caldenia*	↑ during first month	↑ G delay	Seed scarification	↑ G under varying environmental and site conditions	Peinetti *et al.*, 1993
Grizzly bear	Cow parsnip *Heracleum lanatum*	↑ to 65% (C: 51%); frozen: ↑ to 85% (C: 69%)	NT	NT	Dispersal	Applegate *et al.*, 1979

ND, no difference. C, control. G, germination; ↑ Increase; ↓, Decrease; NT, not tested.

seedeaters is less likely than on islands. Relationships among several trophic levels can grow complex. For example, the numbers of agouti and paca on Barro Colorado Island in Panama are over 10 times greater than in a comparable area in Amazonian Peru. Barro Colorado Island is too small ($16\,km^2$) to support large predators such as pumas or jaguars. The lack of predators increased the abundance of some plants, while that of others decreased. Seed predators such as peccaris, pacas, and agoutis will particularly affect numbers of seedlings of large-seeded plants (Terborgh, 1988). In general, passage through vertebrate guts enhances seed germination more in temperate zone plant species than tropical ones, and more often in trees than in herbs and shrubs (Traveset, 1998).

Animals can even affect germination without passing seeds through the digestive system. Heteromyid rodents select certain types of seed, hull them, and collect them in their external, hairy pouches; in *Oryzopsis hymenoides*, a perennial grass of western North America, 31% of seeds germinated after hulling by two *Dipodomys* species whereas only 7% of intact seeds germinated (McAdoo *et al.*, 1983).

The pig and the truffle

The highly priced hypogeous Périgord truffle, *Tuber melanosporum*, occurs in Spain, France, Italy, and Greece. It is commercially grown there in oak and hazelnut woods. The fungus produces an odor that humans can detect as it emanates through cracks in the ground. Traditionally, dogs and female pigs, and sometimes goats, have been used to find the truffles. The truffle produces the steroid androstenol (5α-androst-16-en-3α-ol) (Claus *et al.*, 1981). The same compound occurs in the saliva of the domestic boar and stimulates mating behavior in the sow. Truffles have had a reputation as an aphrodisiac for many years (Harris, 1987). In short, the truffle attracts the female pig with an odor that mimics the sex pheromone of the boar. Since the animal uses the "prey" odor to its advantage, a kairomone is involved. Undoubtedly, the truffle also benefits, probably by spore dispersal. This would make the steroid a synomone. Insects also respond and serve as indicators: Some human truffle hunters find truffles by lying on the ground, and observing the black-and-yellow truffle flies that hover over the invisible truffles, enter the ground, and lay their eggs in mature truffles. Truffles also produce chemicals that kill plants above them, resulting in a patch of "scorched earth," *terre brûlée* in French. It, too, helps in detecting these prized fungi.

Bats, flowers and fruits

Bat-pollinated flowers have "musky" odors, resembling that of bats or fermentation. These flowers contain butyric acid, which smells to humans similar to bat gland odors (Kulzer, 1961). The flower aroma may possibly resemble a bat pheromone, and the bat may be sexually attracted to the flower and pollinate it in the process (Baker, 1963). Fruit bats (Macrochiroptera) use odors to locate and discriminate food. Egyptian fruit bats, *Rousettus aegyptiacus*, discriminated boxes containing 100 mg banana from empty boxes. They also distinguish natural from synthetic banana odor. Frugivorous bats were attracted to odors in a field experiment in Costa Rica's Corcovado National Park. Mist nets were baited with bananas wrapped in plastic, unwrapped, or smeared on leaves. Fourteen species of phyllostomid bats were caught. Smeared banana attracted more bats of the genus *Artibeus* and the family Stenodermatinae. In rain, bats did not discriminate. It was concluded that "olfactory ability may provide a mechanistic link between plant dispersal strategies and chiropteran foraging strategies" (Rieger and Jakob, 1988).

The short-tailed fruit bat, *Carollia perspicillata*, discriminates the odor of ripe from that of green bananas. Odor cocktails of banana odor components were distinguished from the whole odor of banana (Laska, 1990). The two most common esters of banana odor are isopentyl acetate and isobutyl acetate. Nearly all fruit odors contain ethyl acetate. Isopropyl acetate and 2-pentanone are characteristic of putrefying fruits.

Mammals and fungal spores

Small mammals often smell like the fungi they eat. They probably benefit the fungi by dispersing their spores. A strong coevolution between the fungi and these mammalian fungivores has been suggested (Maser *et al.*, 1978).

13

Practical applications of semiochemicals

Give me an ounce of civet,
good apothecary,
to sweeten my imagination

SHAKESPEARE: *King Lear*
Act IV, sc. 6, 1.132

Smelly secretions as repellents: "The devil can be completely undone if you manage to fart into his nostrils." (beating the devil with his own weapons).

MARTIN LUTHER (Erik Erikson: *Young Man Luther*, p. 61–62)

Chemical cues hold considerable promise for manipulating behavior in vertebrates, provided we understand an animal's natural history, biology, and behavior well. However, the development of chemical attractants, stimulants, inhibitors, and repellents for vertebrates has progressed rather slowly for several reasons. First, chemical stimulus and behavior are not connected as rigidly as in insects, for example. Second, the same stimulus may elicit different behaviors, depending on the state of the recipient and the context. Third, chemical cues often are rather complex mixtures of compounds. Fourth, learning, especially early experience plays a major role in vertebrate, notably mammalian behavior. Finally, many behaviors are modulated by several sensory modalities so that chemical stimuli alone trigger only incomplete responses at best.

13.1 Fish

The desire for improved techniques in fishing, aquaculture, and limnological management motivates much of the current basic work in the chemical ecology of fish.

13.1.1 Chemical imprinting

Hatcheries raise fish and release them in lakes and rivers on a large scale. Some species such as salmon imprint early in life on the odor of their home stream. Coho salmon imprint when just over a year old (smolt stage), and sockeye salmon even earlier, right after hatching. Coho salmon prepare to migrate from rivers to the ocean as smolt. Proper imprinting ensures their return as adults to the home stream in order to spawn. Morpholine-imprinted Coho salmon have been successfully attracted to the Mad River in California by adding morpholine to the water (Hassler and Kucas, 1988). Sockeye salmon leave their streams around lakes just a few months after hatching and develop further in a lake where they smolt before they leave for the sea.

The goal is to imprint fish to be released properly so that they do not scatter into other, wild populations and interfere with wild gene pools. Now it is possible to assess whether imprinting has taken place. Fish imprinted on 2-phenylethanol (phenethyl alcohol) have a higher percentage of neurons that respond to this compound in their noses, and these neurons are more sensitive to this compound. Thus, learning leaves its traces in the nose as well as the brain. The sensitive neurons make more cyclic GMP, an intracellular messenger. Changes such as these may be a suitable basis for assays to ascertain whether hatchery-raised fish have been properly imprinted before they are released into the wild (Barinaga, 1999).

13.1.2 Chemical lures, pheromones and feeding attractants and stimulants

Chemical lures, based on feeding stimuli, are now on the market as odor-impregnated artificial bait. They are specific for certain predatory fish and are extremely effective (Schisler and Bergesen, 1996). Efforts to understand improving of feeding attractants and stimulants for economically sound *mariculture* started early (Bardach and Villars, 1974). Another obvious application of fish odors is the use of waterborne male and female pheromones to improve spawning in fish in aquaculture.

13.1.3 Acidification of Water

Acid precipitation and the resulting higher acidity of freshwater bodies may result in severe disruption of feeding and migration, since both behaviors depend on water chemistry. Atlantic salmon (*Salmo salar*) alter responses to amino acids with changes in pH (Royce-Malmgren and Watson, 1987). In the laboratory, the fish are attracted to glycine, but avoid L-alanine. When the pH is lowered from 7.6 to 5.1, the salmon become indifferent to glycine but are attracted to L-alanine. Acidic water also disrupts fish responses to alarm pheromones. At pH 6.0 fish do not respond to the alarm pheromone hypoxanthine 3*N*-oxide. This occurs because the N-oxide group on the molecule is lost through an irreversible covalent change. At pH 8.0 this does not happen (Brown *et al.*, 2000), water pollution by copper affects responses to alarm pheromone in fish such as rainbow trout, chinook salmon, and pikeminnow, *Ptychocheilus lucius* (e.g. Beyers and Farmer, 2001; Scott *et al.*, 2003).

13.1.4 Mosquito control

Mosquito fish, *Gambusia affinis*, are being released into freshwater bodies to control mosquitoes. It appears that chemicals from the fish "warn" the mosquitoes and they avoid laying their eggs in ponds with these fish. The number of mosquito larvae reduced as more fish chemicals (2 liter/day from pools with mosquito fish) were added to outdoor pools in North Carolina. The problem is that if mosquitoes avoid stocked pools, the efforts to control these insects with fish may fail (Angelon and Petranka, 2002).

13.2 Reptiles

13.2.1 Habitat Imprinting

Efforts to apply semiochemicals to manipulate behavior in reptiles are still young. One example is habitat imprinting in sea turtles. Almost all of the endangered Kemp's ridley sea turtles, *Lepidochylus kempi* used to nest on a 15 km beach near Rancho Nuevo, Mexico. In the 1940s, approximately 40 000 females came ashore to lay their eggs. By 1978 there were only 924. Since 1978 a second nesting population 400 km north at Padre Island National Seashore in Texas has slowly been established. The task was to imprint hatchlings to the water of Padre Island so that they return to this new "home" to lay eggs when adult. From 1978 to 1988, 22 507 eggs were collected at Rancho Nuevo and placed in sand and water from Padre Island. They were then brought to Padre Island and hatched there. The juveniles were raised for 9–12 months and then released at the beach where they enter the surf. In multiple-choice experiments, these juveniles spent more time in seawater and sand samples from Padre Island than samples from other areas. For years, none of the released turtles returned to Padre Island and many were found dead at beaches and in fishing nets. Methodological problems, such as not tagging palm-sized hatchlings serving as controls, also occurred (Tauber, 1992). In 1996, this effort at *olfactory imprinting* and "headstarting" bore fruit: two tagged turtles, 10 and 13 years old, landed at the Padre Island beach. Both laid eggs (a total of 590), and 369 hatchlings were released into the ocean. In 1999, 16 nests were found in South Texas. In 2002, 23 of the 40 nests in the USA were found at Padre Island National Seashore. In 2003, 24 sea turtle nests were reported. Of these, 19 were Kemp's ridleys. The hatching success was 90–95%, and 1426 hatchlings were released. In 2004, 3928 eggs were laid in 42 nests and 2608 hatchings were released. That year saw the first *arribada* nesting of more than three females in 1 day. In the 2005 season, 56 clutches had been found by mid-July (www.nps.gov/pis/home). The turtles appear to have established themselves; the headstarting effort has paid off.

13.2.2 Snake control

The brown snake, *Boiga irregularis*, was accidentally introduced on Guam during or shortly after the Second World War. There are now 16–50 snakes per hectare (or as many as 30000 per square mile) on that island. They have caused extinction of native birds, reduced bat and lizard populations, are feeding on chickens and caged birds, and cause frequent power outages. Dispersal to other Pacific islands, notably Hawaii, by commercial traffic is a serious danger. Some snakes have already found their way to Hawaii. To trap brown snakes, the potential of both feeding stimuli from prey (Shivik, 1998) and blood and its components (Chiszar *et al.*, 1988b, 1992, 1993a) and pheromones (Mason and Greene, 2001) are being studied. In the meantime, strict inspections of airplanes from Guam try to prevent accidental spread of the snake.

13.3 Birds

13.3.1 Antifeedants: methyl anthranilate

Feeding repellents for pest birds are the most important application of chemical stimuli to manipulate bird behavior. Methyl anthranilate (Fig. 13.1) and dimethyl anthranilate, esters of benzoic acid, are found in concord grapes and are used as artificial flavorings. Starlings, *Sturnus vulgaris*, have an aversion to methyl anthranilate, which irritates the trigeminal nerve, and they feed less on food flavored with a variety of anthranilates. They avoid the more volatile anthranilates most. The odor is partly responsible for the effect; contact is not necessary. In one particular experiment, only volatile compounds were aversive (Mason and Clark, 1987). If only anthranilate-treated food is offered, the birds will accept more of the flavored food than they do if they offered a choice between

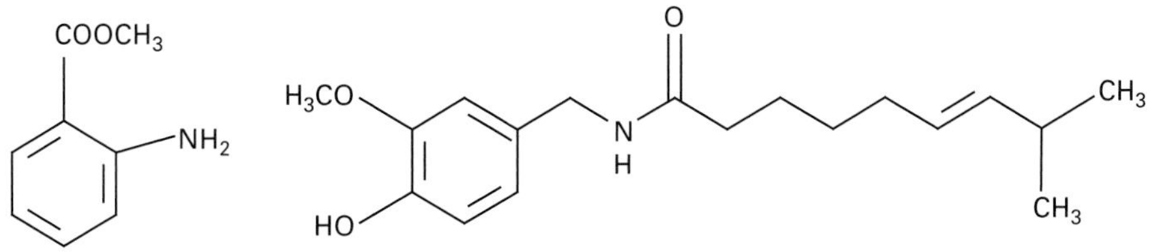

Methyl anthranilate Capsaicin

FIGURE 13.1 Methyl anthranilate deters birds from feeding and stimulates feeding in mammals. Capsaicin deters mammals from feeding but stimulates birds.

treated and untreated food (Mason *et al.*, 1989c). This underscores the need for alternatives for a repellent to work.

Commercially, methyl anthranilate (trade name ReJeX-iT AG-36) is used to repel red-winged blackbirds, *Agelaius phoeniceus*, from feeding on rice seed (Avery *et al.*, 1995). This species avoided methiocarb-poisoned apples; additional chemical (methyl anthranilate) and visual (calcium carbonate) cues enhanced the avoidance response. This way the amount of the toxin methiocarb could be reduced and yet still be equally effective and at lower cost (Mason, 1989). Canada geese (*Branta canadensis*) also feed less on grass treated with methyl anthranilate (Cummings *et al.*, 1995).

13.3.2 Antifeedants: compounds from unripe fruit

Unripe fruits of many plants contain compounds such as quinines that impart an aversive taste and discourage fruit eaters from consuming them before they are ripe and their seeds are mature and ready for dispersal. Unripe fruit of some *Rhamnus* species contain emodin, an anthraquinone that deters birds and mammals from feeding. Anthraquinone reduced red-winged blackbird feeding on rice by 84%, and 71% in boat-tailed grackles (Avery *et al.*, 1997).

13.3.3 Exploitation of automimicry (Batesian mimicry)

Brown-headed cowbirds, *Molothrus ater*, and red-winged blackbirds, *A. phoeniceus*, removed 95–98% less rice seeds sown in flight pens if the seeds had been treated with methiocarb. The effect was the same whether 100% or only 50% of the seeds had been treated. This shows that an entire prey population can be protected even if only a portion is unpalatable. Such automimicry may permit partial treatment of crops with repellents, which would be effective, environmentally sound, and economical (Avery, 1989). Batesian mimicry offers advantages over food aversion learning (Reidinger and Mason, 1983).

Birds are more sensitive to repellents in drinking water than on food. The active molecules in aqueous solution may have better access to the chemoreceptors (Clark *et al.*, 1991).

13.3.4 Food flavor preferences in domestic birds

Domestic birds rely more on odors than initially thought. Chickens are attracted to a sample of familiar odor (soiled substrate, or familiar substrate with orange oil) in an otherwise novel and probably frightening environment (Jones and Gentle, 1985). Chickens reared on mash with a few drops of water added were presented with mash treated with a few drops of orange oil. They showed neophobia (i.e. their latency to feed was longer and they fed less than from the water mash), but only initially. Conversely, chicks raised on orange-scented mash accepted water mash readily. This suggests that diet changes in poultry operations may have little impact on the animals' food intake or well-being (Jones, 1987). Still, farmers in the future may be able to imprint chicks to certain food or coop odors while the chicks are still *in ovo* (Section 9.4).

13.3.5 Non-lethal contact repellents

In the search for environmentally benign chemicals that might deter birds such as starlings, crows, or pigeons from roosting *en masse*, spices and herbs such as rosemary, cumin, and thyme look promising. In some experiments, the birds' feet were immersed in oil extracts of the spices. Starlings also avoided perches treated with starch mixes of (*R*)-limonene, (*S*)-limonene, β-pinene, or methiocarb. The first three occur in rosemary, cumin, and thyme (Clark, 1997).

13.3.6 Bird olfaction and conservation biology

Given how sensitive many birds are to odors, breeders of endangered bird species may be able to use food and environmental odors to imprint young birds on relevant cues of their future habitat (Nevitt, in Malakoff, 1999).

13.3.7 Differences between birds and mammals

Birds and mammals differ in their responses to some flavor compounds

- birds are repelled by methyl anthranilate but accept capsaicin
- mammals are repelled by capsaicin and are accepting of methyl anthranilate.

Mammals avoid capsaicin (Fig. 13.1) from "hot" jalapeño peppers, widely used as dog repellent, but birds eat it and it is used as parrot food (Reidinger and Mason, 1983). Capsaicin is used in commercial feeding repellents for squirrels at bird feeders. Conversely, birds reject methyl anthranilate (Kare and Pick, 1960), but mammals such as livestock and obviously humans, accept it. However, five known bird repellents given in water also reduced drinking in mice. The most active was orthoaminoacetophenone, which eliminated drinking but methyl anthranilate, 2-amino-4′,5′-methoxyacetophenone, 2-methoxyacetophenone, and veratryl amine were also effective (Nolte *et al.*, 1993). Different capsaicin-related compounds repel both starlings and rats. These responses involve the trigeminal nerve. Repellency in birds is enhanced by electron richness of the phenyl ring and basicity and reduced by acidic functionalities. The reverse is true for mammals. Compounds with long side chains (lipophilicity), electron-poor phenyl ring, and acidity repel mammals (Mason *et al.*, 1991; Table 13.1).

Birds may have a trigeminal receptor for *o*-aminoacetophenone and methyl anthranilate, analogous to the capsaicin receptor in mammals (Clark and Shah, 1994).

13.4 Non-human mammals

The bulk of the applications of semiochemicals in vertebrates are aimed at mammals. However, it must be remembered that mammalian behavior is complex and management by chemical cues is more challenging than in insects, for example.

13.4.1 Control of rodent and other herbivore damage

Area repellents

Predator odors

The search for environmentally benign pest repellents has spawned many experiments with excretions and gland secretions from felids, canids, and mustelids.

A predator odor affected a pocket gopher, *Thomomys talpoides*, population in western Canada. First live-trapping removed the gophers from plots of 4 ha area in an orchard. Then synthetic sulfur compounds from stoat, *Mustela erminea* anal glands (1:1 mixture of 2-propylthietane and 3-propyl-1,2-dithiolane) were

Table 13.1 Structure–function relationships in repellents: opposite effects on a bird and a mammal

Compound	Structure	Repellent effect	
		Starling	Rat
Capsaicin	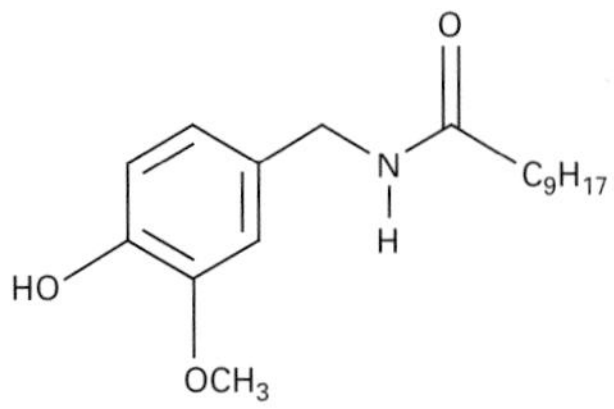 HO, OCH_3, N–H, O, C_9H_{17}	None	Strongest
Methyl capsaicin	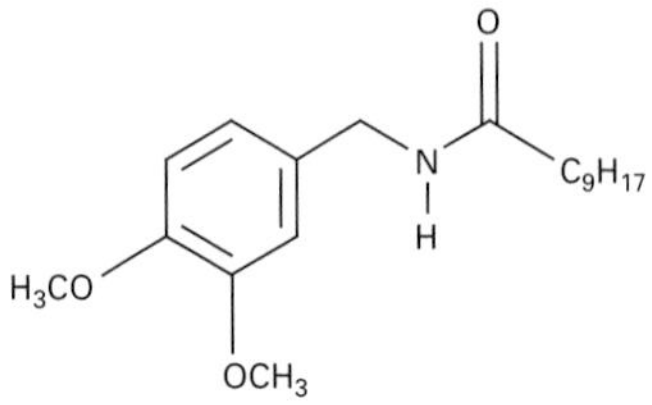 H_3CO, OCH_3, N–H, O, C_9H_{17}	None	Second strongest
Vanillyl acetamide	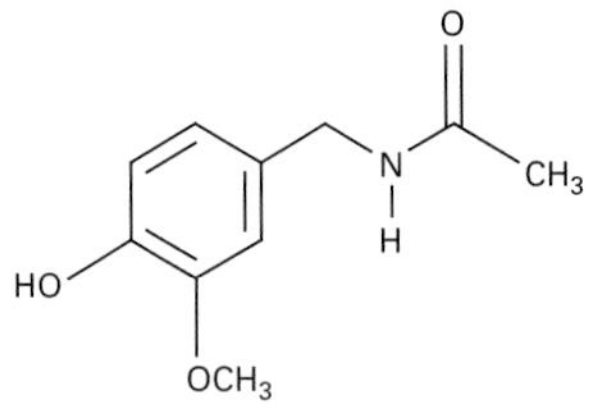 HO, OCH_3, N–H, O, CH_3	None	Medium
Veratryl acetamide	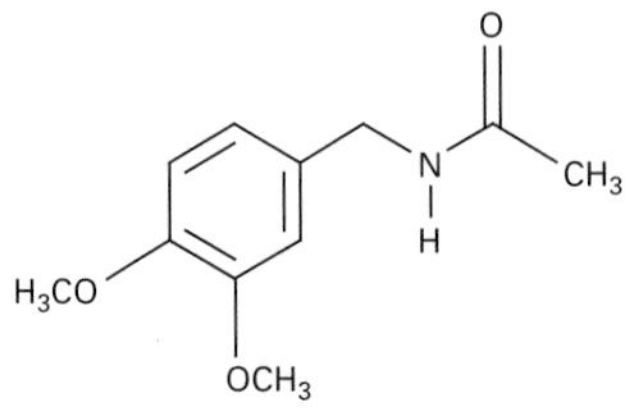 H_3CO, OCH_3, N–H, O, CH_3	Yes	Lowest
Veratryl amine	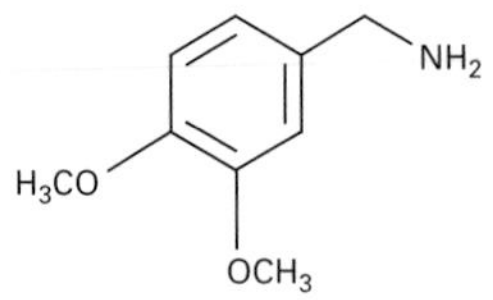 H_3CO, OCH_3, NH_2	Yes	Second lowest

After Mason *et al.*, 1991.

Table 13.2 Some factors that affect chemical repellent efficacy in mammals

Factor	Species	Effect	Reference
Cover	Townsend's vole *Microtus townsendii*	Avoid R in open; ignore R in cover	Merkens *et al.*, 1991
Available choice of food plant species	Many species	R avoided only when food choice is wide	Common experience
Palatability of plant	White-tailed deer *Odocoileus virginianus*	R protects less-preferred plants better	Swihart *et al.*, 1991
Established feeding habits	White-tailed deer	R deters early in growing season but not in summer after having fed in gardens	Müller-Schwarze, 1983
Hunger	Elk *Cervus canadensis*	R ineffective in starving animals	Andelt *et al.*, 1992

R. repellent.

applied on clay pellets inserted into burrows in November. In the following spring (April), recolonization was found to be reduced. The population fell by 50% as judged by the numbers of fresh mounds and did not recover fully, as would happen under usual circumstances (Sullivan *et al.*, 1990).

Predator odors are also effective area repellents for lagomorphs. A rabbit warren sprayed with an extract from lion feces had as many as 80% fewer animals than before the treatment and also fewer than a control warren. Adult rabbits stayed away from the treated warren longer than young ones. The effect lasted up to 5 months (Boag, 1991; Boag and Mlotkiewicz, 1994).

For odors to work, certain environmental conditions have to be met. Townsend's voles, *Microtus townsendii*, avoid a repellent odor if no *cover* is available. With cover present, they feed whether or not the area is scented. Therefore, for effective area repellents, the "pest rodent" should be able to retreat to unscented areas in preferred habitat such as sufficient cover (Merkens *et al.*, 1991; Table 13.2).

Non-target rodent species in the same forests may escape effects of mustelid sulfur compounds. In British Columbia, densities, survival rates, and reproduction of deer mice, *Peromyscus maniculatus*, were little affected when these compounds were used against long-tail voles (*Microtus longicaudus*), meadow voles (*Microtus pennsylvanicus*), and boreal redback voles (*Clethrionomys gapperi*) (Zimmerling and Sullivan, 1994).

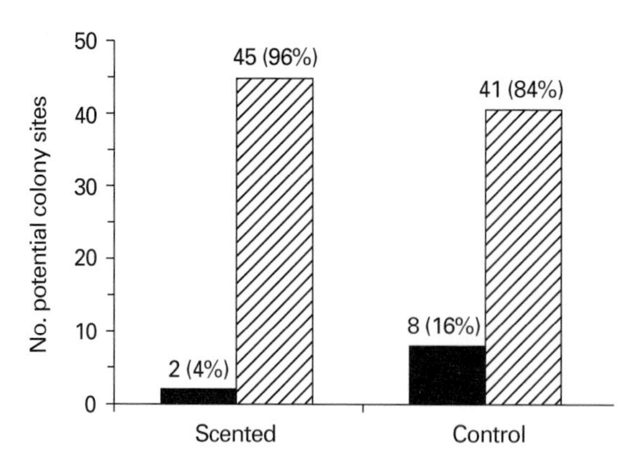

FIGURE 13.2 Beavers colonized only 2 of 47 (4%) experimentally scented potential colony sites (left), but 8 of 49 (16%) unscented control sites (right). Black indicates colonized sites; cross-hatched, sites left vacant. Fulton County Experiment. (From Müller-Schwarze 1990.)

Conspecific odors

Secretions and excretions from the same species can be used as repellents as well as attractants to traps or bait. Repellents are based on territorial exclusion and same-sex competition, while attractants exploit general gregariousness and preference for the opposite sex.

Territorial rodents typically mark their breeding territory with scent gland secretions, urine, feces, or combinations of these. It is possible to mimic occupancy of a vacant area by artificial scenting. North American beaver, *Castor canadensis*, are less likely to colonize available "potential beaver sites" along streams if a mixture of male and female castoreum (from castor sacs) and anal gland secretion is placed as scent marks at "strategic places." These are points of entry along a stream and areas near dam(s) and lodge (Welsh and Müller-Schwarze, 1989; Müller-Schwarze Fig. 13.2).

Adult muskrats, *Ondatra zibethica*, in the dense population at the Montezuma Wildlife Refuge in New York avoided traps scented with musk, the secretion of the preputial glands of this species, whereas juveniles entered clean traps and traps with *Ondatra* musk or a control odor (phenylacetic acid) equally often. This suggests that adults avoid others' territories and their odors, while immature muskrats are tolerated (van den Berk and Müller-Schwarze, 1984). In simultaneous experiments in the Netherlands, however, adult muskrats, especially males, were attracted to musk-scented traps. Here the experimental animals belonged to a colonizing population that was in the process of invading along waterways and were not yet entrenched territorially (Ritter *et al.*, 1982). Both of these responses were noted in a field study of wild red-backed voles,

Clethrionomys clareolus, and wood mice, *Apodemus sylvaticus*, near Munich. Immature animals entered male-scented, female-scented, and control traps equally often, while male odor attracted adult males and female odor attracted adult females (Bäumler and Hock, 1987).

Scents from coyotes, foxes, and other carnivores traditionally have served to attract these animals to traps. Trappers also use conspecific secretions to attract beavers, raccoons, and other fur bearers to traps. The use of scent lures in predator control is discussed below.

Area attractants

The odor of a trap soiled by a previous occupant of the same or different species can affect subsequent trapping success and, therefore, deduced population estimates. Whether rodents prefer dirty to clean traps, the odor of their own species to others, their own odor to other conspecifics, the same or opposite sex, varies between species and studies (reviewed by Gurnell and Little, 1992). For instance, adult California voles, *M. californicus*, prefer vole-scented traps over clean ones only during the breeding season. Juveniles show no preference, whether during the breeding or non-breeding season (Heske, 1987). Short-tailed voles, *M. agrestis*, are also affected by trap odor (Stoddart, 1982) and often the sexes can differ in their responses. In California ground squirrels, *S. b. douglasii*, males are more attracted to male anal gland secretion on traps, but females showed no preferences (Salmon and Marsh, 1989). These responses resemble those by prairie deer mice, *P. m. bairdii*, and house mice, *M. musculus* (Wuensch, 1982).

Wood mice, *A. sylvaticus*, and bank voles, *Clethrionomys glareolus*, differ in their responses to soiled traps. While both species are attracted to conspecific odor, wood mice preferred odor of the opposite sex, while bank voles did not, and wood mice avoided bank vole odor. This study suggests caution when using live-trapping for population censuses of rodents (Tew *et al.*, 1994).

However, several European rodent and shrew species (*A. sylvaticus*, *C. glareolus*, *M. agrestis*, and *Sorex araneus*) entered cleaned (with boiled water or detergent) traps equally often as dirty ones (only scraped after use). This was true for each species and sex. This suggests it may not be necessary to clean traps between uses (Tew, 1987).

Antifeedants

Predator secretions and excretions inhibit feeding in a variety of herbivore species.

Gland secretions

Mountain beaver, *Aplodontia rufa*, a primitive rodent from northwestern North America, fed less from food bowls if the rim was scented with mink anal gland secretion, or urine from mink, bobcat, coyote, or dog. The control odors butyric acid or guinea pig urine had no effect (Epple *et al.*, 1993). Mountain beaver did not respond significantly to mustelid sulfur compounds though, yet these deter other small mammals (Epple *et al.*, 1993).

Excretions

Black-tailed deer, *Odocoileus hemionus columbianus*, feed less on salal leaves (*Gaultheria shallon*), red cedar (*Thuja plicata*), and Douglas fir (*Pseudotsuga menziesii*) in the presence of urine, or feces of large cats and canids (Sullivan *et al.*, 1985b). Browsing by black-tailed deer on salal was most reduced by bobcat feces (by 51%), followed by mountain lion, *Felis concolor* (by 27%), wolf, *Canis lupus*, and coyote, *Canis latrans* (by 8%) (Melchiors and Leslie, 1985). Coyote urine also reduced feeding by tame mule deer on pelleted rations (Andelt *et al.*, 1991).

Predator-derived repellents deterred white-tailed deer, *Odocoileus virginianus*, from gardens after an application in May, the beginning of the growing season, but a second application in August was not effective once the deer had had ample opportunity to sample and utilize the gardens (Müller-Schwarze, 1983). White-tailed deer feed less on Eastern hemlock (*Tsuga canadensis*) and Japanese yew (*Taxus cuspidata*) when these are treated with urine from coyote, *C. latrans*, or bobcat, *Lynx rufus* (Swihart *et al.*, 1991).

The repellent effect depends on hunger, as in female elk (Andelt *et al.*, 1992) and less-preferred plants are better protected by predator urine odors (Swihart *et al.*, 1991).

Dog feces are very effective as feeding repellents for sheep, roedeer, *Cervus capreolus*, and red deer, *Cervus cervus*, while droppings from omnivorous bears had very little repulsive effect (Arnould and Signoret, 1993; Arnould *et al.*, 1993). In Australia, dog urine significantly reduced browsing damage to eucalypt and pine plantations by swamp wallabies, *Wallabia bicolor*, and rabbits (Montague *et al.*, 1990).

The Orkney vole, *Microtus arvalis orcadensis*, removes almost no bark from seedlings of Scots pine, *Pinus sylvestris*, if they are treated with an extract of feces of red fox, *Vulpes vulpes*, even though the two species have been separated for approximately 5500 years (Calder and Gorman, 1991).

Synthetic compounds

The odors of predatory mustelids can be very effective in inhibiting feeding on plants by rodents, lagomorphs, and deer, and, therefore, hold promise

Table 13.3 Predator sulfur compounds active as rodent repellents

Compound	Source animal and secretion	Affected animal	Reference
3,3-Dimethyl-1,2-dithiolane	Ferret AGS Red fox feces	Rat	Vernet-Maury *et al.*, 1984
2-Propylthietane	Stoat AGS	Montane vole	Sullivan *et al.*, 1988
3-Propyl-1,2-dithiolane	Stoat AGS	Montane vole	Sullivan *et al.*, 1988
2,5-Dihydro-2,4,5-trimethylthiazoline	Red fox urine	Montane and meadow vole	Sullivan *et al.*, 1988
3-Methyl-3-butenyl-methyl sulfide	Red fox urine	Snowshoe hare	Sullivan and Crump, 1986a
Trimethylthiazoline	Red fox feces	Rat, mouse	Vernet-Maury *et al.*, 1984; Coulston *et al.*, 1993;

AGS, anal gland secretion

for protecting plantations. Sulfur compounds from mustelid anal gland secretions suppress feeding by snowshoe hares, *Lepus americanus*. The most active compounds were 3-propyl-1,2-dithiolane from the stoat, *M. erminea*, and 2,2-dimethylthietane from stoat and mink, *M. vison*. The effect lasted for up to 38 days (Sullivan and Crump, 1984). Urine, feces and anal gland secretions from various mustelids, canids, and felids suppressed feeding by snowshoe hares (Sullivan *et al.*, 1985a). The most effective constituent of anal gland secretions from the red fox, *V. vulpes*, for reducing feeding on lodgepole pine seedlings was 3-methyl-3-butenylmethyl sulfide (Table 13.3). Other sulfur compounds such as 2-phenylethylmethyl sulfide and 3-methylbutylmethyl sulfide were not active (Sullivan and Crump, 1986b). Odors from wolverines also deter snowshoe hares from feeding (Sullivan, 1986). Sulfur compounds from mustelid anal glands inhibited feeding in a number of rodents such as pocket gophers, *Thomomys talpoides* (Sullivan and Crump, 1986a) and montane, *Microtus montanus*, and meadow, *M. pennsylvanicus*, voles (Sullivan *et al.*, 1988).

A mixture of five sulfur compounds that occur in lion (*Felis leo*) dung proved to be a feeding inhibitor in captive red deer, *Cervus elaphus*. The five compounds are dimethyl disulfide, dimethyl trisulfide, dimethyl tetrasulfide, 3-mercaptopropan-1-ol, and 3-mercaptothiopropan-1-ol (Abbott *et al.*, 1990). This odor also deters rabbits from feeding on carrots for up to 1 month (Boag and Mlotkiewicz, 1994). Sulfur compounds, generated during protein digestion, could be the "general theme" underlying the repellency of predator odors for rodents and other herbivores (Nolte *et al.*, 1994b). Mule deer *Odocoileus hemionus hemionus*, avoided tree seedlings treated with proteins that are high in methionine, a sulfur-containing amino acid, although methionine alone was not effective (Kimball and Nolte, 2004).

Experimentation with sulfur-containing compounds has also utilized egg extract. Such extract, with acrylic adhesive and a red dye, proved effective in repelling sheep from feeding on seedlings of Monterey pine, *Pinus radiata*, in New Zealand. Thiram (tetramethylthiuram disulfide, a fungicide), with or without egg, was not effective (Knowles and Tahan, 1979). Crozier (1991) recommended egg-based repellents, which had proved effective against rabbits, hares, and possums in New Zealand, because they are readily available, easy to apply, and non-toxic to humans. Mixed with acrylic paint adhesives, such repellents stay on tree seedlings for months. A commercial deer and elk repellent containing putrescent whole egg solids has been tested as a spray to change the uneven distribution of steers on their range. The repellent effect in putrefied egg concoctions results from aldehydes and possibly fatty acids (Tiedman *et al.*, 1976).

Plant compounds

"Natural resistance in trees to damage by animals remains essentially unrecognized and unexploited" (Dimock *et al.*, 1976). Despite great interest in chemical plant defenses since the mid-1970s, we still are just beginning to understand these systems well enough to apply them successfully. (Well-defended trees in non-glaciated "resistance centers" and from areas with large wildfires were discussed in Ch. 11, p. 336 and p. 337.)

Over evolutionary time, chemical warfare between plants and herbivores has produced many effective antifeedants that have significant effects on herbivores (Ch. 11). When seeking applications, we can avoid exaggerated expectations by remembering that the vast majority of plants do not totally eliminate herbivory. Instead, they manage to keep it at acceptable levels. It is not surprising that some plant toxins have been used in rodent control. Scilliroside (red squill) is a rodenticidal glycoside from the Mediterranean plant *Urginea maritima*. Quebracho, a condensed tannin, reduces *Microtus* spp. damage in orchards (Lindroth and Batzli, 1984). Intense efforts are underway to harness natural antiherbivore compounds as non-toxic repellents to protect crops and the potential is enormous.

Pine oil is a by-product of the pulp industry. This oily liquid contains a mixture of terpene alcohols and monoterpenes and is available under the trade name Norpine-65. Snowshoe hare, *L. americanus*, and voles, *M. townsendii*, consume less laboratory chow, apples, or carrots if they are treated with pine oil. This response did not habituate within the time of the experiment. When considering the use of pine oil as a herbivore repellent, we have to keep in mind that it is phytotoxic to seedlings if applied topically (Bell and Harestad, 1987). Red deer, *C.*

elaphus, calves avoid monoterpenes from Sitka spruce, *Picea sitchensis*, and lodgepole pine, *Pinus contorta*, even without prior experience. Females are more sensitive than males. Monoterpenes with higher vapor pressure were avoided more, raising the question whether the deer respond to the strength of an odor or to its quality (Elliot and Loudon, 1987). This avoidance is adaptive as monoterpenes inhibit rumen microorganisms (Schwartz *et al.*, 1980). In the search for new and effective deer repellents to protect Douglas fir, *P. menziesii*, fresh leaves of wild ginger, *Asarum caudatum*, deterred black-tailed deer, *O. h. columbianus*, effectively. The deer ate only 2% of a food mix containing 5% shredded wild ginger leaves (Campbell and Bullard, 1972). Poisonous, foxglove, *Digitalis purpurea*, also quickly conditions herbivores to avoid this plant (Campbell, 1987).

In the first practical application of plant antifeedants for forest crop protection and wildlife management, snowshoe hare, *L. americanus*, fed less on coniferous tree seedlings after they had been sprayed with pinosylvin or pinosylvin methyl ether (Sullivan *et al.*, 1992).

Conditioned flavor aversions

"Flavor" consists of taste and odor combined. Animals often avoid tastes after only one experience if delayed illness follows it, but odors are not avoided under such circumstances. If an odor and a taste are presented together before an illness is induced experimentally (e.g. with lithium chloride), animals can learn to avoid the odor after just one such experience. When the odor is intensified in such a combined presentation, only the odor aversion increases, not the taste aversion. If the taste is made more intense, both odor and taste aversions increase. As a rule, the odor aversion strength depends on the taste experience, but not vice versa (Garcia *et al.*, 1986).

Mountain beavers, *A. rufa*, and black-tailed deer, *O. h. columbianus*, severely damage seedlings of forest trees in western North America. These two species learnt to avoid Douglas fir, *P. menziesii*, when mountain beaver were provided with Douglas fir cuttings treated with a putrescent egg preparation, and black-tailed deer were exposed to treated plastic ribbons on seedlings (Campbell, 1987). Damage to Douglas-fir seedlings by mountain beaver can also be reduced if the animals are conditioned first. Sulfur compounds such as the fungicide thiram, dimethyl dithiocarbamate compound are feeding repellents suited for aversion conditioning. Thiram-treated seedlings were placed into burrows and treated and untreated seedlings planted in front of the burrow. After 8 months (note the long-term effect), feeding on treated and untreated seedlings was reduced to 17.4% and 16.0%, respectively. On sites without thiram conditioning, 94.4% of untreated seedlings were cut outside untreated burrows (Campbell and Evans, 1989).

Modifying plants

Lowering or increasing the concentrations of plant compounds that affect palatibilty by selective breeding or bioengineering remains an important goal. Douglas fir, *P. menziesii*, is an example where an increase in secondary compounds would be desirable to render it more resistant to browsing by black-tailed deer, *O. h. columbianus* (Dimock *et al.*, 1976). Sagebrush, *Artemisia tridentata*, poses the opposite problem: selective breeding for *reduced* monoterpene content is needed so that mule deer, *O. h. hemionus*, can better utilize it (Welch and McArthur, 1981).

The levels of antifeedants can be raised safely only in plants such as forest trees or fiber crops, not in food plants for humans or livestock. Such problems have arisen inadvertently. For instance, a new insect-resistant potato cultivar had to be withdrawn from the market because it contained high levels of the carcinogens solanine and chaconine (Renwick *et al.*, 1984). In another example, an insect-resistant celery had 10 times the usual concentration of the carcinogen 8-methoxypsoralen (and related psoralens), which caused dermatitis in produce handlers (Seligman *et al.*, 1987).

Watering the foliage of seedlings with quadrivalent selenium reduced deer browsing of Douglas fir. This produces a garlic-like odor that might be responsible for the effect (Allan *et al.*, 1984).

Feeding stimulants

Conspecific secretions and excretions may stimulate pest rodents to consume bait. An example is the metad, or soft-furred rat, *Rattus meltada pallidior*, of the grasslands and crop fields of India. In a laboratory test, 1% conspecific urine from males or females added to the diet of millet increased food consumption by 32 to 70%. The metads also stayed longer in the half of the cage with the urine-treated food than in that with untreated food, and scent marked more there. The males marked with sebum, the females with urine. Male urine tended to be more effective, and males responded more than females (Soni and Prakash, 1987).

Rats are initially reluctant to take new bait. This bait shyness can be overcome by applying urine of estrous females to attract males, and male urine to attract females, unless pregnant. Males consume more of a feed if female urine odor is present (Gao, 1991). Bank voles are attracted to traps scented with 16-5α-androsten-3α-ol (Bäumler and Haag, 1989).

A novel odor for attracting rats or mice to bait is carbon disulfide. It is found in the breath of rats, serves in communication about food, attracts rodents to bait, and increases bait consumption (Bean *et al.*, 1988). Use of carbon disulfide and carbonyl sulfide in baits attracts mice and rats and prevents bait shyness from developing. It has been patented for that purpose (US Patent No. 4,861,585 to J. R. Mason and B. G. Galef).

13.4.2 Domestic Animals

Predator-derived odor repellents

Alpine goats, *Capra hircus*, fed considerably less from buckets with food (cottonseed hull, corn, and alfalfa) covered with shredded filter paper soaked with extracts of predator feces. The predators were Bengal tiger (*Panthera tigris tigris*), Siberian tiger (*Panthera tigris altaica*), African lion (*Panthera leo*), and brown bear (*Ursus arctos*) (Weldon *et al.*, 1993). Dog odor is also active. A water slurry of dog feces repels sheep, roe-deer, and possibly red deer from feeding on grass, yoke-elm, *Carpinus betulus*, and oak, *Quercus* spp. (Arnould *et al.*, 1993).

Predator odors do not necessarily keep herbivores away from forage. Sheep and cattle, presented with odors from coyote, fox (*V. vulpes*), cougar (*Felis concolor*), or bear (*Ursus americanus*) near their feed rations spent less time feeding there, but they did not stay away from the treated feed (Pfister *et al.*, 1990b).

Similarly, two sulfur compounds from predators, 3,3-dimethyl-1,2-dithiolane and (*E*)-(Z)-2,4,5-trimethyl-Δ^3–thiazoline had no effect on wild roof rats, *Rattus rattus*, in a Hawaiian macadamia orchard (Burwash *et al.*, 1998).

Laboratory and domestic animals may be poor models for avoidance of predator odors. For example, in one experiment, chickpeas were painted with the sulfur compounds *n*-propyldithiolane and *n*-propylthiolane from stoat anal gland secretion and 2,4,5-trimethylthiazoline (Fig. 3.1, p. 37) from fox feces. The chickpeas were planted and wild mice and house mice were tested to see if they would dig up and eat the peas. Wild mice remembered the predator odors better after odor exposure for 1 or 4 weeks and, consequently, may be better than laboratory mice at risk assessment (Coulston *et al.*, 1993).

Conspecific herbivore odors as feeding deterrents

Cattle avoid grazing near cow pats for at least 30 days. The aversive stimulus emanates from the dung and not the plants: cattle feces, 0–35 days old, placed in a feeding trough inhibit feeding. Volatiles extracted from cow dung also reduce feeding. A neutral fraction of a diethyl ether extract was active in one experiment (Dohi *et al.*, 1991). Likewise, horses will not feed near their own droppings (Ödberg and Francis-Smith, 1977). House mice, *M. musculus*, differ. They consume the same amount of food whether it contains fecal pellets or not (Pennycuick and Cowan, 1990).

Plant compounds and food consumption

Secondary plant metabolites can reduce food consumption by livestock. Sheep refuse hay if it contains coumarin, gramine, tannic acid, malonic acid, or glycine. However, butyric acid or amyl acetate stimulate hay consumption. Anosmic and agustatory sheep show the same responses so sensory stimuli are not involved. Tannic acid and gramine lower *in vitro* digestibility, coumarin does so slightly, and the other compounds did not affect digestibility. In general, the compounds affect food intake only if there is an effect on the rumen (Arnold *et al.*, 1980). Some plants are outright toxic to livestock, presenting the chemical ecologist with the twin problems of the effects of plant compounds on herbivores and the need to develop effective repellents to steer grazing animals away from stands of toxic plants. Larkspur (*Delphinium* sp.) contains diterpenoid alkaloids and poses the most serious poisonous plant problem for cattle of the mountain rangeland in the western United States. Cattle are more affected than sheep. During the years 1913–1916, more than 5000 cattle died annually in the national forests of the west from larkspur poisoning (Ralphs *et al.*, 1988).

Olfactory imprinting

Since time immemorial, farmers, breeders, and zookeepers have had to cope with cross-fostering problems. They either have to save orphaned livestock newborns or attach a young of a different species to a lactating female for later interbreeding to produce desirable hybrids. Recent studies have confirmed that in cows, for example, odor transfer between young is an inexpensive, rapid and safe technique to help postpartum beef cows accept alien calves (Dunn *et al.*, 1987). The alien calf is wrapped in a stockinette that the cow's own calf had worn for 48–96 hours and this makes the alien calf acceptable to 9 out of 10 cows. By comparison, only 1 of 12 alien calves was accepted if they wore a jacket with their own odor. Acceptance was defined as lack of aggression and suckling periods lasting at least 20 seconds. Four days after fostering, 100% of the alien calves with the jacket of the cow's own calf were accepted, but only 33% with their own jacket. The highest acceptance rate for this second group was 50% on day 5. All cows accepted calves with transferred jackets, while two (17%) did not accept young with an alien calf's jacket (Dunn *et al.*, 1987).

In sheep too, a ewe whose lamb has died soon after birth will accept an alien lamb if it carries the odor of her own young. Several methods of scenting the new lamb have been compared. First, Hessian coats were fitted on the lambs when they died and left there for 2 to 18 hours; they were then put on the foster lambs. When presented to the ewes whose lambs had died, 73% accepted the new lambs.

Second, dead lambs were smeared with neatsfoot oil (an oil obtained by boiling feet and leg bones of cattle), left with the mother for 2–18 hours and then substituted by neatsfoot oil-treated foster lambs. Here 91% were accepted. The best acceptance rate (100%), however, was achieved by draping the skin of the dead lamb over the alien one. This is the method of choice because it is not only most effective but also saves labor as no treatment is necessary (Alexander *et al.*, 1987). Ewes accept alien lambs if they are rubbed with non-polar oils such as white soft paraffin or liquid paraffin. Polar materials such as polyethylene glycol, glycerol silicone, or diisooctyl phthalate have no effect. Ewes appear to use non-polar volatiles for discrimination (Alexander *et al.*, 1989).

When ewes and lambs become separated during sheep management operations, mothers and lambs first bleat. After approaching one another, the ewe sniffs a lamb at close range. This olfactory test decides whether the mother lets the lamb nurse, or rejects it, possibly by head-butting (Fig. 13.3).

Amniotic fluid

Along with eating the afterbirth, female ungulates lick amniotic fluid during and after giving birth. For cross-fostering experiments, it is important to know whether acceptance of amniotic fluid is species specific. In one experiment, ewes accepted food treated with ovine amniotic fluid, also caprine amniotic fluid (albeit less so) at and after parturition, but always rejected bovine amniotic fluid (Arnould *et al.*, 1991).

Priming pheromones

Priming pheromones hold great promise as means to accelerate puberty or to synchronize and regularize estrus cycles in livestock.

Puberty acceleration

Oronasal application of bull urine accelerated puberty in beef heifers; 67% of the animals reached puberty within 8 weeks of treatment, compared with 32% in water-treated controls. Urine-treated heifers that reached puberty during the experiment calved earlier than water-treated heifers. The calving season also was shorter for urine-treated animals (Izard and Vandenbergh, 1982a). Puberty in female pigs can be accelerated by 1 month if a boar is present (Pearce and Paterson, 1992).

Estrus synchronization

Dairy heifers, after being brought into their follicular phase by injection of prostaglandin $F_{2\alpha}$, were treated with cervical mucus or urine placed on

FIGURE 13.3 Ewes recognize their lambs by first calling and then sniffing lambs after separation during herding operations in paddocks. (Photographs: D. Müller-Schwarze.)

the oronasal area. Cervical mucus had the greatest effect: all heifers in this group were in estrus within 72 hours and the degree of synchrony was high (Izard and Vandenbergh, 1982b).

Signaling pheromones

The boar sex pheromone, consisting of androstenol and androstenone, from the submaxillary salivary gland has been used commercially for artificial insemination of pigs. It was available as a spray under the trade name Suidor (formerly Jeyes Boar Mate) in Europe (Glei *et al.*, 1989). Sows who did not assume the mating stance are sprayed in the face. After this treatment, 60% will assume the mating stance. In addition, fighting among young pigs was reduced by introducing the odor of an adult boar (McGlone, 2002).

13.4.3 Breeding of rare species

Declining species are sometimes hard to breed in captivity. Two techniques have been used to enhance breeding behavior in the harvest mouse, *Micromys minutus*. First, males rejected by females were given scent marks from preferred mates. This stimulated their mating behavior and made them more attractive to females. Second, females were given male scent marks to prefamiliarize them with a male. The females preferred the male odor donors and were less aggressive towards them (Roberts and Gosling, 2004).

13.4.4 Predator control

Gland secretions and urine as scent lures in predator control

Many odor lures have been tried for coyotes, *C. latrans*. The behaviors of coyotes at chemical lures include approach, sniffing, digging, scratching, licking, chewing, biting, rolling, urinating, and defecating. Fagre *et al.* (1983) have listed 23 behavioral and chemical characteristics of an optimal coyote lure. Some effective coyote lures such as a synthetic fermented egg formulation comprise four odor classes: fruity (esters), sulfurous (organosulfurs), sweaty (volatile fatty acids), and fishy (amines) (Bullard *et al.*, 1983). Trimethylamine occurs in anal sacs of coyotes, dogs, and red foxes, valeric acid in coyote urine. In combination with trimethylammonium valerate, trimethylamine is attractive to coyotes. The combination slowly releases trimethylamine and lasts longer than the latter alone (Fagre *et al.*, 1983). Even more attractive is trimethylammonium decanoate, a salt of a C_{10} fatty acid (Fagre *et al.*, 1983).

Both male and female coyotes, *C. latrans*, are attracted to volatile aldehydes from sheep liver and estrous urine of female coyotes. The most active compounds were octanal, nonanal, decanal, and undecanal. The aldehydes released sniffing and rub-rolling, but little lick-chewing and biting. Therefore, these compounds are better suited to attract coyotes to traps than to toxicant-delivery systems that rely on the latter behaviors (Scrivner *et al.*, 1984).

The introduced ferret, *Mustela furo*, in New Zealand destroys native fauna and has to be controlled. For trapping, lures based on conspecific scent gland secretions or urine compounds can be as effective as food bait. Eight constituents of the anal sac secretion have been synthesized and tested for their ability to attract ferrets to traps. Ferrets were most attracted to a mixture of 2-propylthietane and *trans*- and *cis*-2,3-dimethylthietane (Clapperton *et al.*, 1989).

Numbers of visits to scent stations baited with fatty acid tablets or prey odor such as shellfish oil allows estimation of population sizes of free-living red and gray foxes, raccoons, striped skunks, river otter, bobcats, cottontail rabbits, and oppossums. However, Smith *et al.* (1994) showed that the numbers of raccoon visits to scent stations did not reflect the size of their population.

Coyote scent (urine) stations also attracted cottontail rabbits *Sylvilagus floridanus*. In Texas, this behavior has aided rabbit censuses. The scent station count is more accurate than the traditional "headlight count" but more labor intensive (Drew *et al.*, 1988).

Predator repellents

Of many compounds tested, *trans*-cinnamaldehyde was most effective in delaying the approach to food by coyotes (Jankowsky *et al.*, 1974). Other studies found β-chloroacetyl chloride, a strong irritant and lachrymator, an effective olfactory repellent, while cinnamaldehyde was somewhat active in delaying approach to a visible food source (Lehner *et al.*, 1976).

Wildlife managers and conservationists often need carnivore repellents. As an example, in Lothringia, kites' nests had to be moved from barley fields. Red foxes followed the human tracks and preyed on the nestlings (A. Engelhart, personal communication). Measures that discouraged the foxes included chemical repellents, covering the human tracks, and aversive conditioning of the foxes.

Aversive conditioning of predators

To discourage birds and mammals from preying on eggs of the greater sandhill crane, *Grus canadensis*, Nicolaus (1987) injected surrogate crane eggs (turkey eggs) with 2,3,5- and 3,4,5-trimethylphenylmethyl carbamate (UC 27867) and placed them in breeding territories of ravens. Untreated chicken eggs, painted blue–green, were presented in a second experimental nest. At additional control sites, neither the surrogate eggs nor the painted chicken eggs were injected. Significantly more treated eggs (70%) survived intact than did uninjected blue–green eggs (38%). At the control sites, 37% of untreated surrogate eggs and 41% of blue–green chicken eggs survived.

A long line of investigations has tried aversion conditioning for mammalian predators, particularly coyotes. Traditionally, lithium chloride is either injected into the predator after it has consumed a certain food, or it is worked into the bait. A study of aversive conditioning to rabbit meat used three methods: lithium chloride was injected into a coyote after feeding, used to lace rabbit meat wrapped in rabbit fur as a bait package, or used to lace rabbit carcasses. The aversion tended to be specific: the coyotes avoided the bait packages but continued hunting live rabbits (Horn, 1983).

Mongooses, *Herpestes auropunctata*, have been introduced on many tropical islands to fight rats in sugar cane fields. Unfortunately, they also damage or destroy native populations of mammals, birds, reptiles, or amphibians. Mongooses are particularly destructive where they prey on endangered species, such as the eggs of endangered hawksbill, green, and leatherback sea turtles. In a study in the field and in captivity on St. Thomas, American Virgin Islands, mongooses were conditioned with carbachol (carbamyl choline chloride) to avoid eggs. Carbachol acts as a cholinesterase blocker. Unlike the more often used trimethacarb, it is water soluble and it is less detectable by taste or odor. Both are harder to detect than lithium chloride, which is salty-tasting, and they are, therefore, thought to be more suitable for producing conditioned taste aversions. In a T-maze pre-test, the mongooses were exposed to an egg treated with anise extract or one treated with vanilla extract, both without carbachol. Next, during the acquisition phase, each end of the T-maze held an egg treated with anise extract and laced with 12 mg carbachol. Finally, in the post-testing phase, the two eggs smelled as they did in the pretest but the anise egg was injected with 6 mg carbachol. The toxic eggs were not avoided at a distance nor after tasting the outer shell. In the field, the mongoose behaved differently. Here they avoided toxic eggs sprayed with vanilla or mint odor from a distance and 39% of the clutches were left alone (Nicolaus and Nellis, 1987).

Aversive conditioning to deter mammalian predators from preying on waterfowl nests remains elusive. In one experiment, nests with chicken eggs injected with lithium chloride were placed along transects in the Sand Lake National Wildlife Refuge in South Dakota. Later, there was no difference in nest success between treated and control plots. The predators in the area were red fox (*V. vulpes*), raccoon (*Procyon lotor*), striped skunk (*M. mephitis*), mink (*M. vison*), and Franklin's ground squirrel (*Spermophilus franklinii*). The waterfowl nesting in the area were blue-winged teal (*Anas discors*), mallard (*Anas platyrhynchos*), common pintail (*Anas acuta*), gadwall (*Anas strepera*), and northern shoveler (*Anas clypeata*) (Sheaffer and Drobney, 1986).

In a similar study, chicken eggs were injected with emetine dihydrochloride (20–25 mg) and placed along three transects, 0.7 to 1.0 km long. Raccoons, opossums, *Didelphis virginiana*, and striped skunks consumed 75% less treated eggs than untreated controls if the animals were first conditioned and untreated eggs were offered later. Simultaneously offered eggs of both types resulted in both eaten equally often. This suggests that predators should be conditioned before the nesting season, not after egg laying has begun (Conover, 1990).

Aversive conditioning of beeyard-raiding black bears, *Ursus americanus*, using lithium chloride wrapped in honey or brood comb has been inconclusive. Emetics also proved to be "overspecific": bears avoided only the particular foods that had been used as bait. In Yosemite National Park, black bears were successfully discouraged from raiding the foodsacks of backcountry campers by "aversion sacks" that contained balloons filled with ammonium hydroxide (Hastings and Gilbert 1980).

Social learning can modify conditioned aversions. Spotted hyenas conditioned by lithium chloride treatment lost their aversion in the presence of naive conspecifics (Yoerg, 1991). The reverse occurs in rats: a conditioned aversion is transmitted to naive animals (Strupp and Levitsky, 1984).

13.4.5 Use of mammal noses

While the performance of dogs in rescue and police work is legendary, other mammals have been used to detect fungi, explosives, illegal drugs or diseases.

Tracking by dogs

The domestic dog provides a convenient model for wild canids. Like wild dogs, domestic breeds find prey by tracking with their nose near the ground, or by air scenting, picking up odor from the source directly with nose held high. In air scenting, practitioners speak of a "scent cone" spreading from a point source. The dog quarters in a zigzag pattern – like insects – toward the odor source. The age-old practice of training dogs for hunting and finding lost persons acquired a scientific basis during the first three decades of the twentieth century. For training and testing a dog's performance, Löhner (1926) introduced the method of letting the dog fetch a scented piece of wood of standard size. He found that dogs use human sweat, sebum, and dandruff to identify an individual. Dogs are very responsive to body odor components such as fatty acids up to C_8 – formic, acetic, propionic, valeric, caproic, and caprylic acids – plus nitrogen compounds such as ammonia and skatole, and also sulfur compounds. Contact for only 1–2 s between skin and wood left enough scent on the wood for the dog to identify it. Contact with an area as small as a fingertip sufficed. The dogs distinguish whether the odor side on the wood is placed up, sideways, or down when the wood is put on the ground.

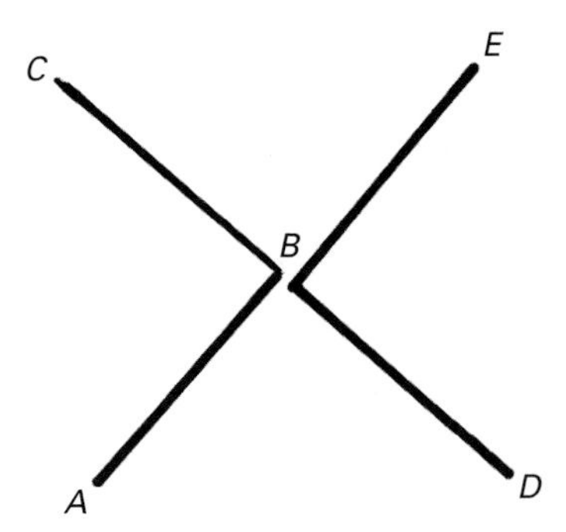

FIGURE 13.4 The track cross was used to test the ability of dogs to follow a track (A to C) that angles back while a second, diversionary track (D to E) leads straight on at point B. (Redrawn from Most and Brückner, 1936.)

Schmid (1935) continued that work. He found that odor from individuals can drift up to 60m from the actual trail in lateral winds. When confronted with diversionary tracks, dogs circle at the intersection with the real track. They are not distracted by diversionary tracks and still can follow individual horse or dog tracks. Most and Brückner (1936) used a "track cross" (Fig. 13.4) to assess track-following ability. The test track veers off at a right angle to the left or right, while a diversionary track is laid as a straight continuation of the test track. A dog will follow the straight line (i.e. the wrong track) while a specially trained dog would double back and was not distracted by the straight diversionary track. The age of the track matters: trained police dogs discriminated tracks that differed only by 3 minutes in age.

Most and Brückner (1936) found human tracks to be complex. A track contains a human species odor and an individual odor, both extruded through boots, but also the odors of crushed plants, disturbed soil, and leather or other shoe material. Dogs still tracked correctly after removal of one or several of these components. To separate the track components, the authors built a chair lift and a track wheel (Fig. 13.5). The rim of the metal wheel (approximately 2m in diameter) carried raised replicas of shoes at stride intervals. Pulled over the terrain, the wheel makes a track of crushed plants and disturbed soil without human scent. The lift consists of a chair suspended on a steel cable about 1.5m above the ground. Although a person riding this chair leaves no foot imprints or disturbed soil, he/she still sheds odorous "rafts" of skin cells. A trained dog uses one or the other of these track components, as terrain and surface change.

As discussed in Ch. 5, dogs are extremely sensitive to odors (Neuhaus, 1953, 1956a,b; Neuhaus and Müller, 1954). A dog can identify a human odor against background odors such as bergamot, clove or oregano oil, tobacco smoke,

FIGURE 13.5 Analysis of tracking components. (*a*) A chairlift. (*b*) The track wheel was used to separate a human track into a person's olfactory footprint and odors emanating from crushed soil and vegetation. The track wheel has wooden footpads on a metal rim, spaced at an average human stride. The wheel produces a track that lacks human scent and consists entirely of odors from disturbed soil and vegetation. (From Most and Brückner, 1936.)

or hay or barn odor. Odor differences between body regions of one individual can be greater than differences between individuals.

Identical twins provide the ultimate test for a dog's prowess in distinguishing individual odors. In his 1875 *Anthropological Miscellany: the History of Twins and the Criterion of their Relevant Powers of Nature and Nurture*, Francis Galton suggested "it would be an interesting experiment for twins who were closely alike to try how far dogs can distinguish between them by scent." H. Kalmus of the Galton Laboratory in London tested seven trained police dogs and two show dogs skilled at retrieving for their ability to discriminate identical twins by smell. He challenged them to single out a handkerchief scented by a particular person, or to follow a person's track among four other confusing tracks. A 3-year old Alsatian bitch was unable to distinguish handkerchiefs scented by two identical twin sisters if the handkerchiefs were laid out in a row, to be smelled one at a time. Tracking experiments, however, permitted the dog to compare the two tracks of the identical twins simultaneously. In this test, a male Alsatian followed the "correct" track of identical twin A among confusing tracks laid by twin B after sniffing a handkerchief scented by A (Kalmus, 1955). Of these two experiments, the tracking test is more relevant to the question of a predator following a prey odor (i.e. a kairomone).

Can dogs, and by extension any mammalian predator, determine the direction of a track? Thesen *et al.* (1993) found that dogs that approached a track of human footprints at a right angle, found the "right" direction after sniffing only two to five footprints in 3–5 seconds. In these experiments, the footprints were 3 or 20 minutes old and had been laid 1 s apart. The dogs apparently were able to use differences in scent concentration in the air above the footprints (Thesen *et al.*, 1993).

Rodents such as gerbils or rats are able to sniff out drugs or explosives. In efforts to find land mines in Mozambique, the Gambian giant pouched rat, *Cricetomys gambianus*, performed well in sniffing out buried TNT scents. (The rats find their buried food stashes by smell.) APOPO, a Belgian organization, trains these giant rats in Tanzania. Trainers condition the animals, harnessed and running along a taut line, for scratching soil at the correct spot to respond to a metallic double click that signals a food reward. Upon this signal, the rat runs back to the handler and receives a food such as banana. These rats found all 20 live mines in a 400 m^2 area. Rats are well suited for this work. They are abundant, cheap, and small enough for easy transport, too light to detonate mines accidentally, and tire less than dogs. Furthermore, human mine specialists are rare, and metal detectors give many false alarms (Mott, 2004; Cox *et al.*, 2004; Wines, 2004).

The mammalian olfactory sense is also being exploited to detect disease. African giant pouched rats can detect tuberculosis in sputum (Holden, 2004) and dogs can discriminate urine from patients with bladder cancel from normal urine, but not consistently (Willis *et al.*, 2004).

13.4.6 Odors, earthquakes, and animals

Wild and domestic birds and mammals, even snakes, might be sensitive to cues preceding earthquakes (Deshpande 1986; Tributsch 1988). If odors are among the pertinent stimuli, sulfurous volatiles are the most likely candidates, as sulfur dioxide is commonly released in earthquakes, along with methane and radon. Methane from the earth migrates to the surface through weak zones. In cold zones it reaches the surface unchanged and could possibly be detected by animals, while in hot volcanic lava it will be oxidized to carbon dioxide (Gold and Soter, 1980). The San Francisco earthquake of October 1989 gave reason to examine anew whether the olfactory sense of animals is reliable enough to predict imminent earthquakes.

13.5 Humans

Odors affect human behavior more than we realize. They are now appreciated as important in human health and disease. Above all, the powerful role of learning is impressive. Odors become associated with pleasant and unpleasant experiences and can retain their hedonic value lifelong. This applies to food, to social and sexual relationships, and to environments such as houses, workplaces, or landscapes. Writers rather than scientists have described such anecdotes. In *Remembrances of Things Past*, Marcel Proust evoked a flood of childhood memories by the taste of a madeleine dipped in lime-blossom tea. Jean-Paul Sartre tells in his autobiography *Les Mots* how the halitosis of his grade-school teacher became to him the odor of authority.

New, powerful techniques in chemistry, odor formulations, bioassays, and olfactometry have supplied us with deeper as well as fresh insights into olfactory effects on our behavior. Medicine, psychology, environmental design, occupational safety, air-quality control, marketing, and advertising now consider and contribute to human chemical ecology.

13.5.1 Diet odors

Odors are important for food discrimination, including detection of spoiled or toxic food, and perhaps surprisingly for stimulating food intake. People who become anosmic in the wake of head injuries often loose their appetite. To them food no longer appears desirable; conversely odors stimulate food intake in obese people (Schecter and Henkin 1974). Reducing stimulating food odors would suggest itself as one measure in treatment of obesity.

Food aversions

Most of us have one or the other food aversion. These aversions develop when we overindulge, are poisoned by a food, or ate the particular food while an illness was developing. It follows that "the time to give children junk food and candy is when they are sick, certainly not when they are healthy and hungry" (Cain, 1982). Drug tolerance and addiction may be acquired in a similar fashion (Siegel, 1979). *Aversion therapy* for overeating, alcoholism, and drug abuse should couple the odor of food, alcoholic beverages, or drugs, respectively, with the condition of being sick or feeling miserable. So far, these efforts have not been successful. The subjects habituate to the odors if no further negative consequences occur. To counteract habituation, several or many different negative odors are needed. Repeated exposure makes pleasant odors less so and unpleasant odors more pleasant. This applies to employees in the perfume industry and candy stores as well as workers in chemical factories, rendering plants, or pigsties.

Unfortunate food aversions can develop in people who receive chemotherapy. Even though they are nauseous, children have to eat regular meals and, subsequently, they may dislike the food because of its association with feeling sick. Even anorexia nervosa may be the result.

13.5.2 Social and sexual odors

Removing or covering body odors is an age-old preoccupation and the basis of lucrative toiletries and perfume industries. Obviously, odors – good and bad – and their control are extremely important to humans. Yet the often-posed question of *human pheromones* has not been answered satisfactorily. It probably never will if we apply standards and definitions that were developed originally for insects – not even for non-human vertebrates. Stories of potent and irresistible sex attractants have remained the province of imaginative writers, as in *Perfume* by Patrick Süsskind, *Come Out To Play* by Alex Comfort, or *Switch Bitch* by Roald Dahl. The much more meager results scientists have reaped so far probably reflect the real biological state of affairs. First, female odors may be important as *primers* that drive – among other stimuli – the ovarian cycles of associating women (McClintock, 1971; Preti *et al.*, 1986). Second, daily nasal application of a chemical analogue of luteinizing hormone-releasing hormone inhibited ovulation in 25 of 27 women (Bergquist *et al.*, 1979). This is the opposite effect to its usual one of increasing fertility. Third, the body odors of men and women may signal attraction or repulsion, based on prior pleasant or negative experiences with certain person(s) and/or odors. Finally, recognition of offspring, mate, sibling, or parent may be aided by body odors.

The influence of odors such as perfumes and fragrances on human behavior is assumed to be acquired, and the responses elicited depend on the often complex previous social experiences. The response will be altered if a laboratory experiment eliminates contextual stimuli (Kirk-Smith and Booth, 1987). Social odors include those of the well-known security blankets in toddlers, familiarly scented bed sheets in new surroundings, and treating insomnia with mother's axillary odor on handkerchief. Removal of bad body odors (diet, metabolism defects) that disrupt interpersonal harmony appears to be universal.

Human odors have been used for sinister purposes. In East Germany, the Stasi (state security police) kept a library of dissidents' smells. Samples of their scarves or dirty underwear were kept in several hundred glass jars. Trained dogs were able to compare the odors of antigovernment materials with those of the standards in this smell library (Rosenberg, 1995). At other times, the persecuted have tried to protect themselves with olfactory countermeasures. During the second World War, the Danish inventor Ernst Morch thwarted the bloodhounds of the Gestapo. The dogs searched for Jews hidden in false bottoms of fishing boats on their way to safety in Sweden. Morch perfected a mixture of rabbit's blood and cocaine that disabled the dogs' noses (Thomas, 1996).

13.5.3 Health and safety hazards

Environmental pollution

The chemical ecology of *Homo sapiens* increasingly includes unpleasant odors from chemical and other factories, oil refineries, rendering plants, sewage plants, car and truck exhausts, lignite fuel burning, dairy or chicken farms, feedlots, silage, liquid manure, or open landfills. These odors may present more than mere esthetic problems by physically irritating or even signaling danger from toxins. Discharges from sulfate cellulose plants (Berglund, 1974) and malodors such as hydrogen sulfide from combustion toilets (Lindvall and Svensson, 1974) are well-investigated examples. Many American cities, such as Jacksonville, Florida or Berlin, New Hampshire have struggled with obnoxious industrial odors, mostly from paper mills (the *Aroma of Tacoma*; Pesce, 1990).

Industrial effluents contain sulfur compounds, with their burnt, pungent quality, and fishy or ruinous-smelling nitrogen compounds. Humans are very sensitive to these odors: hydrogen sulfide has an extremely low recognition threshold of 0.00047 ppm (vol.) (Cain, 1978).

Attempts to label dangerous substances with either characteristic or outright unpleasant odors have not been successful with children, the primary victims of accidental poisonings by toxic household products. Children tolerate odors that adults find unpleasant, such as that of butyric acid. The range between the most pleasant and unpleasant odors is much narrower for children around 4 years of age than for adults, and also much narrower than for taste stimuli. This means that olfactory cues are not suited to produce aversive responses in children (Engen, 1974b, Cain, 1978).

Toxic pollutants may affect the chemical ecology of humans indirectly by altering steroid metabolism and other functions, as animal models indicate. The cancer-causing dioxins found in herbicides, for example, may mimic steroid hormones, altering chemical communication. Hormones affect the expression, development, and functioning of both scent production in glands and chemoreception. Fish, amphibians, and mammals provide examples. In Atlantic salmon, *S. salar*, males lower their response to a priming pheromone from females after exposure to the pesticide cypermethrin (Moore and Waring, 2001). Female newts, *Notophthalmus viridescens*, suffer impaired sexual pheromone communication and lower mating success after exposure to the pesticide endosulfan (Park *et al.*, 2001). Lastly, low doses of pesticides even affect the next generation: male mice scent-marked their territories more if they had been exposed *in utero* to dichlorodiphenyltrichloroethane (DDT) and methoxychlor ingested by their pregnant mothers (vom Saal *et al.*, 1995). Zala and Penn (2004) and Clotfelter *et al.* (2004) have reviewed the effects of endocrine-disrupting chemicals on animal behavior.

13.5.4 Odors in advertising

Much has been expected from odors in advertising and sales, as with fresh bread odor, new car odor, or fragrance samples in magazines. But the last, particularly are controversial: readers of magazines and libraries object to the scent strips that permeate the entire issue. In response to "a couple of hundred letters," the *New Yorker* discontinued running fragrance strips in October 1992 after publishing them for over a year. The media continue to report on novel attempts at using odors to sell products. Popcorn smell in movie theaters is said to lift popcorn sales by 5–15%. There are claims that people spent 45% more money at scented gambling machines. Even the "odor of an honest car salesman" has been tried. Hotels are using floral-citrus fragrances in their lobbies, trying to induce guests to stay longer. Real estate agents advise home sellers to scent the premises with floral-spicy, inviting kitchen odors. In one experiment, people spent more time in textile department stores when odors (at near threshold levels) were added to the fresh air supply; however, sales increased only slightly (Kleinfield, 1992). *Public Citizen* and other research and lobbying groups reject such manipulation. Because the odors being used are kept secret, replication and scientific scrutiny are impossible. Some companies scent invoices with the urinary odor of androstenone. The unpleasant odor is supposed to induce the recipient to "pay fast to get rid of the stink" (Kleinfield, 1992).

Odor and other senses: movies with scent

Visual stimuli may influence the perceived odor intensity. An odorous and an odorless solution that were slightly tinted with yellow were more often considered odorous than the corresponding clear solutions with and without odor (Engen, 1972). Based on these findings, Cain (1978) pointed out that in the realm of air pollution the sight of black smoke from a factory may drastically increase the number of *odor* complaints whether or not the odor is indeed more intense.

Motion pictures still lack the odor dimension. Experimental scent movies are improving. The Science City in Paris features a small theater that screens films with an odor component. Scent cans on a carousel rotate in position, get activated, and an air stream blows odors toward the viewer at the appropriate scene. A more recent computer-controlled mixing apparatus can provide many more odor combinations.

GLOSSARY

active space three-dimensional zone near an odor source where the odor concentration is above the biologically relevant threshold of conspecifics; given air movements, the active space typically extends downwind

aggression aggressive Behaviors such as threat, attack, fighting, or chasing that result in dominance or spacing; injury or death may ensue, but are not the typical outcome (Latin: *aggredior, aggressus sum*, to get ready, to assert oneself)

agonistic interactions between animals that include attack, fighting, defense, or flight (Greek: *agonistes,* combatant)

alarm pheromone animal-produced odors that arouse conspecifics and/or cause them to flee, attack, or defend in the face of danger such as predators

alert odor Animal-produced odor that causes conspecifics to interrupt ongoing behavior and prepare them for evasive action in the face of disturbance or danger

alkaloid A heterogeneous class of compounds that contain nitrogen and often taste bitter (Arabic: *Al-qaliy*: ashes of saltwort *Salsola kali*)

allelochemics Interspecific chemical signals (Greek: *allel-*, mutual)

allomone a chemical substance produced or acquired by an organism, which evokes in the receiver a behavioral or physiological reaction adaptively favorable to the emitter (Greek: *allos*, different)

anadromous (fish) migrating upstream (Greek an-, up, upwards; *dromos*, a running, from *dramein,* to run)

anosmia inability to perceive all or certain specific odors

antifeedants chemicals that inhibit food ingestion, usually by herbivores

apocrine gland a modified sweat gland that produces an aqueous secretion and discharges it into the pilosebaceous canal around a hair shaft; milk glands are a type of apocrine gland (Greek: *apo,* away from, off)

aposematism, aposematic an animal signals distastefulness or other dangerous attributes by visual, olfactory, or auditory cues (e.g. warning colors) to would-be predators (Greek *aposemanein,* to announce by sign)

aversion, primary a non-learned aversion, usually to food

avomic unable to detect chemicals by the vomeronasal organ (Graves and Duvall 1985)

bioassay determination of active compounds by experimental presentation to animals in the field or the laboratory; fractions and compounds to be tested depend on prior responses by the experimental animals (response-guided bioassay)

biotransformation conversion of xenobiotics (compounds ingested with food, but not needed by the organism and potentially harmful) into non-toxic, excretable compounds, usually in the liver

body odor a term used in lieu of pheromone when no clear and compulsory function has been established

chemical ecology the study of interactions among and between organisms and between organisms and their abiotic environment by means of chemical stimuli

chemical signal compound that serves as a messenger between organisms, is processed by the chemical senses, and is usually part of an evolved communication system

communication. To give or exchange information, signals, or messages in any meaningful way; many varied definitions exist depending on the context (Latin: *communicare*, to impart or share)

conditioned aversion avoidance of an originally neutral or attractive food (or other stimulus) by an animal as a consequence of becoming sick (nausea), experiencing pain, or other negative effects after consuming the food; a single experience can have lasting effect

coumarin white, vanilla-flavored crystalline substance for baking, perfumes, etc.; the tree coumarou, *Dipteryx odorata*, yields the sweet-scented tonka bean (native of French Guiana) (French *coumaron*, tonka bean)

detoxication *see* biotransformation

eccrine sweat glands these produce an aqueous secretion, open directly to the skin surface (not associated with hairs), and occur over most of the body in humans being most numerous on the sole of the foot and the palm of the hand (Greek *ex,* out, and *krinein*, to separate; referring to the production of a liquid secretion without removing other cell contents)

exocrine gland a gland secreting to the surface of the body, in contrast to *endocrine* glands that send secretions to other tissues *within* the body

fear substance (schreckstoff) compound(s) released by an animal when encountering a danger, such as a predator or rival, or – in the laboratory – electric shock or other noxious stimuli; also known as *alarm substance*

feeding stimulant chemicals from plants or animals that facilitate ingestion

flavor the combination of odor and taste (of a foodstuff)

holocrine gland sebaceous glands, producing a fatty secretion and opening into the pilosebaceous canal of the hair follicle. From Greek *holos*, whole, and *krinein*, to separate, after the mechanism of secreting whereby the entire cell dissolves and forms the secretions)

home range area utilized, but not necessarily defended, by a resident animal, pair, family, or group of animals; it maybe as large as, or larger than, the defended territory and includes all points the resident(s) visit during the non-migratory (especially breeding) phase of the annual cycle

homing an animal's returning to its site of residence after migration, feeding trips, or artificial displacement; the term does not imply any particular mechanism of reaching the target

information in animal communication and cybernetics, this is the property of a signal or through which message something unpredictable and meaningful to a receiver is conveyed; usually measured in bits

kairomones chemical messengers between species where the adaptive benefit falls on the recipient rather than the emitter; can be non-adaptive or maladaptive to the transmitter (Greek: *kairos*, fitness, opportunity, [by stealth, opportunistic)

macrosmatic endowed with keen sense of smell

message information that is transmitted from one individual to another during communication

***MHC* molecules** all-surface glycoprotien encoded by the major histocompatibility complex of genes; they are involved in immunity and cell recognition

mimicry, Batesian resemblance of one animal (the mimic) to another (the model) to the benefit of the mimic, as when the model is dangerous or inedible (described by Henry Walter Bates 1825–1892)

mimicry, Müllerian the resemblance of two species to their mutual advantage, e.g. if two or more distasteful or toxic species share the same visual or chemical pattern or behavior, a predator needs to learn to avoid only one pattern rather than two or more, which reduces predation on the several prey species in question (described by Fritz Müller, 1821–1897)

navigation finding a target by means other than landmarks, usually over large distances

neophobia avoidance of new food or unfamiliar places.

odor image a biologically relevant odor signal consisting of many chemical compounds, all of which are necessary; reducing the number of components results in loss of activity

odor pattern two or more compounds in certain proportions that are characteristic for a secretion from a particular individual, sex, subspecies, etc.; it is usually used for analytical findings, but the pattern is also thought to be the relevant discrimination cue for the animals themselves

odor profile similar to odor pattern

odor trail chemicals deposited by one individual as it moves along the substrate; it is used as cue by conspecifics, but predators and parasites also cue in on it

odortype genetically based individual odor, as in the urine odors in the laboratory mouse controlled by the major histocompatibility complex

olfaction reception of volatile cues and processing of information about them by means of sensory neurons and higher centers; includes the *main olfactory system* and the *vomeronasal organ*

orientation reaching a target area by means of landmarks

osmetrichia modified hairs involved in chemical communication (scent hairs); they store secretion, release odor, and serve in applying scents to the environment

palatability likelihood that a food will be ingested under specific conditions

phenolics compounds with a benzene ring and a hydroxyl group (French *phene*, for a contaminant that showed up in gas for illumination; Greek: *phenain*: appear, show)

pheromone a chemical released (usually in minute amounts) by one organism into the external environment that is detected by conspecifics resulting in behavioral or physiological changes; these are likely to benefit both individuals (Greek: *pherein*, to carry)

pilosebaceous canal the space around a hair follicle; it receives secretions from sebaceous and apocrine glands and conveys them to the surface of the skin

poison term refers to any substance (of biological or synthetic origin) that is ingested, inhaled, or absorbed and has detrimental systemic effects on respiratory enzymes, neural transmission, or cell-membrane processes; in biology, it is often used for toxic compounds present in potential prey species and it may be associated with visual signals such as "warning colors," as in tropical frogs, or other cues to signal danger to predators (aposematism)

preference a higher likelihood that an animal approaches or stays near one of two or more stimuli (odors or animals), or eats more of a food than from others, under specific circumstances and given a peculiar set of choices; it is a relative measure and characteristic neither of the stimulus nor of the animal

pregnancy block termination of pregnancy upon exposure to an adult male (or his odor) different from the stud male; it occurs typically up to the blastula stage

primary aversion the (usually taste) aversion is not learned

primer (priming) pheromones intraspecific chemical cues that modulate physiological processes, such as growth, development, sexual maturation, estrus cycles, or pregnancy

releaser pheromones intraspecific chemical signals that trigger behavioral responses immediately (i.e. within seconds or minutes of exposure); the original notion of a rigid response being triggered is not easily applicable to mammals

repellent natural repellents are secretions that irritate, are malodorous, or are distasteful to other animals, notably predators, and can be ejected or sprayed; synthetic repellents are compounds or concoctions that reduce damage (by consuming or soiling) inflicted by animals to humans, livestock, crops, forests, stored foods, materials, buildings or any other assets

scent marking depositing glandular secretions, urine, feces, saliva, or their mixtures on the substrate (e.g. soil, rocks, vegetation, or other, often conspicuous features of the environment) or other, conspecifics; the behavior is olfactorily guided as evidenced by sniffing before and after marking and subsequent sniffing by conspecifics and specific, often rigidly elaborated motor patterns are used. The scent mark conveys information on species, sex, age, kinship, reproductive state, social status, or individuality; it facilitates reproductive behavior and regulates social encounters, thereby increasing the inclusive fitness of both marker and addressee

scent matching comparing a scent mark or airborne odor with the odor of a territory owner; this is essential in territorial defense (coined by Gosling, 1981)

semiochemicals chemical compounds or mixtures of compounds that serve as signals in communication. (Greek: *semeion* sign, flag)

sex attractant chemical produced by one sex that causes members of the other sex of the same species to move toward the sender

signal something that conveys information and occasions a certain action or response; it may be a behavior, a chemical, or other communication and it may or may not serve other functions as well

Signaling pheromone provides information to the recipient that may or may not lead to a change in the recipient's behavior; also called *informer pheromone*

skin gland a composite glandular area (macrogland) on the surface of a vertebrate body; it consists mostly of varying proportions of sebaceous and apocrine glands, along with hair, muscles, and blood vessels

stress odor odor emanating from an animal that experiences stress, trauma or danger; in the laboratory, stress odors are stimulated by predators, their odor, or electric foot shock; the odor may transmit information to conspecifics

synergism two or more compounds act together to achieve or enhance a specific biological effect: the total effect is greater than the sum of the component effects and, individually, the compounds may even have no effect

terpene a large class of compounds with the 5-carbon unit isopentane as bulding block; they are classified by multiples of 5-carbon units: monoterpenes (C_{10}), diterpenes (C_{20}), etc. (word derived from *turpentine*)

territory an area defended by an individual, pair, family, clan, or group of animals against conspecifics, and used for breeding, and/or feeding; it should be distinguished from *home range*

threshold smallest amount of a compound detected by a given organism; *detection* threshold and *recognition* threshold may lie at different concentrations

toxin any poison derived from a plant, animal, or microorganism; toxins can cause damage to organisms that feed on, or come in contact with, the producing organism and they are usually considered to be adaptive for the producer as defensive compounds (often also used for specific chemical compounds present in a venom)

trail pheromone chemical(s) deposited on the substrate that conspecifics follow; common in insects but also occurs in snakes and deer

venom toxins produced by animals and *injected* by stings, spines, or specialized mouth parts such as fangs in snakes; an animal may use venom for preying as well as antipredator defense, as in venomous snakes; venoms have a complex composition and often are potent enzymes

volatility ability of compounds to evaporate and disperse through air, thus suited for distance communication between animals (or even plants); a small molecular size is required

vomerolfaction the vomeronasal sense; used analogous to olfaction or gustation

vomodor the largely non-volatile chemical stimuli received by the vomeronasal organ

xenobiotics natural or synthetic compounds that animals ingest but do not need; they are often harmful and have to be detoxified, usually in the liver

REFERENCES

Abbott, D. H., Baines, D. A., Faukes, C. G., *et al.* (1990). A natural deer repellent: chemistry and behavior. In *Chemical Signals in Vertebrates*, vol. 5, ed. D. W. Macdonald, D. Müller-Schwarze, and S. E. Natynczuk, pp. 599–609. Oxford: Oxford University Press.

Adams, M. A. and Johnsen, P. B. (1986). Chemical control of feeding in herbivorous and carnivorous fish. In *Chemical Signals in Vertebrates*, vol. 4, ed. D. Duvall, D. Müller-Schwarze, and R. M. Silverstein, pp. 45–61. New York: Plenum.

Adams, M. A., Teeter, J. H., Katz, Y., and Johnsen, P. B. (1987). Sex pheromones of the sea lamprey, *Petromyzon marinus*: steroid studies. *Journal of Chemical Ecology* **13**, 387–395.

Adams, R., Johnson, J. R., and Wilcox, C. F. (1970). *Laboratory Experiments in Organic Chemistry*, 6th edn. London: MacMillan.

Adron, J. W. and Mackie, A. M. (1978). Studies on the chemical nature of feeding stimulants for rainbow trout, *Salmo gairdneri* Richardson. *Journal of Fish Biology* **12**, 303–310.

Agami, M. and Waisel, Y. (1988). The role of fish in distribution and germination of seeds of the submerged macrophytes *Najas marina* and *Ruppia maritima*. *Oecologia* **76**, 83–88.

Ågren, G., Zhou, Q., and Zhong, W. (1989). Ecology and social behavior of Mongolian gerbils, *Meriones unguiculatus*, at Xilinhot, Inner Mongolia, China. *Animal Behaviour* **37**, 11–27.

Ahn, Y.-J., Lee, S.-B., Okubo, T., and Kim, M. (1995). Antignawing factor of crude oil derived from *Thujopsis dolabrata* S. and Z. var. *hondai* sawdust against mice. *Journal of Chemical Ecology* **21**, 263–271.

Albers, H. E. and Rowland, C. M. (1989). Ovarian hormones influence odor stimulated flank marking behavior in the hamster, *Mesocricetus auratus*. *Physiology and Behavior* **45**, 113–118.

Alberts, A. C. (1989). Ultraviolet visual sensitivity in desert iguanas: implications for pheromone detection. *Animal Behaviour* **38**, 129–137.

(1992a). Constraints on the design of chemical communication systems in terrestrial vertebrates. *American Naturalist* **139**, 562–589.

(1992b). Density-dependent scent gland activity in desert iguanas. *Animal Behaviour* **44**, 774–776.

(1992c). Pheromonal self-recognition in desert iguanas. *Copeia* **1**, 229–231.

(1993). Chemical and behavioral studies of femoral gland secrections in iguanid lizards. *Brain Behavior and Evolution* **41**, 255–260.

Alberts, A. C. and Werner, D. I. (1993). Chemical recognition of unfamiliar conspecifics by green iguanas: functional significance of different signal components. *Animal Behaviour* **46**, 197–199.

Alberts, A. C., Sharp, T. R., Werner, D. I., and Weldon, P. J. (1992). Seasonal variation of lipids in femoral gland secretions of male green iguanas (*Iguana iguana*). *Journal of Chemical Ecology* **18**, 703–712.

Alberts, A. C., Jackintell, L. A., and Phillips, J. A. (1994). Effects of chemical and visual exposure to adults on growth, hormones, and behavior of juvenile green iguanas. *Physiology and Behavior* **55**, 987–992.

Albone, E. S. (1984). *Mammalian Semiochemistry. The Investigation of Chemical Signals between Mammals*. Chichester: Wiley.

Albone, E. S., Blazquez, N. B., French, J., Long, S. E., and Perry, G. C. (1986). Mammalian semiochemistry: issues and futures, with some examples from a study of chemical signalling in cattle. In *Chemical Signals in Vertebrates*, vol. 4, ed. D. Duvall, D. Müller-Schwarze, and R. M. Silverstein, pp. 27–36. New York: Plenum.

Albonetti, B. D. and D'Udine, B. (1986). Social experience occuring during adult life: its effects on socio-sexual olfactory preferences in inbred mice, *Mus musculus*. *Animal Behaviour* **34**, 1844–1847.

Aleksiuk, M. (1968). Scent mound communication, territoriality, and population regulation in the beaver. *Journal of Mammalogy* **49**, 759–762.

Alexander, G., Stevens, D., and Bradley, L. R. (1987). Fostering in sheep V. Use of unguens to foster an additional lamb onto a ewe with a single lamb. *Applied Animal Behaviour Science* **17**, 95–108.

Alexander, G., Goodrich, B. S., Stevens, D., and Bradley, L. R. (1989). Maternal interest in lambs smeared with polar and nonpolar substances. *Australian Journal of Experimental Agriculture* **29**, 513–516.

Ali, M. A. (ed.) (1978). *Sensory Ecology*. New York: Plenum Press.

Allan, G. G., Gustafson, D. I., Mikels, R. A., Miller, J. M., and Neogi, S. (1984). Reduction of deer browsing of Douglas fir (*Pseudotsuga menziesii*) seedlings by quadrivalent selenium. *Forest Ecology and Management* **7**, 161–181.

Allison, M. J. (1978). The role of ruminal microbes in the metabolism of toxic constituents from plants. In *Effects of Poisonous Plants on Livestock*, ed. R. F. Keeler, K. R. Van Kampen, and L. J. James, pp. 101–118. New York: Academic Press.

Alyan, S. H. and Jander, R. (1994). Short-range learning in the house mouse, *Mus musculus*: stages in the learning of directions. *Animal Behaviour* **48**, 285–298.

(1997). Interplay of directional navigation mechanisms as a function of near-goal distance: experiments with the house mouse. *Behavioural Processes* **41**, 245–255.

American Meteorological Association (2000). *Glossary of Meterorology*, 2nd edn.

Ames, B. N. and Gold, L. S. (1990). Too many rodent carcinogens: mitogenesis increases mutagenesis. *Science* **249**, 970–971.

Ames, B. N., Profet, M., and Gold, L. S. (1990). Nature's chemicals and synthetic chemicals: comparative toxicology. *Proceedings of the National Academy of Sciences of the USA* **87**, 7782–7786.

Amoore, J. E., Pelosi, P., and Forrester, L. J. (1977). Specific anosmias to 5α-androst-16-en-3-one and ω-pentadecalactone: the urinous and musky primary odors. *Chemical Senses and Flavour* **2**, 401–425.

Andelt, W. F., Burnham, K. P., and Manning, J. A. (1991). Relative effectiveness of repellents for reducing mule deer damage. *Journal of Wildlife Management* **55**, 341–347.

Andelt, W. F., Baker, D. L., and Burnham, K. P. (1992). Relative preference of captive cow elk for repellent-treated diets. *Journal of Wildlife Management* **56**, 164–173.

Andersen, K. K. and Bernstein, D. T. (1975). Some chemical constituents of the scent of the striped skunk (*Mephitis mephitis*). *Journal of Chemical Ecology* **1**, 493–499.

Andren, C. (1982). The role of the vomeronasal organs in the reproductive behavior of the adder *Viper berus*. *Copeia* **1982**, 148–157.

Angelon, K. A. and Petranka, J. W. (2002). Chemicals of predatory mosquitofish (*Gambusia affinis*) influence selection of oviposition site by *Culex* mosquitoes. *Journal of Chemical Ecology* **28**, 797–806.

Apfelbach, R. (1973). Olfactory sign stimulus for prey selection in polecats. *Zeitschrift für Tierpsychologie* **33**, 270–273.

Apfelbach, V. R., Schutz, S., and Slotnick, B. (1990). Eine verhaltensphysiologische Untersuchung zur Ermittlung olfaktorischer Schwellenwerte bei männlichen Ratten. *Zeitschrift für Säugetierkunde* **55**, 407–412.

Applebaum, S. W. and Birk, Y. (1979). Saponins. In *Herbivores: Their Interaction with Secondary Plant Metabolites,* ed. G. A. Rosenthal and D. H. Janzen, pp. 539–566. New York: Academic Press.

Applegate, R. D., Rogers, L. L., Casteel, D. A., and Novak, J. M. (1979). Germination of cow parsnip seeds from grizzly bear feces. *Journal of Mammalogy* **60**, 655.

Apps, P., Viljoen, H. W., Richardson, P. R. K., and Pretorius, V. (1989). Volatile components of anal gland secretion of aardwolf (*Proteles cristatus*). *Journal of Chemical Ecology* **15**, 1681–1688.

Archer, A. T. and Glen, R. M. (1969). Observations on the behavior of two species of honey-guides *Indicator variegatus* (Lesson) and *Indicator exilia* (Cassin). *Los Angeles County Museum Contributions in Science* **160**, 1–6.

Archibald, J. D. (1996). Fossil evidence for a late Cretaceous origin of 'hoofed" mammals. *Science* **272**, 1150–1152.

Arnold, G. W. (1966). The special senses in grazing animals. II. Smell, taste, and touch and dietary habits in sheep. *Australian Journal of Agricultural Research* **17**, 531–542.

Arnold, G. W. and Hill, J. L. (1972). Chemical factors affecting selection of food plants by ruminants. In *Phytochemical Ecology,* ed. J. B. Harborne, pp. 71–101. London: Academic Press.

Arnold, G. W., de Boer, E. S., and Boundy, C. A. P. (1980). The influence of odour and taste on the food preferences and food intake of sheep. *Australian Journal of Agricultural Research* **31**, 571–587.

Arnold, S. J. (1978). Some effects of early experience on feeding responses in the common garter snake, *Thamnophis sirtalis*. *Animal Behaviour* **26**, 455-462.

(1981a). Behavioral variation in natural populations. I. Phenotypic, genetic and environmental correlations between chemoreceptive responses to prey in the garter snake, *Thamnophis elegans*. *Evolution* **35**, 489–509.

(1981b). Behavioral variation in natural populations. II. The inheritance of a feeding response in crosses between geographic races of the garter snake, *Thamnophis elegans*. *Evolution* **35**, 510–515.

Arnould, C. and Signoret, J. P. (1993). Sheep food repellents: efficacy of various products, habituation, and social facilitation. *Journal of Chemical Ecology* **19**, 225–236.

Arnould, C., Piketty, V., and Levy, F. (1991). Behavior of ewes at parturition toward amniotic fluids from sheep, cows and goats. *Applied Animal Behaviour Science* **32**, 191–196.

Arnould, C., Orgeur, P., Sempéré, A., and Signoret, J.-P. (1993). Repulsion alimentaire chez trois espèces d'ongulés en situation de pâturage: effet des excréments de chien. *Revue Ecologique (Terre Vie)* **48**, 121–132.

Arnould, C., Malosse, C., Signoret, J.-P., and Descoins, C. (1998). Which chemical constituents from dog feces are involved in its food repellent effect in sheep? *Journal of Chemical Ecology* **24**, 559–576.

Arnts, R. R., Peterson, W. B., Seila, R. L., and Gay, B. W., Jr. (1982). Estimates of alpha-pinene emissions from a loblolly pine forest using an atmospheric diffusion model. *Atmospheric Environment* **16**, 2127–2137.

Asa, C. S. (1993). Relative contributions of urine and anal sac secretion in scent marks of large felids. *American Zoologist* **33**, 167–172.

Asa, C. S., Peterson, E. K., Seal, U. S., and Mech, D. L. (1985). Deposition of anal-sac secretions by captive wolves (*Canis lupus*). *Journal of Mammalogy* **66**, 89–93.

Asa, C. S., Mech, L. D., and Seal, U. S. (1990). The influence of social and endocrine factors on urine-marking by captive wolves (*Canis lupis*). *Hormones and Behavior* **24**, 497–509.

Ashkenazy, D., Kashman, Y., Nyska, A., and Friedman, J. (1985). Furocoumarins in shoots of *Pituranthos triradiatus* (Umbelliferae) as protectants against grazing by hyrax (*Procavia capensis syriaca*). *Journal of Chemical Ecology* **11**, 231–239.

Atema, J. (1986). Review of sexual selection and chemical communication in the lobster, *Homarus americanus*. *Canadian Journal of Fisheries and Aquatic Sciences* **43**, 2283–2290.

(1988). Distribution of chemical stimuli. In *Sensory Biology of Aquatic Animals*, ed. J. Atema, A. N. Popper, R. R. Fay, and W. H. Tavolga, pp. 29–56. Heidelberg: Springer-Verlag.

Atema, J., Holland, K., and Ikehara, W. (1980). Olfactory responses of yellow fin tuna (*Thunnus albacares*) to prey odors: chemical search image. *Journal of Chemical Ecology* **6**, 457–465.

Auffenberg, W. (1981). *The Behavioral Ecology of the Komodo Monitor.* Gainesville, FL: University Press of Florida.

Augner, M. and Bernays, E. A. (1998). Plant defence signals and Batesian mimicry. *Evolutionary Ecology* **12**, 667–680.

Austin, P. J., Suchar, L. A., Robbins, C. T., and Hagerman, A. E. (1989). Tannin-binding proteins in saliva of deer and their absence in saliva of sheep and cattle. *Journal of Chemical Ecology* **15**, 1335–1347.

Avery, M. L. (1989). Experimental evaluation of partial repellent treatment for reducing bird damage to crops. *Journal of Applied Ecology* **26**, 433–439.

Avery, M. L., Decker, D. G., and Nelms, C. O. (1992). Use of a trigeminal irritant for wildlife management. In *Chemical Signals in Vertebrates*, vol. 6, ed. R. L. Doty and D. Müller-Schwarze, pp. 319–322. New York: Plenum.

Avery, M. L., Decker, D. G., Humphrey, J. S., *et al.* (1995). Methyl anthranilate as a rice seed treatment to deter birds. *Journal of Wildlife Management* **59**, 50–56.

Avery, M. L., Humphrey, J. S., and Decker, D. G. (1997). Feeding deterrence of anthraquinone, anthracene, and anthrone to rice-eating birds. *Journal of Wildlife Management* **61**, 1359–1365.

Aylor, D. E. (1976). Estimating peak concentrations of pheromones in the forest. In *Perspectives in Forest Entomology*, ed. J. F. Anderson and H. K. Kaya, pp. 177–188. New York: Academic Press.

Badcock, J. (1986). Aspects of the reproductive biology of *Gonostoma bathyphilus*, Gonostomatidae. *Journal of Fish Biology* **29**, 589–604.

Bailey, K. (1978). Flehmen in the ring-tailed lemur (*Lemur catta*). *Behavior* **LXV**, 309–319.

Baker, H. G. (1963). Evolutionary mechanisms in pollination biology. *Science* **139**, 877–883.

Balogh, R. D. and Porter, R. H. (1986). Olfactory preferences resulting from mere exposure in human neonates. *Infant Behavior and Development* **9**, 395–402.

Balthazart, J. and Schoffeniels, E. (1979). Pheromones are involved in the control of sexual behavior in birds. *Naturwissenschaften* **66**, 55–56.

Bang, B. G. (1965). Anatomical adaptations for olfaction in the snow petrel. *Nature* **205**, 513–515.

(1971). *Functional Anatomy of the Olfactory System in Orders of Birds*. Basel: Karger.

Bang, B. G. and Cobb, S. (1968). The size of the olfactory bulb in 108 species of birds. *Auk* **85**, 55–61.

Bardach, J. E. and Villars, T. (1974). The chemical senses of fishes. In *Chemoreception by Marine Organisms*, ed. P. T. Grant and A. M. Mackie, pp. 49–104. New York: Academic Press.

Bardach, J. E., Todd, J. D., and Crickmes, R. (1967). Orientation by taste in fish of the genus *Ictalurus*. *Science* **155**, 1276–1278.

Barinaga, M. (1999). Salmon follow watery odors home. *Science* **286**, 705–706.

Barnard, C. J. and Fitzsimmons, J. (1988). Kin recognition and mate choice in mice: the effects of kinship, familiarity and social interference on intersexual interaction. *Animal Behaviour* **36**, 1078–1090.

Barrette, C. (1977). The social behavior of captive muntjacs (*Muntiacus reevesi*). *Zeitschrift für Tierpsychologie* **43**, 188–213.

Barron, S., Gagnon, W. A., Mattson, S. N., *et al.* (1988). The effects of prenatal alcohol exposure on odor associative learning in rats. *Neurotoxicology and Teratology* **10**, 333–340.

Barrows, E. M. (1995). *Animal Behavior Desk Reference*. Boca Raton, FL: CRC Press.

Barthalmus, G. T. and Zielinski, W. J. (1988). *Xenopus* skin mucus induces oral dyskinesias that promote escape from snakes. *Pharmacology, Biochemistry and Behavior* **30**, 957–959.

Bastakov, V. A. (1986). Preference by young-of-the-year of the edible frog (*Rana esculenta* complex) for their own reservoir ground smell. *Zoologicheskii Zhurnal* **65**, 1864–1868.

Bate-Smith, E. C. (1972). Attractants and repellents in higher animals. *Proceedings of the Phytochemistry Society* **8**, 45–56.

Batzli, G. O. and Pitelka, F. A. (1971). Condition and diet of cycling populations of the California vole, *Microtus californicus*. *Journal of Mammalogy* **52**, 141–163.

Batzli, G. O., Getz, L. L., and Hurley, S. S. (1977). Suppression of growth and reproduction of microtine rodents by social factors. *Journal of Mammalogy* **58**, 583–591.

Baudinette, R. V., Wheldrake, J. F., Hewitt, S., and Hawke, D. (1980). The metabolism of [^{14}C] phenol by native Australian rodents and marsupials. *Australian Journal of Zoology* **28**, 511–520.

Baumann, T. W. and Gabriel, H. (1984). Metabolism and excretion of caffeine during germination of *Coffea arabica*. L. *Plant Cell Physiology* **25**, 1431–1436.

Bäumler, W. and Haag, I. (1989). Sekrete und Delta-16-steroide als Lockmittel für die Rötelmaus (*Clethrionomys glareolus*). *Anzeiger für Schädlingskunde, Pflanzenschutz und Umweltschutz* **62**, 55–59.

Bäumler, W. and Hock, V. (1987). Über den Einfluss von Duftstoffen und Sekreten auf Verhalten und Entwicklung der Rötelmaus (*Clethrionomys clareolus*) und sympatrischer Arten. *Anzeiger für Schädlingskunde, Pflanzenschutz und Umweltschutz* **60**, 105–109.

Bauwens, D., Nuijten, K., van Wezel, H., and Verheyen, R. F. (1987). Sex recognition by males of the lizard *Lacerta vivipara*: an introductory study. *Amphibia-Reptilia* **8**, 49–58.

Beaman, A. L. (1988). A novel approach to estimating the odor concentration distribution in the community. *Atmospheric Environment* **22**, 561–568.

Bean, N. J. (1982). Modulation of agonistic behavior by the dual olfactory system in male mice. *Physiology and Behavior* **29**, 433–437.

Bean, N. J., Galef, B. G., Jr., and Mason, J. R. (1988). The effect of carbon disulfide on food consumption by house mice. *Journal of Wildlife Management* **52**, 502–507.

Bearder, S. W. and Randall, R. M. (1978). The use of fecal marking sites by spotted hyenas and civets. *Carnivore* **1**, 32–48.

Beauchamp, G. K. (1976). Diet influences attractiveness of urine in guinea pigs. *Nature* **263**, 587–588.

Beauchamp, G. K. and Berüter, J. (1973). Source and stability of attractive components in guinea pig (*Cavia porcellus*) urine. *Behavioral Biology* **9**, 43–47.

Beauchamp, G. K., Doty, R. L., Moulton, D. G., and Mugford, R. A. (1976). The pheromone concept in mammalian chemical communication: a critique. In *Mammalian Olfaction, Reproductive Processes and Behavior*, ed. R. L. Doty, pp. 144–160. New York: Academic Press.

Beauchamp, G. K., Wysocki, C. J., and Wellington, J. L. (1985). Extinction of response to urine odor as a consequence of vomeronasal organ removal in male guinea pigs. *Neuroscience* **99**, 950–955.

Beauchamp, G. K., Yamazaki, K., Bard, J., and Boyse, E. A. (1988). Preweaning experience in the control of mating preferences by genes in the major histocompatibility complex of the mouse. [*Symposium of the American Society of Zoology on Kin Recognition in Animals: Empirical Evidence and Conceptual Issues*, New Orleans, December, 1987.] *Behavioral Genetics* **18**, 537–548.

Beauchamp, G. K., Curran, M., and Yamazaki, K. (2000). MHC-mediated fetal odourtypes expressed by pregnant females influence male associative behavior. *Animal Behaviour* **60**, 289–295.

Bechler, D. L. (1986). Pheromone and tactile communication in the subterranean salamander, *Typhlomolge rathbuni*. *Proceedings of the Ninth International Congress of Speleology*, Barcelona, Spain, vol. 2, pp. 120–122.

Becker, J. and van Raden, H. (1986). Meteorologische Gesichtspunkte zur olfaktorischen Navigationshypothese. *Journal of Ornithology* **127**, 1–8.

Begg, C. M., Begg, K. S., Du Toit, J. T., and Mills, M. G. L. (2003). Scent-marking behaviour of the honey badger, *Mellivora capensis* (Mustelidae), in the southern Kalahari. *Animal Behaviour* **66**, 917–929.

Bekoff, M. and Wells, M. C. (1980). The social ecology of coyotes. *Scientific American* **242**, 9–30.

Belcher, A., Epple, G., Küderling, I., and Smith, A. B. III. (1988). Volatile components of scent material from cotton-top tamarin (*Saguinus o. oedipus*): a chemical and behavioral study. *Journal of Chemical Ecology* **14**, 1367–1384.

Belcher, A. M., Epple, G., Greenfield, K. L., *et al.* (1990). Proteins: biologically relevant components of the scent marks of a primate (*Saguinus fuscicollis*). *Chemical Senses* **15**, 431–446.

Belete, H., Tikubet, G., Petros, B., Oyibu, W. A., and Otigbuo, I. N. (2004). Control of human African trypanosomiasis: trap and odour preference of tsetse flies (*Glossina morsitans submorsitans*) in the upper Didessa River Valley of Ethiopia. *Tropical Medicine and International Health* **9**, 710–714.

Bell, C. M. and Harestad, A. S. (1987). Efficacy of pine oil as repellent to wildlife. *Journal of Chemical Ecology* **13**, 1409–1417.

Bell, D. J. (1986). Social effects on physiology in the European rabbit. *Mammal Review* **16**, 131–138.

Bellairs, A. (1970). *The Life of Reptiles*. New York: Universe Books.

Bellringer, J. F., Pratt, H. P. M., and Keverne, E. B. (1980). Involvement of the vomeronasal organ and prolactin in pheromonal induction of delayed implantation in mice. *Journal of Reproduction and Fertility* **59**, 223–228.

Belvedere, P., Colombo, L., Giacoma, C., Malacarne, G., and Andreoletti, G. E. (1988). Comparative ethological and biochemical aspects of courtship pheromones in European newts. *Monitore Zoologica Italiano (Italian Journal of Zoology)* **22**, 397–403.

Benjamini, L. (1987). Primer pheromones in the Levant vole (*Microtus guentheri*): activation of reproduction in the female by male-related stimuli. *Phytoparasitica* **14**, 3–14.

Bennett, M. H. (1968). The role of the anterior limb of the anterior commissure in olfaction. *Physiology and Behavior* **3**, 507–515.

Bennett, N. and Jarvis, J. U. M. (1988). The social structure and reproductive biology of colonies of the mole-rat, *Cryptomys damareasis* (Rodentia, Bathyersidae). *Journal of Mammalogy* **69**, 293–302.

Benson, J. M., Seiber, J. N., Bagley, C. V., *et al.* (1979). Effects on sheep of the milkweeds *Asclepias eriocarpa* and *A. labriformis* and of cardiac glycoside-containing derivative materials. *Toxicon* **17**, 155–165.

Bent, A. C. (1964). *Life Histories of North American Nuthatches, Wrens, Thrashers, and Their Allies.* New York: Dover.

Benvenuti, S., Fiaschi, V., Fiore, L., and Papi, F. (1973). Homing performance of inexperienced and directionally trained pigeons subjected to olfactory nerve section. *Journal of Comparative Physiology* **83**, 81–92.

Benvenuti, S., Ioalé, P., and Papi, F. (1992). The olfactory map of homing pigeons. In *Chemical Signals in Vertebrates*, vol. 6, ed. R. L. Doty and D. Müller-Schwarze, pp. 429–434. New York: Plenum.

Berejikian, B. A., Smith, R, J. F., Tezak, E. B., Schultz, W., and Knudsen, C. M. (1999). Chemical alarm signals and complex hatchery rearing habitats affect antipredator behavior and survival of Chinook salmon (*Oncorhynchus tshawytscha*) juveniles. *Canadian Journal of Fisheries and Aquatic Sciences* **56**, 830–838.

Berenbaum, M. (1981). Patterns of furanocoumarin production and insect herbivory in a population of wild parsnip (*Pastinaca sativa* L.). *Oecologia* **49**, 236–244.

Berenbaum, M. R. (1991). Coumarins. In *Herbivores: Their Interactions with Secondary Plant Metabolites*, ed. G. A. Rosenthal and M. R. Berenbaum, pp. 221-249. San Diego, CA: Academic Press.

Berger, P. J., Sanders, E. H., Gardner, P. D., and Negus, N. C. (1977). Phenolic plant compounds functioning as reproductive inhibitors in *Microtus montanus*. *Science* **195**, 575–577.

Berger, P. J., Negus, N. C., Sanders, E. H., and Gardner, P. D. (1981). Chemical triggering of reproduction in *Microtus montanus*. *Science* **214**, 69–70.

Berger, P. J., Negus, N. C., and Rowsemitt, C. N. (1987). Effect of 6-methoxybenzoazolinone on sex ratio and breeding performance in *Microtus montanus*. *Biology of Reproduction* **36**, 255–260.

Bergeron, J. M., Jodoin, L., and Jean, Y. (1987). Pathology of voles (*Microtus pennsylvanicus*) fed with plant extracts. *Journal of Mammalogy* **68**, 73–79.

Berglund, B. (1974). Quantitative and qualitative analysis of industrial odors with human observers. *Annals of the New York Academy of Sciences* **237**, 35–51.

Bergquist, C., Nillius, S. J., and Wide, L. (1979). Inhibition of ovulation in women by intranasal treatment with a luteinizing hormone-releasing agonist. *Contraception* **19**, 497–506.

Berti, R., Thinès, G., and Lefèvre, B. (1982). Effets des informations chimiques provenant d'un milieu habité par des congéneres su l'orientation topographique du poisson cavernicole *Phreatichthys andruzzii* Vinciguerra (Pisces, Cyprinidae). *International Journal of Speleology* **12**, 103–117.

Beyers, D. W. and Farmer, M. S. (2001). Effects of copper on olfaction of Colorado pikeminnow. *Environmental Toxicology and Chemistry* **20**, 907–912.

Beynon, R. G., Robertson, D. H. L., Hubbard, S. J., Gaskell, S. J., and Hurst, J. L. (1999). The role of protein binding in chemical communication. In *Advances in Chemical Signals in Vertebrates,* vol. 8, ed. R. E. Johnston, D. Müller-Schwarze, and P. W. Sorensen, pp. 137–147. New York: Kluwer Academic/Plenum.

Beynon, R. G., Veggerby, C., Payne, C. E., *et al.* (2002). Polymorphism in major urinary proteins: molecular heterogeneity in a wild mouse population. *Journal of Chemical Ecology* **28**: 1429–1446.

Bierich, J. R. (1981). Puberty. [in German.] *Klinische Wochenschrift* **59**, 985–994.

Billard, R., Bieniarz, K., Popek, W., Epler, P., and Saad, A. (1989). Observations on a possible pheromonal stimulation of milt production in carp (*Cyprinus carpio* L.). *Aquaculture* **77**, 387–392.

Billewicz, W. Z., Fellowes, H. M., and Thomson, A. M. (1981). Menarche in Newcastle upon Tyne, England, UK, girls. *Annals of Human Biology* **8**, 313–320.

Birch, E. J., Knight, T. W., and Shaw, G. J. (1989). Separation of male goat pheromones responsible for stimulating ovulatory activity in ewes. *New Zealand Agricultural Research* **32**, 337–342.

Bird, S. and Gower, D. B. (1982). Axillary 5α-androst-11-en-3-one, cholesterol and squalene in men: preliminary evidence for 5α-androst-16-en-3-one being a product of bacterial action. *Journal of Steroid Biochemistry* **17**, 517–522.

Birke, L. I. A. and Sadler, D. (1987a). Differences in maternal behavior of rats and the sociosexual development of the offspring. *Developmental Psychobiology*. **20**, 85–100.

(1987b). Effects of odor familiarity on the development of systematic exploration in the spiny mouse, *Acomys cahirinus*. *Developmental Psychobiology* **20**, 627–640.

Bjerselius, R. and Olsen, K. H. (1993). A study of the olfactory sensitivity of the crucian carp (*Carassius carassius*) and goldfish (*Carassius auratus*) to 17α,20β-dihydroxy-4-pregnen-3-one. *Chemical Senses* **18**, 427–436.

Bjerselius, R., Li, W., Teeter, J. H., *et al.* (2000). Direct behavioral evidence that unique bile acids released by larval sea lamprey (*Petromyzon marinus*) function as a migratory pheromone. *Canadian Journal of Fisheries and Aquatic Science* **57**, 557–569.

Black, G. A. and Dempson, J. B. (1986). A test of the hypothesis of pheromone attraction in salmon migration. *Environmental Biology of Fishes* **15**, 229–235.

Bland, K. P. and Jubilan, B. M. (1987). Correlation of flehmen by male sheep with female behavior and oestrus. *Animal Behaviour* **35**, 735–738.

Blasquez, N. B., Long, S. E., Perry, G. C., and Matson, E. D. (1987). Effect of estradiol-17β on perineal and neck skin glands in heifer calves. *Journal of Endocrinology* **115**, 43–46.

Blass, E. M. and Teicher, M. H. (1980). Suckling. *Science* **210**, 15–22.

Blass, E. M., Shuleikina-Turpaeva, K., and Luschiekin, V. (1988). Sensory determinants of nipple-attachment behavior in 2- to 4-day-old kittens. *Developmental Psychobiology* **21**, 365–370.

Blaustein, A. R. (1981). Sexual selection and mammalian olfaction. *American Naturalist* **117**, 1001–1010.

Blaustein, A. R. and O'Hara, R. K. (1981). Genetic control for sibling recognition? *Nature* **290**, 246–248.

Blaustein, A. R. and Waldman, B. (1992). Kin recogition in anuran amphibians. *Animal Behaviour* **44**, 207–221.

Block, M., Volpe, V., and Hayes, M. (1981). Saliva as a chemical cue in the development of social behavior. *Science* **211**, 1062–1064.

Bloom, S. J. (1975). Olfaction in children one to five years of age. B.A. Thesis, Brown University, Providence, Rhode Island, USA.

Blum, M. S., Murray, S., Jones, T. H., Rinderer, T. E., and Sylvester, H. A. (1988). Oxygenated compounds in beeswax: identification and possible significance. *Biochemical and Molecular Biology* **91**B, 581–583.

Blytt, H. J., Guscar, T. K., and Butler, L. G. (1988). Antinutritional effects and ecological significance of dietary condensed tannins may not be due to binding and inhibiting digestive enzymes. *Journal of Chemical Ecology* **14**, 1455.

Boag, B. (1991). Evaluation of an odour derived from lion faeces on the behavior of wild rabbits. *Annals of Applied Biology* **118** (Suppl. 12).

Boag, B. and Mlotkiewicz, J. A. (1994). Effect of odor derived from lion feces on behavior of wild rabbits. *Journal of Chemical Ecology* **20**, 631–637.

Böcskei, Z., Groom, C. R., Flower, D. R., *et al.* (1992). Pheromone binding to two rodent urinary proteins revealed by X-ray crystallography. *Nature* **360**, 186–188.

Bogert, C. M. (1941). Sensory cues used by rattlesnakes in their recognition of ophidian enemies. *Annals of the New York Academy of Sciences* **41**, 329–343.

Bolhuis, J. J., Strijkstra, A. M., Moore, E., and van der Lende, K. (1988). Preferences for odors of conspecific non-siblings in the common vole, *Microtus arvalis*. *Animal Behaviour* **36**, 1551–1553.

Bollinger, K. S. (1980). Scent marking behavior of beaver (*Castor canadensis*) M. Sc. Thesis, University of Massachusetts, Amherst.

Bonadonna, F. and Bretagnolle, V. (2002). Smelling home: a good solution for burrow-finding in nocturnal petrels? *Journal of Experimental Biology* **205**, 2519–2523.

Bonadonna, F. and Nevitt, G. A. (2004). Partner-specific odor recognition in an Antarctic seabird. *Science* **306**, 835.

Bonadonna, F., Villafane, M., Bajzak, C., and Jouventin, P. (2004). Recognition of burrow's olfactory signature in blue petrels, *Halobaena caerulea*: an efficient discrimination mechanism in the dark. *Animal Behaviour* **67**, 893.

Booth, W. D. (1984). A note on the significance of boar salivary pheromones to the male-effect puberty attainment in gilts. *Animal Production* **39**, 149–152.

(1987). Factors affecting the pheromone composition of voided boar saliva. *Journal of Reproduction and Fertility* **81**, 427–432.

(1989). Boar pheromones: a model for the integration of endocrinology and exocrinology. *Journal of Endocrinology* **123** (Suppl.), 16.

Bossert, W. H. and Wilson, E. O. (1963). The analysis of olfactory communication in animals. *Journal of Theoretical Biology* **5**, 443-469.

Bounds, D. and Pope, G. S. (1960). Light-absorption and chemical properties of mirestrol, the estrogenic substance of *Pueraria mirafica*. *Journal of the Chemical Society* 3696–3705.

Bouvet, J. F., Godinot, F., Delaleu, J. C., and Holley, A. (1989). Interaction between the olfactory and the trigeminal systems in the frog. *Chemical Senses* **14**, 200.

Bowen, W. D. (1978). Social organization of the coyote in relation to prey size, Ph.D. Thesis, University of British Columbia.

Bowers, J. M. and Alexander, B. K. (1967). Mice: individual recognition by olfactory cues. *Science* **158**, 1208–1210.

Bowers, M. D. and Farley, S. (1990). The behaviour of grey jays, Perisoreus canadensis towards palatable and unpalatable Lepidoptera. *Animal Behaviour* **39**, 699–705.

Bowers, M. D. and Larin, Z. (1989). Acquired chemical defense in the lycaenid butterfly, *Eumaeus atala*. *Journal of Chemical Ecology* **15**, 1133–1146.

Boyd, S. K. and Blaustein, A. R. (1985). Familiarity and inbreeding avoidance in the gray-tailed vole (*Microtus canicaudus*). *Journal of Mammalogy* **66**, 348–352.

Boyle, R. (1999). Folivorous specialization: adaptations in the detoxification of the dietary terpene, *p*-cymene, in Australian marsupial folivores. *American Zoologist* **39**, 102A.

Boyle, R. R. and McLean, S. (2004). Constraint of feeding by chronic ingestion of 1,8-cineole in the brushtail possum (*Trichosurus vulpecula*). *Journal of Chemical Ecology* **30**, 757–775.

Bradbury, J. W. and Vehrencamp, S. L. (1977). Social organization and foraging in emballonurid bats. III–IV. *Behavioral Ecology* **2**, 1–29.

Brannon, E. L. and Quinn, T. B. (1990). Field test of the pheromone hypothesis for homing by Pacific salmon. *Journal of Chemical Ecology* **16**, 603–609.

Braun, J. J. and Marcus, J. (1969). Stimulus generalization among odorants by rats. *Physiology and Behavior* **4**, 245–248.

Bray, R. O., Wamboldt, C. L., and Kelsey, R. G. (1991). Influence of sagebrush terpenoids on mule deer preference. *Journal of Chemical Ecology* **17**, 2053–2062.

Brennan, P., Kaba, H., and Keverne, E. B. (1990). Olfactory recognition: a simple memory system. *Science* **250**, 1223–1226.

Brett, J. R. and MacKinnon, D. (1954). Some aspects of olfactory perception in migrating adult coho and spring salmon. *Journal of the Fisheries Research Board of Canada* **11**, 310–318.

Brett, L. P., Hankins, W. G., and Garcia, J. (1976). Prey-lithium aversions III: Buteo hawks. *Behavioral Biology* **17**, 87–98.

Brightsmith, D. J. (2002). *The Tambopata Macaw Project*. Washington, DC: Earthwatch Institute.

Brightwell, R., Dransfield, R. D., Kyorku, C., Goldes, T. K., and Tarimo, S. A. R. (1987). A new trap for *Glossina pallidipes*. *Tropical Pest Management* **33**, 151–159.

Brinck, C., Erlinge, S., and Sandell, M. (1983). Anal sac secretion in mustelids: a comparison. *Journal of Chemical Ecology* **9**, 727–746.

Brock, O. G. and Meyers, S. N. (1979). Responses of ingestively native *Lampropeltis getulus* (Reptilia, Serpentes, Colubridae) to prey extracts. *Journal of Herpetology* **13**, 209–212.

Brodie, E. D., Jr. (1977). Hedgehogs use toad venom in their own defense. *Nature* **268**, 627–628.

Brodie, E. D., Jr. and Brodie, E. D. III. (1980). Differential avoidance of mimetic salamanders by free-ranging birds. *Science* **208**, 181–182.

Brodie, E. D. Jr., Hensel, J. L., and Johnson, J. A. (1974). Toxicity of the urodele amphibians *Taricha, Notophthalmus, Gynops* and *Paramesotriton*. *Copeia* **2**, 506–511.

Brodie, E. D. Jr., Nowak, R. T., and Harvey, W. R. (1979). The effectivness of anti-predator secrections and behavior of selected salamanders against shrews. *Copeia* **2**, 270–274.

Brodie, E. D., Jr., Ridenhour, B. J., and Brodie, E. D., III. (2002). The evolutionary response of predators to dangerous prey: hotspots and coldspots in the geographic mosaic of coevolution between garter snakes and newts. *Evolution* **56**, 2067–2082.

Bronmark, C. and Miner, J. G. (1992). Predator-induced phenotypical change in body morphology in Crucian carp. *Science* **258**, 1348.

Bronson, F. H. (1979). The reproductive ecology of the house mouse. *Quarterly Review of Biology* **54**, 265–299.

Bronson, F. H. and Coquelin, A. (1980). The modulation of reproduction by priming pheromones in house micee: Speculatons on adaptive function. In *Chemical Signals: Vertebrates and Aquatic Invertebrates*, ed. D. Müller-Schwarze and R. M. Silverstein, pp. 243–265. New York: Plenum.

Brooksbank, B. W. L. (1970). Labelling of steroids in axillary sweat after administration of ^{3}H-5-pregnenolone and ^{14}C-progesterone to a healthy man. *Experientia* **26**, 1012–1014.

Brooksbank, B. W. L., Brown, R., and Gustafsson, J. A. (1974). The detection of 5α-androst-16-en-3α-ol in human male axillary sweat. *Experientia* **30**, 864–865.

Brouette-Lahlou, I., Amouroux, R., Chastrette, F., *et al.* (1991). Dodecyl propionate, attractant from rat preputial gland: characterization and identification. *Journal of Chemical Ecology* **17**, 1343–1345.

Brower, L. P. (1969). Ecological chemistry. *Scientific American*, **220**, 22–29.

Brower, L. P. and Brower, J. V. Z. (1964). Birds, butterflies, and plant poisons: a study in ecological chemistry. *Zoologica* **49**, 137–159.

Brower, L. P., Nelson, C. J., Seiber, J. N., Fink, L. S. and Bond, C. (1988). Exaptation as an alternative to coevolution in the cardenolide-based chemical defense of monarch butterflies (*Danaus plexippus* L.) against avian predators. In *Chemical Mediation of Coevolution*, ed. K. C. Spencer, pp. 447–475. San Diego, CA: Academic Press/Harcourt Brace.

Brown, G. E. and Brown, J. A. (1993). Social dynamics in salmonid fishes: do kin make better neighbours? *Animal Behaviour* **45**, 863–871.

Brown, G. E. and Godin, G. J. (1999). Chemical alarm signals in wild Trinidadian guppies (*Poecilia reticulata*). *Canadian Journal of Zoology* **77**, 562–570.

Brown, G. E., Brown, J., and Crosbie, A. (1993). Phenotype matching in juvenile rainbow trout. *Animal Behaviour* **46**, 1223–1225.

Brown, G. E., Adrian, J. C., Smyth, E., Leet, H., and Brennan, S. (2000). Ostariophysan alarm pheromones: laboratory and field tests of the functional significance of nitrogen oxides. *Journal of Chemical Ecology* **26**, 139–154.

Brown, G. E., Golub, J. L., and Plata, D. L. (2001). Attack cone avoidance during predator inspection visits by wild finescale dace (*Phoxinus neogaeus*): the effect of predator diet. *Journal of Chemical Ecology* **27**, 1657–1666.

Brown, G. E., Adrian, J. C., Jr., Naderi, N. T., Harvey, M. C., and Kelly, J. M. (2003). Nitrogen oxides elicit antipredator responses in juvenile channel catfish, but not in convict cichlids or rainbow trout: conservation of the ostariophysan alarm pheromone. *Journal of Chemical Ecology* **29**, 1781–1796.

Brown, R. E. (1988). Individual odors of rats are discriminable independently of changes in gonadal hormone levels. *Physiology and Behavior* **43**, 359–364.

Brown, R. E. and MacDonald, D. (1985). *Social Odours in Mammals*, vol. 2, p. 635. Oxford: Clarendon Press.

Brown, R. E., Singh, P. B., and Roser, B. (1987a). Both class I and class II regions of the major histocompatibility complex influence the distinctive urinary odors of congenic rat. *American Zoologist* **27**, 47A.

(1987b). The major histocompatibility complex and the chemosensory recognition of individuality in rats. *Physiology and Behavior* **40**, 65–74.

Brown, R. E., Hauschild, M., Holman, S. D., and Hutchison, J. B. (1988). Mate recognition by urine odors in the Mongolian gerbil, *Meriones unguiculatus*. *Behavioral and Neural Biology* **49**, 174–183.

Brown, W. L., Jr., Eisner, T., and Whittaker, R. H. (1970). Allomones and kairomones: transspecific chemical messengers. *BioScience* **20**, 21–22.

Brown, W. S. and MacLean, F. M. (1983). Conspecific scent-trailing by newborn timber rattlesnakes, *Crotalus horridus*. *Herpetologica* **39**, 430–436.

Brownlee, R. G., Silverstein, R. M. Müller-Schwarze, D., and Singer, A. G. (1969). Isolation, identification and function of the chief component of the male tarsal scent in black-tailed deer. *Nature* **221**, 284–285.

Bruce, H. M. (1959). An exteroceptive block to pregnancy in the mouse. *Nature* **184**, 105.

Brundin, A., Andersson, G., Andersson, K., Mossing, T., and Källquist, L. (1978). Short-chain aliphatic acids in the interdigital gland secretion of reindeer (*Rangifer tarandus* L.), and their discrimination by reindeer. *Journal of Chemical Ecology* **4**, 613–622.

Bryant, B. P. and Atema, J. (1987). Diet manipulation affects social behavior of catfish: importance of body odor. *Journal of Chemical Ecology* **13**, 1645–1662.

Bryant, J. P. (1981). Phytochemical deterrence of snowshoe hare browsing by adventitious shoots of four Alaskan trees. *Science* **213**, 889–890.

Bryant, J. P. and Kuropat, P. J. (1980). Selection of winter forage by subarctic browsing vertebrates: the role of plant chemistry. *Annual Review of Ecology* **11**, 261–285.

Bryant, J. P., Wieland, G. D., Reichardt, P. B., Lewis, V. E., and McCarthy, M. C. (1983). Pinosylvin methyl ether deters snowshoe hare feeding on green alder. *Science* **222**, 1023–1025.

Bryant, J. P., Tahvanainen, J., Sulkinoja, M. *et al.* (1989). Biogeographic evidence for the evolution of chemical defense by boreal birch and willow against mammalian browsing. *American Naturalist* **134**, 20–34.

Bryant, J. P., Kuropat, P. J., Reichardt, P. B., and Clausen, T. P. (1991). Controls over the allocation of resources by woody plants to chemical antiherbivore defense. In *Plant Defenses Against Mammalian Herbivory*, ed. R. T. Palo and C. T. Robbins, pp. 83–102. Boca Raton, FL: CRC Press.

Buchsbaum, R., Valiela, I., and Swan, A. (1984). The role of phenolic acids and other plant constituents in feeding of Canada geese in a coastal marsh. *Oecologia* **63**, 343–349.

Buck, L. (1996). Information coding in the vertebrate olfactory system. *Annual Review of Neuroscience* **19**, 517–544.

Buck, L. and Axel, R. (1991). A novel multigene family may encode odorant receptors: a molecular basis for odor recognition. *Cell* **65**, 175–187.

Bucyanayandi, J. D., Bergeron, J.-M., and Menard, H. (1990). Preference of meadow voles (*Microtus pennsylvaticus*) for conifer seedlings: chemical components and nutritional quality of bark of damaged and undamaged trees. *Journal of Chemical Ecology* **16**, 2569–2579.

Buesching, C. D., Waterhouse, J. S., and MacDonald, D. W. (2002a). Gaschromatographic analyses of the subcaudal gland secretion of the European badger (*Meles meles*). Part I: chemical differences related to individual parameters. *Journal of Chemical Ecology* **28**, 41–56.

(2002b). Gaschromatographic analyses of the subcaudal gland secretion of the European badger (*Meles meles*). Part II: time-related variation in the individual-specific composition. *Journal of Chemical Ecology* **28**, 57–69.

Bugloss, A. J., Darling, F. M. C., and Waterhouse, J. S. (1990). Analysis of the anal sac secretion of Hyaenidae. In *Chemical Signals in Vertebrates*, vol. 5, ed. D. W. MacDonald, D. Müller-Schwarze, and S. T. Natynczuk, pp. 65–69. Oxford: Oxford University Press.

Buitron, D. and Nuechterlein, G. L. (1985). Olfactory detection of food caches by black-billed magpies. *Condor* **87**, 92–95.

Bullard, R. W., Turkowski, F. J., and Kilburn, S. R. (1983). Responses of free-ranging coyotes to lures and their modifications. *Journal of Chemical Ecology* **9**, 877–888.

Burger, B. V., Pretorius, P. J., Spies, H. S. C., Bigalke, R. C., and Geierson, G. R. (1990). Mammalian phenomones VIII: chemical characterization of preorbital gland secretion of grey duiker, *Sylvicapra grimmia* (Artiodactyla: Bovidae). *Journal of Chemical Ecology* **16**, 397–416.

Burger, J. (1989). Following of conspecific and avoidance of predator chemical cues by pine snakes (*Pituophis melanoleucus*). *Journal of Chemical Ecology* **15**, 799–806.

(1991). Response to prey chemical cues by hatchling pine snakes (*Pituophis melanoleucus*). Effects of incubation temperature and experience. *Journal of Chemical Ecology* **17**, 1069–1078.

Burger, J. and Gochfeld, M. (1985). A hypothesis on the role of pheromones on age of menarche. *Medical Hypotheses* **17**, 39–46.

Burger, J., Boarman, W., Kurzava, L. and Gochfeld, M. (1991). Effect of experience with pine (*Pituophis melanoleucus*) and king (*Lampropeltis getulus*) snake odors on Y-maze behavior of pine snake hatchlings. *Journal of Chemical Ecology* **17**, 79–87.

Burghardt, G. M. (1966). Stimulus control of the prey attack response in naive garter snakes. *Psychonomic Science* **4**, 37–38.

(1967). Chemical-cue preferences of inexperienced snakes: comparative aspects. *Science* **157**, 718–721.

(1970). Intraspecific geographical variation in chemical food cue preferences of newborn garter snakes (*Thamnophia sirtalis*). *Behaviour* **36**, 246–257.

(1971). Chemical-cue preferences of newborn snakes: influence of prenatal maternal experience. *Science* **171**, 921–923.

(1973). Chemical release of prey attack: extension to naive newly hatched lizards. *Journal of Herpetology* **15**, 77–81.

(1975). Chemical prey preference polymorphism in newborn garter snakes *Thamnophis sirtalis* M. *Behavior* **52**, 202–225.

(1977). The ontogeny, evolution, and stimulus control of feeding in humans and reptiles. *The Chemical Senses and Nutrition*, ed. M. R. Kare and O. Maller, pp. 253–275. In New York: Academic Press.

(1983). Aggregation and species discrimination in newborn snakes. *Zeitschrift für Tierpsychologie* **61**, 89–101.

(1992). Prior exposure to prey cues influences chemical prey preference and prey choice in neonatal garter snakes. *Animal Behaviour* **44**, 787–789.

Burghardt, G. M. and Abeshaheen, J. P. (1971). Responses to chemical stimuli of prey in newly hatched snakes of the genus *Elaphe*. *Animal Behaviour* **19**, 486–489.

Burghardt, G. M. and Hess, E. H. (1966). Food imprinting in the snapping turtle, *Chelydra serpentina*. *Science* **151**, 108–109.

Burghardt, G. M., Wilcoxon, H. C., and Czaplicki, J. A. (1973). Conditioning in garter snakes: aversion to palatable prey induced by delayed illness. *Animal Learning and Behavior* **1**, 317–320.

Burghardt, G. M., Goss, S. E., and Schell, F. M. (1988). Comparison of earthworm- and fish-derived chemicals eliciting prey attack by garter snakes (*Thamnophis*). *Journal of Chemical Ecology* **14**, 855–881.

Burritt, E. A. and Provenza, F. D. (2000). Role of toxins in intake of varied diets by sheep. *Journal of Chemical Ecology* **26**, 1991–2005.

Bursell, E. (1984). Effects of host odour on the behavior of tsetse. *Insect Science and its Application* **5**, 345–349.

Bursell, E., Gough, A. J. E., Beevor, P. S., *et al.* (1988). Identification of components of cattle urine attractive to tsetse flies, *Glossina* spp. (Diptera: Glossinidae). *Bulletin of Entomological Research* **78**, 281–291.

Burwash, M. D., Tobin, M. E., Woolhouse, A. D., and Sullivan, T. P. (1998). Field testing synthetic predator odors for roof rats (*Rattus rattus*) in Hawaiin macadamia nut orchards. *Journal of Chemical Ecology* **24**, 603–630.

Butterstein, G. M., Schadler, M. H., Lysogorski, E., Robin, L., and Sipperly, S. (1985). A naturally occurring compound, 6-methoxybenzoxyzolinone, stimulates reproductive responses in rats. *Biology of Reproduction* **32**, 1018–1023.

Buzzell, G. R., Menendez-Pelaez, A., Chlumecky, V., and Reiter, R. J. (1991). Gender differences and time course of castration-induced changes in porphyrins, indoles, and proteins in the Harderian gland of the Syrian hamster. *Canadian Journal of Physiology and Pharmacology* **69**, 1814–1818.

Caicedo, A. and Roper, S. D. (2001). Taste receptor cells that discriminate between bitter stimuli. *Science* **291**, 1557–1560.

Cain, W. S. (1978). The odoriferous environment and the application of olfactory research. In *Handbook of Perception*, vol. VIA: *Tasting and Smelling*, ed. E. C. Carterette, and M. P. Friedman, pp. 277–304. London: Academic Press.

(1982). Odor identification by males and females: predictions vs. performance. *Chemical Senses* **7**, 129–142.

Caine, N. G. and Weldon, P. J. (1989). Responses by red-bellied tamarins (*Saguinus labiatus*) to fecal scents of predatory and non-predatory Neotropical mammals. *Biotropica* **21**, 186–189.

Calder, C. J. and Gorman, M. L. (1991). The effects of red fox *Vulpes vulpes* faecal odours on the feeding behavior of Orkney voles *Microtus arvalis*. *Journal of Zoology* **224**, 599–606.

Caldwell, J. P. (1996). The evolution of myrmecophagy and its correlates in poison frogs (family Dendrobatidae). *Journal of Zoology* **240**, 75–101.

Camazine, S. M. (1983). Mushroom chemical defense: food aversion learning induced by a hallucinogenic toxin, muscimol. *Journal of Chemical Ecology* **9**, 1473–1481.

(1985). Olfactory aposematism: association of food toxicity with naturally occurring odor. *Journal of Chemical Ecology* **11**, 1289–1295.

Camazine, S. M., Resch, J. F., Eisner, T., and Meinwald, J. (1983). Mushroom chemical defense: pungent sesquiterpenoid dialdehyde antifeedant to opossum. *Journal of Chemical Ecology* **10**, 1439–1447.

Campbell, D. L. (1987). Potential for aversive conditioning in forest animal damage control. In *Proceedings of a Symposium on Animal Damage Management in Pacific Northwest Forests*, March, 1987, pp. 117–118.

Campbell, D. L. and Bullard, R. W. (1972). A preference-testing system for evaluating repellents for black-tailed deer. *Proceedings of the Vertebrate Pest Conference* **5**, 56–63.

Campbell, D. L. and Evans, J. (1989). Aversive conditioning with thiram to reduce mountain beaver damage to Douglas-fir seedlings. *Northwest Science* **63**, 70.

Cantoni, D. and Rivier, L. (1992). Analysis of the secretions from the flank glands of three shrew species and their possible function in a social context. In *Chemical Signals in Vertebrates*, vol. 6, ed. R. L. Doty and D. Müller-Schwarze, pp. 99–106. New York: Plenum.

Carolsfeld, J., Tester, M., Kreiberg, H., and Sherwood, N. M. (1997a). Pheromone-induced spawning of Pacific herring. 1. Behavioral characterization. *Hormones and Behavior* **31**, 256–268.

Carolsfeld, J., Scott, A. P., and Sherwood, N. M. (1997b). Pheromone-induced spawning of Pacific herring. 2. Plasma steroids distinctive to fish responsive to spawning pheromone. *Hormones and Behavior* **31**, 269–276.

Carr, W. E. S. (1976). Chemoreception and feeding behavior in the pigfish, *Orthopristis chrysopterus*: characterization and identification of stimulatory substances in a shrimp extract. *Comparative Biochemical Physiology* **55A**, 153–157.

Carr, W. and Chaney, T. B. (1976). Chemical stimulation of feeding behavior in the pinfish, *Lagodon rhomboides*: characterization and identification of stimulatory substances extracted from shrimp. *Comparative Biochemistry and Physiology* **54A**, 437–441.

Carr, W., Gondeck, A. R., and Delanoy, R. L. (1976). Chemical stimulation of feeding behavior in the pinfish, *Lagodon rhomboides*: a new approach to an old problem. *Comparative Biochemistry and Physiology* **54A**, 161–166.

Carter, C. S. and Roberts, R, L. (1997). The psychobiology of cooperative breeding in rodents. In *Cooperative Breeding in Mammals*, ed. N. G. Solomon and J. French, pp. 231–266. New York: Cambridge University Press.

Carter, C. S., Getz, L. L., Gavish, L., McDermott, J. L., and Arnold, P. (1980). Male-related pheromones and the activation of reproduction in the prairie vole (*Microtus ochrogaster*). *Biology of Reproduction* **23**, 1038–1045.

Carter, C. S., Getz, L. L., and Cohen-Parsons, M. (1986). Relationships between social organization and behavioral endocriology in a monogramous mammal. *Advances in the Study of Behavior* **16**, 109–146.

Cave, A. J. E. (1988). Note on olfactory activity in mysticetes. *Proceedings of the Zoological Society of London*, **214**, 307–311.

Cernoch, J. M. and Porter, R. H. (1985). Recognition of maternal axillary odors by infants. *Child Development* **56**, 1593–1598.

Chabot, D., Gagnon, P., and Dixon, E. A. (1996). Effect of predator odors on heart rate and metabolic rate of wapiti (*Cervus elaphus canadensis*). *Journal of Chemical Ecology* **22**, 839.

Champlin, A. K. (1971). Suppression of oestrus in grouped mice: the effect of various densities and the possible nature of the stimulus. *Journal of Reproduction and Fertility* **27**, 233–241.

Chanin, P. (1985). *The Natural History of Otters*. London: Croom Helm.

Chapple, R. S. and Wodzicka-Tomaszewska, M. (1987). The learning behavior of sheep when introduced to wheat. II. Social transmission of wheat feeding and the role of the senses. *Applied Animal Behaviour Science* **18**, 163–172.

Chelazzi, G. and Delfino, G. (1986). A field test on the use of olfaction in homing by *Testudo hermanni* (Reptilia: Testudinidae). *Journal of Herpetology* **20**, 451–455.

Chen, L. and Martinich, R. L. (1975). Pheromonal stimulation and metabolite inhibition of ovulation in the zebrafish, *Brachydanio rerio*. *NOAA Fisheries Bulletin* **73**, 889–894.

Chernayeva, G. N., Dolgodvorova, S., and Peryshkina, Y. (1982). Seasonal dynamics of tannin content in European white birch bark. *Rastitielnii Resurci* **18**, 63–66.

Chien, A. K. (1973). Reproductive behavior of the angelfish *Pterophyllum scalare* (Pisces: Cichlidae) II. Influence of male stimuli upon the spawning rate of females. *Animal Behaviour* **21**, 457–463.

Chipman, R. K. and Fox, K. A. (1966). Oestrus synchronization and pregnancy blocking in wild house mice (*Mus musculus*). *Journal of Reproduction and Fertility* **12**, 233–236.

Chiszar, D. (1986). Motor patterns dedicated to sensory function, *Chemical Signals in Vertebrates*, vol. 4, ed. D. Duvall, D. Müller-Schwarze, and R. M. Silverstein, pp. 37–44. New York: Plenum.

Chiszar, D. and Scudder, K. M. (1980). Chemosensory searching by rattlesnakes during predatory episodes. In *Chemical Signals: Vertebrates and Aquatic Invertebrates*, ed. D. Müller-Schwarze and R. M. Silverstein, pp. 125–139. New York: Plenum.

Chiszar, D., Radcliff, C. W., Scudder, R. M., and Duvall, D. (1983). Strike-induced chemosensory searching by rattlesnakes: the role of envenomation-related chemical cues in the post-strike environment. In *Chemical Signals in Vertebrates* vol. 3, ed. D.Müller-Schwarze and R. M. Silverstein, pp. 1–24. New York: Plenum.

Chiszar, D., Nelson, P., and Smith, H. M. (1988a). Analysis of the behavioral sequence emitted by rattlesnakes during feeding episodes III: strike-induced chemosensory searching and location of rodent carcasses. *Bulletin of the Maryland Herpetological Society* **24**, 99–108.

Chiszar, D., Kandler, K., Lee, R., and Smith, H. M. (1988b). Stimulus control of predatory attack in the brown tree snake, *Boiga irregularis*. 2. Use of chemical cues during foraging. *Amphibia-Reptilia* **9**, 77–88.

Chiszar, D., Melner, T., and Lee, R. (1990). Chemical cues used by prairie rattlesnakes (*Crotalus viridis*) to follow trails of rodent prey. *Journal of Chemical Ecology* **16**, 79–86.

Chiszar, D., Fox, K., and Smith, H. M. (1992). Stimulus control of predator behavior in the brown tree snake (*Boiga irregularis*). *Behavioral and Neural Biology* **57**, 167–169.

Chiszar, D., Dunn, T. M., and Smith, H. M. (1993a). Response of brown tree snakes (*Boiga irregularis*) to human blood. *Journal of Chemical Ecology* **19**, 91–96.

Chiszar, D., Grant, H., and Hobart, M. (1993b). Prairie rattlesnakes (*Crotalus viridis*) respond to rodent blood with chemosensory searching. *Brain, Behavior and Evolution* **41**, 229–233.

Chivers, D. P. and Smith, R. J. F. (1993). The role of olfaction in chemosensory-based predator recognition in the fathead minnow, *Pimephales promelas*. *Journal of Chemical Ecology* **19**, 623–633.

(1994). The role of experience and chemical alarm signaling in predator recognition by fathead minnows, *Pimephales promelas*. *Journal of Fish Biology* **44**, 273–285.

(1998). Chemical alarm signaling in aquatic predator-prey systems: a review and prospectus. *Ecoscience* **5**, 338-352.

Chivers, D. P., Kiesecker, J. M., Anderson, M. T., and Wildy, B. A. R. (1996). Avoidance response of a terrestrial salamander (*Ambystoma macrodactylum*) alarm cues. *Journal of Chemical Ecology* **22**, 1709–1716.

Chivers, D. P., Kiesecker, J. M., Marco, A., Wildy, E. L., and Blaustein, A. R. (1999). Shifts in life history as a response to predation in Western toads (*Bufo boreas*). *Journal of Chemical Ecology* **25**, 2455-2463.

Chow, Y. S. and Lin, Y. M. (1986). Actinidine, a defensive secretion of stick insect, *Megacrania alpheus*, Westwood (Orthoptera: Phasmatidae). *Journal of Entomological Science* **21**, 97-101.

Ciofi, C. (1999). The Komodo dragon. *Scientific American* **280**, 84–91.

Cipollini, M. L. and Levey, D. J. (1997). Antifungal activity of *Solanum* fruit glycoalkaloids: implications for frugivory and seed dispersal. *Ecology* **78**, 799–809.

Clancy, A. N., Coquelin, A., Macrides, F., Gorski, R. A., and Nobel, E. P. (1984). Sexual behavior and aggression in male mice: involvement of the vomeronasal system. *Journal of Neuroscience* **4**, 2222–2229.

Clapperton, B. K., Minot, E. D., and Crump, D. R. (1988). An olfactory recognition system in the ferret *Mustela furo L.* (Carnivora: Mustelidae). *Animal Behaviour* **36**, 541–553.

(1989). Scent lures from anal sac secretions of the ferret, *Mustela furo* L. *Journal of Chemical Ecology* **15**, 291–308.

Clark, L. (1991). Odor detection thresholds in tree swallows and cedar waxwings. *Auk* **108**, 177–180.

(1997). Dermal contact repellents for starlings: foot exposure to natural plant products. *Journal of Wildlife Management* **61**, 1352–1358.

Clark, L. and Mason, J. R. (1985). Use of nest material as insecticidal and anti-pathogenic agents by the European starling. *Oecologia* **67**, 169–176.

(1987). Olfactory discrimination of plant volatiles by the European starling. *Animal Behaviour* **35**, 227–235.

(1989). Sensitivity of brown-headed cowbirds to volatiles. *Condor* **91**, 922–932.

Clark, L. and Shah, P. S. (1992). Information content of prey odor plumes: what do foraging Leach's storm petrels know? In *Chemical Signals in Vertebrates* vol. 6, ed. R. L. Doty and D. Müller-Schwarze, pp. 421–427. New York: Plenum.

(1994). Tests and refinements of a general structure–activity model for avian repellents. *Journal of Chemical Ecology* **20**, 321–339.

Clark, L. and Smeraski, C. A. (1990). Seasonal shifts in odor acuity by starlings. *Journal of Experimental Zoology* **255**, 22–29.

Clark, L., Shah, P. S. and Mason, J. R. (1991). Chemical repellency in birds: relationship between structure of anthranilate and benzoic acid derivatives and avoidance response. *Journal of Experimental Zoology* **269**, 310–322.

Clark, M. M., Whiskin, E. E., and Galef, B. G. (2003). Mongolian gerbil fathers avoid newborn male pups, but not newborn female pups: olfactory control of early paternal behavior. *Animal Behaviour* **66**, 441–447.

Clark, R. W. 2004. Timber rattlesnakes (*Crotalus horridus*) use chemical cues to select ambush sites. *Journal of Chemical Ecology* **30**, 607–617.

Claus, R. and Hoppen, H. O. (1979). The boar pheromone identified in vegetables. *Experientia* **35**, 1674–1675.

Claus, R., Hoppen, H. O., and Karg, H. (1981). The secret of truffles: a steroidal pheromone? *Experientia* **37**, 1178–1179.

Claus, R., Dehnhard, M., Götz, U., and Lacorn, M. (2001). The pheromone of the male goat: function, sources, androgen dependency and partial chemical characterization. In *Chemical Signals in Vertebrates,* vol. 9, ed. A. Marchlewska-Koj, J. J. Lepri, and D. Müller-Schwarze, pp. 133–140. New York: Kluwer Academic/Plenum.

Clausen, T. P., Reichardt, P. B., and Bryant, J. P. (1986). Pinosylvin and pinosylvin methyl ether as feeding deterrents in green alder. *Journal of Chemical Ecology* **12**, 2117–2131.

Clausen, T. P., Bryant, J. P., and Swihart, R. K. (2004). Has browsing by mammals caused continent-scale variation in the chemical defenses of woody plants? In *Annual Meeting of International Society of Chemical Ecology*, July 2004, Ottawa, Canada.

Clotfelter, E. D., Bell, A. M., and Levering, K. R. (2004). The role of animal behaviour in the study of endocrine-disrupting chemicals. *Animal Behaviour* **68**, 665–676.

Clulow, F. V. and Baddaloo, E. G. Y. (1987). Influence of odors of male organ homogenates on maturation of young female meadow voles, *Microtus pennsylvanicus*. *Behavioral Processes* **14**, 225–228.

Clulow, F. V. and Clarke, J. R. (1968). Pregnancy-block in *Microtus agrestis*, an induced ovulator. *Nature* **219**, 511.

Clulow, F. V. and Langford, P. E. (1971). Pregnancy-block in the meadow vole, *Microtus pennsylvanicus*. *Journal of Reproduction and Fertility* **24**, 275–277.

Cole, K. S. and Smith, R. J. F. (1987). Release of chemicals by prostaglandin-treated female fathead minnows, *Pimephalus promelas*, that stimulate male courtship. *Hormones and Behavior* **21**, 440–456.

(1992). Attraction of female fathead minnows, *Pimephales promelas*, to chemical stimuli from breeding males. *Journal of Chemical Ecology* **18**, 1269–1284.

Coley, P. D., Bryant, J. P., and Chapin, F. S., III (1985). Resource availability and plant antiherbivore defense. *Science* **230**, 895–899.

Colombo, L., Marconato, A., Belvedere, P. C., and Friso, C. (1980). Endocrinology of teleost reproduction: a testicular steroid pheromone in the black gobi, *Gobius jozo* L. *Bolletino di Zoologia* **47**: 355–364.

Commetto-Muñiz, J. E. and Cain, W. S. (1993). Detection thresholds for an olfactory mixture and its three constituent compounds. *Chemical Senses* **18**, 723–734.

Conover, M. R. (1990). Reducing mammalian predation on eggs by using a conditioned taste aversion to deceive predators. *Journal of Wildlife Management* **54**, 360–365.

Cooper, S. M. and Owen-Smith, N. (1985). Condensed tannins deter feeding by browsing ruminants in a South African savanna. *Oecologia* **67**, 142–146.

Cooper, W. E., Jr. (1989a). Strike-induced chemosensory searching occurs in lizards. *Journal of Chemical Ecology* **15**, 1311–1320.

(1989b). Prey odor discrimination in the varanoid lizards: *Heloderma suspectum* and *Varanus exanthematicus*. *Ethology* **81**, 250–258.

(1992). Elevation in tongue-flick rate after biting prey in the broad-headed skink, *Eumeces laticeps*. *Journal of Chemical Ecology* **18**, 455–467.

(1994). Chemical discrimination by tongue-flicking in lizards: a review with hypotheses on its origin and its ecological and phylogenetic relationships. *Journal of Chemical Ecology* **20**, 439–487.

(2000). Responses to chemical cues from plant and animal food by an omnivorous lizard, *Gerrhosaurus validus*. *Journal of Herpetology* **34**, 614–617.

Cooper, W. E., Jr. and Alberts, A. C. (1991). Tongue-flicking and biting in response to chemical food stimuli by an iguanid lizard (*Dipsosaurus dorsalis*) having sealed vomeronasal ducts: vomerolfaction may mediate these behavioral processes. *Journal of Chemical Ecology* **17**, 135–146.

Cooper, W. E., Jr. and Garstka, W. (1987). Discrimination of male conspecific from male heterospecific odors by male scincid lizards, *Eumeces laticeps*. *Journal of Experimental Zoology* **241**, 253–256.

Cooper, W. E. and Trauth, S. E. (1992). Discrimination of conspecific male and female cloacal chemical stimuli by males and possession of a probable pheromone gland by females in a cordylid lizard, *Gerrhosaurus nigrolineatus*. *Herpetologia* **48**, 229–232.

Cooper-Driver, G. A. and Swain, T. (1976). Cyanogenic polymorphism in bracken (*Pteridium*) in relation to herbivore predation (*Schistocerca gregaria*). *Nature* **260**, 604.

Coopersmith, C. B. and Banks, E. M. (1983). Effects of olfactory cues on sexual behavior in the brown lemming, *Lemmus trimucronatus*. *Journal of Comparative Psychology* **97**, 120–126.

Coopersmith, R. and Leon, M. (1986). Neurobehavioral analysis of odor preference development in rodents. In *Ontogeny of Olfaction: Principles of Olfactory Maturation in Vertebrates*, ed. W. Breiphol, pp. 237–242. Irvine, CA: Department of Psychobiology.

Coppola, D. M. (1986). The puberty delaying pheromone of the house mouse: field data and a new evolutionary perspective, In *Chemical Signals in Vertebrates*, vol. 4, ed. D. Duvall, D. Müller-Schwarze, and R. M. Silverstein, pp. 457–462 New York: Plenum.

(2001). The role of the main and accessory olfactory systems in prenatal olfaction. In *Chemical Signals in Vertebrates* 9, ed. A. Marchlewska-Koj, J. J. Lepri and D. Müller-Schwarze, pp. 189–196. New York: Kluwer Academic/Plenum.

Coppola, D. M. and Coltrane, J. A. (1994). Retronasal or internasal olfaction can mediate odor-guided behaviors in newborn mice. *Physiology and Behavior* **56**, 729–736.

Coppola, D. M. and O'Connell, R. J. (1988). Behavioral responses of peripubertal female mice towards puberty-accelerating and puberty-delaying chemical signals. *Chemical Senses* **13**, 407–424.

Coppola, D. M. and Vandenbergh, J. G. (1985). Effect of density, duration of grouping and age of urine stimulus on the puberty delay pheromone in female mice. *Journal of Reproduction and Fertility* **73**, 517.

(1987). Induction of a puberty-regulating chemosignal in wild mouse populations. *Journal of Mammalogy* **68**, 86–91.

Coquelin, A., Clancy, A. N., Macrides, F., Nobel, E. P. and Gorski, R. A. (1984). Pheromonally induced release of luteinizing hormone in male mice: involvement of the vomeronasal system. *Journal of Neuroscience* **4**, 2230–2236.

Cornwell-Jones, C. A., Velasquez, P. Wright, E. L. and McGaugh, J. L. (1988). Early experience influences adult retention of aversively motivated tasks in normal but not DSP4-treated rats. *Developmental Psychobiology* **21**, 177–185.

Costanzo, J. P. (1989). Conspecific scent trailing by garter snakes (*Thamnophis sirtalis*) during autumn. *Journal of Chemical Ecology* **15**, 2531–2538.

Cott, H. B. (1947). The edibility of birds. *Proceedings of the Zoology Society of London* **116**, 371–542.

(1948). Edibility of the eggs of birds. *Nature* **161**, 8–11.

(1952). The palatability of the eggs of birds. *Proceedings of the Zoological Society of London* **122**, 1–54.

(1953). The palatability of the eggs of birds: illustrated by experiments on the food preferences of the ferret (*Putorius furo*) and cat (*Felis cato*); with notes on other egg-eating carnivores. *Proceedings of the Zoological Society of London* **123**, 123–141.

(1955). The palatability of the eggs of birds: mainly based upon observations of an egg panel. *Proceedings of the Zoological Society of London* **124**, 335–463.

Cott, H. B. and Benson, J. M. (1969). The palatability of birds, mainly based upon observations of a tasting panel in Zambia. *Ostrich* **8**, 357–384.

Coulston, S., Stoddart, D. M., and Crump, D. R. (1993). Use of predator odors to protect chick-peas from predation by laboratory and wild mice. *Journal of Chemical Ecology* **19**, 607–612.

Coureaud, G., Schaal, B., Langlois, D., and Perriers, G. (2001). Orientation response of newborn rabbits to odours of lactating females: relative effectiveness of surface and milk cues. *Animal Behaviour* **61**:153–162.

Courtney, A. J. and Masel, J. M. (1997). Spawning stock dynamics of two penaeid prawns, *Metapenaeus bennettae* and *Penaeus esculentus,* in Moreton Bay, Queensland, Australia. *Marine Ecology Progress Series* **148**, 37–47.

Cowles, R. B. and Phelan, R. L. (1958). Olfaction in rattlesnakes. *Copeia* **1958**, 77–83.

Cowley, J. J. and Wise, D. R. (1972). Some effects of mouse urine on neonatal growth and reproduction. *Animal Behaviour* **20**, 499–506.

Cox, C., Weetjens, B., Machangu, R., Billet, M., and Verhagen, R. (2004). Rats for demining: an overview of the APOPO program. www.apopo.org.

Cox, T. P. (1989). Odor-based discrimination between noncontiguous demes of wild *Mus*. *Journal of Mammalogy* **70**, 549–556.

Cox, T. P. and Sacks, O. W. (2002). Cycad neurotoxins, consumption of flying foxes, and ALS-PDC diseases in Guam. *Neurology* **26**, 1664–1665.

Creigh, S. L. and Terman, C. R. (1988). Reproductive recovery of inhibited male prairie deer mice (*Peromyscus maniculatus bairdii*) from laboratory populations by contact with females or their urine. *Journal of Mammalogy* **69**, 603–607.

Cronin, E. W., Jr. and Sherman, P. W. (1976). A resource-based mating system: the orange rumped honeyguide. *Living Bird* **15**, 5–32.

Crozier, E. R. (1991). Practical animal repellents for tree seedlings: a success story. *Forest Research Institute Bulletin* **156**, 172–177.

Cruz, E., Gibson, S., Kandler, K., Sanchez, G., and Chiszar, D. (1987). Strike-induced chemosensory searching in rattlesnakes: a rodent specialist (*Crotalus viridis*) differs from a lizard specialist (*Crotalus pricei*). *Bulletin of the Psychonomic Society* **25**, 136–138.

Cummings, J. L., Pochop, D. A., Davis, J. E., and Krupa, H. W. (1995). Evaluation of ReJeX-iT AG-36 as a Canada goose grazing repellent. *Journal of Wildlife Management* **59**, 47–50.

Cummins, K. A. and Myers, L. J. (1988). Effect of visual and olfactory alterations on social behavior in lactating dairy cows. *Journal of Dairy Science* **71**(Suppl. 1), 189.

Cupp, P. V., Jr. (1988). Avoidance of predators by salamanders through the detection of chemical odors. *American Zoologist* **28**, 156A.

Cushing, B. S. (1983). Responses of polar bears to human menstrual odors. In *International Conference on Bear Research and Management*, pp. 270–274.

(1984). A selective preference by least weasels for oestrus versus dioestrous urine of prairie deer mice. *Animal Behaviour* **32**, 1263–1265.

Cutler, W. B., Preti, G., Huggins, G. R., Garcia, C. R., and Lawley, H. J. (1986). Human axillary secretions influence women's menstrual cycles: the role of donor extract from men. *Hormones and Behavior* **20**, 463–473.

Dagg, A. I. and Windsor, D. E. (1971). Olfactory discrimination limits in gerbils. *Canadian Journal of Zoology* **49**, 283–285.

Daltry, J. C., Wüster, W., and Thorpe, R. S. (1996). Diet and snake venom evolution. *Nature* **379**, 537–540.

Daly, J. W. and Meyers, C. W. (1967). Toxicity of Panamanian poison frogs (*Dendrobates*): some biological and chemical aspects. *Science* **156**, 970–973.

Daly, J. W. and Spande, T. F. (1986). Amphibian alkaloids: chemistry, pharmacology and biology, In *Alkaloids: Chemical and Biological Perspectives*, vol. 4, ed. S. W. Pelletier, pp. 4–274. New York: Wiley.

Daly, J. W., Meyers, C. W., Warnick, J. E., and Albuquerque, E. X. (1980). Levels of bactrachotoxin and lack of sensitivity to its action in poison-dart frogs (*Phyllobates*). *Science* **208**, 1383–1385.

Daly, J. W., Secunda, S. I., Garaffo, H. M., *et al.* (1992). Variability in alkaloid profiles in neotropical poison frogs (Dendrobatidae): genetic versus environmental determinants. *Toxicon* **30**, 887–898.

Daly, J. W., Garraffo, H. M., Spande, T. F., Jaramillo, C., and Stanley, R. A. (1994). Dietary source for skin alkaloids of poison dart frogs (*Dendrobatidae*)? *Journal of Chemical Ecology* **20**, 943–955.

Daly, J. W., Garaffo, H. M., Jain, P., *et al.* (2000). Arthropod–frog connection: decahydroquinoline and pyrrolizidine alkaloids common to microsympatric myrmicine ants and dendrobatid frogs. *Journal of Chemical Ecology* **26**, 73–85.

Daly, M. and Daly, S. (1975). Behavior of *Psammomys obesus* (Rodentia: Gerbillinae) in the Algerian Sahara. *Zeitschrift für Tierpsychologie* **37**, 298–321.

Daly, R. P. (1988). Status of the Arabian oryx in Oman. In *Proceedings of the 63rd Meeting of IUCN Species Survival Commission* San Jose, Costa Rica, January, 1988.

D'Amato, F. and Cabib, S. (1987). Chronic exposure to a novel odor increases pups' vocalizations, maternal care, and alters dopaminergic functions in developing mice. *Behavioral and Neural Biology* **48**, 197–205.

Danell, K., Huss-Danell, K., and Bergström, R. (1985). Interactions between browsing moose and two species of birch in Sweden. *Ecology* **66**, 1867–1878.

Danell, K., Gref, F., and Reza, Y. (1990). Effects of mono- and diterpenes in Scots pine needles on moose browsing. *Scandinavian Journal of Forest Research* **5**, 535–539.

Daranzo, J. P., Sydow, M., and Garris, D. R. (1983). Influence of isolation and training on fighting in mice with olfactory-bulb lesions. *Physiology and Behavior* **31**, 857–860.

Daumae, M. and Kimura, T. (1986). Analysis of urination pattern of male-male counter marking in mice. *Zoological Society of Japan* (Tokyo), **3**, 1103.

David, C. T., Kennedy, J. S., Ludlow, A. R., Perry, J. N., and Wall, C. (1982). A re-appraisal of insect flight towards a distant, point source of wind-borne odor. *Journal of Chemical Ecology* **9**, 1207–1215.

Davis, W. (1996). *One River. Explorations and Discoveries in the Amazon Rain Forest*. New York: Simon and Schuster.

Dawkins, R. and Krebs, J. R. (1979). Arms races between and within species. *Proceedings of the Royal Society of London Series B* **205**, 489–511.

Dawley, E. M. (1984). Recognition of individual, sex and species odours by salamanders of the *Plethodon glutinosus–P. jordani* complex. *Animal Behaviour* **32**, 353–361.

(1987). Salamander vomeronasal systems: why plethodontids smell well. *American Zoologist* **27**, 166A.

(1992). Correlation of salamander vomeronasal and main olfactory system anatomy with habitat and sex: behavioral interpretations. In *Chemical Signals in Vertebrates*, vol. 6, ed. R. L. Doty and D. Müller-Schwarze, pp. 403–409. New York: Plenum.

Dawson, T. M., Arriza, J. L., Jaworsky. D. E., *et al.* (1993). β-Adrenergic receptor kinase-2 and β-arrestin-2 as mediators of odorant-induced desensitization. *Science* **259**, 825–829.

Dearing, M. D. (1997). The manipulation of plant toxins by a food-hoarding herbivore, *Ochotona princeps*. *Ecology* **78**, 774–781.

de Boer, J. N. (1977). The age of olfactory cues function in chemo-communication among male domestic cats. *Behavioral Processes* **2**, 209–225.

De Fanis, E. and Jones, G. (1995). The role of odour in the discrimination of conspecifics by pipistrelle bats. *Animal Behaviour* **49**, 835–839.

De Fraipont, M. (1987). Chemical detection in *Astyanax mexicanus*, Teleostei, Characidae, (cave-dwelling form) as a function of group density. *Annales de la Societé Royale Zoologique de Belgique* **117**, 63–67.

De Fraipont, M. and Thines, G. (1986). Responses of the cavefish *Astyanax mexicanus* (*Anoptichthys antrobius*) to the odor of known or unknown conspecifics. *Experientia* (Basel) **42**, 1053–1054.

De Fraipont, M. and Sorensen, P. W. (1993). Exposure to the pheromone 17α,20 β-dihydroxy-4-pregnen-3-one enhances the behavioral spawning success, sperm production and sperm motility of male goldfish. *Animal Behaviour* **46**, 245–256.

Dehnhard, M. and Claus, R. (1988). Reliability criteria of a bioassay using rats trained to detect estrus-specific odor in cow urine. *Theriogenology* **30**, 1127–1138.

De Lorgeril, M., Salen, P., Paiilard, F., *et al.* (2002). Mediterranean diet and the French paradox: two distinct biogeographic concepts for one consolidated scientific theory on the role of nutrition in coronary heart disease. *Cardiovascular Research* **3**, 503–515.

Del Punta, K., Leinders-Zufall, T., Rodriguez, I., *et al.* (2002). Deficient pheromone responses in mice lacking a cluster of vomeronasal receptor genes. *Nature* **419**, 70–74.

DeMonte, M. and Roeder, J. J. (1990). Responses to inter- and intraspecific scent marks in pine martens (*Martes martes*). *Journal of Chemical Ecology* **16**, 611–618.

Demski, L. S. and Northcutt, R. G. (1983). The terminal nerve: a new chemosensory system in vertebrates? *Science* **220**, 435–437.

Denmead, O. T., Simpson, J. R., and Freney, J. R. (1974). Ammonia flux into the atmosphere from a grazed pasture. *Science* **185**, 609–610.

Deshpande, B. G. (1986). Earthquakes, animals and man: Chapter III: Animal response to earthquakes. *Proceedings of the Indian National Science Academy, Part B, Biological Sciences* **52**, 585–618.

Desjardins, C., Maruniak, J. A., and Bronson, F. H. (1973). Social rank in house mice: differentiation revealed by ultraviolet visualization of urinary marking patterns. *Science* **182**: 939–941.

De Souza, L. L., Ferrari, S. F., Da Costa, M. L., and Kern, D. C. (2002). Geophagy as a correlate of folivory in red-handed howler monkeys (*Alouatta belzebul*) from eastern Brazilian Amazonia. *Journal of Chemical Ecology* **28**, 1613–1621.

Detling, J. K., Dyer, M. I., Procter-Greeg, C., and Winn, D. T. (1980). Plant–herbivore interactions: examination of potential effects of bison saliva on regrowth of *Bouteloua gracilis* (H. B. K.) Lag. *Oecologia* **45**, 26–31.

Detling, J. K., Ross, C. W., Walmsley, M. H., *et al* (1981). Examination of North American bison saliva for potential plant growth regulators. *Journal of Chemical Ecology* **7**, 239–246.

Deutsch, J. D. and Nefdt, R. J. C. (1992). Olfactory cues influence female choice in two lek-breeding antelopes. *Nature* **356**, 596–598.

Devine, G. J., Ingvarsdottir, A., Mordue, W., *et al.* (2000). Salmon lice, *Lepeophtheirus salmonis,* exibit specific chemotactic responses to semiochemicals originating from the salmonid, *Salmo salar. Journal of Chemical Ecology* **26**, 1833–1847.

Devine, M. C. (1977). Copulatory plugs, restricted mating opportunities and reproductive competition among male garter snakes. *Nature* **267**, 345–346.

Dewsbury, D. A. (1981). Effects of novelty on copulatory behavior. *Psychological Bulletin* **89**, 464–482.

Dial, B. E., Weldon, P. J., and Curtis, B. (1989). Chemosensory identification of snake predators (*Phyllorhynchus decurtatus*) by banded geckos (*Coleonyx variegatus*). *Journal of Herpetology* **23**, 224–229.

Dickinson, C. and Keverne, E. B. (1988). Importance of noradrenergic mechanisms in the olfactory bulbs for the maternal behavior of mice. *Physiology and Behavior* **43**, 313–316.

Dickman, C. R. (1992). Predation and habitat shift in the house mouse, *Mus domesticus. Ecology* **73**, 313–322.

Dickman, C. R. and Doncaster, C. P. (1984). Responses of small mammals to red fox (*Vulpes vulpes*) odour. *Journal of Zoology* (London) **204**, 521–531.

Dieterlen, F. (1959). Das Verhalten des Syrischen Goldhamsters. *Zeitschrift für Tierpsychologie* **16**, 47–103.

Dietz, B. A., Hagerman, A. E., and Barrett, G. W. (1994). Role of condensed tannin on salivary tannin-binding proteins, bioenergetics and nitrogen digestibility in *Microtus pennsylvanicus. Journal of Mammalogy* **75**, 880–889.

Dimock, E. J., Silen, R. E., and Allen, V. E. (1976). Genetic resistance on Douglas-fir to damage by snowshoe hare and black-tailed deer. *Forest Science* **22**, 106–121.

Distel, R. A. and Provenza, F. D. (1991). Experience early in life affects voluntary intake of beachbrush by goats. *Journal of Chemical Ecology* **17**, 431–450.

Dixon, M. D., Johnson, W. C., and Adkisson, C. S. (1997). Effects of caching on acorn tannin levels and Blue Jay dietary performance. *Condor* **99**, 756–764.

Dluzen, D. E., Ramirez, V. D., Carter, C. S., and Getz, C. C. (1981). Male vole urine changes luteinizing hormone-releasing hormone and norepinephrine in female olfactory bulb. *Science* **212**, 573–575.

Dmitrieva, T. M. and Ostroumov, V. A. (1986). Role of chemocommunication in the organization of the spawning behavior of the bullhead, *Cottocomephorus grewingki (Dyb.)*. *Biologicheskie Nauki (Moscow)* **10**, 38–42.

Do, J. T., Sullivan, R. M., and Leon, M. (1987). Differential respiration during training is not required for early olfactory learning in infant rats. *Society of Neuroscience Abstracts* **13**, 1402.

Dohi, H., Yamada, A. and Entsu, S. (1991). Cattle feeding deterrents emitted from cattle feces. *Journal of Chemical Ecology* **17**, 1197–1203.

Dominic, C. J. (1964). Source of the male odour causing pregnancy-block in mice. *Journal of Reproduction and Fertility* **8**, 266–267.

Dorries, K. M., Adkins-Regan, E., and Halpern, B. P. (1991). Sex difference in olfactory sensitivity to the boar chemosignal, androstenone, in the domestic pig. *Animal Behaviour* **42**, 403–411.

Dorries, K. M., Adkins-Regan, E., and Halpern, B. P. (1997). Sensitivity and behavioral responses to the pheromone androstenone are not mediated by the vomeronasal organ in domestic pigs. *Brain, Behavior and Evolution* **49**, 53–62.

Doty, R. L. (1972). Odor preferences of female *Peromyscus maniculatus bairdii* for male mouse odors of *P. m. bairdii* and *P. leucopus noveboracensis* as a function of estrous state. *Journal of Comparative and Physiological Psychology* **81**, 191–197.

(1981). Human olfaction. *Chemical Senses* **6**, 351–376.

(1986). Ontogeny of human olfactory function. In *Ontogeny of Olfaction. Principles of Olfactory Maturation in Vertebrates*, ed. W. Breipohl, pp. 3–17. Berlin: Springer-Verlag.

Doty, R. L. and Ferguson-Segall, M. (1989). Influence of adult castration on the olfactory sensitivity of the male rat: a signal detection analysis. *Behavioral Neuroscience* **103**, 691–694.

Doty, R. L., Green, P. A., Ram, C., and Yankell, S. L. (1982). Communication of gender from human breath odors: relationship to perceived intensity and pleasantness. *Hormones and Behavior* **16**, 13–22.

Doty, R. L., Shaman, P., Applebaum, S. L., *et al.* (1984). Smell identification ability: changes with age. *Science* **226**, 1441–1443.

Douglas, H. D., III, Co, J. E., Jones, T. H., and Conner, W. E. (2004). Interspecific differences in *Aethia* spp. auklet odorants and evidence for chemical defense against ectoparasites. *Journal of Chemical Ecology* **30**, 1921–1935.

Døving, K. B. and Pinching, A. J. (1973). Selective degeneration of neurons in the olfactory bulb following prolonged odour exposure. *Brain Research* **52**, 115–129.

Døving, K. B. and Trotier, D. (1998). Structure and function of the vomeronasal organ. *Journal of Experimental Biology*, **201**, 2913–2925.

Døving, K., Selset, B., and Thommsen, R. (1980). Olfactory sensitivity to bile acids in salmonid fishes. *Acta Physiologica Scandinavica* **108**, 123–131.

Døving, K. B., Westerberg, H., and Johnsen, P. B. (1985). Role of olfaction in the behavior and neuronal responses of Atlantic salmon, *Salmo salar*, to hydrographic stratification. *Caradian Journal of Fisheries and Aquatic Science* **42**, 1658–1667.

Døving, K. B., Trotier, D., Rosin, J. F., and Holley, A. (1993). Functional architecture of the vomeronasal organ of the frog (genus *Rana*). *Acta Zoologica* **74**, 173–180.

Downum, K. R., Villegas, S., Rodriguez, E., and Keil, D. J. (1989). Plant photosensitizers: a survey of their occurrence in arid and semiarid plants from North America. *Journal of Chemical Ecology* **15**, 345–355.

Drea, C. M., Vignieri, S. N., Cunningham, S. B., and Glickman, S. E. (2002). Responses to olfactory stimuli in spotted hyenas (*Crocuta crocuta*): I. Investigation of environmental odors and the function of rolling. *Journal of Comparative Psycholology* **116**, 331–341.

Drew, G. S., Fagre, D. B., and Martin, D. J. (1988). Scent-station surveys for cottontail rabbit populations. *Wildlife Society Bulletin* **16**, 396–398.

Drickamer, L. C. (1977). Delay of sexual maturation in female housemice by exposure to grouped females or urine from grouped females. *Journal of Reproduction and Fertility* **51**, 77–81.

(1979). Acceleration and delay of first estrus in wild *Mus musculus*. *Journal of Mammology* **60**, 215–216.

(1982). Acceleration and delay of sexual maturation in female house mice by urinary cues: dose levels and mixing uring from different sources. *Animal Behaviour* **30**, 456–460.

(1984a). Acceleration of puberty by a urinary chemosignal from pregnant and lactating *Mus musculus*. *Journal of Mammalogy* **65**, 697–699.

(1984b). Effects of very small doses of urine on acceleration and delay of sexual maturation in female house mice. *Journal of Reproduction and Fertility* **71**, 475–477.

(1986). Behavioral aspects of rodent urinary chemosignals. *American Zoologist* **26**, 118A.

(1989a). Patterns of deposition of urine containing chemosignals that affect puberty and reproduction by wild stock male and female house mice (*Mus domesticus*). *Journal of Chemical Ecology* **15**, 1407–1421.

(1989b). Odor preferences of wild stock female house mice (*Mus domesticus*) tested at three ages using urine and other cues from conspecific males and females. *Journal of Chemical Ecology* **15**, 1971–1987.

(1989c). Pregnancy block in wild stock house mice, *Mus domesticus*: olfactory preferences of females during gestation. *Animal Behaviour* **37**, 690–698.

(1992). Oestrous female house mice discriminate dominant from subordinate males and sons of dominant from sons of subordinate males by odour cues. *Animal Behaviour* **43**, 868–870.

(2001a). Ecological aspects of house mouse urinary chemosignals. In *Chemical Signals in Vertebrates*, vol. 9, ed. A. Marchlewska-Koj, J. J. Lepri and D. Müller-Schwarze, pp. 35–41. New York: Kluwer Academic/Plenum.

(2001b). Intrauterine position effects on rodent urinary chemosignals. In *Chemical Signals in Vertebrates*, vol. 9, ed. A. Marchlewska-Koj, J. J. Lepri, and D. Müller-Schwarze, pp. 211-216. New York: Kluwer Academic/Plenum.

Drickamer, L. C. and Hoover, J. E. (1979). Effects of urine from pregnant and lactating female house mice on sexual maturation of juvenile females. *Developmental Psychobiology* **12**, 545–551.

Drickamer, L. C. and Lenington, S. (1987). T-locus effects on the male urinary chemosignal that accelerates puberty in female mice. *Animal Behaviour* **35**, 1581–1582.

Drickamer, L. D. and McIntosh, T. K. (1980). Effects of adrenalectomy on the presence of a maturation-delaying pheromone in the urine of female mice. *Hormones and Behavior* **14**, 146–152.

Drickamer, L. C. and Mikesic, D. G. (1990). Urinary chemosignals, reproduction, and population size for house mice (*Mus domesticus*) living in field enclosures. *Journal of Chemical Ecology* **16**, 2955–2968.

Drummond, H. (1985). The role of vision in the predatory behavior of natricine snakes. *Animal Behaviour* **33**, 206–215.

Drummond, H. and Garcia, M. (1995). Congenital responsiveness of garter snakes to a dangerous prey abolished by learning. *Animal Behaviour* **49**, 891–900.

Dubost, G. and Feer, F. (1981). The behavior of the male *Antilope cervicapra L.*, its development and social rank. *Behaviour* **76**, 62–127.

Ducey, P. K. and Ritsema, P. (1988). Intraspecific aggression and responses to marked substrates in *Ambystoma maculatum*, (Caudata: Ambystomatidae). *Copeia* 1988, 1008–1013.

Duchamp-Viret, P., Chaput, M. A. and Duchamp, A. (1999). Odor response properties of rat olfactory receptor neurons. *Science* **284**, 2171–2174.

Duellman, W. E. and Trueb, L. (1986). *Biology of Amphibians*. New York: McGraw Hill.

Dulka, J. G., Stacey N. E., and Sorenson, P. W. (1986). A gonadal sex steroid: 17α-20β-dihydroxy-4-pregnen-3-one, acts as a pheromone to rapidly increase gonadotropin and milt volume in male goldfish *Carassius auratus*. *Society for Neuroscience Abstracts* **12**, 1413.

Dulka, J. G., Stacey, N. E., Sorensen, P. W. and van der Kraak, G. J. (1987). A steroid sex pheromone synchronizes male–female spawning readiness in goldfish. *Nature* **325**, 251–253.

Dumbacher, J. P. (2003). Natural history of chemical defense in New Guinea birds. In *Proceedings of the 10th International Symposium on Chemical Signals in Vertebrates*, July, 2003, Oregon State University, Corvallis.

Dumbacher, J. P., Beehler, B. M., Spande, T. F., Garraffo, H. M., and Daly, W. (1992). Homobatrachotoxin in the genus *Pitohui*: chemical defense in birds? *Science* **258**, 799–801.

Dumbacher, J. P., Spande, T. F., and Daly, J. W. (2000). Batrachotoxin alkaloids from passerine birds: a second toxic bird genus (*Ifrita kowaldi*) from New Guinea. *Proceedings of the National Academy of Sciences of the USA* **97**, 12970–12975.

Dunbar, I. F., Ranson, E., and Buehler, M. (1981). Pup retrieval and maternal attraction to canine amniotic fluids. *Behavioral Processes* **6**, 249–260.

Duncan, M. W. (1991). Role of the cycad neurotoxin BMAA in the amyotrophic lateral sclerosis–parkinsonism dementia complex of the western Pacific. *Advances in Neurology* **56**, 301–310.

Dunn, G. C., Price, E. O., and Katz, L. S. (1987). Fostering calves by odor transfer. *Applied Animal Behaviour Science* **17**, 33–39.

Dusenberry, D. B. 1992. *Sensory Ecology*. New York: W. H. Freeman.

Duvall, D. (1986). Snake, rattle and roll. *Natural History* **95**, 66–73.

Duvall, D. and Chiszar, D. (1990). Behavior and chemical ecology of vernal migration and pre-and poststrike predatory activity in prairie rattlesnakes: field and laboratory experiments. In *Chemical Signals in Vertebrates*, vol. 5, ed. D. W. Macdonald, D. Müller-Schwarze, and S. E. Natynczuk, pp. 539–554 New York: Plenum.

Duvall, D., Scudder, K. M., and Chiszar, D. (1980). Rattlesnake predatory Behavior: mediation of prey discrimination and release of swallowing by cues arising from envenomated mice. *Animal Behaviour* **28**, 674–683.

Duvall, D., Graves, B. M., and Carpenter, G. C. (1987). Visual and chemical composite signaling effects of *Sceloporus* lizard fecal boli. *Copeia* **1987**, 1028–1031.

Dyer, M. I. (1980). Mammalian epidermal growth factor promotes plant growth. *Proceedings of the National Academy of Sciences, USA* **77**, 4836–4837.

Eberhard, I. H., McNamara, J., Pearse, R. N., and Southwell, I. A. (1975). Ingestion and excretion of *Eucalyptus punctata* D. C. and its essential oil by the koala, *Phascolarctas cinereus* (Goldfuss). *Australian Journal of Zoology* **23**, 169–179.

Ebling, F. J. (1977). Hormonal control of mammalian skin glands. In *Chemical Signals in Vertebrates*, vol. 1, ed. D. Müller-Schwarze and M. M. Mozell, pp. 17–33. New York: Plenum.

Egid, K. and Brown, J. L. (1989). The major histocompatibility complex and female mating preference in mice. *Animal Behaviour* **38**, 548–549.

Egid, K. and Lenington, S. (1985). Responses of male mice to odor of females: effects of T- and H2-locus genotype. *Behavior and Genetics* **15**, 287–295.

Ehrenfeld, J. G. and Ehrenfeld, D. W. (1973). Externally secreting glands of freshwater and sea turtles. *Copeia* **1973**, 305–314.

Ehrlich, P. R. and Raven, P. H. (1964). Butterflies and plants: a study in coevolution. *Evolution* **18**, 586–608.

Eibl-Eibesfeldt, I. (1949). Über das Vorkommen von Schreckstoffen bei Erdkrötenquappen. *Experientia* **5**, 236.

Eidelman, A., Katz, M., Good, A., and Roken, A. (1987). Mothers' recognition of their newborns by olfactory cues. *Pediatric Research* **21**, 180A.

Eisenberg, J. F. and Gould, E. (1970). *Smithsonian Contributions to Zoology*, No. 27: *The Tenrecs: A Study in Mammalian Behavior and Evolution*. Washington, DC: Smithsonian Institution Press.

Eisner, T. (1989). Prospecting for nature's chemical riches. *Issues in Science and Technology* **6**, 31-34.

Eisner, T. and Grant, R. P. (1981). Toxicity, odor aversion and "olfactory aposematism." *Science* **213**, 213–476.

Eisner, T., Conner, M. E., Hicks, K., *et al.* (1977). Stink of stinkpot turtle identified: ω-phenylalkanoic acids. *Science* **196**, 1347–1349.

Eleftheriou, B. E., Bronson, F. H., and Zarrow, M. X. (1962). Interaction of olfactory and other environmental stimuli on implantation in the deer mouse. *Science* **137**, 764.

Elkinton, J. S., Cardé, R. T., and Mason, C. J. (1984). Evaluation of time-average dispersion models for estimating pheromone concentrations in a deciduous forest. *Journal of Chemical Ecology* **10**, 1081–1108.

Ellin, R. I., Farrand, R. L., Oberst, F. W., *et al.* (1974). An apparatus for the detection and quantitation of volatile human effluents. *Journal of Chromatography* **100**, 137–152.

Ellingsen, O. F. and Døving, K. B. (1986). Chemical fractionation of shrimp extracts inducing bottom food search behavior in cod (*Gadus morhua* L.). *Journal of Chemical Ecology* **12**, 155–168.

Elliot, E. J. (1986). Chemosensory stimuli in feeding behavior of the leech *Hirudo medicinalis*. *Journal of Comparative Physiology* **159**, 391–401.

Elliot, S. and Loudon, A. (1987). Effects of monoterpene odors on food selection by red deer calves (*Cervus elaphus*). *Journal of Chemical Ecology* **13**, 1343–1349.

Ellis, H. H. (1920). *Sexual Selection in Man*. Philadelphia, PA: F. A. Davis.

El-Sayed, S. Z. (1988). Fragile life under the ozone hole. *Natural History* **97**, 72.

Emlen, S. 1969. Homing ability and orientation in the painted turtle *Chrysemys picta marginata*. *Behaviour* **33**, 58–76.

Endler, J. A. and McLellan, T. (1988). The processes of evolution: toward a newer synthesis. *Annual Review of Ecology and Systematics* 395–421.

Engelhart, A. and Müller-Schwarze, D. (1995). Responses of beavers (*Castor canadensis* Kuhl) to predator chemicals. *Journal of Chemical Ecology* **21**, 1349–1364.

Engen, T. (1972). *The Effect of Expectation on Judgements of Odor*. [*Research Report ICRL-RR-70-11*.] Washington, DC: Injury Control Research Laboratory, US Department of Health, Education and Welfare.

(1974a). Method and theory in the study of odor preferences. In *Human Responses to Environmental Odors*, ed. A. Turk, J. Johnston, and D. Moulton, pp. 121–141. New York: Academic Press.

(1974b). The potential usefulness of sensations of odor and taste in keeping children away from harmful substances. *Annals of New York Academy of Sciences* **237**, 224–228.

(1986). The combined effect of carbon monoxide and alcohol on odor sensitivity. *Environment International* **12**, 207–210.

Enns, H. P. and Hornung, D. E. (1988). Comparisons of the estimate of smell, taste and overall intensity in young and elderly people. *Chemical Senses* **13**, 131–140.

Enomoto, S., Shoji, T., Taniguchi, M., and Kurihara, K. (1992). Role of lipids of receptor membranes in odor reception. In *Chemical Signals in Vertebrates*, vol. 6, ed. R. L. Doty and D. Müller-Schwarze, pp. 55–58. New York: Plenum.

Enserink, M. (2002). What mosquitoes want: secrets of host attraction. *Science* **298**, 90–92.

Epple, G., Belcher, A. M., and Smith, A. B., III. (1986). Chemical signals in callitrichid monkeys: a comparative review. In *Chemical Signals in Vertebrates*, vol. 4, ed. D. Duvall, D. Müller-Schwarze, and R. M. Silverstein, pp. 653–672. New York: Plenum.

Epple, G., Küderling, I. and Belcher, A. (1988). Some communicatory functions of scent marking in the cotton-top tamarin (*Saguinus oedipus oedipus*). *Journal of Chemical Ecology* **14**, 503–515.

Epple, G., Mason, J. R., Nolte, D. L., and Campbell, D. L. (1993). Effects of predator odors on feeding in the mountain beaver (*Aplodontia rufa*). *Journal of Mammology* **74**, 715–722.

Erlinge, S., Sandell, M., and Brinck, C. (1982). Scent-marking and its territorial significance in stoats, *Mustela erminea*. *Animal Behaviour* **30**, 811–818.

Estes, R. D. (1972). The role of the vomeronasal organ in mammalian reproduction. *Mammalia* **36**, 315–342.

Evans, C. M., Mackintosh, J. H., Kennedy, J. T., and Robertson, S. M. (1978). Attempts to characterise and isolate aggression-reducing olfactory signals from the urine of female mice *Mus musculus*. *Physiology and Behavior* **20**, 129–134.

Eyck, G. T. and Halpern, M. (1988). Aggregation in infant corn snakes (*Elaphe guttata*) and garter snakes (*Thamnophis radix*). *Chemical Senses* **13**, 740.

Faak, M. (ed.) (1990). *Die amerikanische Reise. A. v. Humboldt' s Travel Log*, vol. 2. Berlin Akademie-Verlag.

Faeth, S. H. (1992). Do defoliation and subsequent phytochemical responses reduce future herbivory on oak trees? *Journal of Chemical Ecology* **18**, 915–925.

Fagre, D. B., Howard, W. E., Barnum, D. A., *et al.* (1983). Criteria for the development of coyote lures. In *Proceeding of the 4th Symposium of the ASTM STP 812 on Vertebrate Pest Control and Management Materials*, ed. D. E. Kaukeinen, pp. 265–277. Philadelphia, PA: American Society for Testing and Materials.

Farentinos, R. C., Capretta, P. J., Kepner, R. E., and Littlefield, V. M. (1981). Selective herbivory in tassel-eared squirrels: role of monoterpenes in ponderosa pines chosen as feeding trees. *Science* **213**, 1273–1275.

Fares, Y., Sharpe, P. J., and Magnuson, C. E. (1980). Pheromone dispersion in forests. *Journal of Theoretical Biology* **84**, 335–359.

Faulkes, C. G. and Abbott, D. H. (1993). Evidence that primer pheromones do not cause social supression of reproduction in naked mole rats (*Heterocephalus glaber*). *Journal of Reproduction and Fertility* **99**, 225–230.

Feiler, W. and Haas, W. (1988). Trichobilharzia ocellata: chemical stimuli of duck skin for cercarial attachment. *Parasitology* **96**, 507–517.

Feist, J. D. and McCullough, D. (1976). Behavior patterns and communication in feral horses. *Journal of Reproduction and Fertility* **23** (Suppl.), 337–371.

Fergusson, B., Bradshaw, S. D., and Cannon, J. R. (1985). Hormonal control of femoral gland secretion in the lizard, *Amphibolurus ornatus*. *General and Comparative Endocrinology* **57**, 371–376.

Ferkin, M. H. (1988). The effect of familiarity on social interactions in meadow voles, *Microtus pennsylvanicus*: a laboratory and field study. *Animal Behaviour* **36**, 1816–1822.

Ferkin, M. H. and Johnston, R. E. (1995a). Meadow voles, *Microtus pennsylvaticus*, use multiple sources of scent for sex recognition. *Animal Behaviour* **49**, 37–44.

(1995b). Effects of pregnancy, lactation and postpartum oestrus on odor signals and the attraction to odours in female meadow voles, *Microtus pennsylvaticus*. *Animal Behaviour* **49**, 1211–1217.

Ferkin, M. H. and Seamon, J. O. (1987). Odor preference and social behavior in meadow voles, *Microtus pennsylvanicus*: seasonal differences. *Canadian Journal of Zoology* **65**, 2931–2937.

Ferkin, M. H., Sorokin, E. S., Johnston, R. E., and Lee, C. J. (1997). Attractiveness of scents varies with protein content of the diet in meadow voles. *Animal Behaviour* **53**, 133–141.

Ferreira, A., Dahlof, L. G., and Hansen, S. (1987). Olfactory mechanisms in the control of maternal aggression, appetite, and fearfulness: effects of lesions to olfactory receptors, mediodorsal thalamic nucleus, and insular prefrontal cortex. *Behavioral Neuroscience* **101**, 709–717.

Ferris, C. G., Axelson, J. F., Shinto, L. H., and Albers, H. E. (1987). Scent marking and the maintenance of dominant subordinate status in male golden hamsters. *Physiological Behavior* **40**, 661–664.

Filho, O. G. and Mazzafera, P. (2000). Caffeine does not protect coffee against the leaf miner *Perileucoptera coffeella*. *Journal of Chemical Ecology* **26**:1447–1464.

Fillion, T. J. and Blass, E. M. (1986). Infantile experience with suckling odors determines adult sexual behavior in male rats. *Science* **231**, 729–731.

Fine, J. M. and Sorensen, P. W. (2004). Bioassay-guided fractionation demonstrates that the sea lamprey migratory pheromone is a mixture of at least three sulfated steroids. In *Annual Meeting of International Society of Chemical Ecology*, July 2004, Ottawa, Canada.

Fine, J. M., Vrieze, L. A., and Sorensen, P. W. (2004). Evidence that petromyzontid lampreys employ a common migratory pheromone that is partially comprised of bile acids. *Journal of Chemical Ecology* **30**, 2091–2110.

Flannery, T. (1998). *Throwing Way Leg: Tree-kangaroos, Possums, and Penis Gourds: On the track of Unknown Mammals in Wildest New Guinea.* New York: Atlantic Monthly Press.

Fleming, A. S., Vaccarino, F., Tambosso, L., and Chee, P. (1979). Vomeronasal and olfactory system modulation of maternal behavior in the rat. *Science* **203**, 372–374.

Fletcher, K. J. C. and Michener, D. C. (1987). *Kin Recognition in Animals.* New York: Wiley.

Fletcher, T. C. and Lindsay, D. R. (1968). Sensory involvement in the mating behavior of domestic sheep. *Animal Behaviour* **16**, 410–416.

Flier, J. M., Edwards, W., Daly, J. W., and Meyers, C. (1980). Widespread occurrence in frogs and toads of skin compounds interacting with the ovabain site of Na^+, K^+-ATPase. *Science* **208**, 503–505.

Flood, P. F., Abrams, S. R., Muir, G. D., and Rowell, J. E. (1989). Odor of the muskox: a preliminary investigation. *Journal of Chemical Ecology* **15**, 2207–2217.

Flowers, M. A. and Graves, B. M. (1997). Juvenile toads avoid chemical cues from snake predators. *Animal Behaviour* **53**, 641–646.

Fogel, R. and Trappe, J. M. (1978). Fungus consumption (mycophagy) by small animals. *Northwest Science* **52**, 1–31.

Foley, W. J., McLean, S., and Cork, S. J. (1995). Consequences of biotransformation of plant secondary metabolites on acid–base metabolism in mammals: a final common pathway? *Journal of Chemical Ecology* **21**, 721–743.

Fombon, A. M. and Polak, E. H. (1987). Odor similarity between stress-inducing odorants in Wistar rats. *Journal of Chemical Ecology* **13**, 153–166.

Ford, N. B. (1978). Evidence for the species specificity of pheromone trails in two sympatric garter smakes, (*Thamnophis*). *Herpetological Review* **9**, 10.

(1981). Seasonality of pheromone trailing behavior in two species of garter snake, *Thamnophis* (Colubridae). *Southwestern Naturalist* **26**, 385–388.

(1982). Species specificity of sex pheromone trails of sympatric and allopatric garter snakes (*Thamnophis*). *Copeia* **1**, 10–13.

Ford, N. B. and Low, J. R. (1984). Sex pheromone source location by garter snakes: a mechanism for detection of direction in nonvolatile trails. *Journal of Chemical Ecology* **10**, 1193–1199.

Forester, D. C. and Wisnieski, A. (1991). The significance of airborne olfactory cues to the recognition of home area by the dart-poison frog *Dendrobates pumilio*. *Journal of Herpetology* **25**, 502–504.

Formanowicz, D. R. and Brodie, E. D. (1982). Relative palatabilities of members of a larval amphibian community. *Copeia* **1982**, 91–97.

Fornasieri, I. and Roeder, J. J. (1992). Behavioral responses to own and other species' scent marks in *Lemur fulvus* and *Lemur macaco*. *Journal of Chemical Ecology* **18**, 2069–2082.

Fowler, M. E. (1992). *Veterinary Zootoxicology*. Boca Raton, FL: CRC Press.

Fraenkel, G. S. (1959). The raison d'etre of secondary plant substances. *Science* **129**, 1466–1470.

Frank, F. (1954). Beiträge zur Biologie der Feldmaus, *Microtus arvalis* (Pallas). Teil I. Gehegeversuche. *Zoologische Jahrbücher* **82**, 354–404.

Frank, R. A. and Byram, J. (1988). Taste–smell interactions are tastant and odorant dependent. *Chemical Senses* **13**, 445–455.

Franklin, W. L. (1983). Contrasting socioecologies of South America's wild camelids: the vicuña and the guanaco. *American Society of Mammologists, Special Publication* **7**, 573–631.

Freeland, W. J. (1974). Vole cycles: another hypothesis. *American Naturalist* **108**, 238–245.

(1991). Plant secondary metabolites, biochemical coevolution with herbivores. In *Plant Defenses Against Mammalian Herbivory*, ed. R. T. Palo and C. T. Robbins, pp. 61–81. Boca Raton, FL: CRC Press.

Freeland, W. J. and Janzen, D. H. (1974). Strategies in herbivory by mammals: the role of plant secondary compounds. *American Naturalist* **108**, 269–289.

French, J. A., Abbot, D. H., and Snowdon, C. T. (1984). The effect of social environment on estrogen excretion, scent marking, and sociosexual behavior in tamarins (*Saguinus oedipus*). *American Journal of Primatology* **6**, 155–167.

Frey, R. and Hofmann, R. R. (1997). Skull, proboscis musculature and preorbital gland in the saiga antelope and Guenther's dikdik (Mammalia, Artiodactyla, Bovidae). *Zoologischer Anzeiger* **235**, 183–199.

Frid, L. and Turkington, R. (2001). The influence of herbivores and neighboring plants on risk of browsing: a case study using arctic lupine (*Lupinus arcticus*) and arctic ground squirrels (*Spermophilus parryi plesius*). *Canadian Journal of Zoology* **79**, 874–880.

Friedmann, A. (1955). The honey guides. *Bulletin of the United States National Museum* **208**, 1–292.

Friedman, L. and Miller, J. G. (1971). Odor incongruity and chirality. *Science* **172**, 1044–1046.

Frischknecht, P. M., Ulmer-Dufek, J., and Baumann, T. W. (1986). Purine alkaloid formation in buds and developing leaflets of *Coffea arabica*: expression of an optimal defence strategy? *Phytochemistry* **25**, 613–616.

Fry, B. G., Vidal, N., Norman, J. A., *et al.* (2005). Early evolution of the venom system in lizards and snakes. *Nature* advance online publication, 16 November 2005. (doi.10.1038/nature04328).

Fuchs, J. L. and Burghardt, G. M. (1971). Effects of early feeding experience on the responses of garter snakes to food chemicals. *Learning and Motivation* **2**, 271–279.

Fulk, G. W. (1971). *The Behavioral Interactions of the Short-tailed Shrew and the Meadow Vole.* Kingston, RI: University of Rhode Island.

Gagliardo, A., Faschi, V., and Benvenuti, S. (1988). Pigeon homing: olfactory experiments with young inexperienced birds. *Naturwissenschaften* **75**, 211–213.

Galef, B. (1982). Acquisition and waning of exposure-induced attraction to a non-natural odor in rat pups. *Developmental Psychobiology* **15**, 479–490.

Galef, B. G. and Kaner, H C. (1980). Establishment and maintenance of preference for natural and artificial olfactory stimuli in juvenile rats. *Journal of Comparative and Physiological Psychology* **94**, 588–595.

Galef, B. G. and Stein, M. (1985). Demonstrator influence on observer diet preference: analysis of critical social interactions and olfactory signals. *Animal Learning and Behavior* **13**, 31–38.

Galef, B. G., Kennett, D. J., and Stein, M. (1985). Demonstrator influence on observer diet preference: effects of simple exposure and the presence of a demonstrator. *Animal Learning and Behavior* **13**, 25–30.

Galef, B. G., Mason, J. R., Prety, G., and Bean, N. J. (1988). Carbon disulfide: a semiochemical mediating socially-induced diet choice in rats. *Physiology and Behavior* **42**, 119–124.

Gamradt, S. C. and Kats, L. B. (1996). Effect of introduced crayfish and mosquitofish on California newts. *Conservation Biology* **10**, 1155–1162.

Gangrade, B. K. and Dominic, C. J. (1986). Effect of storage and lyophilization on estrus-inducing capacity of male mouse urine. *Indian Journal of Experimental Biology* **24**, 728–729.

Gao, Y. (1991). Behavioral responses of rats to the smell of urine from conspecifics. *Animal Behaviour* **42**, 506–508.

Garcia, J. and Rusiniak, K. W. (1980). What the nose learns from the mouth. In *Chemical Signals in Vertebrates*, vol. 2, ed. D. Müller-Schwarze, and R. M. Silverstein, pp. 141–156. New York: Plenum.

Garcia, J., Holder, M. D., and Yirmiya, R. (1986). Taste and odor interactions in conditioned flavor aversions. *Appetite* **7**, 259.

Garstka, W. R. and Crews, D. (1986). Pheromones and reproduction in garter snakes. In *Chemical Signals in Vertebrates*, vol 4, ed. D. Duvall, D. Müller-Schwarze, and R. M. Silverstein, pp. 243–260. New York: Plenum.

Garton, J. D. and Mushinsky, H. R. (1979). Integumentary toxicity and unpalatability as an antipredator mechanism in the narrow mouthed toad, *Gastrophryne carolinensis*. *Canadian Journal of Zoology* **57**, 1965–1973.

Gause, G. F. (1934). *The Struggle for Existence*. New York: Hafner.

Gauthier, G. and Hughes, R. J. (1995). The palatability of Arctic willow for greater snow geese: the role of nutrients and deterring factors. *Oecologia* **103**, 390–392.

Gauthier-Pilters, H. (1974). The behavior and ecology of camels in the Sahara, with special reference to nomadism and water management. In *The Behavior of Ungulates and its Relation to Management*, ed. V. Geist and F. Walther, pp. 542–551. Morges, Switzerland: International Union for Conservation of Nature and Natural Resources.

Gavish, L., Hofmann, J. E., and Getz, L. L. (1984). Sibling recognition in the prairie vole, *Microtus ochrogaster*. *Animal Behaviour* **32**, 362–366.

Gazdewich, K. J. and Chivers, D. P. (2002). Acquired predator recognition by fathead minnows: influence of habitat characteristics on survival. *Journal of Chemical Ecology* **28**, 439–445.

Gehlbach, F. R., Watkins, J. F., and Reno, H. W. (1968). Blindsnake defensive behavior elicited by ant attacks. *BioScience*. **18**, 784–785.

Gehlbach, F. R., Watkins, J. F., and Kroll, J. C. (1971). Pheromone trail following studies of typhlopid, leptotyphlopid, and colubrid snakes. *Behaviour* **40**, 282–294.

Geiger, R. (1965). *The Climate Near the Ground*. Cambridge, MA: Harvard University Press.

Genna, R. L., Mordue, W., Pike, A. W., and Mordue (Luntz) A. J. (2004). Identification of semiochemicals involved in sea lice host location, and their potential use in pest control. In *Annual Meeting of the International Society of Chemical Ecology*, July 2004, Ottawa, Canada.

George, C. J. W. (1960). Behavioral interaction of the pickerel (*Esox niger* LeSueur and *Esox americanus* LeSueur) and the mosquitofish (*Gambusia patruellis* Bairdad Girard). Ph.D. Thesis, Harvard University Massachussetts.

Gerhart, D. J., Bondura, M. E., and Commito, J. A. (1991). Inhibition of sunfish feeding by defensive steroids from aquatic beetles: structure–activity relationships. *Journal of Chemical Ecology* **17**, 1363–1370.

Gerritsen, A. F. C., van Heezik, Y. M., and Swennen, C. (1983). Chemoreception in two further *Calidris* species (*C. maritima and C. canutus*) with a comparison of the relative importance of chemoreception during foraging in *Calidris* species. *Netherlands Journal of Zoology* **33**, 485–496.

Gervais, R., Holley, A., and Keverne, B. (1988). The importance of central noradrenergic influences on the olfactory bulb in the processing of learned olfactory cues. *Chemical Senses* **13**, 3–12.

Gheusi, G., Goodall, G., and Dantzer, R. (1997). Individually distinctive odors represent individual conspecifics in rats. *Animal Behaviour* **53**, 935–944.

Gilardi, J. D., Duffey, S. S., Munn, C. A., and Tell, L. A. (1999). Biochemical functions of geophagy in parrots: detoxification of dietary toxins and cytoprotective effects. *Journal of Chemical Ecology* **25**, 897–922.

Gilbert, B. K. (1973). Scent marking and territoriality in pronghorn (*Antilocapra americana*) in Yellowstone National Park. *Mammalia* **37**, 25–33.

Gilder, P. M. and Slater, P. J. B. (1978). Interest of mice in conspecific male odours is influenced by degree of kinship. *Nature* **274**, 364–365.

Gleeson, R. A. (1978). Functional adaptation in chemosensory systems. In *Sensory Ecology*, ed. M. A. Ali, pp. 291–317. New York: Plenum.

Glei, M., Schlegel, W., Straube, D., and Blankenger, J. (1989). Untersuchungen zur Beeinflussung des Pubertätseintritts von Jungsauen mittels maskuliner Stimuli. *Archiv für Tierzucht* **32**, 173–179.

Glendinning, J. F., Brower, L. P., and Montgomery, C. A. (1990). Responses of three mouse species to deterrent chemicals in the Monarch butterfly. I. Taste and toxicity tests using artificial diets laced with digitoxin or monocrotaline. *Chemoecology* **1**, 114–123.

Godfrey, J. (1958). The origin of sexual isolation between bank voles. *Proceedings of the Royal Physiological Society Edinburgh* **27**, 47–55.

Golan, L., Radcliffe, C. W., Miller, T., O'Connell, R. J., and Chiszar, D. (1982). Prey trailing by the prairie rattlesnake (*Crotalus v. viridis*). *Journal of Herpetology* **16**, 287–293.

Gold, T. and Soter, S. (1980). The deep-earth-gas hypothesis. *Scientific American* **242**, 154–161.

Goldstein, W. S. and Spencer, K. C. (1985). Inhibition of cyanogenesis by tannins. *Journal of Chemical Ecology* **11**, 847–858.

Goodrich, B. S. and Mykytowycz, R. (1972). Individual and sex differences in the chemical composition of pheromone-like substances from the skin glands of the rabbit, *Oryctolagus cuniculus*. *Journal of Mammalogy* **53**, 540–548.

Goodrich, B. S., Gambale, S., Pennycuik, P. R., and Redhead, T. D. (1990). Volatiles from feces of wild male house mice. Chemistry and effects on behavior and heart rate. *Journal of Chemical Ecology* **16**, 2091–2106.

Gorman, M. L. (1976). A mechanism for individual recognition by odour in *Herpestes auropunctatus* (Carnivora: Viverridae). *Animal Behaviour* **24**, 141–145.

(1984). The response of prey to stoat (*Mustela erminea*) scent. *Journal of Zoology* (London) **202**, 419–423.

Gorman, M. L. and Mills, M. G. L. (1984). Scent marking strategies in hyaenas (Mammalia). *Journal of Zoology* (London) **202**, 535–547.

Gorman, M. L., Jenkins, D., and Cooper, R. J. (1978). The anal scent sacs of the otter (*Lutra lutra*). *Journal of Zoology* **186**, 463–474.

Gosden, P. E. and Ware, G. C. (1976). The aerobic bacterial flora of the anal sac of the red fox. *Journal of Applied Bacteriology* **41**, 271–275.

Gosling, L. M. (1981). Demarkation in a gerenuk territory: an economic approach. *Zeitschrift für Tierpsychologie* **56**, 305–322.

(1982). A reassessment of the function of scent marking in territories. *Zeitschrift für Tierpsychologie* **60**, 89–118.

(1986). Economic consequences of scent marking in mammalian territoriality. In *Chemical Signals in Vertebrates*, vol. 4, ed. D. Duvall, D. Müller-Schwarze, and R. M. Silverstein, pp. 385–395. New York: Plenum.

Gould, S. J. and Vrba, E. S. (1982). Exaptation: a missing term in the science of form. *Paleobiology* **8**, 4–15.

Gove, D. and Burghardt, G. M. (1975). Responses of ecologically dissimilar populations of the water snake, *Natrix s. sipedon*, to chemical cues from prey. *Journal of Chemical Ecology* **1**, 25–40.

Gower, D. B. (1989). Pheromones in the human. *Journal of Endocrinology*. **123**(Suppl.), 13.

Gower, D. B. and Booth, V. (1986). *Ontogeny of Olfaction: Principles of Olfactory Maturation in Vertebrates*, Berlin: Springer-Verlag.

Gower, D. B. and Ruparelia, B. A. (1993). Olfaction in humans with special reference to odourous 16-androstenes: their occurrence, perception and possible social, psychological, and sexual impact. *Journal of Endocrinology* **137**, 167–187.

Gower, D. B., Watkins, J., Mallett, A. I., Rennie, P. J., and Holland, K. T. (1989). Transformations of odorous 16-androstene steroids by human axillary coryneform bacteria. *Chemical Senses* **14**, 208.

Graham, C. A. and McGrew, W. C. (1980). Menstrual synchrony in female undergraduates living in a co-educational campus. *Psychoneuroendocrinology* **5**, 245–252.

Graham, T., Georges, A., and Mcelhinney, N. (1996). Terrestrial orientation by the eastern longnecked turtle, *Chelodina longicollis*, from Australia. *Journal of Herpetology* **30**, 467–477.

Grajal, A. (1995). Structure and function of the digestive tract of the hoatzin (*Opisthocomus hoazin*): a folivorous bird with foregut fermentation. *Auk* **112**, 20–28.

Grajal, A., Strahl, S. D., Parra, R., Dominquez, M. G., and Neher, A. (1989). Foregut fermentation in the hoatzin, a neotropical leaf-eating bird. *Science* **245**, 1236–1238.

Grant, D., Andersen, O., and Twitty, V. (1968). Homing orientation by olfaction in newts (*Taricha rivularis*). *Science* **160**, 1354–1356.

Grassman, M. and Owens, D. (1987). Chemosensory imprinting in juvenile green sea turtles, *Chelonia mydas*. *Animal Behaviour* **35**, 929–931.

Grassman, M. A., Owens, D. W., McVey, J. P., and Marquez, M. R. (1984). Olfactory-based orientation in artificially imprinted sea turtles. *Science* **224**, 83–84.

Grau, H. J. (1982). Kin recognition in the white-footed deer mice (*Peromyscus leucopus*). *Animal Behaviour* **30**, 497–505.

Graves, B. M. and Duvall, D. (1985). Avomic prairie rattlesnakes (*Crotalus viridis*) fail to attack rodent prey. *Zeitschrift für Tierpsychologie* **67**, 161–166.

(1988). Evidence of an alarm pheromone from the cloacal sacs of prairie rattlesnakes. *Southwestern Naturalist* **33**, 339–345.

Graves, B. M. and Halpern, M. (1990). Roles of vomeronasal chemoreception in tongue-flicking, exploratory and feeding behavior of the lizard, *Chalcides ocellatus*. *Animal Behaviour* **39**, 692–698.

(1991). Discrimination of self from conspecific chemical cues in *Tiliqua scincoides* (Sauria: Scincidae). *Journal of Herpetology* **25**, 125–126.

Graves, B. M., Carpenter, G. C. and Duvall, D. (1987). Chemosensory behaviors of neonate prairie rattlesnakes, *Crotalus viridis*. *Southwest Naturalist* **32**, 515–517.

Graves, B. M., Halpern, M., and Gillingham, J. C. (1993). Effects of vomeronasal system deafferentation on home range use in a natural population of eastern garter snakes, *Thamnophis sirtalis*. *Animal Behaviour* **45**, 307–311.

Graziadei, P. P. C. (1977). Functional anatomy of the mammalian chemoreceptor system. In *Chemical Signals in Vertebrates*, vol. 1, ed. R. T. Mason, M. P. LeMaster, and D. Müller-Schwarze, pp. 435–454. New York: Plenum.

Green, G. A. (1988). Living on borrowed turf. *Natural History* **97**, 58–64.

Greenberg, B. (1943). Social behavior of the western banded gecko, *Coleonyx variegatus* Baird. *Physiological Zoology* **16**, 110–122.

Greene, M. J. and Mason, R. T. (2000). Courtship, mating and male combat of the brown tree snake, *Boiga irregularis*. *Herpetologica* **56** 166–175.

(2003). Pheromonal inhibition of male courtship behavior in the brown tree snake, *Boiga irregularis*: a mechanism for the rejection of potential mates. *Animal Behaviour* **65**, 905–910.

Greene, M. J., Stark, S. L., and Mason, R. T. (2002). Predatory response of brown tree snakes to chemical stimuli from human skin. *Journal of Chemical Ecology* **28**: 2465–2473.

Greenwood, D. R., Comeskey, D., Hunt, M. B., and Rasmussen, L. E. L. (2005). Chirality in elephant pheromones. *Nature* **438**, 1097–1098.

Gregory, M. J. and Cameron, G. N. (1989). Scent communication and its association with dominance behavior in the Hispid cotton rat (*Sigmodon hispidus*). *Journal of Mammalogy* **70**, 10–17.

Greig-Smith, P. W. (1988). Bullfinches and ash trees: assessing the role of plant chemicals in controlling damage by herbivores. *Journal of Chemical Ecology* **14**, 1889–1903.

Griffin, R. W. and Beidler, L. M. (1984). Studies in canine olfaction, taste and feeding: a summing up and some comments on the academic–industrial relationship. *Neuroscience and BioBehavioral Reviews* **8**, 261–263.

Griffith, C. R. (1919). A possible case of instinctive behavior in the white rat. *Science* **50**, 166–167.

(1920). The behavior of white rats in the presence of cats. *Psychobiology* **2**, 19–28.

Grimmer, J. L. (1962). Strange little world of the hoatzin. *National Geographic* **122**, 390–401.

Grønneberg, T. Ø. (1978–79). Analysis of a wax ester fraction from anal gland secretion of beaver (*Castor fiber*) by chemical ionization mass spectrometry. *Chemical Scripta* **13**, 56–58.

Grønneberg, T. Ø. and Lee, T. (1984). Lipids of the anal gland secretion of beaver (*Castor canadensis*). *Chemica Scripta* 24, 100-103.

Grubb, J. C. (1973a). Olfactory orientation in breeding Mexican toads, *Bufo valliceps*. *Copeia* **1973**, 490–497.

(1973b). Olfactory orientation in *Bufo woodhousei fowleri, Pseudacris clarki* and *P. streckeri*. *Animal Behaviour* **21**, 726–732.

(1976). Maze orientation by Mexican toads, *Bufo valliceps*, using olfactory and configurational cues. *Journal of Herpetology* **10**, 97–104.

Grubb, T. C., Jr. (1972). Smell and foraging in shearwaters and petrels. *Nature* **237**, 404–405.

(1974). Olfactory navigation to the nesting burrow in Leach's petrel (*Oceanodroma leucorrhoa*). *Animal Behaviour* **22**, 192–202.

Grundvig, J. L., Dustman, R. E., and Beck, E C. (1967). The relationship of olfactory receptor stimulation to stimulus environmental temperature. *Experimental Neurology*. **18**, 416–428.

Gubernick, D. J. (1990). A maternal chemosignal maintains paternal behavior in the biparental California mouse, *Peromyscus californicus*. *Animal Behaviour* **39**, 936–942.

Gubernick, D. J. and Klopfer, P. H. (1980). *Parental Care in Mammals*. New York: Plenum.

Gubernick, D. J. and Nordby, J. C. (1992). Parental influences on female puberty in the monogamous California mouse, *Peromyscus californicus*. *Animal Behaviour* **44**, 259–267.

Gunson, J. R. (1970). Dynamics of the beaver of Saskatchewan's northern forest. M. Sc. Thesis, University of Alberta, Edmonton, Alberta.

Gurnell, J. and Little, J. (1992). The influence of trap residual odor on catching woodland rodents. *Animal Behaviour* **43**, 623–632.

Gustin, M. K. and McCracken, G. F. (1987). Scent recognition between females and pups in the bat *Tadarida brasiliensis mexicana*. *Animal Behaviour* **35**, 13–19.

Gwinner, H., Oltrogge, M., Trost, L., and Nienaber, U. (2002). Green plants in starling nests: effects on nestlings. *Animal Behaviour* **59**, 301–309.

Habermehl, G. G. (1994). *Gift-Tiere und ihre Waffen*, 5th edn. Berlin: Springer-Verlag.

Haftorn, S., Mehilum, F., and Bech, C. (1988). Navigation to nest site in the snow petrel, *Pagodroma nivea*. *Condor* **90**, 484–486.

Hagelin, J. C., Jones, I. L., and Rasmussen, L. E. L. (2003). A tangerine-scented social odour in a monogamous seabird. *Proceedings of the Royal Society of London, Series B* **270**, 1323–1329.

Haigh, G. R. (1987). Reproductive inhibition of female *Peromyscus leucopus*: female competition and behavioral regulation. *American Zoologist* **27**, 867–868.

Haim, A. and Fluxman, S. (1996). Daily rhythms of metabolic rates: role of chemical signals in coexistence of spiny mice of the genus *Acomys*. *Journal of Chemical Ecology* **22**, 223–231.

Haldane, J. B. S. (1955). Animal communication and the origin of human language. *Science Progress* (London) **43**, 385–401.

Hall, D. R., Beevor, P. S., Cork, A., Nesbitt, B. F., and Vale, G. A. (1984). 1-Octen-3-ol. A potent olfactory stimulant and attractant for tsetse isolated from cattle odors. *Insect Science and its Application* **5**, 335–339.

Halpern, M. and Frumin, N. (1979). Roles of the vomeronasal and olfactory systems in prey attack and feeding in adult garter snakes. *Physiology and Behavior* **22**, 1183–1189.

Halpern, M., Scribani, L., and Kubie, J. L. (1985). Vomeronasal stimuli can be reinforcing. *Chemical Senses* **10**, 422.

Halpern, M., Schulman, N., and Kirschenbaum, D. M. (1986). Characteristics of earthworm washings detected by the vomeronasal system of snakes. In *Chemical Signals in Vertebrates*, vol. 4, ed. D. Duvall, Müller-Schwarze, D. and R. M. Silverstein, pp. 63–77, New York: Plenum.

Halpin, Z. T. (1974). Individual difference in the biological odors of the Mongolian gerbil (*Meriones unguiculatus*). *Behavioral Biology* **11**, 253–259.

(1986). Individual odors among mammals: origins and functions. *Advances in the Study of Behavior* **16**, 39–70.

Hanfin, C. T., Brodie, E. D., III, and Brodie, E. D., Jr. (2003). Tetrodotoxin levels in eggs of rough-skin newt, *Taricha granulosa*, are correlated with female toxicity. *Journal of Chemical Ecology* **29**, 1729–1739.

Hansen, L. P., Døving, K.B., and Jonsson, B. (1987). Migration of farmed adult Atlantic salmon with and without olfactory sense, released on the Norwegian coast. *Journal of Fish Biology* **30**, 713–722.

Hansson, L., Gref, R., and Theander, O. (1986). Susceptibility to vole attacks due to bark phenols and terpenes in *Pinus contorta* provenances introduced into Sweden. *Journal of Chemical Ecology* **12**, 1569–1578.

Harada, K. (1989). Role of carboxylic group on feeding repellence of acidic amino acids and organic acids for oriental weatherfish. *Bulletin of the Japanese Society of Scientific Fisheries* **55**, 927.

Harborne, J. B. (1993). *Introduction to Ecological Biochemistry*, 4th edn. London: Academic Press.

Harden-Jones, F. R. (1968). *Fish Migration*. London: Edward Arnold.

Hare, J. F. (1994). Group member discrimination by Columbian ground squirrels via familiarity with substrate-borne chemical cues. *Animal Behaviour* **47**, 803–813.

Hargrove, J. W. and Vale, G. A. (1978). The effect of host odour concentration on catches of tsetse flies (Glossinidae) and other Diptera in the field. *Bulletin of Entomological Research* **68**, 607–612.

Harju, A. (1996). Effect of birch (*Betula pendula*) bark and food protein level on root voles (*Microtus oeconomus*): II. Detoxification capacity. *Journal of Chemical Ecology* **22**, 719–728.

Harrington, F. H. (1981). Urine marking and caching behavior in the wolf. *Behavior* **61**, 82–105.

Harris, B. (1987). *Mushrooms and Truffles: Botany, Cultivation, and Utilization*. Koenigstein: Koeltz Scientific.

Harris, M. A. and Murie, J. O. (1982). Responses to oral gland scents from different males in Columbian ground squirrels. *Animal Behaviour* **30**, 140–148.

Hart, B. L. (1983). Flehmen behavior and vomeronasal organ function. In *Chemical Signals in Vertebrates*, vol. 3, ed. D. Müller-Schwarze and R. M. Silverstein. New York: Plenum.

Hart, B. L. and Leedy, M. G. (1987). Stimulus and hormonal determinants of flehmen behavior in cats. *Hormones and Behavior*. **21**, 44–52.

Hart, B. L., Hart, L. A., and Maina, J. N. (1988). Alteration in vomeronasal system anatomy in alcelaphine antelopes: correlation with alteration in chemosensory investigation. *Physiology and Behavior* **42**, 155–162.

Harvey, S., Jemiolo, B., and Novotny, M. (1989). Pattern of volatile compounds in dominant and subdominant male mouse urine. *Journal of Chemical Ecology* **15**, 2061–2072.

Hasler, A. D. (1954). Odor perception and orientation in fishes. *Journal of the Fisheries Research Board of Canada* **11**, 107–129.

Hasler, A. D. and Scholz, A. T. (1983). *Olfactory Imprinting and Homing in Salmon*. New York: Springer-Verlag.

Hasler, A. D. and Wisby, W. J. (1951). Discrimination of stream odors by fishes and relation to parent stream behavior. *American Naturalist*. **85**, 223–238.

Hasler, A. D., Scholz, A. T., and Horral, R. M. (1978). Olfactory imprinting and homing in salmon. *American Scientist* **66**, 347–355.

Hassanali, A., McDowell, P. G., Owaga, M. L. A., and Saini, R. K. (1986). Identification of tsetse attractants from excretory products of a wild host animal, *Syncerus caffer. Insect Science and its Application* **7**, 5–9.

Hassler, T. J. and Kucas, S. T. (1988). Returns of morpholine – imprinted coho salmon to the Mad River, California. *North American Journal of Fisheries Management* **8**, 356–358.

Hastings, B. C. and Gilbert, B. K. (1980). Aversive conditioning of black bears in the backcountry of Yosemite National Park. *Proceedings of the Second Conference on Scientific Research in National Parks*, vol. 7, pp. 294–303.

Hatanaka, T. and Hanada, T. (1987). Structure of the vomeronasal system and the induced wave in the accessory olfactory bulb of red eared turtle. *Chemical Senses* **12**, 521.

Hayashi, S. (1987). Female mice prefer odors of aggressive males. *Zoological Science* (Tokyo) **4**, 1105.

(1989). Male-mice: social-dominance influenced by strange male odors. *Aggressive Behavior*. **15**, 1–3.

Hayes, R. A., Richardson, B. J., and Wyllie, S. G. (2001). Increased social dominance in rabbits, *Oryctalus cuniculus*, is associated with increased secretion of 2-phenoxyethanol from the chin gland. In *Chemical Signals in Vertebrates*, vol. 9, ed. A. Marchlewska-Koi, J. J. Lepri, and D. Müller-Schwarze, pp. 335–341. New York: Kluwer Academic/Plenum.

(2002). Semiochemicals and social signaling in the wild European rabbit in Australia: I. Scent profiles of chin gland secretion from the field. *Journal of Chemical Ecology* **28**, 363–384.

Hébert, P. and Barrette, C. (1989). Experimental demonstration that scent marking can predict dominance in the woodchuck (*Marmota monax*). *Canadian Journal of Zoology* **67**, 575–578.

Hefetz, A., Ben-Yaacov, R., and Yom-Tov, Y. (1984). Sex specificity in the anal gland secretion of the Egyptian mongoose *Herpestes ichneumon*. *Journal of Zoology* **203**, 205–209.

Hendrichs, H. and Hendrichs, U. (1971). Dikdik und Elefanten: Ökologie und Soziologie zweier afrikanischer Huftiere. In *Studies in Ethology*, ed. W. Wickler. Munich: Piper.

Henessy, F. F. and Owings, D. H. (1979). Snake species discrimination and the role of olfactory cues in the snake-directed behavior of the California ground squirrel. *Behaviour* **65**, 115–124.

Henry, J. D. (1980). The urine marking behavior and movement patterns of red foxes (*Vulpes vulpes*) during a breeding and post-breeding period. In *Chemical Signals in Vertebrates*, vol. 2, ed. D. Müller-Schwarze and R. M. Silverstein, pp. 11–27. New York: Plenum.

Henton, W. W., Smith, J. C., and Tucker, D. (1966). Odor discrimination in pigeons. *Science* **153**, 1138–1139.

Hepper, P. G. (1983). Sibling recognition in the rat. *Animal Behaviour* **31**, 1177–1191.

(1987). The amniotic fluid: an important priming role in kin recognition. *Animal Behaviour* **35**, 1343–1346.

(1988). Adaptive fetal learning, prenatal exposure to garlic affects postnatal preferences. *Animal Behaviour* **36**, 935–936.

(1990). Foetal olfaction. In *Chemical Signals in Vertebrates*, vol. 5, ed. D. W. MacDonald, D. Müller-Schwarze, and S. E. Natynczuk, pp. 282–288. Oxford: Oxford University Press.

Herb, R., Carroll, R., Yoshida, W. Y., Scheuer, P. J., and Paul, V. J. (1990). Polyalkylated cyclopentindoles: cytotoxic fish antifeedants from a sponge, *Axinella sp. Tetrahedron* **46**, 3089–3092.

Herrera, E. A. (1992). Size of testes and scent glands in capybaras, *Hydrochaeris hydrochaeris* (Rodentia: Caviomorpha). *Journal of Mammalogy* **73**, 871–875.

Herz, R. S. and Inzlicht, M. (2002). Sex differences in response to physical and social factors involved in human mate selection: the importance of smell for women. *Evolution and Human Behavior* **23**, 359–364.

Heske, E. J. (1987). Responses of a population of California voles, *Microtus californicus*, to odor-baited traps. *Journal of Mammalogy* **68**, 64–72.

Hews, D. K. (1988). Alarm response in larval western toads, *Bufo boreas*: release of chemicals by a natural predator and it's effect on predator capture efficiency. *Animal Behaviour*. **36**, 125–133.

Hickey, M. B. C. and Fenton, M. B. (1987). Scent-dispersing hairs (Osmetrichia) in some Pteropodidae and Molossidae (Chiroptera). *Journal of Mammalogy* **68**, 381–384.

Hill, J. O., Smith, G. L., III, Pavlik, E. H., Burghardt, G. M., and Coulson, P. B. (1976). Species characteristic responses to catnip by undomesticated felids. *Journal of Chemical Ecology* 2, 239–253.

Hiller, A. and Wand, U. (1984). Radiocarbon dating of breeding places of petrels in the Antarctic. Academy of Sciences, German Democratic Republic. *Zentralinstitut für Isotopen-und Strahlenforschung ZFI-Mitteilungen* **89**, 103–121.

Hirvonen, H., Ranta E., Piironen, J., Laurila, A., and Peuhkuri, N. (2000). Behavioral responses of naive Arctic charr young to chemical cues from salmonid and non-salmonid fish. *Oikos* **88**, 191–199.

Hladick, A. and Hladik, C. M. (1969). Rapports trophiques entre vegetation et primates dans la forêt de Barro Colorado (Panama). *Terre Vie* **116**, 25–117.

Ho, H.-Y. and Chow, Y. S. (1993). Chemical identification of defensive secretion of stick insect, *Megacrania tsudai* Shiraki. *Journal of Chemical Ecology* **19**, 39–46.

Hofmann, J. E., Getz, L. L., and Gavish, L. (1987). Effect of multiple short-term exposures of pregnant *Microtus ochrogaster* to strange males. *Journal of Mammalogy* **68**, 166–169.

Hold, B. and Schleidt, M. (1977). The importance of human odour in non-verbal communication. *Zeitschrift für Tierpsychologie* **43**, 225–238.

Holden, C. (2001). Cyanide named as Kentucky foal killer. *Science* **292**, 1831.

(2004). Ratting out tuberculosis. *Science* **303**, 166.

Holland, K. (1978). Chemosensory orientation to food by a Hawaiian goatfish (*Parupeneus porphyreus*, Mullidae). *Journal of Chemical Ecology* **4**, 173–186.

Holling, C. S. (1958). Sensory stimuli involved in the location and selection of sawfly cocoons by small mammals. *Canadian Journal of Zoology* **36**, 633–653.

Holm, J. C. and Walther, B. T. (1988). Free amino acids in live freshwater zooplankton and dry feed: possible importance for first feeding in Atlantic salmon fry (*Salmo salar*). *Aquaculture* **71**, 223–234.

Holmes, W. G. (1984). Sibling recognition in thirteen lined ground squirrels: effects of genetic relatedness, rearing association and olfaction. *Behavioral Ecology and Sociobiology* 14, 225.

(1986). Kin recognition by phenotype matching in female Belding's ground squirrels. *Animal Behaviour* **34**, 38–47.

(1992). Sternal odors as cues for social discrimination by female Virginia opossums, *Didelphis virginiana*. *Journal of Mammalogy* **73**, 286–291.

Holy, T. E., Dulac, C., and Meister, M. (2000). Responses of vomeronasal organ to natural stimuli. *Science* **289**, 1569–1572.

Horn, S. W. (1983). An evaluation of predatory suppression in coyotes using lithium chloride induced illness. *Journal of Wildlife Management* **47**, 999–1009.

Horne, E. A. and Jaeger, R. G. (1988). Territorial pheromones of female red-backed salamanders. *Ethology* **78**, 143.

Houck, L. D. and Reagan, N. L. (1990). Male courtship pheromones increase female receptivity in a plethodontid salamander. *Animal Behaviour* **39**, 729–734.

Houck, L. D., Bell, A. M., Reagan-Wallin, N. L., and Feldhoff, R. C. (1998). Effects of experimental delivery of male courtship pheromones on the timing of courtship in a terrestrial salamander, *Plethodon jordani* (Caudata: Plethodontidae). *Copeia* **1**, 214–219.

Houlihan, P. W. (1989). Scent mounding by beaver (*Castor canadensis*): functional and semiochemical aspects. M. Sc. Thesis, State University of New York, College of Environmental Science and Forestry, Syracuse, New York.

Houston, D. C. (1984). Does the king vulture *Sarcorhamphus papa* use a sense of smell to locate food? *Ibis* **126**, 67–69.

(1987). Scavenging efficiency of turkey vultures in tropical forests. *Condor* **88**, 318–323.

Howard, R. R. (1971). Avoidance learning of spotted salamanders by domestic chickens. *American Zoologist* **11**, 637.

Howard, W. E. and Cole, R. E. (1967). Olfaction in seed detection by deer mice. *Animal Behaviour* **16**, 13–17.

Howard, W. E., Marsh, R. E., and Cole, R. E. (1968). Food detection by deer mice using olfactory rather than visual cues. *Animal Behaviour* **16**, 13–27.

Hradecky, P. (1989). Possible induction by estrous cows of pheromone production in penmates. *Journal of Chemical Ecology* 15, 1067-1076.

Hubert, H. B., Fabsitz, R. R., Feinleib, M. and Brown, K. S. (1980). Olfactory sensitivity in humans: genetic versus environmental control. *Science* **208**, 607–609.

Huck, U. W., Lisk, R. D., Kim, S., and Evans, A. B. (1989). Olfactory discrimination of estrous condition by the male golden hamster, *Mesocricetus auratus*. *Behavioral and Neural Biology* **51**, 1–10.

Hudson, R. and Altbäcker, V. (1994). Development of feeding and food preference in the European rabbit: environmental and maturational determinants. In *Behavioral Aspects of Feeding: Basic and Applied Research in Mammals*, ed. B. G. Galef, M. Mainardi, and P. Valsecchi, pp. 125–145. Chur: Harwood Academic.

Hudson, R. and Distel, H. (1983). Nipple location by newborn rabbits: behavioral evidence for pheromonal guidance. *Behaviour* **85**, 260–275.

(1986). Pheromonal release of suckling in rabbits does not depend on the vomeronasal organ. *Physiology and Behavior* 37, 123–128.

Hudson, R. and Vodermayer, T. (1992). Spontaneous and odour-induced chin marking in domestic female rabbits. *Animal Behaviour* **43**, 329–336.

Huffman, M. A. and Seifu, M. (1989). Observations on the illness of and consumption of a possibly medicinal plant, *Vernonia amygdalina* (Del.), by a wild chimpanzee in the Mahale Mountains National Park, Tanzania. *Primates* **30**, 51–63.

Hugie, D. M. and Smith, R. J. F. (1987). Epidermal club cells are not linked with an alarm response in reedfish, *Erpetoichthys calabaricus*. *Canadian Journal of Zoology* **65**, 2057–2061.

Hull, M. Q. (1997). The role of semiochemicals in the behavior and biology of *Lepeophtheirus salmonis* (Krøyer 1837): Potential for control? Ph.D. Thesis, University of Aberdeen, Aberdeen, UK.

Humphries, R. E., Robertson, D. H. L., Beynon, R. J., and Hurst, J. L. (1999). Unravelling the chemical basis of competitive scent marking in house mice. *Animal Behaviour* **58**, 1177–1190.

Hurst, J. L. (1987). The functions of urine marking in a population of free-living house mice, *Mus domesticus* Rutty. *Animal Behaviour* **35**, 1433–1442.

(1989). The complex network of olfactory communicationin populations of wild house mice *Mus domesticus* Rutty: urine marking and investigation within family groups. *Animal Behaviour* **37**, 705–725.

(1990a). Urine marking in populations of wild house mice *Mus domesticus* Rutty. I. Communication between males. *Animal Behaviour* **40**, 209–222.

(1990b). Urine marking in populations of wild house mice *Mus domesticus* Rutty. II. Communication between females. *Animal Behaviour* **40**, 223–232.

(1993). The priming effects of urine substrate marks on interactions between male house mice, *Mus musculus domesticus* Schwarz and Schwarz. *Animal Behaviour* **45**, 55–81.

Hurst, J. L. and Rich, T. J. (1999). Scent marks as competitive signals of mate quality. In *Advances in Chemical Signals in Vertebrates*, ed. R. E. Johnston, D. Müller-Schwarze, and P. W. Sorensen, pp. 209–225. New York: Kluwer Academic/Plenum.

Hurst, J. L. and Smith, J. (1995). *Mus spretus* Lataste: a hygienic house mouse? *Animal Behaviour* **49**, 827–834.

Hurst, J. L., Fang, J., and Barnard, C. J. (1993). The role of substrate odours in maintaining social tolerance between male house mice, *Mus musculus domesticus*. *Animal Behaviour* **45**, 997–1006.

Hurst, J. L, Robertson, D. H. L., Tolladay, U., and Beynon, R. J. (1998). Proteins in urine scent marks of male house mice extend the longevity of olfactory signals. *Animal Behaviour* **55**, 1289–1298.

Hurst, J. L., Payne, C. E., Nevison, C. M., *et al.* (2001). Recognition of mice mediated by major urinary protiens. *Nature* **414**, 631–634.

Hutchison, L. V. and Wenzel, B. M. (1980). Olfactory guidance in foraging by procellariiforms. *Condor* **82**, 314–319.

Hutchison, L. V., Wenzel, B. H., Stager, K. E. and Tedford, B. L. (1984). Further evidence for olfactory foraging by Sooty Shearwater and Northern Fulmars. In *Marine Birds: Their Feeding Ecology and Commercial Fisheries Relationships*, ed. D. N. Nettleship, G. A. Sanger, and P. F. Springer, pp. 78–89. Ottawa: Canadian Wildlife Service Special Publication.

Hwang, C. J., Krasner, S. W., and McGuire, M. J. (1984). Determination of subnanogram per liter levels of earth-musty odorants in water by the salted closed-loop stripping method. *Environmental Science and Technology* **18**, 535–539.

Iason, G. R. and Murray, A. H. (1996). The energy costs of ingestion of naturally occurring nontannin plant phenolics by sheep. *Physiological Zoology* **69**, 532–546.

Iason, G. R. and Palo, R. T. (1991). The effects of birch phenolics on a grazing and a browsing mammal: a comparison. *Journal of Chemical Ecology* **17**, 1733–1743.

Idler, D. E., Fagerland, V. H. M., and Mayoh, H. (1956). Olfactory perception in migrating salmon. I. L-Serine, a salmon repellent in mammalian skin. *Journal of General Physiology* **39**, 889–892.

Idris, M. and Prakash, I. (1987). Scent marking activity in the Indian gerbil, *Tatera indica*, in relation to population density. *Animal Behaviour* **35**, 920–941.

Ikeda, H. (1984). Raccoon dog scent marking by scats and its significance in social behavior. *Journal of Ethology* **2**, 77–84.

Illius, A. W. and Jessop, N. S. (1995). Modeling metabolic costs of allelochemical ingestion by foraging herbivores. *Journal of Chemical Ecology* **21**, 693–719.

Izard, M. K. and Vandenbergh, J. G. (1982a). The effects of bull urine on puberty and calving date in crossbred beef heifers. *Journal of Animal Science* **55**, 1160–1168.

(1982b). Priming pheromones from oestrous cows increase synchronization of oestrus in dairy heifers after PGC-Z injection. *Journal of Reproduction and Fertility* **66**, 189–196.

Izhaki, I. and Safiel, U. N. (1990). The effect of some Mediterranean scrubland frugivores upon germination patterns. *Journal of Ecology* **78**, 56–65.

Izhaki, I., Korine, C., and Arad, Z. (1995). The effect of bat (*Rousettus aegyptiacus*) dispersal on seed germination in Eastern Mediterranean habitats. *Oecologia* **101**, 335–342.

Jachmann, H. (1989). Food selection by elephants in the "Miombo" dome, in relation to leaf chemistry. *Biochemical Systematics and Ecology* **17**, 15–24.

Jackson, F. L. C. (1994). Bioanthropological impact of chronic exposure to sublethal cyanides from cassava in Africa. *Actae Horticulturae* **375**, 295–319.

Jacob, J., Balthazart, J., and Schoffeniels, E. (1979). Sex differences in the chemical composition of uropygial gland waxes in domestic ducks. *Biochemical Systems Ecology* **7**, 149–153.

Jaeger, R. G. (1986). Pheromonal markers as territorial advertisement by terrestrial salamanders. In *Chemical Signals in Vertebrates*, vol. 4, ed. D. Duvall, D. Müller-Schwarze, and R. M. Silverstein, pp. 191–203. New York: Plenum.

Jaeger, R. G. and Gergits, W. F. (1979). Intra- and interspecific communication in salamanders through chemical signals on the substrate. *Animal Behaviour* **27**, 150–156.

Jakubas, W. J. and Gullion, G. W. (1990). Coniferyl benzoate in quaking aspen: a ruffed grouse feeding deterrent. *Journal of Chemical Ecology* **16**, 1077–1087.

Jakubas, W. J. and Mason, J. R. (1991). Role of avian trigeminal sensory system in detecting coniferyl benzoate, a plant allelochemical. *Journal of Chemical Ecology* **17**, 2213–2221.

Jakubas, W. J., Gullion, G. W., and Clausen, T. P. (1989). Ruffed grouse feeding behavior and its relationship to secondary metabolites of quaking aspen flower buds. *Journal of Chemical Ecology* **15**, 1899–1917.

Jakubas, W. J., Wentworth, B. C., and Karasov, W. H. (1993). Physiological and behavioral effects of conifer benzoate on avian reproduction. *Journal of Chemical Ecology* **19**, 2353–2377.

James, P. C. (1986). How do Manx shearwaters, *Puffinus puffinus*, find their burrows? *Ethology* **71**, 287–294.

Jameson, D. A. (1964). Forage plant physiology and soil-range relationships. Effect of defoliation on forage plant physiology. *American Society of Agronony, Special Publication* **5**, 67–80.

Jankowsky, M. J., Swanson, V. B., and Cramer, D. A. (1974). Field trials of coyote repellents in western Colorado. *Proceedings of the Annual Meeting of the Western Section of the American Society of Animal Science* **25**, 74–76.

Janzen, D. H. (1978). The ecology and evolutionary biology of seed chemistry as relates to seed predation. In *Biochemical Aspects of Plant and Animal Co-Evolution*, ed. J. B. Harborne, pp. 163–206. London: Academic Press.

(1981). *Enterolobium cyclocarpum* seed passage rate and survival in horses, Costa Rican Pleistocene seed dispersal agents. *Ecology* **62**, 593–601.
Jarvis, J. U. M. (1981). Eusociality in a mammal: cooperative breeding in naked mole-rat. *Science* **212**, 571–573.
Jedrzejewska, B. and Jedrzejewski, W. (1990). Antipredator behaviour of bank voles and prey choice by weasels: enclosure experiments. *Annales Zoologici Fennici* **27**, 321–328.
Jemiolo, B., Xie, T. M., Andreolini, F., Baker, A. E. M., and Novotny, M. (1991). The t complex of the mouse: chemical characterization by urinary volatile profiles. *Journal of Chemical Ecology* **17**, 353–367.
Jobling, M. (1995). *Environmental Biology of Fishes.* London: Chapman and Hall.
Jogia, M. K., Sinclair, A. R. E., and Andersen, R. J. (1989). An antifeedant in balsam poplar inhibits browsing by snowshoe hares. *Oecologia* **79**, 189–192.
Johannesson, B. (1987). Observations related to the homing instinct of Atlantic salmon (*Salmo salar* L.). *Aquaculture* **64**, 339–341.
Johanson, I. B. and Shapiro, E. G. (1986). Intake and behavioral responsiveness to taste stimuli in infant rats from 1 to 15 days of age. *Developmental Psychobiology* **19**, 593–606.
Johns, M. A., Feder, H. H., Komisaruk, B. R., and Mayer, A. D. (1978). Urine-induced reflex ovulation in anovulatory rats may be a vomeronasal effect. *Nature* **272**, 446–447.
Johns, T. (1986). The detoxification function of geophagy and the domestication of the potato. *Journal of Chemical Ecology* **12**, 635–646.
(1990). *With Bitter Herbs They Shall Eat It: Chemical Ecology and Origins of Human Diet and Medicine.* Tucson, AZ: University of Arizona Press.
Johnsen, P. B., Zhou, H., and Adams, M. A. (1988). Olfactory sensitivity of the herbivorous grass carp, *Ctenopharyngodon idella*, to amino acids. *Journal of Fish Biology* **33**, 127–134.
Johnson, A. and Bailey, C. B. (1972). Influence of bovine saliva on grass regrowth in the greenhouse. *Canadian Journal of Animal Sciences* **52**, 573–574.
Johnson, B. A. and Leon, M. (2001). Spatial representations of odorant chemistry in the main olfactory bulb of the rat. In *Chemical Signals in Vertebrates*, vol. 9, ed. A. Marchlewska-Koj, J. J. Lepri, and D. Müller-Schwarze, pp. 85–91. New York: Kluwer Academic/Plenum.
Johnston, R. E. (1983). Mechanisms of individual discrimination in hamsters. In *Chemical Signals in Vertebrates*, vol. 3, ed. D. Müller-Schwarze and R. M. Silverstein, pp. 245–258. New York: Plenum.
(1992). Olfactory and vomeronasal mechanisms of social communication in golden hamsters. In *Chemical Signals in Vertebrates*, vol. 6, ed. R. L. Doty and D. Müller-Schwarze, pp. 515–522. New York: Plenum.
(1993). Memory for individual scents in hamsters (*Mesocricetus auratus*) as assessed by habituation methods. *Journal of Comprehensive Psychology* **107**, 201–207.
(2001). Neural mechanisms of communication: from pheromones to mosaic signals. In *Chemical Signals in Vertebrates*, vol. 9, ed. A. Marchlewska-Koj, J. J. Lepri, and D. Müller-Schwarze, pp. 61–67. New York: Kluwer Academic/Plenum.

(2005). Communication by mosaic signals: Individual recognition and underlying neural mechanisms. In *Chemical Signals in Vertebrates*, vol. 10, ed. R. T. Mason, M. P. LeMaster, and D. Müller-Schwarze, pp. 269–282. New York: Springer.

Johnston, R. E. and Jernignan, P. (1994). Golden hamsters recognize individuals, not just individual scents. *Animal Behaviour* **48**, 129–136.

Johnston, R. E. and Schmidt, T. (1979). Responses of hamsters to scent marks of different ages. *Behavioral and Neural Biology* **26**, 64–75.

Johnston, R. E., Derzie, A., Chiang, G., Jernigan, P., and Lee, H.-C. (1993). Individual scent signatures in golden hamsters: evidence for specialization of function. *Animal Behaviour* **45**, 1061–1070.

Johnston, R. E., Munver, A., and Tung, C. (1995). Scent countermarks: selective memory for the top scent by golden hamsters. *Animal Behaviour* **49**, 1435–1442.

Jones, A. S., Lamont, B. B., Fairbanks, M. M., and Rafferty, C. M. (2003). Kangaroos avoid eating seedlings with or near others with volatile essential oils. *Journal of Chemical Ecology* **29**, 2621–2635.

Jones, D. A. (1998). Why are so many food plants cyanogenic? *Phytochemistry* **47**, 155–162.

Jones, I. L., Hagelin, J. C., Major, H. L., and Rasmussen, L. E. L. (2004). An experimental field study of the function of crested auklet feather odor. *Condor* **106**, 71–78.

Jones, K. A. and Hara, T. J. (1982). Behavioral response by Arctic charr (*Salvelinus alpinus*) to taurocholic acid and L-serine, two putative semio-chemicals. *American Zoologist* **22**, 925.

Jones, R. B. (1987). Food neophobia and olfaction in domestic chicks. *Bird Behavior* **7**, 78–81.

Jones, R. B. and Gentle, M. J. (1985). Olfaction and behavioral modification in domestic chicks (*Gallus domesticus*). *Physiology and Behavior* **34**, 917–924.

Jones, T. H., Gorman, J. S. T., Snelling, R. R., *et al.* (1999). Further alkaloids common to ants and frogs: decahydroquinilines and a quinolizidine. *Journal of Chemical Ecology* **25**, 1179–1193.

Jorgenson, J. W., Novotny, M., Carmak, M., *et al.* (1978). Chemical scent constitutents in the urine of red fox (*Vulpes vulpes*) during the winter season. *Science* **199**, 796–798.

Jouventin, P. (1977). Olfaction in snow petrels. *Condor* **79**, 498–499.

Jouventin, P. and Robin, J. P. (1984). Olfactory experiments on some Antarctic birds. *Emu* **84**, 46–48.

Jung, H. G. and Batzli, G. O. (1981). Nutritional ecology of microtine rodents: effects of plant extracts on the growth of arctic microtines. *Journal of Mammalogy* **62**, 286–292.

Kaba, H., Li, C.-S., Keverne, E. B., Saito, H., and Seto, K. (1992). Physiology and pharmacology of the accessory olfactory system. In *Chemical Signals in Vertebrates*, vol. 6, ed. R. L. Doty and D. Müller-Schwarze, pp. 49–54. New York: Plenum.

Kaiser, H. E. and Bartone, J. C. (1966). The carcinogenic activity of ordinary tea. *Journal of the National Medical Association* **58**, 361.

Kaitz, M., Good, A., Korem, A. H., and Eidelman, A. I. (1987). Mothers' recognition of their newborns by olfactory cues. *Developmental Psychobiology* **20**, 587–592.

Kalmus, H. (1955). The discrimination by the nose of the dog of individual human odours and in particular the odours of twins. *British Journal of Animal Behaviour* **5**, 25–31.

Kano, N. (1976). Experiments on the avoidance by Japanese brown bear (*Ursus arctos*) in Noboribetsu Bear Park. *Higuma* **1**, 16–17.

Kare, M. R. and Pick, H. L. (1960). The influence of the sense of taste on feed and fluid consumption. *Poultry Science* **39**, 697–706.

Kareem, A. M. and Barnard, C. J. (1982). The importance of kinship and familiarity in social interactions between mice. *Animal Behaviour* **30**, 594–601.

Karlson, P. and Lüscher, M. (1959). "Pheromones": a new term for a class of biologically active substances. *Nature* **183**, 55–56.

Kassil, V. G. and Gulina, L. K. (1987). Alteration of signal significance of olfactory stimuli in combining them with negative effects on the organism of puppies. *Fiziologicheskii Zhurnal SSSR* **73**, 246–253.

Kasumyan, A. O. (1994). Olfactory sensitivity of the sturgeon to free amino acids. *Biophysics* **39**, 509–522.

Kats, L. B. (1988). The detection of certain predators via olfaction by small-mouthed salamander larvae, *Ambystoma texanum*. *Behavioral and Neural Biology* **50**, 126–131.

Kats, L. B. and Dill, L. M. (1998). The scent of death: chemosensory assessment of predation risk by prey animals. *Ecoscience* **5**, 361–394.

Kats, L. B., Petranka, J. W., and Sih, A. (1988). Antipredator defenses and the persistence of amphibian larvae. *Ecology* **69**, 1865–1870.

Katsel, P. L., Dmitrieva, T. M., Valeyer, R. B., and Kozlov, Y. P. (1992). Sex pheromones of male yellow fin Baikal sculpin (*Cottocomephorus grewingki*): Isolation and chemical studies. *Journal of Chemical Ecology* **18**, 2003–2010.

Katsir, Z., and Crewe, R. M. (1980). Chemical communication in *Galago crassicaudatus:* investigation of the chest gland secrection. *South African Journal of Zoology* **15**, 249–254.

Kavaliers, M. (1990). Responsiveness of deer mice to a predator, the short-tailed weasel: population differences and neuromodulatory mechanisms. *Physiological Zoology* **63**, 388–407.

Kawamichi, T. and Kawamichi, M. (1979). Spatial organization and territory of tree shrews (*Tupaia glis*). *Animal Behaviour* **27**, 381–393.

Kaye, H., Mackintosh, N. J., Rothschild, M., and Moore, B. P. (1989). Odour of pyrazine potentiates an association between environmental cues and unpalatable taste. *Animal Behaviour* **37**, 563–568.

Keane, B. (1990). The effect of relatedness on reproductive success and mate choice in the white-footed mouse, *Peromyscus leucopus*. *Animal Behaviour* **39**, 264–273.

Keefe, M. L. and Winn, H. E. (1991). Chemosensory attraction to the home stream water and conspecifics by native brook trout *Salvelinus fontinalis,* from two southern New England streams. *Canadian Journal of Fisheries and Aquatic Sciences* **48**, 938–944.

Keeler, R. F. (1986). Teratology of steroidal alkaloids, In *Alkaloids: Chemical and Biological Perspectives*, ed. S. W. Pelletier, pp. 389–425. New York: Wiley.

Keevin, T. M., Zuleyma, T. H., and McCurdy, N. (1981). Individual and sex specific odors in male and female eastern chipmunks (*Tamias striatus*). *Behavioral Biology* **6**, 329–338.

Kelley, M. M. (1988). Chemosensory recognition of conspecifics by striped bass juveniles. *Pacific Science* **42**, 123–124.

Kendall, W. A. and Leath, K. T. (1976). Effect of saponins on palatability of alfalfa to meadow voles. *Agronomy Journal* **68**, 473–476.

Kendall, W. A. and Sherwood, R. T. (1975). Palatability of leaves of tall fescue and reed canarygrass and of some of their alkaloids to meadow voles. *Agronomy Journal* **67**, 667–671.

Kendrick, K. M., Keverne, E. B., Chapman, C., and Baldwin, B. A. (1988). Microdialysis measurement of oxytoxcin, aspartate, gamma-aminobutryic acid and glutamate release from the olfactory bulb of sheep during vaginocervical stimulation. *Brain Research* **442**, 171–174.

Kendrick, K. M., Levy, F., and Keverne, E. B. (1992). Changes in the sensory processing of olfactory signals induced by birth in sheep. *Science* **256**, 833–836.

Kenney, A. M., Evans, R. L., and Dewsbury, D. A. (1977). Postimplantation pregnancy disruption in *Microtus ochrogaster, M. pennsylvanicus* and *Peromyscus maniculatus*. *Journal of Reproduction and Fertility* **49**, 365–367.

Keverne, E. B. (1983). Chemical communication and reproductive behavior; primates. In *Pheromones and Reproduction in Mammals,* ed. J. G. Vandenbergh, pp. 79–92. New York: Academic Press.

(1999). The vomeronasal organ. *Science* **286**, 716–720.

Keverne, E. B. and de la Riva, C. (1982). Pheromones in mice: reciprocal interactions between the nose and brain. *Nature* **296**, 148–150.

Keverne, E. B. and Rosser, A. E. (1986). The evolutionary significance of the olfactory block to pregnancy. In *Chemical Signals in Vertebrates,* vol. 4, ed. D. Duvall, D. Müller-Schwarze, and R. M. Silverstein, pp. 433–439. New York: Plenum.

Khan, T. Y. and Stoddart, D. M. (1986). Demonstration of an odorous intramale primer effect in short-tailed vole, *Microtus agrestis* L. *Journal of Chemical Ecology* **12**, 2097–2106.

Khazenchdari, C., Buglass, A. J., and Waterhouse, J. S. (1996). Anal gland secretion of European mole: volatile constitutuents and significance in territorial maintenance. *Journal of Chemical Ecology* **22**, 383–392.

Kiddy, C. A. and Mitchell, D. S. (1981). Estrus-related odors in cows: time of occurrence. *Journal of Dairy Science* **64**, 267–271.

Kiddy, C. A., Mitchell, D. S., and Hawk, H. W. (1984). Estrus-related odors in body fluids of dairy cows. *Journal of Dairy Science* **67**, 388–391.

Kiepenheuer, J. (1985). Can pigeons be fooled about the actual release site position by presenting them information from another site? *Behavioral Ecology* **18**, 75–82.

(1986). Are site specific airborne stimuli relevant for pigeon navigation only when matched by other release site information? *Naturwissenschaften* **73**, 42–43.

Kiesecker, J. M., Chivers, D. P., Anderson, M., and Blaustein, A. R. (2002). Effect of predator diet on life history shifts of red-legged frogs, *Rana aurora*. *Journal of Chemical Ecology* **28**, 1007–1015.

Kikuyama, S., Toyota, F., Ohmiya, Y., *et al.* (1995). Sodefrin: a female-attracting peptide pheromone in newt cloacal glands. *Science* **267**, 1643–1645.

Kimball, B. A. and Nolte, D. L. (2004). That's disgusting: deer responses to seedlings treated with proteins. In *Proceedings of the Annual Meeting of International Society of Chemical Ecology*, July 2004, Ottawa, Canada.

Kirk-Smith, M. D. and Booth, D. A. (1987). Chemoreception in human behavior: experimental analysis of the social effects of fragrances. *Chemical Senses* **12**, 159–166.

Kirschenbaum, D. M. Schulman, N., and Halpern, M. (1986). Earthworms produce a collagen-like substance detected by the garter snake vomeronasal system. *Proceedings of the National Academy of Sciences, USA* **83**, 1213–1216.

Kita, M., Nakamura, Y., Okumura, Y., *et al.* (2004). Blarinatoxin, a mammalian lethal venom from the short-tailed shrew, *Blarina brevicauda*: isolation and characterization. *Proceedings of the National Academy of Science of the USA* **101**, 7542–7547.

Kleerekoper, H. and Mogensen, J. (1963). Role of olfaction in the orientation of *Petromyzon marinus*. I. Response to a single amine in prey's body odor. *Physiological Zoology* **36**, 347–360.

Kleiman, D. (1966). Scent marking in the Canidae. *Symposium of the Zoological Society of London* **18**, 167–177.

Kleinfield, N. R. (1992). The smell of money. *New York Times*, October 25, Section 9 (Styles), p. 1.

Klemm, W. R., Hawkins, G. N., and De Los Santos, E. (1987). Identification of compounds in bovine cervico-vaginal mucus extracts that evoke male sexual behavior. *Chemical Senses* **12**, 77–88.

Klingel, H. (1974). A comparison of the social behavior of the Equidae. In *New Series 1 Publication 24: The Behavior of Ungulates and Its Relation to Management*, ed. V. Geist and F. Walker, pp. 124–132. Morges, Switzerland: International Union for Conservation of Native and Natural Resources.

Klopfer, P. H. and Gambale, J. (1966). Maternal imprinting in goats: the role of the chemical senses. *Zeitschrift für Tierpsychologie* **23**, 588–592.

Knight, M., Glor, R., Smedley, S. R., *et al.* (1999). Firefly toxicosis in lizards. *Journal of Chemical Ecology* **25**, 1981–1986.

Knight, T. W. and Lynch, P. R. (1980). Source of ram pheromones that stimulate ovulation in the ewe. *Animal Reproduction Science* **13**, 133–136.

Knowles, R. I. and Tahan, F. (1979). A repellent to protect radiata pine seedlings from browsing by sheep. *New Zealand Journal of Forestry Science* **9**, 3–9.

Kobal, G. S. van Toller, S., and Hummel, T. (1989). Is there directional smelling? *Experientia (Basel)* **45**, 130–132.

Kobayashi, T. and Watanabe, M. (1986). An analysis of snake-scent application behavior in Siberian chipmunks (*Eutamias sibiricus asiaticus*). *Ethology* **72**, 40–52.

Korn, H. and Taitt, M. J. (1987). Initiation of early breeding in a population of *Microtus townsendii* (Rodentia) with the secondary plant compound 6-MBOA. *Oecologia* **71**, 593–596.

Krames, L. (1970). Responses of female rats to the individual body odors of male rats. *Psychonomic Science* **20**, 274–275.

Krasnov, B., Khokhlova, I., and Shenbrot, G. (2002). The effect of host density on ectoparasite distribution: an example of a rodent parasitized by fleas. *Ecology* **83**, 164–175.

Krause, J. (1993). The effect of "schreckstoff" on the shoaling behavior of the minnow: a test of Hamilton's selfish herd theory. *Animal Behaviour* **45**, 1019–1024.

Krefting, L. W. and Roe, E. I. (1949). The role of some birds and mammals in seed germination. *Ecological Monographs* **19**, 269–286.

Krestel, D., Passe, D., Smith, J. C., and Jonsson, L. (1984). Behavioral determination of olfactory thresholds to amyl acetate in dogs. *Neuroscience and Biobehavioral Reviews* **8**, 169–174.

Krieger, J., Schmitt, A., Loebel D., *et al.* (1999). Selective activation of G protein subtypes in the vomeronasal organ upon stimulation with urine-derived compounds. *Journal of Biological Chemistry* **274**, 4655–4662.

Kristal, M. B., Whitney, J. F., and Peters, L. C. (1981). Placenta on pup's skin accelerates onset of maternal behavior in non-pregnant rats. *Animal Behaviour* **29**, 82–85.

Kruse, K. C. and Stone, B. M. (1984). Largemouth bass (*Micropterus salmoides*) learn to avoid feeding on toad (*Bufo*) tadpoles. *Animal Behaviour* **32**, 1035–1039.

Kruska, D. and Rohrs, M. (1974). Comparative-quantitative investigations on brains of feral pigs from the Galapagos Islands and of European domestic pigs. *Zeitschrift für Anatomie und Entwicklungsgeschichte* **144**, 61–73.

Kruuk, H. (1972). *The Spotted Hyaena, a Study of Predation and Social Behavior.* Chicago, IL: University of Chicago Press.

(1978). Spatial organization and territorial behavior of the European badger *Meles meles. Journal of Zoology* **184**, 1–19.

(1992). Scent marking by otters (*Lutra lutra*): signaling the use of resources. *Behavioral Ecology* **3**, 133–140.

Kruuk, H., Gorman, M., and Leitch, A. (1984). Scent-marking with the subcaudal gland by the European badger, *Meles meles* L. *Animal Behaviour* **32**, 899–907.

Kubie, J. L. and Halpern, M. (1978). Garter snake trailing behavior: effects of varying prey-extract concentration and mode of prey-extract presentation. *Journal of Comparative and Physiological Psychology* **93**, 362–373.

Kubota, K. and Kobayashi, A. (1988). Identification of unknown methyl ketones in volatile flavor components from cooked small shrimp. *Journal of Agricultural and Food Chemistry* **36**, 121–123.

Kucharski, D. and Hall, W. G. (1987). New routes to early memories. *Science* **238**, 786–788.

Kulzer, E. (1961). Über die Biologie der Nil-Flughunde der Gattung (*Rousettus aegyptiacus*). *Natur und Volk* **91**, 219–228.

Kumari, S. and Prakash, I. (1983). Seasonal variation in the dimension of scent-marking gland of three desert rodents and its possible relationship with their reproductive performance. *Proceedings of the Indian Academy of Sciences, Animal Science* **92**, 299–304.

Kunz, T. H. (1982). Roosting ecology. In *Ecology of Bats*, ed. T. H. Kun, pp. 1–55. New York: Plenum.

Kurland, L. T. (1972). An appraisal of the neurotoxicity of cycad and the etiology of amyotrophic lateral sclerosis on Guam. *Federation Proceedings* **31**, 1540–1542.

Kusnetzov, V. B. (1988). Problem of olfaction reduction in Odontoceti toothed whales. *Zhurnal Obschei Biologii* **49**, 128–135.

Kvitek, R. G. (1991). Sequestered paralytic shellfish poisoning toxins mediate glaucous-winged gull predation on bivalve prey. *Auk* **108**, 381–392.

Labov, C. B. and Wysocki, C. J. (1989). Vomeronasal and social influences on urine marking by male mice. *Physiology and Behavior* **45**, 443–447.

Lacher, T. E. J., Bouchardet de Fonseca, G. A., Alves, C., Jr., and Magalhaes-Castro, B. (1981). Exudate-eating, scent marking and territoriality in wild populations of marmosets. *Animal Behaviour* **29**, 306–307.

Ladewig, J., Price, E. O., and Hart, B. L. (1980). Flehmen in male goats: role in sexual behavior. *Behavioral and Neural Biology* **30**, 312–322.

Laffort, P. and Gortan, C. (1987). Olfactory properties of some gases in hyperbaric atmosphere. *Chemical Senses* **12**, 139–142.

Laing, D. G. (1975). A comparative study of the olfactory sensitivity of humans and rats. *Chemical Senses and Flavour* **1**, 257–269.

Langley, W. M. (1988). Spiny mouse's (*Acomys cahirinus*) use of its distance senses in prey localization. *Behavioral Processes* **16**, 67–73.

Larson, J. K. and McCormick, M. I. (2005). The role of chemical alarm signals in facilitating learned recognition of novel cues in a coral reef fish. *Animal Behaviour* **69**, 51–57.

Larson, R. A. and Berenbaum, M. R. (1988). Environmental phototoxicity. *Environmental Science and Technology* **22**, 354–360.

Laska, M. (1990). Olfactory sensitivity to food odor components in the short-tail fruit bat, *Carollia perspicillata* (Phyllostomatidae, Chiroptera). *Journal of Comparative Physiology* **166**, 395–399.

Laska, M. and Hudson, R. (1991). A comparison of the detection thresholds of odour mixtures and their components. *Chemical Senses* **16**, 651–662.

Laska, M., Rother, G., and Schmidt, V. (1986). Die Beeinflussung des Riechvermögens durch die Luftfeuchte bei *Carollia perspicillata und Phyllostomus discolor* (Chiroptera). *Zeitschrift für Säugetierkunde* **51**, 74–79.

Laska, M., Hudson, R., and Distel, H. (1990). Sensitivity to biologically relevant odours may exceed the sum of component thresholds. *Chemoecology* **1**, 139–141.

Lauman, J., Pern, U., and Blum, V. (1974). Investigations on the function and hormonal regulation of the anal appendices in *Blennius payo* (Risso). *Journal of Experimental Zoology* **190**, 47–56.

Launchbaugh, K. L., Provenza, F. D., and Burritt, E. A. (1993). How herbivores track variable environments: response to variability of phytotoxins. *Journal of Chemical Ecology* **19**, 1047–1056.

Laurila, A. (2000). Responses to predator chemical cues and local variation in antipredator behaviour of *Rana temporaria* tadpoles. *Oikos* **88**, 159–168.

Lavin-Murcio, P., Robinson, B. G., and Kardong, K. (1993). Cues involved in relocation of struck prey by rattlesnakes, *Crotalis viridis oreganus*. *Herpetologica* **49**, 463–469.

Lawless, H. T. and Engen, T. (1977). Associations to odors: interference, mnemonics and verbal labeling. *Journal of Experimental Psychology H*. **3**, 52–59.

Lawton, A. D. (1979). Inhibition of sexual maturation by a urinary pheromone in male prairie deer mice. *Hormones and Behavior* **13**, 128–138.

Leffingwell, J. C. (2001). Olfaction. *Leffingwell Reports*, **1**, 1–24. www.Leffingwell.com/olfaction.htm.

Lehner, P. N., Krumm, R., and Cringa, A. T. (1976). Tests for olfactory repellents for coyotes and dogs. *Journal of Wildlife Management* **40**, 145–150.

Leinders-Zufall, T., Lane, A. P., Puche, A. C., *et al.* (2000). Ultrasensitive pheromone detection in the vomeronasal organ upon stimulation with urine-derived compounds. *Nature* **405**, 792–796.

Leinders-Zufall, T., Brennan, P., Widmayer, P., *et al.* (2004). MHC class I peptides as chemosensory signals in the vomeronasal organ. *Science* **306**, 1033–1037.

LeMaster, M. P. and Mason, R. T. (2002). Variation in a female sexual attractiveness pheromone controls male mate choice in garter snakes. *Journal of Chemical Ecology* **28**, 1269–1285.

(2003). Pheromonally mediated sexual isolation among denning populations of red-sided garter snakes, *Thamnophis sirtalis parietalis*. *Journal of Chemical Ecology* **29**, 1027–1043.

LeMaster, M. P., Moore, I. T., and Mason, R. T. (2001). Conspecific trailing behavior of red-sided garter snakes, *Thamnophis sirtalis parietalis*, in the natural environment. *Animal Behaviour* **61**, 827–833.

Lenington, S. and Egid, K. (1985). Female determination of male odor correlated with male genotype at the T-locus: A response to T-locus or H2 locus variability? *Behavior Genetics* **15**, 53–67.

Leon, M. (1975). Dietary control of maternal pheromone in the learning rat. *Physiology and Behavior* **14**, 311–319.

Leon, M. and Behse, J. H. (1977). Dissolution of the pheromonal bond: waning of approach response by weanling rats. *Physiology and Behavior* **18**, 393.

Leopold, A. S., Erwin, M., Oh, J., and Browning, B. (1976). Phytoestrogens: adverse effects in reproduction in California quail. *Science* **191**, 98–99.

Lepri, J. J. and Wysocki, C. J. (1987). Removal of the vomeronasal organ disrupts the activation of reproduction in female voles. *Physiology and Behavior* **40**, 349–355.

Lepri, J. J., Wysocki, C. J., and Vandenbergh, J. G. (1985). Mouse vomeronasal organ: effects on pheromone production and maternal behavior. *Physiology and Behavior* **35**, 809–814.

Lequette, B., Verheyden, C., and Jouventin, P. (1989). Olfaction in subantarctic sea-birds: its phylogenetic and ecological significance. *Condor* **91**, 732–735.

Levin, D. A. (1971). Plant phenolics: an ecological perspective. *American Naturalist* **105**, 157–181.

Levy, F. and Poindron, P. (1987). The importance of amniotic fluids for the establishment of maternal behavior in experienced and inexperienced ewes. *Animal Behaviour* **35**, 1188–1192.

Lewis, D. M. (1987). Fruiting patterns, seed germination, and distribution of *Sclerocarya caffra* in an elephant-inhabited woodland. *Biotropica* **19**, 50–56.

Leyden, J. J., McGinley, K. J., Hoelzle, E., Labows, J. N., and Kligman, A. M. (1981). The microbiology of the human axillae and its relation to axillary odors. *Journal of Investigative Dermatology* **77**, 413–416.

Li, W., Scott, A. P., Siefkes, M. J., *et al.* (2002). Bile acid secreted by male sea lamprey that acts as a sex pheromone. *Science* **296**, 138–141.

Licht, T. (1989). Discriminating between hungry and satiated predators: the response of guppies (*Poecilia reticulata*) from high and low predation sites. *Ethology* **82**, 238–243.

Liley, N. R. (1982). Chemical communication in fish. *Canadian Journal of Fisheries* **39**, 22–35.

Lincoln, J., Coopersmith, R., Harris, E. W., Cotman, C. W., and Leon, M. (1988). NMDA receptor activation and early olfactory learning. *Developmental Brain Research* **39**, 309–312.

Lindquist, S. B. and Bachman, M. D. (1980). Feeding behavior of the tiger salamander, *Ambystoma tigrinum. Herpetologica* **36**, 144–158.

Lindquist, N., Hay, M. E., and Fenical, W. (1992). Defense of ascidians and their conspicuous larvae: adult vs. larval chemical defenses. *Ecological Monographs* **62**, 547–568.

Lindroth, R. L. (1988). Adaptations of mammalian herbivores to plant chemical defenses. In *Chemical Mediation of Coevolution*, ed. K. C. Spencer, pp. 415–445. San Diego, CA: Academic Press.

Lindroth, R. L. and Batzli, G. O. (1983). Detoxication of some naturally occuring phenolics by prairie voles: a rapid assay of glucuronidation metabolim. *Biochemical Systematics and Ecology* **11**, 405–409.

(1984). Plant phenolics as chemical defenses: effects of natural phenolics on survival and growth of prairie voles (*Microtus ochrogaster*). *Journal of Chemical Ecology* **10**, 229–244.

Lindroth, R. L., Batzli, G. O., and Avildsen, S. I. (1986). *Lespedeza* phenolics and *Penstemon* alkaloids: effects on digestion efficiencies and growth of voles. *Journal of Chemical Ecology* **12**, 713–728.

Lindvall, T. and Svensson, L. T. (1974). Equal unpleasantness matching of malodorous substances in a community. *Journal of Applied Psychology* **59**, 264.

Linn, C. E. Jr., Campbell, M. G., and Roelofs, W. L. (1987). Pheromone components and active spaces: what do moths smell and where do they smell it? *Science* **237**, 650–651.

Lipsitt, L. P., Engen, T., and Kaye, H. (1963). Developmental changes in the olfactory threshold of the neonate. *Child Development* **34**, 371–376.

Löhner, L. (1926). Untersuchungen über die geruchsphysiologische Leistungsfähigkeit von Polzeihunden. *Archiv für die gesamte Physiologie* **121**, 84–94.

Løkkeborg, S., Olla, B. L., Pearson, W. H., and Davis, M. W. (1995). Behavioral responses of sablefish, *Anoplopoma fimbria*, to bait odor. *Journal of Fisheries Biology* **46**, 142–155.

Lomas, D. E. and Keverne, E. B. (1982). Role of vomeronasal organ and prolactin in the acceleration of puberty in female mice. *Journal of Reproduction and Fertility* **66**, 101–107.

Lombardi, J. R. and Vandenbergh, J. G. (1977). Pheromonally induced sexual maturation in females: regulation by the social environment of the male. *Science* **196**, 545–546.

Lombardi, J. R., Vandenbergh, J. G., and Whitsett, J. M. (1976). Androgen control of the sexual maturation pheromone in house mouse urine. *Biology of Reproduction* **15**, 179–186.

Loop, M. S. and Scoville, S. A. (1972). Response of newborn *Eumeces inexpectatus* to prey-object extracts. *Herpetologia* **28**, 254–256.

López, P. and Salvador, A. (1992). The role of chemosensory cues in discrimination of prey odors by the amphisbaenan *Blanus cinereus*. *Journal of Chemical Ecology* **18**, 87–93.

Lorenz, K. (1963). *Das sogenannte Böse: Zur Naturgeschichte der Aggression*. Vienna: Borotha-Schoeler.

Löyttiniemi, K. (1981). Nitrogen fertilization and nutrient contents in Scots pine in relation to browsing preference by moose (*Alces alces*). *Folia Forestalia* **478**, 12–14.

Löyttiniemi, K. and Hiltonen, R. (1978). Monoterpenes in Scots pine in relation to browsing preference by moose. (*Alces alces* L.) *Silvia Fennica* **12**, 85–87.

Lucas, C. E. (1944). Excretions, ecology and evolution. *Nature* **153**, 378–379.

Luschi, P. and Dall'Antonia, P. (1993). Anosmic pigeons orient from familiar sites by relying on the map-and-compass mechanism. *Animal Behaviour* **46**, 1195–1203.

Lydell, K. and Doty, R. L. (1972). Male rat odor preferences for female urine as a function of sexual experience, urine age, and urine source. *Hormones and Behavior* **3**, 205–212.

Lyman, B. J. and McDaniel, M. A. (1986). Effects of encoding strategy on long-term memory for odors. *Quarterly Journal of Experimental Psychology A: Human Experimental Psychology* **38**, 753–766.

Mabry, T. J. and Gill, J. E. (1979). Sesquiterpene lactones and other terpenoids. In *Herbivores: Their Interaction with Secondary Plant Metabolites*, ed. G. A. Rosenthal and D. H. Janzen, pp. 501–537. New York: Academic Press.

MacDonald, D., Kranz, K., and Aplin, R. D. (1984). Behavioral, anatomical and chemical aspects of scent marking amongst capybaras, *Hydrochoerus hydrochaeris* (Rodentia: Caviomorpha). *Journal of Zoology* (London) **202**, 341–360.

MacDonald, K. B. (1980). Activity patterns in a captive wolf pack. *Carnivore* **31**, 62–64.

MacFarlane, A. (1975). Olfaction in the development of social preferences in the human neonate. In *Ciba Foundation Symposium*, No. 33: *Parent–Infant Interaction*, pp. 103–113. New York: Elsevier.

MacGintie, G. E. (1939). The natural history of the blind goby, *Typhlogobius californiensis* Steindachner. *American Midland Naturalist* **21**, 489–505.

MacLain, D. K. and Shure, D. J. (1985). Host plant toxins and unpalatability of *Neacoryphus bicruris* (Hemiptera: Lygaeidae). *Ecological Entomology* **10**, 291–298.

Macrides, F., Bartke, A., and Dalterio, S. (1975). Strange females increase plasma testosterone levels in male mice. *Science* **189**, 1104–1106.

Maderson, P. A. F. (1986). The tetrapod epidermis: a system protoadapted as a semiochemical source. In *Chemical Signals in Vertebrates*, vol. 4, ed. D. Duvall, D. Müller-Schwarze, and R. M. Silverstein, pp. 13–25. New York: Plenum.

Madison, D. M. (1977). Chemical communication in amphibians and reptiles, In *Chemical Signals in Vertebrates*, vol. 1, ed. D. Müller-Schwarze and M. M. Mozell, pp. 135–168. New York: Plenum.

Madison, D. M., Sullivan, A. M., Maerz, J. C., McDarby, J. H., and Rohr, J. R. (2002). A complex, cross-taxon, chemical releaser of antipredator behavior in amphibians. *Journal of Chemical Ecology* **28**, 2271–2282.

Madubunyi, L. C., Hassanali, A., Ouma, W., Nyarango, D., and Kabii, J. (1996). Chemo-ecological role of mammalian urine in host location by tsetse, *Glossina* spp. (Diptera: Glossinidae). *Journal of Chemical Ecology* **22**, 1187–1199.

Magurran, A. E. (1986). Predator inspection behavior in minnow shoals: differences between populations and individuals. *Behavioral Ecology and Sociobiology* **19**, 267–273.

(1989). Acquired recognition of predator odour in the European minnow (*Phoxinus phoxinus*). *Ethology* **82**, 216–223.

Mainardi, D., Marsan, M., and Pasqali, A. (1965). Causation of sexual preferences of the house mouse. The behavior of mice reared by parents whose odor was artificially altered. *Atti della Societa Italiana di Scienze Naturali e el Museo Civico de Storia Naturale di Milano* **104**, 325–338.

Mair, R. G., Bouffard, J. A., Engen, T., and Morton, T. H. (1978). Olfactory, sensitivity during the menstrual cycle. *Sensory Processes* **2**, 90–98.

Malakoff, D. (1999). Following the scent of avian olfaction. *Science* **286**, 704–705.

Malone, N., Payne, C. E., Beynon, R. J., and Hurst, J. L. (2001). Social status, odour communication and mate choice in wild house mice. In *Chemical Signals in Vertebrates*, vol. 9, ed. Marchlewska-Koj, A., Lepri, J. J., and Müller-Schwarze. D., pp. 217–224. New York: Kluwer Academic/Plenum.

Manning, C. J., Wakeland, E. K., and Potts, W. K. (1992). Communal nesting patterns in mice implicate MHC genes in kin recognition. *Nature* **360**, 581–583.

Manteifel, Y. B. and Goncharova, N. V. (1992). Sensitivity and specificity of some L-amino acids perception in a European water turtle *Emys orbicularis*: chemotesting movements and avoidance reaction. *Comparative Biochemistry and Physiology* **102**, 527–531.

Manteifel, Y., Goncharova, N., and Boyo, V. (1992). Chemotesting movements and chemosensory sensitivity to amino acids in the European pond turtle, *Emys orbicularis* L. In *Chemical Signals in Vertebrates*, vol. 6, ed. R. L. Doty and D. Müller-Schwarze, pp. 397–401. New York: Plenum.

Mapert, H. L., Cieslak, A., Alkan, O., *et al.* (1999). The golden hamster *aprodisin* gene. Structure, expression in parotid glands of female animals and comparison with a similar murine gene. *Journal of Biological Chemistry* **274**, 444–450.

Marchisin, A. (1980). Predator–prey interactions between snake-eating snakes and pit vipers. Ph.D. Thesis, Rutgers University, Newark, NJ, USA.

Marchlewska-Koj, A. (1990). Chemical interaction between adult female mice: the role of ovarian and adrenal hormones. In *Chemical Signals in Vertebrates*, vol. 5, ed. D. W. Macdonald, D. Müller-Schwarze, and S. E. Natynczuk, pp. 209–216. New York: Oxford University Press.

Marchlewska-Koj, A. and Kruczek, M. (1986). Female-induced delay of puberty in bank vole and European pine vole females. In *Chemical Signals in Vertebrates*, vol. 4, ed. D. Duvall, D. Müller-Schwarze, and R. M. Silverstein, pp. 551–554 New York: Plenum Press.

Marchlewska-Koj, A., Pochron, E. and Sliwowska, A. (1990). Salivary glands and preputial glands of males as source of estrus-stimulating pheromone in female mice. *Journal of Chemical Ecology* **16**, 2817–2822.

Margolis, F. L., Kudrycki, K., Stein-Izsak, C., Grillo, M., and Akeson, R. (1993). From genotype to olfactory phenotype: the role of the Olf-1-binding site. *Ciba Foundation Symposia* **179**, 3–20.

Marks, D. L., Swain, T., Goldstein, S., Richard, A., and Leighton, M. (1988). Chemical correlates of rhesus monkey food choice: the influence of hydrolyzable tannins. *Journal of Chemical Ecology* **14**, 213–235.

Marlier, L., Schaal, B., Gaugler, C., and Messer, J. (2001). Olfaction in premature newborns: detection and discrimination abilities two months before gestational term. In *Chemical Signals in Vertebrates,* vol. 9, ed. A. Marchlewska-Koj, J. J. Lepri, and D. Müller-Schwarze, pp. 205–209. New York: Kluwer Academic/Plenum.

Marples, N. M. (1993). Do wild birds use size to distinguish palatable and unpalatable prey types? *Animal Behaviour* **46**, 347–354.

Marquis, R. J. and Batzli, G. O. (1989). Influence of chemical factors on palatability of forage to voles. *Journal of Mammalogy* **70**, 503–511.

Marsden, A. M. and Bronson, F. A. (1964). Estrus synchrony in mice: alteration by exposure to male urine. *Science* **144**, 1469.

Marshall, D. A. and Moulton, D. G. (1981). Olfactory sensitivity to α-ionone in humans and dogs. *Chemical Senses* **6**, 53–61.

Martin, G. B., Oldham, C. M., Cognie, Y., and Pearce, D. T. (1986). The physiological responses of anovulatory ewes to the introduction of rams: a review. *Livestock Production Science* **15**, 219–247.

Martin, I. G. and Beauchamp, G. K. (1982). Olfactory recognition of individuals by male cavies (*Cavia aperea*). *Journal of Chemical Ecology* **8**, 1241–1249.

Martin, J. S. and Martin, M. M. (1983). Tannin essays in ecological studies. Precipitation of ribulose-1,5-biphosphate carboxylase/oxygenase by tannic acid, quebracho, and oak foliage extracts. *Journal of Chemical Ecology* **9**, 285–294.

Martin, M. L., Price, E. O., Hallach, J. R., and Dally, M. R. (1987). Fostering lambs by odor transfer: the add-on experiment. *Journal of Animal Science* **64**, 1378–1383.

Martof, B. S. (1962). Some observations on the role of olfaction among salient amphibia. *Physiological Zoology* **35**, 270.

Maruniak, J. A., Silver, W. L., and Moulton, D. G. (1983). Olfactory receptors respond to blood-borne odorants. *Brain Research* **265**, 312–316.

Maruniak, J. A., Wysocki, C. J., and Taylor, J. A. (1986). Mediation of male mouse urine marking and aggression by the vomeronasal organ. *Physiology and Behavior* **37**, 655–657.

Maser, C., Trappe, J. M., and Nussbaum, R. A. (1978). Fungal–small mammal interrelationships with emphasis on Oregon coniferous forests. *Ecology* **59**, 799–809.

Mason, J. R. (1989). Avoidance of methiocarb-poisoned apples by red-winged blackbirds. *Journal of Wildlife Management* **53**, 836–840.

Mason, J. R. and Clark, L. (1986). Chemoreception and the selection of green plants as nest fumigants by starlings. In *Chemical Signals in Vertebrates,* vol. 4, ed. D. Duvall, D. Müller-Schwarze, and R. M. Silverstein, pp. 369–384. New York: Plenum.

(1987). Behavioral assessment of olfactory and trigeminal responsiveness of starlings *Sturnus vulgaris* to nine anthranilates. *Chemical Senses* **12**, 679.

Mason, J. R. and Stevens, D. A. (1981a). Discrimination and generalization among reagent grade odorants by tiger salamanders (*Ambystoma tigrinum*). *Physiology and Behavior* **27**, 647–653.

(1981b). Behavioral determinations of thresholds for *n*-butyl acetate and *n*-butyl alcohol in the tiger salamander (*Ambystoma tigrinum*). *Chemical Senses* **6**, 189–195.

Mason, J. R., Rabin, M. D., and Stevens, D. A. (1982). Conditioned taste aversions: skin secretions used for defense by tiger salamanders (*Ambystoma tigrinum*). *Copeia*. 667–671.

Mason, J. R., Greenspon, J. M. and Silver, W. L. (1987a). Capsaicin and its effects on olfaction and trigeminal chemoreception. *Acta Physiologica Hungarica* **69**, 469–479.

Mason, J. R., Kamalesh, J. and K., Morton, T. H. (1987b). Generalization in olfactory detection of chemical cues containing carbonyl functions by tiger salamanders (*Ambystoma tigrinum*). **13**, 1–18.

Mason, J. R., Bean, N. J., and Galef, B. G., Jr. (1988). Attractiveness of carbon disulfide to wild Norway rats. *Proceedings of Vertebrate Pest Conference*, vol. 13, ed. A. C. Crabb and R. E. Marsh, pp. 95–97.

Mason, R. T., Fales, H. M., Jones, T. H., *et al.* (1989a). Sex pheromones in snakes. *Science* **245**, 290–293.

Mason, J. R., Bullard, R. W., Dolbeer, R. A., and Woronecki, P. P. (1989b). Red-winged blackbird (*Agelaius phoeniceus* L.) feeding responses to oil and anthocyanin levels in sunflower meal. *Crop Protection* **8**, 455–460.

Mason, J. R., Adams, M. A., and Clark, L. (1989c). Anthranilate repellency to starlings: chemical correlates and sensory perception. *Journal of Wildlife Management*. **53**, 55–64.

Mason, J. R., Bean, N. J. Shah, P. S., and Clark, L. (1991). Taxon-specific differences in responsiveness to capsaicin and several analogues: correlates between chemical structure and behavioral aversiveness. *Journal of Chemical Ecology* **17**, 2539–2552.

Mason, J. R., Epple, G., and Nolte, D. L. (1994). Semiochemicals and improvements in rodent control. In *Behavioral Aspects of Feeding: Basic and Applied Research in Mammals,* ed. B. G. Galef, M. Mainardi, and P. Valsecchi, pp. 327–345. Chur, Switzerland: Harwood Academic.

Mason, R. T. (1992). Reptilian pheromones. In *Biology of the Reptilia*, vol. 18, *Hormones, Brain and Behavior*, ed. C. Gans and D. Crews, pp. 114–228. Chicago, IL: University of Chicago Press.

Mason, R. T. and Greene, M. J. (2001). Invading pest species and the threat to biodiversity: pheromonal control of Guam brown tree snakes, *Boiga irregularis*. In *Chemical Signals in Vertebrates,* vol. 9, eds. A. Marchlewska-Koj, J. J. Lepri, and D. Müller-Schwarze, pp. 361–368. New York: Kluwer Academic/Plenum.

Mason, R. T. and Gutzke, W. H. N. (1990). Sex recognition in the leopard gecko, *Eublepharis macularis* (Sauria: Gekkonidae). Possible mediation by skin-derived semiochemicals. *Journal of Chemical Ecology* **16**, 27–36.

Massey, A. and Vandenbergh, J. G. (1980). Puberty delay by a urinary cue from female house mice in feral populations. *Science* **209**, 821–822.

Matochik, J. A. (1988). Role of the main olfactory system in recognition between individual spiny mice. *Physiology and Behavior* **42**, 217–222.

Matsunami, H. and Buck, L. B. (1997). A multigene family encoding a diverse array of putative pheromone receptors in mammals. *Cell* **90**, 775–784.

May, R. M. (1972). Limit cycles in predator–prey communities. *Science* **177**, 900–902.

Mayer, J. J. and Brisbin, I. L., Jr. (1986). A note on the scent-marking behavior of two captive-reared feral boars. *Applied Animal Behaviour Science* **16**, 85–90.

McAdoo, J. K., Evans, C. C., Roundy, B. A., Young, J. A., and Evans, R. A. (1983). Influence of heteromyid rodents on *Oryzopsis hymenoides* germination. *Journal of Range Management* **36**, 61–64.

McArthur, C., Sanson, G. D., and Beal, A. M. (1995). Salivary proline-rich proteins in mammals: roles in oral homeostasis and counteracting dietary tannin. *Journal of Chemical Ecology* **21**, 663–691.

McClenaghan, L. R. (1987). Lack of effect of 6-MBOA on reproduction in *Dipodomys merriami*. *Journal of Mammalogy* **68**, 150–152.

McClintock, M. K. (1971). Menstrual synchrony and suppression. *Nature* **229**, 244–245.

McDonough, L. M., Brown, D. F., and Aller, W. C. (1989). Insect sex pheromones. Effect of temperature on evaporation rates of acetates from rubber septa. *Journal of Chemical Ecology* **15**, 779–790.

McGlone, J. (2002). *Pig Production: Biological Principles and Applications*. Clifton Park, NJ: Delmar Learning.

McGregor, J. H and Teska, W. R. 1989. Olfaction as an orientation mechanism in migrating *Ambystoma maculatum*. *Copeia* **1989**, 779–781.

McKey, D. and Beckerman, S. (1993). Chemical ecology, plant evolution and traditional manioc cultivation systems. In *Man and the Biosphere Series*, vol. 13: *Tropical Forests People and Food*, ed. C. M. Hadlik, A. Hadlik, O. F. Linares, *et al.* pp. 321–338. Paris: UNESCO-MAB/Parthenon Publishing.

McKey, D., Gartlan, J. S., Waterman, P. G., and Choo, G. M. (1981). Food selection by black colobus monkeys (*Colobus satanus*) in relation to plant chemistry. *Biological Journal of the Linnean Society* **16**, 115–146.

McKinnon, J. S. and Liley, N. R. (1987). Asymetric species specificity in response to female pheromone by males of two species of trichogaster (Pisces: Blontidae). *Canadian Journal of Zoology* **65**, 1129–1134.

McLaughlin, S. K., Mckinnon, P. J., Robichon, A., Spickofsky, N., and Margolskee, R. F. (1993). Gustducin and transducin: a tale of two G proteins. *Molecular Basis of Smell and Taste Transduction* **179**, 186–200.

McLean, S., Foley, W. J., Davies, N. W., *et al.* (1993). Metabolic fate of dietary terpenes from *Eucalyptus radiata* in common ringtail possom (*Pseudocheirus peregrinus*). *Journal of Chemical Ecology* **19**, 1625–1643.

McLennan, D. A. (2005). Changes in response to olfactory cues across the ovulatory cycle in brook sticklebacks, *Culaea inconstans*. *Animal Behaviour* **69**, 181–188.

McMahon, T. E. and Tash, J. C. (1979). The use of chemosenses by threadfin shad, *Dorosoma petenense* to detect conspecifics, predators and food. *Journal of Fisheries Biology* **14**, 289–296.

McNaughton, S. J. (1985). Interactive regulation of grass yield and chemical properties by defoliation, a salivary chemical, and inorganic nutrition. *Oecologia* **65**, 478–486.

Mead, R. J., Oliver, A. J., King, D. R., and Hubach, P. H. (1985). The co-evolutionary role of fluoroacetate in plant–animal interactions in Australia. *Oikos* **44**, 55–60.

Mech, L. D. (1970). *The Wolf: The Ecology and Behavior of an Endangered Species*. Minneapolis, MI: University of Minnesota Press.

Mehansho, H., Hagerman, A., Clements, S., *et al.* (1983). Modulation of proline-rich protein biosynthesis in rat parotid glands by sorghums with high tannin levels. *Proceedings of the National Academy of Sciences USA* **80**, 3948–3952.

Mehansho, H., Butler, L. G., and Carlson, D. M. (1987). Dietary tannins and salivary proline-rich proteins: interactions, induction, and defense mechanism. *Annual Review of Nutrition* **7**, 423–440.

Meier, P. T. (1991). Response of adult woodchucks (*Marmota monax*) to oral-gland scents. *Journal of Mammalogy* **72**, 622–624.

Meisami, E. and Wenzel, B. (1987). Is the northern fulmar's large olfactory bulb designed for high sensitivity in odor detection? *Chemical Senses* **12**, 681–682.

Melcer, T. and Chiszar, D. (1989). Striking prey creates a specific chemical search image in rattlesnakes. *Animal Behaviour* **37**, 477–486.

Melchiors, M. A. and Leslie, C. A. (1985). Effectiveness of predator fecal odors as black-tailed deer repellents. *Journal of Wildlife Management* **49**, 358–362.

Melrose, D. R., Reed, H. C. B., and Patterson, R. L. S. (1971). Androgen steroids associated with boar odour as an aid to the detection of oestrus in pig artificial insemination. *British Veterinary Journal* **127**, 497–502.

Mennella, J. A. and Moltz, H. (1988). Infanticide in the male rat: the role of the vomeronasal organ. *Physiology and Behavior* **42**, 303–306.

(1989). Pheromonal emission by pregnant rats protects against infanticide by nulliparous conspecifics. *Physiological Behavior* **46**, 591–596.

Meredith, M. (1983). Sensory physiology of pheromone communication. In *Pheromones and Reproduction in Mammals*, ed. J. G. Vandenbergh, pp. 199–252. New York: Academic Press.

(1986). Vomeronasal organ removal before sexual experience impairs male hamster mating behavior. *Physiological Behavior* **36**, 737–743.

Meredith, M. and O'Connell, R. J. (1979). Efferent control of stimulus access to the hamster vomeronasal organ. *Journal of Physiology* **286**, 301–316.

Merkens, M., Harestad, A. S., and Sullivan, T. P. (1991). Cover and the efficacy of predator-based repellents for Townsend's vole, *Microtus townsendii*. *Journal of Chemical Ecology* **17**, 401–412.

Merkx, J., Slob, A. K., van der Werff, J. J., and ten Bosch, J. J. (1988). Vaginal bacterial flora partially determines sexual attractivity of female rats. *Physiology and Behavior* **44**, 147–149.

Merritt, G. C., Goodrich, B. S., Hesterman, E. R., and Mykytowycz, R. (1982). Microflora and volatile fatty acids present in inguinal pouches of the wild rabbit, *Oryctolagus cuniculus,* in Australia. *Journal of Chemical Ecology* **8**, 1217–1226.

Mester, A. F., Doty, R. L., Shapiro, A., and Frye, R. E. (1988). Influence of body tilt within the sagittal plane on odor identification performance. *Aviation, Space and Environmental Medicine* **59**, 734–737.

Meyer, M. W. and Karasov, W. H. (1989). Antiherbivore chemistry of *Larrea tridentata*: effects on woodrat (*Neotoma lepida*) feeding and nutrition. *Ecology* **70**, 953–961.

Meyers, C. W. and Daly, J. W. (1983). Dart-poison frogs. *Scientific American* **248**, 120–133.

Meyer, M. W. and Karasov, W. H. (1991). Chemical aspects of herbivory in arid and semiarid habitats. In *Plant Defenses Against Mammalian Herbivory*, ed. R.T. Palo and C. T. Robbins, pp. 167–187. Boca Raton, FL: CRC Press.

Millar, J. G. and Haynes, K. F. (1988). *Chemical Ecology Methodology*. New York: Chapman and Hall.

(1998). *Chemical Ecology Methodology*. New York: Chapman & Hall.

Mills, M. G. L., Gorman, M. L., and Mills, M. E. J. (1980). The scent marking behaviour of the brown hyaena *Hyaena brunnea*. *South African Journal of Zoology* **15**: 240–248.

Milton, K. (1979). Factors influencing leaf choice by howler monkeys: a test of some hypotheses of food selection by generalist herbivores. *American Naturalist* **114**, 362–378.

Mínguez, E. (1997). Olfactory nest recognition by British storm-petrel chicks. *Animal Behaviour* **5**, 701–707.

Mirza, R. S. and Chivers, D. P. (2001). Do chemical alarm signals enhance survival of aquatic vertebrates? An analysis of the current research paradigm. In *Chemical Signals in Vertebrates*, vol. 9, ed. A. Marchlewska-Koj, J. J. Lepri, and D. Müller-Schwarze, pp. 19–26. New York: Kluwer Academic/Plenum.

Mitchell, D., Laycock, J. D., and Stevens, W. F. (1977). Motion sickness-induced pica in rats. *American Journal of Clinical Nutrition* **30**, 147–150.

Moen, A. N. (1973). *Wildlife Ecology*. San Francisco, CA: W. H. Freeman.

Molez-Verriere, N. (1988). Detection and capture of prey in two species of Galagidae, Primates. *Biology of Behavior* **13**, 30–42.

Mombaerts, P. (1999). Seven-transmembrane proteins as odorant and chemosensory receptors. *Science* **286**, 707–711.

Moncomble, A.-S., Coureaud, G., Quennedey, B., *et al.* (2005). The mammary pheromone of the rabbit: from where does it come? *Animal Behaviour* **69**, 29–38.

Montagner, H. (1974). Communication non verbale et discrimination olfactive chez le jeune enfant: approche éthologique. In *L' Unite de l' Homme,* ed. E. Morin and M. Piatelli-Palmarini. Paris: Seuil.

Montague, J. C., Pocock, D. C., and Wright, W. (1990). An examination of the animal browsing problem in Australian eucalypt and pine plantations. In *Proceedings of the*

14th Vertebrate Pest Conference, ed. L. R.Davis and R. E. Marsh, pp. 203–208. Davis, CA: University of California at Davis.

Montgomery-St. Laurent, T., Fullenkamp, A. M., and Fischer, R. B. (1988). A role for the hamster's flank gland in heterosexual communication. *Physiology and Behavior* **44**, 759–762.

Moodie, J. D. and Byers, J. A. (1989). The function of scent marking by males on female urine in pronghorns. *Journal of Mammalogy* **70**, 812–814.

Moore, A. and Waring, C. P. (2001). The effects of a synthetic pyrethroid peticide on some aspects of reproduction in Atlantic salmon (*Salmo salar* L.). *Aquatic Toxicology* **52**, 1–12.

Moore, B. D., Wallis, I. R., Palá-Paúl, J., *et al.* (2004). Antiherbivore chemistry of *Eucalyptus*: cues and deterrents for marsupial folivores. *Journal of Chemical Ecology* **30**, 1743–1769.

Moore, C. L. (1984). Maternal contribution to the development of masculine sexual behavior in laboratory rats. *Developmental Psychobiology* **17**, 347–356.

Moore, R. E. (1965). Olfactory discrimination as an isolation mechanism between *Peromyscus maniculatus* and *Peromyscus polionotus*. *American Midland Naturalist* **73**, 85–100.

Moore, W. G. and Marchinton, R. L. (1974). Marking behavior and its social function in white-tailed deer. In *The Behavior of Ungulates and Its Relation to Management*, ed. V. Geist and F. Walter. pp. 447–456. Morges, Switzerland: International Union for the Conservation of Nature and Natural Resources.

Mordue-Luntz, A. J., Ingvarsdöttir, A., Birkett, M. A., *et al.* (2004). The role of pheromones and kairomones in mate location of the salmon louse *Lepeophtheirus salmonis* (Crustacea; Caligidae). In *Proceedings of the Annual Meeting of International Society of Chemical Ecology*, July 2004, Ottawa, Canada.

Mori, K., Nagao, H., and Yoshihara, Y. (1999). The olfactory bulb: coding and processing of odor molecule information. *Science* **286**, 711–715.

Morin, P. P., Dodson, J. J., and Dore, F. Y. (1987). Laboratory identification of a sensitive period for olfactory imprinting in young Atlantic salmon. *American Zoologist* **27**, 77A.

Morris, M. P. and García-Rivera, J. (1955). The destruction of oxalates by the rumen contents of cows. *Journal of Dairy Science* **38**, 1169.

Morrow, J. L. and McGlone, J. J. (1987). Preference of piglets for odors from sow feces. *Journal of Animal Science* **65**(Suppl. 1), 231.

(1988). Sensory systems and nipple attachment behavior in neonatal pigs. *Journal of Animal Science*. **66**(Suppl. 1), 243.

Morrow-Tesch, J. and McGlone, J. J. (1990). Sources of maternal odors and the development of odor preferences in baby pigs. *Journal of Animal Science* **68**, 3563–3571.

Morton, E. S. (1975). Ecological sources of selection on avian sounds. *American Naturalist* **109**, 17–34.

Morton, J. F. (1970). Tentative correlations of plant usage and esophageal cancer zones. *Economic Botany* **24**, 217–226.

(1989). Tannin as a carcinogen in bush-tea: tea, maté, and khat. In *Chemistry and Significance of Condensed Tannins*, ed. R. W. Hemingway and J. J. Karchesy, pp. 403–416. New York: Plenum.

Most, K. and Brückner, G. H. (1936). Über Voraussetzungen und den derzeitigen Stand der Nasenleistungen von Hunden. *Zeitschrift für Hundeforschung* **12**, 9–30.

Mott, M. (2004). Bees, giant African rats used to sniff land mines. *National Geographic News* 10 February 2004.

Mucignat-Caretta, C., Caretta, A., and Cavaggioni, A. (1995). Acceleration of puberty onset in female mice by male urinary proteins. *Journal of Physiology* **486**, 517–522.

Mucignat-Caretta, C., Cavaggioni, A., and Caretta, A. (2004). Male urinary chemosignals differentially affect aggressive behavior in male mice. *Journal of Chemical Ecology* **30**, 777–791.

Müller-Schwarze, D. (1971). Pheromones in black-tailed deer (*Odocoileus heminonus columbianus*). *Animal Behaviour* **19**, 141–152.

(1972). Responses of young black-tailed deer to predator odors. *Journal of Mammalogy* **53**, 393–394.

(1979). Flehmen in the context of mammalian urine communication. In *Chemical Ecology: Odour Communication in Animals*, ed. F. J. Ritter, pp. 85–96. Amsterdam: Elsevier.

(1983). Experimental modulation of behavior of free-ranging mammals by semiochemicals. In *Chemical Signals in Vertebrates*, vol. 3, ed. D. Müller-Schwarze, and R. M. Silverstein, pp. 235–244. New York: Plenum.

(1987). Evolution of cervid olfactory communication. In *Research Symposia of the National Zoological Park: Biology and Management of Cervidae*, ed. C. M. Wemmer, pp. 331–344. Washington DC: Smithsonian Institution Press.

(1990). Leading them by their noses: animal and plant odours for managing vertebrates. In *Chemical Signals in Vertebrates*, vol. 5, ed. D. W. Macdonald, D. Müller-Schwarze, and S. E. Natynczuk, pp. 585–598. Oxford: Oxford University Press.

(1992). Castoreum of beaver (*Castor canadensis*): function, chemistry and biological activity of its components. In *Chemical Signals in Vertebrates*, vol. 6, ed. R. L. Doty, and D. Müller-Schwarze, pp. 457–464. New York: Plenum.

(1998). Signal specialization and evolution in mammals. In *Chemical Signals in Vertebrates*, vol. 8, ed. R. E. Johnston, D. Müller-Schwarze, and P. W. Sorensen, pp. 1–14. New York: Plenum.

(2005). Thirty years on the odor trail: from the *First* to the *Tenth International Symposium on Chemical Signals in Vertebrates*. In *Chemical Signals in Vertebrates*, vol. 10, ed. R. T. Mason, M. P. LeMaster, & D. Müller-Schwarze, pp. 1–6. New York: Springer.

Müller-Schwarze, D. and Giner, J. (2005). Cottontails and gopherweed: antifeedants from a spurge. In *Chemical Signals in Vertebrates*, vol. 10, ed. R. T. Mason, M. P. LeMaster, and D. Müller-Schwarze, pp. 409–412. New York: Springer.

Müller-Schwarze, D. and Heckman, S. (1980). The social role of scent marking in beaver (*Castor canadensis*). *Journal of Chemical Ecology* **6**, 81–95.

Müller-Schwarze, D. and Houlihan, D. (1991). Pheromonal activity of single castoreum constituents in beaver, *Castor canadensis*. *Journal of Chemical Ecology* **17**, 715–734.

Müller-Schwarze, D. and Müller-Schwarze, C. (1971). Olfactory imprinting in a precocial mammal. *Nature* **229**, 55–56.

(1973). Differential predation of South Polar skuas in an Adiplic penguin rookery. *Condor* **75**, 127–131.

(1977). Interactions between South Polar skuas and Adie penguins. In *Adaptations within Antarctic Ecosystems*, ed. G. A. Llano, pp. 619-646. Washington, DC: Smithsonian Institution.

(1985). Behavioral ecology of Labrador caribou. *National Geographic Society Research Reports* **21**, 321–326.

Müller-Schwarze, D., Müller-Schwarze, C., and Franklin., W. F. (1972). Factors influencing scent marking in pronghorn (*Antilocapra americana*). *Verhandlungen der Deutschen Zoologischen Gesellschaft* **66**, 146–150.

Müller-Schwarze, D., Müller-Schwarze, C., Singer, A. G., and Silverstein, R. M. (1974). Mammalian pheromone: identification of active compound in the subauricular scent of the male pronghorn. *Science* **183**: 860–862.

Müller-Schwarze, D., Silverstein, R. M., Müller-Schwarze, C., Singer, A. G., and Volkman, N. J. (1976). Response to a mammalian pheromone and its geometric isomer. *Journal of Chemical Ecology* **2**, 389–398.

Müller-Schwarze, D., Quay, W. B. & Brundin, A. (1977). The caudal gland of *Rangifer*: its behavioral role, histology and chemistry. *Journal of Chemical Ecology* 3, 591–601.

Müller-Schwarze, D., Volkman, N. J., and Zemanek, K. (1977). Osmetrichia: specialized scent hairs in black-tailed deer. *Journal of Ultrastructure Research* **59**, 223–230.

Müller-Schwarze, D., Ravid, U., Claesson, A., *et al.* (1978a). The "deer lactone": source, chiral properties, and responses by black-tailed deer. *Journal of Chemical Ecology* **4**, 247–256.

Müller-Schwarze, D., Källquist, L., Mossing, T., Brundin, A., and Andersson, G. (1978b). Response of reindeer to interdigital secretion of conspecifics. *Journal of Chemical Ecology* **4**, 325–335.

Müller-Schwarze, D., Altieri, R., and Porter, N. (1984). Alert odor from skin gland in deer. *Journal of Chemical Ecology* **10**, 1707–1729.

Müller-Schwarze, D., Brashear, H., Kinnel, R., Hintz, K. A., and Skibo, C. (2001). Food processing by animals: do beavers leach tree bark to improve palatability? *Journal of Chemical Ecology* **27**, 1011–1028.

Muñoz, A. 2004. Chemo-orientation using conspecific chemical cues in the stripe-necked terrapin (*Mauremys leprosa*). *Journal of Chemical Ecology* **30**, 519–530.

Murata, M., Miyagawa-Kohshima, K., Nakanishi, K., and Naya, Y. (1986). Characterization of compounds that induce symbiosis between sea anemone and anemone fish. *Science* **234**, 585–587.

Murphy, B. P., Miller, K. V., and Marchinton, R. L. (1994). Sources of reproductive chemosignals in female white-tailed deer. *Journal of Mammalogy* **75**, 781–786.

Murray, A. H., Iason, G. R., and Stewart, C. (1996). Effect of simple phenolic compounds of heather (*Calluna vulgaris*) on rumen microbial activity *in vitro*. *Journal of Chemical Ecology* **22**, 1493–1504.

Mykytowycz, R. (1965). Further observations on the territorial function and histology of the submandibular cutaneous (chin) glands in the rabbit, *Oryctolagus cuniculus* (L.). *Animal Behaviour* **13**, 400–412.

Mykytowycz, R. (1968). Territorial marking by rabbits. *Scientific American* **218**,116–126.

Mykytowycz, R. and Ward, M. M. (1971). Some reactions of nestlings of the wild rabbit, *Oryctolagus caniculus* (L.), when exposed to natural rabbit odours. *Forma et Functio* **4**, 137–148.

Nathan, J. (1998). Balzac territory. *New York Times*, 7 June 1998, Section 2, p. 14.

Nathanson, J. A. (1984). Caffeine and related methylxanthines: possible naturally occurring pesticides. *Science* **226**, 184–187.

Natoli, E. (1985). Behavioral responses of urban feral cats to different types of urine marks. *Behaviour* **94**, 234–243.

Natynczuk, S. E. and Macdonald, D. W. (1992). Scent communication in the rat. In *Chemical Signals in Vertebrates*, vol. 6, ed. R. L. Doty and D. Müller-Schwarze, pp. 537–542. New York: Plenum.

Negus, N. C., Berger, P. J., and Brown, B. W. (1986). Microtine population dynamics in a predictable environment. *Canadian Journal of Zoology* **64**, 785–792.

Nelson, C. J., Seiber, J. N., and Brower, L. P. (1981). Seasonal and intraplant variation of cardenolide content in the California milkweed, *Asclepias eriocarpa*, and implications for plant defense. *Journal of Chemical Ecology* **7**, 981–1010.

Nelson, K. (1964). Behavior and morphology in the glandulocaudine fishes (Ostariophysi, Characidae). *University of California Publications in Zoology* **75**, 59–152.

Nelson, R. J. and Shiber, J. R. (1990). Photoperiod affects reproductive responsiveness to 6-methoxy-2-benzoxazolinone in house mice. *Biology of Reproduction* **43**, 586–591.

Neuhaus, W. (1953). Über die Riechschärfe des Hundes für Fettsäuren. *Zeitschrift für vergleichende Physiologie* **35**, 527–552.

(1955). Die Unterscheidung von Duftquantitäten bei Mensch und Hund nach Versuchen mit Buttersäure. *Zeitschrift für vergleichende Physiologie* **37**, 234–252.

(1956a). Die Riechschwelle von Duftgemischen beim Hund und ihr Verhältnis zu den Schwellen unvermischter Duftstoffe. *Zeitschrift für vergleichende Physiologie* **38**, 238–258.

(1956b). Die Unterscheidungsfähigkeit des Hundes für Duftgemische. *Zeitschrift für vergleichende Physiologie* **39**, 25–43.

Neuhaus, W., and Müller, A. (1954). Das Verhältnis der Riechzellenzahl zur Riechschwelle beim Hund. *Naturwissenschaften* **41**, 237.

Nevitt, G. A. (1991). Peripheral mechanisms of olfaction in two teleostean fishes. Ph.D. Thesis, DA 908505.

(1995). Antarctic procellariiform seabirds can smell krill. *Antarctic Journal of the USA* **29**, 168–169.

(1999). Foraging by seabirds on an olfactory landscape. *American Scientist* **87**, 46–53.

Nevitt, G. A., Viet, R. R., and Kareiva, P. (1995). Dimethyl sulphide as a foraging cue for Antarctic procellariiform seabirds. *Nature* **376**, 680–682.

Nevo, E., Bodmer, M., and Heth, G. (1976). Olfactory discrimination as an isolating mechanism in speciating mole rats. *Experientia* **32**, 1511–1512.

Newman, K. S. and Halpin, Z. T. (1988). Individual odours and mate recognition in the prairie vole, *Microtus ochrogaster*. *Animal Behaviour* **36**, 1779–1787.

Newton, P. N. and Nishida, T. (1990). Possible buccal administration of herbal drugs by wild chimpanzees, *Pan troglodytes*. *Animal Behaviour* **39**, 798–801.

Nicolaus, L. K. (1987). Conditioned aversions in a guild of egg predators: implications for aposematism and prey defense mimicry. *American Midland Naturalist* **117**: 405–419.

Nicolaus, L. K. and Nellis, D. W. (1987). The first evaluation of the use of conditioned taste aversion to control predation by mongooses upon eggs. *Applied Animal Behaviour Science* **17**, 329–346.

Nicoletto, P. F. (1985). The relative roles of vision and olfaction in prey discrimination by the ground skink, *Scincella lateralis*. *Journal of Herpetology* **19**, 411–415.

Ninomiya, K. and Kimura, T. (1986). Preference of female mice for odors from urine and preputial gland secretion of males. *Zoological Science* **3**, 1103.

(1988). Male odors that influence the preference of female mice: roles of urinary and preputial factors. *Physiology and Behavior* **44**, 791–796.

Nishida, R. and Fukami, H. (1989). Ecological adaptation of an Aristolochiaceae-feeding swallowtail butterfly, *Atrophaneura alcinous*, to aristolochic acids. *Journal of Chemical Ecology* **15**, 2549–2564.

Nitao, J. K. (1988). Artificial defloration and furanocoumarin induction in *Pastinaca sativa* (Umbelliferae). *Journal of Chemical Ecology* **14**, 1515.

Nixon, A., Mallet, A. I., and Gomer, D. B. (1988). Simultaneous quantification of five odorous steroids, 16-androstenes, in the axillary hair of men. *Journal of Steroid Biochemistry* **29**, 505–510.

Njemelä, P. and Danell, K. (1988). Comparison of moose browsing on Scots pine (*Pinus sylvestris*) and lodgepole pine (*P. contorta*). *Journal of Applied Ecology* **25**, 761–775.

Nolte, D. L., Mason, J. R., and Clark, L. (1993). Avoidance of bird repellents by mice (*Mus musculus*). *Journal of Chemical Ecology* **19**, 427–432.

Nolte, D. L., Mason, J. R., and Lewis, S. L. (1994a). Tolerance of bitter compounds by an herbivore, *Cavia porcellus*. *Journal of Chemical Ecology* **20**, 303–308.

Nolte, D. L., Mason, J. R., Epple, G., Aronov, E., and Campbell, D. L. (1994b). Why are predator urines aversive to prey? *Journal of Chemical Ecology* **20**, 1505–1516.

Nordeng, H. (1971). Is the local orientation of anadromous fishes determined by pheromones? *Nature* **233**, 411–413.

(1977). A pheromone hypothesis for homeward migration in anadromous salmonids. *Oikos* **28**, 155–159.

Nordstrom, K. M., Belcher, A. M., Epple, G., *et al.* (1989). Skin surface microflora of the saddle-back tamarin monkey, *Saguinus fuscicollis*. *Journal of Chemical Ecology* **15**, 629–639.

Novotny, M., Jemiolo, B., Harvey, S., Wiesler, D., and Marchlewska-Koj, A. (1986a). Adrenal-mediated endogenous metabolites inhibit puberty in female mice. *Science* **231**, 722–725.

Novotny, M., Harvey, S., Jemiolo, B., and Alberts, A. (1986b). Synthetic pheromones that promote inter-male aggression in mice. *Proceedings of the National Academy of Sciences, USA* **82**, 2059–2061.

Novotny, M. V., Ma, W., Wiesler, D., and Zidek, L. (1999). Positive identification of the puberty-accelerating pheromone of the house mouse: the volatile ligands associating with the major urinary protein. *Proceedings of the Royal Society of London Series B* **266**, 2017–2022.

Nyby, J., Whitney, G., Schmitz, S., and Dizinno, G. (1978). Post pubertal experience establishes signal value of mammalian sex odor. *Behavioral Biology* **22**, 545–552.

Nyström, P. and Abjörnsson, K. (2000). Effects of fish chemical cues on the interactions between tadpoles and crayfish. *Oikos* **88**, 181–190.

O'Connell, R. J., Singer, A. G., Pfaffmann, C., and Agosta, W. C. (1979). Pheromones of hamster vaginal discharge: attraction to femtogram amounts of dimethyl disulfide and to mixtures of volatile compounds. *Journal of Chemical Ecology* **5**, 575–585.

Ödberg, F. O. and Francis-Smith, K. (1977). Studies on the formation of ungrazed eliminative areas in fields used by horses. *Applied Animal Ethology* **3**, 27–34.

O'Donnell, R. P., Ford, N. B., Shine, R., and Mason, R. T. (2004). Male red-sided garter snakes, *Thamnophis sirtalis parietalis*, determine female mating status from pheromone trails. *Animal Behaviour* **68**, 677–683.

Ogurtsov, S. V. and Bastakov, V. A. (2001). Imprinting on native pond odour in the pool frog, *Rana lessonae* CAM. In *Chemical Signals in Vertebrates*, vol. 9, ed. A. Marchlewska-Koj, J. J. Lepri, and D. Müller-Schwarze, pp. 433–438. New York: Kluwer Academic/Plenum.

Oh, H.K., Sakai, T., Jones, M. B., and Longhurst, W. M. (1967). Effect of various essential oils isolated from Douglas fir needles upon sheep and deer rumen microbial activity. *Applied Microbiology* **15**, 777–784.

O'Hara, R. K. and Blaustein, A. R. (1985). *Rana cascadae* tadpoles aggregate with siblings: an experimental field study. *Oecologia* **67**, 44–51.

Ohigashi, H., Huffman, M. A., Izutsu, D., *et al.* (1994). Toward the chemical ecology of medicinal plant use in chimpanzees: the case of *Vernonia amygdalina,* a plant used by wild chimpanzees possibly for parasite-related diseases. *Journal of Chemical Ecology* **20**, 541–553.

Oli, M. K. and Dobson, F. S. (1999). Population cycles in small mammals: the role of age at sexual maturity. *Oikos* **86**, 557–565.

Olivier, R. C. D. and Laurie, W. A. (1974). Habitat utilization by hippopotamus in the Mara River. *East African Wildlife* **12**, 249–271.

Olsén, K. H. (1987). Chemoattraction of juvenile Arctic charr (*Salvelinus alpinus* L.) to water scented by conspecific intestinal content and urine. *Comparative Biochemistry and Physiology* **87**, 641–644.

(1990). Further studies concerning chemoattraction among fry of Arctic charr (*Salvelinus alpinus* L.) to water conditioned by conspecifics. *Journal of Chemical Ecology* **16**, 2084–2090.

Olsén, K. H. and Winberg, S. (1996). Learning and sibling odor preference in juvenile Artic char, *Salvelinus alpinus* (L.). *Journal of Chemical Ecology* **22**, 773–786.

Olsén, K. H., Karlsson, L., and Helandes, A. (1986). Food search behavior in arctic charr, *Salvelinus alpinus* (L.) induced by food extracts and amino acids. *Journal of Chemical Ecology* **12**, 1987–1998.

Olsén, K. H., Bjerselius, R., Petersson, E., *et al.* (2000). Lack of species-specific primer effects of odours from female Atlantic salmon (*Salmo salar L.*) and brown trout (*Salmo trutta L.*). *Oikos* **88**, 213–220.

Olsén, K. A., Johansson, A.-K., Bjerselius, R., Mayer, I., and Kindhal, H. (2002a). Mature Atlantic salmon (*Salmo salar* L.) male parr are attracted to ovulated female urine but not to ovarian fluid. *Journal of Chemical Ecology* **28**, 29–40.

Olsén, K. H., Grahn, M., and Lohm, J. (2002b). Influence of MHC on sibling discrimination in arctic char, *Salvelinus alpinus* (L.). *Journal of Chemical Ecology* **28**, 783–795.

Ophir, D., Guterman, A., and Gross-Isseroff, R. (1988). Changes in smell acuity induced by radiation exposure of the olfactory mucosa. *Archives of Otolaryngology and Head and Neck Surgery* **114**, 853–855.

Or, K. and Ward, D. (2003). Three-way interactions between *Acacia*, large mammalian herbivores and bruchid beetles: a review. *African Journal of Ecology* **41**, 257–265.

Ortmann, R. (1956). Über die Musterbildung von Duftdrüsen in der Sohlenhaut der weissen Hausmaus (*Mus musculus albinus*). *Zeitschrift für Säugetierkunde* **21**, 138–141.

Ouellet, J. P. and Ferron, J. (1988). Scent-marking behavior by woodchucks (*Marmota monax*). *Journal of Mammalogy* **69**, 365–368.

Ovaska, K. (1988). Recognition of conspecific odors by the western red-backed salamander, *Plethodon vehiculum*. *Canadian Journal of Zoology* **66**, 1293–1296.

Over, R., Cohen-Tannoudji, J., Dehnhard, M., Claus, R., and Signoret, J. P. (1990). Effect of pheromones from male goats on LH secretion in anoestrous ewes. *Physiology and Behavior* **48**, 665–668.

Owadally, A. W. (1979). The dodo and the tambalacoque tree. *Science* **203**,1363–1364.

Owaga, M. L. A., Hassanali, A., and McDowell, P. G. (1988). The role of 4-cresol and 3-*n*-propylphenol in the attraction of tsetse flies to buffalo urine. *Insect Science and its Application* **9**, 95–100.

Owaga, M. L. A. (1984). Preliminary observations on the efficacy of olfactory attractants derived from wild hosts of tsetse. *Insect Science and its Application* **5**, 87–90.

Owre, O. T. and Northington, P. O. (1961). Indication of the sense of smell in the turkey vulture, *Cathartes aura* (Linnaeus) from feeding tests. *American Midland Naturalist* **66**, 200.

Page, J. E., Balza, F., and Nishida, T. (1992). Biologically active diterpenes from *Aspilia mossambicensis,* a chimpanzee mecidinal plant. *Phytochemistry* **31**, 3437–3439.

Pages, E. (1972). Comportement interindividual des Pangolins arboricoles du Gabon. *Biologia Gabonica* **1**, 1–120.

Palen, G. F. and Goddard, G. V. (1966). Catnip and estrus behavior in the cat. *Animal Behaviour* **14**, 372–377.

Palo, R. T. (1984). Distribution of birch (*Betula* spp.), willow (*Salix* spp.) and poplar (*Populus* spp.) secondary metabolites and their potential role as chemical defense against herbivores. *Journal of Chemical Ecology* **10**, 499–520.

Palo, R. T., Pehrson, A., and Knutsson, D. G. (1983). Can birch phenolics be of importance in the defense against browsing vertebrates? *Finnish Game Research* **41**, 75–80.

Palo, R. T., Sunnerheim, K., and Theander, O. (1985). Seasonal variation of phenols, crude protein and cell wall content of birch (*Betula pendula* Roth.) in relation to ruminant in vitro digestibility. *Oecologia* **65**, 314–318.

Pandey, S. D., and Pandey, S. C. (1986). Estrus suppression in wild mice: source of pheromonal cue. *Acta Physiologica Hungarica* **67**, 387–392.

Papi, F. (1976). The olfactory navigation system of the pigeon. *Verhandlungen der Deutschen Zoologischen Gesellschaft* **69**, 184–205.

(1990). Olfactory navigation in birds. *Experientia* **46**, 352–363.

Papi, F. and Ioalè, P. (1988). Pigeon navigation: New experiments on interaction between olfactory and magnetic cues. *Comparative Biochemistry and Physiology. A, Comparative Physiology* **91**, 87–90.

Papi, F., Fiore, L., Viaschi, V., and Benvenuti, S. (1972). Olfaction and homing in pigeons. *Monitore Zoologico Italiano* **6**, 85–95.

Papi, F., Fiore, L., Fiaschi, V., and Benvenuti, S. (1973). An experiment for testing the hypothesis of olfactory navigation of homing pigeons. *Journal of Comparative Physiology* **83**, 93–102.

Papi, F., Keeton, W. T., Brown, A. I., and Benvenuti, S. (1978). Do American and Italian pigeons rely on different homing mechanisms? *Journal of Comparative Physiology* **128**, 303–317.

Paquet, P. C. and Fuller, W. A. (1990). Scent marking and territoriality in wolves of Riding Mountain National Park. In *Chemical Signals in Vertebrates*, vol. 5, ed. D. W. Macdonald, D. Müller-Schwarze, and S. E. Natynczuk, pp. 394–400. Oxford: Oxford University Press.

Park, D. and Propper, C. R. (2001). Repellent function of male pheromones in the red-spotted newt. *Journal of Experimental Zoology* **289**, 404–408.

Park, D., Hempleman, S. C., and Propper, C. R. (2001). Endosulfan exposure disrupts pheromonal systems in the red-spotted newt: a mechanism for subtle effects of environmental chemicals. *Environmental Health Perspectives* **109**, 669–673.

Parker, G. H. (1911). The olfactory reactions of the common killifish, *Fundulus heteroclitus* (Linn.). *Journal of Experimental Zoology* **10**, 1–5.

Parkes, A. S. and Bruce, H. M. (1961). Olfactory stimuli in mammalian reproduction. *Science* **134**, 1049–1054.

Parsons, L. M. and Terman, C. R. (1978). Influence of vision and olfaction on the homing ability of the white-footed mouse (*Peromyscus leucopus noveboracensis*). *Journal of Mammalogy* **59**, 761–771.

Paul, V. J. and Pennings, S. C. (1991). Diet-derived chemical defenses in the sea hare *Stylocheirus longicauda* (Quoy et Gaimard 1824). *Journal of Experimental Marine Biology and Ecology* **151**, 227–243.

Pawlik, J. R. (1993). Marine invertebrate chemical defenses. *Chemical Reviews* **93**, 1911–1922.

Pawlik, J. R., McFall, G., and Zea, S. (2002). Does the odor from sponges of the genus *Ircinia* protect them from fish predators? *Journal of Chemical Ecology* **28**: 1103–1115.

Pawson, M. G. (1977). Analysis of a natural chemical attractant for whiting *Merlangus merlangus* L. and cod *Gadus morhua* L. using a behavioral bioassay. *Comparative Biochemical Physiology* **56A**, 129–135.

Payne, R. (2001). The tragedy of the live fish trade: Part 1. www.pbs.org/odyssey/voice/20010606.

Pearce, D. T. and Oldham, C. M. (1988). Ovulation in the merino ewe in the breeding and anoestrous season. *Australian Journal of Biological Science* **41**, 23–26.

Pearce, G. T. and Paterson, A. M. (1992). Physical contact with the boar is required for maximum stimulation of puberty in the gilt because it allows transfer of boar pheromone and not because it induces cortisol release. *Animal Reproduction Science* **27**, 209–224.

Pearl, I. A. and Darling, S. F. (1968). Studies on the hot water extractives of *Populus balsamifera* bark. *Phytochemistry* **7**, 1851–1853.

Peckarsky, B. L. and Penton, M. A. (1988). Why do *Ephemerella* nymphs scorpion posture: a "ghost of predation past"? *Oikos* **53**, 185–193.

Pedersen, P. E. and Blass, E. M. (1982). Prenatal and postnatal determinants of the first suckling episode in albino rats. *Developmental Psychobiology* **15**, 349–355.

Pedersen, P. E., Stewart, W. B., Greer, C. A., and Shepherd, G. M. (1983). Evidence for olfactory function in utero: VNO probably monitors intrauterine environment. *Science* **221**, 478–480.

Pedersen, R. E. and Benson, T. E. (1986). Projection of septal organ receptor neurons to the main olfactory bulb in rats. *Journal of Comparative Neurology* **252**, 555–562.

Peinetti, R., Pereyra, M., Kin, A., and Sosa, A. (1993). Effects of cattle ingestion on viability and germination rate of caldén (*Prosopis caldenia*) seeds. *Journal of Range Management* **46**, 483–486.

Pelletier, S. W. (1983). The nature and definition of an alkaloid. *Alkaloids, Chemical and Biological Perspectives* **1**, 1–31.

Pennycuick, P. R. and Cowan, R. (1990). Odour and food preferences of house mice, *Mus musculus*. *Australian Journal of Zoology* **38**, 241–247.

Perret, M. and Schilling, A. (1987). Intermale sexual effect elicited by volatile urinary ether extract in *Microcebus murinus* (Prosimian, Primates). *Journal of Chemical Ecology* **13**, 495–507.

Perrigo, G. and Bronson, F. H. (1983). Communication disparities between genetically diverging populations of deer mice. In *Chemical Signals in Vertebrates*, vol. 3, ed. D. Müller-Schwarze, and R. M. Silverstein, pp. 195–210. New York: Plenum.

Perry, J. N., and Wall, C. (1984). Orientation of male pea moth, *Cydia nigricana*, to pheromone traps in a wheat crop. *Entomologia Experimentalis et Applicata* **37**, 161–167.

Pesaro, M., Balsamo, M., Gandolfi, G., and Tongiorgi, P. (1981). Discrimination among different kinds of water in juvenile eels, *Anguilla anguilla* (L.). *Monitore Zoologico Italiano* **15**, 183–191.

Pesce, C. (1990). Stanching stench: Florida City a leader. *USA Today*, March 19, 1990.

Peters, R. P. and Mech, L. D. (1975). Scent marking in wolves. *American Scientist* **63**, 628–637.

Petit, C., Hossaert-McKey, M., Perret, P., Blondel, J., and Lambrechts, M. M. (2002). Blue tits use selected plants and olfaction to maintain an aromatic environment for nestlings. *Ecology Letters* **5**, 585.

Petranka, J. W., Kats, L. B., and Sih, A. (1987). Predator–prey interactions among fish and larval amphibians: use of chemical cues to detect predatory fish. *Animal Behaviour* **35**, 420–425.

Pettersson, L. B., Nilsson, P. A., and Bronmark, C. (2000). Predator recognition and defence strategies in crucian carp, *Carassius carassius*. *Oikos* **88**, 200–212.

Pfeiffer, W. (1978). Heterocyclic compounds as releasers of the fright reaction in the giant danio *Danio malabaricus* (Jerdon) (Cyprinidae, Ostariophysi, Pisces). *Journal of Chemical Ecology* **4**, 665–673.

Pfeiffer, W. and Lamour, D. (1976). Die Wirkung von Schreckstoff auf die Herzfrequenz von *Phoxinus phoxinus* (L.) (Cyprinidae, Ostariophysi, Pisces). *Revue Suisse Zoologique* **83**, 861–873.

Pfeiffer, W. and Lemke, J. (1973). Untersuchungen zur Isolierung und Identifizierung des Schreckstoffes aus der Haut der Elritze, *Phoxinus phoxinus* (L.) (Cyprinidae, Ostariophysi, Pisces). *Journal of Comparative Physiology* **82**, 407–410.

Pfennig, D. W. (1997). Kinship and cannibalism. *BioScience* **47**, 662–675.

Pfister, J. A., Provenza, F. D., and Manners, G. D. (1990a). Ingestion of tall larkspur by cattle: separating effects of flavor from post-ingestive consequences. *Journal of Chemical Ecology* **16**, 1697–1705.

Pfister, J. A., Müller-Schwarze, D., and Balph, D. F. (1990b). Effects of predator fecal odors on feed selection by sheep and cattle. *Journal of Chemical Ecology* **16**, 573–583.

Pfister, J. A., Manners, G. D., Gardner, D. R., Price, K. W., and Ralphs, M. H. (1996). Influence of alkaloid concentration on acceptability of tall larkspur (*Delphinium spp.*) to cattle and sheep. *Journal of Chemical Ecology* **22**, 1147–1168.

Pfister, J. A., Provenza, F. D., Manners, G. D., and Ralphs, M. H. (1997). Tall larkspur ingestion: can cattle regulate intake below toxic levels? *Journal of Chemical Ecology* **23**, 759–777.

Phillips, J. A. and Alberts, A. C. (1992). Naive ophiophagus lizards recognize and avoid venomous snakes using chemical cues. *Journal of Chemical Ecology* **18**, 1775–1783.

Pickrell, J. (2003). Cyanide on the side. http://sciencenow.sciencemag.org. 5/30/2003.

Pigozzi, G. (1990). Latrine use and the function of territoriality in the European badger, *Meles meles*, in a Mediterranean coastal habitat. *Animal Behaviour* **39**, 1000–1002.

Placyk, J. S. and Graves, B. M. (2002). Prey detection by vomeronasal chemoreception in a plethodontid salamander. *Journal of Chemical Ecology* **28**, 1017–1036.

Poddar-Sarkar, M. (1996). The fixative lipid of tiger pheromone. *Journal of Lipid Mediators and Cell Signalling* **15**, 89–101.

Poddar-Sarkar, M., Brahmachary, R. L., and Dutta, J. (1991). Short-chain fatty acids as putative pheromones in the marking fluid of the tiger. Indian Chemical Society 68, 255–256.

Poduschka, W. and Firbas, W. (1968). Das Selbstbespeicheln des Igels, *Erinaceus europaeus*, Linne 1758, steht in Beziehung zur Funktion des Jacobson'schen Organes. *Zeitschrift für Säugetierkunde* **33**, 160–172.

Poindron, P., Levy, F., and Krehbiel, D. (1988). Genital, olfactory, and endocrine interactions in the development of maternal behavior in the parturient ewe. *Psychoneuroendocrinology* **3**, 99–125.

Polak, E. H. and Provasi, J. (1992). Odor sensitivity to geosmin enantiomers. *Chemical Senses* **17**, 23–26.

Polkinghorne, C., Olson, J. M., Gallaher, D. G., and Sorensen, P. W. (2001). Larval sea lamprey release two unique bile acids to the water of a rate sufficient to produce detectable riverine pheromone plumes. *Fish Physiology and Biochemistry* **24**, 15–30.

Pontet, A. and Schenk, F. (1988). Effects of conspecific odors on the activity and the ultrasonic vocalizations of the woodmouse, *Apodemus sylvaticus*, in a plus-maze during ontogeny. *Sciences et Techniques de L' Animal de Laboratoire* **13**, 105–109.

Porter, R. H. (1986). Chemical signals and kin recognition in spiny mice (*Acomys cahirinus*). In *Chemical Signals in Vertebrates*. vol. 4, ed. D. Duvall, D. Müller-Schwarze, and R. M. Silverstein, pp. 397–411. New York: Plenum.

Porter, R. H. and Moore, J. D. (1981). Human kin recognition by olfactory cues. *Physiology and Behavior* **27**, 493–495.

Porter, R. H., Cernoch, J. M., and McLaughlin, F. J. (1983). Maternal recognition of neonates through olfactory cues. *Physiology and Behavior* **30**, 151–154.

Porter, R. H., Balogh, R. D., Cernoch, J. F., and Franchi, C. (1986). Recognition of kin through characteristic body odors. *Chemical Senses* **11**, 389–395.

Porter, R. H., McFadyen-Ketchum, S. A., and King, G. A. (1989). Underlying bases of recognition signatures in spiny mice, *Acomys cahirinus*. *Animal Behaviour* **37**, 638–644.

Porter, R. H., Makin, J. W., Davis, L. B., and Christensen, K. M. (1992). Breast-fed infants respond to olfactory cues from their own mother and unfamiliar lactating females. *Infant Behavior and Development* **15**, 85–93.

Powers, J. B. and Winans, S. S. (1975). Vomeronasal organ: critical role in mediating sexual behavior of the male hamster. *Science* **187**, 961–963.

Powers, J. B., Fields, R. B., and Winans, S. S. (1979). Olfactory and vomeronasal system participation in male hamsters' attraction to female vaginal secretion. *Physiology and Behavior* **22**, 77–84.

Prakash, J. and Idris, M. (1982). Scent marking by the female of the Indian gerbil, *Tatera indica*, from two distinct desert habitats during oestrus. *Indian Journal of Experimental Biology* **20**, 915–916.

Prasad, N. L. N. S. (1989). Territoriality in Indian Blackbuck antelope, *Antilope cervicapra* (Linnaeus). *Bombay Natural History Society Journal* **86**, 187–193.

Preti G., Muetterties, E. L. Furman, J. M., Kennelly, J. J., and Johns, B. E. (1976). Volatile constituents of dog (*Canis familiaris*) and coyote (*Canis latrans*) anal sacs. *Journal of Chemical Ecology* **2**, 177–186.

Preti, G., Cutler, W. B., Garcia, C. R., Huggins, G. R., and Lawley, H. J. (1986). Human axillary secretions influence women's menstrual cycles: the role of donor extract from women. *Hormones and Behavior* **20**, 474–482.

Preti, G., Cutler, W. B., Christensen, C. M., *et al.* (1987). Human axillary extracts: analysis of compounds from samples which influence menstrual timing. *Journal of Chemical Ecology* **13**, 717–731.

Preti, G., Zeng, X. N., Leyden, J. J., McGinley, K. J., and Spielman, A. I. (1992). Human axillary odors and their precursors. *Chemical Senses* **17**, 685–686.

Price, E. O., Dunn, G. C., Talbor, J. A., and Dally, M. R. (1984). Fostering lambs by odor transfer: the substitution experiment. *Journal of Animal Science* **59**, 301–307.

Primor, N., Parness, J., and Zlotkin, E. (1978). Paradaxin: the toxic factor from the skin secretion of the flatfish *Paradachirus marmoratus* (Soleidae). In *Toxins: Animals, Plants and Microbial*, ed. P. Rosenberg. Oxford: Pergamon Press.

Probst, B. and Lorenz, M. (1987). Increased scent marking in male Mongolian gerbils by urinary polypeptides of female conspecifics. *Journal of Chemical Ecology* **13**, 851–862.

Provenza, F. D. (1995). Postingestive feedback as an elementary determinant of food preference and intake in ruminants. *Journal of Range Management* **48**, 2–17.

(2004). Linking herbivore experience, varied diets, and plant biochemical diversity. In *Proceedings of the Annual Meeting of International Society of Chemical Ecology*, July 2004, Ottawa, Canada.

Provenza, F. D. and Lynch, J. J. (1994). How goats learn to distinguish between novel foods that differ in postingestive consequences. *Journal of Chemical Ecology* **20**, 609–624.

Provenza, F. D., Kimball, B. A., and Villaba, J. J. (2000). Roles of odor, taste, and toxicity in the food preferences of lambs: implications for mimicry in plants. *Oikos* **88**, 424–432.

Prudente, A. (1963). Patologia geographica do cancer no Brasil. *Boletin de Oncologia* **46**, 281.

Quadagno, D. M., Shubeita, H. E., Deck, J., and Francoeur, D. (1981). Influence of male social contacts, exercise, and all-female living conditions on the menstrual cycle. *Psychoneuroendocrinology* **6**, 239–244.

Quay, W. B. (1970). Histology of the para-anal sebaceous glandular organs of the bat *Eonycteris spelaea* (Chiroptera: Pteropidae). *Anatomical Record* **166**, 189–197.

(1977). Structure and function of skin glands. In *Chemical Signals in Vertebrates,* vol. 1, ed. D. Müller-Schwarze, and M. M. Mozell, pp. 1–16. New York: Plenum.

Quay, W. B. and Müller-Schwarze, D. (1970) Functional histology of integumentary glandular regions in black-tailed deer (*Odocoileus heminonus columbianus*). *Journal of Mammalogy* **51**, 675–694.

Quinn, T. P. (1980). Locomotor responses of juvenile blind cave fish, *Astyanax jordani*, to the odors of conspecifics. *Behavioral and Neural Biology* **29**, 123–127.

Quinn, T. P. and Busack, C. A. (1985). Chemosensory recognition of sibling in juvenile coho salmon (*Oncorhynchus kisutch*). *Animal Behaviour* **33**, 51–56.

Quinn, V. S. and Graves, B. M. (1998). Home pond discrimination using chemical cues in *Chrysema picta*. *Journal of Herpetology* **32**, 457–461.

Radwan, M. A. and Crouch, G. L. (1978). Selected chemical constituents and deer browsing preference of Douglas fir. *Journal of Chemical Ecology* **4**, 675–683.

Rajan, R., Clement, J. P., and Bhalla, U. S. (2006). Rats smell in stereo. *Science* **311**, 666–670.

Rajendren, G. and Dominic, C. J. (1986). Evaluation of involvement of accessory olfactory (vomeronasal) system in estrous cyclicity and mating in female mice. *Indian Journal of Experimental Biology* **24**, 573–577.

(1988). Effect of cyproterone acetate on the pregnancy-blocking ability of male mice and the possible chemical nature of the pheromone. *Journal of Reproduction and Fertility* **84**, 387–392.

Ralphs, M. H., Olsen, J. D., Pfister, J. A., and Manners, G. D. (1988). Plant–animal interactions in larkspur poisoning in cattle. *Journal of Animal Science* **66**, 2334–2342.

Randall, J. A. (1986). Preference for estrous female urine by male kangaroo rats (*Dipodomys spectabilis*). *Journal of Mammalogy* **67**, 736–739.

Rasa, O. A. E. (1973). Marking behavior and its social significance in the African dwarf mongoose *Helogale undulata rufula*. *Zeitschrift für Tierpsychologie* **32**, 293–318.

Rasmussen, L. E. L. (1988). Chemosensory responses in two species of elephants to constituents of temporal gland secretion and musth urine. *Journal of Chemical Ecology* **14**, 1687–1711.

Rasmussen, L. E. L. and Greenwood, D. R. (2003). Frontalin: a chemical message of musth in Asian elephants (*Elephas maximus*). *Chemical Senses* **28**: 433–446.

Rasmussen, L. E. L. and Hultgren, B. (1990). Gross and microscopic anatomy of the vomeronasal organ in the Asian elephant (*Elephas maximus*). In *Chemical Signals in Vertebrates*. vol. 5, eds. D. W. Mac Donald, D. Müller-Schwarze, and S. E. Natynczuk, pp. 154–161. Oxford: Oxford University Press.

Rasmussen, L. E. L. and Perrin, T. E. (1999). Physiological correlates of musth: lipid metabolites and chemical composition of exudates. *Physiology and Behavior* **67**, 539–549.

Rasmussen, L. E. L. and Schulte, B. A. (1998). Chemical signals in the reproduction of Asian (*Elephas maximus*) and African (*Loxodonta africana*) elephants. *Animal Reproduction Science* **53**, 19–34.

Rasmussen, L. E. L., Schmidt, M. J., Henneous, R., Groves, D., and Daves, G. D. (1982). Asian bull elephants: flehmen-like responses to extractable components in female elephant estrous urine. *Science* **217**, 159–162.

Rasmussen, L. E., Buss, C. O., Hess, D. L., and Schmidt, M. J. (1984). Testosterone and dihydrotestosterone concentrations in elephant serum and temporal gland secretions. *Biology of Reproduction* **30**, 352–362.

Rasmussen, L. E. L., Lee, T. D., Roelofs, W. L., Zhang, A., and Daves, G. D., Jr. (1996). Insect pheromone in elephants. *Nature* **379**:684.

Rasmussen, L. E. L., Lee, T. D., Zhang, A., Roelofs, W. L., and Daves, G. D., Jr. (1997). Purification, identification, concentration and bioactivity of (*Z*)-7-dodecen-1-yl acetate: sex pheromone of the female Asian elephant, *Elephas maximus*. *Chemical Senses* **22**, 417–438.

Raymer, J. D., Wiesler, D., Novotny, M., *et al.* (1984). Volatile constituents of wolf (*Canis lupus*) urine as related to gender and season. *Experientia* **40**, 707–709.

Raymer, J., Wiesler, D., Novotny, M., *et al.* (1986). Chemical scent constituents in urine of wolf (*Canis lupus*) and their dependence on reproductive hormones. *Journal of Chemical Ecology* **12**, 297–314.

Reardon, P. O., Leinweber, C. L., and Merrill, J. L. B. (1972). The effect of bovine saliva on grasses. *Journal of Animal Science* **34**, 897–98.

(1974). Responses of sideoats grama to animal saliva and thiamine. *Journal of Range Management* **27**, 400–401.

Reasner, D. S. (1987). Spatially selective alteration of the mitral cell layer: a critical review of the literature. *Chemical Senses* **12**, 365–379.

Reasner, D. S. and Johnston, R. E. (1988). Acceleration of reproductive development in female Djungarian hamsters by adult males. *Chemical Senses* **13**, 729.

Reed, J. P. (1969). Alarm substances and fright reaction in some fishes from the Southeastern United States. *Transactions of the American Fisheries Society* **98**, 664–668.

Reger, R. L., Gerall, A. A., Wysocki, C. J., and Carter, C. S. (1987). LHRH neuronal system in the accessory olfactory bulb of the prairie vole, *Microtus ochrogaster*. In *Proceedings of the Annual Meeting of the Society for Neuroscience*, New Orleans.

Regnier, F. E. and Goodwin, M. (1977). On the chemical and environmental modulation of pheromone release from vertebrate scent marks. In *Chemical Signals in Vertebrates*, vol. 1, ed. D. Müller-Schwarze, and M. M. Mozell, pp. 115–133. New York: Plenum.

Rehnberg, B. G., and Schreck, C. B. (1986). The olfactory L-serine receptor in coho salmon: biochemical specificity and behavioral responses. *Journal of Comparative Physiology* **159**, 61–67.

(1987). Chemosensory detection of predators by coho salmon, *Oncorhynchus kisutch*: behavioral reaction and the physiological stress responses. *Canadian Journal of Zoology* **65**, 481–485.

Rehnberg, B. G., Smith, R. J. F., and Sloley, B. D. (1987). The reaction of pearl dace (Pisces, Cyprinidae) to alarm pheromone: time-course of behavior, brain amines and stress physiology. *Canadian Journal of Zoology* **65**, 2916–2921.

Reichardt, P. B., Bryant, J. P., Clausen, T. P., and Wieland, G. D. (1985). Defense of winter-dormant Alaska paper birch against snowshoe hares. *Oecologia* **65**, 58–69.

Reichardt, P. B., Bryant, J. P., Anderson, B. J., *et al.* (1990a). Germacrone defends Labrador tea from browsing by snowshoe hares. *Journal of Chemical Ecology* **16**, 1961–1970.

Reichardt, P. B., Bryant, J. P., Mattes, B. R., *et al.* (1990b). Winter chemical defense of Alaskan balsam poplar against snowshoe hares. *Journal of Chemical Ecology* **16**, 1941–1959.

Reidinger, R. F., Jr. and Mason, J. R. (1983). Exploitable characteristics of neophobia and food aversions for improvements in rodent and bird control. In *Proceedings of the 4th Symposium of the American Society for Testing and Materials: Vertebrate Pest Control and Management Materials*, ed. D. E. Kaukeinen, D. E., pp. 20–39, Washington, DC: American Society for Testing and Materials.

Reis, P. J., Tunks, D. A., and Chapman, R. E. (1975). Effects of mimosine, a potential chemical defleecing agent, on wool growth and the skin of sheep. *Australian Journal of Biological Sciences* **1**, 69–84.

Renwick, J. H., Claring, W. B., Earthy, M. E., *et al.* (1984). Neural-tube defects produced in Syrian hamsters by potato glycoalkaloids. *Teratology* **30**, 371–381.

Resink, J. W., van den Hurk, R., Peters, R. C., and van Oordt, P. G. W. J. (1987). Steroid glucuronides as sex attracting pheromones in the African catfish, *Clarias gariepinus, Proceedings of the 3rd International Symposium on Reproductive Physiology of Fish*, ed. D. R. Idler, L. W. Crim, and J. M. Walsh, p. 163. St. John's, Newfoundland: Memorial University Press.

Resink, J. W., Schoonen, W. E. G. J., Albers, P. C., *et al.* (1989a). The chemical nature of sex attracting pheromones from the seminal vesicle of the African catfish, *Clarias gariepinus*. *Aquaculture* **83**, 137–151.

Resink, J. W., Voorthuis, P. K., van den Hurk, R., Peters, R. C., and van Oordt, P. G. W. J. (1989b). Steroid glucuronides of the seminal vesicle as olfactory stimuli in African catfish, *Clarias gariepinus*. *Aquaculture* **83**, 153–166.

Resink, J. W., van den Berg, T. W. M., van den Hurk, R., Huisman, E. A., and van Oordt, P. G. W. J. (1989c). Induction of gonadotropin release and ovulation by pheromones in the African catfish, *Clarias gariepinus*. *Aquaculture* **83**, 167–177.

Ressler, K. J., Sullivan, S. L., and Buck, L. B. (1993). A zonal organization of odorant receptor gene expression in the olfactory epithelium. *Cell*, **73**, 597–609.

(1994). Information coding in the olfactory system: evidence for a stereotyped and highly organized epitope map in the olfactory bulb. *Cell* **79** 1245–1255.

Reynolds, J. and Keverne, E. B. (1979). The accessory olfactory system and its role in the pheromonally mediated suppression of oestrus in grouped mice. *Journal of Reproduction and Fertility* **57**, 31–35.

Rhoades, D. F. (1985). Offensive–defensive interactions between herbivores and plants: their relevance in herbivore population dynamics and ecological theory. *American Naturalist* **125**, 205–238.

Richardson, P. R. K. (1990). Scent marking and territoriality in the aardwolf. In *Chemical Signals in Vertebrates*. vol. 5, eds. D. W. MacDonald, D. Müller-Schwarze and S. E. Natynczuk, pp. 378–387. Oxford: Oxford University Press.

Richardson, R. and Campbell, B. A. (1988). Effects of home nest odors on black–white preference in the developing rat: implications for developmental learning research. *Behavioral and Neural Biology* **50**, 361–366.

Rick, C. M. and Bowman, R. I. (1961). Galapagos tomatoes and tortoises. *Evolution* **15**, 407–417.

Rieger, J. F. and Jakob, E. H. (1988). The use of olfaction in food location by frugivorous bats. *Biotropica* **20**, 161–164.

Rieser, J., Yonas, A., and Wikner, K. (1976). Radial localization of odors by newborns. *Child Development* **47**, 856–859.

Risser, J. M. and Slotnick, B. M. (1987). Nipple attachment and survival in neonatal olfactory bulbectomized rats. *Physiology and Behavior* **40**, 545–550.

Rissman, E. F. and Johnston, R. E. (1986). Nutritional and social cues influence the onset of puberty in California voles. *Physiology and Behavior* **36**, 1–5.

Rissman, E. F., Sheffield, S. D., Kretzmann, M. B., Fortune, J. E., and Johnston, R. E. (1984). Chemical cues from families delay puberty in male California voles. *Biology of Reproduction* **31**, 324–331.

Ritter, F. J., Brüggemann, I. E. M., Gut, J., and Persoons, C. J. (1982). Recent pheromone research in the Netherlands on muskrats and some insects pests introduced from America into Europe: the muskrat, *Odatra zibethicus*, the American cockroach, *Periplaneta americana*, and the beet army worm, *Spodoptera exigua*. *American Chemical Society Symposium Series* **190**, 107–130.

Rivard, G. and Klemm, W. R. (1989). Two body fluids containing bovine estrous pheromone(s). *Chemical Senses* **14**, 273–279.

(1990). Sample contact required for complete bull response to oestrous pheromone in cattle. In *Chemical Signals in Vertebrates*, vol. 5, ed. D W. Macdonald, D. Müller-Schwarze, and S. E. Natynczuk, pp. 627–633. Oxford: Oxford University Press.

Robbins, C. T., Hanley, T. A., Hagerman, A. E., *et al.* (1987). Role of tannins in defending plants against ruminants: reduction in protein availability. *Ecology* **68**, 98–107.

Robbins, C. T., Hagerman, A. E., Austin, P. J., McArthur, C., and Hanley, T. A. (1991). Variation in mammalian physiological responses to a condensed tannin and its ecological implications. *Journal of Mammalogy* **72**, 480–486.

Roberts, S. C. and Gosling, L. M. (2001). The economic consequences of advertising scent mark location on territories. In *Chemical Signals in Vertebrates*, vol. 9, ed. A. Marchlewska-Koi, J. J. Lepri, and D. Müller-Schwarze, pp. 11–17. New York: Plenum.

Roberts, S. C. and Gosling, L. M. (2004). Manipulation of olfactory signaling and mate choice for conservation breeding: a case study of harvest mice. *Conservation Biology* **18**, 548–556.

Roberts, S. C., Gosling, L. M., Thornton, E. A., and McClung, J. (2001). Scent-marking by male mice under the risk of predation. *Behavioral Ecology* **12**, 698–705.

Robinson, I. (1990). The effect of mink odour on rabbits and small mammals. In *Chemical Signals in Vertebrates*, vol. 5, ed. D. W. Macdonald, D. Müller-Schwarze, and S. E. Natynczuk, pp. 566–572. New York: Oxford University Press.

Robinson, T. (1979). The evolutionary ecology of alkaloids. In *Herbivores: Their Interaction with Secondary Plant Metabolites*, vol. 1, ed. G. A. Rosenthal, and M. R. Berenbaum, pp. 413–448. San Diego, CA: Academic Press.

Rodriguez, E., Cavin, J. C., and West, J. E. (1982). The possible role of Amazonian psychoactive plants in the chemotherapy of parasitic worms: a hypothesis. *Journal of Ethnopharmacology* **6**, 303–309.

Rodriguez, E., Aregullin, M., Nishida, T., *et al.* (1985). Thiarubrine A, a bioactive constituent of *Aspilia* (Asteraceae) consumed by wild chimpanzees. *Experientia* **41**, 419–420.

Roeder, J. J. (1980). Marking behavior and olfactory recognition in genets (*Genetta genetta* L., Carnivora, Viverridae). *Behaviour* **72**, 200–210.

Roessler, E. S. (1936). Viability of weed seeds after ingestion by California linnets. *Condor* **38**, 62–65.

Rogers, J. G. and Beauchamp, G. K. (1976). Influence of stimuli from populations of *Peromyscus leucopus* on maturation of young. *Journal of Mammalogy* **57**, 320–330.

Rollmann, S. M., Houck, L. D., and Feldhoff, R. C. (1999). Proteinaceous pheromone affecting female receptivity in a terrestrial salamander. *Science* **285**, 1907–1909.

(2000). Population variation in salamander courtship pheromones. *Journal of Chemical Ecology* **26**: 2713–2724.

(2003). Conspecific and heterospecific pheromone effects on female receptivity. *Animal Behaviour* **66**, 857–861.

Romer, A. S. (1959). *Comparative Anatomy of Vertebrates*. [German edition.] Hamburg: Verlag Paul Parey.

Rood, J. P. (1972). Ecological and behavioral comparisons of three genera of Argentine cavies. *Animal Behaviour Monographs* **5**, 1–83.

Ropartz, P. (1968). Contribution à l'étude du déterminisme d'un effet de groupe chez les souris. *Comptes Rendus de l' Academie des Sciences Paris* **262**, 2070–2072.

Roper, T. J., Shepherdson, D. J., and Davies, D. M. (1986). Scent marking with faeces and anal secretion in the European badger (*Meles meles*): seasonal and spatial characteristics of latrine use in relation to territoriality. *Behavior* **97**, 94–117.

Rose, F. L. (1970). Tortoise chin gland fatty acid composition: behavioral significance. *Comparative Biochemistry and Physiology* **32**, 577–580.

Rosell, F. and Bjørkøyli, T. (2002). A test of the dear enemy phenomenon in the Eurasian beaver. *Animal Behaviour* **63**, 1073–1078.

Rosell, F. and Nolet, B. A. (1997). Factors affecting scent marking behavior in Eurasian beaver (*Castor fiber*). *Journal of Chemical Ecology* **23**, 673–689.

Rosell, F. and Sun, L. (1999). Use of anal gland secretion to distinguish the two beaver species *Castor canadensis* and *C. fiber*. *Wildlife Biology* **5**, 119–124.

Rosell, F., Bergan, F., and Parker, H. (1998). Scent marking in the Eurasian beaver (*Castor fiber*) as a means of territory defense. *Journal of Chemical Ecology* **24**, 207–219.

Rosenberg, T. (1995). *The Haunted Land: Facing Europe' s Ghosts After Communism*. New York: Random House.

Rosenthal. G. A. (1991). Nonprotein amino acids as protective allelochemicals. In *Herbivores: Their Interaction with Secondary Plant Metabolites*, vol. 1, ed. G. A. Rosenthal, and M. R. Berenbaum, pp. 1–34. San Diego, CA: Academic Press.

Rosenthal, G. A. and Bell, E. A. (1979). Naturally occurring toxic nonprotein amino acids. In *Herbivores: Their Interaction with Secondary Plant Metabolites*, ed. G. A. Rosenthal, and D. H. Janzen, pp. 353–385. New York: Academic Press.

Ross, P. and Crews, D. (1978). Stimuli influencing mating-behavior in garter snake, *Thamnophis radix*. *Behavioral Ecology* **4**, 133–142.

Ross, C. W. and Detling, J. K. (1983). Investigations of trypsin inhibitors in leaves of four North American prairie grasses. *Journal of Chemical Ecology* **9**, 247–257.

Rosser, A. E. and Keverne, E. B. (1985). The importance of central noradrenergic neurons in the formation of an olfactory memory in the prevention of pregnancy block. *Neuroscience* **15**, 1141–1147.

Rothschild, M. (1965). The rabbit flea and hormones. *Endeavour* **24**, 162–168.

Rousi, M. and Häggman, J. (1984). Relationship between the total phenol content of Scots pine and browsing by the Arctic hare. *Silvae Genetica* **33**, 95–97.

Rowsemitt, C. N. and O'Connor, A. J. (1989). Reproductive function in *Dipodomys ordii* stimulated by 6-methoxybenzoxazolinone. *Journal of Mammalogy* **70**, 805–809.

Roy, J. and Bergeron, J. M. (1990a). Role of phenolics of coniferous trees as deterrents against debarking behavior of meadow voles (*Microtus pennsylvanicus*). *Journal of Chemical Ecology* **16**, 801–808.

(1990b). Branch-cutting behavior by the vole (*Microtus pennsylvanicus*): a mechanism to decrease toxicity in conifers. *Journal of Chemical Ecology* **16**, 735–741.

Royce-Malmgren, C. H., and Watson, W. H., III. (1987). Modification of olfactory-related behavior in juvenile Atlantic salmon by changes in pH. *Journal of Chemical Ecology* **13**, 533–546.

Rozenfeld, F. M. and Denoël, A. (1994). Chemical signals involved in spacing behavior of breeding female bank voles (*Clethrionomys glareolus* Schreber 1780, Microtidae, Rodentia). *Journal of Chemical Ecology* **20**, 803–813.

Rozenfeld, F. M. and Rasmont, R. (1991). Odour cue recognition by dominant male bank voles, *Clethrionomys glareolus*. *Animal Behaviour* **41**, 839–850.

Rozin, P. and Schiller, D. (1980). The nature and acquisition of a preference for chili pepper by humans. *Motivation and Emotion* **4**, 77–100.

Rubin, B. D. and Katz, L. C. (1999). Optical imaging of odorant representation in the mammalian olfactory bulb. *Neuron* **23**, 499–511.

Rubinoff, I. and Kropach, C. (1970). Differential reactions of Atlantic and Pacific predators to sea snakes. *Nature* **228**, 1288–1290.

Russell, G. F. and Hills, J. I. (1971). Odor differences between enantiomeric isomers. *Science* **172**, 1043–1044.

Russell, M. J. (1976). Human olfactory communication. *Nature* **260**, 520–522.

(1983). Human olfactory communications. In *Chemical Signals in Vertebrates*, vol. 3, ed. D. Müller-Schwarze and R. M. Silverstein, pp. 259–273. New York: Plenum.

Russell, M. J., Switz, G. M., and Thompson, K. (1980). Olfactory influences on the human menstrual cycle. *Pharmacology, Biochemistry and Behavior* **13**, 737–738.

Russell, M. J., Mendelson, T., and Peeke, H. V. S. (1983). Mothers' identification of their infant's odors. *Ethology and Sociobiology* **4**, 29–31.

Russell, M. J., Stone, G., and Russell, M. E. (1989). Human infants scent-mark their mothers. In *Proceedings of the Annual Meeting of the American Association for the Advancement of Science*, January 1989, San Francisco p. 209.

Ryan, M. J., Fox, J. H., Wilczynski, R., and Rand, A. S. (1990). Sexual selection for sensory exploitation in the frog *Physalaemus pustulous*. *Nature* **343**, 66–67.

Sachs, B. D. (1999). Airborne aphrodisiac odor from estrous rats: implications for pheromonal classification. In *Advances in Chemical Signals in Vertebrates*, vol. 8, ed. R. E. Johnston, D. Müller-Schwarze, and P. W. Sorensen, pp. 333–342. New York: Kluwer Academic-Plenum.

Saglio, Ph., Fauconneau, B., and Blanc, J. M. (1990). Orientation of carp, *Cyprinus carpio* L., to free amino acids from Tubifex extract in an olfactometer. *Journal of Fisheries Biology* **37**, 887–898.

Saito, T. R. (1986a). Induction of maternal behavior in sexually inexperienced male rats following removal of the vomeronasal organ. *Japanese Journal of Veterinary Science* **48**, 1029–1030.

(1986b). Role of the vomeronasal organ in retrieving behavior in lactating rats. *Zoological Science* **3**, 919–920.

Salamon, M. (1995). Seasonal, sexual and dietary induced variations in the sternal scent secretion in the brushtail possum (*Trichosurus vulpecula*). In *Chemical Signals in Vertebrates*, vol. 7, ed. R. Apfelbach, D. Müller-Schwarze, K. Reutter, and E. Weiler, pp. 211–222. Oxford: Elsevier/Pergamon.

Salmon, T. P. and Marsh, R. E. (1989). California ground-squirrel trapping influenced by anal-gland odors. *Journal of Mammalogy* **70**, 428–431.

Sambraus, H. H. and Waring, G. H. (1975). Der Einfluss des Harns brünstiger Kühe auf die Geschlechtslust von Stieren. *Zeitschrift für Säugetierkunde* **40**, 49–54.

Sananes, C. B., Gaddy, J. R., and Campbell, B. A. (1988). Ontogeny of conditioned heart rate to an olfactory stimulus. *Developmental Psychobiology* **21**, 117–133.

Sanchez-Criado, J. E. (1982). Involvement of the vomeronasal system in the reproductive physiology of the rat. In *Olfaction and Endocrine Regulation*, ed. W. Breipohl, pp. 209–217. London: IRL Press.

Sanders, E. H., Gardner, P. D., Berger, P. J., and Negus, N. C. (1981). 6-Methoxybenzoxazolinone: a plant derivative that stimulates reproduction in *Microtus montanus*. *Science* **214**, 67–69.

San Diego Zoo (2002). Project: Giant panda – Chemical communication in giant pandas. San Diego, CA; China. www.sandiegozoo.org/conservation/fieldproject. Accessed October 28, 2002.

Sawyer, T. G., Marchinton, R. L., and Berisford, C. W. (1982). Scraping behavior in female white-tailed deer. *Journal of Mammalogy*. **63**, 696–697.

Sawyer, T. G., Marchinton, R. L., and Miller, K. V. (1989). Response of female white-tailed deer to scrapes and antler rubs. *Journal of Mammalogy* **70**, 431–433.

Schaal, B. (1988a). Natal discontinuity and chemosensory continuity: animal models and hypotheses for humans. *Année Biologique* **27**, 1–41.

(1988b). Olfaction in infants and children: developmental and functional perspectives. *Chemical Senses* **13**, 145–190.

Schaal, B. and Orgeur, P. (1992). Olfaction *in utero*: can the rodent model be generalized? *Quarterly Journal of Experimental Psychology* **44B**, 245–278.

Schaal, B., Montagner, H., Hertling, E., *et al.* (1980). Les stimulations olfactives dans les relations entre l'enfant et la mère. *Reproduction, Nutrition and Development* **20**, 843–858.

Schaal, B., Orgeur, P., Lecanuet, J.-P., *et al.* (1991). Chimioréception nasale *in utero*: expériences préliminaires chez le foetus ovin. *Comptes Rendus de l' Academie des Sciences, Paris* **313** (Serie III), 319–325.

Schaal, B., Marlier, L., and Soussignan, R. (1998). Olfactory function in the human fetus: evidence from selective neonatal responsiveness to the odor of amniotic fluid. *Behavioral Neuroscience* **112**, 1438–1449.

Schaal, B., Marlier, L., and Soussignan, R. (2000). Human fetuses learn odours from their pregnant mother's diet. *Chemical Senses* **25**, 729–737.

Schaal, B., Coureaud, G., Marlier, L., and Soussignan, R. (2001). Fetal olfactory cognition preadapts neonatal behavior in mammals. In *Chemical Signals in Vertebrates*, Vol. 9, ed. A. Marchlewska-Koj, J. J. Lepri, and D. Müller-Schwarze, pp. 197–209. New York: Kluwer Academic/Plenum.

Schaal, B., Coureaud, G., Langlois, D., *et al.* (2003). Chemical and behavioural characterization of the rabbit mammary pheromone. *Nature* **242**, 68–72.

Schadler, M. H., Butterstein, G. M., Faulkner, *et al.* (1988). The plant metabolite 6-methoxybenzoxazolinone stimulates an increase in secretion of FSH and size of reproductive organs in *Microtus pinetorum*. *Biology of Reproduction* **38**, 817–820.

Schaffer, V. (1940). *Die Hautdrüsenorgane der Säugetiere*. Vienna: Urban and Schwarzenberg.

Schaller, G. B. (1972). *The Serengeti Lion*. Chicago, IL: the University of Chicago Press.

Schank, J. C. (2001). Oestrus and birth synchrony in Norway rats, *Rattus norvegicus*. *Animal Behaviour* **62**, 409–415.

Schecter, P. J. and Henkin, R. I. (1974). Abnormalities of taste and smell after head trauma. *Journal of Neurology, Neurosurgery and Psychiatry* **37**, 802–810.

Schellinck, H. M. and Brown, R. E. (2000). Selective depletion of bacteria alters but does not eliminate odors of individuality in *Rattus norvegicus*. *Physiology and Behavior* **70**, 261–270.

Schilling, A., Serviere, J., Gendrot, G., and Perret, M. (1989). Functional activation of vomeronasal system by urine in a primate (*Microcebus murinus*): a 2-DG study. *Chemical Senses* **14**, 224.

Schisler, G. J. and Bergesen, E. P. (1996). Postrelease hooking mortality of rainbow trout caught on scented artificial baits. *North American Journal of Fish Management* **16**, 570–578.

Schleidt, M. (1980). Personal odor and nonverbal communication. *Ethology and Sociobiology* **1**, 225–231.

Schleidt, M., Hold, B., and Attili, G. (1981). A cross-cultural study on the attitude towards personal odors. *Journal of Chemical Ecology* **7**, 19–31.

Schmid, B. (1935). Über die Ermittlung des menschlichen und tierischen Individualgeruches durch den Hund. *Zeitschrift für vergleichende Physiologie* **22**, 524–538.

Schmidt, H. J. and Beauchamp, G. K. (1988). Adult-like odor preferences and aversions in three-year-old children. *Childhood Development* **59**, 1136–1143.

Schmidt, U. (1975). Vergleichende Riechschwellenbestimmungen bei neotropischen Chiropteren (*Desmodus rotundus, Artibeus literatus, Phyllostomus discolor*). *Zeitschrift fürSäugetierkunde* **40**, 269–296.

Schmidt, U. and Greenhall, A. M. (1971). Untersuchungen zur geruchlichen Orientierung der Vampirfledermäuse (*Desmodus rotundus*). *Zeitschrift für vergleichende Physiologie* **74**, 217–226.

Schmidt, V. U., Schmidt, C., and Wysocki, C. J. (1986). Der Einfluss des Vomeronasalorgans auf das olfaktorisch geleitete Verhalten nestjunger Mäuse. *Zeitschrift fürSäugetierkunde* **51**, 86–90.

Schmidt-Koenig, K. (1987). Bird navigation: has olfactory orientation solved the problem? *Quarterly Review of Biology* **62**, 31–48.

Scholl, J. P., Kelsey, R. G., and Shafizadeh, F. (1977). Involvement of volatile compounds of *Artemisia* in browse preference by mule deer. *Biochemical Systematics and Ecology* **5**, 291–295.

Scholz, A. T., Horrall, R. M., Cooper, J. C., and Hasler, A. D. (1976). Imprinting to chemical cues: the basis for home stream selection in salmon. *Science* **192**, 1247–1249.

Schulte, B. A., Müller-Schwarze, D., and Sun, L. (1995). Using anal gland secretion to determine sex in beaver. *Journal of Wildlife Management* **59**, 614–618.

Schulte, B. A., Bagley, K., Correll, M., *et al.* (2005). Assessing chemical communication in elephants. In *Chemical Signals in Vertebrates*, vol. 10, eds. R. T. Mason, M. P. LeMasters, and D. Müller-Schwarze, pp. 140–151. New York: Springer.

Schultz, T. H., Flath, R. A., Stern, D. J., *et al.* (1988). Coyote estrous urine volatiles. *Journal of Chemical Ecology* **14**, 701–712.

Schultze-Westrum, T. G. (1965). Innerartliche Verständigung durch Düfte beim Gleitbeutler (*Petaurus breviceps papuanus*) Thomas (*Marsupialia Phalangeridae*). *Zeitschrift für vergleichende Physiologie* **50**, 151–220.

Schwanzel-Fukuda, M. and Pfaff, D. (1989). Origin of luteinizing-hormone-releasing hormone neurons. *Nature* **338**, 161–164.

Schwartz, C. C., Regelin, W. L., and Nagy, J. G. (1980). Deer preference for juniper forage and volatile oil treated foods. *Journal of Wildlife Management* **44**, 114–120.

Schwass, D. E. and Finley, J. W. (1985). Overview: the influence of nutrition on xenobiotic metabolism. *American Chemical Society Symposium Series* **277**, 1–10.

Schwende, F. J., Wiesler, D., Jorgenson, J. W., Carmack, M., and Novotny, M. (1986). Urinary volatile constituents of the house mouse, *Mus musculus*, and their endocrine dependency. *Journal of Chemical Ecology*, **12**, 277–296.

Schwenk, K. (1994). Why snakes have forked tongues. *Science* **263**, 1573–1577.

Scott, A. P., Liley, N. R., and Vermeirssen, E. L. M. (1994). Urine of reproductively female rainbow trout, *Oncorhynchus mykiss* (Walbuam), contains a priming pheromone which enhances plasma levels of sex steroids and gonadotropin II in malts. *Journal of Fish Biology* **44**, 131–148.

Scott, G. R., Sloman, K. A., Rouleau, C., and Wood, C. M. (2003). Cadmium disrupts behavioural and physiological responses to alarm substance in juvenile rainbow trout (*Oncorhynchus mykiss*). *Journal of Experimental Biology* **206**, 1779–1790.

Scott, N. L. and Rasmussen, L. E. L. (2005). Chemical communication of musth in captive male Asian elephants, *Elephas maximus*. In *Chemical Signals in Vertebrates*, vol. 10, eds. R. T. Mason, M. P. LeMaster, and D. Müller-Schwarze, pp. 118–127. New York: Springer.

Scott, T. P. and Weldon, P. J. (1990). Chemoreception in the feeding behavior of adult American alligators, *Alligator mississippiensis*. *Animal Behaviour* **39**, 398–399.

Scrivner, J. H., Howard, W. E., and Teranishi, R. (1984). Aldehyde volatiles for use as coyote attractants, In *Proceedings of the 11th Vertebrate Pest Conference*, ed. D. D. Clarke, pp. 157–160. Davis: University of California.

Scudder, K. M., Stewart, N. J., and Smith, H. M. (1980). Response of neonate water snakes (*Nerodia sipedon sipedon*) to conspecific chemical cues. *Journal of Herpetology* **14**, 196–198.

Scudder, K. M., Chiszar, D., and Smith, H. M. (1992). Strike-induced chemosensory searching and trailing behavior in neonatal rattlesnakes. *Animal Behaviour* **44**, 574–576.

Segi, M. (1975). Tea-gruel as a possible factor for cancer for the esophagus. *Gann (Japanese Journal of Cancer Research)* **66**, 199–202.

Seligman, P. J., Mathias, C. G. T., Omalley, M. A., *et al.* (1987). Phytophotodermatitis from celery among grocery store workers. *Archive of Dermatology* **123**, 1478–1482.

Sengupta, S. (1981). Adaptive significance of the use of margosa leaves in nests of the house sparrow (*Passer domesticus*). *Emu* **81**, 114.

Serizawa, S., Miyamichi, K., Nakatani, H., *et al.* (2003). Negative feedback regulation ensures the one receptor–one olfactory neuron rule in mouse. *Science* **302**, 2088–2094.

Sever, D. M. (1988). Male *Rhyacotriton olympicus*, Dicamptodontidae: Urodela, has a unique cloacal vent gland. *Herpetologica* **44**, 274–280.

Shallenberger, R. J. (1975). Olfactory use in the wedge-tailed shearwater (*Puffinus pacificus*) on Manana Island, Hawaii. In *Olfaction and Taste*, vol. 5, ed. D. A. Denton and J. P. Coghlan, pp. 355–359. New York: Academic Press.

Shappira, Z., Terkel, J., Egozi, J., Nyska, A., and Friedman, J. (1990). Reduction of rodent fertility by plant consumption. With particular reference to *Ziziphus spina-christi*. *Journal of Chemical Ecology* **16**, 2019–2026.

Sheaffer, S. E. and Drobney, R. D. (1986). Effectiveness of lithium chloride induced taste aversions in reducing waterfowl nest predation. *Transactions of the Missouri Academy of Science* **20**, 59–64.

Sheffield, L. P., Law, J. M., and Burghardt, G. M. (1968). On the nature of chemical food sign stimuli for newborn garter snakes. *Communications in Behavioral Biology* 7–12.

Shine, R., Langkilde, T. and Mason, R. T. (2003). Confusion within 'mating balls' of garter snakes: does misdirected courtship impose selection on male tactics? *Animal Behaviour* **66**, 1011–1017.

Shivik, J. A. (1998). Brown tree snake response to visual and olfactory cues. *Journal of Wildlife Management* **62**, 105–111.

Shivik, J. A., and Clark, L. (1997). Carrion seeking in brown tree snakes: importance of olfactory and visual cues. *Journal of Experimental Zoology* **279**, 549–553.

Shkolnik, A. (1971). Diurnal activity in a small desert rodent. *International Journal of Biometeorology*, **15**, 115–120.

Shutt, D. A. (1976). The effects of plant estrogens on animal reproduction. *Endeavour* **35**, 110–113.

Siegel, M. A., Richardson, R., and Campbell, B. A. (1988). Effects of home nest stimuli on the emotional response of preweanling rats to an unfamiliar environment. *Psychobiology* **16**, 236–242.

Siegel, S. (1979). The role of conditioning in drug tolerance and addiction. In *Psychopathology in Animals: Research and Clinical Implications*, ed. J. D. Keehn, pp. 143–168. New York: Academic Press.

Signoret, J. A. (1975). Influence of the sexual receptivity of a teaser ewe on the mating preference in the ram. *Applied Animal Ethology* **1**, 229–232.

Signoret, J. P., Cohen-Tannoudji, J., and Gonzalez, R. (1989). Olfactory aspects of sexual partner's effect on endocrine secretions in the domestic sheep. *Journal of Endocrinology* **123**(Suppl.), 15.

Simon, G. S. and Madison, D. M. (1984). Individual recognition in salamanders: cloacal odors. *Animal Behaviour* **32**, 1017–1020.

Sinclair, A. R. E., Jogia, M. K., and Andersen, R. J. (1988). Camphor from juvenile white spruce as an antifeedant for snowshoe hares. *Journal of Chemical Ecology* **14**, 1505.

Singer, A. G. and Macrides, F. (1990). Aphrodisin: pheromone or transducer? *Chemical Senses* **15**, 199–203.

Singer, A. G., Agosta, W. C., O'Connell, R. J., *et al.*, (1976). Dimethyl disulfide: an attractant pheromone in hamster vaginal secretion. *Science* **191**, 948–950.

Singer, A. G., Clancy, A. N., and Macrides, F. (1989). Conspecific and heterospecific proteins related to aphrodisin lack aphrodisiac activity in male hamsters. *Chemical Senses* **14**, 565–575.

Singer, A. G., Tsuchiya, H., Wellington, J. L., Beauchamp, G. K., and Yamazaki, K. (1993). Chemistry of odortypes in mice: fractionation and bioassay. *Journal of Chemical Ecology* **19**, 569–579.

Singh, P. B., Brown, R. E., and Roser, B. (1987). MHC antigens in urine as olfactory cues. *Nature* **327**, 161–163.

Sipos, M. L., Wysocki, C. J., Nygy, J. G., Wysocki, L., and Nemura, T. A. (1995). An ephemeral pheromone of female house mice: perception via the main and accessory systems. *Physiology and Behavior* **58**, 529–534.

Skeen, J. T. and Thiessen, D. D. (1977). Scent of gerbil cuisine. *Physiology and Behavior* **19**, 11–14.

Slotnick, B. H. and Schoonover, F. W. (1984). Redundancy: One-side bulbectomized rats have same absolute threshold and intensity difference threshold as intact rats. *Chemical Senses* **9**, 325–340.

Smale, L., Pedersen, J. M., Block, M. L., and Zucker, I. (1990). Investigation of conspecific male odours by female prairie voles. *Animal Behaviour* **39**, 768–774.

Smallwood, P. D. and Peters. W. D. (1986). Grey squirrel wood preferences: The effects of tannin and fat concentrations. *Ecology* **67**, 168–174.

Smith, A. B. III, Belcher, A. M., Epple, G., Jurs, P. C., and Lavine, B. (1985). Computerized pattern recognition: a new technique for the analysis of chemical communication. *Science* **228**, 175–177.

Smith, B. A. and Block, L. (1990). Preference of Mongolian gerbils for salivary cues: a developmental analysis. *Animal Behaviour* **39**, 512–521.

Smith, D. V. and Margolskee, R. F. (2001). Making sense of taste. *Scientific American* **284**, 32–39.

Smith, G. J. and Spear, N. E. (1978). Effects of the home environment on withholding behaviors and conditioning in infant and neonatal rats. *Science* **202**, 327–329.

(1981). Home environmental stimuli facilitate learning shock escape spatial discrimination in rats 7–11 days of age. *Behavioral and Neural Biology* **31**, 360–365.

Smith, J. L. D., McDougall, C., and Michelle, D. (1989). Scent marking in free-ranging tigers, *Panthera tigris*. *Animal Behaviour* **37**, 1–10.

Smith, P. W., Parks, O. W., and Schwartz, D. P. (1984). Characterization of male goat odours: 6-*trans*-nonenal. *Journal of Dairy Science* **67**, 794–801.

Smith, R. J. F. and Lawrence, B. J. (1989). Behavioral response of solitary fathead minnows, *Pimephalis promelas*, to alarm substance. *Journal of Chemical Ecology* **15**, 209–219.

Smith, S. A. and Paselk, R. A. (1986). Olfactory sensitivity of the turkey vulture (*Cathartes aura*) to three carrion-associated odors *Auk* **103**, 586–592.

Smith, T. E., Faulkes, C. G., and Abbott, D. H. (1997). Combined olfactory contact with the parent colony and direct contact with non breeding animals does not maintain suppression of ovulation in female naked mole-rats (*Heterocephalus glaber*). *Hormones and Behavior* **31**, 277–288.

Smith, T. L. and Kardong, K. V. (2000). Absence of polarity perception by rattlesnakes of envenomated prey trails. *Journal of Herpetology* **34**, 621–624.

Smith, W. P., Borden, D. L., and Endres, K. M. (1994). Scent-station visits as an index to abundance of raccoons: an experimental manipulation. *Journal of Mammalogy* **75**, 637–647.

Smotherman, W. P. (1982). Odor aversion learning by the rat fetus. *Physiology and Behavior* **29**, 769–771.

(1987). Response of the rat fetus to olfactory stimulation presented *in utero*. *Society of Neuroscience* **13**, 8l.

Smotherman, W.P. and Robinson, S. R. (1985). The rat fetus in its environment: behavioral adjustments to novel familiar, aversive and conditional stimuli presented in utero. *Behavioral Neuroscience* **99**, 521–530.

(1987). Prenatal expression of species-typical action patterns in the rat fetus (*Rattus norvegicus*). *Journal of Comparative Psychology* **101**, 190–196.

(1988). Behavior of rat fetuses following chemical or tactile stimulation. *Behavioral Neuroscience* **102**, 24–34.

Smotherman, W. P., Robinson, S. R., La Vallee, P. A. and Hennessy, M. B. (1987). Influences of the early olfactory environment on the survival, behavior and pituitary–adrenal activity of cesarean delivered preterm rat pups. *Developmental Psychobiology* **20**, 415–424.

Snyder, G. K. and Peterson, D. T. (1979). Olfactory sensitivity in the black-billed magpie and in the pigeon. *Comparative Biochemistry and Physiology A* **62**, 921–925.

Snyder, M. A. (1992). Selective herbivory by Abert's squirrel mediated by chemical variability in ponderosa pine. *Ecology* **73**, 1730–1741.

(1993). Interactions between Abert's squirrel and ponderosa pine: the relationship between selective herbivory and host plant fitness. *American Naturalist* **141**, 866–879.

(1998). Abert's squirrel (*Sciurus aberti*) in ponderosa pine (*Pinus ponderosa*) forests: directional selection, diversifying selection. In *Special Publication* 6: *Ecology and Evolutionary Biology of Tree Squirrels*, ed. M. A. Steele, J. F. Merritt, and D. A. Zegers, pp. 195–201. Martinsville, VA: Virginia Museum of Natural History.

Soffie, M. and Lamberty, Y. (1988). Scopolamine effects on juvenile conspecific recognition in rats: possible interaction with olfactory sensitivity. *Behavioral Processes* **17**, 181–190.

Sola, C. (1995). Chemoattraction of upstream migrating glass eels, *Anguilla anguilla* to organic earthy and green odorants. *Environmental Biology of Fishes* **43**, 179–185.

Sola, C. and Tosi L. (1993). Bile salts and taurine as chemical stimuli for glass eels *Anguilla anguilla*: a behavioral study. *Environmental Biology of Fishes* **37**, 197–204.

Soni, G. R. and Prakash, I. (1987). Effect of conspecific urine on behavior of soft-furred field-rat, *Rattus meltada*. *Behavioral Processes* **14**, 175–182.

Sorensen, A. E. (1983). Taste aversion and frugivore preference. *Oeologia* **56**, 117–120.

Sorensen, P. W., and Stacey, N. E. (1990). Identified hormonal pheromones in the goldfish: the basis for a model of sex pheromones function in teleost fish. In *Chemical Signals in Vertebrates*, vol. 6, ed. R. L. Doty and D. Müller-Schwarze, pp. 302–311. New York: Plenum.

(1999). Evolution and specialization of fish hormonal pheromones. In *Chemical Signals in Vertebrates*, vol. 8, ed. R. E. Johnston, D. Müller-Schwarze and P. W. Sorensen, pp. 15–47. New York: Kluwer Academic/Plenum.

Sorensen, P. W., Hara, T. J., Stacey, N. E., and Goetz, F. W. (1988). F prostaglandins function as potent olfactory stimulants that comprise the postovulatory female sex pheromone in goldfish. *Biology of Reproduction* **39**, 1039–1050.

Sorensen, P. W., Hara, T. J., Stacey, N. E., and Dulka, J. G., (1990a). Extreme olfactory specificity of male goldfish to the preovulatory steroidal pheromone $17\alpha, 20\beta$-dihydroxy-4-pregnen-3-one. *Journal of Comparative Physiology A* **166**, 373.

Sorensen, P. W., Stacey, N. E., and Hara, T. J. (1990b). Acute olfactory sensitivity and specificity of mature male goldfish to water borne androgenic steroids: a class of inhibitory pheromones? *Chemical Senses* **15**, 644.

Sorensen, P. W., Irvine, I. A. S., Scott, A. P., and Stacey, N. E. (1992). Electrophysiological measures of olfactory sensitivity suggest that goldfish and other fish use species-specific mixtures of hormones and their metabolites as pheromones. In *Chemical Signals in Vertebrates*, vol. 6, ed. R. L. Doty and D. Müller-Schwarze, pp. 357–354. New York: Plenum.

Sosa, T., Chaves, N., Alias, J. C., Escudero, J. C., Henao, F., and Gutiérrez-Merino, C. (2004). Inhibition of mouth skeletal muscle relaxation by flavonoids of *Cistus ladanifer*

L.: a plant defense mechanism against herbivores. *Journal of Chemical Ecology* **30**, 1087–1101.

Spencer, P., Nunn, P. B., Hugon, J., *et al.* (1987). Guam amyotrophic lateral sclerosis–parkinsonism–dementia linked to a plant excitant neurotoxin. *Science* **237**, 517–522.

Spickett, A. M., Kierans, J. E., Norval, R. A. I., and Clifford, C. M. (1981). *Ixodes matopi*, new species (Acarina: Ixodes): a tick found aggregating on preorbital gland scent marks of the klipspringer in Zimbabwe. *Onderstepoort Journal of Veterinary Research* **48**, 23–30.

Stabell, O. B. (1987). Intraspecific pheromone discrimination and substrate marking by Atlantic salmon parr. *Journal of Chemical Ecology* **13**, 1644.

Stacey, N. E. and Hourston, A. S. (1982). Spawning and feeding behavior of captive Pacific herring, (*Clupea harengus pallasi*). *Canadian Journal of Fisheries* **39**, 489–498.

Stager, K. E. (1964). The role of olfaction in food location by the turkey vulture (*Cathartes aura*). *Los Angeles County Museum Contributions to Science* No. 81.

Stager, K. E. (1967). Avian olfaction. *American Zoologist* **7**, 415–419.

Stahl, E. (1888). Pflanzen und Schnecken. Biologische Studien über die Schutzmittel der Pflanzen gegen Schneckenfrass. *Jenaische Zeitschrift für Medizin und Naturwissenschaft* **22**, 557–684.

Stahlbaum, L. C. and Houpt, K. A. (1989). The role of the flehmen response in the behavioral repertoire of the stallion. *Physiology and Behavior* **45**, 1207–1214.

Stattelman, A. J., Talbot, R. B., and Coulter, D. B. (1975). Olfactory thresholds of pigeons (*Columba livia*), quail (*Colinus virginianus*) and chickens (*Gallus domesticus*). *Comparative Biochemistry and Physiology A* **50**, 807–809.

Staubli, U., Fraser, D., Faraday, R. and Lynch, G. (1987). Olfaction and the "data" memory system in rats. *Behavioral Neuroscience* **101**, 757–765.

Stauffer, H. P. and Semlitch, R. D. (1993). Effects of visual, chemical and tactile cues of fish on the behavioral responses of tadpoles. *Animal Behaviour* **46**, 355–364.

Steel, E. and Hutchinson, J. B. (1987). The aromatase inhibitor, 1,4,6-androstatriene-3,7-dione (ATD) blocks testosterone-induced olfactory behaviour in the hamster. *Physiology and Behavior* **39**, 141–145.

Stehn, R. A. and Richmond, M. E. (1975). Male-induced pregnancy termination in the prairie vole, *Microtus ochrogaster*. *Science* **187**, 1211–1213.

Stein, M., Ottenberg, P., and Roulet, N. (1958). A study of the development of olfactory preferences. *American Medical Association Archives of Neurology and Psychiatry* **80**, 264–266.

Stern, K. and McClintock, M. K. (1998). Regulation of ovulation by human pheromones. *Nature* **392**, 177–179.

Stevens, J. C., and Cain, W. S. (1985). Age-related deficiency in the perceived strength of six odorants. *Chemical Senses* **10**, 517–529.

Stevens, J. C., Cain, W. S., and Burke, R. J. (1988). Variability of olfactory thresholds. *Chemical Senses* **13**, 643–654.

Stewart, P. S., MacDonald, D. W., Newman, C., and Tattersall, F. H. (2002). Behavioral mechanisms of information transmision and reception by badgers, *Meles meles*, at latrines. *Animal Behaviour* **63**, 999–1007.

Stickrod, G., Kimble, D. P., and Smotherman, W. P. (1982). In utero taste/odor aversion conditioning in the rat. *Physiology and Behavior* **28**, 5–7.

Stoddart, D. M. (1976). Effects of weasels (*Mustela nivalis*) on trapped samples of their prey. *Oecologia* **22**, 439–441.

(1979). Specialized scent-releasing hair in the crested rat *Lophiomys imhausi*. *Journal of Zoology* **189**, 551–553.

(1982). Does trap odour influence estimation of population size of the short-tailed vole, *Microtus agrestis*? *Journal of Animal Ecology* **51**, 375–386.

(1983). *The Ecology of Vertebrate Olfaction*. London: Chapman & Hall.

Stoddart, D. M. and Bradley, A. G. (1991). Measurement of short-term changes in heart rate and plasma concentration of cortisol and catecholamine in a small marsupial. *Journal of Chemical Ecology* **17**, 1333–1341.

Stone, R. (1993). Guam: deadly disease dying out. *Science* **261**, 424–426.

(2002). Toxicology. Fruit bats linked to mystery disease. *Science* **296**, 24.

Stonerook, M. J. and Harder, J. D. (1989). Male pheromones advance the age of first estrus in the gray, short-tailed opossum, *Monodelphis domestica*. *Biology of Reproduction* **40**(Suppl.1), 157.

Stralendorff, F. V. (1987). Partial chemical characterization of urinary signaling pheromone in tree shrews (*Tupaia belangeri*). *Journal of Chemical Ecology* **13**, 655–679.

Strier, K. B. and Ziegler, T. E. (1994). Insights into ovarian function in wild *Muriqui* monkeys (*Muriqui arachnoides*). *American Journal of Primatology* **32**, 31–40.

Strom, G. H. (1976). Transport and diffusion of stack effluents. In *Air Pollution*, vol. 1, ed. A. C. Sku, pp. 401–503. New York: Academic Press.

Strupp, B. J. and Levitsky, D. A. (1984). Social transmission of food preferences in adult hooded rat (*Rattus norvegicus*). *Journal of Comparative Psychology* **98**, 257–266.

Sugiyama, T. (1983). Sex odour components in male goat. Estrous goats are interested in 4-ethyl fatty acids secreted by mature male goat. *Kagaku to Seibutsu* (*Chemistry and Biology, Japanese*) **21**, 420–430. [*Chemical Abstracts* 100, 18366s.]

Sugiyama, T., Sasada, H., Masaki, F., and Yamashita, K. (1981). Unusual fatty acids with specific odor from mature male goats. *Agriculture and Biological Chemistry* **45**, 2655–2658.

Sugiyama, T., Matsumura, H., Sasada, H., Masaki, J., and Yamashita, K. (1986). Characterization of fatty acids in the sebum of goats according to sex and age. *Agricultural and Biological Chemistry* **50**, 3049–3052.

Sullivan, R. M. and Leon, M. (1986). Early olfactory learning induces an enhanced olfactory-bulb response in young rats. *Developmental Brain Research* **27**, 278–282.

Sullivan, R. M., Hofer, M. A., and Brake, S. C. (1986). Olfactory-guided orientation in neonatal rats is enhanced by a conditioned change in behavioral state. *Developmental Psychobiology* **19**, 615–624.

Sullivan, T. P. (1986). Influence of wolverine (*Gulo gulo*) odor on feeding behavior of snowshoe hares (*Lepus americanus*). *Journal of Mammalogy* **67**, 385–388.

Sullivan, T. P. and Crump, D. R. (1984). Influence of mustelid scent gland compounds on suppression of feeding by snowshoe hares (*Lepus americanus*). *Journal of Chemical Ecology* **10**, 1809–1821.

(1986a). Avoidance responses of pocket gophers (*Thomomys talpoides*) to mustelid anal gland compounds, In *Chemical Signals in Vertebrates*, vol. 4 , ed. D. Duvall, D. Müller-Schwarze, and R. M. Silverstein. pp. 519–531. New York: Plenum.

(1986b). Feeding responses of snowshoe hares (*Lepus americanus*) to volatile constituents of red fox (*Vulpes vulpes*) urine. *Journal of Chemical Ecology*. **12**, 729–739.

Sullivan, T. P., Nordstrom, L. O., and Sullivan, D. S. (1985a). The use of predator odors as repellents to reduce feeding damage by herbivores. I. Snowshoe hares (*Lepus americanus*) *Journal of Chemical Ecology* **11**, 903–919.

(1985b). The use of predator odors as repellents to reduce feeding damage by herbivores. II. Black-tailed deer (*Odocoileus hemionus columbianus*). *Journal of Chemical Ecology* **11**, 921–935.

Sullivan, T. P., Crump, D. R., and Sullivan, D. S. (1988). Use of predator odors as repellents to reduce feeding damage by herbivores. III. Montane and meadow voles (*Microtus montanus* and *Microtus pennsylvanicus*). *Journal of Chemical Ecology* **14**, 363–377.

Sullivan, T. P., Crump, D. R., Wieser, H. and Dixon, E. A. (1990). Response of pocket gophers (*Thomomys talpoides*) to an operational application of synthetic semiochemicals of stoat (*Mustela erminea*). *Journal of Chemical Ecology* **16**, 941–949.

(1992). Influence of the plant antifeedant, pinosylvin, on suppression of feeding by snowshoe hares. *Journal of Chemical Ecology* **18**, 1573–1561.

Sun, L. and Müller-Schwarze, D. (1997). Sibling recognition in the beaver: a field test for phenotype matching. *Animal Behaviour* **54**, 492–502.

Sun, L., Xiao, B., and Dai, N. (1994). Scent marking behaviour in the male Chinese water deer. *Acta Theriologica* **39**, 177–184.

Sunnerheim, K., Palo, R. T., Theander, O., and Knutsson, P. G. (1988). Chemical defense in birch. Platyphylloside: a phenol from *Betula pendula* inhibiting digestibility. *Journal of Chemical Ecology* **14**, 549.

Sunnerheim-Sjöberg, K. (1992). (1*S*, 2*R*, 4*S*, 5*S*)-Angelicoidenol-2-O-β-D-glucopyranoside: a moose deterrent compound in Scots pine (*Pinus sylvestris* L.). *Journal of Chemical Ecology* **18**, 2025–2039.

Sunnerheim-Sjöberg, K. and Hämäläinen, M. (1992). Multivariate study of moose browsing in relation to phenol patterns in pine needles. *Journal of Chemical Ecology* **18**, 659–672.

Sutter, E. (1946). Das Abwehrverhalten nestjunger Wiedehopfe. *Der Ornithologische Beobachter* **43**, 72–81.

Sutton, O. G. (1953). *Micrometeorology*. New York: McGraw Hill.

Sveinsson, T. and Hara, T. J. (1990). Olfactory receptors in Arctic char (*Salvelinus alpinus*) with high sensitivity and specificity for prostaglandin $F_{2\alpha}$. *Chemical Senses* **15**, 645–646.

Svendsen., G. E. (1980). Patterns of scent-mounding in a population of beaver (*Castor canadensis*). *Journal of Chemical Ecology* **6**, 133–148.

Svendsen, G. E. and Jollick, J. D. (1978). Bacterial contents of the anal and castor glands of beaver (*Castor canadensis*). *Journal of Chemical Ecology* **4**, 563–569.

Svoboda, F. J. and Gullion, G. W. (1972). Preferential use of aspen by ruffed grouse in northern Minnesota. *Journal of Wildlife Management* **36**, 1166–1180.

Swain, T. (1977). Secondary compounds as protective agents. *Annual Review of Plant Physiology* **28**, 479–501.

(1979). Phenolics in the environment. *Recent Advances in Phytochemistry* **12**, 617–640.

Swank, W. (1944). Germination of seeds after ingestion by ring-necked pheasants. *Journal of Wildlife Management* **8**, 223–231.

Swihart, R. K., Pignatello, J. J., and Mattina, M. J. I. (1991). Aversive responses of white-tailed deer, *Odocoileus virginianus*, to predator urines. *Journal of Chemical Ecology* **17**, 767–777.

Tachibana, K., Sakaitanai, M., and Nakanishi, K. (1984). Pavoninins: shark-repelling ichthyotoxins from the defense secretion of the Pacific sole. *Science* **226**, 703–705.

Tahvanainen, J., Helle, E., Julkunen-Tiitto, R., and Lavola, A. (1985). Phenolic compounds of willow bark as deterrents against feeding by mountain hares. *Oecologia* **65**, 319–323.

Tang-Martinez, Z. (2001). The mechanisms of kin discrimination and the evolution of kin recognition in vertebrates: a critical re-evaluation. *Behavioral Processes* **13**, 21–40.

Tang-Martinez, Z., Mueller, L. L., and Taylor, G. T. (1993). Individual odours and mating success in the golden hamster, *Mesocricetus auratus*. *Animal Behaviour* **45**, 1141–1151.

Tattersall, D. B., Bak, S., Jones, P. R., *et al.* (2001). Resistance to an herbivore through engineered cyanogenic glucoside synthesis. *Science* **293**, 1826–1828.

Tauber, G. (1992). A dubious battle to save the Kemp's Ridley sea turtle. *Science* **256**, 614–616.

Teeter, J. (1980). Pheromone communication in sea lampreys (*Petromyzon marinus*): implications for population management. *Canadian Journal of Fisheries and Aquatic Science* **37**, 2123–2132.

Teicher, E. M. and Blass, M. H. (1980). Suckling. *Science* **210**, 15–22.

Teichmann, H. (1957). Das Riechvermögen des Aales (*Anguilla anguilla* L.). *Naturwissenschaften* **44**, 242–246.

(1959). Concerning the power of the olfactory sense of the eel, *Anguilla anguilla* (L.). *Zeitschrift für vergleichende Physiologie* **42**, 206–254.

Temeles, E. J. (1994). The role of neighbours in territorial systems: when are they 'dear enemies'? *Animal Behaviour* **47**, 339–350.

Temple, S. A. (1977). Plant–animal mutualism: coevolution with dodo leads to near extinction of plant (*Calvaria major*). *Science* **197**, 885–886.

Terborgh, J. (1988). The big things that run the world: a sequel to E. O. Wilson. *Conservation Biology* **2**, 402–403.

Terborgh J., Lopez, L., Nuñez P. *et al.* (2001). Ecological meltdown in predator-free forest fragments. *Science* **294**, 1923–1926.

Terrick, T. D., Mumme, R. L. and Burghardt, G. M. (1995). Aposematic coloration enhances chemosensory recognition of noxious prey in the garter snake *Thamnophis radix*. *Animal Behaviour* **49**, 857–866.

Terry, L. M. and Johanson, I. B. (1987). Olfactory influences on the ingestive behavior of infant rats. *Developmental Psychobiology* **20**, 313–332.

Tester, A. L. (1963). Olfaction, gustation and the common chemical sense in sharks. *Sharks and Survival* **8**, 255–282.

Tew, T. (1987). A comparison of small animal responses to clean and dirty traps. *Journal of Zoology (London)* **212**, 361–364.

Tew, T. E., Todd, I. A., and Macdonald, D. W. (1994). Temporal changes in olfactory preferences in murid rodents revealed by live-trapping. *Journal of Mammalogy* **75**, 750–756.

Thesen, A., Steen, J. B., and Døving, K. B. (1993). Behaviour of dogs during olfactory tracking. *Journal of Experimental Biology* **180**, 247–251.

Thibodeaux, L. J. (1979). *Chemodynamics, Environmental Movement of Chemicals in Air, Water, and Soil*. New York: Wiley.

Thieme, H. (1965). Die Phenylglykoside der Salicaceen: Untersuchungen über jahreszeitlich bedingte Veränderungen der Glykosidkonzentrationen, über die Abhängigheit des Glykosidgehalts von der Tageszeit und dem Alter den Pflanzenorgane. *Pharmazie* **20**, 688–691.

Thiessen D. D. and Cocke, R. (1986). Alarm chemosignals in a *Meriones unguiculatus*: prey–predator interactions. In *Chemical Signals in Vertebrates*, vol. 4, ed. D. Duvall, D. Müller-Schwarze and Silverstein, R. M. pp. 507–518. New York: Plenum.

Thiessen, D. D., Clancy, A., and Goodwin, M. (1976). Harderian gland pheromone in the Mongolian gerbil *Meriones unguiculatus*. *Journal of Chemical Ecology* **2**, 231–238.

Thoen, C., Bauwen, D., and Verheyen, R. F. (1986). Chemoreceptive and behavioral responses of the common lizard, *Lacerta vivipara*, to snake chemical deposits. *Animal Behaviour* **34**, 1805–1813.

Thomas, D. W., Samson, C., and Bergeron, J. M. (1988). Metabolic costs associated with the ingestion of plant phenolics by *Microtus pennsylvaticus*. *Journal of Mammalogy* **69**, 512–515.

Thomas, K. J. and Dominic, C. J. (1987a). Evaluation of the role of the stud male in preventing male-induced implantation failure (the Bruce effect) in laboratory mice. *Animal Behaviour* **35**, 1257–1260.

(1987b). Male-induced implantation failure (the Bruce effect) in laboratory mice: investigations on the olfactory memory of the newly inseminated female. *Archives of Biology* **98**, 263–272.

Thomas, R. McG. (1996). Ernst Morch, anesthesiologist and heroic inventor, was 87 (obituary). *New York Times* January 18, p. B11.

Tiedman, G. T., Oh, J. H., Oita, K., and Christoffers, G. W. (1976). Wildlife damage control II: partial identification of the active ingredients in big game repellents derived from fish and eggs. In *Proceedings of the 172nd National Meeting of American Chemical Society* San Francisco.

Tipton, K. W., Floyd, E. H., Marshall, J. C., and McDevitt, J. B. (1970). Resistance of certain grain sorghum hybrids to bird damage in Louisiana. *Agronomy Journal* **62**, 211–213.

Todd, N. (1962). The inheritance of the catnip response in the domestic cat. *Journal of Heredity* **53**, 54–56.

Toftegaard, C. L., McMahon, K. L., Galloway, G. J., and Bradley, A. J. (2002). Processing of urinary pheromones in *Antechinus stuartii* (Marsupialia: Dasyuridae): Functional magnetic resonance imaging of the brain. *Journal of Mammalogy* **83**, 71–80.

Tolhurst, B. E. and Vince, M. A. (1976). Sensitivity to odours in the embryo of the domestic fowl. *Animal Behaviour* **24**, 772–779.

Tomasi, T. E. (1978). Function of venom in the short-tailed shrew, *Blarina brevicauda*. *Journal of Mammalogy* **60**, 751–759.

Tomback, D. F. (1980). How nutcrackers find their seed stores. *Condor* **82**, 10–19.

Tosi, L. and Sola, C. (1993). Role of geosmin, a typical inland water odour, in guiding glass, eel *Anguilla anguilla* (L.) migration. *Ethology* **95**, 177–185.

Tracy, C. R. and Dole, J. W. (1969). Orientation of displaced California toads, *Bufo boreas*, to their breeding sites. *Copeia* **1969**, 693–700.

Traveset, A. (1998). Effects of seed passage through vertebrate frugivores' guts on germination: a review. *Perspectives in Plant Ecology, Evolution and Systematics* **1**, 151–190.

Tributsch, H. (1988). *When the Snakes Awake: Animals and Earthquake Prediction*. Cambridge, MA: MIT Press.

Tristram, D. A. (1977). Intraspecific olfactory communication in the terrestrial salamander (*Plethodon cinereus*). *Copeia* 597–600.

Trowbridge, B. J. (1983). Olfactory communication of the European otter *Lutra lutra* L. Ph.D. Thesis, University of Aberdeen, UK.

Trumler, E. (1958). Beobachtungen an den Böhmzebras des Georg-von-Opel-Freigeheges für Tierforschung e.V., Kronberg im Taunus. J. Das Paarungsverhalten. *Säugetierkundliche Mitteilungen* **6**, 1–18.

Tsuchiya, H., Yamazaki, K., Beauchamp, G. K., and Singer, A. G. (1992). Chemical characterization of MHC-determined body odors. *Chemical Senses* **17**, 593.

Tucker, D. (1963). Physical variables in the olfactory stimulation process. *Journal of General Physiology* **46**, 453–489.

(1971). Non-olfactory responses from the nasal cavity: Jacobson's organ and the trigeminal system. In *Handbook of Sensory Physiology, vol.* 4, ed. L. M. Beidler, pp. 151–181. Berlin: Springer-Verlag.

Twitty, V. C. (1966). *Of Scientists and Salamanders*. San Francisco, CA: W. H. Freeman.

Umemura, K., Sugawara, K., and Ito, I. (1988). The source of the odor emitted from estrous cows. *Japanese Journal of Zootechnical Science* **59**, 779–786.

Utaisincharoen, P., Mackessy, S, P., Miller, R. A., and Ju, A. T. (1993). Complete primary structure and biochemical properties of gilatoxin, a serine protease with kallikrein-like and angiotensin-degrading properties. *Journal of Biological Chemistry* **268**, 21975–21985.

Valderrama, X., Robinson, J. G., Attygalle, A. B., Athula, B., and Eisner, T. (2000). Seasonal anointment with millipedes in a wild primate: a chemical defense against insects? *Journal of Chemical Ecology* **26**, 2781–2790.

Vale, G. A. (1974). The responses of tsetse flies (Diptera: Glossinidae) to mobile and stationary baits. *Bulletin of Entomological Research* **64**, 545–588.

(1980). Field studies of the responses of tsetse flies (Glossinidae) and other Diptera to carbon dioxide, acetone and other chemicals. *Bulletin of Entomological Research* **70**, 563–570.

(1981). An effect of host diet on the attraction of tsetse flies (Diptera: Glosinidae) to host odour. *Bulletin of Entomological Research* **71**, 259–265.

Vale, G. A. and Hall, D. R. (1985). The role of 1-octen-3-ol, acetone and carbon dioxide in the attraction of tsetse flies *Glossina* spp. (Diptera: Glossinidae) to ox odour. *Bulletin of Entomological Research* **75**, 209–217.

Vale, G. A., Flint, S., and Hall, D. R. (1986). The field responses of tsetse flies, *Glossina* spp. (Diptera: Glossinidae), to odours of host residues. *Bulletin of Entomological Research* **76**, 685–693.

Valentincic, T. and Caprio, J. (1992). Gustatory responses of channel catfish to amino acids. In *Chemical Signals in Vertebrates*, vol. 6, ed. R. L. Doty and D. Müller-Schwarze, pp. 365–369 New York: Plenum.

Van Damme, R. and Castilla, A. M. (1996). Chemosensory predator recognition in the lizard *Podarcis hispania*: effects of predation pressure relaxation. *Journal of Chemical Ecology* **22**, 13–22.

Vandenbergh, J. G. (1969). Male odor accelerates female sexual maturation in mice. *Endocrinology* **84**, 658–660.

(1971). The influence of the social environment on sexual maturation in male mice. *Journal of Reproduction and Fertility* **24**, 383.

(1973). Effects of gonadal hormones on the flank gland of the golden hamster. *Hormone Research* **4**, 28–33.

(1987). Regulation of puberty and its consequences on population dynamics in mice. *American Zoologist* **27**, 891–898.

Vandenbergh, J. G. and Hotchkiss, A. K. (2001). Interfetal communication and adult phenotype in mice. In *Chemical Signals in Vertebrates*, vol. 9, ed. A. Marchlewska-Koj, J. J. Lepri and D. Müller-Schwarze, pp. 183–187. New York: Kluwer Academic/Plenum.

Vandenbergh, J. G., Whitsett, J. M., and Lombardi, J. L. (1975). Partial isolation of a pheromone accelerating puberty in female mice. *Journal of Reproduction and Fertility* **43**, 515–523.

Vandenbergh, J. G., Finlayson, J. S., Dobrogosz, W. J., Dills, S. S., and Kost, T. A. (1976). Chromatographic separation of puberty accelerating pheromone from male mouse urine. *Biology of Reproduction* **15**, 260–265.

van den Berk, J. and Müller-Schwarze, D. (1984). Responses of wild muskrats (*Ondatra zibethicus* L.) to scented traps. *Journal of Chemical Ecology* **10**, 1411–1415.

van den Hurk, R. V., Schoonen, W. G. E. J., van Zoelen, G. A., and Lambert, J. G. D. (1987). The biosynthesis of steroid glucuronides of the zebra fish *Brachydanio verio* and their pheromonal function as ovulation inducers. *General and Comparative Endocrinology* **68**, 179–188.

van der Lee, S. and Boot, L. M. (1955). Spontaneous pseudopregnancy in mice. *Acta Physiologica Neerlandica* **4**, 442–446.

Vander Wall, S. B. (1982). An experimental analysis of cache recovery in Clark's nutcracker. *Animal Behaviour* **30**, 84–94.

(1998). Foraging success of granivorous rodents: effects of variation in seed and soil water on olfaction. *Ecology* **79**, 223–241.

(2000). The influence of environmental conditions on cache recovery and cache pilferage by yellow pine chipmunks (*Tamias amoenus*) and deer mice (*Peromyscus maniculatus*). *Behavioral Ecology* **11**, 544–549.

(2003). How rodents smell buried seeds: a model based on the behavior of pesticides in soil. *Journal of Mammalogy* **84**, 1089–1099.

Vander Wall, S. B., Beck, M. J., Briggs, R., *et al.* (2003). Interspecific variation in the olfactory abilities of granivorous rodents. *Journal of Mammalogy* **84**, 847–896.

van Etten, C. H. and Tookey, H. L. (1979). Chemistry and biological effects of glucosinolates. In *Herbivores: Their Interactions with Secondary Plant Metabolites*, ed. G. A. Rosenthal and D. H. Janzen, pp. 471–500. New York: Elsevier.

Van Hoven, W. (1991). Mortalities in kudu (*Tragelaphus strepsiceros*) populations related to chemical defence in trees. *Journal of African Zoology* **105**, 141–145.

Vaughan, T. A. (1978). *Mammalogy*. Philadelphia, PA: Saunders College.

Vaughan, T. A. and Czaplewski, N. J. (1985). Reproduction in Stephen's woodrat: the wages of folivory. *Journal of Mammalogy* **66**, 429–443.

Vermeirssen, E. L. M., Scott, A. P., and Liley, N. R. (1997) Female rainbow trout urine contains a pheromone which causes rapid rise in plasma 17α,20β-dihydroxy-4-pregnen-3-one levels and milt amounts in males. *Journal of Fish Biology* **50**, 107–119.

Vernet-Maury, E. (1980). Trimethylthiazoline in fox feces: a natural alarming substance for the rat. In *Proceedings of the VIIth International Symposium on Olfaction and Taste*, ed. H. van der Starre, p. 407. London: IRL Press.

Vernet-Maury, E., Polak, E. H., and Demael, A. (1984). Structure–activity relationship of stress-inducing odorants in the rat. *Journal of Chemical Ecology* **10**, 1007–1018.

Vickers, N. J. (2000). Mechanisms of animal navigation in odor plumes. *Biological Bulletin of the Marine Biological Laboratory* **198**, 203–212.

Vieuille-Thomas, C. and Signoret, J. P. (1992). Pheromonal transmission of an aversive experience in domestic pig. *Journal of Chemical Ecology* **18**, 1551–1557.

Viitala, J., Korpimäki, E., Palokangas, P., and Kolvula, M. (1995). Attraction of kestrels to vole scent marks visible in ultraviolet light. *Nature* **373**, 425–427.

Vince, M. A. and Billing, A. E. (1986). Infancy in the sheep: the part played by sensory stimulation in bonding between ewe and lamb. In *Advances in Infancy Research*, vol. IV, ed. L. P. Lipsitt and C. Rovee-Collies, pp. 1–37. Norwood, NJ: Ablex.

Vince, M. A. and Ward, T. M. (1984). The responsiveness of newly born Clun forest lambs to odour sources in the ewe. *Behavior* **89**, 117–127.

Vince, M. A., Lynch, J. J., Mottershead, B. E., Green, G. C., and Elwin, R. L. (1987). Interactions between normal ewes and newly born lambs deprived of visual, olfactory and tactile sensory information. *Applied Animal Behaviour Science* **19**, 119–136.

Vittoria, A. and Rendina, N. (1960). Fattori condizionanti la funzionalita tiaminica in piante superiori e cenni sugli effetti della bocca dei ruminanti sulle erbe pascolative. *Acta Medica Veterinaria* **6**, 379–403.

Voigt, C. C. (2002). Individual variation in perfume blending in male greater sac-winged bats. *Animal Behaviour* **63**, 907–913.

Voigt, C. C. and von Helversen, O. (1999). Storage and display of odor by male *Saccopterys bilieata* (Chiroptera; Emballonuridae). *Behavioral Ecology and Sociobiology* **47**, 29–40.

vom Saal, F. S., Nagel, S. C., Palanza, P., Boechler, M., Parmigiani, S. and Welshons, W. V. (1995). Estrogenic pesticides: binding relative to estradiol in MCF-7 cells and effects of exposure during fetal life on subsequent territorial behavior in male mice. *Toxicology Letters* **77** 343–350.

von Frisch, K. (1941). Über einen Schreckstoff der Fischhaut and seine biologische Bedeutung. *Zeitschrift für vergleichende Physiologie* **29** 46–145.

Wabnitz, P. A., Bowie, J. H., Tyler, M. J., Wallace, J. C., and Smith, B. P. (1999). Aquatic sex pheromone from a male tree frog. *Nature* **401**, 444–445.

Waldman, B. (1981). Sibling recognition in toad tadpoles. The role of experience. *Zeitschrift für Tierpsychologie* **56**, 341–358.

Waldvogel, J. A. (1987). Olfactory navigation in homing pigeons: are the current models atmospherically realistic? *Auk* **104**, 369–379.

Waldvogel, J. A., Phillips, J. B., and Brown, A. I. (1988). Changes in the short-term deflector loft effect are linked to the sun compass of homing pigeons. *Animal Behaviour* **36**, 150–158.

Wallace, P. (1977). Individual discrimination of humans by odor. *Physiology and Behavior* **19**, 577–579.

Wallraff, H. G. (1988a). Olfactory deprivation in pigeons: examination of methods applied in homing experiments. *Comparative Biochemistry and Physiology A, Comparative Physiology* **89**, 621–630.

(1988b). Navigation by means of an olfactory map and a sun compass: the homing ability of pigeons. *Naturwissenschaften* **75**, 380–392.

(1989). Simulated navigation based on unreliable sources of information (Models on pigeon homing. Part 1). *Journal of Theoretical Biology* **137**, 1–19.

(1990). Navigation by homing pigeons. *Ethology, Ecology and Evolution* **2**, 81.

(2004). Avian olfactory navigation: its empirical foundation and conceptual state. Animal Behaviour 67, 189–204.

Wallraff, H. G. and Foà, A. (1981). Pigeon navigation: charcoal filter removes relevant information from environmental air. *Behavioral Ecology and Sociobiology* **9**, 67–77.

Wallraff, H. G. and Neuman, M. F. (1990). Contribution of olfactory navigation and non-olfactory pilotage to pigeon homing. *Behavioral Ecology and Sociobiology* **25**, 293–302.

Wallraff, H. G., Papi, F., Ioale, P., and Benvenuti, S. (1986). Magnetic fields affect pigeon navigation only while the birds can smell atmospheric odors. *Naturwissenschaften* **73**, 215–217.

Walls, S. C., Mathis, A., Jaeger, R. G., and Gergits, W. F. (1989). Male salamanders with high-quality diets have faeces attractive to females. *Animal Behaviour* **38**, 546–548.

Walsberg, G. E. (1975). Digestive adaptations of *Phainopepla nitens* associated with the eating of mistletoe berries. *Condor* **77**, 169–174.

Walther, F. R. (1978). Mapping the structure of the marking system of a territory of the Thomson's gazelle. *East African Wildlife Journal* **16**, 167–176.

Wang, D., Chen, P., and Halpern, M. (1987). Further isolation and purification of the garter snake chemoattractant in earthworm wash. *Chemical Senses* **12**, 706.

Wang, J. and Provenza, F. D. (1996). Food deprivation affects preference of sheep for foods varying in nutrients and a toxin. *Journal of Chemical Ecology* **22**, 2011–2021.

Ward, D. and Young, T. P. (2002). Effects of large mammalian herbivores and ant symbionts on condensed tannins of *Acacia drepanolobium* in Kenya. *Journal of Chemical Ecology* **28**, 921–937.

Ward, J. F., MacDonald, D. W., and Doncaster, C. P. (1997). Responses of foraging hedgehogs to badger odour. *Animal Behaviour* **53**, 709–720.

Ware, G. C. and Gosden, P. E. (1980). Anaerobic microflora of the anal sac of the red fox (*Vulpes vulpes*). *Journal of Chemical Ecology* **6**, 97–102.

Warner, T. F. and Azen, E. A. (1988). Tannins, salivary proline-rich proteins and oesophageal cancer. *Medical Hypotheses* **26**, 99–102.

Watkins, J. F., II, Gehlbach, F. R., and Kroll, J. C. (1969). Attractant–repellent secretions of blind snakes (*Leptotyphlops dulcis*) and their army ant prey (*Neivamyrmex nigrescens*). *Ecology* **50**, 1098–1102.

Webb, J. K. and Shine, R. (1992). To find an ant: trail-following in Australian blindsnakes (Typhlopidae). *Animal Behaviour* **43**, 941–948.

Webster, D. B. (1973). Audition, vision, and olfaction in kangaroo rat predator avoidance. *American Zoologist* **13**, 1346.

Webster, D. R. and Weissburg, M. J. (2001). Chemosensory guidance cues in a turbulent chemical odor plume. *Limnology and Oceanography* **46**, 1034–1047.

Weitzman, S. H. and Fink, S. V. (1985). Xenurobryconin phylogeny and putative pheromone pumps in glanduloaudine fishes. *Smithsonian Contributions to Zoology* **421**, 1–121.

Wekesa, K. S. and Lepri, J. J. (1992). Removal of the vomeronasal organ impairs reproduction in male prairie voles. *Chemical Senses* **17**, 718.

Welch, B. L. and McArthur, E. D. (1981). Variation of monoterpenoid content among subspecies and accessions of *Artemisia tridentata* grown in a uniform garden. *Journal of Range Management* **34**, 380–384.

Welch, B. L., McArthur, E. D., and Davis, J. N. (1983). Mule deer preference and monoterpenoids (essential oils). *Journal of Range Management* **36**, 485–487.

Welch, B. L., Pederson, J. C., and Rodriguez, R. L. (1989). Monoterpenoid content of sage grouse ingesta. *Journal of Chemical Ecology* **15**, 961–969.

Weldon, P. J. (1982). Responses to ophiophagous snakes by snakes of the genus *Thamnophis*. *Copeia* **1982**, 788–794.

(2004). Defense anointing: extended chemical phenotype and unorthodox ecology. *Chemoecology* **14**, 1–4.

Weldon, P. J. and Burghardt, G. M. (1979). The ophiophage defensive response in crotaline snakes: extension to new taxa. *Journal of Chemical Ecology* **5**, 141–151.

Weldon, P. J. and Fagre, D. B. (1989). Responses by canids to scent gland secretions of the western diamondback rattlesnake (*Crotalus atrox*). *Journal of Chemical Ecology* **15**, 1589–1604.

Weldon, P. J. and Schell, F. M. (1984). Responses by king snakes (*Lampropeltis getulus*) to chemicals from colubrid and crotaline snakes. *Journal of Chemical Ecology* **10**, 1509–1520.

Weldon, P. J. and Williams J. A. (1988). Rathke's glands: pattern of secretion discharge and tests of antipredator activity. *American Zoologist* **28**, 162A.

Weldon, P. J., Divita, F. M., and Middendorf, G. A., III (1987). Responses to snake odors by laboratory mice. *Behavioral Processes* **14**, 137–146.

Weldon, P. J., Swenson, D. J., Olson, J. K., and Brinkmeier, W. G. (1990). The American alligator detects food chemicals in aquatic and terrestrial environments. *Ethology* **85**, 191–198.

Weldon, P. J., Graham, D. P., and Mears, L. P. (1993). Carnivore fecal chemicals suppress feeding by alpine goats (*Capra hircus*). *Journal of Chemical Ecology* **19**, 2947–2952.

Weldon, P. J., Aldrich, J. R., Klun, J. A., Oliver, J. E., and Debboun, M. (2003). Benzoquinones from millipedes deter mosquitoes and elicit self-anointing in capuchin monkeys (*Cebus* spp.). *Naturwissenschaften* **90**, 301–304.

Weller, L. and Weller, A. (1993). Human menstrual synchrony: a critical view. *Neuroscience and Behavioral Review* **17**, 427–439.

Wells, M. C. and Bekoff, M. (1981). An observational study of scent marking in coyotes, *Canis latrans*. *Animal Behaviour* **29**, 332–350.

Welsh, R. G. and Müller-Schwarze, D. (1989). Experimental habitat scenting inhibits colonization by beaver, *Castor canadensis*. *Journal of Chemical Ecology* **15**, 887–893.

Wenzel, B. M. (1968). Olfactory prowess of the kiwi. *Nature* **220**, 1133–1134.

(1986). The ecological and evolutionary challenges of procellariiform olfaction. In *Chemical Signals in Vertebrates*, vol. 4, ed. D. Duvall, D. Müller-Schwarze and R. M. Silverstein, pp. 357–368. New York: Plenum.

Wenzel, B. M. and Sieck, M. H. (1966). Olfaction. *Annual Review of Physiology* **28**, 381–432.

Werner, D. I., Baker, E. M., Gonzalez, E. D. C., and Sosa, I. R. (1987). Kinship recognition and grouping in hatchling green iguanas. *Behavioral Ecology and Sociobiology* **21**, 83–90.

Wheeler, J. W. (1977). Properties of compounds used as chemical signals. In *Chemical Signals in Vertebrates*, vol. 1, ed. D. Müller-Schwarze and R. M. Silverstein, pp. 61–70. New York: Plenum.

White, P. J., Kreeger, T. J., Tester, J. R., and Seal, V. S. (1989). Anal-sac secretions deposited with feces by captive red foxes (*Vulpes vulpes*). *Journal of Mammalogy* **70**, 814–816.

White, S. M., Flinders, J. T., and Welch, B. L. (1982). Preference of pygmy rabbits (*Brachylagus idahoensis*) for various populations of big sagebrush (*Artemisia tridentata*). *Journal of Range Management* **35**, 724–726.

Whitten, W. K. (1958). Modification of the estrous cycle of the mouse by external stimuli associated with the male. Changes in the estrous cycle determined by vaginal smears. *Journal of Endocrinology* **17**, 307–313.

Whitten, W. K., Wilson, M. C., Wilson, S. R., *et al.* (1980). Induction of marking behavior in wild red foxes (*Vulpes vulpes*) by synthetic urinary constituents. *Journal of Chemical Ecology* **6**, 49–55.

Widowski, T. M., Ziegler, T. E., Elowson, H. M., and Snowdon, C. T. (1990). The role of males in the stimulation of reproductive function in female cotton-top tamarins, *Saguinus o. oedipus*. *Animal Behaviour* **40**, 731–741.

Williams, B. L., Brodie, E. D., Jr., and Brodie E. D., III (2004). A resistant predator and its toxic prey: persistence of newt toxin leads to poisonous (not venomous) snakes. *Journal of Chemical Ecology* **30**, 1901–1919.

Williams, J. R., Slotnick, B. M., Kirkpatrick, B. W., and Carter, S. B. (1992a). Olfactory bulb removal affects partner preference development and estrus induction in female prairie voles. *Physiology and Behavior* **52**, 635–639.

Willliams, J. D., Holland, K. N., Jameson, D. M., and Bruening, R. C. (1992b). Amino acid profiles and liposomes: their role as chemosensory information carriers in the marine environment. *Journal of Chemical Ecology* **18**, 2107–2115.

Williams, J. R. and Lenington, S. (1993). Factors modulating preferences of female house mice for males differing in t-complex genotype: role of t-complex genotype, genetic background, and estrous condition of females. *Behavior Genetics* **23**, 51–58.

Willis, C. M., Church, S. M., Guest, C. M., *et al.* (2004). Olfactory detection of human bladder cancer by dogs: proof of principle study. *British Medical Journal* **329**, 712.

Wilson, E. O. (1970). Chemical communication within animal species. In *Chemical Ecology*, ed. E. Sondheimer and J. B. Simeone, pp. 133–155. London: Academic Press.

Wilson, E. O. and Bossert, W. H. (1963). Chemical communication among animals. *Recent Progress in Hormone Research* **19**, 673–716.

Wilson, H. C. (1987) Female axillary secretions influence women's menstrual cycles: a critique. *Hormones and Behavior* **21**, 536–546.

Wiltschko, R., and Wiltschko, W. (1987). Pigeon homing: olfactory experiments with inexperienced birds. *Naturwissenschaften* **74**, 94–96.

(1992). The effect of temporary anosmia on orientation behavior. In *Chemical Signals in Vertebrates*, vol. 6, ed. R. L. Doty and D. Müller-Schwarze, pp. 435–442. New York: Plenum.

Wiltschko, W., Wiltschko, R. and Mathias, J. (1987a). The orientation behavior of anosmic pigeons in Frankfurt am Main, West Germany. *Animal Behaviour* **35**, 1324–1333.

Wiltschko, W., Wiltschko, R., and Walcott, C. (1987b). Pigeon homing: different effects of olfactory deprivation in different countries. *Behavioral Ecology and Sociobiology* **21**, 333–342.

Wiltschko, R., Wiltschko, W., and Kowalski, U. (1989). Pigeon homing: an unexpected effect of treatment with a local anesthetic on initial orientation. *Animal Behaviour* **37**, 1050–1052.

Winberg, S. and Olsén, H. (1992). The influence of rearing conditions on the sibling odour preference of juvenile Arctic charr, *Salvelinus alpinus* L. *Animal Behaviour* **44**, 157–164.

Wines, M. (2004). For sniffing out land mines, a platoon of twitching noses. *New York Times*, 18 May 2004, pp. A1, A7.

Wingate, L. P. (1956). Mambas strong smell. *African Wildlife* **10**, 256–257.

Wirsig, C. R. and Leonard, C. M. (1985). Terminal nerve damage affects hamster mating behavior. *Chemical Senses* **10**, 423.

Witmer, L. M. (2001). Nostril position in dinosaurs and other vertebrates and its significance for nasal function. *Science* **293**, 850–854.

Wolff, J. O. and Davis-Born, R. (1997). Response of gray-tailed voles to odours of a mustelid predator: a field test. *Oikos* **79**, 543–548.

Wong, G. T. and Gannon K. S. (1996). Transduction of bitter and sweet taste by gustducin. *Nature* **381**, 796–800.

Wood, W. F. (1990). New components in defense secretion of the striped skunk, *Mephitis mephitis*. *Journal of Chemical Ecology* **16**, 2057–2065.

(2004). Straight and branched-chain falty acids in preorbital glands of sika deer, *Cervus nippon*. *Journal of Chemical Ecology* **30**, 479–482.

Wood, W. F., Morgan, G. C., and Miller, A. (1991). Volatile components in defensive spray of the spotted skunk, *Spilogale putorius*. *Journal of Chemical Ecology* **17**, 1415–1420.

Wood, W. F., Fisher, C. O., and Graham, G. A. (1993). Volatile components in defensive spray of the hog-nose skunk, *Conepatus mesoleucus*. *Journal of Chemical Ecology* **19**, 837–841.

Woolhouse, A. D., Weston, R. J., and Hamilton, B. H. (1994). Analysis of secretions from scent-producing glands of brushtail possum (*Trichosurus vulpecula*). *Journal of Chemical Ecology* **20**, 239–253.

Wrangham, R. W. and Nishida, T. (1983). *Aspilia* spp. leaves: a puzzle in the feeding behavior of wild chimpanzees. *Primates* **24**, 276–282.

Wrangham, R. W. and Waterman, P. (1981). Feeding behavior of vervet monkeys on *Acacia tortilis* and *Acacia xanthoploea* with special reference to reproductive strategies and tannin production. *Journal of Animal Ecology* **50**, 715–731.

Wright, J. and Weldon, P. J. (1990). Responses by domestic cats (*Felis catus*) to snake scent gland secretions. *Journal of Chemical Ecology* **16**, 2947–2953.

Wuensch, K. L. (1982). Effect of scented traps on captures of *Mus musculus* and *Peromyscus maniculatus*. *Journal of Mammalogy* **63**, 314–315.

Würdinger, I. (1979). Olfaction and feeding behavior in juvenile geese (*Anser a. anser* and *Anser domesticus*). *Zeitschrift für Tierpsychologie* **49**, 132–135.

Würdinger, V. I. (1990). Die Reaktionen von Zebrafinken (*Taeniopygia guttata*) auf Düfte: eine Pilotstudie. *Die Vogelwarte* **35**, 359–367.

Wyatt, T. D. (2003). *Pheromones and Animal Behaviour*. Cambridge, UK: Cambridge University Press.

Wyatt, T. D., Phillips, D. G., and Grégoire, J.-C. (1993). Turbulence, trees and semiochemicals: windtunnel orientation of the predator, *Rhizophagus grandis*, to its barkbeetle prey, *Dendroctonus micans*. *Physiological Entomology* **18**, 204–210.

Wysocki, C. J. (1989). Vomeronasal chemoreception: its role in reproductive fitness and physiology. In *Neural Control of Reproductive Function*, ed. J. Lakoski, B., Haber, R. Perez-Polo and D. Rossin, pp. 545–566. New York: Alan Liss.

Wysocki, C. J., Beauchamp, G. K., and Erisman, S. (1980). Access of compounds to the vomeronasal organ in pine and meadow voles. *Proceedings of the Fourth Eastern Pine and Meadow Vole Symposium*, February 1980, Hendersonville, New York, pp. 20–23.

Wysocki, C. J., Nyby, J., Whitney, G., Beauchamp, G., and Katz, Y. (1982). The vomeronasal organ: primary role in mouse chemosensory gender recognition. *Physiology and Behavior* **29**, 315–327.

Wysocki, C. J., Katz, Y., and Bernhard, R. (1983). The male vomeronasal organ mediates female-induced testosterone surges. *Biology of Reproduction* **28**, 917–922.

Wysocki, C. J., Beauchamp, G. K., Reidinger, R. R., and Wellington, J. L. (1985). Access of large and nonvolatile molecules to the vomeronasal organ of mammals during social and feeding behaviors. *Journal of Chemical Ecology* **11**, 1147–1160.

Wysocki, C. J., Bean, N. J., and Beauchamp, G. K. (1986). The mammalian vomeronasal system: its role in learning and social behaviors. In *Chemical Signals in Vertebrates*, vol. 4, ed. D. Duvall, D. Müller-Schwarze and R. M. Silverstein, pp. 471–485. New York: Plenum.

Xiao-Nong Zeng, J., Leyden, J., Lawley, H. J., *et al.* (1991). Analysis of characteristic odors from human male axillae. *Journal of Chemical Ecology* 17, 1469–1492.

Xu, Z., Stoddart, M., Ding, H., and Zhang, J. (1995). Self-anointing behavior in the rice-field rat, *Rattus rattoides*. *Journal of Mammalogy* **76**, 1238–1241.

Yahr, P. (1977). Central control of scent marking. In *Chemical Signals in Vertebrates*, vol. 1, ed. D. Müller-Schwarze and M. M. Mozell, pp. 549–562. New York: Plenum Press.

Yamamori, K., Nakamura, M., Matsui, T., and Hara, T. J. (1988). Gustatory responses to tetrodotoxin and saxitoxin in fish: a possible mechanism for avoiding marine toxins. *Canadian Journal of Fisheries and Aquatic Science* **45**, 2182–2186.

Yamamoto, K., Kawai, Y., Hayashi, T., *et al.* (2000). Silefrin, a sodefrin-like pheromone in the abdominal gland of the sword-tailed newt, *Cynops ensicauda*. *FEBS Letters* **472**, 267–270.

Yamazaki, K., Beauchamp, G. K., Matszaki, O., *et al.* (1986). Participation of the murine X and Y chromosomes in genetically determined chemosensory identity. *Proceedings of the National Academy of Sciences, USA* **83**, 4438–4440.

Yamazaki, K., Beauchamp, G. K., Kupniewski, D., *et al.* (1988). Familiar imprinting determines H-2 selective mating preferences. *Science* **240**, 1331–1332.

Yamazaki, K., Beauchamp, G. K., Bard, J., and Boyse, E. A., (1990). Chemosensory identity and the Y chromosome. *Behavior and Genetics* **20**, 157–165.

Yamazaki, K., Beauchamp, G. K., Imai, Y., Bard, J., and Boyse, E. A. (1992a). Expression of urinary H-2 odortypes by infant mice. *Chemical Senses* **17**, 723.

Yamazaki, K., Beauchamp, G. K., Imai, Y., *et al.* (1992b). MHC control of odortypes in the mouse. In *Chemical Signals in Vertebrates*, vol. 6, ed. R. L. Doty and D. Müller-Schwarze, pp. 189–196. New York: Plenum.

Yasumoto, T. and Murata, M. (1993). Marine toxins. *Chemistry Review* **93**, 1897–1909.

Yasumoto, T., Yasumara, D., Yotsu, M., *et al.* (1986). Bacterial production of tetrodotoxin and anhydrotetrodotoxin. *Agricultural and Biological Chemistry* **50**, 793–795.

Yoerg, S. I. (1991). Social feeding reverses learned flavor aversions in spotted hyenas (*Crocuta crocuta*). *Journal of Comparative Psychology* **105**, 185–189.

Yoshihara, Y., Nago, H., and Mori, K. (2001). Sniffing out odors with multiple dendrites. *Science* **291**, 835–837.

Yousem, D. M., Maldjian, J. A., Siddiqui, F., *et al.* (1999). Gender effects on odor-stimulated functional magnetic resonance imaging. *Brain Research* **818**, 480–487.

Yun, S-S., Scott, A. P., and Li, W. (2003). Pheromones of the male sea lamprey, *Petromyzon marinus* L.: structural studies on a new compound, 3-keto allocholic acid, and 3-keto petromyzonol sulfate. *Steroids* **68**, 297–304.

Zala, S. M. and Penn, D. J. (2004). Abnormal behaviors induced by chemical pollution: a review of the evidence and new challenges. *Animal Behaviour* **68**, 649–664.

Zeng, C., Spielman, A. I., Vowels, B. R., *et al.* (1996a). A human axillary odorant is carried by apolipoprotein D. *Proceedings of the National Academy of Sciences of the USA* **93**, 6626–6630.

Zeng, X.-N., Leyden, J. J., Lawley, H. J. *et al.* (1991). Analysis of characteristic odors from human male axillae. *Journal of Chemical Ecology* **17**, 1469–1492.

Zeng, X.-N., Leyden, J. J., Brand, J. G., *et al.* (1992). An investigation of human apocrine gland secretion for axillary odor precursors. *Journal of Chemical Ecology* **18**, 1039–1055.

Zeng, X.-N., Leyden, J. J., Spielman, A. I., and Preti, G. (1996b). Analysis of characteristic human female axillary odors: qualitative comparison to males. *Journal of Chemical Ecology* **22**, 237–257.

Zheng, W., Strobeck, C., and Stacey, N. E. (1997). The steroid pheromone 17α,20β-dihydroxy-4-pregnene-3-one increases fertility and paternity in goldfish. *Journal of Experimental Biology* **200**, 2833–2840.

Ziegler, T. E., Epple, G., Snowdon, C. T., *et al.* (1993). Detection of the chemical signals of ovulation in the cotton-top tamarin, *Saguinus oedipus*. *Animal Behaviour* **45**, 313–322.

Zielinski, B., Arbuckle, W., Belanger, A., *et al.* (2003). Evidence for the release of sex pheromones by male round gobies (*Neogobius melanstomus*). *Fish Physiology and Biochemistry* **28**, 237–239.

Zielinski, W. J. and Barthalmus, G. T. (1989). African clawed frog skin compounds: antipredatory effects on African and North American watersnakes. *Animal Behaviour* **38**, 1038–1086.

Zimmerling, L. L. and Sullivan, T. P. (1994). Influence of mustelid semiochemicals on population dynamics of the deer mouse (*Peromyscus maniculatus*). *Journal of Chemical Ecology* **20**, 667–689.

Zobel, A. M. and Brown, S. A. (1990). Dermatitis-inducing furanocoumarins on leaf surfaces of eight species of rutaceous and umbelliferous plants. *Journal of Chemical Ecology* **16**, 693–700.

Zucker, W. V. (1983). Tannins: does structure determine function? An ecological perspective. *American Naturalist* **121**, 335–365.

INDEX